Crop Biofortification

Crop Biofortification

Biotechnological Approaches for Achieving Nutritional Security Under Changing Climate

Edited by

Adnan Noor Shah
Khwaja Fareed University of Engineering and Information Technology
Punjab
Pakistan

Sajid Fiaz
The University of Lahore
Lahore
Pakistan

Muhammad Aslam
University of Agriculture
Faisalabad
Pakistan

Javed Iqbal
Bacha Khan University
Khyber Pakhtunkhwa
Pakistan

Abdul Qayyum
The University of Haripur
Haripur
Pakistan

WILEY

This edition first published 2025

Registered Offices
John Wiley & Sons, Inc., 111 River Street, Hoboken, NJ 07030, USA

John Wiley & Sons Ltd, New Era House, 8 Oldlands Way, Bognor Regis, West Sussex, PO22 9NQ, UK

For details of our global editorial offices, customer services, and more information about Wiley products visit us at www.wiley.com.

The manufacturer's authorized representative according to the EU General Product Safety Regulation is Wiley-VCH GmbH, Boschstr. 12, 69469 Weinheim, Germany, e-mail: Product_Safety@wiley.com.

Wiley also publishes its books in a variety of electronic formats and by print-on-demand. Some content that appears in standard print versions of this book may not be available in other formats.

Library of Congress Control Number has been applied for:

Hardback ISBN: 9781394273249
ePDF ISBN: 9781394273263
ePub ISBN: 9781394273256
oBook ISBN: 9781394273270

Cover Design: Wiley
Cover Image: © Елена Бутусова/stock.adobe.com

Set in 9.5/12 STIX Two Text by Lumina Datamatics
Printed and bound by CPI Group (UK) Ltd, Croydon, CR0 4YY

C9781394273249_150325

Contents

List of Contributors

Muhammad Aaqil
College of Food Science and Technology
Yunnan Agricultural University
Kunming
China

Banzeer Ahsan Abbasi
Department of Botany
Rawalpindi Women University
Rawalpindi
Pakistan

FNU Abdullah
Department of Soil and Crop Sciences
Texas A&M University
Texas
USA

Muhammad Adnan
Department of Agriculture
University of Swabi
Anbar
Pakistan

Kafeel Ahmad
Department of Botany
University of Sargodha
Sargodha
Pakistan

Mushtaq Ahmad
Horticultural Research Institute
National Agricultural Research Center
Islamabad
Pakistan

Shakeel Ahmad
Institute of Agronomy
Bahauddin Zakariya University
Multan
Pakistan

Department of Soil and Environmental Sciences
Muhammad Nawaz Shareef University of Agriculture
Multan
Pakistan

Zeeshan Ahmed
Xinjiang Institute of Ecology & Geography
Chinese Academy of Sciences
Urumqi
China

Cele National Station of Observation and Research for Desert-Grassland Ecosystems
Chinese Academy of Sciences
Urumqi
China

College of Life Science
Shenyang Normal University
Shenyang
China

Ahmad Ali
Center for Plant Sciences and Biodiversity
University of Swat
Charbagh
Pakistan

Ayesha Ali
Department of Botany
Government College University Lahore
Lahore
Pakistan

Basharat Ali
Department of Agricultural Engineering
Khwaja Fareed University of Engineering and Information Technology
Rahim Yar Khan
Pakistan

Muqarrab Ali
Department of Climate Change
Muhammad Nawaz Shareef University of Agriculture
Multan
Pakistan

Ruqayya Ali
Department of Botany
Government College University Lahore
Lahore
Pakistan

Umed Ali
Department of Agriculture
Mir Chakar-Khan Rind University
Sibi
Pakistan

Muaz Ameen
Department of Botany
University of Agriculture Faisalabad
Faisalabad
Pakistan

Aqsa Anjum
Department of English Literature and Applied Linguistics
Government Post Graduate College
Timergara
Pakistan

Shakeel Ahmad Anjum
Department of Agronomy
University of Agriculture
Faisalabad
Pakistan

Sumera Anwar
Department of Botany
Government College Women University Faisalabad
Faisalabad
Pakistan

Uneebullah Arif
Department of Forestry and Range Management
Gomal University
Dera Ismail Khan
Pakistan

Uzma Arif
Horticultural Research Institute
National Agricultural Research Center
Islamabad
Pakistan

Yunus Emre Arvas
Van Yuzuncu Yil University
Faculty of Science
Department of Molecular Biology and Genetics
Tusba-Van
Turkey

Karadeniz Technical University
Faculty of Science
Department of Biology
Ortahisar- Trabzon
Turkey

Muhammad Ashraf
Institute of Molecular Biology and Biotechnology
The University of Lahore
Lahore
Pakistan

Umair Ashraf
Department of Botany
Division of Science and Technology
University of Education
Lahore
Pakistan

Dilawar Aslam
Department of Plant Breeding and Genetics
University of Agriculture
Faisalabad
Pakistan

Muhammad Aslam
Department of Botany
Division of Science and Technology
University of Education
Lahore
Pakistan

Areeba Azhar
Department of Botany
Government College University Lahore
Lahore
Pakistan

Hafeezullah Babar
Soil Fertility Research Institute
Agriculture Research Center
Tandojam
Pakistan

Mohammad Safdar Baloch
Department of Agronomy
Gomal University
Dera Ismail Khan
Pakistan

Ghazala Batool
Departmemt of Bioinformatics
University of Agriculture
Faisalabad
Pakistan

Mohamed Behnassi
Economic Social & Environmental Council
(ESEC- Rabat);
Ministry of Higher Education
Scientific Research and Innovation
Morocco

Mahpara Khatoon Bhutto
Department of Agronomy
Sindh Agriculture University
Tandojam
Pakistan

Marya Bibi
Department of Agronomy
Horticulture & Plant Science
South Dakota State University
Brookings
USA

Yamin Bibi
Department of Agronomy
Sayed Jamaluddin Afghani University
Salar Bagh
Afghanistan

Noman Ali Buttar
Department of Agricultural Engineering
Khwaja Fareed University of Engineering and Information
Technology
Rahim Yar Khan
Pakistan

Shah Fahad
Department of Agronomy
Abdul Wali Khan University Mardan
Khyber Pakhtunkhwa
Pakistan

Areej Fatima
Department of Botany
Government College University Lahore
Lahore
Pakistan

Sajid Fiaz
Institute of Molecular Biology and Biotechnology
The University of Lahore
Lahore
Pakistan

Muhammad Adeel Ghafar
Department of Botany
Government College University Faisalabad
Faisalabad
Pakistan

Muhammad Salman Hameed
National Key Laboratory of Green Pesticide
International Joint Research Center for Intelligent
Biosensor Technology and Health
Central China Normal University
Wuhan
China

Pat (J S) Heslop Harrison
Key Laboratory of Plant Resources Conservation and Sustainable Utilization/Guangdong Provincial Key Laboratory of Applied Botany
South China Botanical Garden
Chinese Academy of Sciences
Guangzhou
China

Department of Genetics and Genome Biology
University of Leicester
Institute for Environmental Futures
Leicester
United Kingdom

South China National Botanical Garden
Chinese Academy of Sciences
Guangzhou
China

Wiwiek Harsonowati
Research Center for Horticulture
National Research and Innovation Agency (BRIN)
Cibinong
Indonesia

Shah Hassan
Department of Agricultural Extension Education & Communication
The University of Agriculture
Peshawar
Pakistan

Waseem Hassan
Department of Soil and Environmental Sciences
Muhammad Nawaz Shareef University of Agriculture
Multan
Pakistan

Muhammad Baqir Hussain
Department of Soil and Environmental Sciences
MNS University of Agriculture
Multan
Pakistan

Saddam Hussain
Department of Agronomy
University of Agriculture
Faisalabad
Pakistan

Sajjad Hyder
Department of Botany
GC Women University
Sialkot
Pakistan

Shumaila Ijaz
School of Biomedical Engineering
Shenzhen University Medical School
Shenzhen University
Shenzhen
China

Muhammad Imran
Department of Soil and Environmental Sciences
Muhammad Nawaz Shareef University of Agriculture
Multan
Pakistan

Javed Iqbal
Department of Botany
Bacha Khan University
Charsadda
Pakistan

Rashid Iqbal
Department of Agronomy
Faculty of Agriculture and Environment
The Islamia University of Bahawalpur
Bahawalpur
Pakistan

Maham Ishfaq
Institute of Molecular Biology and Biotechnology
The University of Lahore
Lahore
Pakistan

Muhammad Shoaib Ismail
Department of Agronomy
MNS University of Agriculture
Multan
Pakistan

Institute of Plant Nutrition and Soil Science
Kiel University
Kiel
Germany

Ghulam Hussain Jatoi
Department of Plant Pathology
Faculty of Crop Protection
Sindh Agriculture University
Tandojam
Pakistan

Maria Javaid
Department of Botany
University of Peshawar
Peshawar
Pakistan

Maryam Javid
Department of Botany
University of Peshawar
Peshawar
Pakistan

Nitish Joshi
Department of Agronomy
Horticulture & Plant Science
South Dakota State University
Brookings
USA

Sobia Kanwal
Department of Biology and Environmental Sciences
Allama Iqbal Open University
Islamabad
Pakistan

Azizullah Khalili
Department of Agronomy
The University of Haripur
Haripur
Pakistan

Department of Agronomy
Sayed Jamaluddin Afghani University
Salar Bagh
Afghanistan

Asghar Ali Khan
Department of Agronomy
Gomal University
Dera Ismail Khan
Pakistan

Bilal Ahmad Khan
College of Earth and Environmental Sciences
University of the Punjab
Lahore
Pakistan

Imran Khan
Department of Forestry and Range Management
Gomal University
Dera Ismail Khan
Pakistan

Center for Life Science Ventures
Cornell University
Ithaca
USA

Mehnoob Ullah Khan
Department of Botany
Division of Science and Technology
University of Education
Lahore
Pakistan

Muhammad Nasir Rasheed Khan
Department of Agronomy
Abdul Wali Khan University Mardan
Khyber Pakhtunkhwa
Pakistan

Qudrat Ullah Khan
Department of Soil Science
Gomal University
Dera Ismail Khan
Pakistan

Usama Ahmad Khan
Department of Botany
Ghazi University
Dera Ghazi Khan
Pakistan

Zafar Iqbal Khan
Department of Botany
University of Sargodha
Sargodha
Pakistan

Maqsood Ahmed Khaskheli
Department of Plant Pathology
College of Agriculture
Guizhou University
Guiyang
China

Waseem Ahmed Khattak
Department of Plant Sciences
Quaid-I-Azam University
Islamabad
Pakistan

Shiza Ayaz Khilji
Department of Botany
Division of Science and Technology
University of Education
Lahore
Pakistan

Munazza Kiran
Department of Botany
Division of Science and Technology
University of Education
Lahore
Pakistan

Lin Li
School of Breeding and Multiplication
Sanya Institute of Breeding and Multiplication
Hainan University
Sanya
China

Ahmad Mahmood
Department of Soil and Environmental Sciences
Muhammad Nawaz Shareef University of Agriculture
Multan
Pakistan

Athar Mahmood
Department of Agronomy
University of Agriculture Faisalabad
Faisalabad
Pakistan

Sammina Mahmood
Department of Botany
Division of Science and Technology
University of Education
Lahore
Pakistan

Tariq Mahmood
Department of Plant Sciences
Faculty of Biological Sciences
Quaid-i-Azam University
Islamabad
Pakistan

Muhammad Majeed
Department of Botany
University of Gujrat
Gujrat
Pakistan

Konain Zahra Maqbool
Department of Botany
University of Baltistan
Skardu
Pakistan

Sadaf Mehfooz
Department of Botany
Government College University Lahore
Lahore
Pakistan

Iqra Mukhtar
Department of Botany
Government College University Lahore
Lahore
Pakistan

Mamoona Munir
Department of Botany
Rawalpindi Women University
Rawalpindi
Pakistan

Ghulam Murtaza
Faculty of Environmental Science and Engineering
Kunming University of Science and Technology
Kunming
China

Maqzia Mushtaq
Principal Pharmacist
Shifa International Hospital Ltd.
Islamabad
Pakistan

Muhammad Amjad Nadim
Department of Agronomy
Gomal University
Dera Ismail Khan
Pakistan

Fahim Nawaz
Research School of Biology
The Australian National University
Canberra
Australia

Muhammad Nawaz
Department of Agricultural Engineering
Khwaja Fareed University of Engineering and Information Technology
Rahim Yar Khan
Pakistan

Muhammad Amjed Nawaz
National Research Tomsk State University
Tomsk
Russia

Taufiq Nawaz
Department of Biology/Microbiology
South Dakota State University
Brookings
USA

Dillon Nelson
Department of Biology/Microbiology
South Dakota State University
Brookings
USA

Yasir Niaz
Department of Agricultural Engineering
Khwaja Fareed University of Engineering and Information Technology
Rahim Yar Khan
Pakistan

Ijaz Rasool Noorka
Department of Plant Breeding and Genetics
University of Sargodha
Sargodha
Pakistan

Zahra Noreen
Department of Botany
Division of Science and Technology
University of Education
Lahore
Pakistan

Qurban Ali Panhwar
Soil and Environmental Sciences Division
Nuclear Institute of Agriculture
Tandojam
Pakistan

Abdul Qayyum
Department of Agronomy
The University of Haripur
Haripur
Pakistan

Muhammad Tamoor Qureshi
Department of Plant Breeding and Genetics
University of Sargodha
Sargodha
Pakistan

Praviinkumar R
Department of Genetics and Plant Breeding
Faculty of Agriculture
Annamalai University
Chidambaram
India

Arslan Rafique
Department of Botany
University of Gujrat
Gujrat
Pakistan

Tanzeel Ur Rahman
Department of Hotel Management & Tourism
University of Swabi
Anbar
Pakistan

Ambrin Rajput
Soil Fertility Research Institute
Agriculture Research Center
Tandojam
Pakistan

Swaiba Rani
Department of Plant Breeding and Genetics
University of Agriculture
Faisalabad
Pakistan

Madiha Rashid
Department of Botany
Division of Science and Technology
University of Education
Lahore
Pakistan

Nadia Riaz
Department of Botany
Lahore College for Women University
Lahore
Pakistan

Sundas Riaz
Department of Soil and Environmental Sciences
Muhammad Nawaz Shareef University of Agriculture
Multan
Pakistan

Umair Riaz
Department of Soil and Environmental Sciences
Muhammad Nawaz Shareef University of Agriculture
Multan
Pakistan

Muhammad Rizwan
School of Energy Science and Engineering
Central South University
Changsha
China

Nabeel Rizwan
Department of Soil Science
Gomal University
Dera Ismail Khan
Pakistan

Umar Khitab Saddozai
Department of Agronomy
Gomal University
Dera Ismail Khan
Pakistan

Anam Saleem
Department of Biosciences
COMSATS University
Islamabad
Pakistan

Rukhsar Saleem
Department of Agronomy
University of Agriculture
Faisalabad
Pakistan

Shobana Sampath
Department of Biotechnology
Vel Tech Rangarajan Dr Sagunthala R&D Institute
of Science and Technology
Chennai
India

Saira Sattar
Horticultural Research Institute
National Agricultural Research Center
Islamabad
Pakistan

Obaid Ullah Sayal
Department of Plant Breeding and Genetics
Gomal University
Dera Ismail Khan
Pakistan

Fahad Shafiq
Department of Botany
Government College University Lahore
Lahore
Pakistan

Adnan Noor Shah
Department of Agricultural Engineering
Khwaja Fareed University of Engineering and Information
Technology
Rahim Yar Khan
Pakistan

Aqeela Shaheen
Department of Chemistry
Government Sadiq College Women University
Bahawalpur
Pakistan

Muhammad Naveed Shahid
Department of Botany
Division of Science and Technology
University of Education
Lahore
Pakistan

Muhammad Faisal Shahzad
Department of Entomology
Gomal University
Dera Ismail Khan
Pakistan

Tahmina Shar
Agriculture Research Center
Tandojam
Pakistan

Muhammad Asif Shehzad
Institute of Plant Breeding and Biotechnology
Muhammad Nawaz Shareef University of Agriculture
Multan
Pakistan

Minahil Sheikh
Department of Botany
Government College University Lahore
Lahore
Pakistan

Hassan Sher
Center for Plant Sciences and Biodiversity
University of Swat
Charbagh
Pakistan

Kadambot H. M. Siddique
The UWA Institute of Agriculture
The University of Western Australia
Crawley
Australia

Razia Sultana
Wheat Research Institute
Ayub Agricltral Research Institute
Faisalabad
Pakistan

Mehwish Tabassum
Department of Botany
Division of Science and Technology
University of Education
Lahore
Pakistan

Swapan K. Tripathy
Department of Agricultural Biotechnology
College of Agriculture
Bhubaneswar
India

Aasma Tufail
Department of Botany
Division of Science and Technology
University of Education
Lahore
Pakistan

Ikram Ullah
College of Horticulture
Northwest A&F University
Yangling
China

Zahid Ullah
Center for Plant Sciences and Biodiversity
University of Swat
Charbagh
Pakistan

Zakir Ullah
Department of Plant Sciences
Faculty of Biological Sciences
Quaid-i-Azam University
Islamabad
Pakistan

Muhammad Usman
School of Agriculture and Biology
Shanghai Jiao Tong University
Minhang
China

Department of Botany
Government College University
Lahore
Pakistan

Zunaira Yaqoob
Department of Botany
Government College University Lahore
Lahore
Pakistan

Tabassum Yaseen
Department of Botany
Bacha Khan University
Charsadda
Pakistan

Afifa Younas
Department of Botany
Division of Science and Technology
University of Education
Lahore
Pakistan

Iqra Yousaf
Department of Botany
Division of Science and Technology
University of Education
Lahore
Pakistan

Muhammad Zahid
Department of Plant Breeding and Genetics
University of Agriculture
Faisalabad
Pakistan

Preface

In this current era of rapid climate change and increasing food insecurity, challenges in ensuring food and nutritional insecurity have become a research hotspot. World population is expected to surpass 9 billion by 2050 with the increasing global climate change. Nutritional deficiencies, particularly in the developing countries, pose significant health risks, affecting the growth and development of millions of people around the world. Approximately 2 billion people suffer from one or more micronutrient malnutrition. World Health Organization (WHO) and the Food and Agriculture Organization (FAO) of the United Nations reported that 149 million children under 5 years are stunted, 47 million are wasted, and 462 million are underweight. Under such circumstances, strategies such as crop biofortification hold potential to enhance the nutritional quality of staple crops. Biofortification is the combination of different processes by which the nutrient density of food crops is increased through traditional and/or modern plant breeding approaches, improved agronomic practices and/or modern biotechnological techniques without compromising any characteristic favorable to consumers and farming community.

As we navigate through complexities of the food production system under the climate change scenario, this book aims to provide comprehensive understanding about the role of biofortification to achieve nutritional security. In this book, we invited leading researchers, academicians and policymakers to document various approaches: biofortification of food grains, the Golden Rice project, traditional and novel plant breeding approaches, integration of OMICs based technologies, nanotechnology, the global action plan to achieve zero hunger, biofortification of macro- and micronutrients, genetic engineering and marker-assisted selection for biofortification, exploration of genetic diversity and nanofertilizers to mitigate the deficiency of nutrients from crops essential to food security. By exploring the science behind biofortification and its application in diverse agricultural context, we illustrated the pathways to achieve nutritional security. Also, we highlighted the climate change impact on the global food production system, especially the concept of climate smart agriculture through exploring innovative solutions by using cutting-edge technologies to ensure healthier life for everyone and sustainable crop production. The documented knowledge will not only serve as a precious resource for devising precise plant breeding efforts but also help us devise policies and technologies to enrich essential vitamins, minerals and to develop crop resilience to stress conditions posed by climate change.

In this book, we delve into the latest research and breakthroughs in biofortification. We believe this book will be of interest to agronomists, plant breeders, molecular biologists, researchers, and postdoctoral fellows working in related disciplines for developing nutritious and climate-resilient crops.

Adnan Noor Shah, PhD
Department of Agricultural Engineering, Khwaja Fareed University of Engineering and Information Technology, Punjab, Pakistan
Sajid Fiaz, PhD
Institute of Molecular Biology and Biotechnology, The University of Lahore, Lahore, Pakistan
Muhammad Aslam, PhD
Department of Plant Breeding and Genetics, Faculty of Agriculture, University of Agriculture Faisalabad, Pakistan
Javed Iqbal, PhD
Department of Botany, Bacha Khan University, Charsadda, Khyber Pakhtunkhwa, Pakistan
Abdul Qayyum, PhD
Department of Agronomy, The University of Haripur, Pakistan

1

Biofortification of Food Grains in Relation to Food Security

Ijaz Rasool Noorka[1], Muhammad Tamoor Qureshi[1], Zafar Iqbal Khan[2], Kadambot H. M. Siddique[3] and Pat (J S) Heslop Harrison[4,5,6]*

[1] *Department of Plant Breeding and Genetics, University of Sargodha, Pakistan*
[2] *Department of Botany, University of Sargodha, Pakistan*
[3] *The UWA Institute of Agriculture, The University of Western Australia, Crawley, Australia*
[4] *Key Laboratory of Plant Resources Conservation and Sustainable Utilization/Guangdong Provincial Key Laboratory of Applied Botany, South China Botanical Garden, Chinese Academy of Sciences, Guangzhou, China*
[5] *Department of Genetics and Genome Biology, University of Leicester, Institute for Environmental Futures, Leicester, United Kingdom*
[6] *South China National Botanical Garden, Chinese Academy of Sciences, Guangzhou, China*

*Corresponding author: ijazphd@yahoo.com

1.1 Introduction

1.1.1 Micronutrient Malnutrition

The current consumer interest in a healthy diet has forced the development of "super-foods" categorized by health benefiting properties and high nutrition. Undernourishment is determined by the sufficiency of energy intake and it is defined as taking less than required amount of any nutrient of energy (Khush et al. 2012). Micronutrient deficiencies and/or undernourishment hinder development of crops and human beings (Bailey, West, and Black 2015). There are multiple sources of essential mineral nutrients, among which plants are the best source of essential mineral nutrients and are vital to humans for their well-being. Some plants are rich in some minerals and deficient in others; however, no single plant contains all the essential elements important for human health (Hopkins and Hüner 2004). So, it can be concluded that plants are the best source of nutrients for human diet but unfortunately, most of the major crops lack some essential micronutrients, minerals, and vitamins.

Human beings and mammals require more than 25 mineral nutrients, while only 17 have been reported to be essential to plant metabolism for survival, development, and growth (Bouis and Welch 2010). Essential elements are divided into three categories as per their requirement for general growth and development: (1) essential elements that are required by plants and/or animals in small concentrations (micronutrients), (2) those that are required in large amounts (macronutrients), and (3) those required in trace amounts (Hirschi 2009). The term trace elements refer to the elements present in very low concentrations, generally micrograms per kilogram, but which nevertheless remain essential for the proper functioning of the body's various physiological systems (Schulze, Beck, and Hohenstein 2004). O'Dell and Sunde (1997) reported that trace elements are required by adults in the amount of 1 and 100 mg day^{-1}. Essential micronutrients such as foliates, copper, iodine, manganese, selenium, zinc, iron, cobalt, chromium, molybdenum, and vitamins are essential as nutrients for humans and are organic compounds (Bender 2009).

Micronutrient malnutrition is defined as the presence of a limited number of micronutrients that hampers the normal functioning of the body, and mineral deficiency is termed as "hidden hunger" being a major concern worldwide, especially in developing countries (Mayer, Pfeiffer, and Beyer 2008). Micronutrient malnutrition affects more than 50% of the world's population, particularly in developing countries. Malnutrition leads to an obvious increase in mortality and morbidity, poor growth and psychomotor growth in children, declines their academic performance, reduces immunity, leads to infertility, fatigue, irritability, weakness, reduces work efficiency of labors, results in loss of hair, muscle wasting, and stagnation

Crop Biofortification: Biotechnological Approaches for Achieving Nutritional Security under Changing Climate, First Edition. Edited by Adnan Noor Shah et al.

of efforts. National gentrification reduces the lifespan and lifestyle of affected people (Stein 2010). Micronutrients are essential to human life in trace quantities for development, improved nutrition, and growth of physiological and biological functions (Gibson 2007). The health and nutritional status of an individual influence the bioavailability of micronutrients found in foods intended for accumulation in the human body as well as their age, sex, race, genotype, ethnicity, and physiological state. Micronutrient malnutrition is caused by insufficient intake of nutrient-rich food, nutrition loss due to poor diet, diseases, infection, and blood loss due to injuries and during menstruation cycle of women. It is a major worldwide threat to human health in the developing countries particularly after COVID-19, as it became evident that micronutrients play a significant role in providing resistance to respiratory infection (Calder 2020). The cause of this condition is the intake characterized by poor quality of diet of staple foods, low consumption of fish, animal products, fruits, vegetables, legumes, and these types of foods contain a large quantity of essential minerals and vitamins (Bouis, Boy-Gallego, and Meenakshi 2012a).

It is estimated that a large number of people in the developing countries suffer from micronutrient deficiencies and most undernourished people are poor; they cannot afford high-quality food, meat, fish, poultry, vegetables, and fruits, foods rich in micronutrients and some countries cannot grow these quality foods themselves because of limited resources (Bouis, Boy-Gallego, and Meenakshi 2012b). Major population in the developing countries are uniquely at a risk of poor growth and development, decreased skills and ability, disease, spontaneous death due to a diet insufficient to meet their needs and is poor in essential nutrients especially in South and Southeast Asia, sub-Saharan Africa, the Caribbean, and Latin America. For micronutrient-deficient population, cereal-based food is the only source of their everyday diet in a large fraction but unfortunately characteristically deficient in essential nutrients (Cakmak et al. 2010).

Micronutrient malnutrition is a major problem to well-being of humans especially in countries with limited resources (Pfeiffer and McClafferty 2007). For humans, rice, wheat, and corn are major staple food around the world, which are significant sources of calories contributing 23%, 17%, and 10% of total calories intake by humans but they lack sufficient amounts of minerals (iron, zinc, and selenium) and vitamins (A, E, and C) and folate (White and Brown 2010).

Micronutrient malnutrition is very common in areas where the soil has low bioavailability of plant micronutrients. The relationship between the soil and food crops' nutritional status and human health is elucidated by the fact that food products of agricultural origin constitute the main source of human nutrition. The dynamic role of micronutrients is not only for human health but are also important for plant growth and development. In crop sciences, the enormous importance of micronutrients is inevitable since plants are basically dependent on nutrients as they have a profound impact on a range of plant activities. In soil, micronutrients are abundant; however, plants generally require them in trace amounts. Therefore, trace elements such as B, Cu, Fe, Mn, and Zn are considered micronutrients that plants acquire in trace amounts but play a projecting role in plant growth and development (Figure 1.1).

In plants, micronutrients play a major role in morphological, psychological, and biochemical functions including the metabolism of plants, reproductive growth, photosynthesis chlorophyll synthesis, activation of enzymes, defense mechanism of plants, and seed development. So, availability of trace elements promotes health morphological, psychological, biochemical, and metabolic functions in plant, which untimely improves plants growth and development and their absence promotes abnormality in plants. Lintschinger, Fuchs, and Moser (1997) reported that antinutritional components including phytic acid, tannic acid, and other components such as indigestible dietary fibers are the major factors causing low absorption of trace elements in crop plants. However, hydrolysis of these components during the germination process increases the absorption rate of minerals and trace elements (Huertas et al. 2022). Therefore, antinutritional factors must be minimized to counter the negative interference with the nutrient availability (Díaz-Gómez et al. 2017).

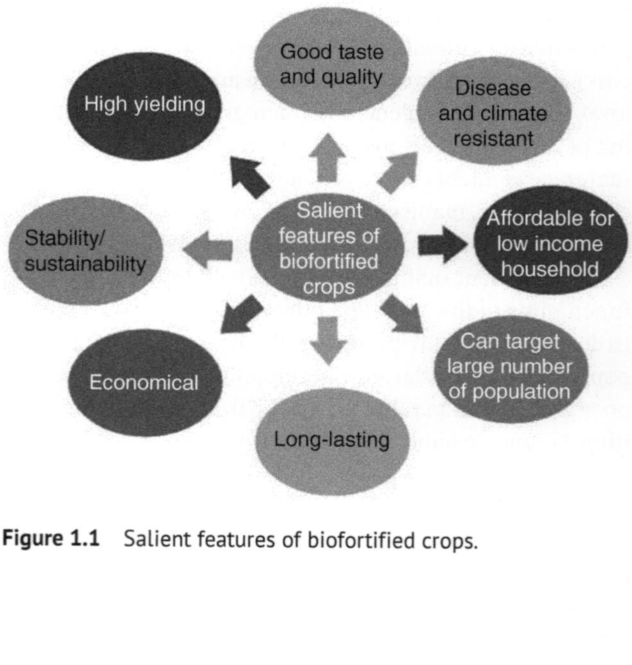

Figure 1.1 Salient features of biofortified crops.

Direct and indirect roles of environmental factors have been reported for micronutrient deficiencies, which is threatening to human health. Soil factors, including pH, humidity, soil type, temperature, high level of CO_2, flooding, high rain, and harmful effects of micronutrients affect the bioavailability of micronutrients and lead to micronutrient deficiency in plants (Neenu and Ramesh 2020). Micronutrient deficiency is a universal phenomenon that exists particularly in soils with anaerobic conditions and soil types ranging from neutral to alkaline in arid regions (Liu et al. 2014). Concentrations of micronutrients in soil vary depending upon various factors such as soil type, soil testing procedure, and soil pH. Some plants can modify the rhizosphere by secreting H^+ ions or organic acids that enhance the bioavailability and accumulation of micronutrients (Giri et al. 2017).

1.1.2 Role of Micronutrients in Humans

For human health, 22 mineral elements are essential (White and Broadley 2009). More than half of the human population is afflicted with the deficiency in iron, zinc, selenium, vitamin A, and folate. The human body cannot synthesize micronutrients, so they should be consumed through diet (Graham, Welch, and Bouis 2001).

1.1.2.1 Iron

Iron (Fe) is significant to growth of plants and humans, so it is considered one of the most critical micronutrients. Fe is the fourth most copious element in the earth crust (Zuo and Zhang 2011). Among the 10 most dominant causes of death, Fe ranks fifth and Zn is ranked sixth in the underdeveloped countries, and half of the world population is affected by insufficiencies in both Fe and Zn (Khush et al. 2012) where poor household and kindergarten children are severely affected due to the high demand for Fe. For plant development and growth, Fe is the most restrictive nutrient. Due to its physicochemical characteristics, Fe plays a key role in redox reactions and several enzymatic activities and acts as a cofactor in the human body (Sheftel, Mason, and Ponka 2012). Insufficient supply of iron may cause restricted mental growth, immune activation, disability, and diseases such as anemia, which rarely cause death (Miller 2013).

1.1.2.2 Zinc

Zinc is an essential micronutrient for both humans and plants because it is a cofactor in more than 300 enzymatic reactions, structural constituent of protein, gene expression regulation, and many biochemical pathways (White and Broadley 2009). Zn deficiency is the most ubiquitous problem in crops. Billions of people in developing countries are at high risk due to high occurrence of zinc insufficiency (Maret and Sandstead 2006). Zn deficiency causes severe immune dysfunctions, stunted and restricted growth, and diarrheal diseases (Hunt 2005); insufficient protein intake (Alloway 2009) is reported to be the cause of Zn insufficiency (Prasad 2013). So, the role of Zn in body function is important, and its deficiency leads to severe consequences. In plants, Zn plays a key role in the photosynthesis process, which is severely affected by Zn deficiency because activity of Rubisco is reduced, and ultimately photosynthesis rate is decreased. Impaired nitrogen metabolism, prolonged growth period, reduced flowering and fruit development, and decreased quality and yield are due to Zn insufficiency (Das and Green 2013).

1.1.2.3 Selenium

Selenium (Se) at very low concentrations has a significant role in humans, plants, and animals. Plants are considered to be a direct dietary source of Se. Se is an important factor in the development of the painful disease Kashin–Beck disease, which significantly affects the human health and ability to work especially in women and children (Yang, Chen, and Feng 2007). Prolonged deficiency of Se affects the cardiovascular system in humans, affects fertility in men, and decreases immunity (Malagoli, Schiavon, and Dall'Acqua 2015). In plants, Se at low concentrations creates resistance by acting as an antioxidant and a pro-oxidant against biotic and abiotic stress, such as extreme temperature, cold, salinity, drought, extreme light, and especially against heavy metals' toxicity, thus improving the plant growth, development, and yield (Feng, Wei, and Tu 2013).

1.1.2.4 Iodine

Iodine is a vital component of the human diet, and iodine deficiency poses a major health problem. The recommended daily intake of iodine is between 90 and 250 μm. Most inland soils are iodine deficient, and crop plants consist of inadequate amounts of iodine to fulfill the daily recommended intake (Gonzali, Kiferle, and Perata 2017). So insufficient intake of iodine is considered one of the leading causes of micronutrient malnutrition. In humans, low intake of I causes iodine deficiency disorders, which leads to goiter (enlarged thyroid gland) due to inadequate thyroid hormones secretion. Iodine insufficiency during pregnancy may impair the growth and neuro-development of offspring, which ultimately affects the quality of life and economic productivity of community (Vasiljev et al. 2022).

1.1.2.5 Vitamin A

Vitamins are a small group of organic compounds that are essentially required for humans. Vitamin A plays a key role in many functions of the human body including growth, development, vision, reproduction, and immune response. Insufficient intake of vitamin A is one of the major globe health risks affecting millions of people in the developing and underdeveloped countries. Vitamin A deficiency causes dry eye or xerophthalmia, night blindness, sight loss, limited growth, increased morality, and corneal ulceration (Dawson 2000) in very young children and women in later adulthood (Stevens et al. 2015). Maximizing the intake of vitamin A is the only way to overcome its deficiency.

1.1.2.6 Folates

Folates are a group of water-soluble vitamin B (vitamin B9) and are essential elements in the human diet. The recommended intake ranges from 400 to 600 μg (Rider et al. 2012). Insufficient intake of folates causes serious health issues and developmental disorders in humans such as anemia and birth defects like neural tube defects (Blancquaert et al. 2014); hence, folate deficiency is one of the major concerns particularly in children and women. Folates play a key role in amino acids, nucleotides biosynthesis and metabolism and regeneration of methionine by homocysteine because they act as donor and acceptor of C1 (Saini et al. 2016).

Naeem et al. (2021) reported that populations around the world are consuming inadequate lithium (Li) compared to daily recommended intake of 1.0 mg day^{-1}. Li has become a beneficial element for human health and it is an effective psychopharmacological agent.

1.1.3 Interventions to Overcome Deficiencies

Hidden hunger and malnutrition of micronutrients have been a serious threat to human life, thereby forcing researchers to develop interventions. To curb hidden hunger and micronutrient malnutrition, four strategies have been devised: dietary diversification, supplementation of mineral elements, food strengthening with micronutrients or food fortification, and biofortification (White and Broadley 2009).

1.1.3.1 Dietary Diversification

Human body needs essential micronutrients for growth and development, and these nutrients are derived from multiple sources including fruits, vegetables, and meat. No single food is capable of fulfilling complete nutrient requirements of vitamins and minerals for humans so a balanced and rich diet is required for adequate intake of micronutrients, and this technique is called dietary diversification (Stein et al. 2005). Basically, dietary diversification is intake of multiple types of foods containing significant amounts of both micro- and macronutrients that are essentially required for human growth and development. It helps alleviate all types of insufficiencies and improve and boost the immune system and is being a culturally suitable and acceptable method. Dietary diversification encourages to consume adequate amount of oils, fats, and vitamins to overcome undernourishment (Mene-Saffrane and Pellaud 2017). Dietary diversification is very difficult to implement in the developing countries because it is difficult to change the dietary pattern and due to lack of nutritional knowledge, and anti-nutritional factors. The major drawback of this approach is it is expensive to purchase high-quality food from diverse sources so practical implementation for a large targeted population is difficult.

1.1.3.2 Supplementation

Micronutrient supplementation is a widely practiced intervention to overcome micronutrient malnutrition of single and multiple deficiencies where high concentrations of micronutrients are consumed orally in the form of capsules, tablets, and syrups particularly in the developing countries where malnutrition is a long-term issue (Allen et al. 2006). Supplementation is the best remedy to curb malnutrition for a large population because it provides the required amount of certain nutrients in absorbable condition. Micronutrient supplementation programs are in place to overcome zinc, iron, vitamin, and calcium deficiencies among the high-risk population. In the developing countries, vitamin supplements are commonly used to overcome hidden hunger. For instance, calcium supplementation during pregnancy reduces the risk of gestational hypertension, and it can be provided in the form of tablets. The best absorbable form of iron includes ferrous gluconate, sulfate, and fumarate, while zinc can be supplied in the form of acetate, sulfate, and gluconate to overcome insufficiency. Supplementation is a direct, short-term, rapid, and controllable technique to tackle malnutrition with acute deficiency in the developing countries. Compared to dietary diversification, supplementation is an effective, quick, and cost-effective approach. Supplementation programs require a well-established network system comprised of awareness, procurement, purchase, and distribution to reach and convince rural population of highly affected. Another major drawback of supplementation is malnutrition owing to reduced intake (Allen et al. 2006). Another disadvantage of direct supplementation is development of toxicity, allergy, and vomit, which has severe effects on health of targeted population.

1.1.3.3 Food Fortification

Food fortification is accumulation of minerals and vitamins in food during processing to upsurge their nutritional value in desirable and commonly consumed foods like oil, flour, salt, and sugar. Food fortification is a cost-effective technique to overcome micronutrient deficiencies in a precise manner and supply the required amount of nutrients (Allen et al. 2006). In industrialized countries through the use of food fortification, micronutrient deficiencies are effectively controlled, for example, iodine, Fe, and vitamins A, B, and D. Food fortification examples include Fe fortification of staple food such as flour to overcome Fe deficiency, addition of folic acid in wheat flour, ionized salt, oil fortification with vitamin A, and folate-fortified rice (Olson et al. 2021). It is no doubt that food fortification is one of the best interventions to overcome hidden hunger and undernourishment, but its major limitation is that it can be implemented in the countries that have well-established resources to develop fortified food, a proper network of distribution, food dispensation equipment, and necessary technical infrastructure and this technique is less common in underdeveloped countries. It is much expensive and not available for the people living in tribal areas. The major drawback of this technique is additives' limited stability in fortified food during processing, dehusking, milling, and consumption (Gómez-Galera et al. 2010). The fortified rice during boiling losses folate, whereas in rice grains, Zn and Fe are removed as they are localized in the outer layer of rice grain (Chugh and Dhaliwal 2013). Even in foods fortified with Fe have the chances to change the taste because they are susceptible to oxidation (Blancquaert et al. 2014). So supplementation and food fortification are not recommended interventions to combat with hidden hunger especially in underdeveloped countries.

1.1.3.4 Biofortification

In contrast to food fortification in which nutrients are added postharvest processing, biofortification is the procedure to enhance the nutritional value of grown food crops by using agricultural techniques (Mayer et al. 2008). Thus, it brings agricultural and nutritional science together to minimize malnutrition and increase the level of desirable nutrients. To tackle malnutrition, biofortification is a novel and complementary approach (Brinch-Pedersen et al. 2007). Development of biofortified varieties holds great promise for well-being and health of human population. Practical examples of biofortification are zinc-biofortified wheat, maize, beans, rice, and sweet potato (Cakmak, McLaughlin, and White 2017); iron-biofortified maize, beans, rice, and sweet potato; provitamin A-biofortified Golden Rice, multivitamin maize (Carolight®), cassava, and orange sweet potato (Brigide et al. 2019). This approach's main advantages are that it is suitable for low-income families from the rural background with limited resources and often relies on subsistence farming. It is a process to improve the desirable amount of nutrients in crops using agricultural management strategies, methods of conventional plant breeding, and biotechnology and genetic engineering (Broadley et al. 2010).

Similar to dietary diversification, in biofortification there is no need to change eating habits because this process is depending on the basic foods of the population and the available food distribution system. Contrary to fortified foods, which requires proper supply chain, market system and technical management system, biofortified crops are easily accessible in the local market because they can be easily grown in any type of soil and environment in targeted areas of highly affected population. Excessive consumption of fortified foods can cause toxicity in humans but biofortified foods pose a lower risk of poisoning. So, in comparison to other intervention of micronutrient malnutrition, biofortification proves to be feasible, cost-effective as it is one-time investment to develop crop, environmentally friendly, and easily adaptive. Biofortification not only improves the nutrient value of food crops but also enhances the productivity of soil by improving the nutrient accumulation, increases germination, develop resistance to diseases, insects, pests, and crop productivity. To address the challenges of hidden hunger, biofortification holds considerable promise.

1.2 Agronomic Biofortification

Agronomic biofortification is supplementation of mineral nutrients that can be easily absorbed by crop plants through foliar application and/or spraying and by basal application of micronutrient-rich fertilizers (Msungu et al. 2022). This technique is recognized to be cheapest, quick, and simple way to increase the nutrition level of crops, and intake of these micronutrient-enriched crops improves human health by reducing mineral deficiency (de Valença and Bake 2016). This method depends on multiple factors including environment factors like soil, pH, humidity, temperature, wind, rainfall, method of fertigation, solubilization of mineral element, target nutrient, stage of application, crop type, plant psychology, and socioeconomic conditions (Sharma, Aggarwal, and Kaur 2017). Agronomic biofortification can be done by basal application and foliar application. Both methods show significant effects on various crops and improve their nutritional status. Agronomic biofortification not only improves the nutritional status of crops but also improves soil fertility and shows influence on the synthesis of beneficial compounds (Khush 2008). Supplementation of Zn, Se, and Fe has been reported to be developing a defense system against biotic stress. Productivity and performance of crop plants is highly affected by the application method, type and formulation of mineral fertilizers essentially required for growth and yield. Micronutrients like Fe, Zn, and Se are essential for human health, but the concentration range of toxic effects and beneficial impact are very narrow, so selected crop type and variety for improving nutritional status need to be studied in depth (Roohani et al. 2013). Bioavailability of micronutrients is largely determined by the formulation of fertilizer, strict selection concentrations, type of fertilizer, and combined application with other fertilizers can have neutral as well as positive and negative effects on the nutrient use efficiencies and growth of crops.

1.2.1 Types of Agronomic Biofortification

Several methods of agronomic biofortification have been attempted worldwide, but the predominant methods are seed dressing, basal and/or soil application, and foliar and/or spraying application. Coating and priming of seeds with micronutrient fertilizers are additional approaches for application of micronutrients that can stimulate plant growth and yields; however, improved nutritional level of grains is seldom reported (Duffner et al. 2014).

1.2.2 Basal Application

Basal application is a type of agronomic biofortification in which nutrient-enriched fertilizers are supplied directly to the soil to improve the micronutrient concentration in the edible parts of the crop and to improve soil fertility as well (de Valença et al. 2017). Basal or soil application aims to improve the basic intake of essential micronutrients, which are deficient in consumers (Broadley et al. 2010). Basal application is less cost-effective, low use efficiency of micronutrients may cause soil pollution (Kanwal et al. 2020).

1.2.3 Foliar Application

Supplying micronutrients by foliar application is one of the common practice used to supply the adequate concentration of the required nutrient, improve the yield, and quality of crop plants as well as nutritional status and concentration in desirable parts of plants. It can be defined as the application and/or spray of essential nutrients above the ground plant parts. This method is a ready-to-use strategy, regulates more nutrient uptake, nutrient use efficiencies, rapid recovery from observed deficiency, and allocation of desirable micronutrients in edible plant parts. The foliar method can also be used for other purposes like alleviating the drastic effects of stress including frost, heat, drought etc., application, spraying pesticides, growth regulators, and amino acids. Foliar application is the best agro technology in delivering Fe, Zn, Se, I, vitamins, and minerals. This technique is considered best suited for arid and semiarid regions where water is limited and fixation of micronutrients in the soil. Immediate uptake of the applied nutrient is the main advantage of this process because immobilization is avoided in the soil. Foliar application is less costly, less chances of fertilizer drainage or loss of micronutrient is less and environmentally friendly because micronutrients are absorbed by plant tissue. Most of the essential nutrients may be applied to the aerial part of the plant directly because they are water soluble and enter the plant through stomata opening or penetrating through cuticles. After entering the plant through leaves, nutrient via plasmodesmata moved to others portion of plants, however, few of the nutrients are not freely mobile (Smoleń 2012). Another major advantage of the foliar method is application of micro- and macrofertilizers together without any phytotoxic effect. Application of micronutrients using foliar application is reported to be best suited for cereals and leafy vegetables. Nutrient absorption depends on many factors such as the thickness of cuticle, age of leaf, permeability of plant tissue, and pH of the foliar solution. Environmental factors such as temperature, humidity, rainfall, and wind speed can also affect this process. Timing of foliar application and/or stage of crop plants is the most important factor responsible for accumulation of nutrients in edible parts of the plant. Alshaal and El-Ramady (2017) reported that Zn fortification at the reproductive stage is better compared with the vegetative stage in terms of high Zn concentration in grains. This finding was supported by Haider et al. (2020). They reported that Zn fortification at the milking stage enhances Zn concentration in grains. In contrast to basal application, uptake of essential nutrients is 8–20 times more efficient (Alshaal and El-Ramady 2017). Therefore, application of micronutrients using foliar application techniques is suggested to be simple, adaptive, less labor intensive, and more efficient. The only downside of foliar application techniques is that fertilizers can be washed out by rain (García-Bañuelos, Sida-Arreola, and Sánchez 2014).

1.2.4 Basal and Foliar Application

Basal and foliar application in combination is considered to be a more effective and better method to improve yield and nutritional status of crops compared with basal and foliar application alone (Phattarakul et al. 2012). Paramesh et al. (2020) recommended that foliar and basal applications of Zn and Fe were promising to improve Fe and Zn concentration in straw and grains. Ramzan et al. (2020) also reported similar findings and they also observed that with the increase of micronutrient status of wheat, quality attributes and yield also improved significantly using basal and foliar application. Chattha et al. (2017) revealed that soil and foliar application of Zn significantly decreased the phytic acid content in grains and improved the grain yield and biofortification. Ramkissoon et al. (2019) uncovered that Se concentration in wheat grain can be improved using agronomic biofortification, i.e. through soil and spraying of Se and with co-application with other macronutrients for better results. For effective agronomic iodine biofortification of Cassava, time of application and type of agronomic biofortification is significant (Ansa, Binang, and Shiyam 2017). Combined application of basal and foliar showed more promising results. Basal and foliar application of Fe and Zn at three different stages of maize and sorghum accessions improves Zn and Fe concentration in grains, respectively (Qureshi et al. 2020, 2021; Table 1.1).

Table 1.1 Advantages, disadvantages, and factors affecting the agronomic biofortification.

Agronomic biofortification		
Types	**Advantages**	**Disadvantages**
Soil/basal application	Improves basic intake of essential micronutrient	Less cost-effective, low use efficiency of micronutrients and with the passage of pollutes soil Maximum chances of fertilizer drainage or loss of micronutrient
Foliar/spray application	Simple, adaptive, less labor, ready-to-use strategy and more efficient Rapid recovery from observed deficiency Regulates more nutrient uptake in edible parts of plant	Can be washed out by rain
Seed dressing/priming of seed	High germination, health seedlings, increase root and shoot growth	Improved nutritional level of grains is seldom reported

Factors affecting method of agronomic biofortification		
	Factors	**References**
Soil	Soil pH, soil characteristics	
Environmental factors	Temperature, humidity, rainfall, wind speed	Ebrahimi et al. (2019)
Fertilizer	Type: organic, inorganic Formulation: granular, powder Dosage and concentration	Sago, Watanabe, and Minami (2018)
Time and stage of crop plant	Before sowing, vegetative stage, reproductive stage	Ramzan et al. (2020)
Type of plant	Crop, variety	Gomez-Coronado et al. (2016)
Type of micronutrient	Fe, Zn, Se, I, vitamins	Turner, Luo, and Buchanan (2020)
Plant morphology	Basal: root length, roots hair Foliar: cuticle thickness, ages of leaf, stomata	Ebrahimi et al. (2019)
Application method	Basal, foliar, and basal + foliar	Gomez-Coronado et al. (2016)

1.3 Conclusion

To tackle malnutrition, biofortification is a novel and complementary approach, which is environmentally friendly, economical, easily accessible, and effective. Among several types of biofortification, agronomic biofortification is highly recommended by researchers because this technique not only improves the nutritional status of crops but also improves soil fertility and exerts influence on the synthesis of beneficial compounds. Agronomic biofortification depends on the plant type (crop and variety), environmental factors (soil type, pH, temperature, humidity, and rainfall), application methods (basal, foliar, and basal + foliar), formulation (granular and powder), and concentration of mineral fertilizer, stage of application (before sowing, vegetative stage, and reproductive stage). Foliar method is an effective method compared to the basal method of application.

References

Allen, L., Benoist, B.D., Dary, O. et al. (2006). Increasing the diversity of foods consumed. In: *Guidelines on Food Fortification with Micronutrients* (eds. L. Allen, B.D. Benoist, O. Dary. et al.), 12. Geneva: World Health Organization, Food and Agriculture Organization of the United Nations.

Alloway, B.J. (2009). Soil factors associated with zinc deficiency in crops and humans. *Environmental Geochemistry and Health* 31 (5): 537–548.

Alshaal, T. and El-Ramady, H. (2017). Foliar application: from plant nutrition to biofortification. *Environment, Biodiversity and Soil Security* 1: 71–83.

Ansa, J.E.O., Binang, W.B. and Shiyam, J.O. (2017). Effect of time of iodine fertilization on effectiveness of agronomic fortification of cassava in Calabar, South-eastern Nigeria. *International Journal of Current Agricultural Sciences* 7 (3): 189–172.

Bailey, R.L., West Jr, K.P. and Black, R.E. (2015). The epidemiology of global micronutrient deficiencies. *Annals of Nutrition and Metabolism* 66 (suppl. 2): 22–33.

Bender, D.A. (2009). *Nutritional Biochemistry of the Vitamins*, 2e. NY: Cambridge University Press.

Blancquaert, D., De Steur, H., Gellynck, X. et al. (2014). Present and future of folate biofortification of crop plants. *Journal of Experimental Botany* 65 (4): 895–906.

Bouis, H.E. and Welch, R.M. (2010). Biofortification—a sustainable agricultural strategy for reducing micronutrient malnutrition in the global south. *Crop Science* 50: S1–S20.

Bouis, H., Boy-Gallego, E. and Meenakshi, J.V. (2012a). Micronutrient malnutrition: causes, prevalence, consequences, and interventions. In: *Fertilizing Crops to Improve Human Health: A Scientific Review* (ed. T.W. Bruulsema, P. Heffer, R.M. Welch. et al.). Norcross, GA: International Plant Nutrition Institute and Paris, France: International Fertilizer Industry Association.

Bouis, H., Boy-Gallego, E. and Meenakshi, J.V. (2012b). Micronutrient malnutrition: causes, prevalence, consequences, and interventions. In *Fertilizing Crops to Improve Human Health: A Scientific Review*. Volume 1: Food and Nutrition Security (ed. T.W. Bruulsema, P. Heffer, R.M. Welch, et al.). Chapter 2, 29–64. https://www.fertilizer.org/resource/fertilizing-crops-to-improve-human-health-a-scientific-review/.

Brigide, P., de Toledo, N.M.V., López-Nicolás, R. et al. (2019). Fe and Zn in vitro bioavailability in relation to antinutritional factors in biofortified beans subjected to different processes. *Food & Function* 10 (8): 4802–4810.

Brinch-Pedersen, H., Borg, S., Tauris, B. et al. (2007). Molecular genetic approaches to increasing mineral availability and vitamin content of cereals. *Journal of Cereal Science* 46 (3): 308–326.

Broadley, M.R., Alcock, J., Alford, J. et al. (2010). Selenium biofortification of high-yielding winter wheat (*Triticum aestivum* L.) by liquid or granular Se fertilisation. *Plant and Soil* 332 (1–2): 5–18.

Cakmak, I., Kalayci, M., Kaya, Y. et al. (2010). Biofortification and localization of zinc in wheat grain. *Journal of Agricultural and Food Chemistry* 58 (16): 9092–9102.

Cakmak, I., McLaughlin, M.J. and White, P. (2017). Zinc for better crop production and human health. *Plant and Soil* 411: 1–4.

Calder, P.C. (2020). Nutrition, immunity and COVID-19. *BMJ Nutrition, Prevention & Health* 3 (1): 74.

Chattha, M.U., Hassan, M.U., Khan, I. et al. (2017). Biofortification of wheat cultivars to combat zinc deficiency. *Frontiers in Plant Science* 8: 281.

Chugh, V. and Dhaliwal, H.S. (2013). Biofortification of staple crops. *Agricultural Sustainability* 177–196.

Das, S. and Green, A. (2013). Importance of zinc in crops and human health. *Journal of SAT Agricultural Research* 11: 1–7.

Dawson, M.I. (2000). The importance of vitamin A in nutrition. *Current Pharmaceutical Design*, 6 (3): 311–325.

Díaz-Gómez, J., Twyman, R.M., Zhu, C. et al. (2017). Biofortification of crops with nutrients: factors affecting utilization and storage. *Current Opinion in Biotechnology* 44: 115–123.

Duffner, A., Hoffland, E., Stomph, T.J. et al. (2014). Eliminating zinc deficiency in rice-based systems (VFRC report/Virtual Fertilizer Research Center; No. 2014/2). VFRC. https://edepot.wur.nl/299383No. 2014/2.

Ebrahimi, N., Stoddard, F.L., Hartikainen, H. et al. (2019). Plant species and growing season weather influence the efficiency of selenium biofortification. *Nutrient Cycling in Agroecosystems* 114: 111–124.

Feng, R., Wei, C. and Tu, S. (2013). The roles of selenium in protecting plants against abiotic stresses. *Environmental and Experimental Botany* 87: 58–68.

García-Bañuelos, M.L., Sida-Arreola, J.P. and Sánchez, E. (2014). Biofortification-promising approach to increasing the content of iron and zinc in staple food crops. *Journal of Elementology* 19 (3): 865–888.

Gibson, R.S. (2007). The role of diet-and host-related factors in nutrient bioavailability and thus in nutrient-based dietary requirement estimates. *Food and Nutrition Bulletin*, 28 (1_suppl 1): S77–S100.

Giri, A., Heckathorn, S., Mishra, S. et al. (2017). Heat stress decreases levels of nutrient-uptake and-assimilation proteins in tomato roots. *Plants* 6 (1): 6.

Gomez-Coronado, F., Poblaciones, M.J., Almeida, A.S. et al. (2016). Zinc (Zn) concentration of bread wheat grown under Mediterranean conditions as affected by genotype and soil/foliar Zn application. *Plant and Soil* 401: 331–346.

Gómez-Galera, S., Rojas, E., Sudhakar, D. et al. (2010). Critical evaluation of strategies for mineral fortification of staple food crops. *Transgenic Research* 19: 165–180.

Gonzali, S., Kiferle, C. and Perata, P. (2017). Iodine biofortification of crops: agronomic biofortification, metabolic engineering and iodine bioavailability. *Current Opinion in Biotechnology* 44: 16–26.

Graham, R.D., Welch, R.M. and Bouis, H.E. (2001). Addressing micronutrient malnutrition through enhancing the nutritional quality of staple foods: principles, perspectives and knowledge gaps. *Advances in Agronomy* 70: 77–142.

Haider, M.U., Hussain, M., Farooq, M. et al. (2020). Zinc nutrition for improving the productivity and grain biofortification of mungbean. *Journal of Soil Science and Plant Nutrition* 20: 1321–1335.

Hirschi, K.D. (2009). Nutrient biofortification of crops. *Annual Review of Nutrition* 29: 401–421.

Hopkins, W.G. and Hüner, N. P. (2004). *Plant Physiology*, 3e. US A: John Willy and Sons.

Huertas, R., Allwood, J.W., Hancock, R.D. et al. (2022). Iron and zinc bioavailability in common bean (*Phaseolus vulgaris*) is dependent on chemical composition and cooking method. *Food Chemistry* 387: 132900.

Hunt, J.M. (2005). The potential impact of reducing global malnutrition on poverty reduction and economic development. *Asia Pacific Journal of Clinical Nutrition* 14: 10–38.

Kanwal, A., Khan, M.B., Hussain, M. et al. (2020). Basal application of zinc to improve mung bean yield and zinc-grains biofortification. *Phyton-International Journal of Experimental Botany* 89 (1): 87–96.

Khush, G.S. (2008). Biofortification of crops for reducing malnutrition. *Plant Biotechnology Reports*.

Khush, G.S., Lee, S., Cho, J.I. and Jeon, J.S. (2012). Biofortification of crops for reducing malnutrition. *Plant Biotechnology Reports* 6: 195–202.

Lintschinger, J., Fuchs, N., Moser, H. et al. (1997). Uptake of various trace elements during germination of wheat, buckwheat and quinoa. *Plant Foods for Human Nutrition* 50: 223–237.

Liu, Z., Cheng, R., Xiao, W. et al. (2014). Effect of off-season flooding on growth, photosynthesis, carbohydrate partitioning, and nutrient uptake in Distylium chinense. *PLoS One* 9 (9): e107636.

Malagoli, M., Schiavon, M. and Dall'Acqua, S. (2015). Effects of selenium biofortification on crop nutritional quality. *Frontiers in Plant Science* 6: 280.

Maret, W. and Sandstead, H.H. (2006). Zinc requirements and the risks and benefits of zinc supplementation. *Journal of Trace Elements in Medicine and Biology* 20 (1): 3–18.

Mayer, J.E., Pfeiffer, W.H. and Beyer, P. (2008). Biofortified crops to alleviate micronutrient malnutrition. *Current Opinion in Plant Biology* 11 (2): 166–170.

Mene-Saffrane, L. and Pellaud, S. (2017). Current strategies for vitamin E biofortification of crops. *Current Opinion in Biotechnology* 44: 189–197.

Miller, J.L. (2013). Iron deficiency anemia: a common and curable disease. *Cold Spring Harbor Perspectives in Medicine* 3 (7): a011866.

Msungu, S.D., Mushongi, A.A., Venkataramana, P.B. et al. (2022). A review on the trends of maize biofortification in alleviating hidden hunger in sub-Sahara Africa. *Scientia Horticulturae* 299: 111029.

Naeem, A., Aslam, M. and Mühling, K.H. (2021). Lithium: perspectives of nutritional beneficence, dietary intake, biogeochemistry, and biofortification of vegetables and mushrooms. *Science of the Total Environment* 798: 149249.

Neenu, S. and Ramesh, K. (2020). Weather–micronutrient interactions in soil and plants—a critical review. *Chemical Science Review Letters* 9 (33): 205–219.

O'Dell, B.L. and Sunde, R.A. (1997). *Handbook of Nutritionally Essential Mineral Elements*. NY: Marcel Decker Inc.

Olson, R., Gavin-Smith, B., Ferraboschi, C. et al. (2021). Food fortification: the advantages, disadvantages and lessons from sight and life programs. *Nutrients* 13 (4): 1118.

Paramesh, V., Dhar, S., Dass, A. et al. (2020). Role of integrated nutrient management and agronomic fortification of zinc on yield, nutrient uptake and quality of wheat. *Sustainability* 12 (9): 3513.

Pfeiffer, W.H. and McClafferty, B. (2007). HarvestPlus: breeding crops for better nutrition. *Crop Science* 47: S1–88.

Phattarakul, N., Rerkasem, B., Li, L.J. et al. (2012). Biofortification of rice grain with zinc through zinc fertilization in different countries. *Plant and Soil* 361: 131–141.

Prasad, A.S. (2013). Discovery of human zinc deficiency: its impact on human health and disease. *Advances in Nutrition* 4 (2): 176–190.

Qureshi, M.T., Ahmad, M.F., Iqbal, N. et al. (2021). Agronomic bio-fortification of iron, zinc and selenium enhance growth, quality and uptake of different sorghum accessions. *Plant, Soil and Environment* 67 (10): 549–557.

Qureshi, M.T., Iqbal, N., Noorka, I.R. et al. (2020). Bio-fortification of iron and zinc improves the biomass, uptake, distribution and yield of different maize varieties. *Journal of Plant Nutrition* 44 (1): 120–129.

Ramkissoon, C., Degryse, F., da Silva, R.C. et al. (2019). Improving the efficacy of selenium fertilizers for wheat biofortification. *Scientific Reports* 9 (1): 19520.

Ramzan, Y., Hafeez, M.B., Khan, S. et al. (2020). Biofortification with zinc and iron improves the grain quality and yield of wheat crop. *International Journal of Plant Production* 14 (3): 501–510.

Rider, K.S., Yang, T.P., Berry, R.J. et al. (2012). Folate and DNA methylation: a review of molecular mechanisms and the evidence for folate's role. *Advances in Nutrition* 3 (1): 21–38.

Roohani, N., Hurrell, R., Kelishadi, R. et al. (2013). Zinc and its importance for human health: an integrative review. *Journal of Research in Medical Sciences: The Official Journal of Isfahan University of Medical Sciences* 18 (2): 144.

Sago, Y., Watanabe, N. and Minami, Y. (2018). Zinc biofortification of hydroponic baby leaf lettuce grown under artificial lighting with elevated wind speed and root zone temperature. *Journal of Agricultural Meteorology* 74 (4): 173–177.

Saini, R.K., Nile, S.H. and Keum, Y.S. (2016). Folates: chemistry, analysis, occurrence, biofortification and bioavailability. *Food Research International* 89: 1–13.

Schulze, E.D., Beck, E. and Hohenstein, M.K. (2004). *Plant Physiology*. Berlin Heidelberg New York: Springer-Verlag.

Sharma, P., Aggarwal, P. and Kaur, A. (2017). Biofortification: a new approach to eradicate hidden hunger. *Food Reviews International* 33 (1): 1–21.

Sheftel, A.D., Mason, A.B. and Ponka, P. (2012). The long history of iron in the Universe and in health and disease. *Biochimica et Biophysica Acta (BBA) – General Subjects* 1820 (3): 161–187.

Smoleń, S. (2012). Foliar nutrition: current state of knowledge and opportunities. *Advances in Citrus Nutrition* 41–58.

Stein A.J. (2010). Global impacts of human malnutrition. *Plant Soil* 335 (1): 133–154.

Stein, A.J., Meenakshi, J.V., Qaim, M. et al. (2005). *Health Benefits of Biofortification: An Ex-Ante Analysis of Iron-Rich Rice and Wheat in India*. RI, USA: Paper presented at the American Agricultural Economics Association Annual Meeting, Providence.

Stevens, G.A., Bennett, J.E., Hennocq, Q. et al. (2015). Trends and mortality effects of vitamin A deficiency in children in 138 low-income and middle-income countries between 1991 and 2013: a pooled analysis of population-based surveys. *The Lancet Global Health* 3 (9): e528–e536.

Turner, E.R., Luo, Y. and Buchanan, R.L. (2020). Microgreen nutrition, food safety, and shelf life: a review. *Journal of Food Science* 85 (4): 870–882.

de Valença, A.W. and Bake, A. (2016). Micronutrient management for improving harvests, human nutrition, and the environment. In: *Scientific Project, Assigned by Food & Business Knowledge Platform*. Netherlands, 24.

de Valença, A.W., Bake, A., Brouwer, I.D. et al. (2017). Agronomic biofortification of crops to fight hidden hunger in sub-Saharan Africa. *Global Food Security* 12: 8–14.

Vasiljev, V., Subotić, A., Marinović Glavić, M. et al. (2022). Overview of iodine intake. *Southeastern European Medical Journal: SEEMEDJ* 6 (1):12–20.

White, P.J. and Broadley, M.R. (2009). Biofortification of crops with seven mineral elements often lacking in human diets—iron, zinc, copper, calcium, magnesium, selenium and iodine. *New Phytologist* 182 (1): 49–84.

White, P.J. and Brown, P.H. (2010). Plant nutrition for sustainable development and global health. *Annals of Botany* 105: 1073–1080.

Yang, X.E., Chen, W.R. and Feng, Y. (2007). Improving human micronutrient nutrition through biofortification in the soil–plant system: China as a case study. *Environmental Geochemistry and Health* 29 (5): 413–428.

Zuo, Y. and Zhang, F. (2011). Soil and crop management strategies to prevent iron deficiency in crops. *Plant Soil* 339 (1–2): 83–95.

2

Golden Rice Project and Its Impact on Global Nutritional Security

Yunus Emre Arvas[1,2*]

[1] *Van Yuzuncu Yil University, Faculty of Science, Department of Molecular Biology and Genetics, Tusba-Van, Turkey*
[2] *Karadeniz Technical University, Faculty of Science, Department of Biology, Ortahisar Trabzon, Turkey*

*Corresponding author: yunusearvas@gmail.com

2.1 Introduction

From past to present, human beings have based their lives on activities such as nutrition, clothing, and shelter. These three activities are basically dependent on the soil and will remain dependent on the soil throughout humanity and will continue to live in a way that will be intertwined with agriculture (Dhaliwal et al. 2019; Zou et al. 2019). Factors such as the uncontrolled expansion of settlement areas due to the increasing population in the world and the increase in industrial activities cause agricultural areas to gradually decrease and damage. Approximately 97% of the world population, which is expected to exceed eight and a half billion in 2030, is expected to be in developing countries (Anonymous 2017).

According to these data, rapid population growth will be observed in the world. Rapid population growth will bring along a number of problems. It is known that the most important of these problems is the adequate, balanced, and healthy nutrition of the growing population. In order to solve the problem of feeding the ever-increasing world population, it is important to increase the amount of product obtained from unit areas rather than expanding the arable areas, which have already reached the limit.

All living organisms need microelements as well as macroelements in order to sustain their metabolic activities, and this need is met by nutrition (Welch and Graham 2004). The world population is approaching 8 billion in the first quarter of 2020, and 1 out of every 7 people in the world is struggling with hunger (Dubock 2019; Peters et al. 2024).

Biofortification utilizes modern biotechnology techniques, traditional plant breeding practices, and some agricultural practices to produce plants with high bioavailability and nutrient content. The Food and Agriculture Organization (FAO) of the United Nations reported that 792.5 million people in the world are undernourished, of which 780 million are living in developing countries (McGuire 2015; FAO 2017). However, despite increases in crop yields worldwide, it is known that approximately 2 billion people face hidden hunger due to inadequate intake of essential micronutrients (Muthayya et al. 2013; Hodge 2016; Gould 2017).

Although there are different strategies to increase the intake of microelements in the human diet, such as dietary diversification, mineral element supplements, and postharvest enrichment of plants with mineral elements, such practices are expensive and do not last long (White and Broadley 2009; Sharma et al. 2020). Biofortification is a good opportunity to increase the mineral element concentration and bioavailability of foods, especially edible parts.

This study aims to examine the Golden Rice project, which will address vitamin A deficiency (VAD) and affect approximately 250 million children and lactating women, and its impact on global food security. VAD can lead to many diseases such as death, diarrhea, measles, meningitis, lower respiratory tract infection, malaria, bitot spot, night blindness, and xerophthalmia. This project, which will directly and indirectly affect millions of people, will investigate whether the implementation of this project to prevent an important deficiency will have effects on global food security.

Crop Biofortification: Biotechnological Approaches for Achieving Nutritional Security under Changing Climate, First Edition. Edited by Adnan Noor Shah et al.

2.2 Rice

The rice plant was first domesticated thousands of years ago in China and India, making it one of the oldest cultivated plants (Poehlman and Sleper 1995). Rice, wheat, and maize are among the world's three most important crops. While maize and wheat are primarily processed for use as feed or food, rice is predominantly consumed directly without humans needing industrial processing. Rice is a monocotyledonous plant that ranks second globally in crop production and cultivation area, serving as a staple food source for more than half of the world's population (Anonymous 2017). Rice is the only cereal crop that can grow in water and benefit from dissolved oxygen through its roots. A rice grain, known as a karyopsis, consists of the endosperm (enclosed by the hull without adhering to it) and the lemma. The product with the hull intact is called paddy rice, while the product with the hull removed and subjected to polishing is referred to as rice. Unlike other cereal products, rice can be consumed at home without further processing after removing the hull. The sole purpose of polishing is to enhance its visual appeal (Turan 2013). Unlike other starchy foods, rice does not require lengthy or tedious cooking steps to become palatable. Thus, rice remains the most valuable and will continue to be the most valuable food product to support the rapidly growing world population. Rice is highly significant in human nutrition among cereals. Rice ranks second in global production volume among cereals, following wheat. Approximately 60% of the world's population's daily caloric needs are met by rice (Akay 2010; Kaya et al. 2017; Arvas et al. 2022). Rice can be cultivated in many countries worldwide, but approximately 90% of rice production occurs in Asia, where 60% of the world's population resides (Mosleh et al. 2015).

2.3 The Place of Rice (*Oryza sativa* L.) in Human Nutrition

China, India, and Indonesia are the top three rice-producing nations. Approximately 160 million hectares of land are used for rice cultivation worldwide. Approximately 11% of all cultivated land is used for rice production, which produces over 700 million tons of product yearly (Al Mamun et al. 2024). About 21% of the world's population gets their calories from rice, while for those in the Far East, rice is their primary source of 40–70% of their daily caloric intake (De Nardi et al. 2024). However, given that by 2030, there will be more than 8.5 billion people on the planet and that 4.3 billion of them will depend on rice for food (Thamarai et al. 2024), a 25% increase in rice production must be achieved while utilizing less water, land, and agricultural chemicals (Wani and Sah 2014). It has taken over 30 years to see any appreciable increase in rice yield, even with enhanced varieties and technologies. Thus, biotechnology – especially genetic engineering (GE) – offers many benefits to improve rice productivity and plant protection, allow rice to thrive in saline and drought-prone environments, and provide more nutrient-dense rice to reduce hunger (Datta et al. 2003; Ranasinghe and Mapa 2024). There are two strategies for raising rice yield. The first is to increase the area under cultivation for rice, which presents difficulties because of rising urbanization and population. The second strategy focuses on creating cultivars that boost production per hectare by combining conventional breeding techniques with biotechnology approaches. In addition to increasing productivity, biotechnological techniques successfully mitigate a range of biotic and abiotic stress factors, minimizing yield losses (Gelvin 2010; Ozawa and Takaiwa 2010; Mahmood-ur-Rahman et al. 2014; Oyebamiji et al. 2024).

Agriculture has changed with the introduction of transgenic plant production using biotechnology techniques (Şahin and Gözükırmızı 2012). Transgenic crop farming has expanded substantially since it started in 1996 and currently occupies 17 685 million hectares of planting area (ISAAA 2022). The significant increase in productivity and yield provided by transgenic crops has positively impacted farm incomes, leading to widespread acceptance among farmers. Because of its superior techniques, affordability, and ecologically friendly agricultural practices. GE technology has been applied to improve socio-economic and environmental conditions (Arvas 2017). By inserting one or more genes, it is now feasible to alter the genomes of essential plants using biotechnological techniques (Arvas and Kaya 2019). In contrast to conventional breeding techniques, these transferred features may come from different or the same species (Kaya et al. 2013b). Genetically modified organisms (GMOs) are individuals with a gene or genes combined from another organism or the same species inserted into them by recombinant DNA technology. This process results in GMOs. GMOs do not have limited gene transfer as in traditional breeding methods; the transferred gene can be from the same species or different species and genera. For instance, a gene from bacteria can be transferred to the tobacco plant to create genetically engineered tobacco (Kaya et al. 2013a). Nowadays, many genetic modification studies have been carried out on the rice plant. While some of them have been commercialized, the majority of them are still undergoing laboratory studies. Most studies focus on abiotic stress factors such as drought, salinity, herbicide, and pesticide resistance (Table 2.1). However, none of these studies had been as famous as Golden Rice obtained through biofortification. Because Golden Rice has a quality that can directly affect the lives of millions of people.

Table 2.1 Regulatory elements used in transgenic rice production.

S. no.	Plant/variety	Transformation method	Transferred structure	References
1.	Yamahoushi	Electroporation	prolC-gus-nost	Matsuki et al. (1989)
2.	Nipponbare	Electroporation	pADH1-i(ADH1)-gus-nost	Kyozuka et al. (1991)
3.	Nipponbare	Electroporation	pLHCP-gus-nost	Tada et al. (1991)
4.	Radon	Electroporation	pRTBV-gus-nost	Bhattacharyya-Pakrasi et al. (1993)
5.	Nipponbare	Electroporation	priceRBCS-gus-nost	Kyozuka et al. (1993)
6.	Nipponbare, Kinuhikari	Electroporation	pH3-gus-nost	Terada et al. (1993)
7.	Taipei 309	Electroporation	pPINII-gus-PINIIt, pPINII-i(ACT1)-gus-PINIIt	Xu et al. (1993)
8.	Kinuhikari	Electroporation	pLTP2-gus-nost	Kalla et al. (1994)
9.	Nipponbare	Electroporation	pCAB-gus-nost	Yamamoto et al. (1994)
10.	Yuukara	Electroporation	pWX-gus-nost	Hirano et al. (1995)
11.	Nipponbare, Kinuhikari, Toride-1	Electroporation	pRAMY1A-gus-nost	Itoh et al. (1995)
12.	Taipei 309	Ag	pCVMV-gus-nost	Verdaguer et al. (1996)
13.	Yamahoushi	Ag	pOSG6B-gus-nost	Yokoi et al. (1997)
14.	Kenfong	Bombardment	ABRC1-pACT1- HVA22(1) gus-PINIIIt, 4ABRC1-p ACT1-HVA22(1)-gus-PINIIIt pMtENOD12-gus-35St	Reddy et al. (1998)
15.	IRAT349	Bombardment	pPURO-B-gus-nost	Digeon et al. (1999)
16.	Taipei 309	Bombardment	pRTBV-i(RTBV)-gus-35St-p35S-i(RTBV)-gus-35St	Klöti et al. (1999)
17.	Kitaake	Ag	pricePDK-gus nost pmaizePDK-gus-nost	Nomura et al. (2000)
18.	Kitaake	Ag	pGT1-gus-HSV/his6-t	Choi et al. (2000)
19.	Nipponbare	Electroporation	pPOXA/N-gus- nost	
20.	Dongjin	Ag	pOSTUBA1-i1(OSTUBA1)-gus-nost	Ito et al. (2000)
21.	ITA212	Bombardment	pMSV-CP-gus-SIR-MSV C2 ORF	Mazithulela et al. (2000)
22.	M12	Bombardment	pUBI1-i(UBI1)- ADC-nost	Bassie et al. (2000)
23.	Nipponbare, Kinuhikari	Electroporation	pWX-i1(WX)-AsWX- i1(WX) nost, pADH1-i1(ADH1)- AsWX i1(WX)-nost	Terada et al. (2000)
24.	Kitaake	Ag	pGLB1O-gus- nost	Wu Gang et al. (2001)
25.	Nipponbare	Bombardment	pLTP1-gus-nost pCHI26-gus-nost	Hwang et al. (2002)
26.	Kitaake	Ag	pRISBZ1-gus- nost	Onodera et al. (2001)
27.	Ariete	PEG	pC1-i(MPI)-gus-nost pC1-i(MPI)-cry1B-nost	Breitler et al. (2001)
28.	Taipei 309	Bombardment	pREB-REB- REBt, pGLB-LYS-t	Yang et al. (2001)
29.	Nipponbare	Bombardment	pGLB-LYSOZYME-nost	Hwang et al. (2001)
30.	Tainung 67	Ag	pACT1u-nost,SR/ ARS-pACT13-luc-nost, (RS/G,S-p35S,luc-nost	Chen et al. (2002)
31.	ITA212	Bombardment	pUBI-aphIV-soyt, p35S-gus-nost, Rb7MAR-p35S-gus-nost-Rb7MAR, ARS1MAR-p35S-gus- nost-ARS1MAR	Vain et al. (2002)

(*Continued*)

Table 2.1 Regulatory elements used in transgenic rice production. *(continued)*

S. no.	Plant/variety	Transformation method	Transferred structure	References
32.	Nakdong	Ag	pOsCC1-sGFP-PINIIt pACT1-sGFP-PINIIt pRBCS-sGFP- PINIIt	Jang et al. (2002)
33.	Taipei 309	Bombardment	pUBI1-uORF- gus-nost	Locatelli et al. (2002)
34.	Taipei 309	Bombardment	pOLTP2-gus-t	Guiderdoni et al. (2002)
35.	Notohikari	Ag	pOSEM-gus-nost	Miyoshi et al. (2002)
36.	Taipei 309	Ag	pOSKN2-eGFP-gus-nost, pOsKN3-eGFP-gus-nost	Postma-Haarsma et al. (2002)
37.	-	Ag	pCMP-gus-nost	Stavolone et al. (2003)
38.	Taipei 309	Bombardment	pOUBI2-gus-nost	Wang and Oard (2003)
39.	Zhonghua 11	Ag	pOSBP73-gus-nost	Chen et al. (2003)
40.	Nipponbare	Ag	pMT-gus-nost	Fukuzawa et al. (2004)
41.	Bengal	Bombardment	pα-tubulin-variable-UTR- gus-nost	Fiume et al. (2004)
42.	Nipponbare	Ag	pOCATB-gus-nost	Iwamoto et al. (2004)
43.	Taipei 309	Ag	pOGIII-SP-gus-nost	Li et al. (2005)
44.	Nipponbare	Ag	pcAAT-gus-nost, pmAAT-gus-nost, pmAAT-i(AAT)-gus-nost	Nomura et al. (2005a)
45.	Nipponbare	Ag	pPCK-gus-nost, pME-gus-nost, pME-i(ME)-gus-nost	Nomura et al. (2005b)
46.	Nipponbare	Ag	pHEV1.1-gus-nost, pHEV2.1-gus-nost	Pujade-Renaud et al. (2005)
47.	Nipponbare	Ag	pONAC300-gus-nost	Kusano et al. (2005)
48.	Nipponbare, Chiyohonami	Ag	pMC8-gus-nost, pNCR-gus-nost, pPCNA-gus-nost, p35S-i(CAT)-gus-nost	Shirasawa-Seo et al. (2005)
49.	Taipei 309	Bombardment	pCDPK2-gus-nost, pCDPK2-gus- 3jUTR(CDPK2)-nost	Morello et al. (2006)
50.	Kasalath	*Agrobacterium*-mediated transformation	*CaMV35S:gus CaMV35S:sgfp CaMV35S:luc*	Saika and Toki (2010)
51.	MR219	*Agrobacterium*-mediated transformation	*CaMV35S:hpt CaMV35S:nptII*	Rahman et al. (2011)
52.	BR29, IR68899B	*Agrobacterium*-mediated transformation	*rd29:hpt*	Datta et al. (2012)
53.	BPT5204	*Agrobacterium*-mediated transformation	*CaMV35S:hpt CaMV35S:nptII*	Manimaran et al. (2013)
54.	Samba Mahsuri (BPT-5204)	*Agrobacterium*-mediated transformation	CaMV35S:rd29A CaMV35S:hpt	Ravikumar et al. (2014)
55.	MR219	*Agrobacterium*-mediated transformation	*mGFP:hptII*	Fook et al. (2015)
56.	JK1044R, JKRH401	*Agrobacterium*-mediated transformation	*CaMV35S:hpt*	Chakraborty et al. (2016)
57.	Nipponbare	*Agrobacterium*-mediated transformation	*Ubi:hpt*	Xu et al. (2017)
58.	*Citrus sinensis* L. Osbeck cv.		*pCAM1300-VvMyb, 35S*	Dutt et al. (2018)
59.	Sambha Mahsuri, Cotton Sannula	*Agrobacterium*-mediated transformation	*CaMV35S:nptII*	Reddy et al. (2019)
60.	no specific name given		pDR5:RUBY construction	He et al. (2020)
61.	Hashemi		pUhCPR, CaMV35S: hpt	Chamani Mohasses et al. (2024)

2.4 Biofortification

Biofortification is a method used to improve the microelement content of foods, which is of great importance in the healthy nutrition of the human population (Díaz-Gómez et al. 2017; Amerian et al. 2024). Since biofortification is used to increase the bioavailable microelement content in foods, it can also be referred to as "biofortification." Biofortification is a sustainable method that can be used to prevent the deficiency of essential microelements such as vitamin A in humans. Some agricultural approaches, such as VAD, breeding, and fertilization, have been identified as important tools of biofortification (Tanin et al. 2024).

Many of the nutrients that play an important role in human nutrition are of plant origin, but not all plant products are rich enough in some microelements, vitamins, etc. that are essential for human nutrition (Welch 2005; Zhu et al. 2007; Mayer et al. 2008; Sharma et al. 2024; Sonawane et al. 2024).

Cereals such as wheat, rice, and maize are an important source of energy in the human diet, accounting for about 17%, 23%, and 10% of total energy globally, respectively (Rawat et al. 2013; Sá and House 2024). However, the endosperm lacks sufficient amounts of vitamins (especially A, E, C, and folic acid) and nutrients (especially Fe, Zn, and Se) (Gómez-Galera et al. 2010; Chaudhary et al. 2024; Laryea et al. 2024). Iron and Zn deficiency affects more than 50% of the human population, resulting in poor growth and development, impaired immune system, fatigue, muscle loss, infertility, and even death (Maret and Sandstead 2006; Padoan et al. 2024; Shi et al. 2024; Zeidan et al. 2024). Zinc deficiency is as important as iron and VAD in humans, and is one of the most important health problems, ranking eleventh among risk factors worldwide (WHO 2002).

Cereals are used as a staple food stock in many countries around the world. Therefore, *Oryza sativa* L. and *Oryza glaberrima* L. genotypes are grown on a large scale for use in the nutrition of millions of people around the world and are cultivated in more than a hundred countries (WHO 2004; Mohidem et al. 2022). While 21% of people's caloric needs worldwide are met by paddy rice, it meets approximately 76% of the daily calorie needs of Far East Asians (Ma et al. 2007; Fitzgerald et al. 2009; Kaya et al. 2017).

2.4.1 Biofortification with Plant Breeding

Current rice varieties are not capable of accumulating optimum levels of beta carotene in the grain and have narrow genetic variation in vitamin A uptake. Compared to wild wheat varieties, cultivated rice varieties have genetic characteristics such as lower amounts of β-carotene intake. Traditional rice varieties do not naturally contain β-carotene, so they do not provide a direct source of vitamin A. As a result, Golden Rice has the potential to help combat VAD in populations where rice is an important dietary component. However, it is important to note that Golden Rice is not intended to replace a varied diet or other sources of vitamin A, but rather to complement existing strategies to address malnutrition. Additionally, Golden Rice is nutritionally similar to other rice varieties in terms of macronutrients such as carbohydrates, protein, and fat, except for the addition of β-carotene (Raiola et al. 2014; Sharma et al. 2020; Amerian et al. 2024).

2.4.2 Agricultural Biofortification

Agricultural biofortification is based on the application of inorganic fertilizers to increase the concentration of mineral elements in the edible parts of plants and/or to improve the movement and solubility of mineral elements in soil. By 2050, the global population is expected to exceed 9 billion. To meet the increasing food requirements of a growing population, it will be necessary to develop new crop varieties with higher yields (FAO 2024). To meet these needs, crops such as wheat, soybeans (*Glycine max*), and maize (*Zea mays*) will require a 70–100% increase in yield (Ewing et al. 2024).

Rice will require 25% more than is currently produced. There is a need to improve the quality of nutrition by maximizing the efficiency of agricultural biotechnology to respond to societal needs while maximizing the yields from arable land (Wani and Sah 2014). Golden Rice represents an important innovation in agricultural biotechnology with several potential implications and avenues for further development. Golden Rice studies have demonstrated the potential to increase the nutritional content of staple crops through VAD methods. This work has opened up new avenues for further studies and biotechnological endeavors, with many innovations and alternatives. For example, it could be extended to other crops to address various nutrient deficiencies such as iron, zinc, and vitamin E in different populations around the world. Overall,

Golden Rice is a pioneering example of how agricultural biotechnology can be used to address global challenges such as malnutrition and food insecurity, paving the way for further innovations in crop improvement and sustainable agriculture (Dubock 2019; Wu et al. 2021; IRRI 2023; 2024).

2.4.3 Genetic Biofortification

Biotechnology and VAD applications are useful options to ameliorate some of the physiological and genetic limitations that prevent mycoelement bioaccumulation in consumable plant parts such as seeds, fruits, tubers, and roots. This technology allows the expression of specific genes responsible for the expression of some plant characters specific to essential biological functions, namely, being a source of vitamin A in case of excess nutrient supply, translocation mechanism, and plant tolerance mechanism (Akhtar et al. 2019; Akhtar 2019).

β-carotene is synthesized by an enzyme called phytoene synthase. This enzyme catalyzes the condensation of two molecules of geranylgeranyl pyrophosphate to form phytoene, a precursor in the biosynthesis of carotenoids, including β-carotene. In the case of Golden Rice, genetic modification introduced a phytoene synthase gene from another plant species into the rice genome, allowing rice to produce β-carotene (Tian 2015; Zhou et al. 2022).

The phytoene synthase gene plays a crucial role in the biosynthesis of carotenoids in plants. Carotenoids are a class of pigments responsible for the yellow, orange, and red colors in many fruits, vegetables, and flowers. They also have important functions in photosynthesis, protection from light, and as precursors in the synthesis of hormones and other compounds. Specifically, phytoene synthase catalyzes the condensation of two molecules of geranylgeranyl pyrophosphate (GGPP), an isoprenoid precursor, to form phytoene. Phytoene is a colorless intermediate in the biosynthesis of carotenoids. Subsequent enzymatic steps convert phytoene into various carotenoids, including β-carotene, lycopene, and others, depending on the plant species and environmental factors (Zhou et al. 2022). The incorporation of the phytoene synthase gene from other organisms into plants engineered to produce β-carotene, such as Golden Rice, enables the synthesis of β-carotene, which has important implications in improving the nutritional quality of staple crops and addressing VAD in human populations (Sahoo et al. 2021; Sonawane et al. 2024).

2.5 Golden Rice

Golden Rice represents a novel rice variant enriched with β-carotene, a precursor to vitamin A, crucial for human health. Upon consumption, the body converts β-carotene into vitamin A, imparting the rice its distinctive golden hue (Figure 2.1). Developed through genetic research, Golden Rice emerges as a promising addition to combat VAD, aiming to address 30–50% of the average daily vitamin A requirement. Through genetic manipulation, scientists identified two novel enzymes facilitating β-carotene production within the rice grain. Cultivating Golden Rice mirrors the process of traditional rice varieties, with GR2E Golden Rice trait varieties exhibiting comparable yield and agronomic performance (Bharti et al. 2024). The β-carotene synthesized in Golden Rice grains mirrors the compound found in green leaves, various yellow vegetables, and orange fruits, as well as synthetic β-carotene formulations (Stein et al. 2006; Bin et al. 2023; IRRI 2023; 2024). Worldwide, approximately 250 million children suffer from VAD. This number corresponds to approximately one-third of preschool-age children worldwide. VAD causes xerophthalmia, making the eyes dry and more prone to infections for basic reasons. Millions of women and children lose their lives or their vision each year (Al-Babili and Beyer 2005; Potrykus 2012).

Research suggests that supplementing the inadequately consumed vitamin A could avert the deaths of 1.3–2.5 million newborns and young children. This figure exceeds the annual average of children killed by tetanus, pertussis, and measles. It is anticipated that rice with a higher β-carotene concentration will offer a remedy for this significant issue. Commercialization of Golden Rice is still pending. In the Philippines, field tests are still under progress. It is expected to be completed in the near future, and all legal and formal procedures are expected to be completed and introduced to the markets (Beyer 2010; Potrykus 2012; Tang et al. 2012).

2.5.1 Golden Rice Project

Scientists have conducted various studies on rice varieties produced worldwide due to VAD, which affects low-income communities relying on rice as a primary food source, but have failed to identify any variety rich in β-carotene content.

Figure 2.1 The genetically modified rice variety known as Golden Rice appears golden in color due to its increased β-carotene content. *Source:* International Rice Research Institute; photo licensed under CC BY 2.0.

Therefore, Golden Rice has not been developed using traditional breeding methods. Instead, scientists have enabled the production of β-carotene in rice grains by using recombinant DNA techniques, incorporating genes from corn (daffodil gene encoding phytoene synthase [psy]) and a common soil microorganism (Erwinia uredovora carotene desaturase [crtI]) (Zhou et al. 2022; IRRI 2024). Thus, the production of Golden Rice has been achieved. Due to its high content of β-carotene, Golden Rice appears golden in color. Therefore, this rice variety developed is referred to as "Golden Rice." β-carotene, the pigment responsible for the golden hue of rice and the vibrant colors of fruits and vegetables like pumpkin, papaya, and carrots, serves as a precursor to vitamin A within the body (Potrykus 2012; IRRI 2024). Tang et al. (2009) demonstrated that a modest daily intake of Golden Rice, as little as one cup (approximately 150 g of uncooked rice), fulfills around 50% of the recommended daily allowance of vitamin A for an adult individual (Tang et al. 2009; Wu et al. 2021).

2.5.2 History of the Golden Rice Project

Potrykus and Beyer, two professors, created Golden Rice. The Golden Rice prototype was created and made public in 2000 when it was published in the prestigious international scientific magazine "Science" (Ye et al. 2000). Then, a public-private collaboration was carried out on the Golden Rice research because a private corporation wanted to create and spread this new kind. Private sector scientists conducted additional laboratory tests, greenhouse experiments, and field trials to increase the β-carotene content of Golden Rice, which helped to develop the crop. This research was published in "Nature Biotechnology" in 2005 (Grusak 2005). As per the findings of this study, a new variety of Golden Rice was created, which yields roughly 23 times more β-carotene than the 2000 prototype. Because of this characteristic, it has been observed that Golden Rice has even higher β-carotene content than spinach (Tang et al. 2012). The commercial corporation then gave IRRI (International Rice Research Institutes) and other nonprofit public organizations royalty-free access to several critical technologies employed in Golden Rice, allowing them to create "Golden Rice" varieties. To further create and promote Golden Rice types and make them easily accessible to anyone seeking to work on rice research, the Golden Rice Humanitarian Board and the Golden Rice Network were later founded. Since 2006, IRRI has been developing Golden Rice in collaboration with its national and international partners, taking on the Golden Rice Network coordinator role. Given that Golden Rice is predicted to provide 30–50% of the projected average daily requirement for vitamin A, IRRI seeks to supplement current approaches (IRRI 2023; 2024).

2.5.3 Gene Transfer for Golden Rice

Through backcrossing with DNA marker genes, the gene transfer process is carried out to produce Golden Rice from local variations. The use of marker genes accelerates backcrossing. Developing new types can take two to three years instead of five to six years when this strategy is used. The plants will resemble the original genome after six generations of visual selection combined with backcrossing. The utilization of marker-supported genes in backcrossing is recommended due to its dual benefits, which include improving the accuracy and efficiency of conventional backcrossing. To start, it reduces unfavorable links from unimproved sources. Second, it expedites the entire undertaking. Therefore, after two or three backcrossings, the employment of marker-supported genes allows about 100% of the genome to be reconstructed. The same outcomes can be achieved from the genome after three backcrosses when marker-supported marker genes are used, unlike traditional breeding, where a small linkage is still dragged from the donor chromosome even after 20 backcrosses. It can take two to three years instead of five to six years to produce new kinds when marker-supported marker genes are used in breeding research. Backcrossing using marker-supported marker genes is a promising way to quickly convert local population diversity or mega-diversity into a new variety that shares traits with its parent gene. The use of DNA marker genes on rice has also contributed to creating this new product in rice. The new generation of DNA markers are called single nucleotide polymorphisms (SNPs) (Potrykus 2008; Beyer 2010; Potrykus 2012).

2.6 Malnutrition

Nutrition can generally be defined as the utilization of food. In other words, nutrition is the use of nutrients for growth, maintenance of life, and protection of health. All edible plant and animal tissues are called nutrients (food) (Tripathi 2024). Nutrients (nutrients) are the six groups of protein, carbohydrate, fat, vitamins, minerals, and water found in foods. Each of the nutrients must be taken into the body regularly every day as needed (Amawi et al. 2024; Li and Wu 2024).

Healthy nutrition is the intake of all nutrients in the amounts required for the individual and the maintenance of ideal body weight by enjoying the food eaten and eating a varied and balanced diet. Adequate and balanced nutrition is defined as adequate and balanced nutrition if each of the energy and nutrients are taken in adequate and balanced amounts and used appropriately for the growth and development of the body, regeneration, and functioning of the tissues; malnutrition is defined as inadequate nutrition if less than the required amount is taken. Unbalanced nutrition is the disruption of the ratio of nutrients taken. The goal of healthy nutrition is to ensure adequate and balanced nutrition. Inadequate nutrition problems mostly occur as "protein-energy malnutrition," whereas unbalanced nutrition is usually seen as a higher intake than required and results in obesity (Amawi et al. 2024; Tripathi 2024).

Malnutrition refers to a persistent imbalance between the intake of nutrients (protein, energy, and other nutrients) consumed and the satisfaction of changing metabolic needs. This results in loss of body mass and organ-system dysfunction. Malnutrition traditionally refers to protein and/or calorie (protein-energy) malnutrition (Hegazi et al. 2024). Sometimes it may also refer to caloric excess or vitamin toxicity, such as obesity. There are currently no generally accepted international criteria for the diagnosis of malnutrition (Santangeli et al. 2024).

In general, malnutrition leads to impaired wound healing, suppression of the immune system, decreased striated muscle mass, atrophy of the intestinal mucosa, development of generalized edema, decline in cognitive function, growth retardation in children, and a general decline in functional capacity (Boccardi and Marano 2024). Physiologic losses due to PEM involve the gastrointestinal system, cardiovascular system, immune functions, respiratory system, endocrine system, wound healing, skin, hair, kidney, and bone marrow (Padoan et al. 2024; Qin et al. 2024; Yan et al. 2024).

Biofortification should include micronutrient fortification to address malnutrition, integration of fortified crops into existing farming systems, improvement of traditional varieties to preserve their taste, appearance, and agronomic characteristics, cultural acceptance and integration into local diets, and food preferences. It also has the potential to reduce the risk of micronutrient deficiencies, leading to improved associated health outcomes such as slowed growth, impaired cognitive development, and increased susceptibility to infectious diseases. Overall, biofortification contributes to the fight against malnutrition by increasing the availability and consumption of nutrient-rich foods, promoting sustainable agricultural practices, and improving health outcomes, particularly in resource-limited settings where access to diverse diets and nutrition interventions is limited (Welch 2005; White and Broadley 2009; Rawat et al. 2013; Sharma et al. 2020; Amerian et al. 2024).

However, it is already recognized that while Golden Rice is a valuable tool to address VAD, it cannot alone provide a solution to malnutrition (Li et al. 2024; Zulfiqar et al. 2024). This project has the potential to be integrated into broader strategies to promote diversified diets, food fortification, and other interventions aimed at improving overall nutrition and food security, which will, directly and indirectly, affect millions of people, including children and lactating women, by addressing VAD (IRRI 2024). As a global project in its own right, Golden Rice could be a great stimulus for many global projects to follow and contribute to the fight against global malnutrition (Rao and Annadana 2017).

2.6.1 What About Global Malnutrition, According to the World Health Organization?

A recent World Health Organization (WHO) report highlights a staggering statistic: over 2 billion individuals worldwide suffer from deficiencies in vital vitamins and minerals, notably vitamin A, iodine, iron, and zinc (WHO 2006). This deficiency disproportionately affects populations in low-income countries, where access to nutrient-rich foods like fruits, vegetables, animal products, and fortified foods is often limited due to economic constraints or local unavailability (Newby 2018; Jumrani and Rai 2020). Such deficiencies heighten susceptibility to infectious diseases, amplifying the risks of ailments like diarrhea, measles, malaria, and pneumonia, which rank among the top 10 global health challenges. Among the most vulnerable groups are pregnant and breastfeeding women, along with young children, who have increased nutrient requirements and are particularly susceptible to the adverse effects of deficiencies. For pregnant women, these effects include elevated risks during childbirth and the potential for delivering underweight or mentally impaired infants (FAO et al. 2021). The micronutrient status of breastfeeding mothers profoundly impacts the health and development of their infants, especially during the critical first six months of life. Similarly, micronutrient deficiencies in young children heighten mortality risks from infectious diseases and impair both physical and cognitive development (WHO 2006).

One way to meet the recommended daily micronutrient intake is to provide micronutrient-enriched foods (WHO 2006). During emergencies, fortified foods like corn-soy mixtures, biscuits, vitamin A–enhanced vegetable oils, and iodized salt are frequently distributed as part of emergency relief provisions. These efforts aim to stave off or alleviate micronutrient deficiencies within affected populations. However, it's essential that such fortified foods are carefully formulated, considering that other non-fortified dietary sources may contribute to meeting certain micronutrient requirements. Despite these efforts, certain nutritionally sensitive subgroups, such as pregnant and breastfeeding women and young children, may still face challenges in obtaining adequate nutrition solely from fortified foods. In recognition of this, UNICEF and WHO have collaborated to develop a daily multi-micronutrient formula, outlined in Table 2.2, specifically tailored to meet the recommended nutrient intake (RNI) for these vulnerable groups during emergencies (WHO 2006).

2.6.2 What Has Happened Since the Golden Rice Work Began?

Ever since Potrykus and Beyer initiated the first studies of Golden Rice in 1999 to address VAD, it has faced constant criticism that has undoubtedly delayed the progress of what is undoubtedly a major humanitarian project for humanity (Wesseler and Zilberman 2014; Amna et al. 2020; Wu et al. 2021; De Steur et al. 2022). A major obstacle is thought to be the controversy surrounding transgenic studies.

Hundreds of crop varieties obtained through traditional biofortification are already on the market. But the world's first transgenic crop to be commercialized through biofortification is Golden Rice. In July 2021, the authorization to grow Golden Rice in the Philippines took this issue one step further (Birol et al. 2015; CAST 2020; Matacic 2021; De Steur et al. 2022). Golden Rice was previously declared safe for consumption in Australia, New Zealand, Canada, and the United States. It is expected to be authorized in Bangladesh in 2017 (Dubock 2019; Amna et al. 2020).

2.7 Golden Rice Project and Its Impact on Global Nutritional Security

The development of Golden Rice began in the 1990s as a joint project between the IRRI and the Swiss Federal Institute of Technology to provide a sustainable and cost-effective solution to VAD. The project involved inserting genes from maize and a common soil bacterium into the rice genome to enable the production of β-carotene (IRRI 2023; 2024).

Vitamin A, vital for human health, is exclusively sourced from dietary intake, with oily fish being its richest natural reservoir. Additionally, milk fat, eggs, and liver are notable sources of this essential nutrient. While plants themselves

Table 2.2 The composition of multiple micronutrient supplements for pregnant women, lactating women, and children from 6 to 59 months of age, designed to provide the daily recommended intake of each nutrient (one RNI) (UNICEF/UNU/WHO, 1999; FAO/WHO, 2005).

Micronutrients	Pregnant women[a]	Children (6–59 months)[a]
Vitamin A (μg)	800.0	400.0
Vitamin D (μg)	5.0	5.0
Vitamin E (mg)	15.0	5.0
Vitamin C (mg)	55.0	30.0
Thiamine (vitamin B1) (mg)	1.4	0.5
Riboflavin (vitamin B2) (mg)	1.4	0.5
Niacin (vitamin B3) (mg)	18.0	6.0
Vitamin B6 (mg)	1.9	0.5
Vitamin B12 (μg)	2.6	0.9
Folic acid (μg)	600.0	150.0
Iron (mg)	27.0[b]	10.0
Zinc (mg)	10.0	4.1
Copper (mg)	1.15[c]	0.56[c]
Selenium (μg)	30.0	17.0
Iodine (μg)	250.0[d]	90.0

a. See ref. 3.
b. See ref. 5.
c. See ref. 13.
d. See ref. 14.

don't contain vitamin A, many fruits and vegetables harbor β-carotene and similar pigments that the body converts into vitamin A. However, in low-income and developing regions, access to these nutrient-rich foods is often restricted, heightening susceptibility to diseases linked with VAD. This deficiency poses significant health risks, including developmental complications and mortality (Sonawane et al. 2024). The WHO underscores the severity of this issue, estimating that approximately 250 million individuals in developing nations suffer from VAD, with children under the age of five constituting 40% of this population (WHO/NMH/NHD/EPG 2014). Childhood blindness, largely attributable to VAD, underscores its gravity, with half of the affected children succumbing to the condition within a year of losing their sight (Zhao et al. 2022; Dubock 2024).

Golden rice, a genetically modified variant of rice (*Oryza sativa* L.), is engineered to synthesize β-carotene, a precursor to vitamin A. This pigment, also responsible for the vibrant orange hues of carrots and certain plants, is amplified within rice grains, imparting them with their distinctive golden color. Originally conceived to address prevalent vitamin A deficiencies in rice-dependent low-income nations, particularly among children, Golden Rice has sparked substantial controversy as a transgenic crop.

While genetically modified crops are powerful tools to increase yields, improve crop quality, reduce pesticide use, combat micronutrient deficiencies, and adapt crops to climate change, there is still considerable controversy over them (Arvas 2017; Arvas and Kaya 2019; Arvas and Kocacaliskan 2020; Bharti et al. 2024).

To date, the Golden Rice project has been mainly the domain of plant scientists. It is still cultivated and seed production continues in areas controlled by IRRI. In all countries, including Bangladesh, the permission of regulatory authorities is required for the commercialization of this crop and access to farmers in the countries where it will be cultivated. Once legal permission is obtained, this product, Golden Rice, will need to take its place in the existing food system. That is, it needs to be taken up by value chains and integrated into food environments and consumer diets. It is important to emphasize that this crop is a transgenic crop obtained through biofortification to end micronutrient deficiencies. Improving the micronutrient status of poor populations of developing countries is again of particular importance, as it will also contribute to stronger immune systems in general and increase the resilience of a large population to VAD, albeit with a poor dietary diversity (WHO 2006; Hodge 2016; Li et al. 2024).

The dissemination of Golden Rice is expected to be a valuable and cost-effective complementary intervention to win the fight against VAD in particular and malnutrition in general. However, this intervention is still not enough and it would be better to move the perspective forward and adopt an understanding of how to achieve sustainable and healthy diets by communicating the technological aspects of individual crops such as Golden Rice (Swamy et al. 2019; Willett et al. 2019; De Steur et al. 2022).

In order to achieve healthy diets for healthy individuals and societies worldwide, it has been suggested that the adequacy of most micronutrients in low-income countries needs to be increased by increasing dietary diversity, at least through increased consumption of fruits and vegetables, which are an essential source. In this case, by increasing sustainable dietary diversity, the need for vitamin A would also be met and there would be less need for such searches.

Moreover, even if food transfers are achieved by increasing sustainable dietary diversity, it will still not be enough to eliminate these problems fundamentally. This is because around 3 billion people globally today cannot afford the minimum cost of a healthy diet recommended by national governments (Herforth et al. 2020). While it is very good that diseases caused by dietary deficiencies such as Golden Rice are at least minimized, it is of great importance that it is delivered to consumers under very favorable economic conditions. Otherwise, the same order will continue with the current problems. It means that consumers in poor countries with no purchasing power will not be able to access it for the foreseeable future as it will exceed their disposable daily income. Interventions that increase the affordability of nutritious foods are vital until economic growth and widespread consumption of more diverse diets and healthy diets are achieved, with access to vital commodities such as Golden Rice (Fan et al. 2021).

Of course, Golden Rice will not jeopardize global food security or end poverty and dietary deficiencies on its own, but it has a valuable role to play as it will impact the lives of millions of people. Golden rice, as an ingredient rich in provitamin A, will support consumers' diets better than traditional rice, enabling access to more nutrient-rich diets and shedding light on promising further work in societies that rely on rice as a traditional and accepted staple (Custodio et al. 2023).

As a result, when Golden Rice becomes commercially available, it will need to be at least the same as other preferred rice in terms of various sensory properties such as aroma, taste, color, etc. in the consumer diet, and in this way, it can be included in the diet as a consumption preference. This is expected to be an adaptation process according to consumer culture, socio-economic status, and the environments they are exposed to. So let us ask the question clearly: what solutions will Golden Rice offer to global hunger problems now and in the future?

1. Golden Rice is a special project designed primarily to address VAD rather than hunger in general. It is important to note that malnutrition, which includes micronutrient deficiencies such as VAD, is an important component of food security, and is a major problem. While Golden Rice alone cannot solve global hunger problems, it will contribute to increasing the need for micronutrient nutrition in societies that rely heavily on rice as a staple food.
2. Golden Rice will increase vitamin A supplementation and help relieve many diseases. However, combating hunger and malnutrition requires a broad dietary program. When Golden Rice is planted in existing agricultural areas, it will become part of daily diets and perhaps the first step in a broader strategy to work towards biofortification. It contributes to strategic biofortification efforts to address VAD as well as deficiencies in other essential micronutrients such as iron, zinc, and vitamin D.
3. Golden Rice will not only be a sustainable solution to the VAD of millions of people, but they will also continue to grow a crop that already meets most of their daily nutritional needs.

2.7.1 How Can Malnutrition Be Prevented in the Future? What Should Be Done?

Preventing malnutrition undoubtedly requires a comprehensive and multifaceted approach that addresses the various factors contributing to undernutrition, overnutrition, and micronutrient deficiencies. Here are the main strategies that can be implemented to prevent malnutrition in the future.

1. Complete efforts to increase the nutrient content of staple crops and continue to improve other agricultural practices.
2. Support biofortification programs that focus on the development of crops with high levels of essential nutrients, such as vitamin A-rich crops like Golden Rice.
3. Transgenic studies to obtain vitamins, which are significantly deficient in malnutrition, from other crops should be supported, and necessary legislative work should be carried out.

In conclusion, by focusing on nutrient-rich crops, sustainable agricultural practices, and promoting agricultural diversity, it is possible to contribute significantly to global efforts to combat malnutrition. Preventing undernutrition requires a coordinated effort involving governments, civil society organizations, health professionals, communities, and individuals. By implementing a combination of nutrition-specific and nutrition-sensitive interventions, it is possible to make significant progress in preventing malnutrition and improving overall health outcomes.

2.8 Conclusion

Micronutrient deficiencies are the direct result of poor diets, where vitamins and minerals are below the threshold for maintaining a healthy life. Through biofortification, there is potential to improve the bioavailability of important nutrients in new crops. Current approaches represent a significant and sustained financial output to artificially meet the micronutrient requirements of large populations, both by adding them to food after processing and by delivering them in pill form. Through manufacturing or VAD, the need to provide nutritional supplements to large populations over a long period of time can be reduced. Previous studies have shown the manipulation of the expression levels of genes for the metabolism of provitamin A, folate, and vitamin C, resulting in an increase in these micronutrients and combining these manipulations in multivitamin maize.

The most important part of the Golden Rice project is to make it available to those who want to grow or consume it. The availability of Golden Rice in the daily diet is essential for realizing the potential health and welfare benefits.

However, because Golden Rice is a transgenic crop, it is only safe to grow and consume after official approval from the relevant country's law-making authorities that it is safe to grow and consume. If approval is obtained, only then can farmers in that country access the seed and grow the crop.

They will also want to grow it if it is as productive as the conventional rice they already grow, as we mentioned in the previous section. While Golden Rice and similar biofortified crops can contribute to the fight against malnutrition, it is important to remember that they are part of a broader set of solutions needed to comprehensively tackle hunger and food security. When Golden Rice is planted in existing agricultural areas, it will become part of daily diets and perhaps the first step in a broader strategy for biofortification efforts. It contributes to strategic biofortification efforts to address VAD as well as deficiencies in other essential micronutrients such as iron, zinc, and vitamin D. Increased efforts should also be directed towards ensuring and promoting agricultural diversification, ensuring dietary diversity, improving access to nutritious daily foods, and supporting initiatives that contribute to hunger. There should also be a focus on addressing socio-economic factors and ensuring sustainable and resilient agricultural practices. Finally, initiatives such as Golden Rice with strategic biofortification should recognize that local context, cultural acceptance, and community engagement are critical to the successful implementation of a nutrition intervention and should be strategized accordingly.

References

Akay, H. (2010). *Çeltilte (Oryza sativa L.) Farklı Somatik Eksplantlardan Kallus Oluşumunun ve Bitki Elde Etme Potansiyelinin Belirlenmesi*. Samsun: Ondokuz Mayıs Üniversitesi.

Akhtar, M., Yousaf, S., Sarwar, N. et al. (2019). Zinc biofortification of cereals – role of phosphorus and other impediments in alkaline calcareous soils. *Environmental geochemistry and health* 41 (5): 2365–2379.

Akhtar, M.S. (2019). *Salt Stress, Microbes, and Plant Interactions: Mechanisms and Molecular Approaches*. Singapore: Springer.

Al-Babili, S. and Beyer, P. (2005). Golden Rice – five years on the road – five years to go? *Trends in Plant Science* 10 (12): 565–573.

Al Mamun, M.A., Nihad, S.A.I., Sarker, M.R. et al. (2024). Spatiotemporal mapping of rice acreage and productivity growth in Bangladesh. *PLoS One* 19 (3): e0300648.

Amawi, A., AlKasasbeh, W., Jaradat, M. et al. (2024). Athletes' nutritional demands: A narrative review of nutritional requirements. *Frontiers in Nutrition* 10: 1331854.

Amerian, M., Palangi, A., Gohari, G. et al. (2024). Enhancing salinity tolerance in cucumber through Selenium biofortification and grafting. *BMC Plant Biology* 24 (1): 24.

Amna, Qamar, S., Tantray, A.Y., Bashir, S.S. et al. (2020). Golden rice: Genetic engineering, promises, present status and future prospects. *Rice Research for Quality Improvement: Genomics and Genetic Engineering: Volume 2: Nutrient Biofortification and Herbicide and Biotic Stress Resistance in Rice* 581–604.

Anonymous. (2017). *(USDA) United States of Department Agriculture. Rice Yearbook* [Online]. Available: https://www.ers.usda.gov/webdocs/DataFiles/53146/World_Rice_trade_Exports_and_imports_of_selected_countries_or_regions.xls?v=43189 (Erişim Tarihi 09.06.2018).

Arvas, Y.E. (2017). *Genetiği Değiştirilmiş Bitkiler ve Tanısı*. Germany: LAP Lambert Academic Publishing.

Arvas, Y.E. and Kaya, Y. (2019). Genetiği Değiştirilmiş Bitkilerin Biyolojik Çeşitliliğe Potansiyel Etkileri. *Yüzüncü Yıl Üniversitesi Tarım Bilimleri Dergisi* 29 (1): 168–177.

Arvas, Y.E. and Kocacaliskan, İ. (2020). Genetiği değiştirilmiş bitkilerin biyogüvenlik riskleri. *Türk doğa ve fen dergisi* 9 (2): 201–210.

Arvas, Y.E., Yolci, M.S. and Marakli, S. (2022). Investigation the effect of drought stress on tolerant and resistant rice (*Oryza sativa L.*) varieties in terms of morphologic and genetic characters. *Genetic Resources and Crop Evolution* 70 (2): 427–436. doi: 10.1007/s10722-022-01436-0.

Bassie, L., Noury, M., Lepri, O. et al. (2000). Promoter strength influences polyamine metabolism and morphogenic capacity in transgenic rice tissues expressing the oat adc cDNA constitutively. *Transgenic research* 9 (1): 33–42.

Beyer, P. (2010). Golden Rice and 'Golden' crops for human nutrition. *New Biotechnology* 27 (5): 478–481.

Bharti, U., Gaur, P., Kaur, K. et al. (2024). Tracking genetically modified (GM) rice ingredients in samples of packed rice and food products from the marketplace in India: A pilot study for regulatory compliance. *Journal of Environmental Health Science and Engineering* 22 (1): 263–269.

Bhattacharyya-Pakrasi, M., Peng, J., Elmer, J.S. et al. (1993). Specificity of a promoter from the rice tungro bacilliform virus for expression in phloem tissues. *The Plant Journal* 4 (1): 71–79.

Bin Rahman, A.R. and Zhang, J. (2023). Trends in rice research: 2030 and beyond. *Food and Energy Security* 12 (2): e390.

Birol, E., Meenakshi, J., Oparinde, A. et al. (2015). Developing country consumers' acceptance of biofortified foods: A synthesis. *Food Security* 7 (3): 555–568.

Boccardi, V. and Marano, L. (2024). Improving geriatric outcomes through nutritional and immunonutritional strategies: Focus on surgical setting by a comprehensive evidence review. *Ageing Research Reviews* 96: 102272.

Breitler, J.C., Cordero, M.J., Royer, M. et al. (2001). The− 689/+ 197 region of the maize protease inhibitor gene directs high level, wound-inducible expression of the *cry1B* gene which protects transgenic rice plants from stemborer attack. *Molecular Breeding* 7: 259–274.

CAST (2020). *Food Biofortification: Reaping the Benefits of Science to Overcome Hidden Hunger: A Paper in the Series on the Need for Agricultural Innovation to Sustainably Feed the World by 2050*. Ames: CAST Issue Paper 69.

Chakraborty, M., Sairam Reddy, P., Laxmi Narasu, M. et al. (2016). *Agrobacterium*-mediated genetic transformation of commercially elite rice restorer line using *npt II* gene as a plant selection marker. *Physiology and Molecular Biology of Plants* 22 (1): 51–60.

Chamani Mohasses, F., Mousavi Pakzad, S.M., Moatamed, E. et al. (2024). Efficient genetic transformation of rice using *Agrobacterium* with a codon-optimized chromoprotein reporter gene (*ChromoP*) and introducing an optimized iPCR method for transgene integration site detection. *Plant Cell, Tissue and Organ Culture (PCTOC)* 156 (1): 5.

Chaudhary, A., Kumari, M. and Vyas, S. (2024). Strategies to combat iron deficiency anemia among lactating women in India: A Review. *Food and Humanity* 2 (1): 100253.

Chen, P.-W., Lu, C.-A., Yu, T.-S. et al. (2002). Rice α-amylase transcriptional enhancers direct multiple mode regulation of promoters in transgenic rice. *Journal of Biological Chemistry* 277 (16): 13641–13649.

Chen, S., Jin, W., Wang, M. et al. (2003). Distribution and characterization of over 1000 T-DNA tags in rice genome. *The Plant Journal* 36 (1): 105–113.

Choi, S.-B., Wang, C., Muench, D.G. et al. (2000). Messenger RNA targeting of rice seed storage proteins to specific ER subdomains. *Nature* 407 (6805): 765–767.

Custodio, M.C., Demont, M. and De Steur, H. (2023). Market intelligence for guiding crop improvement: A systematic review of stakeholder preference studies in the rice sector in the Global South and beyond. *Comprehensive Reviews in Food Science and Food Safety* 22 (6): 4404–4432.

Datta, K., Baisakh, N., Ganguly, M. et al. (2012). Overexpression of *Arabidopsis* and rice stress genes' inducible transcription factor confers drought and salinity tolerance to rice. *Plant Biotechnology Journal* 10 (5): 579–586.

Datta, K., Baisakh, N., Oliva, N. et al. (2003). Bioengineered 'golden' indica rice cultivars with β-carotene metabolism in the endosperm with hygromycin and mannose selection systems. *Plant Biotechnology Journal* 1 (2): 81–90.

De Nardi, S., Carnevale, C., Raccagni, S. et al. (2024). Data-driven models to forecast the impact of temperature anomalies on rice production in Southeast Asia. *Forecasting* 6 (1): 100–114.

De Steur, H., Stein, A.J. and Demont, M. (2022). From Golden Rice to golden diets: How to turn its recent approval into practice. *Global Food Security* 32: 100596.

Dhaliwal, S., Ram, H., Shukla, A. et al. (2019). Zinc biofortification of bread wheat, triticale, and durum wheat cultivars by foliar zinc fertilization. *Journal of Plant Nutrition* 42 (8): 813–822.

Díaz-Gómez, J., Twyman, R.M., Zhu, C. et al. (2017). Biofortification of crops with nutrients: Factors affecting utilization and storage. *Current Opinion in Biotechnology* 44: 115–123.

Digeon, J.F., Guiderdoni, E., Alary, R. et al. (1999). Cloning of a wheat puroindoline gene promoter by IPCR and analysis of promoter regions required for tissue-specific expression in transgenic rice seeds. *Plant Molecular Biology* 39 (6): 1101–1112.

Dubock, A. (2019). Golden rice: To combat vitamin A deficiency for public health. *Vitamin A* 1.

Dubock, A. (2024). Prejudice, against GMO crops and Golden Rice, in US academia drove unethical behaviour, with global and detrimental consequences for vitamin A deficiency alleviation. *Medical Research Archives* 12 (2).

Dutt, M., Zambon, F.T., Erpen, L. et al. (2018). Embryo-specific expression of a visual reporter gene as a selection system for citrus transformation. *PLoS One* 13 (1): e0190413.

Ewing, P.M., Chim, B.K., Lehman, R.M. et al. (2024). Diversified grain rotations can be highly and reliably productive in unstable climates. *Field Crops Research* 310: 109361.

Fan, S., Headey, D., Rue, C. et al. (2021). Food systems for human and planetary health: Economic perspectives and challenges. *Annual Review of Resource Economics* 13 (1): 131–156.

FAO (2017). Statistical pocketbook world food and agriculture. Food and Agriculture Organization of the United Nations [Online]. http://www.fao.org/land-water/databases-and-software/en/ (accessed: 30 October 2018).

FAO (2024). Plant production and protection. About FAO's work on plant production and Protection [Online]. https://www.fao.org/plant-production-protection/about/en (accessed: 26 April 2024).

FAO, I. et al. (2021). *The State of Food Security and Nutrition in the World 2021. Transforming Food Systems for Food Security, Improved Nutrition and Affordable Healthy Diets for All.* Rome: FAO.

FAO/WHO. (2005). *Vitamin and Mineral Requirements in Human Nutrition*. Geneva: FAO/WHO.

Fitzgerald, M.A., McCouch, S.R. and Hall, R.D. (2009). Not just a grain of rice: The quest for quality. *Trends in Plant Science* 14 (3): 133–139.

Fiume, E., Christou, P., Gianì, S. et al. (2004). Introns are key regulatory elements of rice tubulin expression. *Planta* 218 (5): 693–703.

Fook, C.W.K., Lai, K.S., Wong, M.Y. et al. (2015). Efficient regeneration and Agrobacterium-mediated transformation protocol for recalcitrant indica rice (*Oryza sativa L.*). *Emirates Journal of Food and Agriculture* 27 (2): 837–848.

Fukuzawa, H., Yu, L.H., Umeda-Hara, C. et al. (2004). The rice metallothionein gene promoter does not direct foreign gene expression in seed endosperm. *Plant Cell Reports* 23 (4): 231–235.

Gelvin, S.B. (2010). Plant proteins involved in *Agrobacterium*-mediated genetic transformation. *Annual Review of Phytopathology* 48 (1): 45–68.

Gómez-Galera, S., Rojas, E., Sudhakar, D. et al. (2010). Critical evaluation of strategies for mineral fortification of staple food crops. *Transgenic Research* 19 (2): 165–180.

Gould, J.F. (2017). Complementary feeding, micronutrients and developmental outcomes of children. In: *Complementary Feeding: Building the Foundations for a Healthy Life*, 13–28. Basel: Karger Publishers.

Grusak, M.A. (2005). Golden rice gets a boost from maize. *Nature Biotechnology* 23 (4): 429–430.

Guiderdoni, E., Cordero, M.J., Vignols, F. et al. (2002). Inducibility by pathogen attack and developmental regulation of the rice Ltp1 gene. *Plant Molecular Biology* 49 (6): 679–695.

He, Y., Zhang, T., Sun, H. et al. (2020). A reporter for noninvasively monitoring gene expression and plant transformation. *Horticulture Research* 7 (1):152.

Hegazi, R., Miller, A. and Sauer, A. (2024). Evolution of the diagnosis of malnutrition in adults: A primer for clinicians. *Frontiers in Nutrition* 11: 1169538.

Herforth, A., Bai, Y., Venkat, A. et al. (2020). *Cost and Affordability of Healthy Diets Across and Within Countries: Background Paper for The State of Food Security and Nutrition in the World 2020. FAO Agricultural Development Economics Technical Study No. 9*. Rome: Food & Agriculture Org.

Hirano, H.Y., Tabayashi, N., Matsumura, T. et al. (1995). Tissue-dependent expression of the rice *wx+* gene promoter in transgenic rice and *Petunia*. *Plant and Cell Physiology* 36 (1): 37–44.

Hodge, J. (2016). *Hidden Hunger: Approaches to Tackling Micronutrient Deficiencies*. Washington, D.C.: International Food Policy Research Institute.

Hwang, Y.-S., Nichol, S., Nandi, S. et al. (2001). Aleurone-and embryo-specific expression of the β-glucuronidase gene controlled by the barley *Chi26* and *Ltp1* promoters in transgenic rice. *Plant Cell Reports* 20: 647–654.

Hwang, Y.-S., Yang, D., McCullar, C. et al. (2002). Analysis of the rice endosperm-specific globulin promoter in transformed rice cells. *Plant Cell Reports* 20 (9): 842–847.

IRRI. (2023). Golden rice [Online]. International Rice Research Institute (IRRI). https://www.irri.org/golden-rice-faqs (accessed 26 April 2024).

IRRI. (2024). Golden Rice FAQs [Online]. International Rice Research Institute (IRRI). https://www.irri.org/golden-rice (accessed 26 April 2024).

ISAAA. (2022). *Global Status of Commercialized Biotech/GM Crops in: Biotech Crop Adoption Surges as Economic Benefits Accumulate in 22 Years*. Ithaca, NY: The International Service for the Acquisition of Agri-biotech Applications (ISAAA).

Ito, H., Hiraga, S., Tsugawa, H. et al. (2000). Xylem-specific expression of wound-inducible rice peroxidase genes in transgenic plants. *Plant Science* 155 (1): 85–100.

Itoh, K., Yamaguchi, J., Huang, N. et al. (1995). Developmental and hormonal regulation of rice [alpha]-amylase (*RAmy1A*)-gusA fusion genes in transgenic rice seeds. *Plant Physiology* 107 (1): 25–31.

Iwamoto, M., Higo, H. and Higo, K. (2004). Strong expression of the rice catalase gene *CatB* promoter in protoplasts and roots of both a monocot and dicots. *Plant Physiology and Biochemistry* 42 (3): 241–249.

Jang, I.-C., Choi, W.-B., Lee, K.-H. et al. (2002). High-level and ubiquitous expression of the rice cytochrome c gene *OsCc1* and its promoter activity in transgenic plants provides a useful promoter for transgenesis of monocots. *Plant Physiology* 129 (4): 1473–1481.

Jumrani, J. and Rai, V.N. (2020). The unacceptable status quo: Malnutrition challenges of the developed and developing world. In: *Wheat and Barley Grain Biofortification*, 1–25. Amsterdam: Elsevier.

Kalla, R., Shimamoto, K., Potter, R. et al. (1994). The promoter of the barley aleurone-specific gene encoding a putative 7 kDa lipid transfer protein confers aleurone cell-specific expression in transgenic rice. *The Plant Journal* 6 (6): 849–860.

Kaya, Y., Kuyumcu, G., Karakütük, S. et al. (2017). Kır Çeltik Bitkisi. *Yyü Tar Bil Derg* 27 (1): 151–156.

Kaya, Y., Marakli, S., Gozikirmizi, N. et al. (2013a). Herbicide tolerance genes derived from bacteria. *The Journal of Animal and Plant Sciences* 23 (1): 85–91.

Kaya, Y., Yilmaz, S., Gozukirmizi, N. et al. (2013b). Evaluation of transgenic Nicotiana tabacum with *dehE* gene using transposon based IRAP markers. *American Journal of Plant Sciences* 4 (08): 41.

Klöti, A., Henrich, C., Bieri, S. et al. (1999). Upstream and downstream sequence elements determine the specificity of the rice tungro bacilliform virus promoter and influence RNA production after transcription initiation. *Plant Molecular Biology* 40 (2): 249–266.

Kusano, H., Asano, T., Shimada, H. et al. (2005). Molecular characterization of ONAC300, a novel NAC gene specifically expressed at early stages in various developing tissues of rice. *Molecular Genetics and Genomics* 272 (6): 616–626.

Kyozuka, J., Fujimoto, H., Izawa, T. et al. (1991). Anaerobic induction and tissue-specific expression of maize Adh1 promoter in transgenic rice plants and their progeny. *Molecular and General Genetics MGG* 228: 40–48.

Kyozuka, J., McElroy, D., Hayakawa, T. et al. (1993). Light-regulated and cell-specific expression of tomato *rbcS-gusA* and rice *rbcS-gusA* fusion genes in transgenic rice. *Plant Physiology* 102 (3): 991–1000.

Laryea, D., Mills, S.T.E., Dapuliga, C.C. et al. (2024). Millets: Food diversification and processing techniques. In: *Sustainable and Functional Foods from Plants*, 3–24. Florida: Apple Academic Press.

Li, G., Xu, X., Xing, H. et al. (2005). Insect resistance to Nilaparvata lugens and Cnaphalocrocis medinalis in transgenic indica rice and the inheritance of *gna+ sbti* transgenes. *Pest Management Science: formerly Pesticide Science* 61 (4): 390–396.

Li, J., Martin, C. and Fernie, A. (2024). Biofortification's contribution to mitigating micronutrient deficiencies. *Nature Food* 5 (1):19–27.

Li, P. and Wu, G. (2024). Characteristics of nutrition and metabolism in dogs and cats. In: *Nutrition and Metabolism of Dogs and Cats*, 55–98. Berlin: Springer.

Locatelli, F., Magnani, E., Vighi, C. et al. (2002). Inhibitory effect of myb7 uORF on downstream gene expression in homologous (rice) and heterologous (tobacco) systems. *Plant Molecular Biology* 48 (3): 309–318.

Ma, H., Chong, K. and Deng, X.W. (2007). Rice research: Past, present and future. *Journal of Integrative Plant Biology* 49 (6): 729–730.

Mahmood-ur-Rahman, M., Qasim, M., Bukhari, S.A. et al. (2014). *Bt Crops: A Sustainable Approach Towards Biotic Stress Tolerance*. Amsterdam: Academic Press.

Manimaran, P., Kumar, G.R., Reddy, M.R. et al. (2013). Infection of early and young callus tissues of indica rice BPT 5204 enhances regeneration and transformation efficiency. *Rice Science* 20 (6): 415–426.

Maret, W. and Sandstead, H.H. (2006). Zinc requirements and the risks and benefits of zinc supplementation. *Journal of Trace Elements in Medicine and Biology* 20 (1): 3–18.

Matacic, C. (2021). Golden Rice to Sprout in the Philippines. New York Ave, NW, Washington, DC 20005 USA: American Association for the Advancement of Science 1200.

Matsuki, R., Onodera, H., Yamauchi, T. et al. (1989). Tissue-specific expression of the rolC promoter of the Ri plasmid in transgenic rice plants. *Molecular and General Genetics MGG* 220: 12–16.

Mayer, J.E., Pfeiffer, W.H. and Beyer, P. (2008). Biofortified crops to alleviate micronutrient malnutrition. *Current Opinion in Plant Biology* 11 (2): 166–170.

Mazithulela, G., Sudhakar, D., Heckel, T. et al. (2000). The maize streak virus coat protein transcription unit exhibits tissue-specific expression in transgenic rice. *Plant Science* 155 (1): 21–29.

McGuire, S. (2015). FAO, IFAD, and WFP. The state of food insecurity in the world 2015: Meeting the 2015 international hunger targets: taking stock of uneven progress. Rome: FAO, 2015. *Advances in Nutrition* 6 (5): 623–624.

Miyoshi, K., Kagaya, Y., Ogawa, Y. et al. (2002). Temporal and spatial expression pattern of the *OSVP1* and *OSEM* genes during seed development in rice. *Plant and Cell Physiology* 43 (3): 307–313.

Mohidem, N.A., Hashim, N., Shamsudin, R. et al. (2022). Rice for food security: Revisiting its production, diversity, rice milling process and nutrient content. *Agriculture* 12 (6): 741.

Morello, L., Bardini, M., Cricrì, M. et al. (2006). Functional analysis of DNA sequences controlling the expression of the rice *OsCDPK2* gene. *Planta* 223: 479–491.

Mosleh, M.K., Hassan, Q.K. and Chowdhury, E.H. (2015). Application of remote sensors in mapping rice area and forecasting its production: A review. *Sensors* 15 (1): 769–791.

Muthayya, S., Rah, J.H., Sugimoto, J.D. et al. (2013). The global hidden hunger indices and maps: An advocacy tool for action. *PLoS One* 8 (6): e67860.

Newby, P. (2018). *Food and Nutrition: What Everyone Needs to Know®*. Oxford University Press.

Nomura, M., Higuchi, T., Ishida, Y. et al. (2005a). Differential expression pattern of C4 bundle sheath expression genes in rice, a C3 plant. *Plant and Cell Physiology* 46 (5): 754–761.

Nomura, M., Higuchi, T., Katayama, K. et al. (2005b). The promoter for C4-type mitochondrial aspartate aminotransferase does not direct bundle sheath-specific expression in transgenic rice plants. *Plant and Cell Physiology* 46 (5): 743–753.

Nomura, M., Sentoku, N., Nishimura, A. et al. (2000). The evolution of C4 plants: Acquisition of cis-regulatory sequences in the promoter of C4-type pyruvate, orthophosphate dikinase gene. *The Plant Journal* 22 (3): 211–221.

Onodera, Y., Suzuki, A., Wu, C.-Y. et al. (2001). A rice functional transcriptional activator, RISBZ1, responsible for endosperm-specific expression of storage protein genes through GCN4 motif. *Journal of Biological Chemistry* 276 (17): 14139–14152.

Oyebamiji, Y.O., Adigun, B.A., Shamsudin, N.A.A. et al. (2024). Recent advancements in mitigating abiotic stresses in crops. *Horticulturae* 10 (2): 156.

Ozawa, K. and Takaiwa, F. (2010). Highly efficient *Agrobacterium*-mediated transformation of suspension-cultured cell clusters of rice (*Oryza sativa L.*). *Plant Science* 179 (4): 333–337.

Padoan, F., Guarnaroli, M., Brugnara, M. et al. (2024). Role of nutrients in pediatric non-dialysis chronic kidney disease: From pathogenesis to correct supplementation. *Biomedicines* 12 (4): 911.

Peters, K., Fleuren, H. and Cruijssen, F. (2024). *The Role of Analytics in Achieving the Sustainable Development Goal of Zero Hunger*. Tilburg: Center for Economic Research.

Poehlman, J. and Sleper, D. (1995). *Breeding Field Crops*. Ames, IA: Iowa State University Press.

Postma-Haarsma, A.D., Rueb, S., Scarpella, E. et al. (2002). Developmental regulation and downstream effects of the *knox* class homeobox genes *Oskn2* and *Oskn3* from rice. *Plant Molecular Biology* 48: 423–441.

Potrykus, I. (2008). Golden rice – from idea to reality. In: *Bertebos Conference: Golden Rice and other Biofortified Food Crops for Developing Countries – Challenges and Potential*. Falkenberg, Sweden: Åke Barklund, Secretary General and Managing Director, KSLA.

Potrykus, I. (2012). Golden rice, a GMO-product for public good, and the consequences of GE-regulation. *Journal of Plant Biochemistry and Biotechnology* 21 (Suppl 1): 68–75.

Pujade-Renaud, V., Sanier, C., Cambillau, L. et al. (2005). Molecular characterization of new members of the *Hevea brasiliensis* hevein multigene family and analysis of their promoter region in rice. *Biochimica et Biophysica Acta (BBA)-Gene Structure and Expression* 1727 (3): 151–161.

Qin, Y., Pillidge, C., Harrison, B. et al. (2024). Pathways in formulating foods for the elderly. *Food Research International* 186: 114324.

Rahman, Z.A., Seman, Z.A., Basirun, N. et al. (2011). Preliminary investigations of *Agrobacterium*-mediated transformation in indica rice MR219 embryogenic callus using *gusA* gene. *African Journal of Biotechnology* 10 (40): 7805–7813.

Raiola, A., Rigano, M.M., Calafiore, R. et al. (2014). Enhancing the health-promoting effects of tomato fruit for biofortified food. *Mediators of Inflammation* 2014 (1): 139873.

Ranasinghe, P. and Mapa, M.S. (2024). A comprehensive approach to develop rice varieties to cope with climate change impacts. *YSF Thematic Publication* 2024: 128.

Rao, C.K. and Annadana, S. (2017). Nutrient biofortification of staple food crops: Technologies, products and prospects. *Phytonutritional Improvement of Crops* 113–183.

Ravikumar, G., Manimaran, P., Voleti, S. et al. (2014). Stress-inducible expression of AtDREB1A transcription factor greatly improves drought stress tolerance in transgenic indica rice. *Transgenic Research* 23: 421–439.

Rawat, N., Neelam, K., Tiwari, V.K. et al. (2013). Biofortification of cereals to overcome hidden hunger. *Plant Breeding* 132 (5): 437–445.

Reddy, P.M., Ladha, J.K., Ramos, M.C. et al. (1998). Rhizobial lipochitooligosaccharide nodulation factors activate expression of the legume early nodulin gene *ENOD12* in rice. *The Plant Journal* 14 (6): 693–702.

Reddy, S.S.S., Singh, B., Peter, A.J. et al. (2019). Genetic transformation of indica rice varieties involving *Am-SOD* gene for improved abiotic stress tolerance. *Saudi Journal of Biological Sciences* 26 (2): 294–300.

Sá, A.G. and House, J.D. (2024). Protein quality of cereals: Digestibility determination and processing impacts. *Journal of Cereal Science* 117:103892.

Sahoo, J.P., Samal, K.C. and Mishra, A.P. (2021). Provitamin-A biofortified Golden Rice Approved for commercial cultivation in Philippines: A breakthrough in fighting hidden hunger. *Biotica Research Today* 3 (7): 648–650.

Saika, H. and Toki, S. (2010). Mature seed-derived callus of the model indica rice variety Kasalath is highly competent in *Agrobacterium*-mediated transformation. *Plant Cell Reports* 29: 1351–1364.

Santangeli, E., Abbati, C., Chen, R. et al. (2024). Pathophysiological-based nutritional interventions in cirrhotic patients with sarcopenic obesity: A state-of-the-art narrative review. *Nutrients* 16 (3): 427.

Sharma, D., Ghimire, P., Bhattarai, S. et al. (2020). Biofortification of wheat: Genetic and agronomic approaches and strategies to combat iron and zinc deficiency. *International Journal of Environment, Agriculture and Biotechnology* 5 (4): 1077–1088.

Sharma, V., Ahmad, S., Mohan, C. et al. (2024). Food products of non–plant origin to combat the problem of nutritional deficiency. In: *Food Production, Diversity, and Safety Under Climate Change*, 141–151. Berlin: Springer.

Shi, Y., Hao, R., Ji, H. et al. (2024). Dietary zinc supplements: Beneficial health effects and application in food, medicine and animals. *Journal of the Science of Food and Agriculture* 104 (10): 5660–5674.

Shirasawa-Seo, N., Sano, Y., Nakamura, S. et al. (2005). The promoter of milk vetch dwarf virus component 8 confers effective gene expression in both dicot and monocot plants. *Plant Cell Reports* 24: 155–163.

Sonawane, V.N., Surana, K.R., Ahire, E.D. et al. (2024). Vitamins as nutraceuticals for human nutrition. In: *Preventive and Therapeutic Role of Vitamins as Nutraceuticals*, 281–302. Florida: Apple Academic Press.

Stavolone, L., Kononova, M., Pauli, S. et al. (2003). Cestrum yellow leaf curling virus (CmYLCV) promoter: A new strong constitutive promoter for heterologous gene expression in a wide variety of crops. *Plant Molecular Biology* 53: 703–713.

Stein, A.J., Sachdev, H. and Qaim, M. (2006). Potential impact and cost-effectiveness of Golden Rice. *Nature Biotechnology* 24 (10): 1200–1201.

Swamy, B.M., Samia, M., Boncodin, R. et al. (2019). Compositional analysis of genetically engineered GR2E Golden Rice in comparison to that of conventional rice. *Journal of Agricultural and Food Chemistry* 67 (28): 7986–7994.

Tada, Y., Sakamoto, M., Matsuoka, M. et al. (1991). Expression of a monocot LHCP promoter in transgenic rice. *The EMBO Journal* 10 (7): 1803–1808.

Tang, G., Hu, Y., Yin, S.-a. et al. (2012). β-Carotene in Golden Rice is as good as β-carotene in oil at providing vitamin A to children. *The American Journal of Clinical Nutrition* 96 (3): 658.

Tang, G., Qin, J., Dolnikowski, G.G. et al. (2009). Golden Rice is an effective source of vitamin A. *The American Journal of Clinical Nutrition* 89 (6): 1776–1783.

Tanin, M.J., Saini, D.K., Kumar, P. et al. (2024). Iron biofortification in wheat: Past, present, and future. *Current Plant Biology* 38: 100328.

Terada, R., Nakajima, M., Isshiki, M. et al. (2000). Antisense waxy genes with highly active promoters effectively suppress waxy gene expression in transgenic rice. *Plant and Cell Physiology* 41 (7): 881–888.

Terada, R., Nakayama, T., Iwabuchi, M. et al. (1993). A wheat histone H3 promoter confers cell division-dependent and independent expression of the *gus A* gene in transgenic rice plants. *The Plant Journal* 3 (2): 241–252.

Thamarai, P., Deivayanai, V., Saravanan, A. et al. (2024). Carbon mitigation in agriculture: Pioneering technologies for a sustainable food system. *Trends in Food Science & Technology* 147: 104477.

Tian, L. (2015). Carotenoids, genetically modified foods, and vitamin A nutrition. In: *Genetically Modified Organisms in Food: Production, Safety, Regulation and Public Health*, 353–360. Amsterdam: Elsevier.

Tripathi, P. (2024). *Food Nutrition in Human Science*. Noida: Blue Rose Publishers.

Turan, A. (2013). *Çimlendirilmiş Esmer Pirinç Keki Üretiminin Biyoaktif Bileşenlere Etkisi*. Yüksek Lisans Tezi: Hacettepe Üniversitesi.

UNICEF/UNU/WHO. (1999). Composition of a multi-micronutrient supplement to be used in pilot programmes among pregnant women in developing countries. New York: UNICEF/UNU/WHO.

Vain, P., James, V., Worland, B. et al. (2002). Transgene behaviour across two generations in a large random population of transgenic rice plants produced by particle bombardment. *TAG Theoretical and Applied Genetics* 105 (6): 878–889.

Verdaguer, B., de Kochko, A., Beachy, R.N. et al. (1996). Isolation and expression in transgenic tobacco and rice plants, of the cassava vein mosaic virus (CVMV) promoter. *Plant Molecular Biology* 31: 1129–1139.

Wang, J. and Oard, J. (2003). Rice ubiquitin promoters: Deletion analysis and potential usefulness in plant transformation systems. *Plant Cell Reports* 22: 129–134.

Wani, S. and Sah, S. (2014). Biotechnology and abiotic stress tolerance in rice. *Journal of Rice Research* 2: e105.

Welch, R.M. (2005). Biotechnology, biofortification, and global health. *Food and Nutrition Bulletin* 26 (4_suppl3): S304–S306.

Welch, R.M. and Graham, R.D. (2004). Breeding for micronutrients in staple food crops from a human nutrition perspective. *Journal of Experimental Botany* 55 (396): 353–364.

Wesseler, J. and Zilberman, D. (2014). The economic power of the Golden Rice opposition. *Environment and Development Economics* 19 (6): 724–742.

White, P.J. and Broadley, M.R. (2009). Biofortification of crops with seven mineral elements often lacking in human diets – iron, zinc, copper, calcium, magnesium, selenium and iodine. *New Phytologist* 182 (1): 49–84.

WHO (2002). World health report: 2002 reducing risks, promoting healthy life [Online]. Switzerland: WHO library cataloguing in publication data. https://www.who.int/publications/i/item/9241562072 (accessed 26 April 2024).

WHO (2004). Vitamin and mineral requirements in human nutrition [Online]. Bangkok, Thailand: WHO library cataloguing-in-publication data. https://iris.who.int/bitstream/handle/10665/42716/9241546123.pdf?sequence=1 (accessed 26 April 2024).

WHO (2006). Preventing and controlling micronutrient deficiencies in populations affected by an emergency. World Health Organization, World Food Programme, United Nations Children's Fund [Online]. Nutrition and Food Safety (NFS). https://www.who.int/publications/m/item/WHO-WFP-UNICEF-statement-micronutrients-deficiencies-emergency (accessed 26 April 2024).

WHO/NMH/NHD/EPG (2014). Xerophthalmia and night blindness for the assessment of clinical vitamin A deficiency in individuals and populations. Vitamin and mineral nutrition information system [Online]. Geneva: World Health Organization. http://apps.who.int/iris/bitstream/10665/133705/1/WHO_NMH_NHDEPG_14.4_eng.pdf?ua=1 (accessed 27 April 2024).

Willett, W., Rockström, J., Loken, B. et al. (2019). Food in the Anthropocene: The EAT–Lancet commission on healthy diets from sustainable food systems. *The Lancet* 393 (10170): 447–492.

Wu, F., Wesseler, J., Zilberman, D. et al. (2021). Allow Golden Rice to save lives. *Proceedings of the National Academy of Sciences* 118 (51): e2120901118.

Wu Gang, W.G., Cui HaiRui, C.H., Shu QingYao, S.Q. et al. (2001). Expression patterns of *cry1Ab* gene in progenies of Kemingdao and the resistance to striped stem borer. *Scientia Agricultura Sinica* 34 (5): 496–501.

Xu, D., McElroy, D., Thornburg, R.W. et al. (1993). Systemic induction of a potato pin2 promoter by wounding, methyl jasmonate, and abscisic acid in transgenic rice plants. *Plant Molecular Biology* 22: 573–588.

Xu, R., Qin, R., Li, H. et al. (2017). Generation of targeted mutant rice using a CRISPR-Cpf1 system. *Plant Biotechnology Journal* 15 (6): 713–717.

Yamamoto, Y., Rikiishi, S., Chang, Y.-C. et al. (1994). Quantitative estimation of aluminium toxicity in cultured tobacco cells: Correlation between aluminium uptake and growth inhibition. *Plant and Cell Physiology* 35 (4): 575–583.

Yan, H., Wang, P., Yang, F. et al. (2024). Anticancer therapy-induced adverse drug reactions in children and preventive and control measures. *Frontiers in Pharmacology* 15: 1329220.

Yang, D., Wu, L., Hwang, Y.-S. et al. (2001). Expression of the REB transcriptional activator in rice grains improves the yield of recombinant proteins whose genes are controlled by a Reb-responsive promoter. *Proceedings of the National Academy of Sciences* 98 (20): 11438–11443.

Ye, X., Al-Babili, S., Kloti, A. et al. (2000) Engineering the provitamin A (beta-carotene) biosynthetic pathway into (carotenoid-free) rice endosperm. *Science* 287 (5451), 303–305.

Yokoi, S., Tsuchiya, T., Toriyama, K. et al. (1997). Tapetum-specific expression of the Osg6B promoter-β-glucuronidase gene in transgenic rice. *Plant Cell Reports* 16: 363–367.

Zeidan, R.S., Martenson, M., Tamargo, J.A. et al. (2024). Iron homeostasis in older adults: Balancing nutritional requirements and health risks. *The Journal of Nutrition, Health and Aging* 28 (5): 100212.

Zhao, T., Liu, S., Zhang, R. et al. (2022). Global burden of vitamin A deficiency in 204 countries and territories from 1990–2019. *Nutrients* 14 (5): 950.

Zhou, X., Rao, S., Wrightstone, E. et al. (2022). Phytoene synthase: The key rate-limiting enzyme of carotenoid biosynthesis in plants. *Frontiers in Plant Science* 13: 884720.

Zhu, C., Naqvi, S., Gomez-Galera, S. et al. (2007). Transgenic strategies for the nutritional enhancement of plants. *Trends in Plant Science* 12 (12): 548–555.

Zou, C., Du, Y., Rashid, A. et al. (2019). Simultaneous biofortification of wheat with zinc, iodine, selenium, and iron through foliar treatment of a micronutrient cocktail in six countries. *Journal of Agricultural and Food Chemistry* 67 (29): 8096–8106.

Zulfiqar, U., Khokhar, A., Maqsood, M.F. et al. (2024). Genetic biofortification: Advancing crop nutrition to tackle hidden hunger. *Functional & Integrative Genomics* 24 (2): 34.

3

Biofortification of Cereals and Pulses Using New Breeding Techniques

Shumaila Ijaz[1†], Javed Iqbal[2*†], Banzeer Ahsan Abbasi[3], Zakir Ullah[4], Tabassum Yaseen[2], Ghulam Murtaza[5], Rashid Iqbal[6], Sajjad Hyder[7], Wiwiek Harsonowati[8], Sobia Kanwal[9], Tariq Mahmood[4]*

[1] *School of Biomedical Engineering, Shenzhen University Medical School, Shenzhen University, Shenzhen, China*
[2] *Department of Botany, Bacha Khan University, Charsadda, Khyber Pakhtunkhwa, Pakistan*
[3] *Department of Botany, Rawalpindi Women University, 6th Road, Satellite Town, Rawalpindi, Pakistan*
[4] *Department of Plant Sciences, Faculty of Biological Sciences, Quaid-i-Azam University Islamabad, Pakistan*
[5] *Faculty of Environmental Science and Engineering, Kunming University of Science and Technology, Kunming, China*
[6] *Department of Agronomy, Faculty of Agriculture and Environment, The Islamia University of Bahawalpur, Bahawalpur, Pakistan*
[7] *Department of Botany, GC Women University Sialkot, Pakistan*
[8] *Research Center for Horticulture, National Research and Innovation Agency (BRIN), Jl. Raya Bogor Km. 46, Cibinong, Bogor, West Java, Indonesia*
[9] *Department of Biology and Environmental Sciences, Allama Iqbal Open University, Islamabad-, Pakistan*
† *Equal contribution*

*Correspondence: shumailaijaz1996@yahoo.com; javed89qau@gmail.com

3.1 Introduction

Cereal and pulse products are basic meals in both industrialized and developing countries, which contain minerals and provide nutritional energy. Cereals supply around 50% of the world's energy terms and are typically made up of roughly 75% carbs (mostly starch) and 6–15% proteins. The availability of cereals, which amounts to 2600 million tons annually, is a main factor in determining the degree of global food security, which enables the assessment of relevance of cereals (Laskowski et al. 2019). With the growing need for crops that can endure extreme weather, such as heat stress and drought, and that can provide human beings with a balanced diet, the changing climate is placing significant pressure on agricultural output (Jamil et al. 2021; Shahzad et al. 2021). In these conditions, pulses have become an important part of the food chain and can offer a sustainable supply of lipids, proteins, and minerals. For human nutrition, pulses are an excellent source of dietary proteins, complex carbs, vitamins, and minerals. They are widely used as traditional diets in many regions of the world due to their high protein content, balanced amino acid profile, and delayed carbohydrate digestion because they provide balanced meals, proteins, and micronutrients to the general public in an economical way. If biofortified for nutritional components, pulses and grains provide a complete diet (Ahmad et al. 2022; Jha and Warkentin 2020).

In 2015, 17 sustainable development goals were established by the United Nations. SDG 3 is centered around the promotion of well-being and the assurance of healthy lives for individuals of all ages. The foundation of good health is nutrition; yet, humans cannot survive, develop, fight off disease, or lead fulfilling lives without consuming a regular and healthy diet. According to research by the European Food Safety Authority, micronutrient deficiencies result from not achieving the daily dietary reference values of several nutritional components for adults. Depending on the gender, these values vary from 8 to 11 mg for zinc, 8–18 mg for iron, and 750 mg for calcium. In a quiet pandemic, micronutrient deficiencies retard intellectual and physical development, impair the immune system over time, and can even be fatal (Ahmad et al. 2022; Gorji and Ghadiri 2020). The phenomenon referred to as hidden hunger, or micronutrient insufficiency, is widespread, affecting about 2 billion people worldwide (Muthayya et al. 2013). This deficiency increases the risk of infectious diseases and death from malaria, measles, pneumonia, and diarrhea in many low-income countries. One industrialized answer to this worrying nutritional deficit is food fortification by crop biofortification and supplementation (Majumder et al. 2019). By targeting and modifying movement routes, such as the root absorption phenomenon, conveying, remobilizing, storing, and increasing the bioavailability of the minerals, biofortification is a micronutrient-enriching technique. The four primary

Crop Biofortification: Biotechnological Approaches for Achieving Nutritional Security under Changing Climate, First Edition. Edited by Adnan Noor Shah et al.

approaches commonly used in crop biofortification are as follows: The main goals of traditional plant breeding, genetic engineering, and gene editing are to enhance cultivars in terms of micronutrient accumulation, bioavailability, and anti-nutrient suppression (Riaz et al. 2020). Agronomic biofortification, on the other hand, focuses on improving mineral solubilization and mobilization.

Originally, the goal of biofortification was to address the shortcomings and downsides of supplementing (Zulfiqar et al. 2020). Thus, direct gene transfer in elite genotypes was a contemporary weapon against the mineral deficit. Transgenic crops improved and modified the genotypes of carbohydrates, proteins, lipids, vitamins, minerals, and other secondary metabolites by using genetic engineering techniques to target specific metabolic pathways of the plant (Malik and Maqbool 2020). The primary approaches commonly used in crop biofortification are as follows: in terms of micronutrient accumulation, bioavailability, and anti-nutrient suppression, the main goals of traditional plant breeding, genetic engineering, and gene editing are cultivar improvement (Riaz et al. 2020). Agronomic biofortification, on the other hand, focuses on improving mineral solubilization and mobilization. Originally, the goal of biofortification was to address the shortcomings and downsides of supplementing (Zulfiqar et al. 2020). Thus, direct gene transfer in elite genotypes was a contemporary weapon against the mineral deficit. Transgenic crops improved and modified the genotypes of carbohydrates, lipids, vitamins, proteins, minerals, and other secondary metabolites by using genetic engineering techniques to target specific metabolic pathways of the plant. In accordance with the most recent advancements, additional regulatory elements of crops generated by NBT and future possibilities were also covered (Ijaz et al. 2024).

3.2 Malnutrition a Hidden Hunger

Globally, as the population grows, malnutrition is becoming a more serious concern. It is projected that the current 7.8 billion people on the planet will give way to 8.3 billion people by 2030. About 792.5 million people worldwide, a greater proportion of whom are in developing nations, suffer from malnutrition, according to UNFAO. Hunger and nutrient-poor food are common problems for the majority of people living in underdeveloped nations. Rising to 24 000 deaths worldwide each day is the result of hunger (Ahmad et al. 2024a; Sedeek et al. 2019). The percentage of people experiencing "hidden hunger" is about one-third. Some micro- and macronutrients, such as zinc, folic acid, iron, iodine, lysine, selenium, vitamin A, vitamin C, vitamin B12, vitamin D, and others, are either completely absent from their diets or just partially present (Malik and Maqbool 2020).

These vitamins and minerals are essential for the proper growth, development, and upkeep of the human body. Short-term deficiencies in these micronutrients don't do much damage, but long-term deficiencies can lead to a number of diseases, including scurvy (vitamin C deficiency), anemia (fertility), pellagra (niacin deficiency), rickets (vitamin D deficiency), and beriberi (vitamin B deficiency) (Figure 3.1). Anemia affects around 25% of people worldwide, with a shortage in iron accounting for 50–60% of cases. In areas of the world where rice was consumed, 50% of child mortality in the early 1920s was attributed to beriberi (Blicharz et al. 2020). A significant number of groups worldwide are at risk from rickets: 3% of people in China, 13% in Bangladesh, 25.6% in Magnolia, 13% in India, and 10% in the Middle East are afflicted (Bouillon and Antonio 2020). These data emphasize how crucial crop biofortification is to providing nutritious food to the global population and resolving these problems.

Cereals and pulses are what most people eat to meet their nutritional needs. Cereals and pulse-based foods provide around 60% of the dietary calorie needs in the underdeveloped world and 30% in the developed world (Awika 2011). The main reason of malnutrition, however, is that cereal lacks a stable nutritional profile, which includes iron, zinc, and other vital elements. The population of developing nations is very susceptible to malnourishment; over 50% of neonates do not obtain a balanced diet and experience delayed intellectual development. Malnutrition affects almost 40% of children in South Africa and Asia. One in 20 children are overweight, 1 in 14 children are wasted, and 45% of mortality in children under five years is linked to eating a diet low in nutrients. These facts are related to a baby and its nursing mother eating a diet low in nutrients (Adepoju and Allen 2019; Adeyeye et al. 2021). Half of the world's population relies mostly on white rice due to its flavor, texture, and palatability. For this reason, using brown rice is highly advised in order to get the most nutritional advantages (Saleh et al. 2019). The most vital component for health is iron (Fe), as a shortage of it can cause severe anemia. Similarly, deficiencies in β-carotene and niacin cause dementia, diarrhea, pellagra (a condition in which a person has oral sores), and infantile blindness (Majumder et al. 2019).

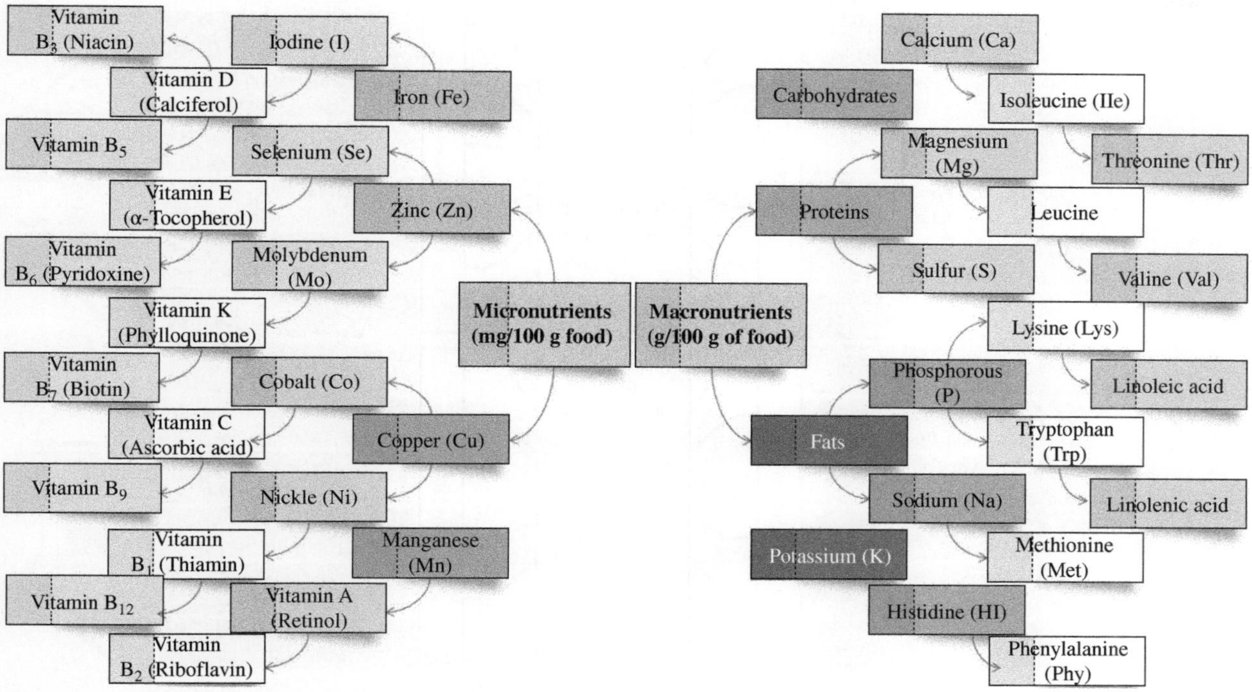

Figure 3.1 Vital macro- and micronutrients needed for optimal human health.

3.3 What Has to Be Biofortifying?

Maximum mineral absorption, their transfer to edible portions, and their bioavailability are all guaranteed by biofortification. To provide the general public with a balanced diet, it is imperative to biofortify grains and pulses with vital micronutrients (Jha and Warkentin 2020; Kumar and Pandey 2020). Several crops, including pulses and cereals, that require biofortification are briefly discussed in the section that follows.

3.3.1 Cereals

Cereal crops including wheat, rice, barley, sorghum, oats, and millets are all often grown. According to reports, rice lacks provitamin A, zinc, and iron, which can contribute to anemia, night blindness, and vision loss (Majumder et al. 2019). For wheat to effectively provide zinc, iron, grain anthocyanins, selenium, provitamin A, and vital amino acids, biofortification is required. In a similar vein, certain antinutrient compounds, such as phytic acid, are present in excessive amounts and need to be checked (Saini et al. 2020). In addition to strong antioxidants, vitamin A, anti-nutrient elements, vitamin C, and high-quality protein, maize is shown to be lacking in vitamin E (tocopherol, tocotrienol) (Prasanna et al. 2020). Enhancing the tryptophan and lysine levels is necessary to increase the nutritional quality of essential zein proteins. Lysine and tryptophan concentrations in maize may rise in the opaque-2 (o2) mutant type. Compared to wild-type variations, o2 allele-carrying cultivars had comparatively higher tryptophan and lysine levels (Sultana et al. 2019; Grover et al. 2020). Barley is also said to be lacking in micronutrients such as zinc, iron, polyunsaturated fatty acids, stearidonic acid, and the health-promoting γ-linolenic acid. The phytase gene (*HvPAPhy a*) in barley seeds is in charge of the increased phytase activity. The bioavailability of iron and zinc is improved by this elevated phytase activity. The β-glucan concentration rises as a result of cellulose synthase-like gene (HvClF) overexpression. These are the dietary fibers that are most responsible for lowering the risk of cardiovascular disease and type II diabetes (Ahmad et al. 2022). Along with iodine, selenium, magnesium, calcium, iron, and zinc deficiencies, sorghum is also said to be low in these elements (Debelo et al. 2020).

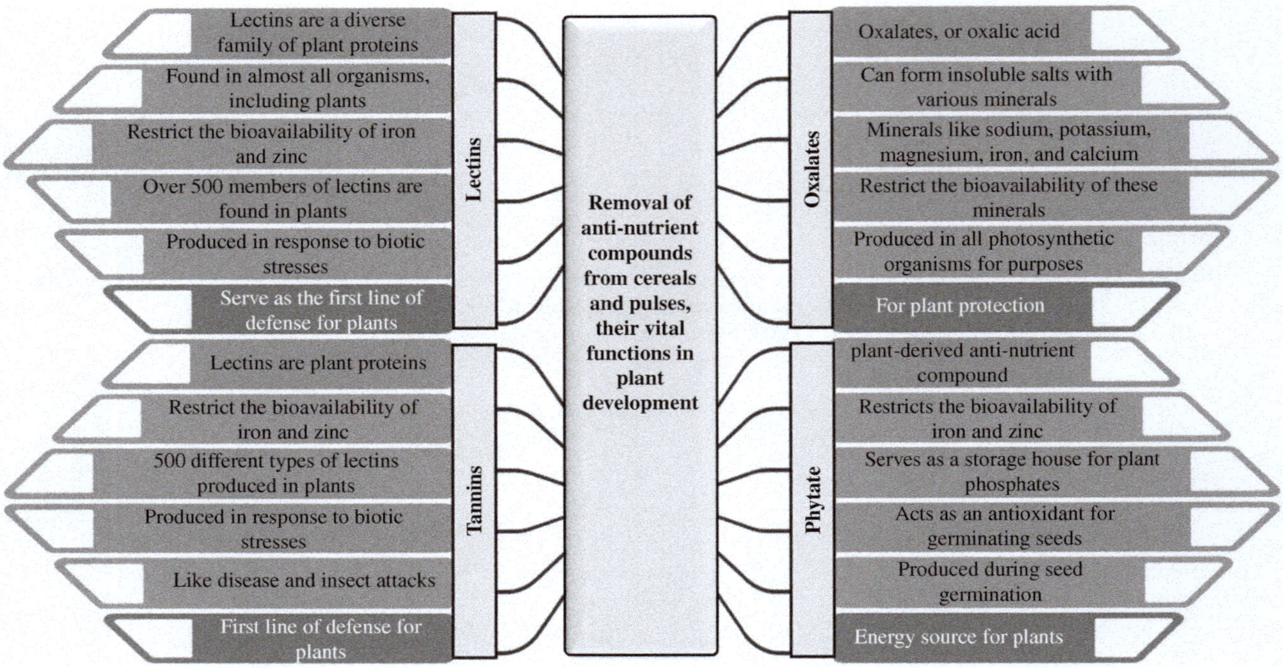

Figure 3.2 Elimination of anti-nutrient substances from pulses and grains, which are essential for plant growth.

3.3.2 Pulses

Only 20 of the approximately 1000 pulses known to science are grown for human consumption. Some of these include pigeon pea (*Cajanus cajan* L.), cowpea (*Vigna unguiculata*), chickpea (*Cicer arietinum* L.), mung bean (*Vigna radiata* L.), urban (*Vigna mungo* L.), French bean (*Phaseolus vulgaris*), moth bean (*Vigna aconitifolia*), horse gram (*Macrotyloma uniflorum*), lentil (*Lens culinaris Medik*), soybeans (*Glycine max*), field pea (*Pisum sativum* L.), and Lathyrus (*Lathyrus sativus* L.). Similar to most cereals, pulses are low in zinc and iron (Kumar and Pandey 2020). They are devoid of dietary fiber, selenium, and amino acids that include sulfur, such as methionine and cysteine (Kaur et al. 2020). While pulses are abundant in micronutrients and proteins, they also contain certain anti-nutrients that limit the bioavailability of these substances. Iron, magnesium, calcium, zinc, and other essential minerals are also not absorbed because of these anti-nutrients (Sandarani and Kulathunga 2019). Understanding the metabolic route involved in the production of these substances and the disruption of crucial genes responsible for their decreased synthesis is essential (Figure 3.2).

3.4 Methods to Address Hunger

A range of direct and indirect methods are used to increase the concentration of vital minerals and vitamins in grains and other plant components in order to fight malnutrition (Garg et al. 2018). Malnutrition is combated via dietary variety, artificial supplementation, and biofortification. To lessen this nutrient shortage, biofortification is a practical, long-term, and affordable option. Biofortified crops include all the vitamins, minerals, and fatty acids required for an individual to achieve their nutritional requirements (Ahmad et al. 2024b; Haider et al. 2022b; Sanjeeva Rao et al. 2020). Biofortified crops are also environmentally benign as they promote soil health, increase nutrient absorption from the soil, and are sustainable (Waters and Sankaran 2011; Singla and Grover 2017). Various agronomic, transgenic, traditional breeding, and gene editing techniques are employed for biofortification, as outlined in the following sections.

3.4.1 Conventional Approaches

Agronomic biofortification is a cheap and simple technique, but it involves careful consideration of the type of nutrients to be used, how they are applied, and the surrounding conditions. This method concentrates on improving the nutrients' availability to the plants, their efficient uptake and movement within the plants, and a rise in microbial activity to support this uptake. To improve nutrient availability and absorption, microbes such as rhizobium, azotobacter, bacillus, and some fungal strains, such as *Piriformospora indica*, are used (Grover et al. 2020). Mineral nutrients have great potential for biofortification through foliar and soil application. Based on nitrogen, phosphorus, and potassium, which are vital for human as well as plant health, fertilizer is most commonly utilized. Crops also require a number of additional micro-minerals, including iron, copper, zinc, iodine, manganese, nickel, and molybdenum.

Most of these micronutrients are readily taken by plants and integrated into the food chain in the soil, where they are present in abundance. However, there are instances when they are provided as mineral nourishment to plants since they are unable to absorb them (Kaur et al. 2020a). Conventional breeding, aside from agronomic biofortification, depends on the genetic diversity of the trait of interest within the gene pool (Garg et al. 2018). Using traditional crossing approaches, the required genes are pyramided, after which segregated populations undergo extensive screening. Biofortified wheat cultivars with increased iron and zinc concentrations, called "Zincol" and "Akbar-2019," were introduced by the Wheat Research Institute (WRI), Faisalabad, and the National Agriculture Research Council (NARC), Islamabad, Pakistan, in 2015 and 2019, respectively (Rashid et al. 2020; Lowe et al. 2018). Essential micronutrients come from the crucial genes carried by crop wild relatives (CWRs). To achieve traditional nutritional concentrations and protein demands, genetic diversity from CWRs will inevitably make its way into current crops (Kumar et al. 2016). The most successful example of wild hybridization is the chromosomal replacement method used to introduce the Gpc-B1 locus from wild emmer into bread wheat. This introgression proved to be highly successful since there was no yield penalty and wheat crops with an improved nutritional profile (Mn, Fe, and Zn) were generated (Engels and Thormann 2020).

3.4.2 HarvestPlus Program

It is important to highlight the role played by the HarvestPlus initiative, which aims to eradicate hidden hunger and micronutrient deficiencies worldwide by collaborating with agriculture and nutrition. It aims to manage deficits of micronutrients by giving a food-based solution to the most disadvantaged groups. The World Health Organization (WHO) has identified zinc, iron, and vitamin A, as the three minerals that are most deficient in diets worldwide. WHO is collaborating with its partners to create agricultural types that are high in these elements (Fatiukha et al. 2020). It is aiming to use conventional breeding techniques to create nutritionally enhanced crops that might be supported by evidence, affordable, and sustainable. By 2030, the program hopes to provide a supply of biofortified crops to 1 billion people.

At the moment, HarvestPlus is focused on the following areas: Nigeria, Rwanda, Uganda, Zambia, Zimbabwe, the Democratic Republic of the Congo, Bangladesh, India, Nigeria, Latin America, and the Caribbean. The program was split into three stages: a feasibility study, a pilot project to supply provitamin-enhanced sweet potatoes in Africa, and a discovery phase (2003–2008) that focused on identifying high-risk people, their dietary patterns, and plant breeding resources. During the development phase (2009–2013), in order to provide biofortified crops, researchers established country-specific teams after evaluating the nutritional profile of produced varieties. In order to reach the largest population, the delivery phase (starting in 2014) was focused on creating consumer desires for biofortified crops. Researchers were attempting to ensure the long-term viability of biofortified crops by estimating the area covered by them (Ijaz et al. 2023; Bouis and Saltzman 2017).

3.5 New Breeding Techniques

Genome editing (GE), transgenic breeding, and RNA interference (RNAi) are examples of innovative breeding techniques that are significantly contributing to biofortification of pulses and cereals crops by providing new pathways for the emergence of unique genetic diversity not found in the gene pool (Van Der Straeten et al. 2020).

3.5.1 Transgenic Breeding

3.5.1.1 Cereals

Combating malnutrition may be done effectively, sustainably, and economically by using transgenic breeding to biofortified food. Transgenic crops are produced when there is any genetic heterogeneity in the gene pool. Transgenic breeding facilitates biofortification by adding genes that either decrease anti-nutrient compounds or increase micronutrient accumulation and bioavailability (Figure 3.3) (Malik and Maqbool 2020).

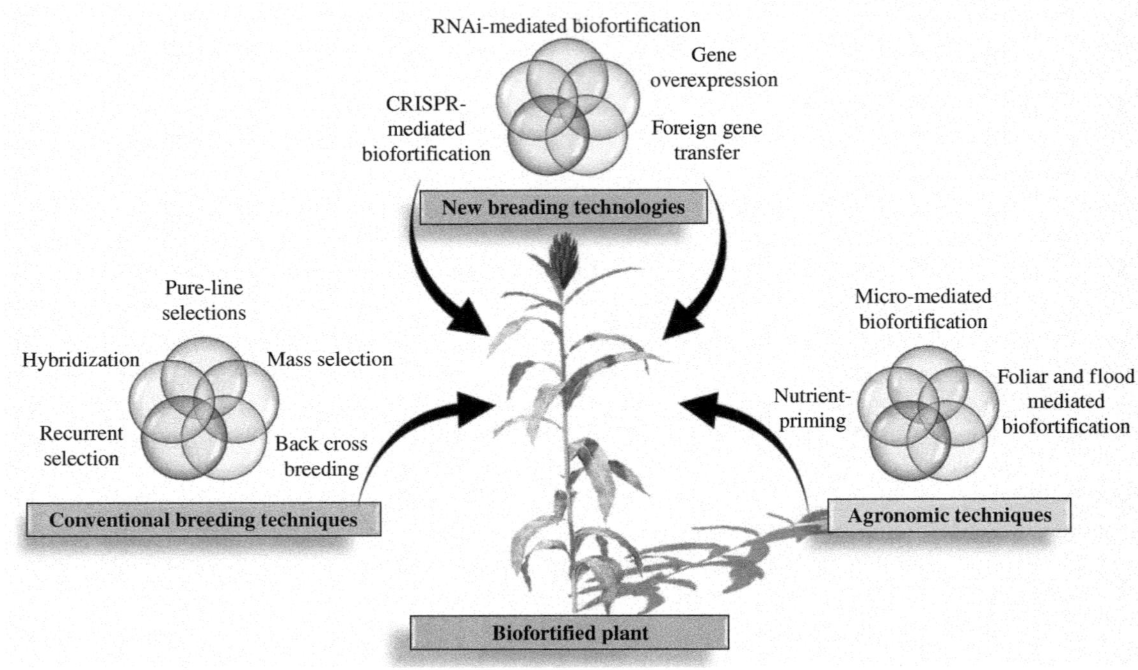

Figure 3.3 Various approaches of crop biofortification.

Genes relating to micronutrients such as iron, zinc, vitamin A, and proteins might be overexpressed in cereal grains to compensate for their shortages. For instance, ectopic overexpression of *OsNAS2* and *TaFer1-A* genes resulted in substantial iron accumulation in wheat grain, increasing iron content to 80 and 85 $\mu g\ g^{-1}$ in grains, respectively, although *TaFer1-A* expression varies throughout generations (Singh et al. 2017). In a similar vein, overexpression of *TaVIT* increased the amount of iron present in wheat grain, albeit unevenly across the grain (Kumar et al. 2019). Tyrosine, lysine, cysteine, and methionine deficiencies are addressed by overexpression of the Amaranthus albumin-based *Ama1* gene (Xu et al. 2017). While silencing the *TaABC13* transporter gene and the phytochrome (*PhyA*) gene reduced the formation of phytic acid (PA; 18–19%) and increased the bioavailability of iron and zinc, overexpression of the *PhyA* gene increased the production of phytase and anti-nutrient activity (Abid et al. 2017). The *SBEIIA* gene was silenced, increasing the concentration of less digestible amylose starch to combat obesity and overnutrition.

The percentage of vitamin A was similarly increased by the overexpression of the bacterial *PSY, CrtI*, and *CrtB* carotene desaturase genes (Kumar et al. 2019). Using the bacterial *Crtl* genes and the maize *PSY1* gene producing phytoene synthase, a greater quantity of vitamin A, 4.96 $\mu g\ g^{-1}$ dried weight (DW), was obtained (Abid et al. 2017). Provitamin A accumulation was similarly enhanced by *CrtB* or *Crtl*, reaching a maximum of 3.21 $\mu g\ g^{-1}$ of DW.

Antioxidant activity of rice is associated with flavonoids, which may be improved by overexpressing the genes for phenylalanine ammonia lyase, chalcone synthase, maize C1, GmFAD3, and flavone ammonia lyase. This would lead to an increased buildup of linoleic acid and flavonoids in rice (Zhu et al. 2021). The overexpression of the bacterial genes *CrtB* and *CrtI*, which are controlled by the endosperm-specific super gema zein promotor, resulted in an increase in vitamin A accretion in maize endosperm up to 9.8 $\mu g\ g^{-1}$ of DW. In a similar vein, maize's storage of β-carotene was similarly enhanced by overexpressing the *Psy1* gene. Increased vitamin C accumulation in maize was achieved by overexpressing the dehydroascorbate reductase gene, which converted ascorbic acid from its oxidizing to its reduced form (Foyer et al. 2020). Iron bioavailability in maize was increased by overexpressing the soybean ferrite gene, *Aspergillus phytase*, and *Aspergillus niger phy2*. The amount of lysine in maize grain increased due to the overexpression of the potato *sb401* gene. Tryptophan and lysine accretion were boosted in maize by antisense dsRNA targeting α-zeins. The quantities of tocopherol and tocotrienol in maize were increased by overexpressing the homogentisic acid geranylgeranyl transferase (*HGGT*) gene, leading to the production of biofortified maize (Figure 3.4).

Moreover, the overexpression of *PSY1* maize verified a greater vitamin A content of 59.32 $\mu g\ g^{-1}$ of DW (Hossain et al. 2019; Thanh 2019). The transgenic line homo188 of sorghum has a high nutritional value and distinct genes, such as *PSY-1,*

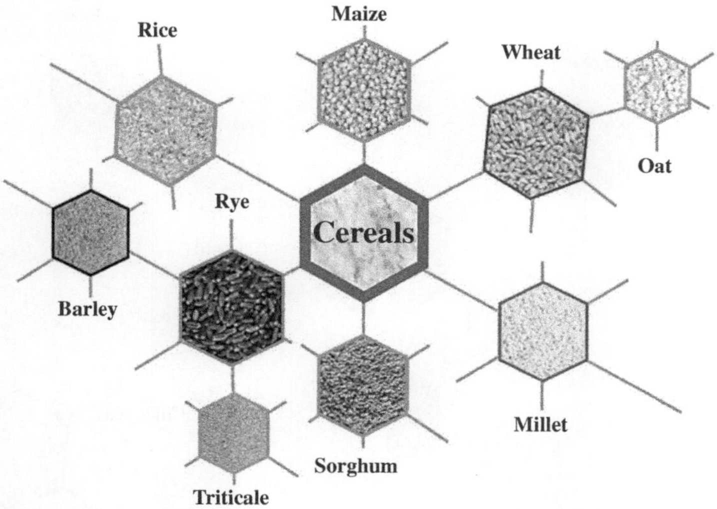

Figure 3.4 Cereals are consumed all around the world because of their high protein level and vitamin richness. They are basic cereals that offer both a filling meal and vital nutrients. Their adaptability and health advantages make them a favorite in many different cuisines around the globe, whether they are used as the main source of protein or as a side dish. Major cereals have a high value since they may provide a healthy, well-balanced diet from Asia to the Americas.

CRT-I, PMI, and *LPA-1*, for increased provitamin A content. The quantity of zinc was increased in barley by overexpressing the genes encoding zinc transporters. In barley, the expression of the *DHPS* and phytase genes increases the bioavailability of zinc and lysine. It has been observed that vitamin E in barley seeds was increased by the co-expression of the 2-methyl-6-phytyl benzoquinol methyl transferase *AtVTE-4* and *AtVTE3* genes. Similarly, the *D6D* gene's activation increased barley's concentration of linolenic and stearidonic acids, both of which are vital to human health. A small increase in the concentration of dietary fibers such as β-glucans was also seen with the overexpression of the cellulose synthase-like (*HvCslF*) gene (Ahmad et al. 2022). Reports state that zinc transporters in transgenic millet contributed to its higher zinc content, while transcriptomics showed that calcium sensor genes involved in calcium absorption, translocation, and accumulation in finger millet were expressed at higher levels. Moreover, it has been discovered that the genome-edited transgenic millet contains less antinutrients including phytic acid, tannins, and polyphenols (Vinoth and Ravindhran 2017).

3.5.1.2 Pulses

Not much research has been done on transgenic breeding for biofortification in pulses, and only a small number of pulses such as lupines, common beans, soybeans, and chickpeas are adapted to withstand malnutrition (Tan et al. 2017) (Figure 3.5). The primary objectives of biofortifying pulses by transgenic breeding are the increase of essential amino acids, iron and zinc fortification, and reduction of anti-nutrient compounds. The deficiency of these amino acids was compensated for by the overexpression of heterologous proteins rich in sulfur-rich amino acids. To increase the nutritional value of cysteine amino acid, a protein-rich in cysteines termed 27 kDa γ-zein, which is generated from maize, was introduced and overexpressed in a number of pulse crops.

In a similar vein, methionine content rose in lupin and narbon beans when S-rich protein overexpression occurred. Brazil nuts had four times more methionine accumulated in the seed as a result of overexpressing the 2S albumin storage protein and aspartate kinase (AK) using a promoter unique to the seed. After transformation, the adzuki bean's free tryptophan accumulation was boosted by the rice OASA1D transgene (Kaur et al. 2020b).

A accretion in soybeans (Qin et al. 2017). Soybeans have less α-linolenic acid because the *ω-3 FAD3* gene was silenced by siRNA-mediated silencing. Parallel to this, overexpression of the 16-desaturase gene resulted in an increase in the conversion of α-linolenic acid to ω-3 fatty acids and its stable form, γ-linolenic acid. Soybean isoflavone production was enhanced by the induction of gene activation by maize C1 and R transcription factors (Shahzad et al. 2021; Garg et al. 2018). Common beans now contain more methionine thanks to the transformation of methionine-rich storage albumin from Brazil nuts. In a similar vein, lupins' S-rich amino acid profile was elevated upon the transformation of the equivalent gene from sunflower and albumin (Garg et al. 2018).

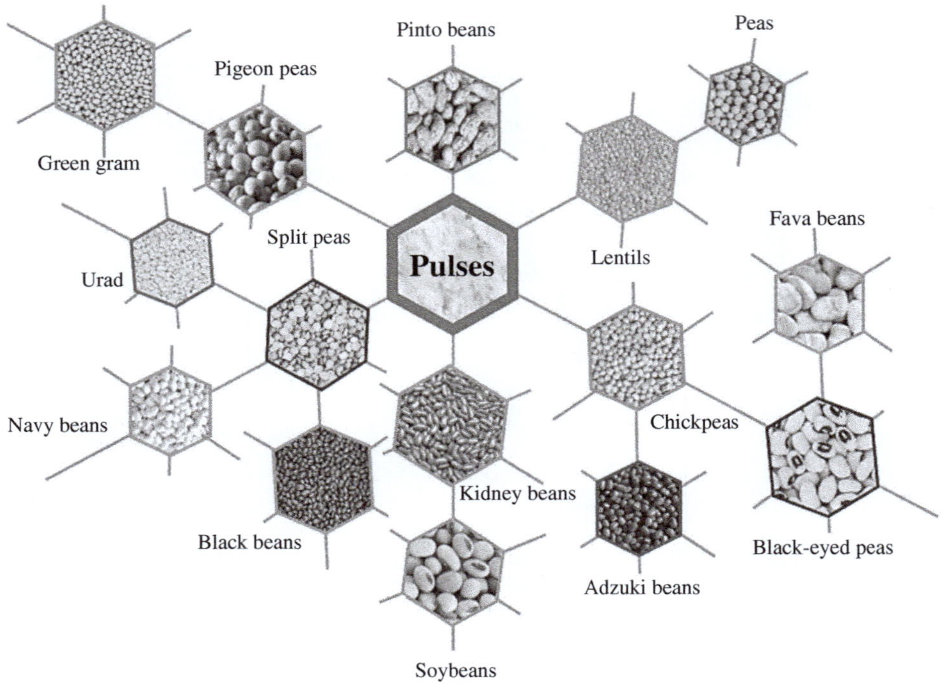

Figure 3.5 The high protein level and vitamin richness of pearls make them a popular food across the world. They are a basic diet pulses that offer important nutrients. Because of their adaptability and health advantages, they are used in many different types of cuisine all over the world, either as the main source of protein or as a complement to other recipes. To increase iron bioavailability, glycine max ferritin and chickpea *NAS2* genes were overexpressed. The overexpression of the cystathionine γ-synthase gene resulted in an increase in the methionine content in soybeans. Overexpression of the maize zein protein resulted in an increase in the cysteine and methionine content in soybean seeds. Similar to this, overexpression of the O-acetylserine sulfhydrylase gene increased the quantity of cysteine in seeds. In transgenic soybeans, the quantity of lysine in the seeds was enhanced by upregulating the genes for aspartokinase and dihydrodipicolinic acid synthase (Garg et al. 2018). Overexpression of the genes for carotene desaturase, bacterial *PSY*, *bkt1*, *crtW*, and *crtB* increased provitamin.

3.5.1.3 RNA Interference

RNAi is a sequence-specific gene control mechanism that is powered by double-stranded RNA (dsRNA) molecules. This procedure represents a major breakthrough in genetic technology as it inhibits the transcription or translation of a specific gene. The discovery of RNAi has brought about a revolution in crop development since it offers an accurate, reliable, effective, and superior substitute for antisense technology. Through techniques like biofortification and bio-elimination, RNAi provides a strong foundation for incorporating biotic and abiotic stress tolerance into crops and accelerating the supply of high-quality food. According to academic literature, its extensive use includes removing pollutants and allergies from food as well as improving the nutritional value of crops (Rajam 2020; Ahmad et al. 2021). PA is a well-known anti-nutrient found in grains and pulses that is recognized for its ability to bind micronutrients and prevent the absorption of vital minerals. RNAi was used to target *inositol pentakisphosphate kinase* (*TaIPK1*), a key gene in the PA production pathway in wheat. The resultant homozygous transgenic wheat lines at the T4 stage showed reduced expression of *TaIPK1*, which resulted in enhanced levels of grain phosphate, zinc, and iron as well as a significant decrease in PA accumulation (26–58%) (Waters and Sankaran 2011). Because they are precursors to provitamin A, carotenoids are essential to human nutrition.

The carotenoid-cleavage dioxygenases (CCDs) genes, which are in charge of carotenoid degradation, were disrupted by RNAi in rice, resulting in an increase in the carotenoid content. Rice mutants created by RNAi showed enhanced carotenoid synthesis in comparison to wild types by targeting genes including OsCCD1, OsCCD4a, and OsCCD4b. This demonstrated the effectiveness of this strategy in nutritional improvement (Ko et al. 2018). Among the necessary amino acids found in rice, lysine stands out as being particularly important for energy and nourishment. Dihydrodipicolinate synthase (DHDPS) and AK are enzymes that are essential for the biosynthesis of lysine and are tightly regulated by feedback. By effectively inhibiting the AK and DHDPS enzymes in rice through the use of RNAi, scientists were able to significantly raise the lysine content of mutant rice lines.

The ability of RNAi to improve nutritional profiles was demonstrated when both mutations were combined in a single genotype, resulting in an astounding 58.5-fold increase in lysine levels over wild types (Yang et al. 2021). Provitamin A and β-carotene deficiencies in maize endosperm lead to widespread vitamin A insufficiency. To increase the quantities of β-carotene in maize endosperm, RNAi was used to target the maize genes *ZmBCH2* and *ZmBCH1*, which are involved in the hydroxylation reaction pathway. Significant increases in β-carotene content were seen in mutants missing the *ZmBCH2* gene, indicating the gene's critical function in carotenoid conversion mechanisms (Berman et al. 2017).

Because of this protein, sorghum, which has γ-kafirin, is found to be less digestible. The digestibility of sorghum was significantly enhanced by the targeted suppression of γ-kafirin 1A, γ-kafirin 1, and γ-kafirin 2, using RNAi silencing (Garg et al. 2018). This effective use of RNAi to improve sorghum's digestibility highlights how adaptable it is in addressing dietary issues in a variety of crop species. To sum up, the creative application of RNAi technology has great potential to transform agricultural development techniques, make it possible to produce nutrient-rich meals, and support international initiatives to fight hunger and malnutrition.

There is a huge promise for increasing the nutritional content of pulses using this technique, even if the role of RNAi in pulse biofortification has not been well addressed. Research findings reveal that phytate, a substance included in pulses, impedes the bioavailability of vital micronutrients by forming complexes with iron, calcium, magnesium, and copper, so diminishing their solubility properties. In addition, it has been shown that certain substances, for example, prebiotics such as inulin and fructans, sequester zinc, calcium, and iron, hence reducing the action of phytic acid. Furthermore, it has been demonstrated that β-carotene helps legumes, peas, and chickpeas absorb iron and zinc. Consequently, it may be possible to overexpress the genes that encode these advantageous substances, improving the absorption of micronutrients.

Furthermore, studies have demonstrated that selenium raises the bioavailability of iodine in legumes, such as lentils, chickpeas, and peas. However, there are still some inhibitors that prevent bioavailability and need to be investigated further. To generate nutritionally enhanced pulse crops, it is imperative to explore the biochemical processes involved in the generation of anti-nutrients and identify important genes that may be targeted for silencing or deletion using RNAi (Uppal et al. 2021). Plant biotechnology relies heavily on GE techniques, especially in the area of plant GE, where transgene-free plants with stable genetic traits are produced by using sequence-specific nucleases (SSNs) to modify specific genes in crops. Plant GE has made use of a variety of SSNs, such as ZFNs, TALENs, and the CRISPR-Cas system. For example, the CRISPR-Cas system uses RNA molecules to guide the Cas9/13 enzymes, which are used to insert specific alterations in the plant genome. The precise genome editing approach involves the formation of a complex at the target location. While GE has been used sparingly for biofortification in pulses and grains, it has produced encouraging results in some cases (Ijaz et al. 2024; Jamil et al. 2020).

3.5.1.4 Cereals

Using GE methods, the *OsVIT2* gene in rice has been effectively edited to increase iron availability in the setting of cereals. In a similar vein, a carotenogenesis cassette was inserted into rice using CRISPR-Cas9, leading to an endosperm accumulation of β-carotene that was on par with conventional breeding techniques. Simultaneous mutation of many rice carotenoid catabolic genes was attempted to raise β-carotene levels; however, the results were not totally satisfactory. Furthermore, modifying genes such as *OsWaxy* in rice decreased the amount of amylose, but focusing on genes associated with nitrogen absorption raised the protein content in rice grains. Zinc content in wheat was increased using GE to target the *TaVIT2* gene (Ullah et al. 2023a; Riaz et al. 2020; Guo et al.2021).

Furthermore, through a variety of regulatory mechanisms, GE has been used to increase the nutritional value of maize. In an effort to speed up the synthesis process, genetic alterations were introduced to *ZmALS1* and *ZmALS2* genes of the acetolactate synthase gene, increasing the synthesis of branched-chain amino acids such as leucine, aline, and isoleucine. Using the CRISPR-Cas9 type II system, the phytoene synthase gene linked to the biosynthesis of carotenoid was edited, leading to a stable transformation in progeny. However, the functional examination of mutants with respect to the concentration of carotenoids is still in progress.

3.5.1.5 Pulses

There are not enough examples in the corpus of literature currently available to clarify how GE is used to biofortify pulses. However, there is a chance to improve the iron and zinc biofortification of pulses by applying GE methods. The uptake of iron and zinc in all pulse crops can be increased by modifying genes such as zinc-regulated transporters, YELLOW STRIPE 1-like, ferric-chelate reductase oxidase, iron-regulated transporter, natural resistance-associated macrophage protein, and zinc-iron regulated transporters like protein (ZIP) (Tan et al. 2017). Targeting anti-nutrient genes that impede the bioavailability of micronutrients is another method of pulse biofortification using GE. One such class of chemicals is saponins, which are thought to be helpful at low concentrations but anti-nutrients at greater ones (Kumar and Pandey 2020). To successfully minimize the production of saponins, it is imperative to identify important regulatory genes in the route of saponin biosynthesis (Tibbs Cortes, Zhang, and Yu 2021). To reduce the formation of anti-nutrient compounds, genes linked to the synthesis of other anti-nutrients such as α-amylase, protease inhibitors, and lathyrogens should also be identified and changed (Ahmad et al. 2022).

3.6 Role of Genome-wide Association Studies

It is impossible to overstate the importance of genome-wide association studies (GWASs) in the field of biofortification. In developmental biology, a thorough understanding of the genetics underpinning complex phenotypes is crucial. Numerous association studies have been initiated as a result of the long-standing fascination of plant scientists with the differences in features across different genetic origins. GWAS has been made possible by the ongoing developments in genome sequencing technology, high-throughput phenotyping platforms, and high-density genotyping arrays, (Ijaz et al. 2024; Haider et al. 2021a). Generally, the GWAS procedure starts with genotyping using various types of next-generation sequencing technologies like whole-genome sequencing or genotyping by sequencing, which is followed by careful phenotyping in a variety of environmental settings.

After processing the genotyping data to find meaningful single nucleotide polymorphisms (SNPs), each trait is submitted to GWAS analysis with the goal of determining SNP-trait relationships and predicting potential genes controlling certain characteristics (Suwarno et al. 2015) (Figure 3.3). With the use of GWAS, scientists have been able to discover the genes that control complex traits, opening up new possibilities for crop development. Numerous GWASs have been conducted on pulses and grains to identify the genes implicated in micronutrient accumulation. Using 476 000 SNP markers and 380 genotypes of maize from the CIMMYT carotenoid association mapping panel, GWAS was used in maize research to identify genes associated with the carotenoid biosynthetic pathway.

It has been successfully determined which genes are important for the carotenoid biosynthesis pathways in maize. These genes include *GGPS1, DXS1*, and *GGPS2*, which are in charge of the accumulation of precursor isoprenoids, and downstream genes *CCD1, HYD5*, and *ZEP1*, which are involved in carotenoid degradation and hydroxylation (Bhatta et al. 2018). A total of 35 648 SNPs and 123 wheat genotypes was used in a GWAS of wheat to identify genetic areas associated with 10-grain minerals, such as Ni, Cd, Li, Ca, Mn, Co, Fe, Cu, Mg, and Zn. Of the 92 SNPs linked to attributes that were found, 60 were found to be new, and 40 were found to be inside genes. Twenty of the 40 genes have functional annotations indicating how important they are for grain mineral accumulation.

The D-genome was found to include the bulk of the SNPs, suggesting that it plays a critical role in controlling the variety of grain minerals in wheat (Ozkuru et al. 2019). Understanding the genetic composition of seed selenium (Se) and molybdenum (Mo) in wild-type and farmed chickpea was the goal of another GWAS project. A total of 180 individuals were used in this work; 107 were wild (*C. reticulatum*) and 73 were farmed (*C. arietinum*). The individuals were phenotyped during a two-year period at two distinct sites and surveyed using 121 840 SNP markers. The association between 16 SNPs and the Mo and Se contents of seeds highlights the effectiveness of GWAS in examining the genetic basis of complex characteristics (Haider et al. 2021b; Ahmar et al. 2020).

3.7 Speed Breeding's Part in the Slow Development of Biofortified Crops

The duration of the breeding cycle is an important obstacle to many crop breeding programs. To produce genetically stable homozygous lines for field testing, parent plant selection, and crossbreeding usually require four to six generations. The breeding cycle is extended by an additional two to three generations by subsequent field evaluation, making the overall duration of the cycle around six to seven years for crops with a single growing season and three to four years for those with two growing seasons (Kim et al. 2020). This extended timeframe calls for the investigation of novel technologies to accelerate crop growth and shorten the breeding season. To reduce the length of the crop breeding cycle, breeders have already embraced techniques like shuttle breeding and double haploid breeding (Haider et al. 2021b; Bhatta et al. 2021).

However, because of the inherent limits of these methods, speed breeding has emerged as a potential strategy that breeders are beginning to take notice of. Speed breeding is a method that was developed in 2003 by the University of Queensland.

It involves modifying the environment to stimulate early blooming and reduce generation time (Jähne et al. 2020). In contrast to double haploid and shuttle breeding methods, which require labor-intensive tasks like in vitro culturing or moving plants across different climates to achieve multiple generations in a year, speed breeding is applicable to a wide range of germplasm (Jähne et al. 2020).

Plant development is significantly influenced by a wide range of internal and external parameters, including photoperiod, temperature, planting density, light intensity, and light quality. By tampering with different biological systems in plants, synthetic biology (SB) interferes with these organic processes and speeds up generation augmentation. Comprehending these basic mechanisms is essential to creating effective SB platforms customized for particular crops (Jähne et al. 2020). Through the use of photoreceptors to sense changes in the external environment, photoperiod is essential in the transition to reproductive improvement. As a result, when a species is synchronized with changes in environmental stimuli, its reproductive success increases.

Three groups of plants are identified: day-neutral plants, which are not influenced by night duration for the commencement of blooming, long-day plants, which require shorter nights than a critical length, and short-day plants (Alahmad et al. 2018). In a similar vein, a plant's reproductive development is greatly influenced by planting density, light intensity, and ambient temperature. To produce four to five generations of a species at the same site in a year, these environmental signals are carefully controlled. Breeding cycles for several cereals and pulses, including wheat (Rana et al. 2019), rice (Cazzola et al. 2020), peas (Samineni et al. 2020), chickpeas (Lassoued et al. 2018), and many more crops, have been successfully shortened by using this strategy. The above-discussed results suggest that the production of biofortified crops for consumers can be accelerated by employing NBTs to change genes or plant processes and by increasing the generation speed through methods such as speed breeding. The necessity of quick generation augmentation, like speed breeding, is also emphasized in this section as a means of guaranteeing food security going forward.

3.8 NBT-developed Varieties' Regulatory Aspects

Within the field of plant breeding, the regulatory elements of varieties created by NBTs have gained significant attention. The increasing understanding of genomics and its use in breeding procedures has led to a substantial increase in the usage of modern plant biotechnology to improve crop yield and quality. The precise and accurate creation of desirable mutants with specified targets is made possible by new plant breeding methods (e.g. NBTs; Figure 3.6). By modifying gene segments by insertion or deletion, these methods allow for direct crop development in a time- and money-efficient manner while achieving desired plant traits without compromising other qualities. Because of their distinct and persistent mutations, NBTs differ from conventional genetically modified (GM) crops. Despite their effectiveness, there is continuous debate over acceptance of these methods in many nations (Rehman et al. 2019). Policies pertaining to the consumption and cultivation of GM crops differ between the United States and the European Union (EU), mostly due to popular opinions on these goods. In general, Americans are more accepting of agricultural biotech goods than Europeans are, as the latter are worried about the uncertain hazards attached to GM crops (Sedeek et al. 2019). As a result, scientists, regulators, and legislators are debating how to govern crops that have had their genomes altered.

Their main point of contention is whether or not crops that have undergone GE should be governed by the same laws that govern GM organisms. In order to categorize genome-edited crops, the conversation revolves around the development of character characteristics, the different paths used for enhancement, the methods utilized, and the possible hazards connected with the final products of crops. In 2012, the US Department of Agriculture declared that plants modified with zinc finger nucleases (ZNFs) and meganucleases should not be governed by GM organisms. This ruling cleared the path for the production and marketing of CRISPR-edited waxy maize and mushrooms without going through the required genetically modified organism (GMO) regulatory processes. The US regulatory strategy is product-focused, whereas the EU has embraced a process-focused technique.

However, some anti-GMO organizations are making the absurd argument that genome-edited crops are unnatural and might damage the environment in an effort to categorize them as subject to GMO rules. The European Court of Justice's order that CRISPR-modified crops comply with GMO legislation has hampered the commercialization and trade ties between European markets and other nations. Therefore, nongovernmental organizations, regulators, and consumers' acceptance of technology is just as crucial to its success as scientific evidence and methodologies. The public acceptance of these technologies is poor, and the production and sale of transgenic crops are subject to onerous, costly, and time-consuming regulatory processes. For example, the development of Golden Rice was a major biotechnology breakthrough,

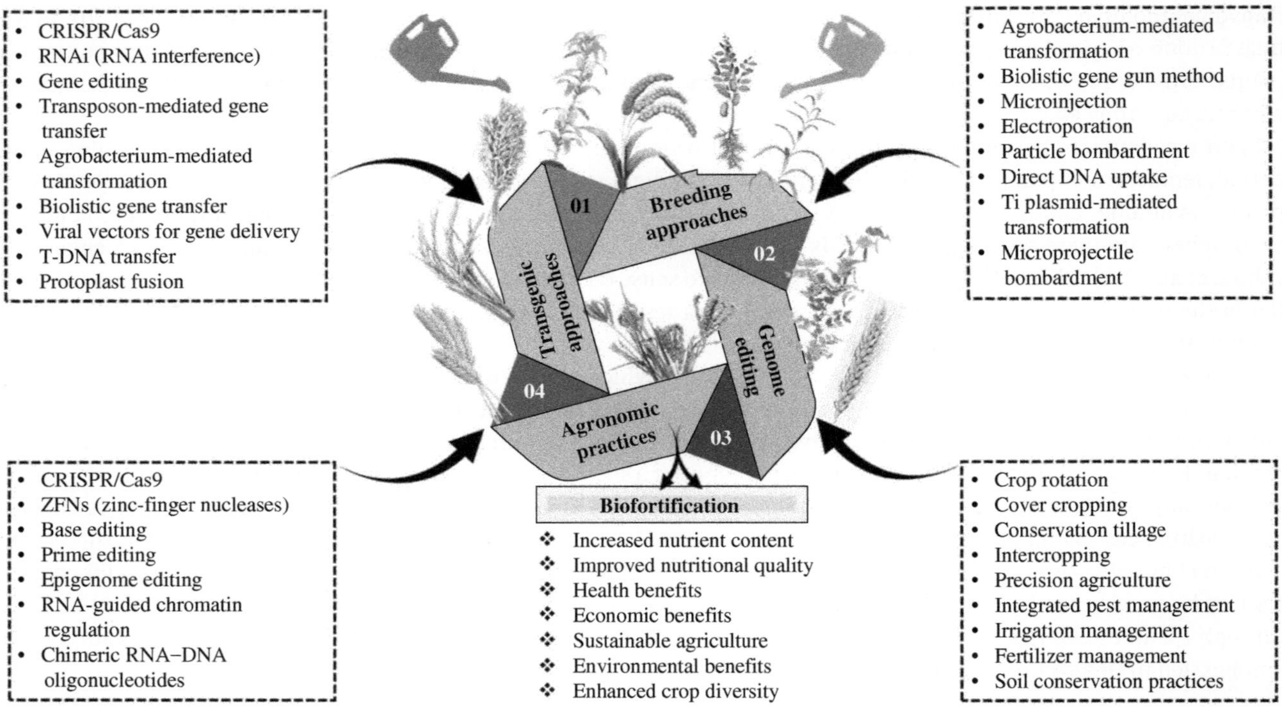

Figure 3.6 Methods for biofortifying crops with agricultural plants. Utilizing techniques like GE, agronomic methods, transgenic plants, and plant breeding, biofortification may be enhanced. The transgenic technique of overexpressing zinc and iron transporters in crops may be advantageous to enrich edible grains with these elements. The breeding strategy facilitates the development of biofortified crop varieties through the use of clonal propagation, marker-assisted breeding, backcrossing, and other procedures. Agronomic techniques including administering growth hormones and fertilizer at the proper dosage, foliar spraying fertilizer based on zinc and iron, and employing microorganism-based biofertilizer can all help increase the bioavailability of zinc and iron in crops.

but its commercial production has been hindered by inconsistent yields, which has led to government opposition. Similar to this, farmers, scientists, and anti-GMO organizations opposed the production of Bt brinjal, which resulted in a brief prohibition. Four further types were approved and released for commercial production after further testing. Therefore, public acceptance of these technologies is closely related to their endorsement.

3.9 Conclusion and Future Perspectives

The Sustainable Development Goals (SDGs) are a collection of 17 objectives that member nations of the United Nations must accomplish by 2030. The third SDG, which focuses on improving health and well-being for all age groups, emphasizes the critical role that nutrition and the intake of nutrient-dense foods play in this regard. Still, the fact that people of all ages in underdeveloped nations are disproportionately exaggerated by hidden hunger presents a serious threat to world health. The enduring worldwide shortage of micronutrients can be ascribed to the extensive usage of cereals worldwide, leading to an imbalanced intake of nutrients. When eaten in the right amounts, pulses and cereals together can provide a well-rounded nutritional profile and satisfy all dietary requirements (Ullah et al. 2023b).

For example, one important barrier to traditional breeding is the absence of genetic variety in the gene pool, which can occasionally be addressed by crossing distant relatives to introduce a trait of interest (TOI). Generally speaking, nevertheless, it is challenging to identify the preferred characteristic in distant relatives. Therefore, it is nearly hard to enhance such characteristics via traditional breeding, that is, enhancing the Se levels in wheat and the linoleic and oleic acid in soybeans (Yeom et al. 2020; Anzalone et al. 2020). It is safe for biological systems and incredibly precise and successful at causing targeted alteration in the gene of interest (Ijaz et al. 2024). The most recent developments in GE, such as base editing that involves converting a base irreversibly at the target site without the need for donor templates, double-strand breaks, or non-homologous end joining (NHEJ) and homology-directed repair (HDR) prime editing, and GE with rice zygote which solves the challenge of delivering macromolecules to host cells/tissues and simplifies transformation and regeneration have given biotechnologists new avenues to pursue their research (Ahmad et al. 2021). Though there is still discussion in the EU, the United States and Canada have already proclaimed GE crops to be exempt from GM laws (Haider et al. 2022a; Haider et al. 2022b).

Since GE crops do not contain transgenes, they should be controlled internationally just like traditional breeding products and should be distributed the most benefits of technology in the shortest amount of time (Ahmad et al. 2021). GE is where biofortification is headed. The engineering of genes for maximal translocation to the seed, enhanced bioavailability, and enhanced absorption of micronutrients from the soil is the targeted area that requires attention. When biofortified with iron, sulfur-rich amino acids, zinc, and anti-nutrient genes eliminated, cereals and pulses provide a complete diet (Perera et al. 2018). The other area that requires extra care is crop engineering to minimize the buildup of anti-nutrients in pulses and grains. In this case, rice's PA content is the emphasized example. The most prevalent form of phosphorus stored in plants is called PA, which functions as an anti-nutrient by chelating metal ions and converting them to phytate. In order to minimize PA buildup and maximize the bioavailability of micronutrients, it is necessary to repress a number of genes (Jamil et al. 2019) involved in the formation of PA. However, anti-nutrient molecules are essential for biotic and abiotic stress tolerance, among other plant activities; therefore, eliminating them would also affect those processes, necessitating trade-offs (Ahmad et al. 2024b).

To understand the genetic complexities underpinning these techniques, it will be helpful to investigate the genes and metabolic pathways involved (Van Der Straeten et al. 2020). The use of CWRs as source material for biofortification is a crucial component of crop biofortification. Studies have shown that CWRs have a higher nutritional value than cultivated crops; for this reason, careful thought must be given to genetically modify them to transmit desirable genes (Jamil et al. 2019; Yadava et al. 2018).

A number of transgenic biofortified crops may be found at research facilities. Due to regulatory concerns and limited consumer acceptance, the majority are not commercialized. Furthermore, a major barrier to the commercialization of GM crops is the requirement by regulatory authorities and biosafety assessment organizations for GM crops to invest an equivalent amount of cash as was spent in their research. Due to the presence of bacterial marker genes in GM crops, consumers are also concerned about antibiotic resistance. These problems might be resolved by using GE techniques to create transgene-free plants or by creating marker-free transgenic plants. Except for the EU, other nations have laxer regulations governing GE crops (Ahmad et al. 2021). A significant complication to the biofortification of pulses is the lack of research on genetic variety in terms of starch and protein composition, as well as the amount of micro- and macronutrients. Thus, in order to successfully use global pulse germplasm for pulse biofortification, large-scale research to comprehend the genetic diversity of significant macro and micronutrients must be initiated (Ahmad et al. 2022). Along with raising the concentration of micronutrients, the main goal should be to increase their bioavailability. In order to do this, the content of promoter substances that boost mineral absorption must be raised while the concentration of anti-nutrients should be lowered. Vitamins A, C, D, and E, choline, and niacin are promoter chemicals that promote the absorption of tryptophan, methionine, Ca, Fe, Zn, and P (Jha and Warkentin 2020).

Safer food delivery to consumers is a major challenge facing postharvest crop management. The general public is ignorant about the dietary and nutritional value of various plant components. For instance, most cereals' milled grain is consumed globally. While the majority of important elements, such as Se and S, are present in larger concentrations in germ, bran contains abundant amounts of other critical elements, including copper, zinc, and iron, which are removed during milling and inaccessible to humans (Wang et al. 2020). The majority of micronutrients are present in the cereals' aleurone layers, which are removed along with the bran during the milling process. These are therefore not fit for ingestion. This is a problem that requires particular attention and has two possible approaches. The first alternative that will improve the nutritional content of items obtained later is whole-grain processing. The second choice may involve designing biosynthetic pathways to change how micronutrients are deposited from the aleurone layer to the endosperm, preventing their loss during processing and grinding.

References

Abid, N., Khatoon, A., Maqbool, A. et al. (2017). Transgenic expression of phytase in wheat endosperm increases bioavailability of iron and zinc in grains. *Transgenic research* 26: 109–122. doi: 10.1007/s11248-016-9983-z.

Adepoju, A.A. and Allen, S. (2019). Malnutrition in developing countries: nutrition disorders, a leading cause of ill health in the world today. *Paediatrics and Child Health* 29: 394–400. doi: 10.1016/j.paed.2019.06.005.

Adeyeye, S.A.O., Ashaolu, T.J., Bolaji, O.T. et al. (2021). Africa and the Nexus of poverty, malnutrition and diseases. *Critical Reviews in Food Science and Nutrition.* 63 (5): 641–6560. doi: 10.1080/10408398.2021.1952160.

Ahmad, S., Shahzad, R., Jamil, S. et al. (2021). Regulatory aspects, risk assessment, and toxicity associated with RNAi and CRISPR methods. In: *CRISPR and RNAi Systems* (eds. K. Abd-Elsalam and L. Ki-Taek), 687–721. Amsterdam: Elsevier.

Ahmad, S., Tang, L., Shahzad, R. et al. (2021). CRISPR-based crop improvements: a way forward to achieve zero hunger. *Journal of Agricultural and Food Chemistry* 69: 8307–8323. doi: 10.1021/acs.jafc.1c02653.

Ahmad, Y., Haakim, Z., Iqbal, J. et al. (2024b). *Technological Innovations for Abiotic Stress Resistance in Horticultural Crops.* Tuskegee: OMICs-based Techniques for Global Food Security, 233–244.

Ahmad, Y., Haider, S., Iqbal, J. et al. (2022). The mechanisms of genome editing technologies in crop plants. In: *Principles and Practices of OMICS and Genome Editing for Crop Improvement*, 295–313. Cham: Springer International Publishing.

Ahmad, Y., Haider, S., Iqbal, J. et al. (2024a). In-silico analysis and transformation of OsMYB48 transcription factor driven by CaMV35S promoter in model plant–Nicotiana tabacum L. conferring abiotic stress tolerance. *GM Crops & Food* 15 (1): 130–149.

Ahmar, S., Gill, R.A., Jung, K-H. et al. (2020). Conventional and molecular techniques from simple breeding to speed breeding in crop plants: Recent advances and future outlook. *International Journal of Molecular Sciences* 21: 2590. doi: 10.3390/ijms21072590.

Alahmad, S., Dinglasan, E., Leung, K.M. et al. (2018). Speed breeding for multiple quantitative traits in durum wheat. *Plant Methods.* 14: 1–15. doi: 10.1186/s13007-018-0302-y.

Anzalone, A.V., Koblan, L.W. and Liu, D.R. (2020). Genome editing with CRISPR–Cas nucleases, base editors, transposases and prime editors. *Nature biotechnology* 38: 824–844. doi: 10.1038/s41587-020-0561-9.

Awika, J.M. (2011). Major cereal grains production and use around the world. In: *Advances in Cereal Science: Implications to Food Processing and Health Promotion* (eds. J.M. Awika, V. Piironen and S. Bean.), 1–13.Washington, DC: ACS Publications.

Berman, J., Zorrilla-López, U., Sandmann, G. et al. (2017). The silencing of carotenoid β-hydroxylases by RNA interference in different maize genetic backgrounds increases the β-carotene content of the endosperm. *International Journal of Molecular Sciences* 18: 2515. doi: 10.3390/ijms18122515.

Bhatta, M., Baenziger, P.S., Waters, B.M. et al. (2018). Genome-wide association study reveals novel genomic regions associated with 10 grain minerals in synthetic hexaploid wheat. *International Journal of Molecular Sciences* 19: 3237. doi: 10.3390/ijms19103237.

Bhatta, M., Sandro, P., Smith, M.R. et al. (2021). Need for speed: manipulating plant growth to accelerate breeding cycles. *Current Opinion in Plant Biology* 60: 101986. doi: 10.1016/j.pbi.2020.101986.

Blicharz, A., Marzeda, M., Drozd, M. et al. (2020). Beriberi disease–a picture of thiamine deficiency in underdeveloped and highly developed countries. *Journal of Education, Health and Sport* 10: 407–414. doi: 10.12775/JEHS.2020.10.09.048.

Bouillon, R. and Antonio, L. (2020). Nutritional rickets: historic overview and plan for worldwide eradication. *The Journal of steroid biochemistry and molecular biology* 198: 105563. doi: 10.1016/j.jsbmb.2019.105563.

Bouis, H.E. and Saltzman, A. (2017). Improving nutrition through biofortification: A review of evidence from HarvestPlus, 2003 through 2016. *Global Food Security* 12: 49–58. doi: 10.1016/j.gfs.2017.01.009.

Cazzola, F., Bermejo, C.J., Guindon, M.F. et al. (2020). Speed breeding in pea (Pisum sativum L.) an efficient and simple system to accelerate breeding programs. *Euphytica.* 216: 1–11. doi: 10.1007/s10681-020-02715-6.

Debelo, H., Albertsen, M., Simon, M. et al. (2020). Identification and characterization of carotenoids, vitamin E and minerals of biofortified sorghum. *Current Developments in Nutrition* 4: 1792. doi: 10.1093/cdn/nzaa067_019.

Engels, J.M. and Thormann, I. (2020). Main challenges and actions needed to improve conservation and sustainable use of our crop wild relatives. *Plants* 9: 968. doi: 10.3390/plants9080968.

Fatiukha, A., Filler, N., Lupo, I. et al. (2020). Grain protein content and thousand kernel weight QTLs identified in a durum × wild emmer wheat mapping population tested in five environments. *Theoretical and Applied Genetics* 133: 119–131. doi: 10.1007/s00122-019-03444-8.

Foyer, C.H., Kyndt, T. and Hancock, R.D. (2020). Vitamin C in plants: novel concepts, new perspectives, and outstanding issues. *Antioxidants & Redox Signaling* 32: 463–485. doi: 10.1089/ars.2019.7819.

Garg, M., Sharma, N., Sharma, S. et al. (2018). Biofortified crops generated by breeding, agronomy, and transgenic approaches are improving lives of millions of people around the world. *Frontiers in Nutrition* 5:12. doi: 10.3389/fnut.2018.00012.

Gorji, A. and Ghadiri M.K. (2020). The potential roles of micronutrient deficiency and immune system dysfunction in COVID-19 pandemic. *Nutrition* 82: 111047. doi: 10.1016/j.nut.2020.111047.

Grover, K., Arora, S. and Choudhary, M. (2020). Development of quality protein product using biofortified maize to combat malnutrition among young children. *Cereal Chemistry* 97: 1037–1044. doi: 10.1002/cche.10326.

Guo, C., Lyu, W. and Geng, Y. (2021). Bioinformatics-based screening of key genes for saponin metabolism in quinoa. *Research Square [Preprint]*. doi: 10.21203/rs.3.rs-139481/v1.

Haider, S., Iqbal, J., Naseer, S. et al. (2021a). Molecular mechanisms of plant tolerance to heat stress: current landscape and future perspectives. *Plant Cell Reports* 40(12): 2247-2271.

Haider, S., Iqbal, J., Naseer, S. et al. (2022b). Unfolding molecular switches in plant heat stress resistance: A comprehensive review. *Plant Cell Reports* 41 (3): 775–798.

Haider, S., Iqbal, J., Shaukat, M. et al. (2021b). The epigenetic chromatin-based regulation of somatic heat stress memory in plants. *Plant Gene* 27: 100318.

Haider, S., Raza, A., Iqbal, J. et al. (2022a). Analyzing the regulatory role of heat shock transcription factors in plant heat stress tolerance: A brief appraisal. *Molecular Biology Reports* 49 (6): 5771–5785.

Hossain, F., Muthusamy, V., Zunjare, RU. et al. (2019) Biofortification of maize for protein quality and provitamin-A content. In: *Nutritional Quality Improvement in Plants* (eds. P.K. Jaiwal, A.K. Chhillar, D. Chaudhary. et al.), 115–136. New York, NY: Springer.

Ijaz, S., Iqbal, J., Abbasi, B.A. et al. (2023). Biofortification: Lessons from the Past and Strategies for Future Food Security. In: *Legumes Biofortification*, 521–545. Cham: Springer International Publishing.

Ijaz, S., Iqbal, J., Abbasi, B.A. et al. (2024). Investigation of bioactive constituents and evaluation of different in vitro antimicrobial, antioxidant, and cytotoxicity potentials of different Portulacaria afra extracts. *Journal of King Saud University-Science* 36 (2): 103033.

Ijaz, S., Iqbal, J., Abbasi, B.A. et al. (2024). *Role of OMICS-Based Technologies in Plant Sciences*, 45–66. Tuskegee: OMICs-based Techniques for Global Food Security.

Jähne, F., Hahn, V., Würschum, T. et al. (2020). Speed breeding short-day crops by LED-controlled light schemes. *Theoretical and Applied Genetics* 133:2335–2342. doi: 10.1007/s00122-020-03601-4.

Jamil, S., Arshad, S., Kanwal, S. et al. (2019). Impact of transgenic crops on global food security: A review. *Journal of Agricultural Research* 57: 245–258. https://apply.jar.punjab.gov.pk/upload/1584357010_138_3._1442.pdf.

Jamil, S., Shahzad, R., Ahmad, S. et al. (2020). Role of genetics, genomics and breeding approaches to combat stripe rust of wheat. *Frontiers in Nutrition* 7: 173. doi: 10.3389/fnut.2020.580715.

Jamil, S., Shahzad, R., Ahmad, S. et al. (2021). Climate change and role of genetics and genomics in climate-resilient sorghum. In: *Developing Climate-Resilient Crops* (eds. S. Fahad, O. Sonmez, S. Saud. et al), 111–138. Boca Raton, FL: CRC Press.

Jha, AB. and Warkentin, TD. (2020). Biofortification of pulse crops: Status and future perspectives. *Plants* 9:73. doi: 10.3390/plan ts9010073.

Kaur, S., Kumari, A., Singh, P. et al. (2020a). Biofortification in pulses. In: *Advances in Agri-Food Biotechnology* (eds. S. Tilak Raj, D. Rupesh, and S. Humira), 85–103. New York, NY: Springer.

Kaur, T., Rana, K.L., Kour, D. et al. (2020b). Microbe-mediated biofortification for micronutrients: Present status and future challenges. In: *Trends of Microbial Biotechnology for Sustainable Agriculture and Biomedicine Systems: Perspectives for Human Health* (eds. A.A. Rastegari, N. Yadav and A.N. Yadav), 1–17. Amsterdam: Elsevier.

Kim, J-J., Jeong, E-G., Jeong, J-M. et al. (2020). Development of cold tolerance genetic resource 'Jungmo1022'through shuttle breeding in rice. *Korean Journal of Breeding Science* 52: 206–211. doi: 10.9787/KJBS.2020.52.2.206.

Ko, M.R., Song, M-H., Kim, J.K. et al. (2018). RNAi-mediated suppression of three carotenoid-cleavage dioxygenase genes, OsCCD1, 4a, and 4b, increases carotenoid content in rice. *Journal of Experimental Botany* 69: 5105–5116. doi: 10.1093/jxb/ery 300.

Kumar, J., Gupta, D.S., Kumar, S. et al. (2016). Current knowledge on genetic biofortification in lentil. *Journal of Agricultural and Food Chemistry* 64: 6383–6396. doi: 10.1021/acs.jafc.6b02171.

Kumar, S. and Pandey, G. (2020). Biofortification of pulses and legumes to enhance nutrition. *Heliyon* 6: e03682. doi: 10.1016/j.h eliyon.2020.e03682.

Kumar, S., Palve, A., Joshi, C. et al. (2019). Crop biofortification for iron (Fe), zinc (Zn) and vitamin A with transgenic approaches. *Heliyon* 5:e01914. doi: 10.1016/j.heliyon.2019.e01914.

Laskowski, W., Górska-Warsewicz, H., Rejman, K. et al. (2019). How important are cereals and cereal products in the average Polish diet? *Nutrients* 11:679. doi: 10.3390/nu11030679.

Lassoued, R., Smyth, S.J., Phillips, P.W. et al. (2018). Regulatory uncertainty around new breeding techniques. *Frontiers in Plant Science* 9: 1291. doi: 10.3389/fpls.2018.01291.

Lowe, N.M., Khan, M.J., Broadley, M.R. et al. (2018). Examining the effectiveness of consuming flour made from agronomically biofortified wheat (Zincol-2016/NR-421) for improving Zn status in women in a low-resource setting in Pakistan: Study protocol for a randomised, double-blind, controlled cross-over trial (BiZiFED). *BMJ Open* 8: e021364. doi: 10.1136/bmjopen-2017-021364.

Majumder, S., Datta, K. and Datta, S.K. (2019). Rice biofortification: High iron, zinc, and vitamin-A to fight against "hidden hunger". *Agronomy* 9:803. doi: 10.3390/agronomy9120803.

Malik, K.A. and Maqbool A. (2020). Transgenic crops for biofortification. *Frontiers in Sustainable Food Systems* 4: 182. doi. 10.3389/fsufs.2020.571402.

Muthayya, S., Rah, JH., Sugimoto, JD. et al. (2013). The global hidden hunger indices and maps: An advocacy tool for action. *PLoS One* 8: e67860. doi: 10.1371/journal.pone.0067860.

Ozkuru, E., Ates, D., Nemli, S. et al. (2019). Genome-wide association studies of molybdenum and selenium concentrations in C. arietinum and C. reticulatum seeds. *Molecular Breeding* 39: 1–15. doi: 10.1007/s11032-019-0947-x.

Perera, I., Seneweera, S. and Hirotsu N. (2018). Manipulating the phytic acid content of rice grain toward improving micronutrient bioavailability. *Rice* 11: 1–13. doi: 10.1186/s12284-018-0200-y.

Prasanna, B.M., Palacios-Rojas, N., Hossain, F. et al. (2020). Molecular breeding for nutritionally enriched maize: Status and prospects. *Frontiers in genetics* 10: 1392. doi: 10.3389/fgene.2019.01392.

Qin, Y., Park, S-Y., Oh, S-W. et al. (2017). Nutritional composition analysis for beta-carotene-enhanced transgenic soybeans (*Glycine max* L.). *Applied Biological Chemistry* 60: 299–309. doi: 10.1007/s13765-017-0282-z.

Rajam, M.V. (2020). RNA silencing technology: A boon for crop improvement. *Journal of biosciences* 45: 1–5. doi: 10.1007/s12038-020-00082-x.

Rana, M.M., Takamatsu, T., Baslam, M. et al. (2019). Salt tolerance improvement in rice through efficient SNP marker-assisted selection coupled with speed-breeding. *International Journal of Molecular Sciences* 20: 2585. doi: 10.3390/ijms20102585.

Rashid, A., Ram, H., Zou, C. et al. (2020). Simultaneous biofortification of rice with zinc, iodine, iron and selenium through foliar treatment of a micronutrient cocktail in five countries. *Frontiers in Plant Science* 11: 589835. doi: 10.3389/fpls.2020.589835.

Rehman, H.M., Cooper, J.W., Lam, H.M. et al. (2019). Legume biofortification is an underexploited strategy for combatting hidden hunger. *Plant, cell & environment* 42: 52–70. doi: 10.1111/pce.13368.

Riaz, U., Aziz, H., Anum, W. et al. (2020) Biofortification technologies used in agriculture in relation to micronutrients. In: *Plant Micronutrients* (eds. T.A. Khalid and R. Hakeem), 225–239. New York, NY: Springer.

Saini, D.K., Devi, P. and Kaushik, P. (2020). Advances in genomic interventions for wheat biofortification: A review. *Agronomy* 10: 62. doi: 10.3390/agronomy10010062.

Saleh, A.S., Wang, P., Wang, N. et al. (2019). Brown rice versus white rice: Nutritional quality, potential health benefits, development of food products, and preservation technologies. *Comprehensive Reviews in Food Science and Food Safety* 18: 1070–1096. doi: 10.1111/1541-4337.12449.

Samineni, S., Sen, M., Sajja, S.B. et al. (2020). Rapid generation advance (RGA) in chickpea to produce up to seven generations per year and enable speed breeding. *The Crop Journal* 8: 164–169. doi: 10.1016/j.cj.2019.08.003.

Sandarani, M. and Kulathunga, K. (2019). A brief review: Lectins, protease inhibitors and saponins in cereals and legumes. *Asian Food Science Journal* 10: 1–4. doi: 10.9734/afsj/2019/v10i430044.

Sanjeeva Rao, D., Neeraja, C., Madhu Babu, P. et al. (2020). Zinc biofortified rice varieties: Challenges, possibilities, and progress in India. *Frontiers in Nutrition* 7: 26. doi: 10.3389/fnut.2020.00026.

Sedeek, K E., Mahas, A. and Mahfouz, M. (2019). Plant genome engineering for targeted improvement of crop traits. *Frontiers in Plant Science* 10: 114. doi: 10.3389/fpls.2019.00114.

Shahzad, R., Shakra Jamil, S A., Nisar, A. et al. (2021). Harnessing the potential of plant transcription factors in developing climate resilient crops to improve global food security: Current and future perspectives. *Saudi Journal of Biological Sciences* 28: 2323. doi: 10.1016/j.sjbs.2021.01.028.

Singh, S.P., Keller, B., Gruissem, W. et al. (2017). Rice NICOTIANAMINE SYNTHASE 2 expression improves dietary iron and zinc levels in wheat. *Theoretical and Applied Genetics* 130: 283–292. doi: 10.1007/s00122-016-2808-x.

Singla, P. and Grover, K. (2017). Biofortified cereal crops: A sustainable approach for food and nutritional security. *Current Journal of Applied Science and Technology* 24: 1–13. doi: 10.9734/CJAST/2017/37865.

Sultana, R., Jamil, S., Aslam, M. et al. (2019). Overview of quality protein maize and molecular breeding approaches for its development. *International Journal of Biosciences* 14: 533–542. doi: 10.12692/ijb/14.1.533-542.

Suwarno, W.B., Pixley, K.V., Palacios-Rojas, N. et al. (2015). Genome-wide association analysis reveals new targets for carotenoid biofortification in maize. *Theoretical and Applied Genetics* 128: 851–864. doi: 10.1007/s00122-015-2475-3.

Tan, G.Z., Das Bhowmik, S.S., Hoang, T.M. et al. (2017). Finger on the pulse: Pumping iron into chickpea. *Frontiers in Plant Science* 8: 1755. doi: 10.3389/fpls.2017.01755.

Thanh, N.D. (2019). Provitamin A biofortification in maize through genetic engineering and marker-assisted selection. *Academia Journal of Biology* 41: 37–53. doi: 10.15625/2615-9023/v41n4.13804.

Tibbs Cortes, L., Zhang, Z. and Yu, J. (2021). Status and prospects of genome-wide association studies in plants. *Plant Genome* 14: e20077. doi: 10.1002/tpg2.20077.

Ullah, R., Ullah, Z., Iqbal, J. et al. (2023a). Aspartic acid-based nano-copper induces resilience in zea mays to applied lead stress via conserving photosynthetic pigments and triggering the antioxidant biosystem. *Sustainability* 15 (16): 12186.

Ullah, Z., Iqbal, J., Abbasi, B.A. et al. (2023b). How Genome Editing Can Be Helpful in the Biofortification of Legumes. In: *Legumes Biofortification*, 207–232. Cham: Springer International Publishing.

Ullah, Z., Iqbal, J., Abbasi, B.A. et al. (2023c). Assessment of Gus expression induced by anti-sense *OsPPO* gene promoter and antioxidant enzymatic assays in response to drought and heavy metal stress in transgenic *Arabidopsis thaliana*. *Sustainability* 15 (17): 12783.

Uppal, C., Kaur, A. and Sharma, C. (2021). Genome engineering for nutritional improvement in pulses. In: *Genome Engineering for Crop Improvement* (ed. S.K. Upadhyay), 157–180. Hoboken, NJ: Wiley.

Van Der Straeten, D., Bhullar, N.K., De Steur, H. et al. (2020). Multiplying the efficiency and impact of biofortification through metabolic engineering. *Nature Communications* 11: 1–10. doi: 10.1038/s41467-020-19020-4.

Vinoth, A. and Ravindhran, R. (2017). Biofortification in millets: A sustainable approach for nutritional security. *Frontiers in Plant Science* 8: 29. doi: 10.3389/fpls.2017.00029.

Wang, M., Kong, F., Liu, R. et al. (2020). Zinc in wheat grain, processing, and food. *Frontiers in Nutrition* 7: 124. doi: 10.3389/fnut.2020.00124.

Waters, B.M. and Sankaran, R.P. (2011). Moving micronutrients from the soil to the seeds: Genes and physiological processes from a biofortification perspective. *Plant Science* 180: 562–574. doi: 10.1016/j.plantsci.2010.12.003.

Xu, M., Zhao, S., Zhang, Y. et al. (2017). Production of marker-free transgenic rice (*Oryza sativa* L.) with improved nutritive quality expressing AmA1. *Iranian Journal of Biotechnology* 15: 102. doi: 10.15171/ijb.1527.

Yadava, D.K., Hossain, F. and Mohapatra, T. (2018). Nutritional security through crop biofortification in India: Status & future prospects. *Indian Journal of Medical Research* 148:621. doi: 10.4103/ijmr.IJMR_1893_18.

Yang, Q.Q., Yu, W.H., Wu, H.Y. et al. (2021). Lysine biofortification in rice by modulating feedback inhibition of aspartate kinase and dihydrodipicolinate synthase. *Plant Biotechnol J* 19: 490–501. doi: 10.1111/pbi.13478.

Yeom, W.W., Kim, H.J., Lee, K-R. et al. (2020). Increased production of α-linolenic acid in soybean seeds by overexpression of lesquerella FAD3-1. *Frontiers in Plant Science* 10: 1812. doi: 10.3389/fpls.2019.01812.

Zhu, Q., Tan, J., Wang, B. et al. (2021). Genetic engineering for increasing antioxidant content in rice: Recent progress and future perspectives. In: *Molecular Breeding for Rice Abiotic Stress Tolerance and Nutritional Quality* (eds. M.A. Hossain, L. Hassan, K. Iftekharuddaula. et al.), 358–381. Hoboken, NJ: Wiley.

Zulfiqar, U., Maqsood, M. and Hussain, S. (2020). Biofortification of rice with iron and zinc: Progress and prospects. In: *Rice Research for Quality Improvement: Genomics and Genetic Engineering* (ed. R. Aryadeep). New York, NY: Springer.

4

Crops Biofortification through OMICs based Knowledge

Imran Khan[1], Uneebullah Arif[1], Mohammad Safdar Baloch[2], Asghar Ali Khan[2], Muhammad Faisal Shahzad[3], Qudrat Ullah Khan[4], Muhammad Amjad Nadim[2], Umar Khitab Saddozai[2], Sajid Fiaz[5], Adnan Noor Shah[6,]*

[1] *Department of Forestry and Range Management, Gomal University, Dera Ismail Khan, Pakistan*
[2] *Department of Agronomy, Gomal University, Dera Ismail Khan, Pakistan*
[3] *Department of Entomology, Gomal University, Dera Ismail Khan, Pakistan*
[4] *Department of Soil Science, Gomal University, Dera Ismail Khan, Pakistan*
[5] *Institute of Molecular Biology and Biotechnology, The University of Lahore, Lahore, Pakistan*
[6] *Department of Agricultural Engineering, Khwaja Fareed University of Engineering and Information Technology, Rahim Yar Khan, Pakistan*

*Corresponding author: adnan.shah@kfueit.edu.pk

4.1 Introduction

Omics technologies have revolutionized the landscape of healthcare and beyond by providing a comprehensive understanding of biological systems at various levels, including genomics, transcriptomics, proteomics, metabolomics, and epigenomics. These technologies have revolutionized biological research, providing vast amounts of data that hold the promise of unraveling complex biological processes and advancing personalized medicine. However, the adoption of omics approaches presents numerous challenges, ranging from technical hurdles in data integration to profound ethical considerations. This comprehensive review delves into the multifaceted landscape of omics adoption, examining the current challenges and potential solutions across various domains, including data integration, analytical techniques, regulatory frameworks, and ethical guidelines. By elucidating these challenges and proposing strategies to address them, this review aims to guide researchers, policymakers, and stakeholders in navigating the complexities of omics technologies and harnessing their full potential for scientific advancement and societal benefits (Lane et al. 2014; Kim et al. 2014; Wilhelm et al. 2014; Low et al. 2013).

Omics technologies, encompassing genomics, transcriptomics, proteomics, metabolomics, and beyond, have emerged as indispensable tools in biological research, offering unprecedented insights into the molecular underpinnings of life. These approaches enable comprehensive analysis of biological molecules on a genome-wide scale, facilitating the exploration of gene expression patterns, protein dynamics, metabolic pathways, and more. The wealth of data generated by omics technologies holds immense potential for elucidating disease mechanisms, identifying biomarkers, and guiding personalized therapeutic interventions. However, the widespread adoption of omics approaches is beset by a myriad of challenges, spanning technical, regulatory, and ethical dimensions. This review seeks to examine these challenges comprehensively, with a focus on data integration, analytical methodologies, regulatory frameworks, and ethical considerations (Aebersold et al. 2013; Wilhelm et al. 2014; Low et al. 2013).

Biofortification, the process of increasing the nutritional value of crops through conventional breeding or biotechnology, holds immense promise for addressing global health and malnutrition challenges. Omics technologies – including genomics, transcriptomics, proteomics, and metabolomics – have revolutionized our understanding of plant biology and offer unprecedented opportunities to enhance biofortification efforts. This summary explores the potential of omics technologies in advancing biofortification strategies for improving global health and nutrition (Singh et al. 2016).

Improving micronutrients in various diets on a large basis to enhance biotic health is quite crucial. Numerous methods, including conventional breeding, agronomic breeding, and biotechnology, help accomplish food biofortification. The

Crop Biofortification: Biotechnological Approaches for Achieving Nutritional Security under Changing Climate, First Edition. Edited by Adnan Noor Shah et al.

majority of these practices are accepted on global basis; therefore, it should be a primary practice in various research efforts. It is believed that micronutrient deficiencies are known as hidden hunger affecting approximately more than 2 billion communities, with most of the cases originated in Asia, Latin America, and Africa. The most well-known micronutrient deficiencies concern iron, iodine, zinc, and vitamin A. Normally, micronutrient deficiency can have extreme well-being repercussions, such as effect on mental health, stunted growth, perinatal complexities, and sometime sudden deaths. Furthermore, vitamin A deficiency is a leading cause of blindness during childhood, i.e. xerophthalmia, and various researchers have been depicted that approximately 190 million children are affected by micronutrient deficiencies (Locato, Cimini, and Gara 2013; Paine et al. 2005).

4.1.1 Navigating the Current Challenges in Omics Adoption: From Data Integration to Ethical Considerations

4.1.1.1 Technical Challenges in Omics Adoption

4.1.1.1.1 Data Integration

One of the primary challenges in omics research lies in integrating heterogeneous datasets generated from different omics platforms and experimental sources. Each omics technology produces vast quantities of data with distinct formats, measurement units, and quality metrics, complicating efforts to merge and analyze them cohesively. Data integration requires sophisticated bioinformatics tools and computational algorithms capable of harmonizing disparate datasets, resolving inconsistencies, and extracting meaningful insights. Moreover, ensuring data interoperability and standardization is essential for facilitating cross-study comparisons and meta-analyses. Advances in data integration methodologies, such as machine learning (ML) algorithms and semantic web technologies, hold promise in overcoming these challenges and enhancing the utility of omics data for scientific discovery (Ruggieri et al. 2016; Chen et al. 2012; Kato, Takahashi and Saito 2011).

4.1.1.2 Analytical Techniques

In addition to data integration, omics researchers face challenges in selecting appropriate analytical techniques to extract relevant information from high-dimensional datasets. Traditional statistical methods may be ill-suited for omics data due to its complexity, nonlinearity, and inherent noise. Consequently, there is a growing need for advanced analytical approaches, including ML, network analysis, and deep learning, capable of uncovering hidden patterns, identifying biomarkers, and predicting biological outcomes. However, the application of these techniques requires careful validation, robust model interpretation, and consideration of potential biases to ensure the reliability and reproducibility of results. Collaborative efforts between bioinformaticians, statisticians, and domain experts are essential for developing and validating analytical frameworks tailored to omics data analysis (Thavarajah, Thavarajah, and Gupta 2014; Chen et al. 2012; Kato, Takahashi, and Saito 2011).

4.1.1.3 Regulatory Frameworks and Standards

4.1.1.3.1 Data Privacy and Security

As omics technologies generate vast amounts of sensitive biological data, ensuring privacy and security is paramount to protect individuals' confidentiality and prevent unauthorized access or misuse. Regulatory frameworks, such as the General Data Protection Regulation in Europe and the Health Insurance Portability and Accountability Act in the United States, impose stringent requirements for the collection, storage, and sharing of personal health information, including genomic data. Compliance with these regulations necessitates robust data encryption, access controls, and anonymization techniques to safeguard genomic privacy while facilitating data sharing for research purposes. Moreover, ethical guidelines and institutional review processes play a crucial role in ensuring that genomic research adheres to principles of beneficence, autonomy, and justice, thereby mitigating potential risks and ensuring responsible conduct (Baky et al. 2022).

4.1.1.4 Reproducibility and Transparency

Another regulatory challenge in omics research pertains to the reproducibility and transparency of findings, which are essential for validating scientific discoveries and informing evidence-based decision-making. Concerns have been raised about the lack of reproducibility in omics studies due to issues such as inadequate sample sizes, incomplete reporting of methods and results, and publication bias. To address these challenges, journals, funding agencies, and scientific communities have adopted initiatives to promote open science practices, such as preregistration of studies, sharing of raw data and code, and replication studies. By enhancing transparency and promoting data sharing, these initiatives aim to foster collaborative research environments and improve the robustness of omics research findings (Sauer et al. 2017).

4.1.1.5 Ethical Considerations

4.1.1.5.1 Informed Consent and Participant Autonomy

Ethical considerations loom large in omics research, particularly concerning informed consent and participant autonomy. Obtaining informed consent for genomic studies entails adequately informing participants about the nature of the research, potential risks and benefits, and the implications of genetic testing or data sharing. However, communicating complex genetic concepts to participants poses challenges due to the inherent uncertainty and multifactorial nature of genomic information. Furthermore, ensuring voluntary participation and respecting participants' right to withdraw from research are essential principles that must be upheld to protect individual autonomy and prevent coercion or undue influence (Ruggieri et al. 2016; Chen et al. 2012; Micheel, Nass, and Omenn 2012).

4.1.1.6 Equity and Justice

Omics research has the potential to exacerbate existing health disparities if not conducted with attention to equity and justice. Biases in sample recruitment, data representation, and interpretation can lead to underrepresentation or misrepresentation of marginalized populations, perpetuating inequities in healthcare and exacerbating social injustices (Table 4.1). To address these concerns, omics researchers must adopt inclusive recruitment strategies, engage with diverse communities, and consider sociocultural factors that may influence health outcomes. Moreover, efforts to democratize access to genomic technologies and ensure equitable distribution of benefits are essential for promoting social justice and reducing health disparities (Kato, Takahashi, and Saito 2011; Thavarajah, Thavarajah, and Gupta 2014; Chakraborty et al. 2018; Manzoni et al. 2018).

4.2 Advancements in Omics Technologies

Omics technologies have revolutionized biological and medical research by offering comprehensive insights into the molecular underpinnings of life processes. Recent advancements in this field have propelled the understanding of genomics, transcriptomics, proteomics, metabolomics, and beyond. High-throughput sequencing techniques, such as next-generation sequencing, have significantly enhanced genomic and transcriptomic studies, enabling the analysis of entire genomes and transcriptomes with unprecedented speed and precision. Moreover, advancements in mass spectrometry and chromatography have revolutionized proteomics and metabolomics, allowing for the identification and quantification of proteins and metabolites in complex biological samples with remarkable sensitivity and specificity. Additionally, the integration of multi-omics approaches has emerged as a powerful strategy to unravel intricate biological networks and biomolecular interactions, paving the way for breakthroughs in personalized medicine, disease diagnostics, and drug discovery. As omics technologies continue to evolve, they hold immense promise for driving scientific discoveries and transforming healthcare in the future (Dai and Shen 2022). Advancements in omics technology focus on the following credentials.

Table 4.1 Outlining the current challenges faced in the adoption of omics technologies to enhance biofortification.

Current challenges	
Challenges	**Description**
Genetic complexity	Genetic basis of traits is often complex, requiring sophisticated analyses.
Phenotypic variation	Omics must correlate genotype with phenotype, which can be influenced by environment.
Data integration	Integrating multi-omics data poses challenges in interpretation and analysis.
Regulatory hurdles	Regulatory approval for genetically modified biofortified crops can be lengthy.
Accessibility	Omics technologies can be costly and require specialized expertise.

4.2.1 Genomics

The study of an individual's complete set of DNA, enabling personalized medicine, predictive diagnostics, and targeted therapies.

4.2.2 Transcriptomics

Analysis of RNA transcripts to understand gene expression patterns, identify biomarkers, and uncover disease mechanisms (Manzoni et al. 2018).

4.2.3 Proteomics

Examination of proteins to elucidate their functions, interactions, and role in disease progression, aiding drug discovery and development (Veenstra 2021).

4.2.4 Metabolomics

Profiling metabolites to assess metabolic pathways, identify disease biomarkers, and optimize treatment strategies.

4.2.5 Epigenomics

Investigation of epigenetic modifications to understand gene regulation, environmental influences, and disease susceptibility (Swamy et al. 2018).

4.2.6 Integration of Omics Data

4.2.6.1 Multi-omics Approaches

Combining data from multiple omics platforms to gain a holistic view of biological systems, enabling more accurate disease diagnosis, prognosis, and treatment.

4.2.6.2 Systems Biology

Utilizing computational models and network analyses to integrate omics data with clinical parameters, environmental factors, and lifestyle information for personalized healthcare interventions (Manzoni et al. 2018; Kaushik et al. 2020a).

4.2.6.3 Big Data Analytics

Leveraging advanced computational algorithms and ML techniques to analyze vast omics datasets, extract meaningful insights, and identify novel therapeutic targets (Kaushik et al. 2020a; Kaushik et al. 2020b).

4.3 Applications in Healthcare

4.3.1 Precision Medicine

Tailoring healthcare interventions based on an individual's genetic makeup, molecular profile, and lifestyle factors to optimize treatment outcomes and minimize adverse effects (Cakmak 2008; Lung'aho et al. 2011).

4.3.2 Disease Biomarkers

Identification of molecular biomarkers for early disease detection, patient stratification, and monitoring treatment response, facilitating timely interventions and personalized therapies (Yang and Römheld 1999).

4.3.3 Drug Discovery and Development

Utilizing omics technologies to identify druggable targets, predict drug efficacy and toxicity, and streamline the drug development process, leading to the discovery of novel therapeutics and repurposing existing drugs (Sauer et al. 2017).

4.3.4 Predictive Diagnostics

Development of omics-based tests for predicting disease risk, progression, and response to treatment, enabling proactive healthcare management and preventive interventions (McConnell et al. 2014).

4.4 Challenges and Opportunities

4.4.1 Data Standardization and Integration

Establishing standardized protocols for omics data generation, storage, and analysis to ensure interoperability, reproducibility, and data sharing across research and clinical settings (Kaushik et al. 2020b).

4.4.2 Ethical and Regulatory Considerations

Addressing ethical issues related to privacy, consent, and data security, while adhering to regulatory frameworks governing omics research and clinical applications (Veenstra 2021).

4.4.3 Cost and Accessibility

Mitigating the high cost of omics technologies and ensuring equitable access to these tools and services, particularly in resource-limited settings, to realize their full potential in improving healthcare outcomes (Baky et al. 2022).

4.5 Future Directions

4.5.1 Single-cell Omics

Advancing single-cell omics technologies to unravel cellular heterogeneity, dynamics, and interactions within complex tissues and ecosystems, opening new avenues for understanding disease pathogenesis and developing targeted therapies (Sauer et al. 2017).

4.5.2 Multi-omics Integration

Further integrating data from diverse omics platforms with clinical, environmental, and behavioral data to enhance predictive modeling, biomarker discovery, and treatment optimization in precision medicine (Manzoni et al. 2018; Kaushik et al. 2020a).

4.5.3 AI and Deep Learning

Harnessing the power of artificial intelligence (AI) and deep learning algorithms to analyze omics data, identify patterns, and generate actionable insights for personalized healthcare delivery, drug discovery, and disease management (Kaushik et al. 2020b).

4.6 Genomics in Biofortification

Genomics plays a pivotal role in identifying genes associated with desirable nutritional traits in crops. Through genome sequencing and association studies, researchers can pinpoint genetic variations responsible for traits, such as increased micronutrient content, enhanced yield, and improved stress tolerance. Utilizing genomic information, breeders can employ marker-assisted selection to accelerate the development of biofortified crop varieties with targeted nutritional traits. Additionally, genome editing techniques such as CRISPR/Cas9 offer precise means to introduce beneficial genetic modifications, further enhancing biofortification efforts (Shin et al. 2006; Bouis et al. 2010).

4.6.1 Biofortification Constraints

Biofortification is considered as the idea of various crops breeding to enhance their nutritional value. However, adopting these methods in an underdeveloped country is hellacious. It requires huge funding and a time span to undertake these techniques and implementation on farmland. For genetic engineering, high-class laboratories will be needed and specific instrument and experienced labor. Besides this, biofortification efforts are yet to be taken for numerous crops (Ray et al. 2016; Rehman et al. 2020).

4.7 Transcriptomics and Proteomics

Transcriptomics and proteomics enable researchers to elucidate the molecular mechanisms underlying nutrient accumulation and metabolism in plants. By analyzing gene expression patterns and protein profiles, scientists can identify key regulatory genes and proteins involved in nutrient uptake, transport, and storage. This knowledge facilitates the manipulation of metabolic pathways to enhance nutrient bioavailability and accumulation in edible plant parts. Integration of transcriptomic and proteomic data also enables the identification of biomarkers for selecting superior biofortified crop varieties with optimized nutritional traits (Hegde, White, and Debouck 2003; Voelckel, Gruenheit, and Lockhart 2017).

4.8 Metabolomics for Nutrient Profiling

Metabolomics allows comprehensive profiling of small molecule metabolites in crops, providing insights into metabolic pathways related to nutrient composition and quality. By comparing metabolite profiles between conventional and biofortified crops, researchers can identify metabolic signatures associated with enhanced nutritional content. Metabolomics also facilitates the assessment of nutritional quality and safety of biofortified crops by detecting potential anti-nutritional compounds or contaminants. Furthermore, metabolomics data can guide targeted breeding strategies to optimize the accumulation of specific nutrients while minimizing undesirable compounds in biofortified crops (Baky et al. 2022; Menni et al. 2013).

4.9 Integration of Omics Data and Multi-omics Approaches

Integration of genomics, transcriptomics, proteomics, and metabolomics data enables a holistic understanding of plant nutritional traits and their underlying molecular mechanisms. Multi-omics approaches allow researchers to unravel complex interactions between genes, proteins, and metabolites influencing nutrient metabolism and biofortification outcomes (Table 4.2).

Table 4.2 Recent advancements in the adoption of omics technologies to enhance biofortification.

Recent advancements	
Advancements	**Description**
CRISPR-Cas9	Precise gene editing allows targeted modifications for enhanced biofortification.
Multi-omics integration	Advanced algorithms enable integration of genomics, transcriptomics, proteomics, and metabolomics data for comprehensive analysis.
Machine learning	Machine learning (ML) algorithms predict phenotype from genomic data, aiding in crop breeding.
High-throughput	Automated platforms accelerate data generation for large-scale phenotypic analysis.
Public–private partnerships	Collaborations facilitate sharing of resources and expertise, advancing research.

By combining omics datasets, scientists can identify key genetic targets for engineering biofortified crops with improved nutritional content, yield, and agronomic performance. Furthermore, systems biology approaches enable predictive modeling of plant metabolic networks, facilitating the design of tailored biofortification strategies for specific nutritional requirements and environmental conditions (Pineda et al. 2015; Skol et al. 2020).

4.9.1 Challenges and Future Perspectives

Despite the tremendous potential of omics technologies in biofortification, several challenges remain to be addressed. High-throughput omics data generation and analysis require sophisticated computational tools and bioinformatics infrastructure, posing challenges for researchers in developing countries with limited resources. Furthermore, ethical and regulatory considerations surround the deployment of genetically modified biofortified crops, necessitating robust risk assessment and regulatory frameworks. Collaboration between academia, industry, and government agencies is essential to overcome these challenges and translate omics-based discoveries into tangible solutions for global health and nutrition (Singh et al. 2016; Okwuonu et al. 2021).

4.10 Importance of Biofortification

Food fortification, supplementation, and dietary diversification initiatives have effectively tackled micronutrient malnutrition in numerous countries. However, biofortification offers distinct advantages over other interventions. It is a cost-effective and sustainable method, particularly beneficial for economically disadvantaged households and subsistence farmers (Nestel et al. 2006). Moreover, the initial investment is primarily required for research and development, making subsequent implementation self-sustaining (Gómez-Galera et al. 2010).

In various regions worldwide, food crops often lack essential micronutrients due to factors such as low native soil micronutrient levels, cultivation of high-yielding varieties requiring increased micronutrient quantities, soil depletion over time, acidic soil conditions exacerbated by liming practices, and limited micronutrient availability in sandy and calcareous soils (Fageria and Baligar 2008). Additionally, continuous application of high-analysis fertilizers with minimal micronutrient content and limited use of organic amendments further contribute to this deficiency (Fageria et al. 2012). Thus, enhancing micronutrient density and improving their bioavailability in major food crops emerges as a promising strategy to combat micronutrient malnutrition.

The primary focus of such programs is to enhance the levels of crucial micronutrients in staple foods (e.g., rice and potato) widely consumed by a significant portion of the global population. Even marginal improvements in bioavailable micronutrient levels in targeted food crops can have a significant impact on alleviating malnutrition, particularly in developing nations (Sperotto et al. 2012). Additionally, an optimized soil–plant system ensures nutrient cycling and promotes ecological sustainability (Yang, Chen, and Feng 2007).

In the long term, the biofortification approach is expected to gain prominence as a strategy with lower infrastructure dependence and compliance requirements (Gómez-Galera et al. 2010).

4.11 Conclusion

Omics technologies hold immense promise in transforming healthcare and beyond by providing unprecedented insights into the molecular mechanisms underlying health and disease. Embracing emerging trends such as multi-omics integration, AI-driven analytics, and single-cell profiling will accelerate the translation of omics research into clinical practice, paving the way for more personalized, predictive, and preventive healthcare strategies. However, addressing challenges related to data standardization, ethical considerations, and accessibility is crucial to realizing the full potential of omics in improving patient outcomes and population health.

Omics technologies offer unprecedented opportunities to enhance biofortification efforts and address global health and malnutrition challenges. Genomics, transcriptomics, proteomics, and metabolomics provide valuable insights into the molecular mechanisms underlying nutrient metabolism in plants, facilitating the development of biofortified crop varieties with improved nutritional content and agronomic traits. Integration of omics data and multi-omics approaches enables a holistic understanding of plant biology and inform targeted breeding strategies for enhancing nutrient bioavailability and accumulation. Despite challenges, continued investment in omics research holds great promise for advancing biofortification as a sustainable solution to improve global health and nutrition.

The adoption of omics technologies holds immense promise for advancing our understanding of human biology and improving healthcare outcomes. However, realizing this potential requires addressing a myriad of challenges, ranging from technical hurdles in data integration and analysis to profound ethical considerations surrounding privacy, consent, and equity. By elucidating these challenges and proposing strategies to overcome them, this review aims to guide researchers, policymakers, and stakeholders in navigating the complexities of omics adoption and harnessing its transformative power for scientific discovery and societal benefit. Through interdisciplinary collaboration, transparent communication, and ethical stewardship, we can navigate the current challenges in omics adoption and chart a course toward a more equitable, responsible, and impactful era of genomic medicine.

References

Aebersold, R., Bader, G.D., Edwards, A.M. et al. (2013). The biology/disease-driven human proteome project (B/D-HPP): enabling protein research for the life sciences community. *Journal of Proteome Research* 12 (1): 23–27.

Baky, M.H., Shamma, S.N., Xiao, J. et al. (2022). Comparative aroma and nutrients profiling in six edible versus non-edible cruciferous vegetables using MS-based metabolomics. *Food Chemistry* 383: 132374.

Cakmak, I. (2008). Enrichment of cereal grains with zinc: agronomic or genetic biofortification? *Plant & Soil* 302: 1–17.

Chakraborty, S., Hosen, M.I., Ahmed, M. et al. (2018). Onco-multi-OMICS approach: a new frontier in cancer research. *Biomed Research International* 2018 (1): 9836256. doi: 10.1155/2018/9836256.

Chen, R., Mias, G.I., Li-Pook-Than, J. et al. (2012). Personal omics profiling reveals dynamic molecular and medical phenotypes. *Cell* 148 (6): 1293–1307.

Dai, X. and Shen, L. (2022). Advances and trends in omics technology development. *Frontiers in Medicine* 9: 911861.

Fageria, N.K. and Baligar, V.C. (2008). Ameliorating soil acidity of tropical oxisols by liming for sustainable crop production. *Advances in Agronomy* 99: 345–399.

Fageria, N.K., Moraes, M.F., Ferreira, E.P.B. et al. (2012). Biofortification of trace elements in food crops for human health. *Communications in Soil Science and Plant Analysis* 43 (3): 556–570.

Gómez-Galera, S., Rojas, E., Sudhakar, D. et al. (2010). Critical evaluation of strategies for mineral fortification of staple food crops. *Transgenic Research* 19: 165–180.

Hegde, P.S., White, I.R., and Debouck, C. (2003). Interplay of transcriptomics and proteomics. *Current Opinion in Biotechnology* 14 (6): 647–651.

Kato, H., Takahashi, S. and Saito, K. (2011). Omics and integrated omics for the promotion of food and nutrition science. *Journal of Traditional Complement Medicines* 1 (1): 25–30. https://www.sciencedirect.com/science/article/pii/S2225411016300530https://www.sciencedirect.com.

Kaushik, A.C., Mehmood, A., Dai, X. et al. (2020a). WeiBI (web-based platform): enriching integrated interaction network with increased coverage and functional proteins from genome-wide experimental OMICS data. *Scientific Reports* 10 (1): 5618.

Kaushik, A.C., Mehmood, A., Upadhyay, A.K. et al. (2020b). CytoMegaloVirus infection database: a public omics database for systematic and comparable information of CMV. *Interdisciplinary Sciences: Computational Life Sciences* 12: 169–177.

Kim, M.S., Pinto, S.M., Getnet, D. et al. (2014). A draft map of the human proteome. *Nature* 509 (7502): 575–581.

Lane, L., Bairoch, A., Beavis, R.C. et al. (2014). Metrics for the Human Proteome Project 2013–2014 and strategies for finding missing proteins. *Journal of Proteome Research* 13 (1): 15–20.

Locato, V., Cimini, S. and Gara, L.D. (2013). Strategies to increase vitamin C in plants: from plant defense perspective to food biofortification. *Frontiers in Plant Science* 4: 152.

Low, T.Y., van Heesch, S., van den Toorn, H. et al. (2013). Quantitative and qualitative proteome characteristics extracted from in-depth integrated genomics and proteomics analysis. *Cell Reports* 5 (5): 1469–1478.

Lung'aho, M.G., Mwaniki, A.M., Szalma, S.J. et al. (2011). Genetic and physiological analysis of iron biofortification in maize kernels. *PLoS One* 6 (6): e20429.

Manzoni, C., Kia, D.A., Vandrovcova, J. et al. (2018). Genome, transcriptome and proteome: the rise of omics data and their integration in biomedical sciences. *Briefings in Bioinformatics* 19 (2): 286–302. https://academic.oup.com/bib/article/19/2/286/2562648.

McConnell, E.R., Bell, S.M., Cote, I. et al. (2014). Systematic Omics Analysis Review (SOAR) tool to support risk assessment. *PLoS One* 9: e110379.

Menni, C., Zhai, G., MacGregor, A. et al. (2013). Targeted metabolomics profiles are strongly correlated with nutritional patterns in women. *Metabolomics* 9: 506–514.

Micheel, C.M., Nass, S.J. and Omenn, G.S. (2012). Committee on the Review of Omics-based Tests for Predicting Patient Outcomes in Clinical Trials: Board on Health Care Services; Board on Health Sciences Policy; Institute of Medicine. In: *Evolution of Translational Omics: Lessons Learned and the Path Forward*. Washington, DC: National Academies Press.

Nestel, P., Bouis, H.E., Meenakshi, J.V. et al. (2006). Biofortification of staple food crops. *The Journal of Nutrition* 136 (4): 1064–1067.

Okwuonu, I.C., Narayanan, N.N., Egesi, C.N. et al. (2021). Opportunities and challenges for biofortification of cassava to address iron and zinc deficiency in Nigeria. *Global Food Security* 28: 100478.

Paine, J.A., Shipton, C.A., Chaggar, S. et al. (2005). Improving the nutritional value of Golden Rice through increased pro-vitamin A content. *Nature Biotechnology* 23 (4): 482–487.

Pineda, S., Gomez-Rubio, P., Picornell, A. et al. (2015). Framework for the integration of genomics, epigenomics and transcriptomics in complex diseases. *Human Heredity* 79 (3–4): 124–136.

Ray, P., Datta, S.P., Rakshit, R. et al. (2016). Agronomic bio-fortification of food crops with zinc and iron for ameliorating their deficiencies in humans: constraints and possibilities. *Indian Journal of Fertilisers* 12 (7): 28–35.

Rehman, A., Farooq, M., Ullah, A. et al. (2020). Agronomic biofortification of zinc in Pakistan: status, benefits, and constraints. *Frontiers in Sustainable Food Systems* 4: 591722.

Ruggieri, V., Bostan, H., Barone, A. et al. (2016). Integrated bioinformatics to decipher the ascorbic acid metabolic network in tomato. *Plant Molecular Biology* 91: 397–412.

Sauer, U.G., Deferme, L., Gribaldo, L. et al. (2017). The challenge of the application of 'omics technologies in chemicals risk assessment: background and outlook. *Regulatory Toxicology and Pharmacology* 91: S14–S26.

Shin, Y.M., Park, H.J., Yim, S.D. et al. (2006). Transgenic rice lines expressing maize C1 and R-S regulatory genes produce various flavonoids in the endosperm. *Plant Biotechnology Journal* 4 (3): 303–315.

Singh, U., Praharaj, C.S., Chaturvedi, S.K. et al. (2016). Biofortification: introduction, approaches, limitations, and challenges. *Biofortification of Food Crops* 3–18.

Skol, A.D., Jung, S.C., Sokovic, A.M. et al. (2020). Integration of genomics and transcriptomics predicts diabetic retinopathy susceptibility genes. *elife* 9: e59980.

Sperotto, R.A., Ricachenevsky, F.K., de Abreu Waldow, V. et al. (2012). Iron biofortification in rice: it's a long way to the top. *Plant Science* 190: 24–39.

Swamy, B.M., Descalsota, G.I.L., Nha, C.T. et al. (2018). Identification of genomic regions associated with agronomic and biofortification traits in DH populations of rice. *PLoS One* 13 (8): e0201756.

Thavarajah, D., Thavarajah, P. and Gupta, D.S. (2014). Pulses biofortification in genomic era: multidisciplinary opportunities and challenges. *Legumes in the Omic Era*. 207–220. doi: 10.1007/978-1-4614-8370-0_10.

Veenstra, T.D. (2021). Omics in systems biology: current progress and future outlook. *Proteomics* 21 (3–4): 2000235.

Voelckel, C., Gruenheit, N. and Lockhart, P. (2017). Evolutionary transcriptomics and proteomics: insight into plant adaptation. *Trends in Plant Science* 22 (6): 462–471.

Wilhelm, M., Schlegl, J., Hahne, H. et al. (2014). Mass-spectrometry-based draft of the human proteome. *Nature* 509 (7502): 582–587.

Yang, X.E., Chen, W.R. and Feng, Y. (2007). Improving human micronutrient nutrition through biofortification in the soil–plant system: China as a case study. *Environmental Geochemistry and Health* 29: 413–428.

5

Current Challenges and Recent Advancements in the Adoption of Omics to Enhance Biofortification

Zakir Ullah[1†], Javed Iqbal[2*†], Banzeer Ahsan Abbasi[3], Shumaila Ijaz[4], Aqsa Anjum[5], Tabassum Yaseen[2], Ghulam Murtaza[6], Rashid Iqbal[7], Sajjad Hyder[8], Sobia Kanwal[9], Tariq Mahmood[1]*

[1] *Department of Plant Sciences, Faculty of Biological Sciences, Quaid-i-Azam University, Islamabad, Pakistan*
[2] *Department of Botany, Bacha Khan University, Charsadda, Khyber Pakhtunkhwa, Pakistan*
[3] *Department of Botany, Rawalpindi Women University, 6th Road, Satellite Town, Rawalpindi, Pakistan*
[4] *School of Biomedical Engineering, Shenzhen University Medical School, Shenzhen University, Shenzhen, China*
[5] *Department of English Literature and Applied Linguistics Government Post Graduate College Timergara Dir Lower*
[6] *Faculty of Environmental Science and Engineering, Kunming University of Science and Technology, Kunming, China*
[7] *Department of Agronomy, Faculty of Agriculture and Environment, The Islamia University of Bahawalpur, Bahawalpur, Pakistan*
[8] *Department of Botany, GC Women University Sialkot, Pakistan*
[9] *Department of Biology and Environmental Sciences, Allama Iqbal Open University, Islamabad, Pakistan*
[†] *Equal contribution*

*Correspondence: zakirullah@bs.qau.edu.pk; javed89qau@gmail.com

5.1 Introduction

5.1.1 Biofortification

Biofortification aims to increase food crops' nutrient content and bioavailability during the growing season (Rana et al. 2020). This process seeks to improve crops' micronutrient levels and maintains desirable agronomic traits like high yield and disease resistance (Ghorai et al. 2022). Biofortification adds nutrients to crops before harvest, as opposed to food fortification, which does it during the postharvest processing of food products (Ullah et al. 2023a). Food fortification requires repeated nutrient additions, while biofortification occurs once per crop variety (Pathak et al. 2018). The most long-term remedy for concealed hunger and malnutrition is thought to be biofortification (Singh et al. 2020a). Major initiatives, including Harvest Plus, the Biocassava project, and the National Agricultural Research Organization (NARO), focus on developing biofortified crop varieties to improve nutritional security (Sheoran et al. 2022). Established in 2003, the Harvest Plus Program aims to enhance the nutritional value of food crops through biofortification, targeting countries in Asia and Africa to ensure high-quality biofortified staples are accessible and nutrient-rich after consumption (Ijaz et al. 2023a). Traditional interventions like food supplementation and industrial food fortification mainly benefit developed countries, whereas biofortification specifically targets developing and rural areas, though it also benefits the developed world (Bohra et al. 2016).

Evaluating nutrient bioavailability, establishing targeted nutrient levels, comprehending the population's nutritional demands, and maximizing nutrient absorption and retention during processing and storage are all necessary for the successful use of biofortification (Shahzad et al. 2021). Thirty three million people in Africa, Asia, Latin America, and the Caribbean consumed biofortified crops in 2017 (Gupta et al. 2023). Orange-fleshed sweet potatoes (OFSPs), Golden Rice, vitamin A-enriched yellow and orange maize, and zinc- and iron-fortified rice and wheat are a few examples of biofortified crops (Kiran et al. 2022). Biofortification programs have shown positive results. For instance, a six-month study in Nigeria found that preschool children who consumed biofortified yellow cassava had significantly improved vitamin A status compared to those who ate white cassava. Similarly, in Rwanda, reproductive women who ate iron-biofortified beans showed increased levels of hemoglobin, serum ferritin, and body iron (Nasar et al. 2024). Biofortification offers several advantages over other nutritional interventions: it addresses the needs of both urban and rural populations, can be implemented at low costs after initial development, allows for multiple nutrients to be biofortified without increasing food prices, and has minimal impact on the sensory properties of food (Haas et al. 2016). However, it has a limitation in that it cannot quickly rectify severe nutrient deficiencies in populations (Singh et al. 2020b).

Crop Biofortification: Biotechnological Approaches for Achieving Nutritional Security under Changing Climate, First Edition. Edited by Adnan Noor Shah et al.

5.1.2 Biofortification of Plant-based Foods

There has been a notable increase in the production, consumption, and sales of plant-based foods (PBFs) worldwide. Having the ability to be grown and consumed locally, vegetables are essential in the fight against hunger, poverty, and undernutrition (Bohra et al. 2015). Many consumers prefer exclusive PBFs due to their associations with health improvements, reduced environmental impact, and enhanced food security (Bohra et al. 2015). Increased fiber, vitamins K and C, folate, magnesium, beta-carotene, and potassium are among the nutritional advantages of PBFs (Husaini 2022). A high-vegetable diet also provides nutrients including calcium, iodine, and selenium, which support hormone production, blood pressure regulation, bone strength, heart health, and mental wellness (Singh et al. 2022). Nonetheless, strict plant-based diets need to be followed to prevent shortages in some micronutrients, especially vitamin B12, which is mostly found in animal products unless it is supplemented or fortified (Ashoka et al. 2023). High concentrations of antinutritional elements such as tannins and phytates, which prevent minerals from being absorbed in the intestine, can also be found in PBFs (Kumar et al. 2019). Cereals can lose even more of their nutritious content when they are polished, milled, or pearled (Kumar et al. 2021a).

For populations where traditional fortification methods and supplementation prove difficult, PBF biofortification is a viable alternative. By improving nutrient bioaccessibility in the soil, uptake, and transportation through plant tissues, as well as accumulating in nontoxic amounts in edible portions, it entails raising nutrient levels and bioavailability (Tayade et al. 2023). This tackles the problems of certain nutrients not being able to be synthesized by plants and the uneven distribution of nutrients in various plant sections (Susmitha et al. 2024). For instance, biofortifying rice with pro-vitamin A aims to enhance its presence and bioavailability in the grains, which are the consumed portion of the plant (Gupta et al. 2022). Steps for biofortifying PBFs are outlined in the Harvest Plus Program and are divided into four categories: impact and socioeconomics, breeding, nutrition and food technology, and consumer response (Gupta et al. 2022a). While it is important to increase the concentration of nutrients, it is also important to minimize overaccumulation, as this can have negative effects on the crop's edible portion and result in the accumulation of unwanted metals that pose health hazards (Amah et al. 2019). For instance, adults should consume no more than 0.055 mg of selenium per day, as it is required in trace amounts (Ijaz et al. 2024). Overconsumption of selenium can result in toxicity, which can cause digestive problems, cardiovascular illnesses, painful muscles, and even death. Therefore, biofortified foods ought to assist in fulfilling dietary requirements without posing any toxicity hazards (Rehman et al. 2020).

The percentage of nutrients available for physiological processes, bioavailability, and the percentage possibly absorbed, or bioaccessibility are essential in PBFs (Rao and Annadana 2017). Studies have shown significant improvements in the content, bioaccessibility, and bioavailability of iron and zinc in biofortified cowpea cultivars (Thakur et al. 2022). For instance, iron bioaccessibility in these cultivars was 2.5 times higher than in common beans, with bioavailability also significantly higher. Similarly, selenium biofortification in wheat enhanced both its content and bioaccessibility, although increased selenate doses did not consistently improve bioavailability (Bashir et al. 2023). Nutrient bioavailability in biofortified crops can be influenced by several factors, including nutritional, physiological, human, and genetic factors (Kathuria et al. 2024). The presence of antinutrients like phytic acid, tannins, lectins, saponins, and oxalic acid can limit the absorption of essential nutrients (Zulfiqar et al. 2024). Phytic acid, found in wheat, rice, barley, and beans, forms complexes with minerals, reducing their solubility and absorption (Ingle et al. 2023). Tannins, common in legumes and berry fruits, form complexes with iron, decreasing its bioavailability (Kumar et al. 2021b). Lectins, found in legumes, cereals, and fruits, can damage gut epithelial cells, limiting nutrient absorption (Gedil et al. 2024). Saponins, present in legumes and oats, can form complexes with sterols, affecting the absorption of fat-soluble vitamins. Oxalic acid binds with calcium and other minerals, preventing their absorption (Younas et al. 2023).

Pre-processing and processing techniques such as soaking, germination, heating, extrusion, milling, and chemical treatments can lower the antinutrient concentrations in food to lessen these effects (Chandra et al. 2020). Despite their negative impact on nutrient absorption, antinutrients have health-promoting properties (Kadapa et al. 2023). For instance, phytotate contains anti-inflammatory, anticarcinogenic, and hypoglycemic qualities. Polyphenols have been shown to reduce low-density lipoproteins and eliminate free radicals (Naik et al. 2024). Lectins inhibit cancer cells by promoting cell division. Saponins lower the risk of heart disease and cancer because they have antibacterial, cancer-preventing, and cholesterol-lowering qualities. Thus, achieving a balance between these antinutrients and bioactive substances can be advantageous (Rakkammal et al. 2022).

The food environment, which includes the matrix in which micronutrients are embedded, can influence micronutrient bioavailability (Knez and Stangoulis 2021). The presence of fats in the diet enhances the absorption of fat-soluble vitamins, while dietary fibers can bind minerals and reduce their bioavailability (Roy et al. 2022). Human factors such as age, health conditions, and genetic variations also impact nutrient absorption (Haq et al. 2024). For instance, micronutrient absorption can increase during pregnancy, lactation, and breastfeeding to meet the nutritional needs of the infant (Kumar et al. 2022). Health conditions like diabetes, obesity, celiac disease, and chronic pancreatitis can limit nutrient absorption (Lal et al. 2020). Genetic variations among different ethnic groups can affect micronutrient bioavailability (Mishra et al. 2022).

Agronomic intervention, traditional plant breeding, and genetic engineering are the three primary methods for biofortifying PBFs (Joshi et al. 2023). These tactics have been used on fruits, vegetables, legumes, oilseeds, and cereals; the greatest number of biofortified varieties are found in grains. The transgenic technique is frequently the most appropriate since oilseeds have low genetic diversity (Rasheed et al. 2024). In summary, while the global production and consumption of PBFs continue to rise, biofortification presents a sustainable solution to enhance their nutritional value and address micronutrient deficiencies, especially for exclusive plant-based diet consumers (Jha et al. 2022). Balancing the benefits and challenges of biofortification and antinutritional factors can help optimize the health benefits of PBFs for diverse populations (Shoeb et al. 2024).

5.2 Omics Technologies

The development of food crops in Africa has been transformed by omics technologies (OTs), which include proteomics, metabolomics, and genomics (Shoeb et al. 2024). The consequences of climate change, such as droughts and irregular rainfall, illnesses, pests, and inadequate infrastructure are major barriers to African agriculture (Sharma et al. 2021a). Through the study of an organism's whole gene pool, a process known as genomics, scientists can find genes associated with desired characteristics such as caloric content, disease resistance, and yield (Shoeb et al. 2024). This genetic information can be utilized in breeding programs to develop improved crop varieties (Jha et al. 2022). Understanding plant protein expression and function, as well as how they react to different stresses, is the main goal of proteomics. It also aims to uncover possible genetic targets for improvement (Ahmad et al. 2024b). To understand metabolic pathways and find molecules linked to crop performance and stress tolerance, metabolomics looks at a crop's whole set of metabolites (Ullah et al. 2023a).

OTs facilitate high-throughput screening and analysis, which enables researchers to identify genetic and molecular markers associated with important traits (Kadirimangalam et al. 2022). By analyzing the safety, usability, and environmental effects of genetically modified (GM) crops, these technologies also aid in the responsible development and evaluation of GM crops (Jha et al. 2022). This knowledge is crucial for creating GM crop varieties that are resistant to the biotic and abiotic stresses that are typical in Africa. OTs (Sharma et al. 2021a) were used by Kadirimangalam et al. (2022) to develop transgenic African rice (*Oryza glaberrima*) cultivars with improved tolerance to ultraviolet-B (UVB) exposure. Access to omics data can also foster collaboration and data sharing among researchers, leading to the development of large-scale databases and platforms for agricultural improvement (Jha et al. 2022). By applying OTs to produce crop varieties with improved nutritional profiles, higher yields, and increased resilience to climate change, Africa might potentially address its agricultural challenges and enhance food security (Sharma et al. 2021) (Figure 5.1).

5.2.1 Dietary Diversification

Micronutrients like iron (Fe), zinc (Zn), iodine (I_2), and various vitamins are often limited in staple foods but are present in certain other foods (Singh et al. 2020a; Khandagale et al. 2023). Thus, concealed hunger may result from a restricted diet consisting mainly of staple foods and insufficient amounts of foods high in micronutrients. This emphasizes how crucial it is to diversify your food to fight malnutrition and hidden hunger (Reddy et al. 2021). Dietary diversification or modification involves consuming a variety of foods that collectively provide adequate amounts of macro- and micronutrients to meet the recommended daily allowance (RDA) for a specific time frame (Sheoran et al. 2022; Kiran et al. 2022; Bohra et al. 2015).

In contrast to other strategies for reducing malnutrition and hidden hunger, dietary variety is culturally acceptable, boosts the immune system, and helps address all forms of nutrient shortages (Rana et al. 2020; Sharma et al. 2021b). However, this tactic comes with several serious drawbacks. These include the requirement for dietary pattern modifications, correct food data, and the existence of antinutrients in food that may interfere with the absorption of nutrients (Sheoran et al. 2022). Additionally, dietary diversification requires financial investment to purchase and produce high-quality

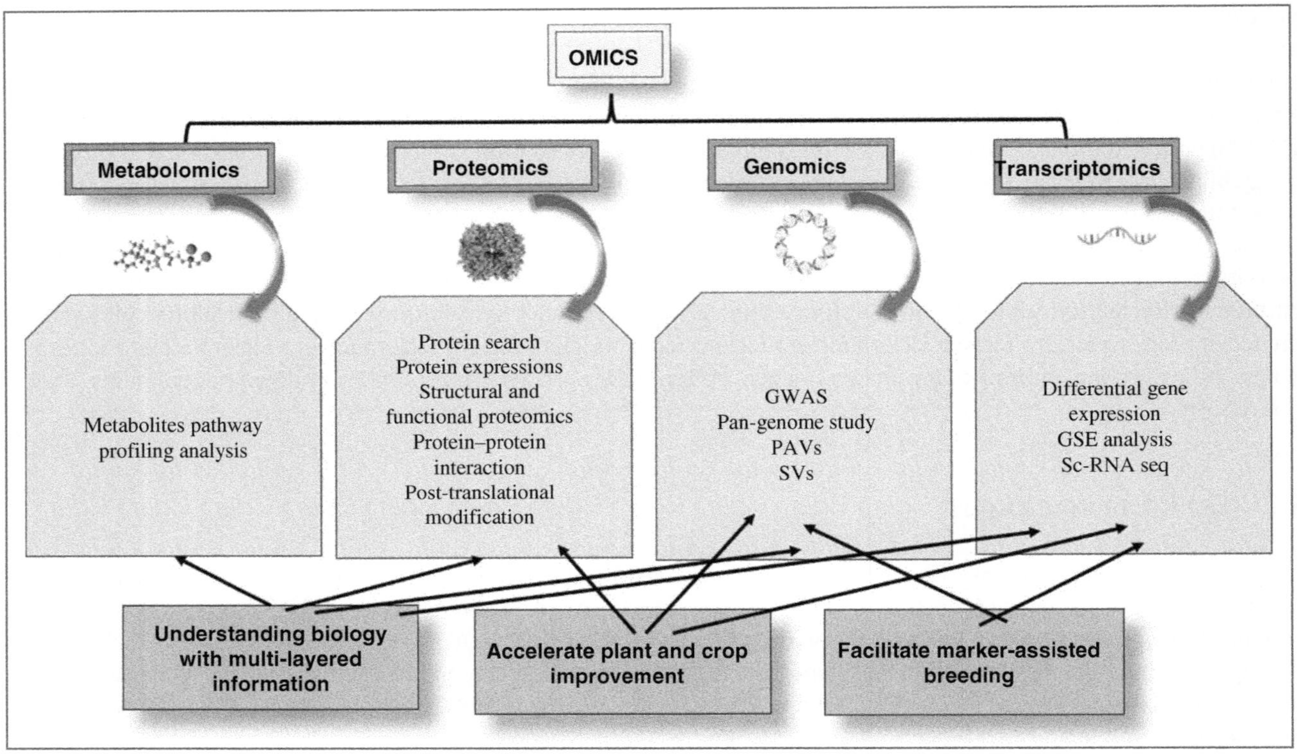

Figure 5.1 Advancing plant breeding efficiency through the multi-omics revolution.

food varieties (Singh et al. 2020a; Sharma et al. 2021; Hafeez et al. 2023). It is especially difficult to implement dietary diversity in impoverished nations, which makes people prefer other interventions such as food fortification, biofortification, and supplementation (Singh et al. 2020a; Sheoran et al. 2022).

5.2.2 Food Supplementation

To complement the nutrients gained from food and achieve the required dietary allowance (RDA), food supplementation entails taking extra nutrients in the form of capsules, syrups, or tablets (Sheoran et al. 2022; Singh et al. 2024). Vitamins, proteins, amino acids, minerals, fiber, vitamins, and calorie supplements are some of these supplements (Singh et al. 2024; Devi and Aglawe 2021). While mineral supplements such as zinc (Zn), iron (Fe), and amino acid supplements are less often employed, vitamin supplements are the most often used treatment for concealed hunger in many developing countries (Singh et al. 2020a; Sheoran et al. 2022; Benavente and Giménez 2021). In impoverished nations, food supplements are usually directed toward small populations with severe dietary shortages. This approach is direct, short-term, manageable, and adaptable to the requirements of a certain group of people. When compared to other interventions like dietary diversification, it is more cost-effective and produces good outcomes more quickly (Sheoran et al. 2022; Singh et al. 2024). In Ceará, Brazil, for example, a population-based study examined the relationship between vitamin A supplementation and child development using a sample of 1232 children from 8000 homes, ages 0–35 months (Siddique and Aley 2024). The study found that children who received vitamin A supplements had over 40% lower probability of delays in cognitive and motor development compared to those who did not receive the supplement. This highlights the positive impact of vitamin A supplementation on child development, demonstrating that food supplementation can improve the nutritional status of populations with micronutrient deficiencies.

It has been demonstrated that zinc supplementation during infancy improves certain growth outcomes, particularly after age two (Bohra et al. 2015). However, the absence of trustworthy diagnostic methods makes it difficult to identify persons who may be at risk of zinc insufficiency. Furthermore, it can be challenging to reach rural people because ongoing supplement provision is necessary (Sharma et al. 2021). Additional drawbacks of food supplementation include its high unsustainability, especially in developing nations, its reliance on the targeted population's compliance, and the requirement for well-defined procedures for successful implementation (Rana et al. 2020; Singh et al. 2020a; Bohra et al. 2015). Moreover, food supplements may result in toxicity, which could have a serious negative impact on the targeted populations' health (Devi and Aglawe 2021).

5.2.3 Food Fortification

To improve food's nutritional value and assist consumers in meeting the recommended daily allowance (RDA) for nutrients, food fortification entails adding such elements to meals, whether they are naturally present (Sheoran et al. 2022; Wang et al. 2023). Although increasing the nutritional value of food is the goal of both food enrichment and fortification, they are very different (Sharma et al. 2024). While food fortification replenishes depleted nutrients and supplements inadequately present nutrients, food enrichment replenishes nutrients lost during manufacturing. Four types of fortification can be applied to food: mass, target, voluntarily, and required (Ullah et al. 2023a). When food processing businesses decide to fortify processed foods with nutrients without being required by law, this is known as voluntary fortification (Ullah et al. 2023a). Examples include the addition of iron and vitamin A to wheat flour and morning cereals in nations such as the United Arab Emirates, Qatar, and the Gambia (Ullah et al. 2023a; Ahmad et al. 2024a). By adding nutrients to foods that are frequently consumed by a population, mass fortification seeks to prevent specific nutrient deficits. For example, rice is fortified in Asian nations like China (Wang et al. 2023; Sen et al. 2022; Verma et al. 2021). Government regulations manage mandatory fortification, which calls for the inclusion of nutrients such as iodine in salt, iron and folic acid in wheat flour, and vitamin A in edible oils (Ullah et al. 2023a; Sen et al. 2022). Target fortification focuses on specific groups within a population to address nutrient deficiencies, such as adding iron to infant formulas (Sen et al. 2022; Verma et al. 2021).

Food fortification has become increasingly relevant in both developing and developed countries due to changing dietary patterns and the rise in processed food consumption (Ullah et al. 2023a; Sheoran et al. 2022). Extensive food processing and storage can deplete nutrients like water-soluble vitamins and minerals (Wairich et al. 2022). Food fortification restores these lost nutrients and complements those that are present in insufficient quantities. Iodized salt, wheat fortified with iron and folic acid, milk fortified with calcium and vitamin D, and rice and edible oil enhanced with vitamin A are common examples (Ullah et al. 2023a; Sheoran et al. 2022; Ahmad et al. 2024a). However, the absence of well-organized processing and distribution networks may restrict the use of food fortification (Singh et al. 2020a; Bohra et al. 2015; Benavente and Giménez 2021). Additionally, food fortification often benefits urban areas more than rural ones, where higher socioeconomic status and better health education are prevalent (Bohra et al. 2015; Sharma et al. 2021). In developed nations, food fortification is quantifiable, sustainable, and inexpensive to apply (Sheoran et al. 2022; Ahmad et al. 2024b). Fortifying foods with nutrients like iron, iodine, and vitamins has significantly reduced the prevalence of nutrient deficiency-related diseases in developing countries (Sagare et al. 2023). However, in some areas, consumers may find fortified foods less enticing due to their increased costs (Sharma et al. 2021; Dharani et al. 2024). Therefore, biofortification is viewed as an additional intervention in underdeveloped nations to reduce malnutrition and hidden hunger.

5.3 Approaches for Biofortification

5.3.1 Agronomic Biofortification

Applying mineral fertilizers to soil or crops to increase the concentration and bioavailability of nutrients in the crops is known as agronomic biofortification (Husaini 2022). These methods were first intended to increase crop health and productivity. However, the focus has expanded to improve the nutritional quality of crops due to increasing awareness of nutrition's importance (Singh et al. 2020a; Reddy et al. 2021; Sagare et al. 2023). Climate change and soil nutrient depletion further necessitate expanding agronomic practices to enhance crop nutrition (Hafeez et al. 2023). Enhancing the solubilization and mobilization of minerals is the goal of agronomic biofortification (Kadirimangalam et al. 2022; Sharma et al. 2021b). The solubility and mobility of minerals, crop absorption capacity, soil composition, and the build-up of bioavailable minerals at nontoxic levels in edible portions all affect how successful it is (Gupta et al. 2022a; Kadirimangalam et al. 2022; Dharani et al. 2024). Since vitamins are made inside the crops, this method mostly addresses minerals. Thus, to successfully address micronutrient deficiencies, agronomic biofortification should be used in conjunction with other techniques (Rana et al. 2020; Singh et al. 2020a; Kadirimangalam et al. 2022). It is important to apply fertilizer correctly because incorrect application can damage crops and the environment. Ecological sustainability and economic benefits result from balanced fertilizing (Kadirimangalam et al. 2022). Soil microorganisms, essential for soil health, are sensitive to fertilization practices (Razzaq et al. 2019). While uneven fertilization can have long-term negative consequences on soil health, insufficient fertilization can cause nutrient deficits and change the microbial population in the soil (Kadirimangalam et al. 2022; Sharma et al. 2021b). Usually, mineral fertilizers are sprayed directly onto leaves or mixed into the soil. When higher concentrations of nutrients are required, soil application is more frequently utilized (Razzaq et al. 2019). When crop indications of nutrient deficiencies are apparent, foliar treatment is employed because it is more cost-effective (Gupta et al. 2022a; Benavente and Giménez 2021). Because foliar application enhances micronutrient content rather than just yield, it is frequently more effective than soil application (Gupta et al. 2022a; Lal et al. 2020). The type of fertilizer, crop characteristics, application timing, and environmental factors all affect the effectiveness of foliar application (Devi and Aglawe 2021).

Adjustments are necessary for agronomic biofortification with minerals like iron (Fe) and zinc (Zn), which have low mobility (Ishfaq et al. 2022). Metal chelators must be added to the fertilizer. $FeSO_4$ used topically is beneficial for Fe biofortification (Sharma et al. 2021b). Potassium iodate has been beneficial for iodine (Khan et al. 2024), as demonstrated in China (Singh et al. 2020a; Sharma et al. 2021b). Zinc biofortification can benefit from the use of inorganic fertilizers such as $ZnSO_4$, ZnO, and Zn-oxy-sulfate (Devi and Aglawe 2021). Zn chelators, like ZnEDTA, work quite well when applied topically (Nasar et al. 2024). Agronomically, selenium (Se) is reinforced as selenate, which the crop then transforms into organic selenomethionine (Siddique and Aley 2024). Depending on the type of soil and when it is applied, both foliar and soil treatments are appropriate for Se biofortification (Ayoub et al. 2020). On the other hand, foliar treatments can be expensive and quickly removed by rain (Ayoub et al. 2020). Leaf characteristics play a significant role in nutrient absorption during foliar applications (Dharani et al. 2024). The plasmodesmata allow nutrients to pass through the cuticle of the leaf and reach other areas. The permeability, age, and structure of the leaf all influence how well nutrients are absorbed (Bollinedi et al. 2022). Foliar application is most effective during the flowering and early milk phases of crop development, as these are the stages when nutrient absorption for fruit formation begins (Razzaq et al. 2019). This effectiveness was observed during the Zn biofortification of wheat, attributed to enhanced phloem mobility and nutrient remobilization (Roorkiwal et al. 2020). The efficiency of foliar application is further influenced by environmental factors such as temperature, humidity, wind speed, and time of day (Jaiwal et al. 2019). Nutrient permeability is enhanced by warm, humid weather in the early morning and late evening, but it might be decreased by low humidity and high temperatures, which enable the sprayed solution to evaporate (Kumar et al. 2017).

Agronomic biofortification techniques also involve covering and priming seedlings with mineral fertilizers. These practices have little impact on the quality of the food but increase crop output and development (Dar 2014). Many nations have successfully employed agronomic biofortification to address micronutrient shortages and increase agricultural productivity. For example, in Ghana, increased potassium (K) fertilizer application improved the weight of fruits and vegetables and significantly increased fruit K concentration (Aggarwal et al. 2022). In South Africa, fertilizer application increased the yield and β-carotene content of OFSP (Manzoor et al. 2023). While agronomic biofortification is simple and yields rapid results (Ahmad et al. 2022a), it can be costly, increasing the prices of biofortified crops and making them less accessible to poorer populations (Anushree et al. 2023). The success of agronomic biofortification depends heavily on farmer compliance (Ahmad et al. 2022). Repeated fertilizer application can lead to mineral accumulation and potential toxicity (Patel et al. 2023). Moreover, the increasing demand for minerals like Se can lead to resource depletion and environmental impact (Xia et al. 2020).

5.3.2 Plant Breeding

The process of creating new or genetically unique crop types with improved critical micronutrients is known as plant breeding (Singh and Prabha 2019). By utilizing genetic variations among species, plant breeding for biofortification seeks to increase the concentration and bioavailability of minerals in crops (Sagare et al. 2023). Plant breeding initially concentrated on increasing output and improving agronomic qualities; however, more recent methods seek to increase agronomic traits as well as nutritional quality (Aggarwal et al. 2022). To improve biofortification efficiency, effective plant breeding should concentrate on genotypes that improve mineral uptake, transport, and redistribution (Manzoor et al. 2023). Since the phloem vessels are the site of mineral redistribution and remobilization, it is imperative to improve mineral mobility there (Ahmad et al. 2022a). Unfortunately, compared to leaves or roots, the edible portions of crops have lower zinc (Zn) contents due to the poor mobility of Zn in phloem vessels (Patel et al. 2023). Conventional, molecular, and mutant breeding are the three primary methods of plant breeding (Kadirimangalam et al. 2022).

5.3.3 Conventional Breeding

The most widely used and recognized technique for biofortification is conventional breeding (Singh et al. 2020b; Kadirimangalam et al. 2022). It improves crop nutrition without sacrificing other agronomic characteristics (Ullah et al. 2023b; Kadirimangalam et al. 2022; Gupta, Verma, and Ravi 2023). To create new varieties with superior features, this strategy entails crossing crops with high nutrient density and other desired agronomic qualities (Singh et al. 2020b). The procedure includes choosing crop types that are biodiverse, evaluating attributes and nutrient levels, and figuring out how growing conditions affect these qualities (Singh et al. 2024). Through conventional breeding, around 299 biofortified crop types have been released in over 30 countries to date (Aggarwal et al. 2022). Quality protein maize (QPM) and OFSP biofortified with provitamin A are two examples (Ashoka et al. 2023; Kathuria et al. 2024; Haq et al. 2024). Additional instances include provitamin A-biofortified cassava and maize, Fe-biofortified beans, and biofortified wheat cultivars including "Zincol" and "Akbar-2019" (Ashoka et al. 2023; Ingle et al. 2023; Kadirimangalam et al. 2022).

5.3.4 Mutation Breeding

Through chemical treatments or physical means like irradiation, mutations are induced, leading to genetic variations through mutation breeding (Gupta et al. 2022a; Kadirimangalam et al. 2022). Using this method, the Indian Agricultural Research Institute (IARI) has biofortified chickpea mutants Pusa-408, Pusa-413, Pusa-417, and Pusa-547. Pusa-547 has been improved with a thin testa, bold and appealing seeds, higher yield performance, and improved cooking quality (Aggarwal et al. 2022).

5.3.5 Molecular Breeding

Finding the location of genes that control nutritional quality and closely related indicators is a key component of molecular breeding (Sagare et al. 2023). Conventional approaches can be used to breed desirable features into the crop utilizing these markers (Rana et al. 2020; Kadirimangalam et al. 2022). Compared to conventional breeding approaches, this methodology is faster because it can identify the presence of desirable features during developmental stages (Gupta et al. 2022a; Kadirimangalam et al. 2022). Improved provitamin A content in maize cultivars has been achieved using molecular or marker-assisted breeding, which can fulfill 25–50% of women's and children's vitamin A needs (Sharma et al. 2021b). Nigeria, India, and Zambia have all received these types of releases (Sharma et al. 2021b). Furthermore, several rice cultivars with enhanced agronomic features and elevated Fe and Zn levels have been bred (Sagare et al. 2023).

Compared to other biofortification techniques, plant breeding is more affordable and sustainable, requiring primarily research and development expenditures (Rana et al. 2020; Singh et al. 2020b). It has less of an influence on the environment than agronomic biofortification (Ashoka et al. 2023). Conventionally bred biofortified crops are usually accepted by consumers and are easier to get regulatory approval for than GM foods (Ashoka et al. 2023). To create cultivars with desired nutritional densities and agronomic characteristics, conventional breeding is labor-intensive and takes longer (Ashoka et al. 2023; Tayade et al. 2023). Insufficient genetic diversity across crops can impede the process of biofortification through plant breeding (Singh et al. 2020b; Ingle et al. 2023). Furthermore, this approach is not effective for biofortifying all nutrients. For instance, it was previously difficult to develop rice types with increased vitamin A content, but advancements in OTs have made this possible (Razzaq et al. 2019). Vegetatively propagated crops like bananas are not suitable for conventional breeding (Razzaq et al. 2019).

In summary, plant breeding for biofortification involves developing crops with enhanced micronutrient content by leveraging genetic differences (Manzoor et al. 2023). While it is sustainable and generally accepted, it is labor-intensive, time-consuming, and sometimes limited by genetic variation and crop propagation methods (Kumar et al. 2022). Advances in molecular techniques have improved the speed and efficiency of this process, offering promising solutions to micronutrient deficiencies (Kumar et al. 2022).

5.4 Genomics in Biofortification

Genetic variety across crops is crucial for plant breeding, and low variation can impede biofortification efforts (Sagare et al. 2023). Genetic engineering provides a workable answer to this issue since it is not restricted to related species, unlike traditional plant breeding (Singh et al. 2020b; Reddy et al. 2021). Crops like rice and bananas, resistant to conventional breeding methods, can be biofortified by genetic engineering with success (Razzaq et al. 2019; Ishfaq et al. 2022). This approach allows for the introduction of new nutrients or agronomic traits into specific crop varieties by applying principles of plant breeding and biotechnology (Rao and Annadana 2017; Thakur et al. 2022). By identifying and characterizing suitable genes, genetic engineering can introduce these genes into crops to enhance their nutritional qualities (Singh et al. 2020b). This method makes use of genes from many different species, such as fungi, bacteria, and other living things. For example, it is possible to engineer crops to improve nutrient absorption, transportation, and concentration by introducing genes from microbes that boost nutrient uptake (Gupta et al. 2022a; Benavente and Giménez 2021). Gene editing, commonly referred to as genome editing, allows DNA sequences in different cells and species to be added, changed, or removed (Singha et al. 2022). By developing GMOs without transgenes, this technique can overcome the regulatory issues surrounding transgenic crops (Sharma et al. 2021b). Using techniques such as mega-nucleases, zinc-finger nucleases (ZFNs), transcription activator-like effector nucleases (TALENs), and clustered regularly interspaced short palindromic repeats (CRISPR/Cas9), β-carotene biofortified rice and zinc-rich wheat varieties have been produced (Reddy et al. 2021; Devi and Aglawe 2021). CRISPR/Cas9 is precise, adaptable, and less expensive than most genome editing techniques, which are labor-intensive, complicated, and expensive. On the other hand, it occasionally results in unwanted alterations in untargeted genomic regions (Reddy et al. 2021; Siddique and Aley 2024). The dimeric nuclease technique, which is extremely specific and accurate, can help to alleviate this problem (Sagare et al. 2023). Research indicates that a lack of efficient consumer communication of scientific findings is the cause of low public awareness of gene editing (Dharani et al. 2024; Razzaq et al. 2019).

In affluent nations, transgenic biofortification has been well-studied (Ishfaq et al. 2022). One prominent example is Golden Rice, which is biofortified with provitamin A through the expression of genes that are essential for the β-carotene pathway, such as phytoene synthase and carotene desaturase (Singh et al. 2020a; Sharma et al. 2021b). Provitamin A levels in Golden Rice increased to 1.6–3.7 $\mu g\ g^{-1}$ DW because of this genetic alteration (Bertola et al. 2021). Furthermore, a 37-fold rise in the Fe concentration of the storage roots was seen in cassava when the *Arabidopsis thaliana* vacuolar Fe transporter VIT1 was overexpressed (Aggarwal et al. 2022). Zn levels and bioavailability were increased by overexpressing the genes encoding phytase activity and zinc transporters in barley (Kadirimangalam et al. 2022; Sharma et al. 2021b) (Figure 5.2).

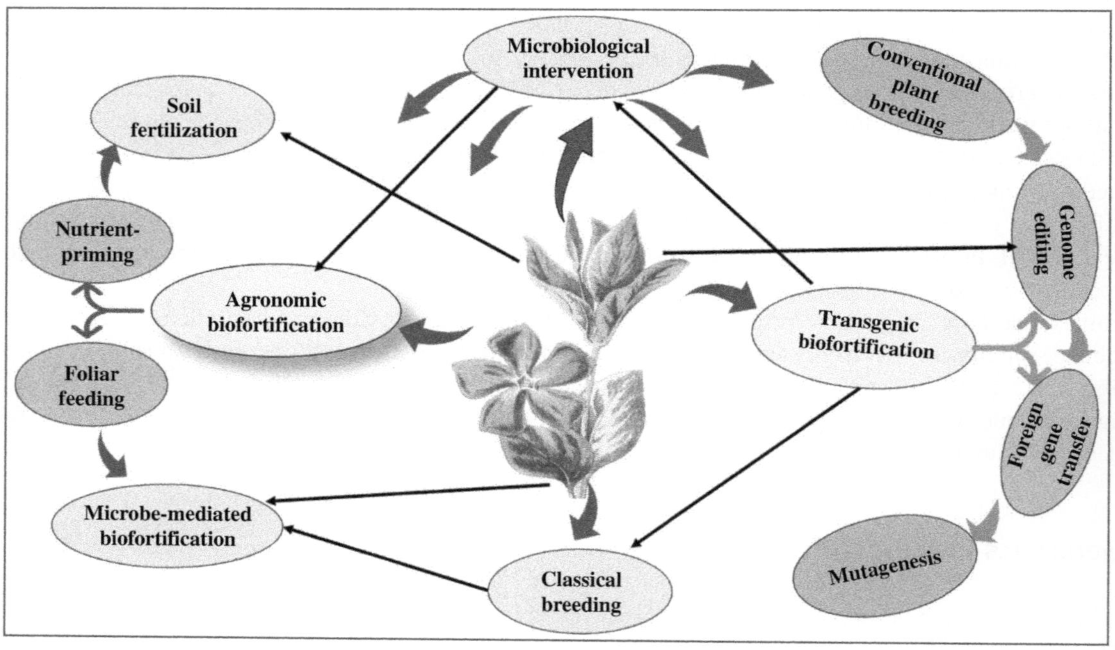

Figure 5.2 Various breeding approaches in crops and their consequences.

5.5 Regulations, Consumer Acceptance, Opportunities, and Prospects

The adoption of biofortified crops by farmers and consumers is a critical factor in the success of biofortification projects (Nasar et al. 2024). Acceptance levels for GM crops differ significantly between these two groups. Generally, consumers exhibit lower acceptance due to skepticism about the risks and benefits associated with GM crops and foods (Ahmad et al. 2022a). A multitude of elements, including knowledge, trust, beliefs, and perceptions of risks and advantages, influence consumer attitudes. Concerns have been expressed regarding the implications of gene flow and transfer on human health, environmental concerns such as the emergence of resistant plants and crops, and the repercussions on traditional farming practices (Sharma et al. 2021b). Furthermore, concerns exist regarding the synthetic origin, toxicity, and allergic properties of GM crops (Kumar et al. 2021c). Despite scientific proof that GM plants are safe, these worries frequently result in the rejection of these plants and their products (Ahmad et al. 2022). To address these issues, it is crucial to implement informational programs that emphasize the benefits of biofortified GM plants and dispel misconceptions about GM crops (Ahmad et al. 2022a). In regions where biotech seeds are available, farmers generally embrace them due to the tangible benefits they provide (Anushree et al. 2023).

Various rules apply to GM crops worldwide (Patel et al. 2023). The acceptance and marketing of GM crops are greatly impacted by these laws (Xia et al. 2020). Over 40 countries have enacted strict labeling rules for GM crops (Singha et al. 2022). For instance, the European Union has implemented a stringent authorization system for GM crops and foods, requiring labeling based on the process, even if no genetic modification traces are detectable in the final product (Sagare et al. 2023). This regulatory environment has made it difficult to cultivate GM crops in many European countries. In contrast, US legislation is more lenient, not requiring labels for novel GM foods if they are compositionally like non-GM foods (Aggarwal et al. 2022). Comparably, China has required food generated from GM crops to be labeled for more than 10 years (Kumar et al. 2021c). It can take up to 13 years to produce and commercialize GM crops, and the authorization process can cost up to $35 million (Ahmad et al. 2022a). Additionally, many agri-businesses' companies patent new GM crops, monopolizing their commercialization (Ahmad et al. 2022). This has sparked debates about whether GM crops are developed for privatization and profit-making or to promote food security (Kumar et al. 2021c). As a result, to promote the adoption of GM crops, laws and regulations controlling them must be changed to be less onerous, more economical, and more time-efficient, especially in developing nations (Satyavathi et al. 2021).

5.5.1 Recent Advancements

Since the early days of crop domestication, scientists and breeders have been striving to enhance various traits in crops, including growth, yield, quality, and resilience to environmental factors such as pests, pathogens, drought, and salinity. Initially, favorable traits were selected based on observation, without understanding the genetic basis of these differences (Haider et al. 2021a). However, Gregor Mendel's groundbreaking discoveries in the 1800s established the fundamental laws of inheritance, revealing that traits in crops are controlled by genes inherited from both parents (Satyavathi et al. 2021). Mendel's work laid the foundation for traditional breeding methods and later advancements in molecular and transgenic breeding techniques (Haider et al. 2021a; Murphy 2014). Traditionally, breeders employed crossbreeding and backcrossing to combine desirable traits in crops (Satyavathi et al. 2021). With the development of gene cloning and transformation technologies, transgenic approaches emerged as a method to introduce foreign genes for improving specific traits in plants, particularly insect- and herbicide-resistant traits (Aggarwal et al. 2022). However, the random insertion of foreign genes into the plant genome posed challenges, including potential gene silencing or altered expression of other genes (Sagare et al. 2023). Thus, there is a need for more precise methods to edit genes for both functional studies and crop improvement (Aggarwal et al. 2022).

A novel technique for precisely and correctly changing certain sequences inside a genome for a variety of reasons is genome editing (Satyavathi et al. 2021). Genome editing approaches, in contrast to conventional transgenic techniques, target genome sequences (Haider et al. 2021a). Restriction nucleases, which identify and cut DNA sequences, are used in this procedure (Murphy 2014). The four primary types of nucleases utilized for genome editing are meganucleases, zinc finger nucleases (ZFNs), transcription activator-like effector-based nucleases (TALENs), and clustered regularly interspaced short palindromic repeats associated nucleases (CRISPR/Cas) (Satyavathi et al. 2021). CRISPR/Cas systems, initially discovered as a defense mechanism in bacteria and Archaea against viral infections, have garnered significant attention for genome editing applications (Haider et al. 2021a). CRISPR sequences are short palindromic repeats found in prokaryotes, which bind to foreign DNA, such as viral DNA, enabling Cas nucleases to cleave the invading DNA (Murphy 2014). When CRISPR-associated protein (Cas) genes and their significance in bacterial defense systems were discovered, the purpose of the CRISPR sequences that were initially cloned from bacteria in the late 1980s became evident (Hossain et al. 2021). Advancements in 2012 showed that CRISPR/Cas systems might be used for genome editing (Satyavathi et al. 2021). Researchers successfully reconstructed CRISPR/Cas systems in vitro and showed their ability to target and cut specific DNA sequences (Haider et al. 2021a). These studies paved the way for using CRISPR/Cas as a genome editing tool (Murphy 2014). Subsequent research further optimized CRISPR/Cas systems for efficient genome editing in mammalian cells, enabling the knockout of multiple genes simultaneously and demonstrating the versatility and flexibility of the technology (Hossain et al. 2021).

Following these pioneering studies, CRISPR/Cas systems have been widely adopted by the scientific community for editing genes in various organisms, including bacteria, animals, and plants (Jadhav et al. 2020). The simplicity and efficiency of CRISPR/Cas systems have revolutionized genome editing, opening new possibilities for functional genomics, crop improvement, and biomedical research (Kaur et al. 2023) (Figure 5.3).

5.5.2 Genome-wide Association Studies

By comparing the allele frequencies of genetic variations in people who are phenotypically different but share a common ancestor, genome-wide association studies (GWAS) aim to find associations between genotypes and phenotypes (Yağız et al. 2022). Single-nucleotide polymorphisms (SNPs) are the most researched genetic variants in GWAS; however, they can also include copy-number variants and sequence differences in the human genome (Raj and Nadarajah 2022). Typically, GWAS identifies blocks of correlated SNPs, known as genomic risk loci, all showing statistically significant associations with the trait under investigation (Singh et al. 2024). Many repeated genetic risk loci linked to characteristics and illnesses have been found by GWAS during the past 15 years, including PTPN22 for autoimmune disorders and FTO for obesity (Haider et al. 2021b). These findings have provided valuable insights into disease biology, informing subsequent clinical trials for targeted treatments (Graw et al. 2021). The results from GWAS have diverse applications (Hossain et al. 2021). For instance, trait-associated genetic variants can serve as control variables in epidemiological studies to adjust for confounding genetic differences among groups (Jadhav et al. 2020). Additionally, these results enable the prediction of an individual's risk for various diseases based on their genetic profile (Kaur et al. 2023). Recent studies have demonstrated that genomic risk prediction using genome-wide polygenic risk scores (PRSs) can rival monogenic risk prediction strategies based on rare mutations, suggesting potential clinical utility as a stratification tool and genetic biomarker (Yağız et al. 2022). With

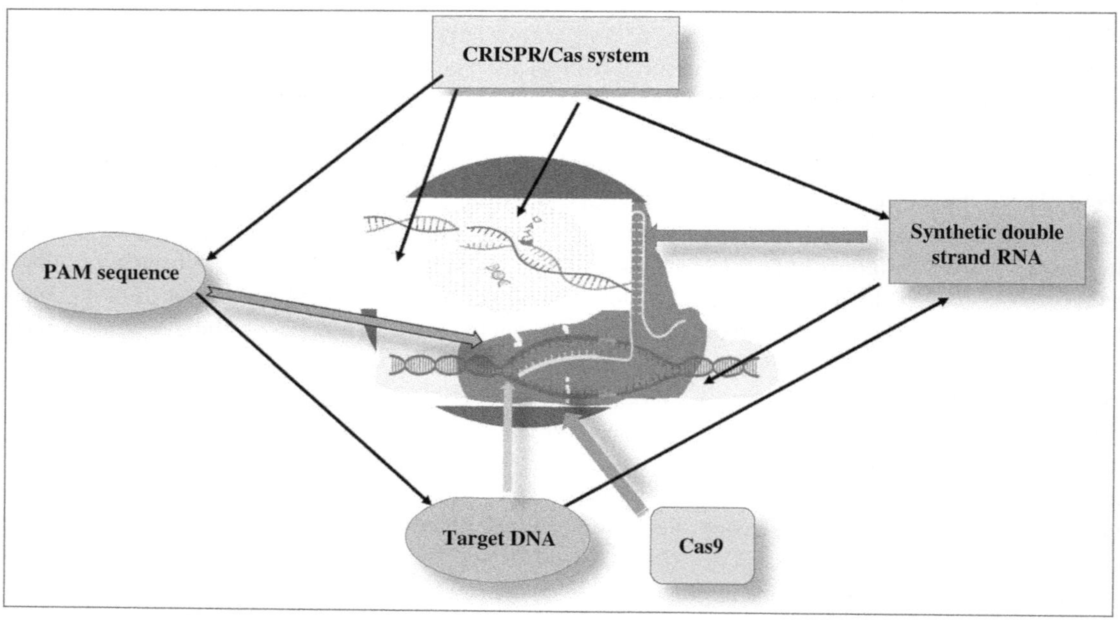

Figure 5.3 CRISPR-associated protein (Cas) genes and their role in defense mechanisms.

over 5700 GWAS conducted for more than 3300 traits, the focus has shifted toward achieving greater statistical power by increasing sample sizes to well beyond a million participants (Singh et al. 2024). This has led to the identification of numerous associated and replicable variants for many heritable traits (Haider et al. 2021b).

Interpreting these correlations within a biological and genetic context is now the difficult part. According to earlier GWAS, hundreds of causal variants impact most characteristics, each of which has a negligible risk on its own and is frequently linked to many additional traits (Graw et al. 2021). Direct comparisons between demographic groupings are further complicated by the possibility that genetic correlations differ among ancestries. These difficulties restrict the usefulness of GWAS results for mechanistic insights or medication development, and they make it difficult to draw unambiguous conclusions about the biological implications of the data (Vasaikar et al. 2018; Kiran et al. 2022; Hasin et al. 2017).

5.6 Transcriptomics in Biofortification

The dynamic nature of transcriptomes refers to the constantly changing profile of RNA transcripts within a cell, tissue, or organism (Canzler et al. 2020). This dynamism is influenced by a myriad of factors, making it a complex task to identify stable markers for processes such as biofortification. Here is a detailed look at this concept:

5.6.1 Influences on the Transcriptome

Gene expression in plants is intricately influenced by a multitude of environmental factors, biotic interactions, developmental stages, and circadian rhythms (Fernandez-Banet et al. 2016). Temperature fluctuations can induce the upregulation or downregulation of specific genes, such as the activation of stress-response genes during extreme heat or cold (Vasaikar et al. 2018). Light exposure significantly impacts gene expression related to photosynthesis, circadian rhythms, and growth. Water availability, whether through drought or excess, triggers changes in genes involved in water transport, stress response, and metabolic adjustments (Canzler et al. 2020). Similarly, the presence or absence of vital nutrients like nitrogen and phosphorus can dramatically alter the expression of genes responsible for nutrient uptake and metabolism (Planell et al. 2021).

Biotic factors also play a crucial role; for instance, pathogen infections or pest attacks activate defense-related genes while suppressing other nonessential pathways (Vasaikar et al. 2018). In contrast, symbiotic relationships, such as those with mycorrhizal fungi or nitrogen-fixing bacteria, can modify gene expression to enhance nutrient exchange and support mutualistic interactions (Hasin et al. 2017). Developmental stages from seedling to fruiting exhibit distinct transcriptomic profiles, with specific sets of genes being activated or repressed to facilitate appropriate growth and development at each stage (Vasaikar et al. 2018). Additionally, the internal biological clock, or circadian rhythm, regulates the timing of stress, leading to daily fluctuations in the plant's transcriptome. Collectively, these factors create a dynamic and responsive gene expression landscape essential for plant adaptation, survival, and growth (Hasin et al. 2017).

5.7 Multi-omics Platforms

Systems biology approaches have become indispensable for modeling the intricate interactions among genes, proteins, and metabolites, offering valuable insights into biofortification (Hasin et al. 2017). Large-scale omics datasets (genomic, transcriptomics, proteomics, metabolomics, metagenomics, phenomics, etc.) are now widely accessible (Figure 5.4), which has transformed biology and made it possible for systems-based methods to enhance our comprehension of biological processes (Canzler et al. 2020). Integrating omics data offers researchers from a variety of fields both great prospects and major problems, as the time and cost associated with producing these datasets decrease (Reddy et al. 2021). Sagare et al. (2023) introduced a comprehensive analytical approach known as the trans-omics concept of dynamic networks, which integrates multiple layers of omics datasets, including transcriptomics, proteomics, and metabolomics, with more recent datasets like phosphoproteomics, protein–protein interactions, DNA–protein interactions, and allosteric regulation (Aggarwal et al. 2022). Our method highlights significant components of dynamic biological networks when omics data are adequately integrated, as demonstrated by example studies in bacteria and rats (Zulfiqar et al. 2024). Furthermore, research on the connections between phenotypes and omics networks representing genetic and environmental factors is presented, including trans-ome-wide association (Trans-OWAS), pathway-wide association, and phenome-wide association (Phenome-wide

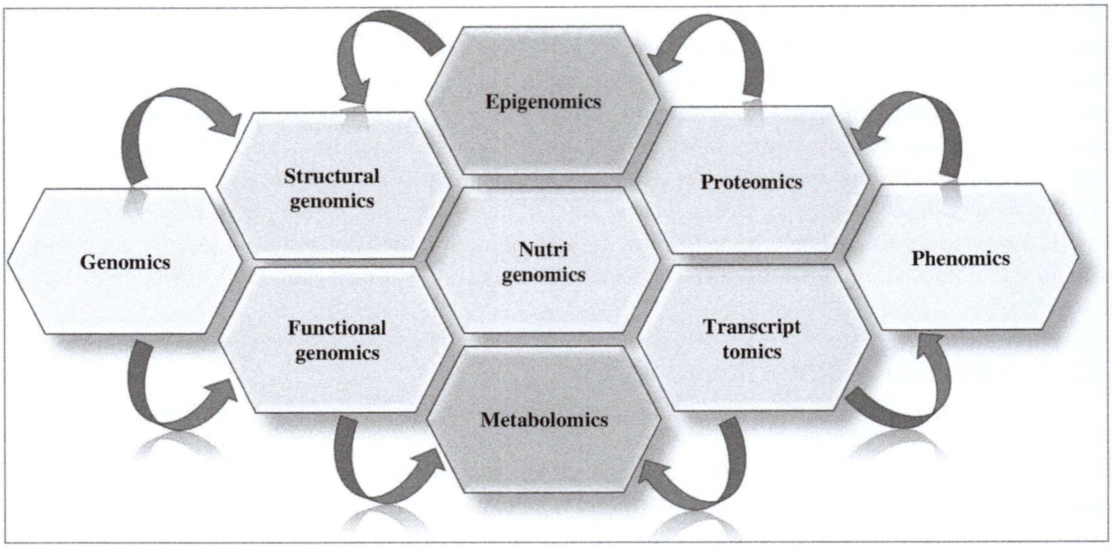

Figure 5.4 Multi-omics platform and technologies.

association) (Kadirimangalam et al. 2022; Sharma et al. 2021). However, visually representing these multi-layered, multifactorial approaches poses computational challenges due to their complexity. This is further complicated by extra information gleaned from interactions between proteins, RNA, DNA, and microRNA (Manzoor et al. 2023). While various tools and resources are available for analyzing and visualizing omics data individually, a thorough summary of methods for combining three or more omics datasets is still absent (Sagare et al. 2023; Dharani et al. 2024). Although the integration of omics data across several study domains is acknowledged as crucial, there are still a few instances when more than two omics datasets are successfully implemented (Kumar et al. 2021c; Satyavathi et al. 2021). Understanding the statistical behavior of readouts from each omics regime separately, identifying subtle connections across omics regimes in the context of biology, and utilizing time resolution in omics data, such as time course studies, are among the challenges (Canzler et al. 2020). Although techniques for integrating genomes and proteomics datasets have been presented in recent reviews, methods for integrating metabolomics datasets have not been sufficiently covered (Yang et al. 2023). Moreover, there are difficulties in merging high-dimensional datasets with over 1000 variables, such as the "curse of dimensionality," which causes sample variances to become sparse and enormous, making cluster analysis useless (Canzler et al. 2020). In summary, integrated omics approaches not only offer significant potential for advancing our understanding of biological systems, but they also pose considerable challenges that must be addressed to fully capitalize on the wealth of omics data available (Hossain et al. 2021; Jadhav et al. 2020).

5.7.1 The Advantages and Disadvantages of Individual Omics

Transcriptomics and genomics have been very useful in many domains, such as the pharmaceutical business, diagnostics, therapeutics, gene therapy, pharmacogenomics, disease prevention, developmental biology, evolutionary genomics, and comparative genomics (Planell et al. 2021). Consequently, biological investigators now need to manage and analyze various kinds of data (Fernandez-Banet et al. 2016). Improvements in next-generation sequencing (NGS) technologies and genomic data analysis have made it easier to generate high-throughput data for genomes (including single nucleotide polymorphisms (SNPs), copy number variants (CNVs), loss of heterozygosity variants, genomic rearrangements, and rare variants), epigenomes (like DNA methylation, histone modifications, chromatin accessibility, and transcription factor (TF) binding), and transcriptomes (like gene expression, alternative splicing, long noncoding RNAs, and small RNAs like microRNAs).

The five main steps in nucleic acid-based omics approaches for data generation are usually as follows: sample collection, high-quality nucleic acid extraction, library preparation, clonal amplification, and sequencing (using methods such as pyrosequencing, sequencing-by-ligation, or sequencing-by-synthesis) (Jadhav et al. 2020; Yağız et al. 2022; Singh et al. 2024). The desired downstream use determines the strategy for each stage (Graw et al. 2021; Vasaikar et al. 2018; Fernandez-Banet et al. 2016). Data cleaning, filtering, assembly, alignment (de novo or reference-based), variant calling, annotation, and functional predictions are steps in the pipeline that follow sequencing (Hasin et al. 2017; Canzler et al. 2020; Planell et al. 2021). Furthermore, to offer biological context, route and/or network studies are frequently used (Satyavathi et al. 2021). Because distinct datasets differ in their approaches to data reduction, data normalization, quality assurance, and quality control, heterogeneous datasets provide issues (Raj and Nadarajah 2022). For instance, short RNA-Seq datasets often contain fewer than 2000 micro RNAs, whereas RNA-Seq datasets typically consist of tens of thousands of transcripts (Seneviratne et al. 2020). As a result, the normalization and scaling of small RNA-Seq data differ from those of RNA-Seq data. With the rapid advancements in single-cell sequencing technology and longer-read sequencing methods, new difficulties are developing, such as appropriate sequence coverage and statistical interpretation of single-cell data (Ullah et al. 2023d). An overview of genomics applications and technologies is accessible (Canzler et al. 2020) for those who wish to learn more, and the NIH National Cancer Institute (Planell et al. 2021) offers best practices for DNA-seq pipelines. To answer a wide range of biological problems, sequencing-based technologies provide exceptional chances to gather high-quality data from tiny volumes of tissues or individual cells. In terms of analytical instruments, public databases for data sharing, and laboratory reagents for standardized protocols, these technologies are the most sophisticated OTs available (Ullah et al. 2023d; Sun and Hu 2016).

5.7.2 Proteomics

Using both shotgun and targeted methods, proteomics is essential for measuring proteins in a variety of sample types (Eddy et al. 2020). Mass spectrometry (MS) has made tremendous strides recently, increasing sensitivity while lowering the sample volume needed for high-throughput investigations (Canzler et al. 2020). These advancements make it possible to identify posttranslational changes, detect subtle variations in protein abundances, and perform other tasks on a variety of samples and tissues (Planell et al. 2021). Choosing between a chemically labeled and an unlabeled quantitative proteomic approach involves six main steps: selecting the right samples, extracting proteins, breaking down proteins into peptides using enzymes, separating/fractionating using liquid chromatography (LC) techniques, identifying and quantifying peptides and proteins using MS analysis, and performing additional bioinformatics analyses such as pathway and network analyses (Yang et al. 2023; Seneviratne et al. 2020).

The field has progressed from traditional 2D-PAGE-based methods for protein spot extraction (using dye/fluorescence labeling) to more comprehensive screening strategies that include quantitative steps, like matrix-assisted laser desorption/ionization time-of-flight (MALDI-TOF) MS characterization or LC-MS (Ullah et al. 2023d; Sun and Hu 2016). Tandem mass tags (TMT), ^{18}O stable isotope labeling, isobaric tagging for relative and absolute quantitation (iTRAQ), isotope-coded affinity tagging, and stable isotope labeling with amino acids in cell culture (SILAC) are a few of the label-free or label-based techniques used in these procedures (Lloyd-Price et al. 2019; Ahmed 2020). Promising outcomes have been obtained using label-free techniques (Lloyd-Price et al. 2019) as well as label-based techniques such as TMT proteomics from different biological matrices (Ahmed 2020). An agreement on data formatting, cleaning, and normalization has not yet been reached by the community, despite continuous attempts through projects such as the Proteomics Standards Initiative (Eddy et al. 2020; Ahmed 2020). This covers discussions on whether to employ peptide-to-spectrum matching or ion intensity. Nevertheless, proteomics keeps expanding our knowledge in the field of biomedical research, with implications for treatments, protein-based biomarker creation, and diagnostics (Reska et al. 2021).

5.7.3 Metabolomics

As the products of complex biochemical pathways that connect the proteome, transcriptome, and genome to phenotype, metabolites are essential resources for determining the genetic underpinnings of metabolic diversity (Reska et al. 2021; Chakraborty et al. 2024). With the use of metabolomics, one can ascertain the relative and absolute concentrations of a wide range of substances, including organic acids, sugars, lipids, amino acids, steroids, drugs, nucleotides, and environmental constituents from a variety of sample types, such as primary cells, tissues, biofluids, organisms, and different geo-climatic environments (Agamah et al. 2022). The methods used in metabolomics include gas chromatography MS (GC-MS) for gas phase analysis, liquid chromatography MS (LC-MS), and capillary electrophoresis MS (CE-MS) for liquid samples, and solid-state NMR for solid samples (Reska et al. 2021; Chakraborty et al. 2024). These techniques record information on tiny molecules using spectroscopy (e.g. NMR) and MS (e.g. LC/GC-MS or tandem MS) (Chakraborty et al. 2024; Agamah et al. 2022). Optimal metabolite extraction and reconstitution, metabolism quenching, appropriate sample collection methods, MS or NMR analysis, optional chemical derivatization, experimental design, and subsequent data analysis that includes alignment, filtering, imputation, statistical analysis, annotation, and pathway/network analysis are all crucial steps in metabolomics analyses (Bibi et al. 2023).

Depending on the platform being utilized for sample analysis, every stage in the metabolomics workflow might vary greatly (Rai et al. 2016). Moreover, each data type and instrument may require a different approach for data structure, imputation techniques, metabolite identification, standardization, scaling, and transformation (Reska et al. 2021). Different approaches are utilized for targeted and untargeted studies; untargeted analyses are often used to provide data that is hypothesis-generating and discovery-oriented, whereas targeted analyses are used to assess specific hypotheses (Bibi et al. 2023). Combining LC-MS with complementary GC-MS can cover most of the chemical space presented by a given biofluid or tissue sample, providing comprehensive insights into metabolic patterns (Rai et al. 2016).

5.7.4 Exclusive Experiments to Specific Omics Platforms

Because of their respective advantages and disadvantages, omics platforms each present different obstacles. Since the complexity and completeness of each data type vary, they are crucial to comprehend when creating strategies and methodologies for integrating omics data (Ma et al. 2020).

5.7.4.1 Linking Genotype to Phenotype

Efficient amplification of nucleic acids from tiny biological samples, together with precise quantification and identification of molecules based on sequence identity, is critical to the success of high-throughput genomics and transcriptomics research (Gurke et al. 2022). The methods for sample preparation now in use may be used to evaluate any DNA or RNA discovered in a biological sample, including coding and noncoding RNAs (Brönstrup 2004). However, interpreting transcriptome and genome data in the context of biological function is a major difficulty, especially when trying to figure out how certain polymorphisms affect phenotypic variance (Rodríguez-Suárez and Whetton 2013). The combination of proteomics and metabolomics data with genomes and transcriptomics provides a solution to this barrier by providing molecular insights that connect genetic and epigenetic modifications with phenotypic changes (Reska et al. 2021; Bibi et al. 2023).

5.7.5 Quantification of the Proteome

Understanding biological function may be gained from proteomics data, particularly concerning posttranslational modifications and isoform variation (Ma et al. 2020). However, the lack of protein amplification techniques means that proteomics approaches still need large sample sizes (Gurke et al. 2022). There are still issues with identifying low-abundance proteins, separating membrane proteins, and studying insoluble proteins (Brönstrup 2004). For example, whereas nuclei enrichment is usually required for nuclear protein representation in proteomic datasets, untargeted proteomic approaches could not include information for every protein in a sample (Ahmed 2020; Reska et al. 2021). In both top–down and bottom–up proteomics, chromatography is used to separate complicated chemistries, such as various charged states and posttranslational modifications, which leads to variability in protein quantification (Agamah et al. 2022). Variations in peptide structure, charge, and hydrophobicity affect their detectability and identification by NMR or MS, leading to variability in peptide identification (Rai et al. 2016). Pipelines for proteomic data processing must handle normalization, absolute versus relative quantification, and missing data (Rai et al. 2016). Peptide and protein quantification accuracy and repeatability using MS have been significantly improved by recent improvements in instrument sensitivity and the creation of useful isotopic tagging techniques for tissue samples (Gurke et al. 2022). These developments allow for the efficient measurement of peptides in complicated mixtures, such as tissue lysates, by applying techniques such as isobaric tags for quantification, peak intensity or area quantification, and peptide spectral match counts (Brönstrup 2004).

5.7.6 Quantification of the Metabolome

The relationship between genetic and proteomic variation and functional alterations is mostly dependent on metabolomics data, which provide insights into the metabolic, regulatory, and signaling processes that occur inside cells or tissues (Rodríguez-Suárez and Whetton 2013). However, metabolites, like proteins, cannot be amplified, and only 15–30% of the mass spectra are recognizable and measurable (Brönstrup 2004). The enormous volume of information collected is limited in its usefulness by this constraint (Reska et al. 2021). Furthermore, the score-based spectrum annotation of molecules presents a hurdle in terms of false positives. Significant hurdles include the unpredictability of sample processing, platform variations, chemical variety among tiny compounds, different quantification techniques, and lack of standards for data formats and analytic pipelines (Agamah et al. 2022). Large-scale initiatives to resolve these issues are currently being aggressively pursued by the metabolomics research community, with a particular emphasis on standardization, metabolite annotation, protocol and technique compatibility, and statistical deliberations (Bibi et al. 2023).

5.7.7 Concerns Revealed Among the Omics Programs

Most omics methodologies necessitate proficiency in managing large datasets, annotating biomolecules within a dataset, assessing the balance between sample size and the number of quantified biomolecules, distinguishing between relevant and irrelevant biomolecules (signal versus noise), evaluating the quality of output, and facilitating data sharing due to the volume and complexity of data (Lloyd-Price et al. 2019; Ahmed 2020).

5.7.7.1 Data Handling

Regardless of the kind of omics data, data handling must address issues such as scaling, imputation, transformation, and normalization, as well as data filtering and cleaning (which is like data wrangling in data science) (Chakraborty et al. 2024; Agamah et al. 2022). Regretfully, no omics data type has a workflow that is considered the "gold standard" by all parties (Bibi et al. 2023). Varied analytic pipelines or tools can yield varied findings, even while genomics and sequencing procedures frequently follow well-accepted standards for activities like variant calling, quality control, and sequence alignment (Al Shweiki et al. 2017). Furthermore, workflows are continually evolving with the development and adoption of new computational tools (Rauniyar and Yates III 2014). Consequently, it is crucial to thoroughly document every analysis pipeline, including the software versions utilized at each step (version control), and justify the chosen parameters (Hultin-Rosenberg et al. 2013).

5.7.7.2 Annotation

Annotation of biomolecules in omics datasets presents significant challenges (Hultin-Rosenberg et al. 2013). While standard model organisms like flies, nematodes, mice, nonhuman primates, and humans have well-annotated genomes, transcriptomes, and proteomes, nonstandard model organisms lack such comprehensive annotations (Zhou et al. 2021). Consequently, working with nonstandard organisms requires comparative approaches to prevent the loss of extensive data (Cox et al. 2014). Using a reiterative policy to interpret transcripts in nonstandard model species, for example, greatly increases the number of annotated transcripts by using the species genome as the main annotation source and aligning unannotated transcripts against numerous additional genomes. Furthermore, peptide annotation may be significantly enhanced by building peptide reference libraries with RNA-Seq transcript sequence data relevant to person and species (Hultin-Rosenberg et al. 2013). An investigation into the baboon liver proteome, for instance, revealed unexpected, unannotated splice variants, and distinctive peptides that are overlooked by conventional reference databases. The chemical catalog of an organism is frequently unknown in metabolomics data processing, which makes the identification of molecules much more difficult in the absence of information about their transcriptomes and proteomes (Zhou et al. 2021).

Study design and analytical statements play crucial roles in enhancing inferential robustness and reproducibility in large omics studies (Cox et al. 2014). A well-planned research design and subject/sample recruitment that is in line with the study design enables the clear testing of a priori hypotheses and agnostic investigations (Reska et al. 2021). Although biases not seen in formal randomized trials might make convenience samples valuable, these biases should be addressed through appropriate research design (Hultin-Rosenberg et al. 2013). Preventing pseudoreplication requires an understanding of the independence of experimental units, particularly when numerous measurements are performed on independent experimental units (Reska et al. 2021). Analytical approaches, whether classical statistical or machine-learning methods, must align with the study design and question, with underlying assumptions verified. Proper analysis also involves distinguishing whether the omics variable is a predictor or an outcome, aligning the analytic approach accordingly, and ensuring that assumptions like normality and constant variance are met (Lloyd-Price et al. 2019; Ahmed 2020; Al Shweiki et al. 2017). To fulfill these assumptions, it could be essential to modify the outcome variables, and for the variables that are difficult to change, it might be necessary to use alternative models (Ghorai et al. 2022; Kumar et al. 2019; Kadirimangalam et al. 2022). By minimizing false discoveries and enhancing actual discoveries, these strategies improve repeatability and bring down the total cost of scientific research (Tayade et al. 2023; Haq et al. 2024; Sharma et al. 2021b).

5.7.8 Statistical Power

Omics data pose significant statistical challenges, including the balance between sample size and the number of molecules quantified, which can lead to both false positives and true negatives (Lloyd-Price et al. 2019; Al Shweiki et al. 2017). Other significant issues include processing data for targeted and untargeted techniques and analyzing time series data, especially when separating actual biological signals from noise. The number of samples in each group varies depending on the OT; transcriptomics and genomes allow for high multiplexing, proteomics allows for moderate multiplexing, and metabolomics does not have multiplexing procedures (Hultin-Rosenberg et al. 2013). When integrating several omics datasets, there is an additional challenge due to the notable variation in the number of observations per sample. Genomes frequently contain a variety of features, such as a transcriptome with tens of thousands of quantified molecules, a small transcriptome with less than 2000 molecules, millions of variants, and proteomes and metabolomes with thousands of quantified molecules (Rodríguez-Suárez and Whetton 2013; Al Shweiki et al. 2017).

5.7.9 Data Archiving and Sharing

The absence of defined terminology, data formats, and public availability is a problem for many omics datasets (Zhou et al. 2021). Other omics data lack this uniformity, but transcriptomic, proteomic, and genomic data have mostly solved these issues by being placed in public databases upon journal publication (Brunner et al. 2022). Establishing a central public database for these datasets requires defining data standards to enable re-analysis of deposited data, akin to the Minimum Information About a Microarray Experiment (MIAME) standards developed for microarray data (Reska et al. 2021; Zhou et al. 2021). Several initiatives have tackled this need for standardization, including Minimum Information About a Genome Sequence (MIGS) for genome sequences, Minimum Information About a Proteomics Experiment (MIAPE) for proteomics, Mass Spectrometry Imaging (MSI) for metabolomics, Minimum Information About a Study of a Genomic sequence (MISAG) for genomics, and Minimum Information About a Metagenome Sequence (MIMAG) for metagenomics (Zhou and Yin 2016). Following these minimal requirements for information will encourage protocols that are both scalable and interoperable, which will make it easier to create datasets that can be replicated (Rochat 2016). This will make it possible to compare omics datasets from different biological samples, analytical systems, and research facilities all around the world.

5.7.10 Tools Accessible for Incorporation of Multi-omics Data

Each omics experiment involves hundreds of observations, which creates computational complexity and makes it difficult to identify significant correlations and interactions (Hultin-Rosenberg et al. 2013). It can be challenging to distinguish between genuine biological signals and random noise in biological systems due to their frequent multifactor joint effects and nonlinear interactions (Zhou et al. 2021). This noise can be generated by biological systems, unrelated analytical platforms, and other data-specific processing techniques (Zhou and Yin 2016). Variability in gene, protein, or metabolite abundances across cell types, tissues, and organs adds further complexity to extracting useful data from high-dimensional datasets. To detect and predict health status or consequences from treatments, a variety of omics methods, including fluxomics, ionomics, microbiomics, and glycomics, are being integrated with biological information (Lelli et al. 2021). Normalization is necessary before merging omics-scale data, particularly considering the usage of various technologies. Multi-omics analysis frequently makes use of statistical and machine-learning technologies, which make it easier to do dimension reduction, grouping, correlation with clinical measurements, and illness prediction (Haider et al. 2022a).

Various integrative frameworks and tools are available, ranging from simplistic descriptive approaches like principal component analysis (PCA) to more complex methods such as canonical correlation analysis (CCA), sparse CCA, multiple factor analysis, and multivariate partial least square regression analysis (Hultin-Rosenberg et al. 2013). While several tools exist for integrating genomic, proteomic, and metabolomic datasets, user-friendliness remains a challenge, with many requiring advanced computational skills. Integrated omics analyses for cancer research face additional challenges due to sample heterogeneity (Lelli et al. 2021). Fuzzy logic modeling frameworks, clustering strategies, and proteomics-first approaches have been proposed to address these challenges. Weighted gene coexpression network analysis (WGCNA) is a commonly used method for the objective integration of transcriptomic and genomic data in network creation techniques, which seek to identify new functions and connections among molecules (Haider et al. 2022a). Tools for integrating omics data range from web-based platforms like PaintOmics and Galaxy to more versatile tools like IntegrOmics and MixOmics

(Haider et al. 2022a; Giavalisco et al. 2009). While user-friendly tools are accessible, caution is advised in their blind application without understanding the underlying methods. For advanced users, tools like SteinerNet and Omics Integrator offer more customization and control over data analyses, catering to specific research needs (Röhnisch et al. 2018).

5.8 Immunity and Infection

One individual's genomes, transcriptomics, proteomics, metabolomics, and autoantibody profiles were combined for 14 months in the innovative integrated personal omics profile (iPOP) research (Ullah et al. 2023c). This novel method considered the structure and longitudinal design of the pathways while computing pathways enriched for molecules that were differently expressed at each time point (Trammell and Brenner 2013). Similarly, multi-omic integrated tools customized for specific organs have been developed; the renal and urinary route knowledge base for kidney illnesses is one such example (Reska et al. 2021). For instance, this integrated database facilitated the rapid generation of ideas by identifying calreticulin as a key protein in human interstitial fibrosis and tubular atrophy during chronic kidney transplant rejection (Hultin-Rosenberg et al. 2013). The significance of this protein was subsequently confirmed in vivo and in vitro (Zhou et al. 2021). Another work used an integrated omics method combining proteome, metabolomic, and lipidomic datasets to investigate how ferrets responded to pandemic H1N1 influenza virus infection (Fernandez-Banet et al. 2016). According to this study, pro-inflammatory lipid precursors significantly influence the pathogenesis of viruses (Lloyd-Price et al. 2019). These illustrations highlight how integrated omics techniques are effective in identifying new compounds that affect infection and immune response (Reska et al. 2021).

5.8.1 Cancer

Trammell and Brenner (2013) used both an integrated and a nonintegrated approach to examine the NCI-60 cancer cell line panel to identify potential molecular mechanisms that are dysregulated in cancer (Røst et al. 2020). They examined the transcriptome, proteome, and tiny transcriptome (miRNAs) together using factor analysis and linear discriminant analysis (LDA). The findings demonstrated that the integrated method functions as a surrogate marker of melanoma development, offering a more comprehensive knowledge of miRNA/gene interactions within the Wnt signaling pathway (Lloyd-Price et al. 2019). In follow-up research, Brönstrup (2004) used information from transcriptome, small transcriptome, genomic methylation, and genomic CNV databases to identify hepatocellular carcinoma subtypes (Trammell and Brenner 2013). After analyzing 256 samples of hepatocellular carcinoma, they discovered five separate subgroups, each of which had a varied survival rate and a distinct genetic signature (Obrenovich et al. 2018). Similar integrated omics analyses have been applied to various cancer types, yielding high quality and comprehensive molecular insights (Haider et al. 2022a).

Moreover, MiRbooking algorithms have contributed valuable insights into integrating miRNA–mRNA interactions within a given cellular condition (Giavalisco et al. 2009). Using an integrated omics approach in cancer research has the potential to improve carcinoma subtype detection and etiology knowledge. Label-free and focused proteomics, lipidomics, and metabolomics analyses were recently (Bolten et al. 2007) carried out in human blood samples from patients with metastatic melanoma, and data integration was then conducted (Trammell and Brenner 2013). This study proposed an explanation for how platelet-activating factors are formed from long-chain polyunsaturated phosphatidylcholines under oxidative conditions, and how this reprogramming of organ functions is caused by metastatic melanoma (Røst et al. 2020).

5.9 Host Microbiome Interactions

used a multi-omics approach to combine metagenomic, metatranscriptomic, and metaproteomic data from the gut microbiota to investigate intra- and inter-individual variation in people with type 1 diabetes mellitus (T1DM) (Haider et al. 2022a). Their research revealed that several microbial communities are responsible for the functional variations seen in T1DM patients (Bolten et al. 2007). Ullah et al. (2024) conducted integrated omics studies combining transcriptome, methylome, metagenome, and metabolome data with imaging data (Haider et al. 2022a) to quantify the global programming of host circadian transcriptional, epigenetic, and metabolic oscillations by the intestinal microbiota. Research has demonstrated a direct relationship between the host's circadian activities and the gut microbiome. Additionally, disruptions to the microbiome's rhythmicity have been found to impact the host's normal transcriptional oscillations in the gut and liver, as well as the host genome and epigenome (Ullah et al. 2023c). The complex relationships between the gut microbiota and the host, as well as how these interactions affect host metabolism, are becoming clearer thanks to these integrated omics investigations (Ullah et al. 2024). Ijaz et al. (2023b) developed an integrated omics workflow that includes LC-MS/MS metabolomics, inferred gene function profiles, and 16S rRNA gene sequencing for the analysis of human and environmental omic materials (Rauniyar and Yates III 2014). They accomplished this in less than 48 hours using Qiita, Phylogenetic Investigation of Communities by Reconstruction of Unobserved States (PICRUSt), and Global Natural Product Social Molecular Networking (GNPS) pipelines (Ijaz et al. 2023b). It is possible to assess human health status using an omics method within a period that is commensurate with traditional clinically relevant culture-based methodologies, according to their research (Haider et al. 2022b). These kinds of investigations have the potential to replace the necessity for microbe culture by offering more practical and precise techniques for identifying microorganisms in clinical settings (Bolten et al. 2007).

5.10 Statistical Methods for Present Challenges

The mathematical techniques utilized in network analysis across many scientific domains, such as the integration of omics datasets, are based on two theories: holistic general system theory, which emerged in the 1950s, and graph theory, which emerged in the 1930s (Al Shweiki et al. 2017). According to graph theory, a graph is made up of nodes that are paired together and connected by edges. An unordered pair consists of two nodes connected to each edge (Giavalisco et al. 2009). Nonetheless, a system is an entity composed of interdependent and connected elements, and altering one of these parts has predictable repercussions on other parts of the system, according to holistic general system theory (Trammell and Brenner 2013). Since its inception, this concept has been central to large-scale, high-throughput, high-dimensional omics studies (Hultin-Rosenberg et al. 2013).

5.11 Sample Number Versus Molecule Numbers

Effective statistical analyses are pivotal within the computational framework for integrating omics data. Each layer of omics, along with its associated analytical methodology, exhibits varying levels of noise (Brunner et al. 2022). The choice of statistical tools is influenced by the sampling process, which needs to be defined before conducting a study (Lelli et al. 2021). Even with small sample sizes, Bayesian network-based methods have shown value in successfully integrating several high-dimensional datasets (Haider et al. 2022a). Penalized regression techniques, such as the elastic net (ENET) (Hultin-Rosenberg et al. 2013) and the least absolute shrinkage and selection operator (LASSO) (Giavalisco et al. 2009), can model various omics data types after proper standardization, addressing issues like multicollinearity and the "n \ p" problem (Haider et al. 2022a), in which the number of independent samples (n) is significantly less than the number of measurements per sample (p). These problems can be solved with statistical techniques like multivariate regression methods, regularized generalized canonical correlation analysis (RGCCA), extensions of PCA (e.g. Structuring of Tables by Inertia [STATIS], Dual-Structuring of Tables by Inertia [dual-STATIS], Dynamic and Interactive Structuring of Tables by Inertia [DISTATIS], Anisotropic Structuring of Tables by Inertia [ANISOSTATIS]), multiblock redundancy analysis (mbRA), and multiblock continuum redundancy (MCR; Ijaz et al. 2023b).

5.11.1 Dimension Reduction

Handling high-dimensional multi-omics datasets can be computationally demanding, prompting the need for dimension reduction techniques to alleviate this burden while attending to multiple testing matters (Reska et al. 2021). Only a small number of dimension reduction methods currently in use allow for data variability (Tayade et al. 2023; Kiran et al. 2022). Common approaches to data dimensionality reduction lack some elements, such as value ratio calculations, low variance and high correlation filters, and methods like random forest, PCA, and backward or forward feature removal. Hultin-Rosenberg et al. (2013) have noted that among them, PCA is the most often used dimension reduction technique for omics investigations. Dimension reduction techniques for integrative analysis employ a variety of techniques, including partial triadic analysis, higher-order generalizations of SVD and PCA (CANDECOMP/PARAFAC/Tucker3), componential analysis (CIA; statico), regularized generalized canonical correlation analysis (rGCCA), sparse generalized canonical correlation analysis (sGCCA), multiple coinertia analysis (MCIA), and the X-STATIS family of methods (STATIS) (Bolten et al. 2007).

Extracting linear connections that most clearly depict variability within and between observations, as well as related structures across datasets, is the aim of these methods, which are available at Comprehensive R Archive Network (CRAN) as R-packages. They are also capable of identifying anomalies or batch effects in a dataset (Ullah et al. 2024). For a more in-depth examination of predictive modeling and analytics methods, readers are referred to recent research (Haider et al. 2022a; Ijaz et al. 2023b; Haider et al. 2022b; Lordan et al. 2019; Bosviel et al. 2016).

5.12 Recent Challenges and Viewing to the Future

Efficient sample preparation poses significant challenges across diverse samples for individual omics studies, spanning genomics, transcriptomics, proteomics, and metabolomics. Numerous reviews have highlighted these challenges, addressing sample preparation for various organisms, including plants, animals, and microbes. Specialized efforts have been made for specific types of omics studies, such as fecal metabolomics, lipidomics, and single-cell genomics. However, in the context of multi-omics, sample quantity emerges as a major bottleneck, compounded by the need for unified extraction strategies capable of simultaneously extracting nucleic acids, proteins, and metabolites without significant loss. This need is addressed by proposed single-tube extraction techniques, which enable multiphasic extraction of all three classes of biomolecules. The simultaneous metabolite, protein, and lipid extraction (SIMPLEX) process and metabolite, protein, and lipid extraction (MPLEx) protocol are two examples of protocols and kits designed for integrated omics analysis that have been created by academic and commercial initiatives. These techniques provide viable paths for obtaining large amounts of data; some are designed to provide multi-omics data quickly. Nevertheless, unified sample preparation procedures are still in their infancy; they frequently produce samples of differing quality and need to be further improved before they can be used universally for a wide range of biological matrices.

5.12.1 Adjusting, Supporting, and Sharing Workflows

Integrated omics processes require reliable experimental design and execution, including the best possible methods for sample collection and preparation, to generate multiple omics datasets from a given material in a single step. This method reduces batch effects and technical variation problems that are frequently present in high-dimensional data creation procedures while improving comparability across datasets. It is essential to document every stage of the data handling procedure, from the creation of separate omics datasets to integration. Multi-omics data handling, storage, and analysis are computationally demanding processes that result in many output files at every stage of the analytical pipeline. The kinds and quantity of data processed and saved are impacted by decisions made on which output files to share and save. When implementing data analytical pipelines, local hardware and software or cloud infrastructures must be chosen for data processing and storage.

Various workflow management systems facilitate omics data integration and streamline processes. Tools for data preparation, statistical analysis, and visualization of omics scale data are available from the Konstanz Information Miner (KNIME), BioMart, Taverna, and the Bioinformatics Institute of Singapore (BII) Infrastructure. These platforms support the development of modular environments for omics workflow construction. Furthermore, collaborative environments like my experiment allow scientists to publish and share workflows, fostering reproducibility and collaboration in the research community. Data processing in integrated omics workflows involves critical decisions, such as whether to integrate raw datasets or first process individual omics datasets before merging significant results for interpretation. Analysis tools chosen for integration efforts significantly impact outcomes. Even while techniques for imputation on a single-omics scale are well-established, handling missing data in multi-omics datasets is still difficult. One recent development is a multi-omics imputation method that uses an iterative algorithm to combine estimates of missing values from several datasets. Because transcriptome, proteomic, and metabolomic alterations respond differently, time course investigations provide additional complications. Existing tools do not adequately compensate for these differences, posing limitations in understanding integrated network dynamics. Normalization, transformation, and scaling approaches vary across different omics platforms due to unique data characteristics. Addressing missing values requires tailored strategies for each omics data type. Integration issues arise from differences in numerical scales, dynamic ranges, and time scales across genomics, transcriptomics, proteomics, and metabolomics platforms. Furthermore, the process of accounting for false positives and unknowns becomes more difficult when data from several sources are integrated. Less strict techniques involve physiologically guided integration via route or ontology enrichment analysis, whereas stricter methods rely on statistical models and data structure. While some investigators have adapted transcriptomics/microarray analysis methods like weighted gene co-expression network analysis (WGCNA) for integrated omics workflows, this approach does not fully address the unique data structures of different omics datasets. Recent developments, such as the MultiDataSet R-package, aim to encapsulate multiple datasets with applications for omics data integration.

Identifying unknown molecules remains a challenge in omics analyses, with limited consensus on annotation methods across different omics platforms. Efforts to generate and share "gold standard" datasets aim to address this issue and improve the reproducibility and utility of omics datasets. Data archiving and sharing are essential for reproducible research. Existing public databases like GEO, TCGA, and PRIDE DB archive individual omics datasets, but integrated omics datasets lack dedicated repositories. Cloud computing technologies facilitate dataset sharing and accessibility for postpublication analyses. Initiatives like DRYAD and Fig Share enable the sharing of large datasets within the research community. Standard databases allowing submission and retrieval of integrated omics datasets are needed to advance data archiving and sharing practices. In summary, integrated omics workflows require careful consideration of experimental design, data handling, processing, and sharing. Addressing challenges in sample preparation, data processing, integration, and data archiving is crucial for advancing reproducible and collaborative omics research.

5.12.3 Challenges in Using Multi-omics Approaches for Clinical Diagnostics and Prognostics

In addition to numerous genetic, protein, and metabolite changes, complex illnesses including cancer, cardiovascular diseases, and neurodegenerative disorders are also impacted by environmental variables like nutrition and lifestyle choices. Most of the research has concentrated on genetic and epigenetic changes alone, frequently in single-omics datasets, despite the Human Genome Project's promise to uncover genetic variants causing complex disorders. However, this approach has not fully elucidated the complexities of disease etiology, as additional biological variations, such as posttranslational modifications of proteins and metabolite abundance, also play significant roles. To address this gap, there is a need not only to quantify various biological variants, but also to integrate these data to enhance our understanding of health and disease. Currently, single metabolite tests like glucose and creatinine are routinely used in clinical settings, but capturing a broader range of biomolecules for clinical applications presents several challenges. Standardized sample collection, processing, and storage methods vary across omics analyses, making it impractical for clinicians to rapidly process samples for multiple analyses. Moreover, limited sample availability may hinder comprehensive omics analysis from a single sample. Simplifying procedures, lowering technical variance, enhancing repeatability, standardizing analytical techniques, and cutting down on sample analysis expenses and time are all necessary to overcome these obstacles. With examples of small molecule assays already in use and genetic testing becoming a standard procedure in clinical practice, it will soon be necessary to address these challenges to enable newer, more comprehensive omics platforms to contribute crucial clinical data for disease prognosis and diagnosis. The biggest challenge for every omics dataset, though, is figuring out how to interpret the data to get useful information. Multidimensional omics techniques aim to uncover biomarkers, and their use in biological research and clinical translation depends on their sensitivity and specificity. It takes a lot of time and computing power to interpret intricate multilayered networks, and it also requires an in-depth understanding of the biological system being studied.

Research that uses omics methodologies without utilizing biological expertise frequently leads to the identification of molecules and networks for hypothesis testing that lack biological significance. Therefore, biologically informed nomination of candidate molecules is crucial for efficient validation and inclusion of relevant networks in further analyses. In summary, integrating omics datasets holds great promise for elucidating the molecular mechanisms underlying complex diseases. Overcoming technical and analytical challenges, standardizing methods, and leveraging biological knowledge are essential steps toward translating integrated omics approaches into clinical practice for improved disease diagnosis and prognosis (Figure 5.5).

5.13 Conclusion

In conclusion, the adoption of OTs to enhance biofortification presents both challenges and opportunities in the pursuit of improving global nutrition and health outcomes. While significant advancements have been made in recent years, particularly in the development of high-throughput omics platforms and computational tools for data analysis, several challenges persist. One of the primary challenges lies in the integration and interpretation of multi-omics datasets, which are

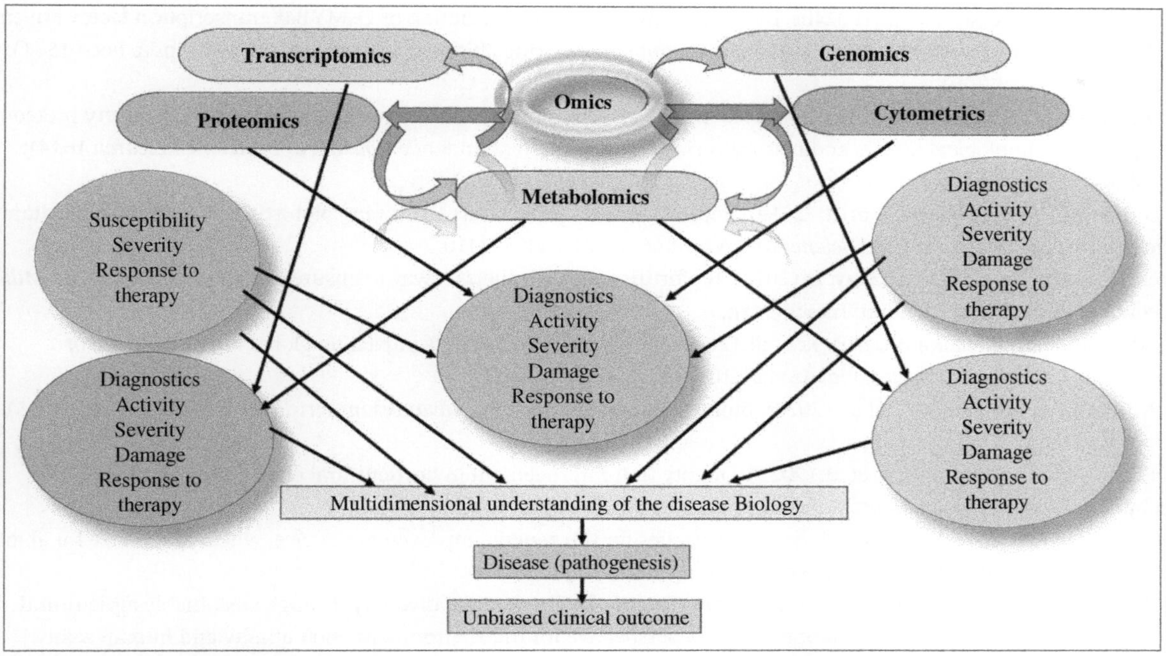

Figure 5.5 A multi-omics approach implemented in the clinic for diagnostic/prognostic purposes.

inherently complex and high-dimensional. Developing robust analytical methods and computational algorithms capable of effectively integrating genomics, transcriptomics, proteomics, and metabolomics data remains a pressing need. Additionally, addressing issues related to data standardization, normalization, and validation is paramount to ensure the reliability and reproducibility of omics-based findings. Furthermore, the translation of omics-based discoveries into actionable strategies for biofortification programs requires interdisciplinary collaboration among researchers, breeders, agronomists, and nutritionists. Bridging the gap between omics research and agricultural practices necessitates the development of innovative strategies for crop improvement, including the identification of key genes and pathways underlying nutrient accumulation and bioavailability in crops. Recent advancements in OTs, such as the application of CRISPR-Cas9 gene editing for targeted crop improvement and the use of advanced imaging techniques for spatial omics analysis, offer promising avenues for addressing these challenges. Moreover, the integration of omics data with other data modalities, such as environmental and socio-economic data, holds great potential for optimizing biofortification strategies tailored to specific agro-ecological contexts and population needs. In summary, while the adoption of OTs in biofortification presents challenges related to data integration, interpretation, and translation, recent advancements offer unprecedented opportunities for accelerating progress in global efforts to combat malnutrition and improve food security. By harnessing the power of omics-based approaches in conjunction with traditional breeding methods and agronomic practices, we can pave the way for the development of nutrient-rich crops that address the nutritional needs of vulnerable populations worldwide.

References

Agamah, F.E., Bayjanov, J.R., Niehues, A. et al. (2022). Computational approaches for network-based integrative multi-omics analysis. *Frontiers in Molecular Biosciences* 9: 967205.

Aggarwal, P.R., Pramitha, L., Choudhary, P. et al. (2022). Multi-omics intervention in Setaria to dissect climate-resilient traits: progress and prospects. *Frontiers in Plant Science* 13: 892736.

Ahmed, Z. (2020). Practicing precision medicine with intelligently integrative clinical and multi-omics data analysis. *Human Genomics* 14 (1): 35.

Ahmad, Y., Haider, S., Iqbal, J. et al. (2022). The mechanisms of genome editing technologies in crop plants. In: *Principles and Practices of OMICS and Genome Editing for Crop Improvement*, 295–313. Cham: Springer International Publishing.

Ahmad, Y., Haakim, Z., Iqbal, J. et al. (2024a). Technological innovations for abiotic stress resistance in horticultural crops. In: *OMICs-based Techniques for Global Food Security*, 233–244. Wiley.

Ahmad, Y., Haider, S., Iqbal, J. et al. (2024b). In-silico analysis and transformation of OsMYB48 transcription factor driven by CaMV35S promoter in model plant – *Nicotiana tabacum L.* conferring abiotic stress tolerance. *GM Crops & Food* 15 (1): 130–149.

Al Shweiki, M.R., Mönchgesang, S., Majovsky, P. et al. (2017). Assessment of label-free quantification in discovery proteomics and impact of technological factors and natural variability of protein abundance. *Journal of Proteome Research* 16 (4): 1410–1424.

Amah, D., van Biljon, A., Brown, A. et al. (2019). Recent advances in banana (*Musa spp.*) biofortification to alleviate vitamin a deficiency. *Critical Reviews in Food Science and Nutrition* 59 (21): 3498–3510.

Anushree, R.K., Durgapal, S. and Yadav, L. (2023). Biofortification of millets: A way to ensure nutritional security. In: *Millets-Rediscover Ancient Grains*. London: IntechOpen.

Ashoka, P., Spandana, B., Saikanth, D.R. K. et al. (2023). Bio-fortification and its impact on global health. *Journal of Experimental Agriculture International* 45 (10): 106–115.

Ayoub, A., Mushtaq, M., Mir, Z.A. et al. (2020). Biofortification in fruits. In: *Advances in Agri-Food Biotechnology*, 131–151. Singapore: Springer.

Bashir, T., Ul Haq, S.A., Masoom, S. et al. (2023). Quality trait improvement in horticultural crops: OMICS and modern biotechnological approaches. *Molecular Biology Reports* 50 (10): 8729–8742.

Benavente, E. and Giménez, E. (2021). Modern approaches for the genetic improvement of rice, wheat and maize for abiotic constraints-related traits: A comparative overview. *Agronomy* 11 (2): 376.

Bertola, M., Ferrarini, A. and Visioli, G. (2021). Improvement of soil microbial diversity through sustainable agricultural practices and its evaluation by-omics approaches: A perspective for the environment, food quality and human safety. *Microorganisms*, 9 (7): 1400.

Bibi, S., Ullah, R., Burni, T. et al. (2023). Impact of resorcinol and biochar application on the growth attributes, metabolite contents, and antioxidant systems of tomato (*Lycopersicon esculentum Mill.*). *ACS Omega* 8 (48): 45750–45762.

Bohra, A., Jha, U.C. and Kumar, S. (2016). Enriching nutrient density in staple crops using modern "-Omics" tools. *Biofortification of Food Crops* 85–103.

Bohra, A., Sahrawat, K.L., Kumar, S. et al. (2015). Genetics-and genomics-based interventions for nutritional enhancement of grain legume crops: Status and outlook. *Journal of Applied Genetics* 56: 151–161.

Bollinedi, H., Singh, A.K., Vinod, K.K. et al. (2022). Genetic fortification of rice to address hidden hunger: Progress and prospects. In: *Biofortification of Staple Crops*, 63–103. Singapore: Springer.

Bolten, C.J., Kiefer, P., Letisse, F. et al. (2007). Sampling for metabolome analysis of microorganisms. *Analytical Chemistry* 79 (10): 3843–3849.

Bosviel, R., Joumard-Cubizolles, L., Chinetti, G. et al. (May 2016). Comparison of the anti-inflammatory properties of selected DHA-derived oxylipins and insights into their mechanism of action. In: *LIPID MAPS Annual Meeting 2016: Lipidomics Impact on Metabolic, Cancer, Cardiovascular and Inflammatory Diseases*, 46. France: Chamonix-Mont Blanc.

Brönstrup, M. (2004). Absolute quantification strategies in proteomics based on mass spectrometry. *Expert Review of Proteomics* 1 (4): 503–512.

Brunner, A.D., Thielert, M., Vasilopoulou, C. et al. (2022). Ultra-high sensitivity mass spectrometry quantifies single-cell proteome changes upon perturbation. *Molecular Systems Biology* 18 (3): e10798.

Canzler, S., Schor, J., Busch, W. et al. (2020). Prospects and challenges of multi-omics data integration in toxicology. *Archives of Toxicology*, 94: 371–388.

Chakraborty, S., Sharma, G., Karmakar, S. et al. (2024). Multi-OMICS approaches in cancer biology: New era in cancer therapy. *Biochimica et Biophysica Acta (BBA) – Molecular Basis of Disease* 1870 (5): 167120.

Chandra, A.K., Kumar, A., Bharati, A. et al. (2020). Microbial-assisted and genomic-assisted breeding: A two way approach for the improvement of nutritional quality traits in agricultural crops. *3 Biotech* 10 (1): 2.

Chaudhary, N., Salgotra, R.K. and Chauhan, B.S. (2023). Genetic enhancement of cereals using genomic resources for nutritional food security. *Genes* 14 (9): 1770.

Cox, J., Hein, M.Y., Luber, C.A. et al. (2014). Accurate proteome-wide label-free quantification by delayed normalization and maximal peptide ratio extraction, termed MaxLFQ. *Molecular & Cellular Proteomics* 13 (9): 2513–2526.

Dar, W.D. (2014). Introduction: Biotechnological interventions for crop improvement: Answers to global challenges. In: *Plant Biotechnology: Experience and Future Prospects*, 1–10. Cham: Springer International Publishing.

Devi, S.R. and Aglawe, S.B. (2021). Social acceptance and regulatory prospects of genomics in addressing food security. *Omics Technologies for Sustainable Agriculture and Global Food Security* 2: 329–351.

Dharani, E., Shricharan, S., Balaji, B. et al. (2024). Nutritional revolution: Biotechnological strategies for fortified crop produce. In: *Phytoremediation and Biofortification*, 355–371. Palm Bay, FL: Apple Academic Press.

Dwivedi, S.L., Garcia-Oliveira, A.L., Govindaraj, M. et al. (2023). Biofortification to avoid malnutrition in humans in a changing climate: Enhancing micronutrient bioavailability in seed, tuber, and storage roots. *Frontiers in Plant Science* 14: 1119148.

Eddy, S., Mariani, L.H. and Kretzler, M. (2020). Integrated multi-omics approaches to improve classification of chronic kidney disease. *Nature Reviews Nephrology* 16 (11): 657–668.

Fernandez-Banet, J., Esposito, A., Coffin, S. et al. (2016). OASIS: Web-based platform for exploring cancer multi-omics data. *Nature Methods* 13 (1): 9–10.

Gedil, M., Mengesha, W., Ilesanmi, O. et al. (2024). Advances in genetic enhancement of nutritional quality of tropical maize in West and Central Africa. *Agriculture* 14 (4): 577.

Ghorai, M., Kumar, V., Kumar, V. et al. (2022). Beneficial role of selenium (Se) biofortification in developing resilience against potentially toxic metal and metalloid stress in crops: Recent trends in genetic engineering and omics approaches. *Journal of Soil Science and Plant Nutrition* 22 (2): 2347–2377.

Giavalisco, P., Köhl, K., Hummel, J. et al. (2009). 13C isotope-labeled metabolomes allowing for improved compound annotation and relative quantification in liquid chromatography-mass spectrometry-based metabolomic research. *Analytical Chemistry* 81 (15): 6546–6551.

Graw, S., Chappell, K., Washam, C.L et al. (2021). Multi-omics data integration considerations and study design for biological systems and disease. *Molecular Omics* 17 (2): 170–185.

Gupta, A., Sharma, T., Singh, S.P. et al. (2023). Prospects of microgreens as budding living functional food: Breeding and biofortification through OMICS and other approaches for nutritional security. *Frontiers in Genetics* 14: 1053810.

Gupta, O.P., Singh, A.K., Singh, A. et al. (2022a). Wheat biofortification: Utilizing natural genetic diversity, genome-wide association mapping, genomic selection, and genome editing technologies. *Frontiers in Nutrition* 9: 826131.

Gupta, S., Verma, R. and Ravi, R.K. (2023). Multiomics approach for crop improvement under climate change. In: *Sustainable Agriculture in the Era of the OMICs Revolution*, 17–36. Cham: Springer International Publishing.

Gupta, S.M., Gupta, S. and Kumar, A. (2022b). Genetic transformation for crop improvement and biofortification. In: *The Finger Millet Genome*, 239–259. Cham: Springer International Publishing.

Gurke, R., Bendes, A., Bowes, J. et al. (2022). Omics and multi-omics analysis for the early identification and improved outcome of patients with psoriatic arthritis. *Biomedicines* 10 (10): 2387.

Haas, J.D., Luna, S.V., Lung'aho, M.G. et al. (2016). Consuming iron biofortified beans increases iron status in Rwandan women after 128 days in a randomized controlled feeding trial. *The Journal of Nutrition* 146 (8): 1586–1592.

Hafeez, U., Ali, M., Hassan, S.M. et al. (2023). Advances in breeding and engineering climate-resilient crops: A comprehensive review. *International Journal of Research and Advances in Agricultural Sciences* 2 (2): 85–99.

Haider, S., Iqbal, J., Naseer, S. et al. (2021a). Molecular mechanisms of plant tolerance to heat stress: current landscape and future perspectives. *Plant Cell Reports* 40: 2247–2271.

Haider, S., Iqbal, J., Naseer, S. et al. (2022b). Unfolding molecular switches in plant heat stress resistance: A comprehensive review. *Plant Cell Reports* 41 (3): 775–798.

Haider, S., Iqbal, J., Shaukat, M. et al. (2021b). The epigenetic chromatin-based regulation of somatic heat stress memory in plants. *Plant Gene* 27: 100318.

Haider, S., Raza, A., Iqbal, J. et al. (2022a). Analyzing the regulatory role of heat shock transcription factors in plant heat stress tolerance: A brief appraisal. *Molecular Biology Reports* 49 (6): 5771–5785.

Haq, S.A.U., Bashir, T., Roberts, T.H. et al. (2024). Ameliorating the effects of multiple stresses on agronomic traits in crops: Modern biotechnological and omics approaches. *Molecular Biology Reports* 51 (1): 41.

Hasin, Y., Seldin, M. and Lusis, A. (2017). Multi-omics approaches to disease. *Genome Biology* 18: 1–15.

Hossain, A., Islam, M.T., Maitra, S. et al. (2021). Neglected and underutilized crop species: are they future smart crops in fighting poverty, hunger and malnutrition under changing climate? In: *Neglected and Underutilized Crops – Towards Nutritional Security and Sustainability*, 1–50. Singapore: Springer.

Hultin-Rosenberg, L., Forshed, J., Branca, R.M. et al. (2013). Defining, comparing, and improving iTRAQ quantification in mass spectrometry proteomics data. *Molecular & Cellular Proteomics* 12 (7): 2021–2031.

Husaini, A.M. (2022). High-value pleiotropic genes for developing multiple stress-tolerant biofortified crops for 21st-century challenges. *Heredity* 128 (6): 460–472.

Ijaz, S., Iqbal, J., Abbasi, B.A. et al. (2023a). Biofortification: Lessons from the past and strategies for future food security. In: *Legumes Biofortification*, 521–545. Cham: Springer International Publishing.

Ijaz, S., Iqbal, J., Abbasi, B.A. et al. (2023b). Rosmarinic acid and its derivatives: Current insights on anticancer potential and other biomedical applications. *Biomedicine & Pharmacotherapy* 162: 114687.

Ijaz, S., Iqbal, J., Abbasi, B.A. et al. (2024). Heavy metal–polluted arable land and its consequences: A global scenario. In: *Biochar-Assisted Remediation of Contaminated Soils Under Changing Climate*, 71–99. Amsterdam: Elsevier.

Ingle, K.P., Suprasanna, P., Narkhede, G.W. et al. (2023). Biofortified foxtail millet: Towards a more nourishing future. *Plant Growth Regulation* 99 (1): 25–34.

Ishfaq, M., Kiran, A., ur Rehman, H. et al. (2022). Foliar nutrition: Potential and challenges under multifaceted agriculture. *Environmental and Experimental Botany* 200: 104909.

Jadhav, P.V., Magar, S., Thakur, P. et al. (2020). Biofortified fodder crops: An approach to eradicate hidden hunger. In: *Advances in Agri-Food Biotechnology*, 173–198. Singapore: Springer.

Jaiwal, P.K., Chhillar, A.K., Chaudhary, D. et al. (eds.) (2019). *Nutritional Quality Improvement in Plants*. Singapore: Springer Nature.

Jha, U.C., Nayyar, H., Parida, S.K. et al. (2022). Ensuring global food security by improving protein content in major grain legumes using breeding and 'Omics' tools. *International Journal of Molecular Sciences* 23 (14): 7710.

Joshi, B.K., Shrestha, H.K. and Ayer, D.K. (2023). Crop breeding and biotechnological advances towards nutrition and environment security. In: *Emerging Solutions in Sustainable Food and Nutrition Security*, 255–285. Cham: Springer International Publishing.

Kadapa, S., Gunturi, A., Gundreddy, R. et al. (2023). *Agronomic Biofortification of Millets: New Way to Alleviate Malnutrition*. London: IntechOpen.

Kadirimangalam, S.R., Bagudam, R., Variath, M.T. et al. (2022). Prospects of biofortification in groundnut using modern breeding approaches. In: *Biofortification of Staple Crops*, 345–377. Singapore: Springer.

Kathuria, S., Anshu, R.G., Bhogal, A.S. et al. (2024). Omics-based applications for sustainable agriculture. In: *Sustainable Agriculture: Nanotechnology and Biotechnology for Crop Production and Protection*, 255. Boston: De Gruyter.

Kaur, S., Dwibedi, V., Sahu, P.K. et al. (Eds.). (2023). *Metabolomics, Proteomes and Gene Editing Approaches in Biofertilizer Industry*. Singapore: Springer Nature.

Khan, A.A., Iqbal, B., Jalal, A. et al. (2024). Advanced molecular approaches for improving crop yield and quality: A review. *Journal of Plant Growth Regulation*, 43 (7): 1–13.

Khandagale, K., Shirsat, D. and Ade, A. (2023). Biofortification of barley for nutritional security. In: *Biofortification in Cereals: Progress and Prospects*, 235–258. Singapore: Springer Nature Singapore.

Kiran, A., Wakeel, A., Mahmood, K. et al. (2022). Biofortification of staple crops to alleviate human malnutrition: Contributions and potential in developing countries. *Agronomy* 12 (2): 452.

Knez, M. and Stangoulis, J.C. (2021). Calcium biofortification of crops – challenges and projected benefits. *Frontiers in Plant Science* 12: 669053.

Kumar, A., Anju, T., Kumar, S. et al. (2021a). Integrating omics and gene editing tools for rapid improvement of traditional food plants for diversified and sustainable food security. *International Journal of Molecular Sciences* 22 (15): 8093.

Kumar, A., Anju, T., Kumar, S. et al. (2021b). Linking omics and gene editing tools for rapid improvement of traditional food plants for diversified foods and sustainable food security. *International Journal of Molecular Sciences* 22: 8093.

Kumar, P., Kumar, A., Dhiman, K. et al. (2022). Recent progress in cereals biofortification to alleviate malnutrition in India: An overview. In: *Agricultural Biotechnology: Latest Research and Trends*, 253–280. Singapore: Springer.

Kumar, R., Bohra, A., Pandey, A.K. et al. (2017). Metabolomics for plant improvement: status and prospects. *Frontiers in Plant Science* 8: 271676.

Kumar, R., Nizampatnam, N.R., Alam, M. et al. (2021c). Genome editing technologies for plant improvement: Advances, applications and challenges. *Omics Technologies for Sustainable Agriculture and Global Food Security* 1: 213–240.

Kumar, S., Palve, A., Joshi, C. et al. (2019). Crop biofortification for iron (Fe), zinc (Zn) and vitamin a with transgenic approaches. *Heliyon* 5 (6): e01872.

Lal, M.K., Kumar, A., Kardile, H.B. et al. (2020). Biofortification of vegetables. In: *Advances in Agri-Food Biotechnology*, 105–129. Singapore: Springer.

Lelli, V., Belardo, A. and Timperio, A.M. (2021). From targeted quantification to untargeted metabolomics. In: *Metabolomics – Methodology and Applications in Medical Sciences and Life Sciences*, 10. London: IntechOpen.

Lloyd-Price, J., Arze, C., Ananthakrishnan, A.N. et al. (2019). Multi-omics of the gut microbial ecosystem in inflammatory bowel diseases. *Nature* 569 (7758): 655–662.

Lordan, R., Tsoupras, A. and Zabetakis, I. (2019). The potential role of dietary platelet-activating factor inhibitors in cancer prevention and treatment. *Advances in Nutrition* 10 (1): 148–164.

Ma, A., McDermaid, A., Xu, J. et al. (2020). Integrative methods and practical challenges for single-cell multi-omics. *Trends in Biotechnology* 38 (9): 1007–1022.

Manzoor, Z., Batool, M., Ali, M.F. et al. (2023). Biomolecular aspects of plant nutrition related to food biofortification. In: *Legumes Biofortification*, 465–479. Cham: Springer International Publishing.

Mishra, G.P., Dikshit, H.K., Priti. et al. (2022). Historical overview of biofortification in crop plants and its implications. In: *Biofortification of Staple Crops*, 31–61. Singapore: Springer.

Murphy, D.J. (2014). Using modern plant breeding to improve the nutritional and technological qualities of oil crops. *Oilseeds and Fats, Crops and Lipids* 21 (6): D607.

Naik, B., Kumar, V., Rizwanuddin, S. et al. (2024). Biofortification as a solution for addressing nutrient deficiencies and malnutrition. *Heliyon* 10 (9): e30595.

Nasar, S., Muhammad, M., Siddiqi, E.H. et al. (2024). OMICS-based knowledge for achieving food and nutritional security. In: *OMICs-Based Techniques for Global Food Security*, 67–90. Hoboken, NJ: Wiley.

Obrenovich, M.E., Donskey, C.J., Schiefer, I.T. et al. (2018). Quantification of phenolic acid metabolites in humans by LC–MS: A structural and targeted metabolomics approach. *Bioanalysis* 10 (19): 1591–1608.

Patel, A., Miles, A., Strackhouse, T. et al. (2023). Methods of crop improvement and applications towards fortifying food security. *Frontiers in Genome Editing* 5: 1171969.

Pathak, R.K., Baunthiyal, M., Pandey, D. et al. (2018). Augmentation of crop productivity through interventions of omics technologies in India: Challenges and opportunities. *3 Biotech* 8 (11): 1–28.

Planell, N., Lagani, V., Sebastian-Leon, P. et al. (2021). STATegra: Multi-omics data integration – A conceptual scheme with a bioinformatics pipeline. *Frontiers in Genetics* 12: 620453.

Rai, V., Karthikaichamy, A., Das, D. et al. (2016). Multi-omics frontiers in algal research: Techniques and progress to explore biofuels in the postgenomics world. *OMICS: A Journal of Integrative Biology* 20 (7): 387–399.

Raj, S.R. G. and Nadarajah, K. (2022). QTL and candidate genes: Techniques and advancement in abiotic stress resistance breeding of major cereals. *International Journal of Molecular Sciences* 24 (1): 6.

Rakkammal, K., Priya, A., Pandian, S. et al. (2022). Conventional and omics approaches for understanding the abiotic stress response in cereal crops – an updated overview. *Plants* 11 (21): 2852.

Rana, N., Rahim, M.S., Kaur, G. et al. (2020). Applications and challenges for efficient exploration of omics interventions for the enhancement of nutritional quality in rice (*Oryza sativa L.*). *Critical Reviews in Food Science and Nutrition* 60 (19): 3304–3320.

Rao, C.K. and Annadana, S. (2017). Nutrient biofortification of staple food crops: Technologies, products and prospects. *Phytonutritional Improvement of Crops* 113–183.

Rasheed, A.A., Raza, Q., Waqas, M. et al. (2024). Biofortification of chickpea: Genetics, genomics, and breeding perspectives. In: *Biofortification of Grain and Vegetable Crops*, 139–159. Cambridge: Academic Press.

Rauniyar, N. and Yates III, J.R. (2014). Isobaric labeling-based relative quantification in shotgun proteomics. *Journal of Proteome Research* 13 (12): 5293–5309.

Razzaq, A., Sadia, B., Raza, A. et al. (2019). Metabolomics: A way forward for crop improvement. *Metabolites* 9 (12): 303.

Reddy, A.R., Tripura, U. and Shirisha, G. (2021). Biofortification of various vegetable crops – A review. *Journal of Emerging Technologies and Innovative Research* 8 (5): d395–d405.

Rehman, A., Almas, H.I., Mazhar, K. et al. (2020). Breeding approaches to generate biofortified rice for nutritional enhancement. *Rice Research for Quality Improvement: Genomics and Genetic Engineering* 2: 509–540.

Reska, D., Czajkowski, M., Jurczuk, K. et al. (2021). Integration of solutions and services for multi-omics data analysis towards personalized medicine. *Biocybernetics and Biomedical Engineering* 41 (4): 1646–1663.

Rochat, B. (2016). From targeted quantification to untargeted metabolomics: Why LC-high-resolution-MS will become a key instrument in clinical labs. *TrAC Trends in Analytical Chemistry* 84: 151–164.

Rodríguez-Suárez, E. and Whetton, A.D. (2013). The application of quantification techniques in proteomics for biomedical research. *Mass Spectrometry Reviews* 32 (1): 1–26.

Röhnisch, H.E., Eriksson, J., Müllner, E. et al. (2018). AQuA: An automated quantification algorithm for high-throughput NMR-based metabolomics and its application in human plasma. *Analytical Chemistry* 90 (3): 2095–2102.

Roorkiwal, M., Bharadwaj, C., Barmukh, R. et al. (2020). Integrating genomics for chickpea improvement: Achievements and opportunities. *Theoretical and Applied Genetics*, 133; 1703–1720.

Røst, L.M., Brekke Thorfinnsdottir, L., Kumar, K. et al. (2020). Absolute quantification of the central carbon metabolome in eight commonly applied prokaryotic and eukaryotic model systems. *Metabolites* 10 (2): 74.

Roy, C., Kumar, S., Ranjan, R.D. et al. (2022). Genomic approaches for improving grain zinc and iron content in wheat. *Frontiers in Genetics* 13: 1045955.

Sagare, D.B., Shetti, P., Yankanchi, S. et al. (2023). Maize nutraceutomics: Genomics, biotechnology, and nanotechnology. In: *Compendium of Crop Genome Designing for Nutraceuticals*, 85–113. Singapore: Springer Nature.

Satyavathi, C.T., Ambawat, S., Khandelwal, V. et al. (2021). Pearl millet: a climate-resilient nutricereal for mitigating hidden hunger and provide nutritional security. *Frontiers in Plant Science* 12: 659938.

Sen, A., Das, B., Roy, S. et al. (2022). Phytoremediation and biofortification: Contrasting yet similar approaches of manipulating plant metal (loid) homeostasis for societal benefit. In: *Plant Metal and Metalloid Transporters*, 407–454. Singapore: Springer Nature.

Seneviratne, C.J., Suriyanarayanan, T., Widyarman, A.S. et al. (2020). Multi-omics tools for studying microbial biofilms: current perspectives and future directions. *Critical Reviews in Microbiology* 46 (6): 759–778.

Shahzad, R., Jamil, S., Ahmad, S. et al. (2021). Biofortification of cereals and pulses using new breeding techniques: Current and future perspectives. *Frontiers in Nutrition* 8: 721728.

Sharma, N., Pradhan, S. and Majumder, S. (2024) Proteomics-based strategies to develop food crops with enhanced nutritional quality. In: *Plant Proteomics*, 226–242. Boca Raton, FL: CRC Press.

Sharma, V., Choudhary, M., Kumar, P. et al. (2021a). Harnessing the wild relatives and landraces for Fe and Zn biofortification in wheat through genetic interventions – A review. *Sustainability* 13 (23): 12975.

Sharma, V., Gupta, P., Kagolla, P. et al. (2021b). Metabolomics intervention towards better understanding of plant traits. *Cells* 10 (2): 346.

Sheoran, S., Kumar, S., Ramtekey, V. et al. (2022). Current status and potential of biofortification to enhance crop nutritional quality: An overview. *Sustainability* 14 (6): 3301.

Shoeb, E., Venkataraman, S., Badar, U. et al. (2024). Genome editing for crop biofortification. In: *Applications of Genome Engineering in Plants*, 91–121.

Siddique, A. and Aley, P. (2024). Nano-omics approaches for studying plant growth, yield, and nutritional value. In: *Harnessing NanoOmics and Nanozymes for Sustainable Agriculture*, 19–41. Hershey, PA: IGI Global.

Singh, D.P. and Prabha, R. (2019). Omics data integration in microbial research for agricultural and environmental applications. In: *Microbial Interventions in Agriculture and Environment* 3: 461–491. Singapore: Soil and Crop Health Management.

Singh, N., Bhatt, V., Rana, N. et al. (2020a). Advances of next-generation sequencing (NGS) technologies to enhance the biofortifications in crops. *Advances in Agri-food Biotechnology* 2020: 427–450.

Singh, N., Rai, V. and Singh, N.K. (2020b). Multi-omics strategies and prospects to enhance seed quality and nutritional traits in pigeonpea. *The Nucleus* 63 (3): 249–256.

Singh, N., Ujinwal, M. and Singh, A. (2022). Advances in agricultural bioinformatics: An outlook of multi "omics" approaches. In: *Bioinformatics in Agriculture*, 3–21. Cambridge: Academic Press.

Singh, S., Chaudhary, R., Lokya, V. et al. (2024). Genome editing based trait improvement in crops: current perspective, challenges and opportunities. *The Nucleus* 67 (1): 97–126.

Singh, V.K., Thakur, R., Sharma, J. et al. (2024). Future prospects: High-altitude crop improvement via genomics. In: *Genetics and Genomics of High-Altitude Crops*, 217–255. Singapore: Springer Nature.

Singha, D.L., Das, D., Paswan, R.R. et al. (2022). Novel approaches and advanced molecular techniques for crop improvement. In: *Plant–Microbe Interactions*, 1–27. Boca Raton, FL: CRC Press.

Sun, Y.V. and Hu, Y.J. (2016). Integrative analysis of multi-omics data for discovery and functional studies of complex human diseases. *Advances in Genetics* 93: 147–190.

Susmitha, P., Kapoor, M., Sanjay, M. et al. (2024). Unlocking nutritional potential: Multi-omics strategies for enhancing millet nutritional traits. *Journal of Advances in Biology & Biotechnology* 27 (6): 131–149.

Tayade, R., Kabange, R., Ali, M.W. et al. (2023). Biofortification of wheat using current resources and future challenges. In: *Biofortification in Cereals: Progress and Prospects*, 173–208. Singapore: Springer Nature.

Thakur, V., Sharma, A., Sharma, P. et al. (2022). Biofortification of vegetable crops for vitamins, mineral and other quality traits. *The Journal of Horticultural Science and Biotechnology* 97 (4): 417–428.

Trammell, S.A. and Brenner, C. (2013). Targeted, LCMS-based metabolomics for quantitative measurement of NAD+ metabolites. *Computational and Structural Biotechnology Journal* 4 (5): e201301012.

Ullah, Z., Iqbal, J., Abbasi, B.A. et al. (2023a). How genome editing can be helpful in the biofortification of legumes. In: *Legumes Biofortification*, 207–232. Cham: Springer International Publishing.

Ullah, R., Ullah, Z., Iqbal, J. et al. (2023b). Aspartic acid-based nano-copper induces resilience in Zea mays to applied lead stress via conserving photosynthetic pigments and triggering the antioxidant biosystem. *Sustainability* 15 (16): 12186.

Ullah, Z., Gul, F., Iqbal, J. et al. (2023c). Biogenic synthesis of multifunctional silver oxide nanoparticles (Ag_2ONPs) using *Parieteria alsinaefolia* delile aqueous extract and assessment of their diverse biological applications. *Microorganisms* 11 (4): 1069.

Ullah, Z., Iqbal, J., Abbasi, B.A. et al (2023d). Assessment of GUS expression induced by anti-sense Os PPO gene promoter and antioxidant enzymatic assays in response to drought and heavy metal stress in transgenic *Arabidopsis thaliana*. *Sustainability* 15 (17): 12783.

Ullah, Z., Iqbal, J., Gul, F. et al. (2024). Biogenic synthesis, characterization, and in vitro biological investigation of silver oxide nanoparticles (AgONPs) using *Rhynchosia capitata*. *Scientific Reports* 14 (1): 10484.

Vasaikar, S.V., Straub, P., Wang, J. et al. (2018). Linked omics: Analyzing multi-omics data within and across 32 cancer types. *Nucleic Acids Research* 46 (D1): D956–D963.

Verma, V., Vishal, B., Kohli, A. et al. (2021). Systems-based rice improvement approaches for sustainable food and nutritional security. *Plant Cell Reports* 40 (11): 2021–2036.

Wairich, A., Ricachenevsky, F.K. and Lee, S. (2022). A tale of two metals: Biofortification of rice grains with iron and zinc. *Frontiers in Plant Science* 13: 944624.

Wang, Q., Hu, J., Lou, T. et al. (2023). Integrated agronomic, physiological, microstructure, and whole-transcriptome analyses reveal the role of biomass accumulation and quality formation during Se biofortification in alfalfa. *Frontiers in Plant Science* 14: 1198847.

Xia, H., Wang, L., Qiao, Y. et al. (2020). Elucidating the source–sink relationships of zinc biofortification in wheat grains: A review. *Food and Energy Security* 9 (4): e243.

Yağız, A.K., Yavuz, C., Naeem, M. et al. (2022). Genome editing for nutrient use efficiency in crops. In: *Principles and Practices of OMICS and Genome Editing for Crop Improvement*, 347–383. Cham: Springer International Publishing.

Yandeau-Nelson, M.D., Lauter, N. and Zabotina, O.A. (2016). Advances in metabolomic applications in plant genetics and breeding. *CABI Reviews* (2015): 1–15.

Yang, Z., Wang, S., Wei, L. et al. (2023). BnIR: A multi-omics database with various tools for *Brassica napus* research and breeding. *Molecular Plant* 16 (4): 775–789.

Younas, A., Yousaf, Z., Riaz, N. et al. (2023). Nutritional enrichment in legumes through omics approaches. In: *Legumes Biofortification*, 177–206. Cham: Springer International Publishing.

Zhou, J. and Yin, Y. (2016). Strategies for large-scale targeted metabolomics quantification by liquid chromatography-mass spectrometry. *Analyst* 141 (23): 6362–6373.

Zhou, Y., Tan, Z., Xue, P. et al. (2021). High-throughput, in-depth and estimated absolute quantification of plasma proteome using data-independent acquisition/mass spectrometry ("HIAP-DIA"). *Proteomics* 21 (5): 2000264.

Zulfiqar, U., Khokhar, A., Maqsood, M.F. et al. (2024). Genetic biofortification: Advancing crop nutrition to tackle hidden hunger. *Functional & Integrative Genomics* 24 (2): 34.

6

Role of Nanoparticles in Improving Biofortification: An Overview

Muhammad Zahid[1], Basharat Ali[2,], Dilawar Aslam[1], Swaiba Rani[1], Aqeela Shaheen[3], Saddam Hussain[4], Noman Ali Buttar[2], Yasir Niaz[2], Shakeel Ahmad[5], Javed Iqbal[2,*]*

[1] *Department of Plant Breeding and Genetics, University of Agriculture Faisalabad, Pakistan*
[2] *Department of Agricultural Engineering, Khwaja Fareed University of Engineering and Information Technology, Rahim Yar Khan, Pakistan*
[3] *Department of Chemistry, Government Sadiq College Women University, Bahawalpur, Pakistan*
[4] *Department of Agronomy, University of Agriculture, Faisalabad, Pakistan*
[5] *Institute of Agronomy, Bahauddin Zakariya University, Multan, Pakistan*

*Corresponding author: dr.javed@kfueit.edu.pk; dr.basharat@kfueit.edu.pk

6.1 Introduction

Globally, more than 2 billion people (one in three) suffer from micronutrient deficiencies, also known as a "hidden hunger," which is particularly common in women, children, and the elderly (Szerement et al. 2022; Karthika et al. 2023). According to a United Nations report on food and agriculture, over 735 million people had experienced malnutrition worldwide in 2023. Bailey et al. (2015) reported that a deficiency of micronutrients is a common problem in Sub-Saharan Africa and Southeast Asia. The deficiency of micronutrients including vitamins (vitamin A and B9) and minerals, such as iron, zinc, and iodide, in the diet negatively influences human health, which leads to decreased growth, perinatal complications, dementia, and increased mortality. Biofortification is a technique that reduces micronutrient deficiencies by improving the nutritional value of crops like maize, rice, barley, wheat, and vegetables by application of nanoparticles (NPs), mineral fertilization, agronomic practices, conventional breeding, and transgenic approaches. Moreover, this technique not only enhances crop productivity, resistance against insects and pests, postharvest losses, and quality of the product, but also fulfills the nutritional requirement of people who live in rural areas and depend on cereals crops as a staple food mostly in developing countries. Biofortified crops improve the nutritional values of consumers, and by developing their markets, these crops could be easily available to consumers. Therefore, nutrient deficiencies in consumers can be minimized by including biofortified crops in their diet; however, this technique cannot eliminate the deficiencies completely (Signorell et al. 2019). However, in urban areas, biofortified crops are readily available in the local markets providing direct access to a large population. Unfortunately, the same cannot be considered for rural areas due to the expensive cost associated with these nutritious foods. The biofortification process offers numerous benefits in developing countries, using various strategies including the engineering of staple crops (Gani et al. 2018), such as agronomic, plant breeding, and transgenic biofortification methods (Jha and Warkentin 2020). Adu et al. (2018) reported that agronomic biofortification, which was applied through mineral fertilizers to the soil or directly on plants, improves the nutritional values of the crops. Genetic biofortification leverages genetic variation among crops to improve nutritional values (Marques et al. 2021).

However, the lack of genetic differences between crops makes it challenging to enhance the nutritional values of the crops through conventional plant breeding alone. In such cases, transgenic approaches become essential to identify and characterize specific genes for transferring into desired crops thereby achieving biofortification. Biofortified crops including vegetables, cereals, fruits, legumes, and oilseeds have been developed by using these three approaches. Among them,

Crop Biofortification: Biotechnological Approaches for Achieving Nutritional Security under Changing Climate, First Edition. Edited by Adnan Noor Shah et al.

cereals stand out as the most prominent biofortified crops. In oilseed crops, where minimal genetic differences exist, transgenic approaches have gained attention for biofortification (Garg et al. 2018). Marker-assisted selection (MAS) has also played a role in producing micronutrients enriched across different crops (Mayer et al. 2008; Singh et al. 2017). Introducing a new dimension, "Nanotechnology" has emerged as a technique that increases resistance against abiotic stresses, contributing to sustainable agriculture globally. NPs have paved the way for promoting sustainable agriculture due to their unique physiochemical parameters, such as greater mobility, particle shape, large surface area, and reactivity. Over the past decade, many scientists have noted a significant increase in the application of nanotechnology in agricultural practices (Lowry et al. 2019). This technology employs environment-friendly techniques and increases the availability of nutrients for plants and ultimately contributing to nutrient enrichment. Nano-fertilizer is an effective approach that improves the fertility of the soil and decreases agroecological degradation (Manjunatha et al. 2016). Sarlak and Taherifar (2017) concluded that nanofertilizers play a crucial role in the process of biofortification by increasing nutrient availability, reducing the need for chemical fertilizers, and fostering the development of environmentally friendly crops with resistance against biotic and abiotic stresses. This approach addresses nutritional gaps in plants by enhancing the uptake of nutrients. Information in this chapter will enlighten the information on the efficient biofortification approaches, which can be used to improve the crop yield, quality, and nutrient's concentration in different crops (Figure 6.1).

6.2 Biofortification of Food Crops: Tackling Malnutrition and Hidden Hunger

This is attributed to their reliance on a variety of biofortified crops rather than single staples ensuring they meet their nutritional needs (Weisstaub and Araya 2008). The problems of hidden hunger are more common in developing countries due to the inability to access adequate foods because of poor economic and political status, changes in dietary patterns and overconsumption of staple foods (Gödecke et al. 2018). Mantadakis et al. (2020) concluded that micronutrient deficiencies including iron, iodine, and vitamins D and E are more common in developing countries. Gani et al. (2018) reported that the problems of hidden hunger and malnutrition are continuously drastic, and their effects are more in the children; approximately 1.1 million children are dying annually due to the deficiency of micronutrients. Recognizing the severity of these situations, governments and stakeholders worldwide are compelled to develop strategies to overcome these problems (Sriatmi et al. 2021). To tackle these challenges, the concept of biofortification has developed through the intervention of food supplementation and dietary diversification (Bouis et al. 2019). Food fortification increases the particular nutrients in food, which are naturally available in foods or introduced from other crops into the selected crops to meet the recommended dietary allowance (RDA) for consumers (Lowe 2021). Biofortification is a type of food fortification that increases nutrient availability during the vegetative growth of food crops (Lowe 2021). Bouis et al. (2019) reported that the HarvestPlus Program began in 2003 to target African and Asia countries to increase nutrient accessibility through the development of high-quality biofortified cultivars of staple grains. The most common examples of biofortified crops by this program comprise maize, Golden Rice, and orange-fleshed sweet potato biofortified with vitamin A; rice and wheat biofortified with Fe and Zinc; and beans biofortified with Fe. In short, plant biofortification plays a vital role in addressing hidden hunger and

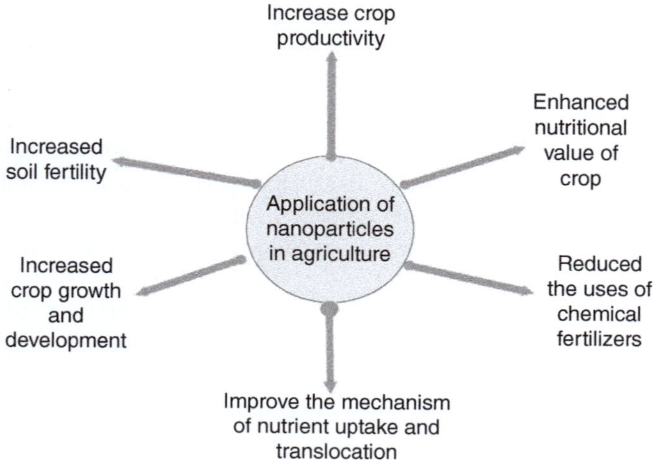

Figure 6.1 Application of nanoparticles in agriculture.

malnutrition globally, making it a practical approach to improving micronutrient consumption alongside fortification, diet diversification, and supplementation strategies.

6.2.1 Food Fortification

Food fortification includes the addition of specific nutrients in the food that are not naturally present. The basic aim of food fortification is to improve the nutritional value of foods to fulfill the RDA of consumers in terms of those specific nutrients (Das et al. 2019). Several types of food fortification involve target fortification, voluntary fortification, mass fortification, and mandatory fortification (Olson et al. 2021). Mass fortification addresses nutrient deficiencies in staple foods, and the most common example of mass fortification is the biofortification of rice, which is mostly consumed on a large scale in Asian countries such as China, Japan, and the Philippines (Das et al. 2019). Voluntary fortification refers to the addition of micronutrients like iron and vitamin A in the staple cereals, mainly wheat flour, in countries such as the United Arab Emirates, Pakistan, Qatar, and Gambia (Osendarp et al. 2018). Mandatory fortification, driven by government laws and regulations, includes the addition of nutrients like vitamin A in edible oils, folic acid in the flour of wheat, and iodine in salts (Younis et al. 2015). Target fortification contains the addition of targeted nutrients in the foods to overcome the deficiency of nutrients in a targeted population. A common example of target fortification is the addition of nutrients such as zinc to increase the growth of children (Jan et al. 2019).

Food fortification has gained importance due to the changes in dietary patterns and the increase in the consumption of processed foods in developing and developed countries (Olson et al. 2021). Large-scale processing of food and storage conditions tends to decrease the nutrients such as minerals and water-soluble vitamins (Martínez et al. 2020). The most common examples of food-fortified processes include vitamin-A-fortified oil and rice, iodized salts, calcium- and vitamin-D-fortified milk, folic acid, and iron-fortified wheat flour (Olson et al. 2021). Food fortification is more common in urban areas as compared to rural areas because of higher levels of education, health, and socioeconomic status (Kiran et al. 2022). It is quantifiable, cost-effective, and sustainable, particularly in developed countries (Gani et al. 2018). The major benefits of fortified foods with nutrients such as iron, iodine, and vitamins include reduced occurrence of diseases, which is correlated with nutrient deficiencies in most developing countries (Mkambula et al. 2020). Therefore, biofortification is considered a complementary intervention to reduce hidden hunger and malnutrition in developing countries (Figure 6.2).

6.2.2 Food Supplementation

Food supplementation includes the consumption of additional nutrients in the form of tablets, capsules, and syrups to meet the RDA of nutrients (Tam et al. 2020). These supplements include calorie supplements, vitamins, fiber, amino acids, proteins, minerals, and essential fatty acids (Ghosh et al. 2018). Buturi et al. (2021) reported that in developing countries

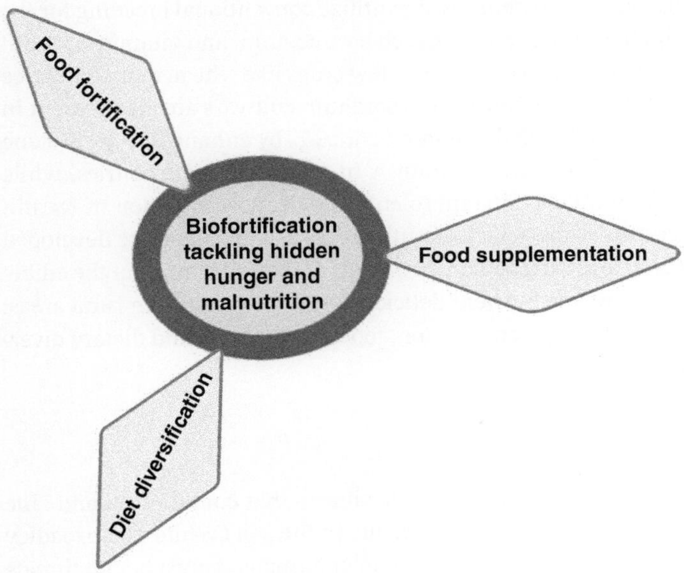

Figure 6.2 Strategies to improve micronutrient consumption in human diet.

vitamin supplements are commonly used to address hidden hunger, in contrast to mineral supplements like amino acids, zinc, and iron supplements. Food supplements cater to the nutritional requirements of small populations in developing countries offering a short-term, direct, and controllable strategy that produces effective and positive results, outperforming alternatives such as dietary diversification (Tam et al. 2020). A study conducted in Brazil on 1232 children of 8000 households between the ages of 0 and 35 months to observe the relationship between child development and vitamin A supplementation (Correia et al. 2019) revealed that children supplemented with vitamin A showed a 40% lower probability each for delay in the development of cognitive and motor abilities in comparison to children without vitamin A supplements. Vitamin A supplementation significantly impacted the developmental process, demonstrating that food supplementation with vitamins can increase the nutritional value of populations against nutrient deficiencies. Bouis et al. (2019) concluded that zinc supplementation after the age of two years to children improves their growth. However, it is reported that observing the effects of zinc deficiency remains challenging due to the absence of a suitable diagnostic technique. Food supplements are becoming a major challenge for availability in rural areas as compared to urban areas (Kiran et al. 2022). An additional disadvantage is that their applicability is limited for a small population and also requires a well-developed structure for successful implementation in developing countries (Bouis et al.2019). The health of targeted populations is also affected due to the toxic effect of food supplementation (Ghosh et al. 2018).

6.2.3 Diet Diversification

Micronutrients including vitamins, iodine, zinc, and iron are present in staple foods in limited quantities (Ul-Allah 2018). Dietary diversification is a strategy to combat micronutrient deficiencies, providing different types of food with sufficient quantities of micronutrients and Macronutrients to fulfill the RDAs of nutrients (Malik and Maqbool 2020; Bouis 2018). Dietary diversification helps to combat micronutrient and macronutrient deficiencies by improving the immune system of humans and is an effective and sustainable strategy to reduce the problem of hidden hunger and malnutrition. However, drawbacks include the presence of antinutrients in consumed foods, the need for nutrition education, proper food data, and changes in dietary patterns that can affect nutrient absorption (Gani et al. 2018). Another disadvantage is the financial obligation involved in producing and purchasing nutrient-enriched foods for consumers. However, the implementation of dietary diversification in developing countries is very hard, so other strategies like food fortification, food supplementation, and biofortification of crops are possible in developing countries (Wakeel et al. 2018).

6.3 Strategies for Crop Biofortification

Biofortification is an effective approach to combat nutrient deficiencies, which are threatening human health. Conventional breeding of plants depends on the manipulation of two plant genomes for the improvement of micronutrients. Many research programs have been conducted like BioCassava Plus and HarvestPlus, which utilize conventional breeding for the development of biofortified crops. Biofortified crops, with higher nutrient values such as zinc, iron, and vitamin A, fulfill the daily requirement of nutrients and boost the immune systems of people relying on few crops like wheat, maize, and rice (Garg et al. 2018). However, zinc and vitamin A-biofortified wheat, rice, maize, and sorghum cultivars are also present in developing countries. For instance, many scientists developed a biofortified *Ipomoea batatas* L. by enhancing a β-carotene content through conventional plant breeding to decrease the deficiency of vitamin A in some African countries, while biofortified rice with iron was also developed through the HarvestPlus program to enhance the accumulation of ferritin proteins in Filipino women (Low et al. 2017). Biofortified cereals crops, which are utilized as a staple food, are developed through agronomic practices and transgenic and plant breeding to attain the standard levels of micronutrients in the edible parts of the plant. Various strategies have been developed to overcome nutrient deficiencies in poor people in rural areas. Many similar strategies have also been practiced by increasing food supplementation, food fortification, and dietary diversity for poultry and livestock (Lyons et al. 2004; Haug et al. 2007).

6.3.1 Agronomic Strategies

Biofortification through agronomic strategies depends on the treatment of mineral fertilizers that contain iron and zinc mineral fertilizers through soil application, which increases the essential micronutrients in the soil (White and Broadley 2009). Boosting zinc nutrients in wheat and rice cultivars strengthens the immune system of humans against health threats and viral infections. The deficiencies of micronutrients in the soil can be addressed by applying fertilizers either through foliar or soil treatments (Imtiaz et al. 2010). The foliar application with mineral fertilizers to the plants is the most effective and environment-friendly agronomic strategy of biofortification (Rawat et al. 2013). The micronutrient content in edible parts of plants can be improved with mineral fertilizers, such as selenium, zinc, nickel, and iodine (Cakmak 2008). The foliar application with zinc nutrients to the pea plants improved the amount of zinc in the seeds of the pea plant. Cakmak (2008) reported that the application of zinc fertilizer to the soil for wheat plants increased the uptake and sequestration of zinc up to 60 mg kg^{-1} in grains of wheat. The foliar application of zinc fertilizer can be absorbed through the epidermis surface of the leaf and translocate into the whole plant through phloem tissue, which increases the zinc in the edible parts of the plants (Figure 6.3). The application of selenium fertilizer in maize plants in the form of aqueous Na_2SeO_4, which contains 41.8% selenium, increased selenium content in maize grain by 19.7 μg per kg. Similarly, calcium ammonium nitrate with 0.005% selenium increased selenium content in the maize grain by 14.8 μg per kg. Also, nitrogen, phosphorus, and potassium (NPK) with 0.0012% selenium increased the content of selenium by 20.7 μg selenium per kg in the grains of maize (Chilimba et al. 2012).

Before selenium fertilization, the average concentrations of selenium in all grain crops were less than 10 μg kg^{-1} through dry weight. After selenium fertilization, the concentrations of selenium in Finnish grain crops increased significantly reaching up to 250 μg kg^{-1} in spring wheat, 40 μg kg^{-1} in the rye, 180 μg kg^{-1} in bread wheat, 50 μg kg^{-1} in winter wheat, and 170 μg kg^{-1} in wheat flour (Eurola et al. 1990). Pooniya and Shivay (2015) reported that foliar application of zinc-enriched urea with $ZnSO_4$ improved the uptake of zinc in rice plants, increased their production, and enhanced the qualitative attributes. Therefore, proper application of zinc concentration in cereal crops, which are grown in the soil with high cadmium concentration decreases the accumulation and uptake of cadmium and protects the plants from the adverse effect of this heavy metal (Shahzad et al. 2014).

Additionally, the iodine deficiency can be addressed through the application of potassium iodide (KIO_3) in plants like rice, maize, and wheat. Like iron and zinc biofortification research, the foliar treatment of iodine is more effective in comparison to the application of soil (Budke et al. 2020). Pal et al. (2021) reported that the foliar application of potassium iodate

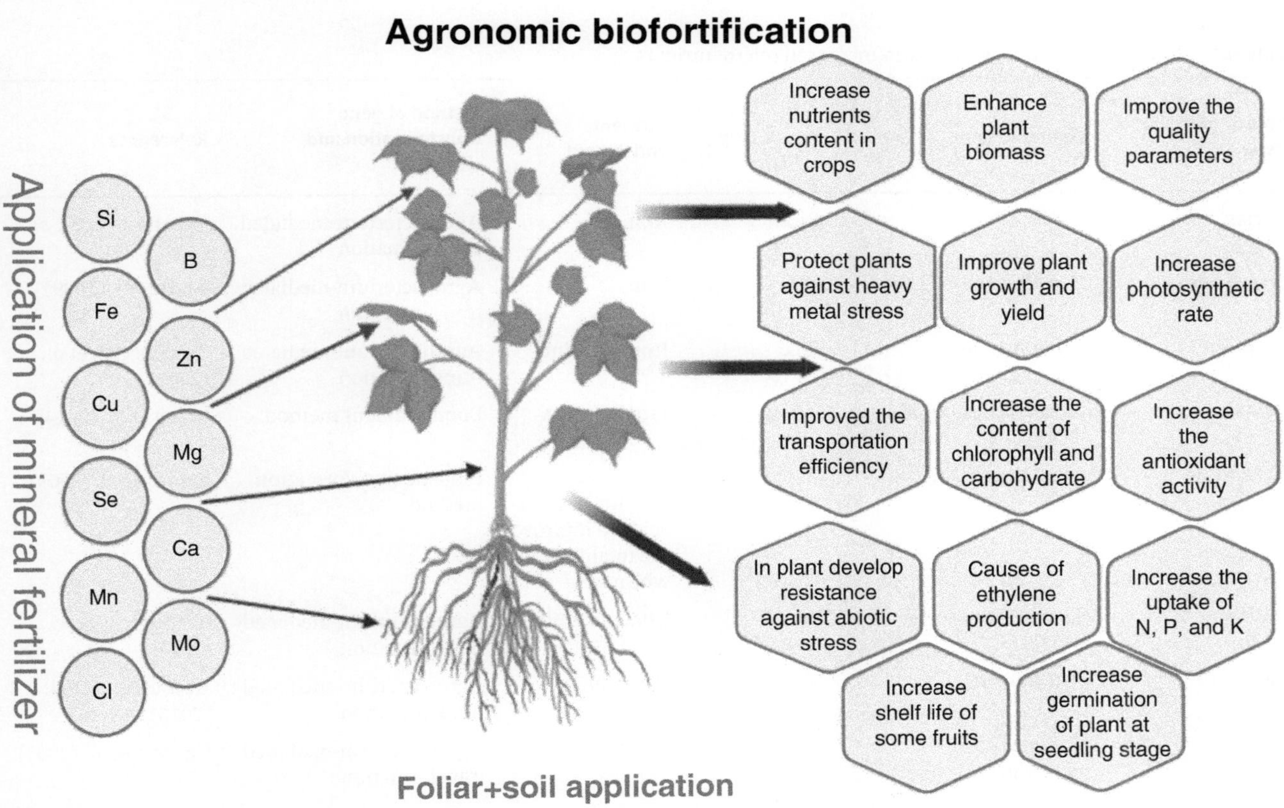

Figure 6.3 Schematic diagram of agronomic biofortification through soil and foliar application for the improvement of plant.

is effective for the biofortification of iodine. It is also, concluded that foliar applications with $FeSO_4$ (iron sulfate) was effective for iron biofortification (Pal et al. 2021). Naz et al. (2015) reported that application of iron and zinc at 4 mg kg^{-1} soil improved the nutrient availability to the wheat plants, with the maximum concentration of zinc and iron observed in the grains and straws as compared to the controlled treatment.

6.3.2 Genetic Strategies

Biofortification relies on a two-step strategy: (1) breeding, if possible, and (2) modifying, if necessary (White and Broadley 2011). The applications of micronutrients, through either leaves or soil, increase the concentration of micronutrients in the edible parts like bulbs, fruits, and seeds of plants. However, some factors like the limited movement of micronutrients, different soil compositions, and restricted movements within plants can increase the concentration of micronutrients in the nonedible parts such as leaves and straws of the plants (Kalra et al. 2020). While the use of an agronomic strategy to improve nutrient concentrations in plants is a temporary solution to the biofortification process, it can be expensive for farmers in developing countries due to the high cost of foliar fertilizers (Jewell et al. 2020).

However, conventional breeding faces challenges in the biofortification of crops due to the low heritability and lack of genetic diversity (Marques et al. 2021). This method needs a long-term inclusive selection to introduce a desirable nutrient into a crop. To overcome these types of problems, many genetic applications have been developed, such as chromosomal mapping, genomic selection, MAS, and quantitative trait loci (QTL) (Table 6.1). These approaches are cost-effective for the selection of targeted plants and biofortification of peas, wheat, carrot, rice, barley, and maize (Koc and Karayigit 2022). MAS determines the plant that carries the genomic regions essential for the expression of a trait and whether it possesses the desired trait (Lema 2018). However, a significant number of molecular markers and statistical techniques are needed to identify QTL. A good understanding of the genetic basis of micronutrients at the molecular level and identification of the important effects of QTLs help in the production of biofortified varieties through marker-assisted breeding. MAS has been used for the development of maize cultivars with pro-vitamin A from the germplasm with a large concentration of β-carotenoid (Masuka et al. 2017).

Table 6.1 Transgenic biofortified crops to increased micronutrients.

Source of desirable gene	Desirable gene	Target crops	Nutrients enrichment	Method of gene transformation and silencing	References
Barley	HvNAAT	Rice	Iron	Agrobacterium-mediated transformation	Takahashi et al. (2001)
Soya bean	Ferritin gene	Rice	Iron	Agrobacterium-mediated transformation	Goto et al. (1999)
Soya bean	Ferritin gene	Rice	Iron and zinc	Agrobacterium-mediated transformation	Vasconcelos et al. (2003)
Bacteria	Carotene desaturase and PSY	Wheat	Provitamin A	Bombardment method	Wang et al. (2014)
Amaranth	Albumin gene (AmA1)	Wheat	Increased essential amino acids to improve the quality of wheat	Biolistic transformation method	Tamas et al. (2009)
Barley	Sucrose transporter (HvSUT1)	Wheat	Iron and zinc	Agrobacterium-mediated transformation	Saalbach et al. (2014)
Bacteria	psy1 and crtI	Maize	Carotenoids	Agrobacterium-mediated transformation	Decourcelle et al. (2015)
Barley	Homogentisic acid geranylgeranyl transferase (HGGT)	Maize	Vitamin E	Agrobacterium-mediated transformation	Cahoon et al. (2003)

(Continued)

Table 6.1 Transgenic biofortified crops to increased micronutrients. *(continued)*

Source of desirable gene	Desirable gene	Target crops	Nutrients enrichment	Method of gene transformation and silencing	References
Potato	sb401	Maize	Lysine and total protein	Agrobacterium-mediated transformation	Yu et al. (2005)
Arabidopsis	Zinc transporter gene (AtZIP1)	Barley	Zinc and iron	Clone DNA	Ramesh et al. (2004)
Colonic bacteria	Starch branching enzymes (SBE I, SBE Iia and SBE Iib)	Barley	Starch	RNAi	Carciofi et al. (2012)
Bacteria	Dihydrodipicolinate synthase (DHPS)	Barley	Essential amino acid	Agrobacterium-mediated transformation	Ohnoutkova et al. (2012)
Wild sorghum	Homo188-A	Cultivated sorghum	Provitamin A	Agrobacterium-mediated transformation	Lipkie et al. (2013)
Wild sorghum	γ-kafirin	Sorghum	Seed storage protein	RNAi	Elkonin et al. (2016)
Pineapple	Seed-specific bacterial phytoene synthase gene	Soya bean	Provitamin A	Agrobacterium-mediated transformation	Schmidt et al. (2015)
Arabidopsis	Arabidopsis cystathionine γ-synthase (AtD-CGS)	Soya bean	Methionine	Agrobacterium-mediated transformation	Song et al. (2013)
French bean	CRC gene	Soya bean	Isoflavone	Particle gun bombardment	Yu et al. (2003)
Bacteria	crtB gene	Potato	Carotenoid	Agrobacterium-mediated transformation	Ducreux et al. (2005)
Strawberry	GalUR	Potato	Vitamin C	Agrobacterium-mediated transformation	Upadhyaya et al. (2009)
Bacteria	Cystathionine γ-synthase (CgS)	Potato	Methionine	Agrobacterium-mediated transformation	Dancs et al. (2008)
Yeast	PSY	Cassava	Zinc, iron, and beta-carotene		Welsch et al. (2010)
Arabidopsis	CAX1 (Cation exchanger 1)	Carrot	Calcium	Agrobacterium-mediated transformation	Park et al. (2004)
Arabidopsis	Calcium transporter (sCAX1)	Carrot	Calcium	Agrobacterium-mediated transformation	Morris et al. (2008)
Bacteria	PSY	Canola	alpha and beta-carotenes		Newell-McGloughlin (2008)
	Lycopene ε-cyclase (ε-CYC)	Canola	β-carotene	RNAi	Yu et al. (2008)
	DE-ETIOLATED1 (DET1)	Canola	carotenoid and flavonoid	RNAi	Wei et al. (2009)
Yeast	Pythium irregulare Δ6 desaturase (PiD6)	Mustard	γ-Linolenic acid	Agrobacterium-mediated transformation	Hong et al. (2002)
Arabidopsis thaliana	3-hydroxymethylglutaryl CoA (hmgr-1)	Tomato	Isoprenoid and sterol	*Agrobacterium tumefaciens* transformation	Enfissi et al. (2005)
Onion	GDP-mannose 3′,5′-epimerase (*SlGME1, SlGME2*)	Tomato	Ascorbic acid	Electroporation	Zhang et al. (2011)

The biofortified maize grain has been developed through MAS, with provitamin A concentrations improving from 1.60 to 5.25 μg g^{-1} and from 1.80 to 8.14 μg g^{-1} (Goredema-Matongera et al. 2021). From all strategies like QTL mapping and genomic selection, single nucleotide polymorphism (SNPs) are widely used for the identification of genomic regions that disturb characteristics of nutritional value. For example, 20 SNPs were identified that have a direct effect on the accumulation of zinc concentration in maize (Hindu et al. 2018). Additionally, 80 QTLs were observed to increase the efficiency of breeding programs for the improvement of zinc and iron concentration, which are mapped on the 12th chromosomes of rice (Sharma et al. 2019). To improve manganese content in cereal crops, many QTLs have been identified and mapped on the 3, 7, and 8 chromosomes of rice (Liu et al. 2017). Liu et al. (2017) observed that one major QTL and gene (*OsNRAMP5*) caused the enhancement of manganese accumulation in the grain of rice. A study at seedling stage of wheat identified 16 QTLs that were involved in the enhancement of selenium concentration (Wang et al. 2017). Genome editing is an effective transgenic approach used to develop genetically modified organisms for the desirable traits (Kumar et al. 2020). Many approaches, such as mega-nucleases, transcription activator-like effector nuclease, zinc-finger nucleases, and clustered regularly interspaced short palindromic repeats (CRISPR/Cas9), have been utilized in gene editing for the development of zinc-rich wheat and β-carotene-biofortified rice cultivars (Jaganathan et al. 2018).

The biofortification processes have occurred primarily in most developed countries through transgenic techniques (Figure 6.4). An example is Golden Rice, developed through a gene editing technique (Jha and Warkentin 2020). The expression of the transgenic gene in Golden Rice increased 1.6–3.7 μg g^{-1} pro-vitamin A (Siwela et al. 2020). The over-expression of vacuolar iron transporter VIT1 of *Arabidopsis thaliana* in cassava increases the concentration of iron in

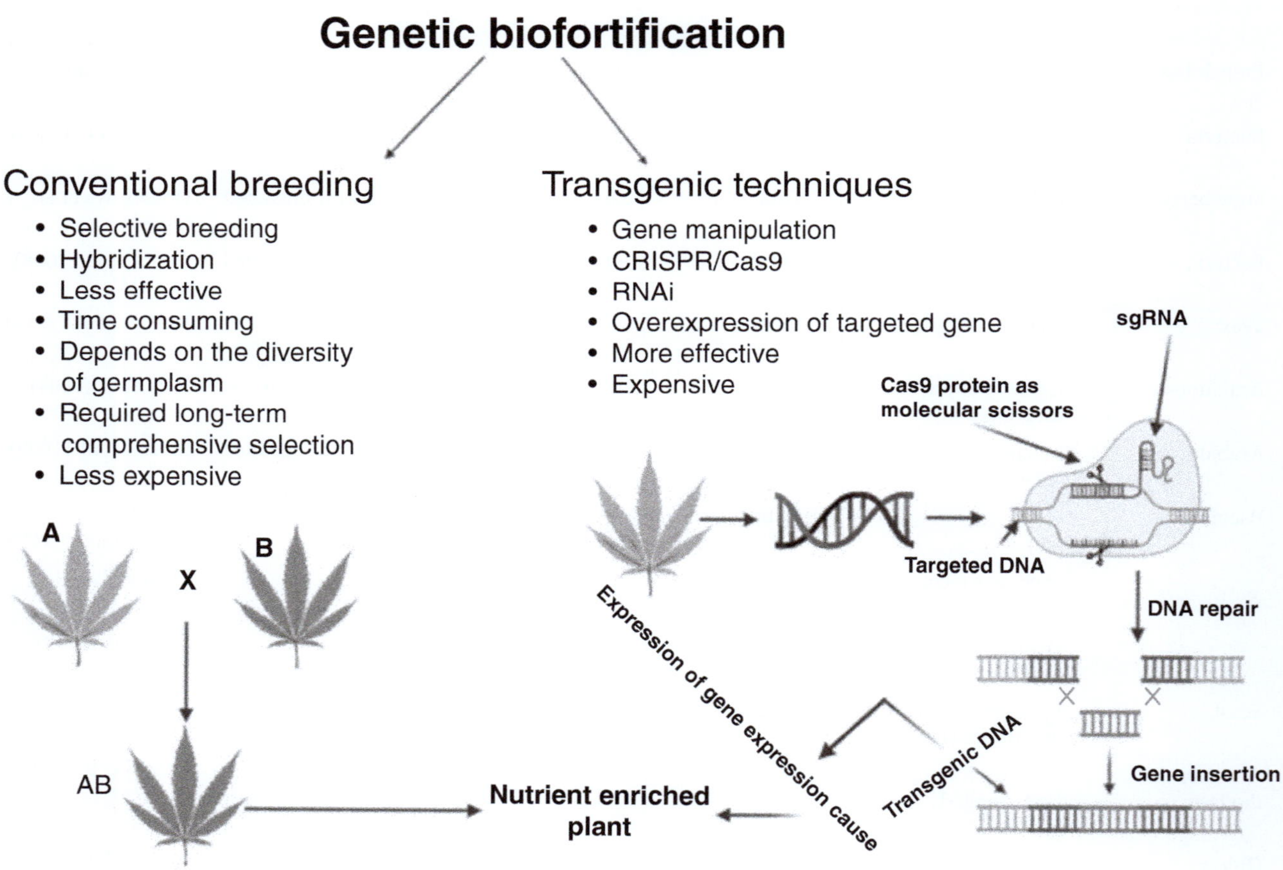

Figure 6.4 Genetic and conventional biofortification for the improvement of nutrients in plants.

the storage organs of the root (Narayanan et al. 2019). The overexpression of zinc transporters and expression of these genes responsible for the activity of phytase in barley increased the bioavailability and concentration of zinc (Sheoran et al. 2022). Ducreux et al. (2005) reported that the expression of the Erwinia uredovora crtB gene, which is encoded with phytoene synthase, increased the concentration of carotenoid in potato tubers. While the overexpression of tissue-specific genes such as bacterial genes CrtY, CrtB, and CrtI developed the Golden potato tubers or β-carotene-enriched potato containing up to 47 $\mu g\ g^{-1}$ dry weight of β-carotene (Diretto et al. 2007). Recently, omics technology has been used to improve the process of biofortification for the enhancement of nutrients (Riaz et al. 2020). Omics technology determines the relationships among nutrients, genes, metabolites, proteins, and transcripts (Carvalho and Vasconcelos 2013). Specific genes regulate the bioavailability of nutrients, uptake, translocation, and accumulation of nutrients through crops. The omics technologies have also been utilized to study the biofortification of vitamin C, lysine, iron, calcium, and zinc in biofortified crops like tomatoes, maize, wheat, and finger millet (Jain et al. 2019; Garg et al. 2018). The transgenic approach is considered rapid and sustainable for introducing desirable traits into the targeted crops (Mir et al. 2020).

6.4 Nanotechnology-based Approaches for Crop Biofortification

Nanotechnology is a new and environmentally friendly technique that enables plants to utilize nutrients efficiently and enhances the concentration of nutrients in food crops, overcoming the problem of malnutrition. This approach has both advantages and disadvantages, including the use of NP-based nutrients or fertilizers (El-Ramady et al. 2020a). The most important advantage of nanotechnology is its ability to boost crop productivity when applied through water and soil. However, the major drawback lies in the potential for toxicity and contamination when NP concentrations increase in the soil (Rizwan et al. 2021). The mechanism of NP synthesis plays a crucial role in human health due to its physical and chemical properties, as it proves to be more effective, less expensive, and environment-friendly (Stephen et al. 2021). Mostly, metal and metal oxide NPs, extracted from various sources like black and green tea, basil eucalyptus leaves, and mint including copper, sulfur, and iron, are commonly used in nanotechnology (Ragab and Saad-Allah 2020). The incorporation of zeolites and nano clays enhances the efficiency and performance of NPs, ultimately contributing to nutritional improvements in plants, plant growth, and soil fertility. The most common NPs that are being utilized for the improvement of nutritional values are Se, Zn, Cu, and Fe (Prasad et al. 2017). Nickel oxide and zinc oxide NPs are derived from the leaves of *Nilgirian tuscíliantus* and seeds of black cumin by using an environment-friendly approach (Resmi et al. 2021). Using NPs is an effective approach, as these are controlled and slowly release nutrients for plant growth, development, and nutritional improvement e.g. nano-based fertilizers, which overcome the problem of malnutrition and deficiency of micronutrients.

6.5 Nutrient-based Nanoparticles in Food Crops and Human Health

Understanding the role of NPs and their interactions in the human body is essential, considering the potential risk associated with their accumulation of degradable and non-degradable NPs in the organs of the human body. The process of phagocytosis, responsible for destroying the foreign invaders in the human body, faces challenges when interfered, activating the stress reaction that leads to inflammation, and compromises the body defense system. NPs, during exposure to fluid and tissue, accumulate various macromolecules on their surfaces, affecting the enzyme and protein regulatory mechanisms. In addition, the properties of NPs, such as surface charge, aggregation and solubility, chemical composition, functional groups, and surface structure induce the toxicity in human body.

The interaction of NPs with the environment and humans may be hazardous. Many scientists conducted an experiment and their results showed that when rats consume the TiO_2 and Fe-based NPs, then these NPs damage their lungs. Similarly, the utilization of fullerene and TiO_2-based NPs causes brain damage in fishes and dogs. TiO_2 and silver-based NPs showed genotoxic and cytotoxic effects on human and mice cells. Their effects are due to the production of reactive oxygen species which cause cell proliferation and DNA damage (Wang et al. 2012).

The application of CuO and ZnO NPs through soil has been reported to have toxic effects on human health. These NPs were passed by the pathway of biochemical reactions that blocked the biological nitrogen fixation and damaged the plant cells, while CuO and ZnO NPs cause a serious threat to the health of humans. Notably, copper oxide NPs have exhibited cytotoxic effects on human lymphocytes due to the oxidative stress. However, addressing these concerns, researchers recommended the toxicological risk assessment studies and safety evaluation, involving specified exposure concentrations of the ZnO and CuO NPs. It is worth mentioning that NP-based nutrients, such as selenium and zinc oxide, have demonstrated protective effects against various diseases, particularly in the defense system of the human body in contradiction of COVID-19 (He et al. 2019; Gatadi et al. 2021). The most-studied nano-fertilizers such as Se, Zn, Cu, and Fe are involved in crop biofortification (Prasad, Bhattacharyya, and Nguyen 2017). Hence, the efficiency of NP-based fertilizers against plant growth is based on the physical and chemical properties of NPs (composition, concentration, shape, and size), growth conditions, environmental factors, and crop species.

6.5.1 Zn-based Nanoparticles

Zinc oxide and zinc sulfate are two important sources of zinc-based NPs. ZnO NPs improve the concentration of micronutrients, making them easily stored, absorbed, and translocated within plants due to their lower volatilization and large surface area (Milani et al. 2015). Many scientists concluded that foliar application of zinc oxide NPs increases the Zn concentration to 82% in maize (Umar et al. 2020). ZnO NPs were applied through various methods such as soil, seed-priming, and foliar application, which increased the protein content in the seeds, decreased heavy metal, such as cadmium, concentrations, and improved the concentration of zinc in the grain in comparison to the chemical fertilizer (Hussain et al. 2018; Sheoran et al. 2021). The utilization of ZnO-based nano-fertilizers has proven effective in the process of biofortification by enhancing the nutritional value of the plants (Ivanov et al. 2021). Many scientists reported that soil application of ZnO NPs with a concentration of 5 mg kg^{-1} increased the concentration of zinc in the grain of sorghum and finger millet up to 94%, while priming of seed with ZnO NPs (5 ppm) elevated zinc levels in grains up to 13.96% in comparison to the nontreated plant with ZnO NPs (Kumar et al. 2021; Dimkpa et al. 2019).

6.5.2 Fe-based Nanoparticles

The large population of the world is facing the problem of iron deficit, which is a highly essential micronutrient element for growth, development, and metabolism of plants and humans (Chugh et al. 2022). The deficiency of iron in the diet affects 14% of the total world population (Matres et al. 2021). Rice and wheat are important sources of staple food in the world, but during the processing, the outer layers of bran are removed which ultimately causes iron deficiency (Zhoa et al. 2020; Fakharzadeh et al. 2020). Chandra et al. (2021) concluded that seed-priming with iron oxide NPs with a concentration of 100 ppm enhances iron content in the grains of finger millet in comparison to iron sulfate. However, soil application of iron-based NPs with concentrations of 50 and 500 mg kg^{-1} increased biomass, and enhanced iron and amino acid concentrations such as cysteine and tyrosine in the grains of wheat crop (Wang et al. 2019a).

The priming of seed with the concentration of 25 ppm iron oxide NPs increased the iron content in the grain of wheat cultivar IITR26 up to 45.7% and in wheat cultivar WL711 up to 26.8% in comparison to nontreated seed (Sundaria et al. 2019). However, it is demonstrated that foliar spray of iron oxide NPs with levels of 1 and 10 mM increased the iron content in the leaves of wheat (70–75%) in comparison to nontreated treatments of wheat (Zimbovskaya et al. 2020). The wheat seed-priming with iron NPs at a concentration of 5–20 mg L^{-1} enhanced the content of iron in the grain by 20–121% (Rizwan et al. 2019). Therefore, soil treatment of iron oxide NPs with a concentration of 25–100 mg kg^{-1} increased the uptake of nutrients, enhanced the nutrient concentrations, particularly, N, P, and K 33%, 35%, and 32.7%, respectively, and decreased cadmium uptake by up to 72.5% in wheat crop (Manzoor et al. 2021). These studies collectively indicate that the application of iron-based NPs represents an environmentally friendly and new technique for addressing iron deficiency in crops and accomplishing effective biofortification.

6.5.3 Se-based Nanoparticles

Selenium is an important micronutrient crucial for animal and human health. Selenium is present in the selenozyme, acids, selenium nucleic acid, oxyhalides, selenides, oxides, oxyacids, and halides in both plants and soil (Skalickova et al. 2017). Selenium-based NPs function as nano-fertilizers with high bioavailability for plants due to their large surface area-to-volume ratio, resulting in increased anti-cancer, antioxidant, and antioxidant properties (Hosnedlova et al. 2018). One billion people in the world are facing the problem of selenium deficiency (Haug et al. 2007). The major functions of selenium NPs in agriculture are to bind with ligands, detoxify the heavy metals from the soil, and protect the plant from the effects of heavy metals (Zoidis et al. 2018). Mateus et al. (2021) concluded that the foliar treatment of selenium-based NPs with concentrations of 10–160 mg L^{-1} enhanced the selenium content in the coffee grains (4.84–5.82 mg kg^{-1}) as compared to the application of sodium selenite, which increased the selenium content in coffee grains in the range of 0.116–4.47 mg kg^{-1}.

Shalaby et al. (2021) investigated the impacts of foliar treatments of selenium-based NPs on cucumber at 25 mg L^{-1} concentration and reported that these nanofertilizers increased the uptake of plant nutrients, enhanced the concentration of nitrogen, N, P, and K in the plant body, and increased production of cucumber. Selenium-based NPs at concentrations of 10 and 30 μm were applied on rice seedlings under a hydroponic condition and their results increased the selenium concentration in a range of 6.7–20.4 as compared to inorganic selenium concentration (Wang et al. 2020). Many scientists concluded that foliar application of selenium NPs (25–100 μm L^{-1}) on brown rice enabled selenium improvement with a range of 218.9–1096.6 μg kg^{-1} (Wang et al. 2021). However, selenium NPs are less expensive and efficient micronutrients for crop biofortification and are included in the improvement of biochemical production of glucosinolates, proteins, phenolics, and amino acids (Carvalho et al. 2003).

6.5.4 Cu-based Nanoparticles

Copper (Cu) is a vital micronutrient that plays an important function against diabetes, metabolic syndrome, and cardiovascular disease of humans due to their oxidative properties and also enhances the inflammatory response in human beings. Xiong et al. (2017) reported that copper oxide-based NPs are used as a plant growth regulator and for the remediation of soil. The foliar spraying ranging from 10 to 250 mg L^{-1} of copper-based NPs in tomato increased the vitamin C and β-carotenoid content by up to 36% (Hernández-Hernández et al. 2019). Yusefi-Tanha et al. (2020) reported that soil-applied copper NPs (50–500 mg kg^{-1}) on soybean plants play a significant role in increasing the growth of agronomic traits and improving the concentration of copper in the seed, which was 1.8 times higher as compared to the nontreated copper plant (Table 6.2). The copper NPs that are applied through soil at a concentration of 40–60 mg kg^{-1} enhanced the copper content (74.5%) in sugarcane (Tamez et al. 2019). Moreover, the soil treatment of copper oxide-based NPs with 75–600 mg kg^{-1} concentration increased the 56–187% leaf allicin content in onion and also enhanced magnesium, copper, and calcium concentration in the bulb of onion (Wang et al. 2020).

6.6 Mechanism of Nanoparticle Uptake and Translocation in Plants

In recent years, extensive research has been conducted to determine the mechanism of NP uptake and transport in plants. Despite significant progress, some plants remain unexplored in terms of NP interactions. The plant and NP interaction consists of three basic steps, such as uptake, absorption, and translocation. However, the absorption and translocation of NPs occur through the leaf, stem, and root surface, while the uptake of NPs occur through the stomata, cuticle, and epidermis, which further translocate through the phloem and xylem. NPs can be applied to plants through various methods such as seed-priming, foliar spray, and soil treatment. The toxicity and bioavailability of NPs are determined through a series of bio- or geotransformations that occur in the soil after the NPs translocate to aerial portions of the plants, interact with the root of the plant, and accumulate in tissue, cellular, and subcellular organelles (Nair et al. 2010). Wang et al. (2013) reported the correlation between NP shape and surface area, accumulation, and reactivity on the surface of cells or inside structures of plants. The surface of NPs is measured through its morphology and area to determine the specific zone of interaction among the structure of plant cells and NPs. The NPs are attached

Table 6.2 Different concentrations of nanoparticles for micronutrient enrichment and their effect on crops.

Target crops	Concentration	Type of nanoparticles	Type of application method	Micronutrient enrichment	Positive effects on crops	References
Maize	80 mg kg^{-1}	ZnO	Foliar	82% more zinc in a grain of maize	Foliar spray of ZnO nanoparticles (NPs) increases the concentration of zinc in the grain of maize and improves their growth and yield.	Umar et al. (2021)
Rice	25 and 100 mg kg^{-1}	ZnO and $ZnSO_4$	Foliar	Application of ZnO NPs increased the 13.5–39.4% zinc in the brown rice.	ZnO NPs increased the NPK concentration in total biomass (6.8–7.6%), panicle number (3.8–10.3%), and spikelet number per panicle (2.2–4.7%), which ultimately increased the yield of rice.	Yang et al. (2021)
Wheat	100 mg L^{-1}	ZnO	Foliar	The Zn content in the grain increased by 45 mg kg^{-1} and 100–150 mg kg^{-1} in the root–shoot tissue of wheat	Foliar application of ZnO NPs increases the Zn concentration in various parts of wheat crop and increases the yield.	Hussain et al. (2021)
Wheat	25–600 ppm	FeO	Seed-priming	Enhancement of iron by 25 ppm in the grain of wheat	Seed-priming with FeO NPs increase the accumulation of iron in different parts of wheat, especially the pigment strand, aleurone layer, nucellar projection, and endosperm.	Khan et al. (2021)
Sorghum	5 mg kg^{-1}	ZnO	Soil treatment	Enhancement of Zn content by 94% in the grain of sorghum	Improved the transportation efficiency of nitrogen and phosphorus in the plant, which caused an increase in growth and yield	Dimkpa et al. (2019)
Finger millet	5 and 100 ppm	ZnO and Fe_3O_4	Foliar treatment	The Zn concentration in the finger millet grain increased by 13.96 %.	Foliar application of ZnO and Fe_3O_4 improves the nutrient uptake mechanism of plants and increases their growth	Chandra et al. (2021)
Mungbean	0, 10, 20, 40, 80, and 160 ppm	ZnO	Murashige and Skoog media culture	Increased the Zn concentration by 106.41 μg g^{-1} in the dry weight of mung bean under controlled conditions and by 4600 μg g^{-1} under 80 and 160 ppm levels of ZnO	ZnO NPs increased the growth and photosynthetic activity in the mung bean	Sorahinobar et al. (2023)
Wheat	25–100 mg kg^{-1}	FeO	Soil treatment	Increased the uptake of Iron	Decreased the cadmium effect on wheat up to 72.5% and enhanced the uptake of NPK	Manzoor et al. (2021)
Wheat	0.08986 and 0.8986 g L^{-1}	Fe $(OH)_2$	Foliar	Enhanced the concentration of iron by up to 75%	A foliar spray of iron hydroxide on wheat plant increases the movement of iron into the leaf.	Zimbovskaya et al. (2020)

(Continued)

Table 6.2 Different concentrations of nanoparticles for micronutrient enrichment and their effect on crops. *(continued)*

Target crops	Concentration	Type of nanoparticles	Type of application method	Micronutrient enrichment	Positive effects on crops	References
Wheat	10, 20, and 30 mg L^{-1}	Fe NPs	Foliar and seed-priming	Increased the concentration of iron in the wheat plant	Seed-priming and foliar spray on wheat plants increase the concentration of chlorophyll a, b, carbohydrate, fibers, lipids, proline, and protein.	Hassan et al. (2023)
Soyabean	50–500 mg kg^{-1}	CuO	Soil treatment	1.8 times more concentration of copper in the seed	Enhanced the sequestration of copper in the seed, leaf, stem, and root	Yusefi-Tanha et al. (2020)
Sugarcane	20, 40, and 60 mg kg^{-1}	Cu	Soil treatment	Increased the copper content by 73% in the sugarcane by the soil treatment of copper NP	Increased the tolerance in sugar cane against heavy metal stress. Enhanced the uptake of micronutrients and increased the concentration of chlorophyll	Tamez et al. (2019)
Wheat	50 and 500 mg kg^{-1}	CuO, Fe_2O_3 and TiO_2	Soil treatment	Increased the copper concentration from 18.84% to 30.45% in the grain of wheat	Increased the growth of wheat and concentration of amino acid in the grain	Wang et al. (2019)
Onion	75–600 mg kg^{-1}	CuO	Soil treatment	Increased the concentration of copper in the bulb of the onion	Increased the allicin content by 56–187% in leaf	Wang et al. (2020)
Rice	0.7896 and 2.3688 g L^{-1}	Se	Hydroponics	Increased the selenium content up to 20.4%	Increased the growth of rice seedlings	Wang et al. (2020)
Faba bean	1.27 mM solution	Sodium selenite and sodium selenate	Foliar	Increased the Se concentration by 1.5–2.3 times in the faba bean	Increased the accumulation of selenium in the seed of faba bean and increased their growth	Sindireva et al. (2023)
Red pitaya	3 and 5 mg L^{-1}	Se	Foliar and soil treatment	Enriched the pitaya plant with selenium	Treatment of pitaya plants with Se NPs increased their shelf life by 4–8 days as compared to non-treated plants	Yu et al. (2023)
Tomato	10 mg L^{-1}	Sodium selenate and Se NPs	Foliar	No significant effects of selenium enrichment	Caused ethylene production, which increased the ripening of tomato and accumulation of chlorogenic acid and naringenin, and reduced concentration of coumaric acid	Shiriaev et al. (2023)
Cucumber	25 mg^{-1}	Se	Foliar	Increased the uptake of selenium	Enhanced the concentration of N, P, K, yield, and growth of cucumber plant	Shalaby et al. (2021)
Coffee	10–160 mg L^{-1}	Se	Foliar	Increased the selenium concentration in coffee by 4.84–5.82 mg Kg^{-1}	The antioxidant activity and yield increased in the Se-treated plant	Mateus et al. (2021)

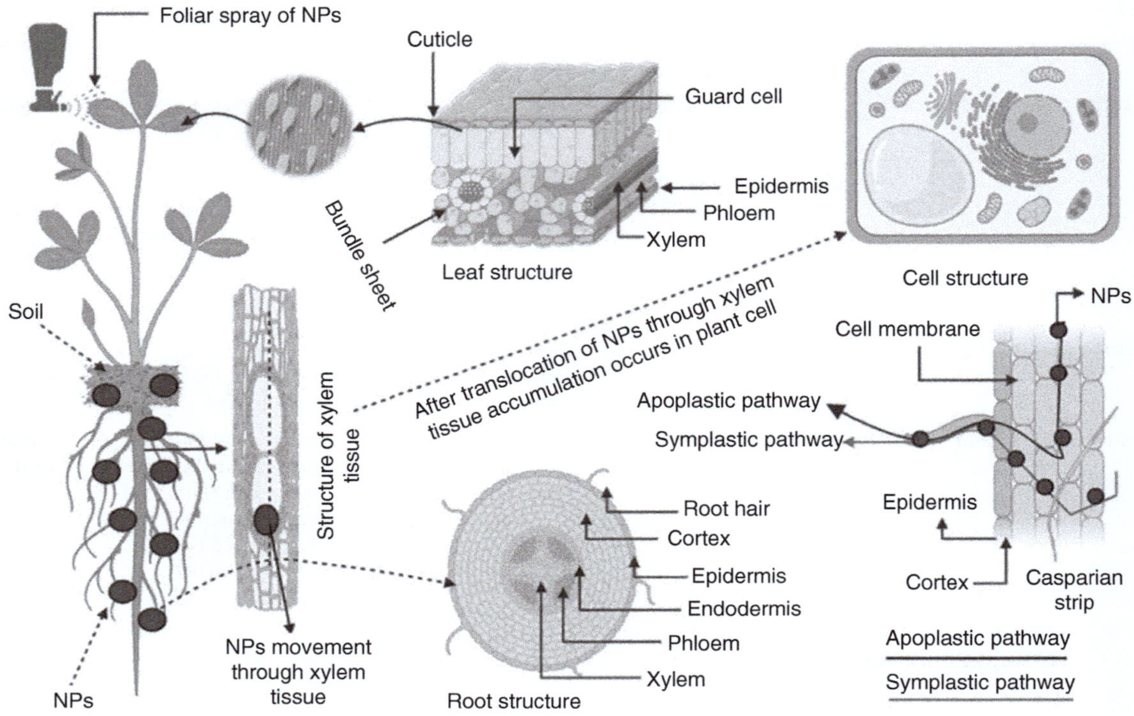

Figure 6.5 Schematic diagram mechanism of nanoparticle uptake, translocation, and accumulation in plants.

to the cell wall of the plant cell structure due to the negative charge of the cell wall of the plant (Figure 6.5). However, hydrophobicity, particle size, and charge play an important role in the process of uptake and translocation (Kaphle et al. 2018).

The basic structure of NPs is necessary for the assessment of their impact on NP uptake, accumulation, and translocation in plants (Raliya et al. 2016). It has been observed by many scientists that small NPs with a diameter range of 3–5 nm easily penetrate plant roots through capillary forces and osmotic pressure, and transit directly through the epidermal cells of the root. However, the semipermeable epidermal cell wall inhibits the movement of NPs into the cell. Some NPs induce new pores into the epidermal cell of the root cell wall, which allows the movement of large NPs (Lin and Xing 2008; Du et al. 2011). After passing through the root epidermal cell wall, the NPs are transported into extracellular spaces through the apoplastic pathway and reach the vascular system of the plant. The central vascular cylinder allows the xylem tissue to transport NPs in the upper portion of the plant, while the symplast pathway is crucial for crossing the Casparian strip barrier for entrance into the vascular system of the plant. This process occurs through interaction with the endodermal cell membrane's carrier proteins through transport, endocytosis, and pore formation. The NPs move through plasmodesmata (pathways that allow direct entry of materials into the cytoplasm) from one cell to another. While, the NPs cannot traverse the Casparian strip, reaching the central vascular cylinder enables xylem tissue to transport them to shoots, with phloem tissue returning NPs to the roots of the plant (Pérez-de-Luque 2017).

Nanoparticles are taken up through the epidermal cell wall of the plant, nuclei, and cortical cell cytoplasm (Wang et al. 2012). The NPs applied through the foliar sprays enter the leaves through the cuticles or stomata. The cuticle acts as a primary leaf barrier, which allows the entry of NPs with a diameter range of less than 5 nm. However, the entry through the stomata occurs through the apoplastic and symplastic pathways into the central vascular cylinder (Ruttkay-Nedecky et al. 2017). The NPs with a diameter range of 10–50 are transferred by the adjacent cell's cytoplasm or plasmodesmata (symplastic pathway), while large NPs with a diameter range of 50–200 nm are translocated among the cells through the apoplastic pathway (Tripathi et al. 2017). The apoplastic pathway is also known as nonselective movement with less resistance. However, many researchers confirmed that numerous nonessential metal complexes and water nutrients favor the apoplastic pathway for translocation (Banijamali et al. 2019). The maximum accumulation of zinc happens in the shoots and roots of maize under the hydroponic condition when treated with zinc oxide NPs, and also due to enhanced dissolution in the rhizosphere, translocation of zinc in its ionic form can increase plant absorption (Zhang et al. 2012). The translocation of copper oxide

NPs has been observed in maize from roots to shoots through xylem tissue and further translocation happens from shoots to roots facilitated by the phloem (Wang et al. 2012). Larue et al. (2014) highlighted the significant role of TiO_2-based NPs in the translocation of Ti NPs in wheat. They reported that NPs with a diameter of 140 nm or above cannot accumulate in the root, while NPs with a diameter of 36 nm can accumulate in the root parenchyma but cannot translocate into the shoot of wheat.

6.6.1 Accumulation of Nanoparticles in Plants and Their Effects

The uptake of NPs from the external environment to plant tissue follows different mechanism to enter into the plant cell, including organic matter assistance, ion channel transport, pore formation, passive transport, endocytosis, transport with water molecules through aquaporins and carrier protein assistance (Yanga et al. 2017). Among all crops, the maximum toxicity, uptake, and accumulation of NPs have been reported in rice, onion (Rajeshwari et al. 2015), lettuce (Margenot et al. 2018), wheat (Gao et al. 2018), soybean (Rezaei et al. 2015), cucumber (García-Gómez et al. 2018b), pumpkin, and tomato (Raliya et al. 2015). Upon reaching the vascular system of the plant, NPs accumulate in many subcellular locations such as walls, vacuoles, pericycle, cytoplasm, chloroplast, nucleus, cellular membranes, tonoplast, endodermis, cortex, mitochondria, and many plant organs like fruits, stem nodes, flowers, and foliage (Rajput et al. 2018b). It is reported that generally, NPs accumulate at various locations within plant, while vascular tissue is responsible for their translocation among plant organs (Pradas del Real et al. 2017; Lv et al. 2019). When NPs reach the vascular tissue of the plant such as the xylem, their translocation toward the upper parts of the plant becomes rapid. The NPs also sequestrate in the fruits due to phloem tissue. Some examples reported that cerium dioxide NPs not only accumulate in the shoots of tomato plants, but also accumulate in the fruits of tomato with lower concentrations (Wang et al. 2012c). This indicates that NPs with specific size can pass and transport on phloem tissues, the only passage entering fruit tissues. Priester et al. (2017) reported that the soybean plant, which was grown under hydroponic conditions with metal and metal-oxide NPs, had accumulated these nanoparticles in pods, roots, stems, and nodules.

Therefore, different NPs perform differently in the vascular tissue of the plant and accumulate in the tissue of plant roots. For instance, CeO_2-based NPs accumulate in the root system of three cereal plants, including wheat, rice, and barley, without affecting the elongation of roots and germination (Rico et al. 2015). However, some molecular changes were assessed in rice crops. In contrast, copper-based NPs were accumulated 10–20 times more in the cucumber plant compared to non-treated plants. The sequestration rate of NPs through the plant root can be obstructed through the properties of NPs and environmental conditions. Many scientists conducted an experiment and reported that the application of silver sulfide NPs with ammonium thiosulfate and potassium chloride increased the concentration of silver NPs in the roots and shoots of the sativa plant (Doolette et al. 2015). However, it is reported that organic matter in soil significantly reduced the cerium dioxide NPs in the roots of maize (Zhao et al. 2012).

6.7 Challenges of Nanoparticle-induced Biofortification

There are many challenges of NPs for the improvement of biofortification. First, nano-based food fortification is unable to be utilized on a large scale in rural areas of developing countries due to expensive cost (Sahoo et al. 2021). Second, the application of NPs at higher concentrations has demonstrated a toxic effect on plant at certain levels. The continuous application of NPs can lead to accumulation in the soil, posing a risk when plants take up these NPs in large amounts. Nano-based nutrients' transfer from leaves to roots and vice versa can result in plant toxicity, leading to the accumulation of nano-based nutrients in plant by-products, which ultimately causes problems for humans. Therefore, the toxicity of NPs is based on the coating, shape, base materials, size, and concentration. There is a wide range of NP by-products used in the process of biofortification where they can be broken down into different compounds, such as semiconductor materials, metals, and metal-oxides, which produce harmful pollutants in the environment (Rajput et al. 2021). Third, many scientists reported that fortified foods that are developed through NPs are less adaptive by the consumers due to the effects of nano-agrochemical applications such as nanopesticides and nano-fertilizers (Giles et al. 2015). Moreover, many studies are needed to observe the interaction between plants and nanoparticles to understand the mechanism of NP uptake and translocation in plants (Nair et al. 2010; Lowry et al. 2019). The major challenge of NPs for the process of biofortification is to introduce this technology in developed and developing countries. It needs some cautionary measures to make the technology more effective and accurate, and proper training is required for the farmers to operate this technology and handle the associated risk that developed through nano-based nutrients to overcome the problems of hidden hunger across the world (Mukhopadhyay 2014).

6.8 Conclusions

Micronutrient deficiency is a common problem all over the world, which causes malnutrition and affects both plant and human growth and development. The development of nutrient-enriched plant through environmentally friendly techniques is a key approach to global food security. The soil contains essential micronutrients like Zn, Fe, Mo, Cu, Mg, Mn, and Se. As a result, plant uptake these essential micronutrients in low quantitates. This, in turn, leads to lower food production and food with inadequate nutrient levels, contributing to malnutrition. The nano-based biofortification is an important strategy to overcome the problem of malnutrition. Nanotechnology is an effective and useful approach for the improvement of essential micronutrients. The applications of nanotechnology through foliar and soil treatment increase the nutrients in the soil and plant, and improve the nutrient uptake, translocation, and accumulation processes. This, in turn, indirectly improves the nutrients in the foods and reduces the nutrient-related health problems in humans. Moreover, nanotechnology increases the quality attributes and shelf life of different fruits, while enhancing the nutrient concentration in the grains of cereal crops like wheat, maize, rice, and barley, which are used as staple foods. Nanotechnology-based biofortification is a green technology that increases agriculture productivity and minimizes the problem of hidden hunger. This technology showed an economical and sustainable approach. Hence, using nanomaterials for biofortification in an area of agricultural food crops might enhance sustainable food availability and nutritional health in both developing and developed countries.

References

Adu, M.O., Asare, P.A., Yawson, D.O. et al. (2018). Agronomic biofortification of selected underutilised solanaceae vegetables for improved dietary intake of potassium (K) in Ghana. *Heliyon* 4 (8): e00750. doi: 10.1016/j.heliyon.2018.e00750.

Bailey, R.L., West Jr, K.P. and Black, R.E. (2015). The epidemiology of global micronutrient deficiencies. *Annals of Nutrition and Metabolism* 66 (2): 22–33. doi: 10.1159/000371618.

Banijamali, S.M., Feizian, M., Alinejadian Bidabadi, A. et al. (2019). Evaluation uptake and translocation of iron oxide nanoparticles and its effect on photosynthetic pigmentation of chrysanthemum (*Chrysanthemum morifolium*) 'Salvador'. *Journal of Ornamental Plants* 9 (4): 245–258. doi: 10/667495?jid=667495.

Bouis, H. (2018). Reducing mineral and vitamin deficiencies through biofortification: progress under HarvestPlus. In: *Hidden Hunger: Strategies to Improve Nutrition Quality*, 118: 112–122. Switzerland: Karger Publishers. doi: 10.1159/000484342.

Bouis, H.E., Saltzman, A. and Birol, E. (2019). Improving nutrition through biofortification. In: *Agriculture for Improved Nutrition: Seizing the Momentum*. Wallingford: *IFPRI* 47–57. doi: 10.1079/9781786399311.0047.

Budke C, Mühling K.H. and Daum, D. (2020). Iodine uptake and translocation in apple trees grown under protected cultivation. *Journal of Plant Nutrition and Soil Science* 183 (4): 468–481. doi: 10.1002/jpln.202000099.

Buturi, C.V., Mauro, R.P., Fogliano, V. et al. (2021). Mineral biofortification of vegetables as a tool to improve human diet. *Foods* 10 (2):223. doi: 10.3390/foods10020223.

Cahoon, E.B., Hall, S.E., Ripp, K.G. et al. (2003). Metabolic redesign of vitamin E biosynthesis in plants for tocotrienol production and increased antioxidant content. *Nature Biotechnology* 21 (9): 1082–1087. doi: 10.1038/nbt853.

Cakmak, I. (2008). Enrichment of cereal grains with zinc: agronomic or genetic biofortification? *Plant and Soil* 302: 1–17. doi: 1 0.1007/s11104-007-9466-3.

Carciofi, M., Blennow, A., Jensen, S.L. et al. (2012). Concerted suppression of all starch branching enzyme genes in barley produces amylose-only starch granules. *BMC Plant Biology* 12 (1): 1–6. doi: 10.1186/1471-2229-12-223.

Carvalho, K.M., Gallardo-Williams, M.T., Benson, R.F. et al. (2003). Effects of selenium supplementation on four agricultural crops. *Journal of Agricultural and Food Chemistry* 51 (3): 704–709. doi: 10.1021/jf0258555.

Carvalho, S.M. and Vasconcelos, M.W. (2013). Producing more with less: strategies and novel technologies for plant-based food biofortification. *Food Research International* 54 (1): 961–971. doi: 10.1016/j.foodres.2012.12.021.

Chandra, A.K., Pandey, D., Tiwari, A. et al. (2021). Metal based nanoparticles trigger the differential expression of key regulatory genes which regulate iron and zinc homeostasis mechanism in finger millet. *Journal of Cereal Science* 100: 103235. doi: 10.101 6/j.jcs.2021.103235.

Chilimba, A.D., Young, S.D., Black, C.R. et al. (2012). Agronomic biofortification of maize with selenium (Se) in Malawi. *Field Crops Research* 125: 118–128. doi: 10.1016/j.fcr.2011.08.014.

Chugh, G., Siddique, K.H. and Solaiman, Z.M. (2022). Iron fortification of food crops through nanofertilisation. *Crop Pasture Science* 5: 471. doi: 10.1071/CP21436.

Correia, L.L., Rocha, H.A.L., Campos, J.S. et al. (2019). Interaction between vitamin A supplementation and chronic malnutrition on child development. *Ciência & Saúde Coletiva* 24: 3037–3046. doi: 10.1590/1413-81232018248.22242017.

Dancs, G., Kondrak, M. and Banfalvi, Z. (2008). The effects of enhanced methionine synthesis on amino acid and anthocyanin content of potato tubers. *BMC Plant Biology* 8: 65. doi: 10.1186/1471-2229-8-65.

Das, J.K., Salam, R.A., Mahmood, S.B. et al. (2019). Food fortification with multiple micronutrients: impact on health outcomes in general population. *Cochrane Database of Systematic Reviews* 12: 12CD011400. doi: 10.1002/14651858.CD011400.pub2.

Decourcelle, M., Perez-Fons, L., Baulande, S., et al. (2015). Combined transcript, proteome, and metabolite analysis of transgenic maize seeds engineered for enhanced carotenoid synthesis reveals pleotropic effects in core metabolism. *Journal of Experimental Botany* 66 (11): 3141–3150. doi: 10.1093/jxb/erv120.

Dimkpa, C.O., Singh, U., Bindraban, P.S. et al. (2019). Zinc oxide nanoparticles alleviate drought-induced alterations in sorghum performance, nutrient acquisition, and grain fortification. *Science of the Total Environment* 688: 926–934. doi: 10.1016/j.scitotenv.2019.06.392.

Diretto, G., Al-Babili, S., Tavazza, R. et al. (2007). Metabolic engineering of potato carotenoid content through tuber-specific overexpression of a bacterial mini-pathway. *PLoS One* 2 (4): e350. doi: 10.1371/journal.pone.0000350.

Doolette, C.L., McLaughlin, M.J. and Kirby, J.K. (2015). Bioavailability of silver and silver sulfide nanoparticles to lettuce (*Lactuca sativa*): effect of agricultural amendments on plant uptake. *Journal of Hazardous Materials* 300: 788–795. doi: 10.1016/j.jhazmat.2015.08.012.

Du, W., Sun, Y., Ji, R., et al. (2011). TiO2 and ZnO nanoparticles negatively affect wheat growth and soil enzyme activities in agricultural soil. *Journal of Environmental Monitoring* 13 (4): 822-828. doi: 10/2011/em/c0em00611d/unauth.

Ducreux, L.J., Morris, W.L., Hedley, P.E. et al. (2005). Metabolic engineering of high carotenoid potato tubers containing enhanced levels of β-carotene and lutein. *Journal of Experimental Botany* 56 (409): 81–89. doi: 10.1093/jxb/eri016.

Elkonin, L.A., Italianskaya, J.V., Domanina, I.V. et al. (2016). Transgenic sorghum with improved digestibility of storage proteins obtained by Agrobacterium-mediated transformation. *Russian Journal of Plant Physiology* 63: 678–89. doi: 10.1134/S1021443716050046.

El-Ramady, H., Brevik, E., Amer, MM. et al. (2020a). Soil and air pollution in the era of COVID-19: a global issue. *Egyptian Journal of Soil Science* 60 (4): 437–448. doi: 10.21608/ejss.2020.49996.1411.

Enfissi, E.M., Fraser, P.D., Lois, L.M. et al. (2005). Metabolic engineering of the mevalonate and non-mevalonate isopentenyl diphosphate-forming pathways for the production of health-promoting isoprenoids in tomato. *Plant Biotechnology Journal* 3 (1): 17–27. doi: 10.1111/j.1467-7652.2004.00091.

Eurola, M., Ekholm, P.A., Ylinen, M.A. I.J.A. et al. (1990). Effects of selenium fertilization on the selenium content of cereal grains, flour, and bread produced in Finland. *Cereal Chemical* 67 (4): 334–337. doi: 10.1079/NRR200255.

Fakharzadeh, S., Hafizi, M., Baghaei, M.A. et al. (2020). Using nano chelating technology for biofortification and yield increase in rice. *Scientific Reports* 10 (1): 1–9. doi: 10.1038/s41598-020-60189-x.

Gani, G., Beenish, O.B., Tashooq, A.B. et al. (2018). Hidden hunger and its prevention by food processing: a review. *International Journal of Unani Integrative Medicine* 2 (3): 1–10. doi: 10.3389/fnut.2022.1043655.

Gao, X., Avellan, A., Laughton, S. et al. (2018). CuO nanoparticle dissolution and toxicity to wheat (*Triticum aestivum*) in rhizosphere soil. *Environmental Science and Technology* 52 (5): 2888–2897. doi: 10.1021/acs.est.7b05816.

García-Gómez, C., Obrador, A., González, D. et al. (2018). Comparative study of the phytotoxicity of ZnO nanoparticles and Zn accumulation in nine crops grown in a calcareous soil and an acidic soil. *Science of the Total Environment* 644: 770–780. doi: 10.1016/j.scitotenv.2018.06.356.

Garg, M., Sharma, N., Sharma, S. et al. (2018). Biofortified crops generated by breeding, agronomy, and transgenic approaches are improving lives of millions of people around the world. *Frontiers in Nutrition* 5: 12. doi: 10.3389/fnut.2018.00012.

Gatadi, S., Madhavi, Y.V. and Nanduri, S. (2021). Nanoparticle drug conjugates treating microbial and viral infections: a review. *Journal of Molecular Structure* 1228: 129750. doi: 10.1016/j.molstruc.2020.129750.

Ghosh, S., Pahari, S. and Roy, T. (2018). An updated overview on food supplement Asian. *Journal of Research in Chemistry* 11 (3): 691–697. doi: 10.5958/0974-4150.2018.00122.0.

Giles, E.L., Kuznesof, S., Clark, B. et al. (2015). Consumer acceptance of and willingness to pay for food nanotechnology: a systematic review. *Journal of Nanoparticle Research* 17: 464. doi: 10.1007/s11051-015-3270-4.

Gödecke, T., Stein, A.J. and Qaim, M. (2018). The global burden of chronic and hidden hunger: trends and determinants. *Global Food Security* 17:21–9. doi: 10.1016/j.gfs.2018.03.004.

Goredema-Matongera, N., Ndhlela, T., Magorokosho, C. et al. (2021). Multinutrient biofortification of maize (*Zea mays* L.) in Africa: current status, opportunities and limitations. *Nutrients* 13 (3): 1039. doi: 10.3390/nu13031039.

Goto, F., Yoshihara, T., Shigemoto, N. et al. (1999). Iron fortification of rice seed by the soybean ferritin gene. *Nature Biotechnology* 17 (3): 282–286. doi: 10.1038/7029.

Hassan, U.U., Khan, M., Younas, Z. et al. (2023). Effect of phytogenic iron nanoparticles on the bio-fortification of wheat varieties. *Green Processing and Synthesis* 12 (1): 20238002. doi: 10.1515/gps-2023-8002.

Haug, A., Graham, R.D., Christophersen, O.A. et al. (2007). How to use the world scarce selenium resources efficiently to increase the selenium concentration in food. *Microbial Ecology in Health and Disease* 19 (4): 209–228. doi: 10.1080/089106007 01698986.

He, L., Zhao, J., Wang, L. et al. (2019). Using nano-selenium to combat coronavirus disease (COVID-19)? *Nano Today* 36: 101037. doi: 10.1016/j. nantod.2020.101037.

Hernández-Hernández, H., Quiterio-Gutiérrez, T., Cadenas-Pliego, G. et al. (2019). Impact of selenium and copper nanoparticles on yield, antioxidant system, and fruit quality of tomato plants. *Plants* 8 (10): 355. doi: 10.3390/plants8100355.

Hernandez-Viezcas, J.A., Castillo-Michel, H., Joy Cooke, A. et al. (2013). In situ synchrotron X-ray fluorescence mapping and speciation of CeO_2 and ZnO nanoparticles in soil cultivated soybean (*Glycine max*). *ACS Nano* 7 (2): 1415–1423. doi: 10.1021/ nn305196q.

Hindu, V., Palacios-Rojas, N., Babu, R. et al. (2018). Identification and validation of genomic regions influencing kernel zinc and iron in maize. *Theoretical and Applied Genetics*. 131: 1443–1457. doi: 10.1007/s00122-018-3089-3.

Hong, H., Datla, N., Reed, D.W. et al. (2002). High-level production of γ-linolenic acid in *Brassica juncea* using a $\Delta 6$ desaturase from pythium irregulare. *Plant Physiology* 129 (1): 354–362. doi: 10.1104/pp.001495.

Hosnedlova, B., Kepinska, M., Skalickova, S. et al. (2018). Nano-selenium and its nanomedicine applications: a critical review. *International Journal of Nanomedicine* 13: 2107–2128. doi: 10.2147/IJN.S157541.

Hussain, A., Ali, S., Rizwan, M. et al. (2018). Zinc oxide nanoparticles alter the wheat physiological response and reduce the cadmium uptake by plants. *Environmental Pollution* 242: 1518–1526. doi: 10.1016/j.envpol.2018.08.036.

Hussain, A., Rizwan, M., Ali, S. et al. (2021). Combined use of different nanoparticles effectively decreased cadmium (Cd) concentration in grains of wheat grown in a field contaminated with Cd. *Ecotoxicology and Environmental Safety* 215: 112139. doi: 10.1016/j.ecoenv.2021.112139.

Imtiaz, M., Rashid, A., Khan, P. et al. (2010). The role of micronutrients in crop production and human health. *Pakistan Journal of Botany* 42 (4): 2565–2578. doi: 10.30848/PJB2020-6(5).

Ivanov, K., Vasilev, A., Mitkov, A. et al. (2021). Application of Zn containing foliar fertilisers for recovery of the grain productivity potential of Zn deficient maize plants. *Italian Journal of Agronomy* 16 (2): 1759. doi: 10.4081/ija.2021.175.

Jaganathan, D., Ramasamy, K., Sellamuthu, G. et al. (2018). CRISPR for crop improvement: an update review. *Frontiers in Plant Science* 9: 985. doi: 10.3389/fpls.2018.00985.

Jain, S., Rustagi, A., Kumar, D. et al. (2019). Meeting the challenge of developing food crops with improved nutritional quality and food safety: leveraging proteomics and related omics techniques. *Biotechnology Letters* 41: 471–81. doi: 10.1007/s10529-01 9-02655-9.

Jan, Y., Malik, M., Yaseen, M. et al. (2019). Vitamin D fortification of foods in India: present and past scenario. *The Journal of Steroid Biochemistry and Molecular Biology* 193:105417. doi: 10.1016/j.jsbmb.2019.105417.

Jewell, C.P., Zhang, S.V., Gibson, M.J.S. et al. (2020). Intraspecific genetic variation underlying postmating reproductive barriers between species in the wild tomato clade (*Solanum sect. Lycopersicon*). *Journal of Heredity* 111 (2): 216–226. doi: 10.1101/718544.

Jha, A.B. and Warkentin, T.D. (2020). Biofortification of pulse crops: status and future perspectives. *Plants* 9 (1): 73. doi: 10.3390 /plants9010073.

Kalra, T., Tomar, P.C. and Arora, K. (2020). Micronutrient encapsulation using nanotechnology: nanofertilizers. *Plant Archives* 20 (2): 1748–1753. doi: 10.1186/s12870-022.

Kaphle, A., Navya, P.N., Umapathi, A. et al. (2018). Nanomaterials for agriculture, food and environment: applications, toxicity and regulation. *Environmental Chemistry Letters* 16: 43–58. doi: 10.1007/s10311-017-0662-y.

Karthika, K.S., Rashmi, I., Neenu, S. et al. (2023). Agronomic biofortification: an effective tool for alleviating nutrient deficiency in plants and human diet. In: *Mineral Biofortification in Crop Plants for Ensuring Food Security*, 1–35. Singapore: Springer Nature. doi: 10.1007/978-981-99-4090-5_1.

Khan, M.K., Pandey, A., Hamurcu, M. et al. (2021). Insight into the Prospects for nanotechnology in wheat biofortification. *Biology* 10 (11): 1123. doi: 10.3390/biology10111123.

Kiran, A., Wakeel, A., Mahmood, K. et al. (2022). Biofortification of staple crops to alleviate human malnutrition: contributions and potential in developing countries. *Agronomy* 12 (2):452. doi: 10.3390/agronomy12020452.

Koc, E. and Karayigit, B. (2022). Assessment of biofortification approaches used to improve micronutrient-dense plants that are a sustainable solution to combat hidden hunger. *Journal of Soil Science and Plant Nutrition* 22 (1): 475–500. doi: 10.1007/s42729-021-00663-1.

Kumar, Ajay, Pandey, D., Tiwari, A. et al. (2021). Metal based nanoparticles trigger the differential expression of key regulatory genes which regulate iron and zinc homeostasis mechanism in finger millet. *Journal of Cereal Science* 100 (2021): 103235. doi: 10.1016/j.jcs.2021.103235.

Kumar, K., Gambhir, G., Dass, A. et al. (2020). Genetically modified crops: current status and future prospects. *Planta* 251 (4): 91. doi: 10.1007/s00425-020-03372-8.

Larue, C., Castillo-Michel, H., Sobanska, S. et al. (2014). Foliar exposure of the crop *Lactuca sativa* to silver nanoparticles: evidence for internalization and changes in Ag speciation *Journal of Hazardous Materials* 264: 98–106. doi: 10.1016/j.jhazmat.2013.10.053.

Lema, M. (2018). Marker assisted selection in comparison to conventional plant breeding. *Agricultural Research Technology* 14: 555914. doi: 10.19080/ARTOAJ.2018.14.555914.

Lin, D. and Xing, B. (2008). Root uptake and phytotoxicity of ZnO nanoparticles *Environmental Science & Technology* 42 (15): 5580–5585. doi: 10.1021/es800422x.

Lipkie, T.E., De-Moura, F.F., Zhao, Z.Y. et al. (2013). Bioaccessibility of carotenoids from transgenic provitamin A biofortified sorghum. *Journal of Agricultural and Food Chemistry* 61 (24): 5764–71. doi: 10.1021/jf305361s.

Liu, C., Chen, G., Li, Y. et al. (2017). Characterization of a major QTL for manganese accumulation in rice grain. *Scientific Reports* 7 (1): 1–17. doi: 10.1038/s41598-017-18090-7.

Low, J.W., Mwanga, R.O.M., Andrade, M. et al. (2017). Tackling vitamin, A deficiency with biofortified sweet potato in sub-Saharan Africa. *Global Food Security* 14: 23–30. doi: 10.1016/j.gfs.2017.01.004.

Lowe, N.M. (2021). The global challenge of hidden hunger: perspectives from the field. *Proceedings of the Nutrition Society* 80 (3): 283–9. doi: 10.1017/S0029665121000902.

Lowry, G.V., Avellan, A. and Gilbertson, L.M. (2019). Opportunities and challenges for nanotechnology in the agri-tech revolution. *Nature Nanotechnology* 14 (6): 517–522. doi: 10.1038/s41565-019-0461-7.

Lv, J., Christie, P. and Zhang, S. (2019). Uptake, translocation, and transformation of metal-based nanoparticles in plants: recent advances and methodological challenges. *Environmental Science: Nano* 6 (1): 41–59. doi: 10.1039/C8EN00645H.

Lyons, G.H., Lewis, J., Lorimer, M.F. et al. (2004). High-selenium wheat: agronomic biofortification strategies to improve human nutrition. *Food Agriculture and Environment* 2 (1): 171–178. doi: 10.1186/s12870-018-1231-7.

Malik, K. and Maqbool, A. (2020). Transgenic crops for biofortification. *Frontiers in Sustainable Food Systems* 4: 571402. doi: 10.3389/fsufs.2020.571402.

Manjunatha, S., Biradar, D. and Aladakatti, YR. (2016). Nanotechnology and its applications in agriculture: a review. *Journal Farm Science* 29 (1): 1–13. doi: 10.113/jfc2913.

Mantadakis, E., Chatzimichael, E. and Zikidou, P. (2020). Iron deficiency anemia in children residing in high and low-income countries: risk factors, prevention, diagnosis and therapy. *Mediterranean Journal of Hematology and Infectious Diseases* 12 (1):e2020041. doi: 10.4084/mjhid.2020.041.

Manzoor, N., Ahmed, T., Noman, M. et al. (2021). Iron oxide nanoparticles ameliorated the cadmium and salinity stresses in wheat plants, facilitating photosynthetic pigments and restricting cadmium uptake. *Science of the Total Environment*769: 145221. doi: 10.1016/j.scitotenv.2021.145221.

Margenot, A.J., Rippner, D.A., Dumlao, M.R. et al. (2018). Copper oxide nanoparticle effects on root growth and hydraulic conductivity of two vegetable crops. *Plant and Soil* 431: 333–345. doi: 10.1007/s11104-018-3741-3.

Marques, E., Darby, H.M. and Kraft, J. (2021). Benefits and limitations of non-transgenic micronutrient biofortification approaches. *Agronomy* 11 (3): 464. doi: 10.3390/11030464.

Martínez, S., Armesto, J., Gómez-Limia, L. et al. (2020). Impact of processing and storage on the nutritional and sensory properties and bioactive components of Brassica spp. A review. *Food Chemistry* 313: 126065. doi: 10.1016/j.foodchem.2019.126065.

Masuka, B., Magorokosho, C., Olsen, M. et al. (2017). Gains in maize genetic improvement in eastern and southern Africa: II.CIMMYT open-pollinated variety breeding pipeline. *Crop Science* 57 (1): 180–191. doi: 10.2135/cropsci2016.05.0408.

Mateus, M.P., de, B., Tavanti, R.F.R. et al. (2021). Selenium biofortification enhances ROS scavenge system increasing yield of coffee plants. *Ecotoxicology and Environmental Safety* 209: 111772. doi: 10.1016/j.ecoenv.2020.111772.

Matres, J.M., Arcillas, E., Cueto-Reaño, M.F. et al. (2021). Biofortification of rice grains for increased iron content. In: *Rice Improvement Physiological*, 1–506. Cham: Springer. doi: 10.1007/978-3-030-66530-2.

Mayer, J.E., Pfeiffer, W.H. and Beyer, P. (2008). Biofortified crops to alleviate micronutrient malnutrition. *Current Opinion in Plant Biology* 11 (2): 166–170. doi: 10.1016/j.pbi.2008.01.007.

Milani, N., Hettiarachchi, G.M., Kirby, J.K. et al. (2015). Fate of zinc oxide nanoparticles coated onto macronutrient fertilizers in an alkaline calcareous soil. *PLoS One* 10 (5): 1–16. doi: 10.1371/journal.pone.0126275.

Mir, Z., Yadav, P., Ali, S. et al. (2020). Transgenic biofortified crops: applicability and challenges. In: *Advances in Agri-food Biotechnology* (ed. TR. Sharma, R. Deshmukh, H. Sonah), 153–172. Singapore: Springer. doi: 10.1007/978-981-15-2874-37.

Mkambula, P., Mbuya, M.N.N., Rowe, L.A. et al. (2020). The unfinished agenda for food fortification in low- and middle-income countries: quantifying progress, gaps and potential opportunities. *Nutrients* 12 (2): 354. doi: 10.3390/nu12020354.

Morris, J., Hawthorne, K.M., Hotze, T. et al. (2008). Nutritional impact of elevated calcium transport activity in carrots. *Proceedings of the National Academy of Sciences*. 105 (5): 1431–1435. doi: 10.1073/pnas.0709005105.

Mukhopadhyay, S.S. (2014) Nanotechnology in agriculture: prospects and constraints. *Nanotechnology, Science and Applications* 7: 63–71. doi: 10.2147/NSA.S39409.

Nair, R., Varghese, S.H., Nair, B.G. et al. (2010). Nanoparticulate material delivery to plants. *Plant Science* 179 (3): 154–163. doi: 10.1016/j.plantsci.2010.04.012.

Narayanan, N., Beyene, G., Chauhan, R.D. et al. (2019). Biofortification of field-grown cassava by engineering expression of an iron transporter and ferritin. *Nature Biotechnology* 37 (2): 144–51. doi: 10.1038/s41587-018-0002-1.

Naz, S., Yousaf, B., Tahir, M.A. et al. (2015). Iron and zinc bio-fortification strategies in wheat crop by exogenous application of micronutrients. *Food Science and Quality Management* 35: 49–54. doi: 10.2224/05058fsqm6088.

Newell-McGloughlin, M. (2008). Nutritionally improved agricultural crops. *Plant Physiology* 147 (3): 939–953. doi: 10.1104/pp.108.121947.

Ohnoutkova, L., Zitka, O., Mrizova, K. et al. (2012). Electrophoretic and chromatographic evaluation of transgenic barley expressing a bacterial dihydrodipicolinate synthase. *Journal of Electrophoresis* 33 (15): 2365–73. doi: 10.1002/elps.201200033.

Olson, R., Gavin-Smith, B., Ferraboschi, C. et al. (2021). Food fortification: The advantages, disadvantages and lessons from sight and life programs. *Nutrients* 13 (4): 1118. doi: 10.3390/nu13041118.

Osendarp, S.J.M., Martinez, H., Garrett, G.S. et al. (2018). Large-scale food fortification and biofortification in low-and middle-income countries: a review of programs, trends, challenges, and evidence gaps. *Food and Nutrition Bulletin* 39 (2): 315–31. doi: 10.1177/0379572118774229.

Pal, V., Singh, G. and Dhaliwal, S.S.A. (2021). New approach in agronomic biofortification for improving zinc and iron content in chickpea (*Cicer arietinum* L.) grain with simultaneous foliar application of zinc sulphate, ferrous sulphate and urea. *Journal of Soil Science and Plant Nutrition* 21: 883–96. doi: 10.1007/s42729-021-00408-0.

Park, S., Kim, C.K., Pike, L.M. et al. (2004). Increased calcium in carrots by expression of an Arabidopsis H^+/Ca^{2+} transporter. *Molecular Breeding* 14: 275–282. doi: 10.1023/B:MOLB.0000047773.20175.ae.

Pérez-de-Luque, A. (2017). Interaction of nanomaterials with plants: what do we need for real applications in agriculture. *Frontiers in Environmental Science* 5: 12. doi:10.3389/fenvs.2017.00012.

Pooniya, V. and Shivay, Y.S. (2015). Influence of green manuring and zinc fertilization on quality parameters of Basmati rice. *Communications in Soil Science and Plant Analysis* 46 (3): 382–392. doi: 10.1080/00103624.2014.981275.

Pradas del Real, A.E., Vidal, V., Carrière, M. et al. (2017). Silver nanoparticles and wheat roots: a complex interplay. *Environmental Science & Technology* 51 (10): 5774–5782. 10.1021/acs.est.7b00422.

Prasad, R., Bhattacharyya, A. and Nguyen, Q.D. (2017). Nanotechnology in sustainable agriculture: recent developments, challenges, and perspectives. *Frontiers in microbiology* 8: 1014. doi: 10.3389/fmicb.2017.01014.

Priester, J.H., Moritz, S.C., Espinosa, K. et al. (2017). Damage assessment for soybean cultivated in soil with either CeO_2 or ZnO manufactured nanomaterials. *Science of the Total Environment* 579: 1756–1768. doi: 10.1016/j.scitotenv.2016.11.149.

Ragab, G.A. and Saad-Allah, K.M. (2020). Green synthesis of sulfur nanoparticles using *Ocimum basilicum* leaves and its prospective effect on manganese-stressed *Helianthus annuus* (L.) seedlings. *Ecotoxicology and Environmental Safety* 191: 110242. doi: 10.1016/j.ecoenv.2020.110242.

Rajeshwari, A., Kavitha, S., Alex, S.A. et al. (2015). Cytotoxicity of aluminum oxide nanoparticles on *Allium cepa* root tip – effects of oxidative stress generation and biouptake. *Environmental Science and Pollution Research* 22: 11057–11066. doi: 10.1007/s11356-015-4355-4.

Rajput, V.D., Minkina, T., Sushkova, S. et al. (2018). Effect of nanoparticles on crops and soil microbial communities *Journal of Soils and Sediments* 18: 2179–2187. doi: 10.1007/s11368-017-1793-2.

Rajput, V.D., Singh, A., Minkina, T. et al. (2021). Nano-enabled products: challenges and opportunities for sustainable agriculture. *Plants* 10 (12): 2727. doi: 10.3390/plants10122727.

Raliya, R., Franke, C., Chavalmane, S. et al. (2016). Quantitative understanding of nanoparticle uptake in watermelon plants. *Frontiers in Plant Science* 7: 198920. doi: 10.3389/fpls.2016.01288.

Raliya, R., Nair, R., Chavalmane, S. et al. (2015). Mechanistic evaluation of translocation and physiological impact of titanium dioxide and zinc oxide nanoparticles on the tomato (*Solanum lycopersicum* L.) plant. *Metallomics* 7 (12): 1584–1594. doi: 10.1039/c5mt00168d.

Ramesh, S.A., Choimes, S. and Schachtman, D.P. (2004). Over-expression of an Arabidopsis zinc transporter in *Hordeum vulgare* increases short-term zinc uptake after zinc deprivation and seed zinc content. *Plant Molecular Biology* 54: 373–385. doi: 10.1023/B:PLAN.0000036370.70912.34.

Rawat, N., Neelam, K., Tiwari, V.K. et al. (2013). Biofortification of cereals to overcome hidden hunger. *Plant Breeding* 132 (5): 437–445. doi: 10.1111/pbr.12040.

Resmi, R., Yoonus, J. and Beena, B. (2021). A novel greener synthesis of ZnO nanoparticles from *Nilgiri Antusciliantus* leaf extract and evaluation of its biomedical applications. *Materials Today: Proceedings* 46: 3062–3068. doi: 10.1016/j.matpr.2021.02.498.

Rezaei, F., Moaveni, P. and Mozafari, H. (2015). Effect of different concentrations and time of nano TiO_2 spraying on quantitative and qualitative yield of soybean (*Glycine max* L.) at Shahr-e-Qods, Iran. *Biology Forum: International Journal* 7: 957–964. doi: 10.5555/20153255017.

Riaz, U., Aziz, H. and Anum, W. (2020). *Plant Micronutrients*, 225–239. Cham: Springer. doi: 10.1007/978-3-030-49856-6_9.

Rico, C.M., Peralta-Videa, J.R. and Gardea-Torresdeya, J.L. (2015). Differential effects of cerium oxide nanoparticles on rice, wheat, and barley roots: a Fourier transform infrared (FT-IR) microspectroscopy study. *Applied Spectroscopy* 69 (2): 287–295. doi: 10.1366/14-07495.

Rizwan, M., Ali, S., Ali, B. et al. (2019). Zinc and iron oxide nanoparticles improved the plant growth and reduced the oxidative stress and cadmium concentration in wheat. *Chemosphere* 214: 269–277. doi: 10.1016/j.chemosphere.2018.09.12.

Rizwan, M., Ali, S., Rehman, M.Z., et al. (2021). Effects of nanoparticles on trace element uptake and toxicity in plants: a review. *Ecotoxicology and Environmental Safety* 221: 112437. doi: 10.1016/j.ecoenv.2021.112437.

Ruttkay-Nedecky, B., Krystofova, O., Nejdl, L. et al. (2017). Nanoparticles based on essential metals and their phytotoxicity. *Journal of Nanobiotechnology* 15: 1–19. doi: 10.1186/s12951-017-0268-3.

Saalbach, I., Mora-Ramirez, I., Weichert, N. et al. (2014). Increased grain yield and micronutrient concentration in transgenic winter wheat by ectopic expression of a barley sucrose transporter. *Journal of Cereal Science* 60 (1): 75–81. doi: 10.1016/j.jcs.2014.01.017.

Sahoo, M., Vishwakarma, S., Panigrahi, C. et al. (2021). Nanotechnology: current applications and future scope in food. *Food Frontiers* 2 (1): 3–22. doi: 10.1002/fft2.58.

Sarlak, N. and Taherifar, A. (2017). Encapsulation of nanomaterials and production of nanofertilizers and nanopesticides: insecticides for agri-food production and plant disease treatment. In: *Nanoscience and Plant–Soil Systems*, 481–498. Cham: Springer. doi: 10.1007/978-3-319-46835-8_18.

Schmidt, M.A., Parrott, W.A., Hildebrand, D.F. et al. (2015). Transgenic soya bean seeds accumulating β-carotene exhibit the collateral enhancements of oleate and protein content traits. *Plant Biotechnology Journal* 13 (4): 590–600. doi: 10.1111/pbi.12286.

Shahzad, Z., Rouached, H. and Rakha, A. (2014). Combating mineral malnutrition through iron and zinc biofortification of cereals. *Comprehensive Reviews in Food Science and Food Safety*13 (3): 329–346. doi: 10.1111/1541-4337.12063.

Shalaby, T.A., Abd-Alkarim, E., El-Aidy, F. et al. (2021). Nano-selenium, silicon and H_2O_2 boost growth and productivity of cucumber under combined salinity and heat stress. *Ecotoxicology and Environmental Safety* 212: 111926. doi: 10.1016/j.ecoenv.2021.111962.

Sharma, V., Saini, D.K., Kumar, A. et al. (2019). Review of important QTLs for biofortification traits in rice. *Preprints* 2019:2019120158. doi: 10.20944/preprints201912.0158.v1.

Sheoran, P., Grewal, S., Kumari, S. et al. (2021). Enhancement of growth and yield, leaching reduction in *Triticum aestivum* using biogenic synthesized zinc oxide nanofertilizer. *Biocatalysis and Agricultural Biotechnology* 32: 101938. doi: 10.1016/j.bcab.2021.101938.

Sheoran, S., Kumar, S., Ramtekey, V. et al. (2022). Current status and potential of biofortification to enhance crop nutritional quality: an overview *Sustainability* 14 (6): 3301. doi: 10.3390/su14063301.

Shiriaev, A., Brizzolara, S., Sorce, C. et al. (2023). Selenium biofortification impacts the tomato fruit metabolome and transcriptional profile at ripening. *Journal of Agricultural and Food Chemistry* 71 (36): 13554–13565. doi: 10.1021/acs.jafc.3c02031.

Signorell, C., Zimmermann, M.B., Cakmak, I. et al. (2019). Zinc absorption from agronomically biofortified wheat is similar to post-harvest fortified wheat and is a substantial source of bioavailable zinc in humans. *Journal of Nutrition* 149 (5): 840–846. doi: 10.1093/jn/nxy328.

Sindireva, A., Golubkina, N., Bezuglova, H. et al. (2023). Effects of high doses of selenate, selenite and nano-selenium on biometrical characteristics, yield and biofortification levels of vicia faba *L. Cultivars. Plants* 12 (15): 2847. doi: 10.3390/plants12152847.

Singh, R., Govindan, V. and Andersson, M.S. (2017). *Zinc-Biofortified Wheat: Harnessing Genetic Diversity for Improved Nutritional Quality. No. 2187-2019-666.* Germany: CIMMYT, HarvestPlus, and the Global Crop Diversity Trust. Bonn. doi: 10 02117agedu2017.

Siwela, M., Pillay, K., Govender, L. et al. (2020). Biofortified crops for combating hidden hunger in South Africa: availability, acceptability, micronutrient retention and bioavailability. *Foods* 9 (6): 815. doi: 10.3390/foods9060815.

Skalickova, S., Milosavljevic, V., Cihalova, K. et al. (2017). Selenium nanoparticles as a nutritional supplement. *Nutrition* 33: 83–90. doi: 10.1016/j.nut.2016.05.001.

Song, S., Hou, W., Godo, I., et al. (2013) Soybean seeds expressing feedback-insensitive cystathionine γ-synthase exhibit a higher content of methionine. *Journal of Experimental Botany* 64(7): 1917–1926.

Sorahinobar, M., Deldari, T., Nazem Bokaeei, Z. et al. (2023). Effect of zinc nanoparticles on the growth and biofortification capability of mungbean (*Vigna radiata*) seedlings. *Biologia* 78 (4): 951–960. doi: 10.1007/s11756-022-01269-3.

Sriatmi, A., Jati, SP., Suryoputro, A. et al. (2021). Stakeholder mapping analysis on the scaling-up nutrition movement during the 1000 days of life between the urban and rural government areas. *Unnes Journal of Public Health* 10 (1): 68–77. doi: 10.15294/ujph.v10i1.38029.

Stephen, B.J., Sharma, M.M., Jain, D. et al. (2021). Biogenic synthesized nanoparticles a boon to human health. *Materials Today: Proceedings* 43: 3211–3213. doi: 10.1016/j.matpr.2021.01.869.

Sundaria, N., Singh, M., Upreti, P. et al. (2019). Seed priming with Iron oxide nanoparticles triggers iron acquisition and biofortification in wheat (*Triticum aestivum* L.) grains. *Journal of Plant Growth Regulation* 38: 122–131. doi: 10.1007/s00344-018-9818-7.

Szerement, J., Szatanik-Kloc, A., Mokrzycki, J. et al. (2022). Agronomic biofortification with Se, Zn, and Fe: an effective strategy to enhance crop nutritional quality and stress defense – a review. *Journal of Soil Science and Plant Nutrition* 22 (1): 1129–1159. doi: 10.1007/s42729-021-00719-2.

Takahashi, M., Nakanishi, H., Kawasaki, S. et al. (2001). Enhanced tolerance of rice to low iron availability in alkaline soils using barley nicotianamine aminotransferase genes. *Nature Biotechnology* 19 (5): 466–469. doi: 10.1038/88143.

Tam, E., Keats, E.C., Rind, F. et al. (2020). Micronutrient supplementation and fortification interventions on health and development outcomes among children under-five in low-and middle-income countries: a systematic review and meta-analysis. *Nutrients* 12 (2): 289. doi: 10.3390/nu12020289.

Tamas, C., Kisgyorgy, B.N., Rakszegi, M. et al. (2009). Transgenic approach to improve wheat (*Triticum aestivum* L.) nutritional quality. *Plant Cell Reports* 28 (2009):1085–1094. doi: 10.1007/s00299-009-0716-0.

Tamez, C., Morelius, E.W., Hernandez-Viezcas, J.A. et al. (2019). Biochemical and physiological effects of copper compounds/nanoparticles on sugarcane (*Saccharum officinarum*). *Science of the Total Environment* 649: 554–562. doi: 10.1016/j.scitotenv.2018.08.337.

Tripathi, D.K., Singh, S., Singh, V.P. et al. (2017). Silicon nanoparticles more effectively alleviated UV-B stress than silicon in wheat (*Triticum aestivum*) seedlings. *Plant Physiology and Biochemistry* 110: 70–81. doi: 10.1016/j.plaphy.2016.06.026.

Ul-Allah, S. (2018). Combating hidden hunger in agriculture perspective. *Strategies to Improve Nutrition Quality* 118: 161–166. doi: 10.1159/000484511.

Umar, W., Hameed, M.K., Aziz, T. et al. (2021). Synthesis, characterization and application of ZnO nanoparticles for improved growth and Zn biofortification in maize. *Archives of Agronomy and Soil Science* 67(9): 1164–1176. doi: 10.1080/03650340.2020.1782893.

Upadhyaya, C.P., Young, K.E., Akula, N. et al. (2009). Over-expression of strawberry D-galacturonic acid reductase in potato leads to accumulation of vitamin C with enhanced abiotic stress tolerance. *Plant Science* 177 (6):659–667. doi: 10.1016/j.plantsci.2009.08.004.

Vasconcelos, M., Datta, K., Oliva, N. et al. (2003). Enhanced iron and zinc accumulation in transgenic rice with the ferritin gene. *Plant Science* 164 (3): 371–-378. doi: 10.1016/S0168-9452(02)00421-1.

Wakeel, A., Farooq, M., Bashir, K. et al. (2018). Micronutrient malnutrition and biofortification: recent advances and future perspectives. In: *Plant Micronutrient Use Efficiency: Molecular and Genomic Perspectives in Crop Plants* (eds. MA. Hossain, T. Kamiya, DJ. Burritt, et al.), 225–243. Amsterdam: Elsevier Inc. doi: 10.1016/B978-0-12-812104-7.00017-4.

Wang, C., Zeng, J., Li, Y., et al. (2014). Enrichment of provitamin A content in wheat (*Triticum aestivum* L.) by introduction of the bacterial carotenoid biosynthetic genes CrtB and CrtI. *Journal of Experimental Botany* 65 (9): 2545–2556. doi: 10.1093/jxb/eru138.

Wang, K., Wang, Y., Li, K. et al. (2020). Uptake, translocation and biotransformation of selenium nanoparticles in rice seedlings (*Oryza sativa* L.). *Journal of Nanobiotechnology* 18: 1–15. doi: 10.1186/s12951-020-00659-6.

Wang, P., Wang, H., Liu, Q. et al. (2017). QTL mapping of selenium content using a RIL population in wheat. *PLoS One* 12 (9):e0184351. doi: 10.1371/journal.pone.0184351.

Wang, W.N., Tarafdar, J.C. and Biswas, P. (2013). Nanoparticle synthesis and delivery by an aerosol route for watermelon plant foliar uptake. *Journal of Nanoparticle Research* 15: 1–13. doi: 10.1007/s11051-013-1417-8.

Wang, X., Liu, L., Zhang, W. et al. (2021). Prediction of plant uptake and translocation of engineered metallic nanoparticles by machine learning. *Environmental Science & Technology* 55 (11): 7491–7500. doi: 10.1021/acs.est.1c01603.

Wang, Y., Deng, C., Cota-Ruiz, K. et al. (2020). Improvement of nutrient elements and allicin content in green onion (*Allium fistulosum*) plants exposed to CuO nanoparticles. *Science of the Total Environment* 725: 138387. doi: 10.1016/j.scitotenv.2020.138387.

Wang, Y., Jiang, F., Ma, C. et al. (2019). Effect of metal oxide nanoparticles on amino acids in wheat grains (*Triticum aestivum*) in a life cycle study. *Journal of Environmental Management* 241: 319–327. doi: 10.1016/j.jenvman.2019.04.041.

Wang, Z., Xie, X., Zhao, J. et al. (2012). Xylem-and phloem-based transport of CuO nanoparticles in maize (*Zea mays* L.). *Environmental Science & Technology* 46 (8): 4434–4441. doi: 10.1021/es204212z.

Wei, S.H., Li, X., Gruber, M.Y. et al. (2009). RNAi-mediated suppression of DET1 alters the levels of carotenoids and sinapate esters in seeds of *Brassica napus*. *Journal of Agricultural and Food Chemistry* 57 (12): 5326–5333. doi: 10.1021/jf803983w.

Weisstaub, G. and Araya, M. (2008). Acute malnutrition in Latin America. The challenge of ending avoidable death. *Journal of Pediatric Gastroenterology and Nutrition* 47: S10–S14. doi: 10.1097/MPG.0b013e3181818e78.

Welsch, R., Arango, J., Bär, C. et al. (2010). Provitamin A accumulation in cassava (*Manihot esculenta*) roots driven by a single nucleotide polymorphism in a phytoene synthase gene. *The Plant Cell* 22 (10): 3348–3356. doi: 10.1105/tpc.110.077560.

White, P.J. and Broadley, M.R. (2011). Physiological limits to zinc biofortification of edible crops. *Frontiers in Plant Science* 80: 1–11. doi: 10.3389/fpls.2011.00080.

White, P.J. and Broadley, M.R. (2009). Biofortification of crops with seven mineral elements often lacking in human diets: iron, zinc, copper, calcium, magnesium, selenium and iodine. *New Phytologist* 182 (1): 49–84. doi: 10.1111/j.1469-8137.2008.02738.x.

Xiong, T., Dumat, C., Dappe, V. et al. (2017). Copper oxide nanoparticle foliar uptake, phytotoxicity, and consequences for sustainable urban agriculture. *Environmental Science & Technology* 51: 5242–5251. doi: 10.1021/acs.est.6b05546.

Yang, G., Yuan, H., Ji, H. et al. (2021). Effect of ZnO nanoparticles on the productivity, Zn biofortification, and nutritional quality of rice in a life cycle study. *Plant Physiology and Biochemistry* 163: 87–94. doi: 10.1016/j.plaphy.2021.03.053.

Yanga, J., Cao, W. and Rui, Y. (2017). Interactions between nanoparticles and plants: phytotoxicity and defense mechanisms. *Journal of Plant Interactions* 12 (1): 158–169. doi: 10.1080/17429145.2017.1310944.

Younis, K., Ahmad, S. and Badpa, A. (2015). Malnutrition: causes and strategies. *Journal of Food Processing & Technology* 6:434. doi: 10.4172/2157-7110.1000434.

Yu, B., Lydiate, DJ., Young, L.W. et al. (2008). Enhancing the carotenoid content of *Brassica napus* seeds by downregulating lycopene epsilon cyclase. *Transgenic Research* 17: 573–585. doi: 10.1007/s11248-007-9131-x.

Yu, H., Miao, P., Li, D. et al. (2023). Improving red pitaya fruit quality by nano-selenium biofortification to enhance phenylpropanoid and betalain biosynthesis. *Ecotoxicology and Environmental Safety* 267: 115653. doi: 10.1016/j.ecoenv.2023.115653.

Yu, J., Peng, P., Zhang, X. et al. (2005). Seed-specific expression of the lysine-rich protein gene sb401 significantly increases both lysine and total protein content in maize seeds. *Food and Nutrition Bulletin* 26 (4_suppl 3): S312–S316. doi: 10.1177/15648265050264S311.

Yu, O., Shi, J., Hession, A.O. et al. (2003). Metabolic engineering to increase isoflavone biosynthesis in soybean seed. *Phytochemistry* 63 (7): 753–63. doi: 10.1016/S0031-9422(03)00345-5.

Yusefi-Tanha, E., Fallah, S., Rostamnejadi, A. et al. (2020). Root system architecture, copper uptake and tissue distribution in soybean (*Glycine max* (L.) *Merr.*) grown in copper oxide nanoparticle (CuONP)-amended soil and implications for human nutrition. *Plants* 9 (10): 1326. doi: 10.3390/plants9101326.

Zhang, C., Liu, J., Zhang, Y., et al. (2011). Overexpression of SlGMEs leads to ascorbate accumulation with enhanced oxidative stress, cold, and salt tolerance in tomato. *Plant Cell Reports* 30: 389–98. doi:10.1007/s00299-010-0939-0.

Zhang, Z., Kong, F., Vardhanabhuti, B. et al. (2012). Detection of engineered silver nanoparticle contamination in pears. *Journal of Agricultural and Food Chemistry* 60 (43): 10762–10767. doi: 10.1021/jf303423q.

Zhao, L., Peralta-Videa, J.R., Ren, M. et al. (2012). Transport of Zn in a sandy loam soil treated with ZnO NPs and uptake by corn plants: electron microprobe and confocal microscopy studies. *Chemical Engineering Journal* 184: 1–8. doi: 10.1016/j.cej.2012.01.041.

Zhao, L., Peralta-Videa, J.R., Varela-Ramirez, A. et al. (2012). Effect of surface coating and organic matter on the uptake of CeO_2 NPs by corn plants grown in soil: insight into the uptake mechanism. *Journal of Hazardous Materials* 225: 131–138. doi: 10.1016/j.jhazmat.2012.05.008.

Zhao, M., Lin, Y. and Chen, H. (2020). Improving nutritional quality of rice for human health. *Theoretical and Applied Genetics* 133 (5):1397–1413. doi: 10.1007/s00122-019-03530-x.

Zimbovskaya, M.M., Polyakov, A.Y., Volkov, D.S. et al. (2020). Foliar application of humic-stabilized nanoferrihydrite resulted in an increase in the content of iron in wheat leaves. *Agronomy* 10 (12): 1891. doi: 10.3390/agronomy10121891.

Zoidis, E., Seremelis, I., Kontopoulos, N. et al. (2018). Selenium-dependent antioxidant enzymes: actions and properties of selenoproteins. *Antioxidants* 7 (5): 1–26. doi: 10.3390/antiox7050066.

7

Role of Seed-priming in Biofortification

Qudrat Ullah Khan[1], Mohammad Safdar Baloch[2], Nabeel Rizwan[1], Asghar Ali Khan[2], Muhammad Amjad Nadim[2], Umar Khitab Saddozai[2], Obaid Ullah Sayal[3], Adnan Noor Shah[4,]*

[1] *Department of Soil Science, Gomal University, Dera Ismail Khan, Pakistan*
[2] *Department of Agronomy, Gomal University, Dera Ismail Khan, Pakistan*
[3] *Department of Plant Breeding and Genetics, Gomal University, Dera Ismail Khan, Pakistan*
[4] *Department of Agricultural Engineering, Khwaja Fareed University of Engineering and Information Technology, Rahim Yar Khan, Pakistan*

*Corresponding author: adnan.shah@kfueit.edu.pk

7.1 Introduction

Seed-priming is the pregermination soaking of seed in water or any other low osmotic potential solutions to a level where the controlled metabolic activity of a seed begins, but the emergence of the radical does not occur. The moisture in this process is sufficient to initiate the pregerminative activities of a seed but not enough to permit the radicle protrusion. Seeds that are primed before sowing have the ability to germinate even under unsuitable conditions. This process enhances the performance of crops and increases crop yield (Marthandan et al. 2020).

Other workers have reported that seed-priming ensures a uniform and increased rate of germination by reducing the time for imbibition (Brocklehurst and Dearman 1983). It also increases the enzyme activation and metabolite production (Hussain et al. 2016), repairs DNA damage (Farooq et al. 2009), and regulates osmosis.

7.1.1 Historical Perspective of Seed Priming

The history of seed-priming is as old as agriculture itself. Farmers after cultivating their lands soon realized that some of the seeds do not germinate properly and lack uniformity in the rate of germination. The ancient civilization, Greek Theophrastus (372–287 BC) initiated work on the physiology of seed and soon became aware about the interruption of seeds during germination (Evenari 1984). In 1664, Roman farmers started soaking of seeds prior to sowing to increase and synchronize the seed germination. The seed-priming techniques were later on practiced by the French Agronomist Olivier de Serres (1539–1619). Evelyn (1729) revealed that there is an impact of temperature on the seed germination process. In 1779, Ingenhousz found that the emergence of seed is influenced by light (Ingenhousz 1779). Later on in the nineteenth century, a number of botanists studied and described the morphology of seeds (Amici 1830; Sachs 1859). The word seed-priming was first coined by Heydecker and his coworkers. They successfully used this technique by using the osmotic agent polyethylene glycol (PEG) for seed-priming, which was presented in symposium in 1977 (Heydecker et al. 1973). In the following years, seed vigor was associated with seed-priming. This concept was adopted by the scientific community and the method was improved, developed, and commercialized (Paparella et al. 2015; Lutts et al. 2016; Paul et al. 2022; Waterworth et al. 2019; Harris et al. 2001).

In order to comprehend the seed-priming process, it is inevitable to understand the seed structure, germination of seed, and seed physiology.

Crop Biofortification: Biotechnological Approaches for Achieving Nutritional Security under Changing Climate, First Edition. Edited by Adnan Noor Shah et al.

7.2 Seed Structure

Seed structure is important to understand the process of germination.

7.2.1 Viable Seed

A seed is the matured fertilized ovule that has a viable seed coat for the protection of embryonic plant and food storage for germination. Seed is the reproductive structure of all the flowering plants. The seed has the genetic capability to propagate the flowering plants. The parts of seed include as follows:

a. **Testa:** The outermost protective covering of a seed that protects the embryo from adverse conditions, i.e. temperature, injury, etc. The seed coat may be soft in some seeds, such as beans while it may be hard in others, such as coconut.
b. **Endosperm:** The endosperm are tissues produced inside the seed in most of the flowering plants following double fertilization. It surrounds and nourishes the embryo. The endosperm has an important role in nurturing the embryo by providing nutrients, controlling its growth, and protecting the embryo during the development stages. The endospermic tissues are rich in carbohydrates, proteins, and lipids.
c. **Aleuron layer**: This is an important layer that exits between the seed coat and the embryo of the monocotyledonous seeds. This layer is one-celled in most of the cereal crops, such as wheat, rice, maize, millet, sorghum, and rye. However, in barley, it is a three-celled thick layer. It is rich in protein. During the germination, as the seed imbibes water, the embryo releases the hormone gibberellin, which activates the aleuron layer to produce enzyme α-amylase. This enzyme further hydrolyses the stored starch and proteins, including proteases. The breakdown of starch into sugar initiates the growth of root and shoot.
d. **Embryo**: An embryo is *a fertilized ovule*, an immature plant from which a new plant grows under favorable conditions. It is composed of the following parts:
   Embryonal axis: The embryonic axis having one end is called the **plumule**, which develops into the shoot as it grows upward, while the second axis is the **radicle**, which grows downward as root.
   The embryonic axis does not includes the cotyledon. The portion between the cotyledon attachment point and radicle is called **hypocotyl** and the portion between the cotyledon attachment point and the plumule is called the **epicotyl**.

7.2.2 Process of Seed Germination

Seed germination and dormancy are the aspect of seed quality which result in the growth, vigorousness, and yield of crops. The rate of seed germination vary from crop to crop also due to the abiotic factor.

A viable seed requires moisture, temperature, and oxygen to germinate. Moisture is the driving force that enables quick germination as it imbibes the seed and regulates temperature and oxygen. Sufficient moisture should be present for the seed to germinate. Research has revealed that limiting the moisture content in soil or its unavailability to seed restricts the seed to germinate.

Szczerba et al. (2021) reported that water has a significant role in the enzymatic activity of seeds and it also mobilizes the reserves stored in seeds including carbohydrates, lipids, and proteins. Water shortage inhibits the enzymes responsible for the seed to hydrolyze the endospermic starch (Muscolo et at. 2014). Also, an oxidative stress arises from the water stress by the seed as it increases the reactive oxygen species (ROSs) in the seed (Kovács et al. 2022, Kunos et al. 2022). As the embryo of the seed grows, the seed germination starts to sprout. However, germination of seed requires both internal and external conditions to be favorable (Figure 7.1).

a. **Imbibition**: Water is an important driving force in the process of seed germination. As the seed absorbs water to initiate germination, this process is termed as seed imbibition. The absorption of water produces a growth regulator gibberellic acid (GA_3); the GA3 stimulates the germination by two means, first by releasing the enzymatic enzymes, and second, by enhancing the growth potential of embryo (Kucera et al. 2005; Finch-Savage et al. 2006). During the imbibition process, the seed swells.
b. **Enzymatic and respiratory activities**: As the seed is rehydrated, the metabolic activities in the seed start. During the anaerobic respiration, the bacteria utilizes the oxygen from glycolysis to produce energy prior to the aerobic respiration. As the seed is mostly sown at surface, the macrospores provide oxygen to the seed. Also, the water activates the enzymes in the seed, allowing for the breakdown of stored food and produce energy.

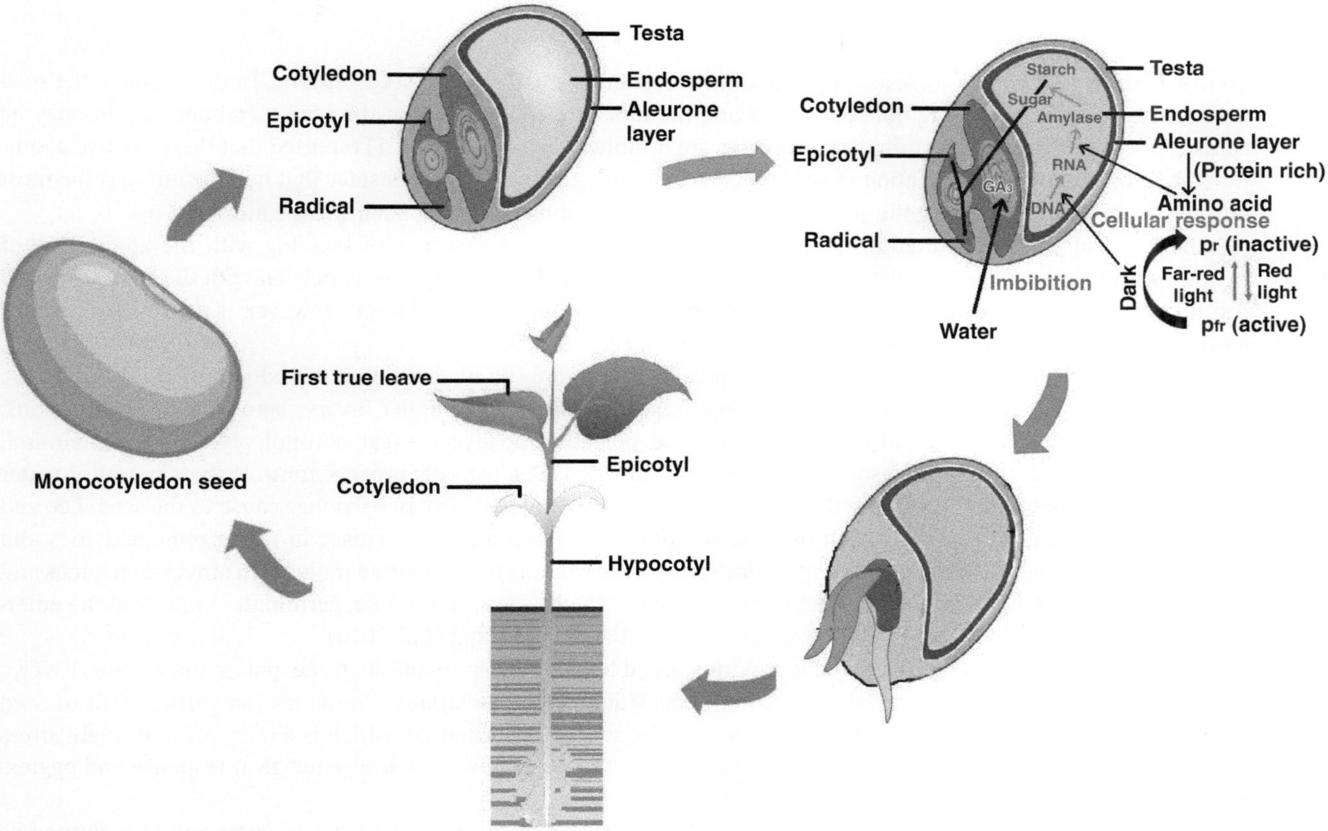

Figure 7.1. Germination cycle of monocotyledon seed.

c. **Impact of light on seed germination:** Seed germination to light varies greatly. The response of seed to light is referred to as photoblastic. The seed may be positive photoblastic, negative photoblastic, and nonphotoblastic. Photoblastic seeds do not germinate in darkness; they require exposure to light for germination. This exposure may be for a short period. Negative photoblastic seeds grow better under darkness and their growth is inhibited by longer exposure to sunlight. The nonphotoblastic seed may grow under exposure and absence of light.

d. **Phytochrome and red and far-red light impact on seed germination:** The phytochrome is the pigment that controls and regulates the development of light-dependent seed processes. This phytochrome pigment that absorbs the light exists in two inert convertible forms, i.e. P_r, which is metabolically inactive, and P_{fr}, which is active. The P_r absorbs the red light of wavelength 660 nm and converts into P_{fr}. The P_{fr} absorbs the far-red light of wavelength 730 nm and reverts back to P_r. The absorption of red light by the pigment stimulates seed germination, while on the other hand the far-red light exhibits an inhibitory effect. The exposure of seed to the quality of light is an important thing in the process of germination. The application of plant growth regulator, i.e. gibberellin and cytokinine, may replace the requirement of seed exposure to light.

e. **Mobilization of reserve food in the endosperm:** The reserve food present in the endosperm of the seed in the form of starch, protein, and fat has to be digested during the metabolic activity of the seed. The aleurone layer secretes certain enzymes, i.e. amylase and proteases, which causes the digestion of the reserved food in the endosperm. The gibberellic acid produced during the process of imbibition plays an important role in the hydrolyzing of the enzymes. The digested food is converted into sugars and amino acids, which are diluted by water, and it is passed to the embryonic parts, such as hypocotyl, epicotyl, plumule, and radicle, through the cotyledon. As the energy assimilates into these parts, the growing organs initiate growth and soon a shape is assumed.

Growth: After accumulation of energy by the seed, the cell in the growing region becomes more active and the growth is started. The stem starts to grow upward to the surface from the plumule and the root moves downward to the soil from the radicle. The seed keeps growing in this manner until it becomes a full-fledged plant.

7.3 Methods of Seed-priming

a. **Hydro-priming**: This is the oldest, simple, and most cost-effective method of seed-priming. In this method, the seed is soaked followed by redrying the seed to the original moisture content so that the seed metabolic activity may be started, but the subsequent germination processes are inhibited. Ermiş et al. (2021) reported that there are two conditions that influence the germination of seed, i.e. drought and salt stress. They consider that hydropriming is the most effective method for enhancing the seed quality and rate of germination under such stress conditions.

   Damalas et al. (2019) reported an increase in seed germination, emergence of seedling with the application of hydropriming methods used for faba bean. But the response to hydro-priming was associated with the environmental condition, especially when the soil moisture is limited after sowing of primed seed. However, if rainfall occurs after sowing, the beneficial effects of primed seed may be masked.

b. **Osmo-priming**: The use of hydro-priming has been considered to be nonamiable for seed germination due to the higher potential of pure water. So the hydro-priming may not be effective under diverse agro-ecological conditions. Osmo-priming refers to the use of osmotic solution, i.e. polyethylene glycol, sugar, sorbitol, glycerol, and mannitol. After treatment with the osmotic solution, the seed is air-dried. The osmosis process controls the amount of water absorbed during imbibition, lowering the ROS for protection against the oxidative damage cause to the seed. Lee and Kim (2000) reported that priming with osmotic solution PEG showed quick increase in the enzyme activities and enhanced the glucose and fructose content of rice seed. Other workers have reported increase in amylase enzymes and breakdown of the starch that has led to an increase in seed viability, respiration rate, germination rate, and the emergence of seedlings as compared to the unprimed seed (Li et al. 2018; Zhang et al. 2010).

c. **Halo-priming**: Halo-priming refers to the soaking of seed by different inorganic salts, i.e. potassium nitrate (KNO_3), sodium chloride (NaCl), calcium sulfate ($CaSO_4$), etc. This priming technique stimulates the germination of seed under saline stress conditions. The use of salts lowers the seed water potential, which is analogous to drought stress conditions (Patade et al. 2009). Also, due to halo-priming, the seed shows quick physiological response and aggressively responds to the imminent stress (Jisha et al. 2013).

d. **Hormonal-priming**: The use of growth regulator solution for soaking the seed prior to germination is termed as hormonal-priming. During the germination of seed, the gibberellic acid produced regulates and triggers the germination process. The use of artificial plant growth regulators, such as auxins, gibberellins, cytokinins, abscisic acid, and ethylene, plays a crucial role in regulating the germination process of seeds. Seed-priming with indole acetic acids has been found to be an effective tool in reducing the influence of different abiotic stresses (Akbari et al. 2007). Gibberellic acids have been also reported to protect seeds of different crops from abiotic stress when used as a seed-priming tool (Guangwu and Xuwen 2014)

e. **Bio-priming**: In bio-priming, the seed is inoculated with beneficial microorganisms to promote growth and development. It is used as an alternative method for controlling soil-borne disease. It is effective in improving stress tolerance and developing a resistance mechanism. This method is considered to be cost-effective and environment-friendly when compared with the chemical methods of seed-priming. Reddy (2013) reported that bio-priming with fungicides and biocontrol agent is an effective method for uniform seed emergence and reducing the disease. Shaffique et al. (2022) reported that the use of microbes as a biostimulant agent has played an important role in the growth of seeds under drought stress conditions.

f. **Nutri-priming**: Nutri-priming is one of the newest methods of seed-priming. In this method, the seed is treated with a solution of micro- and macronutrients. Nutri-priming has become an effective method to enhance seed germination, growth under stress condition, and seed vigor (Farooq et al. 2019). The nutri-priming has been most commonly used for micronutrients, i.e. copper, zinc, manganese, and molybdenum. Also, among the macronutrients, phosphorus is used as a seed primer. The effect of nutri-priming has been influencing the germination, seedling growth, and root system. In recent years, there has been an increased focus on nutri-priming due to its effectiveness in biofortifying crops that are deficient in essential nutrients.

7.3.1 Seed-priming as an Effective Method of Nutrient Biofortification

Biofortification is nutrient enrichment in grains through breeding and agronomic methods. This is the most cost-effective and simple method for increasing the concentration of nutrients that are deficient in crops. The breeding techniques of biofortification focus on the selection of potential varieties of greater nutrient content, and this trait may be transferred to the improved varieties by selective breeding or genetic modification and biotechnological techniques.

Rehman et al. (2018) and Ludwig and Slamet-Loedin (2019) reported that breeding approach used for attaining varieties with high-nutrient content required initial valuable genetic variation among the germplasm for accumulation of a specific nutrient. Also, there might be low heritability. Advanced genetic modification and biotechnological methods are considered the most effective approaches when specific nutrients are not produced in a crop. In such cases, transgenic methods may be the only option (Pérez-Massot et al. 2013). Among the agronomic biofortification methods, the cultural practices including application of soil and foliar fertilizers and biofertilizers are commonly used. The new and most effective methods of biofortification include the use of seed-priming technique.

Poudel et al. (2023) reported that the alternative agronomic biofortification by the use of soaking of seeds with plant nutrients is an easy, simple, and cost-effective technique. Biofortification through the seed treatment can be achieved by two methods. First, by seed coating in which a layer of finely ground solid or liquid is covered all over the seed, and second, through the seed-priming that involves soaking of seed in a nutrient solution for a specific time (Farooq et al. 2019). The seed nutri-priming helps in meeting the crop micronutrient and also in achieving the quantitative micronutrients of grains (Haider et al. 2020). Recently, nutri-priming with nanoparticles has proven to be the most effective approach to address zinc malnutrition due to a number of advantages and is gaining popularity (Veena and Pathur 2022).

Ikram et al. (2023) primed the tomato seed with iron sulfate (1–15 mg L^{-1}). The iron content of tomato showed increased Fe content of 2.04 and 2.37 ppm in the treatment where 10 mg Fe L^{-1} was applied over the control where the Fe content was measured to be 1.47 ppm. Also, the growth and yield traits were significantly enhanced. Treated the seed of barley and wheat with the solution containing nutrients, i.e. phosphorus, copper, and Zn in different concentrations for 10 hours. The seed was again washed with pure water and dried again to the initial moisture content. The results of different nutrient contents measured in the plants showed a 1.61, 10.5, and 13.6-fold increase in P, Cu, and Zn content, respectively.

Ali et al. (2018) carried out an experiment using the bread wheat seed treated with nutrient solution of zinc (Zn), boron (B), and manganese (Mn) alone, and in combination. The seed was soaked with 0.5 M Zn, 0.01 M B, and 0.1 M Mn for 12 hours and then aerated. The result of the study showed that the use of nutrient solution alone and in combination increased the nutrient content of respective elements in the wheat grain. It was further concluded that the nutrient seed-priming showed the highest benefit–cost ratio. The application of Zn + Mn was the most effective treatment in the experiment.

7.4 Conclusion

Seed is the basic input for agriculture crops. The quality of seed determines the growth of plant and ultimately contributes to the yield. Often during the adverse stress conditions, the germination of seed is affected. Seed-priming is the most effective seed treatment used for enhancing the seed germination, seed vigor, seedling emergence, growth, yield, and nutrient content of plant. Hence, it is important to understand the seed structure, morphology, and stages of germination. Seed-priming may be classified into different types; however, among the different priming methods, the nutri-priming, seed soaked with the macro- and micronutrient solution, is the most effective method, especially for biofortification of nutrients that are deficient in plants and grains.

References

Akbari, G., Sanavy, S.A. and Yousefzadeh, S. (2007). Effect of auxin and salt stress (NaCl) on seed germination of wheat cultivars (*Triticum aestivum* L.). *Pakistan Journal of Biological Sciences* 10 (15): 2557–2561.

Ali, N., Farooq, M., Hassan, M.A. et al. (2018). Micronutrient seed priming improves stand establishment, grain yield and biofortification of bread wheat. *Crop & Pasture Science* 69 (5): 479–487.

Brocklehurst, P.A. and Dearman, J. (1983). Interactions between seed priming treatments and nine seed lots of carrot, celery and onion. II. Seedling emergence and plant growth. *Annals of Applied Biology* 102 (3): 585–593.

Damalas, C.A., Koutroubas, S.D. and Fotiadis, S. (2019). Hydro-priming effects on seed germination and field performance of faba bean in spring sowing. *Agriculture* 9 (9): 201.

Ermiş, S., Öktem, G., Gökdaş, Z. et al. (2021). Effect of hydro-priming on seed germination and early seedling growth in three cucurbit rootstock cultivars under salt and osmotic stresses. *Journal of Agricultural Biotechnology* 2 (1): 1–5.

Evenari, M. (1984). Seed physiology: its history from antiquity to the beginning of the 20th century. *The Botanical Review* 50: 119–142.

Farooq, M., Usman, M., Nadeem, F. et al. (2019). Seed priming in field crops: potential benefits, adoption and challenges. *Crop and Pasture Science* 70 (9): 731–771.

Farooq, M., Wahid, A., Kobayashi, N.S.M.A. et al. (2009). Plant drought stress: effects, mechanisms and management. *Sustainable Agriculture* 29: 153–188.

Finch-Savage, W.E. and Leubner-Metzger, G. (2006). Seed dormancy and the control of germination. *New Phytologist* 171 (3): 501–523.

Guangwu, Z. and Xuwen, J. (2014). Roles of gibberellin and auxin in promoting seed germination and seedling vigor in *Pinus massoniana*. *Forest Science* 60 (2): 367–373.

Haider, M.U., Hussain, M., Farooq, M. et al. (2020). Optimizing zinc seed priming for improving the growth, yield and grain biofortification of mungbean (*Vigna radiata* (L.) wilczek). *Journal of Plant Nutrition* 43 (10): 1438–1446.

Harris, D., Pathan, A.K., Gothkar, P. et al. (2001). On-farm seed priming: using participatory methods to revive and refine a key technology. *Agricultural Systems* 69 (1–2): 151–164.

Heydecker, W., Higgins, J. and Gulliver, R.L. (1973). Accelerated germination by osmotic seed treatment. *Nature* 246 (5427): 42–44.

Hussain, S., Khan, F., Hussain, H.A. et al. (2016). Physiological and biochemical mechanisms of seed priming-induced chilling tolerance in rice cultivars. *Frontiers in Plant Science* 7: 178194.

Ikram, N.A., Ghaffar, A., Khan, A.A. et al. (2023). Optimizing iron seed priming for enhanced yield and biofortification of tomato. *Journal of Plant Nutrition* 46 (12): 2796–2810.

Jisha, K.C., Vijayakumari, K. and Puthur, J.T. (2013). Seed priming for abiotic stress tolerance: an overview. *Acta Physiologiae Plantarum* 35 (5): 1381–1396.

Kovács, Z., Bedő, J., Pápai, B. et al. (2022). Ripening-induced changes in the nutraceutical compounds of differently coloured pepper (*Capsicum annuum* L.) breeding lines. *Antioxidants* 11 (4): 637.

Kucera, B., Cohn, M.A. and Leubner-Metzger, G. (2005). Plant hormone interactions during seed dormancy release and germination. *Seed Science Research* 15 (4): 281–307.

Kunos, V., Cséplő, M., Seress, D. et al. (2022). The stimulation of superoxide dismutase enzyme activity and its relation with the *Pyrenophora teres f. teres* infection in different barley genotypes. *Sustainability* 14 (5): 2597.

Lee, S.S. and Kim, J.H. (2000). Total sugars, -amylase activity, and germination after priming of normal and aged rice seeds. *Korean Journal of Crop Science* 45 (2): 108–111.

Li, Z., Song, Z., Yan, Z. et al. (2018). Silicon enhancement of estimated plant biomass carbon accumulation under abiotic and biotic stresses. A meta-analysis. *Agronomy for Sustainable Development* 38: 1–19.

Ludwig, Y. and Slamet-Loedin, I.H. (2019). Genetic biofortification to enrich rice and wheat grain iron: from genes to product. *Frontiers in Plant Science* 10: 450726.

Lutts, S., Benincasa, P., Wojtyla, L. et al. (2016). Seed priming: new comprehensive approaches for an old empirical technique. *New Challenges in Seed Biology-Basic and Translational Research Driving Seed Technology* 46 (10.5772): 64420.

Marthandan, V., Geetha, R., Kumutha, K. et al. (2020). Seed priming: a feasible strategy to enhance drought tolerance in crop plants. *International Journal of Molecular Sciences* 21 (21): 8258.

Muscolo, A., Sidari, M., Anastasi, U. et al. (2014). Effect of PEG-induced drought stress on seed germination of four lentil genotypes. *Journal of Plant Interactions* 9 (1): 354–363.

Paparella, S., Araújo, S.S., Rossi, G. et al. (2015). Seed priming: state of the art and new perspectives. *Plant Cell Reports* 34 (8): 1281–1293.

Patade, V.Y., Bhargava, S. and Suprasanna, P. (2009). Halopriming imparts tolerance to salt and PEG induced drought stress in sugarcane. *Agriculture, Ecosystems & Environment* 134 (1–2): 24–28.

Paul, S., Dey, S. and Kundu, R. (2022). Seed priming: an emerging tool towards sustainable agriculture. *Plant Growth Regulation* 97 (2):215–234.

Pérez-Massot, E., Banakar, R., Gómez-Galera, S. et al. (2013). The contribution of transgenic plants to better health through improved nutrition: opportunities and constraints. *Genes & Nutrition* 8 (1): 29–41.

Poudel, P., Di Gioia, F., Lambert, J.D. et al. (2023). Zinc biofortification through seed nutri-priming using alternative zinc sources and concentration levels in pea and sunflower microgreens. *Frontiers in Plant Science* 14: 1177844.

Reddy, P.P. (2013). Bio-priming of seeds. In: *Recent Advances in Crop Protection*, 83–90. New Delhi: Springer. https://doi.org/10.1007/978-81-322-0723-8_6.

Rehman, A., Farooq, M., Ozturk, L. et al. (2018). Zinc nutrition in wheat-based cropping systems. *Plant and Soil* 422: 283–315.

Sachs, J. (1859). About treatments modifying germination strength in seeds *Botanical Zeitung* 17: 177–188.

Shaffique, S., Khan, M.A., Imran, M. et al. (2022). Research progress in the field of microbial mitigation of drought stress in plants. *Frontiers in Plant Science* 13: 870626.

Szczerba, A., Płażek, A., Pastuszak, J. et al. (2021). Effect of low temperature on germination, growth, and seed yield of four soybean (*Glycine max* L.) cultivars. *Agronomy* 11 (4): 800.

Veena, M. and Puthur, J.T. (2022). Seed nutripriming with zinc is an apt tool to alleviate malnutrition. *Environmental Geochemistry and Health* 44 (8): 2355–2373.

Waterworth, W.M., Bray, C.M. and West, C.E. (2019). Seeds and the art of genome maintenance. *Frontiers in Plant Science* 10: 458877.

Zhang, H., Dou, W., Jiang, C.X. et al. (2010). Hydrogen sulfide stimulates β-amylase activity during early stages of wheat grain germination. *Plant Signaling & Behavior* 5 (8): 1031–1033.

8

Selenium Biofortification in Wheat: A Way Forward Toward Nutritional Security

Muhammad Shoaib Ismail[1,2,*], *Fahim Nawaz*[3], *Muhammad Asif Shehzad*[4], *Waseem Hassan*[5], *Rukhsar Saleem*[6], *Adnan Noor Shah*[7], *Muhammad Baqir Hussain*[5]

[1] *Department of Agronomy, MNS University of Agriculture, Multan, Pakistan*
[2] *Institute of Plant Nutrition and Soil Science, Kiel University, Kiel, Germany*
[3] *Research School of Biology, The Australian National University, Canberra, Australia*
[4] *Institute of Plant Breeding and Biotechnology, MNS University of Agriculture, Multan, Pakistan*
[5] *Department of Soil and Environmental Sciences, MNS University of Agriculture, Multan, Pakistan*
[6] *Department of Agronomy, University of Agriculture, Faisalabad, Pakistan*
[7] *Department of Agricultural Engineering, Khwaja Fareed University of Engineering and Information Technology, Rahim Yar Khan, Pakistan*

*Corresponding author: shoaibismail007@gmail.com

8.1 Introduction

Selenium (Se) is considered as an essential mineral element, and its optimum concentration has several health impacts on human health (Schwarz and Foltz 1957). Its presence is confirmed in several kingdoms, including cyanobacteria, humans, and some plants (Germ and Stibilj 2007), as it prevents plants from several kinds of stress, thereby promoting growth and development (Pennanen et al. 2002). The distribution of Se throughout the globe varies, as it is present in the lithosphere layer of Earth's crust as well as in the open environment, soil, and water (Hasanuzzaman et al. 2020). Soil type is the factor that describes the concentration of Se by nature (Yasin et al. 2015); however, the majority of soil falls in the range of 0.01 and 2.0 mg Se kg^{-1} of soil with the global mean of 0.4 mg Se kg^{-1} of soil (Shahid et al. 2018). In Pakistan, soil Se concentration is unknown in many places due to very little work on it; however, the soils of southern Punjab are considered as Se deficient with the soil Se status of 0.05–0.07 mg kg^{-1} (Khan et al. 2008). The existence of Se in soil is in various forms including selenate (Se^{6+}), selenite (Se^{4+}), elemental selenium (Se^0), selenide (H_2Se), and organic selenide, and among these forms most available is selenate followed by selenite and organic selenide, respectively (Zhao et al. 2005).

8.2 Essentiality of Selenium for Animal Health

Globally, it is estimated that around one billion people are deficient in Se intake in their diet (Zhang et al. 2019a). According to World Health Organization (WHO 2004), the recommended daily intake of Se is 50–55 μg day^{-1}. The deficiency of Se in the daily diet led to more than 40 types of health issues and diseases, and among these diseases, top of the list are Kashin–Beck and Keshan disease, while other health problems include liver dysfunction, cancer, cataracts, cardiovascular issues, and many more (Feng et al. 2013). In various parts of the world, Se is deficient in the daily diets, especially in those areas where the soil Se levels are below average causing lower assimilation in the plants of edible species (Schiavon et al. 2016). Since humans and other organisms cannot synthesize Se autonomously, Se intake can be usually through the consumption of Se-enriched food, which is alternately achieved through value addition and other strategies like biofortification to make the staple foods rich in Se (Wang et al. 2021; Zhou et al. 2019).

Selenium (Se) is a vital component in the building blocks of biological life. Many living forms, especially humans, require minute quantities of selenium as it is an essential component of the amino acid selenocysteine (SeCys). SeCys is sometimes referred to be the 21st proteinogenic amino acid due to its scarcity. The majority of the species that require Se also contain

Crop Biofortification: Biotechnological Approaches for Achieving Nutritional Security under Changing Climate, First Edition. Edited by Adnan Noor Shah et al.

proteins between 1 and 30 with residue of SeCys on the active sites. Likewise, humans also have 25 selenoproteins which are involved in several health functions, and their deficiency causes severe concerns. Glutathione peroxidases (GPx) have the properties of scavenging free radicles causing negative effects. Due to this nature of Se, selenoproteins are involved in various kinds of health benefits, including thyroid protection, improved immunological function, proper spermatogenesis, and gastrointestinal functioning (Rayman 2012). Selenium insufficiency, like other micronutrient deficits, is subtle. It has been related to an elevated likelihood of infections (such as COVID-19 and HIV), malignancies (particularly colorectal), and other disorders. Many researchers have reported that selenoprotein phospholipid hydroperoxide glutathione peroxidase (PHGPx) is directly involved in sperm motility and hence improves male functions. Moreover, selenium shortage can resemble iodine deficiency, resulting in hypothyroidism. Se and other deficiency states of micronutrients are known as "hidden hunger" due to their cryptic yet severe effects.

The contents of Se in various foods vary according to the nature of Se assimilation. Organ meats and seafood include 0.4–1.5 μg Se g^{-1}, muscle meats contain 0.1–0.4 μg Se g^{-1}, most agricultural crops have <1 μg Se g^{-1} dry weight and cereals (grains) have <0.1–0.8 μg Se g^{-1}, dairy products have 0.1–0.3 μg Se g^{-1}, and fruits and vegetables have <0.1 μg Se g^{-1} (Rayman 2008). In case of humans, the absorption of Se from plant-based food is easier than that from animal-derived items. That's why researchers have their main concern with investigating the Se speciation in plant-derived Se fortified foods (D'Amato et al. 2020). Enhancing Se uptake and assimilation by various food crops during growth through biofortification can provide additional Se supplementation in the human diet (Newman et al. 2019). Se concentrations in foods, such as rice and wheat grain, can vary substantially by country and area. Therefore, to steer clear the toxicity and deficiency of Se, it is much important to optimize and check the amount of Se accumulated in various food crops (Zhu et al. 2009).

8.3 Role of Selenium in Plant Growth and Development

When applied in optimum concentration, Se boosts the plant growth and physiological functions and also acts as an anti-senescent agent. This helps maintain its cellular constituents, thus improving the overall function, growth, and development of plants (Hasanuzzaman et al. 2020). Plants uptake Se in the form of selenate and latterly translocate it into the chloroplast, where it follows the sulfur (S) assimilation pathway and further reduced to other forms (Finley 2005). Although the essentiality of Se in lower plants has been confirmed, it is still unclear for higher plants (El-Ramady et al. 2020). Se has multiple physiological functions in plants, especially the tendency to boost antioxidative metabolites and enzymes that have the functions of scavenging free radicles and reactive oxygen species (ROS) ultimately improving plant growth even under any kind of stress (Pilon-Smits 2019).

In some cases, Se accumulation in plants may act as a protective agent against several pathogens, which are Se-sensitive, thus preventing them from attacking such plants and also has allelopathic effects on plant community (Galić et al. 2021). Se assimilation in plants was also found to be beneficial in regulating stomatal regulations by maintaining the levels of K^+ in plants, which also increases chlorophyll a and b and total chlorophyll contents up to 65% as the uptake and assimilation of Se was linked with the increase in uptake of other minerals like Fe and Mg, which are key components in chlorophyll structure (Hasanuzzaman et al. 2020). Studies revealed multiple positive physiological alterations in plants by Se applications including the inhibition of uptake and translocation of heavy metals (HMs), changes in the speciation of HMs, regulation of the antioxidants and ROS activities, recovery of photosynthetic system under stress, and enhancement of chloroplast structure and cell membrane stability (Feng et al. 2013). Addition of Se to the growing media or direct application of Se on plants as foliar spray helps in increasing the net photosynthesis rate, transpiration rate, and gaseous exchange rate in several plants (Hasanuzzaman et al. 2020).

Variation in response of different plants to Se application has been observed in terms of growth and development, uptake, and assimilation of other minerals that interact with the uptake of Se in plants (El-Ramady et al. 2015). Se at low levels resulted in improved plant growth and development. It is revealed that Se application at 5 μM resulted in enhanced relative water contents by 13% compared with the non-Se-treated plants (Shekari et al. 2016). Similarly, the leaf area was observed to increase by 25% under both 3 and 5 μM Se application which in the end contributes to the overall growth and total biomass yield of plant (Chauhan et al.2017). Se also acts as an anti-senescent agent, which maintains the cellular structure and helps in improving the growth and development of plants (Kaur et al. 2014). Soil Se application is a convenient method of Se delivery to the plants, which helps maintain physiological functions and production of plants (Wang, Yu, and Li 2017; Hemmati et al. 2019). Moreover, Se application to the plants directly or in growth media also helps in the uptake and translocation of other minerals which are required by plants' internal functions and thus contribute to the growth and development

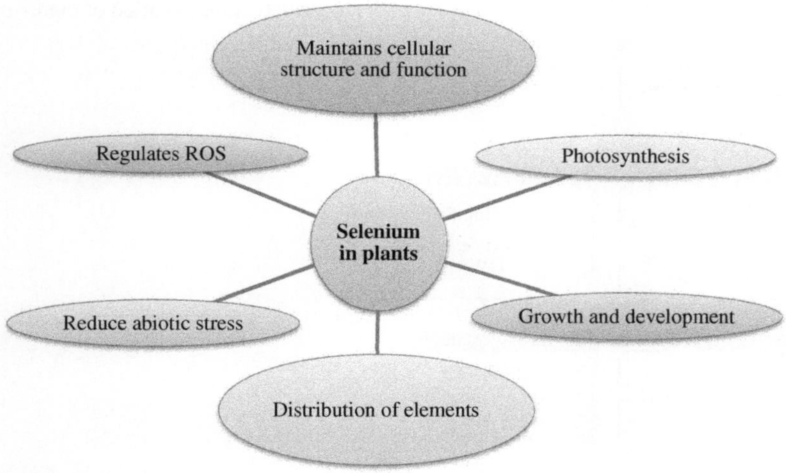

Figure 8.1 Role of Se in plant physiology.

of plants (Gupta and Gupta 2017). Se application has also been reported to improve the nutritional value of wheat in terms of mineral contents and protein fractions (Ismail et al. 2024). One of the main reasons behind the improved growth and performance of plants even under stressful circumstances is the key role of Se as a protective agent, which acts as an antioxidant, causing the plants to grow better under production of ROS in stress conditions. The Se-induced increase in the activities of the antioxidant defense system improves the plant's physiological status, which exerts favorable effects on the yield attributes of crops (Hasanuzzaman et al. 2020). A significant increment in the osmoprotectant accumulation in wheat and enhanced activity of antioxidant enzymes activity were observed under exogenous Se supply, which ultimately improves the wheat production and grain quality even under water stress conditions (Figure 8.1; Nawaz et al. 2015).

8.4 Uptake, Absorption, and Assimilation Dynamics of Selenium in Plants

Many variables influence plant Se absorption. The method of treatment and the form of Se are the two most critical aspects in increasing Se content in grains, while systematic research on these topics is still lacking (Deng et al. 2017). Plants absorb Se more rapidly when administered foliarly compared to soil treatment, with the added benefit of minimal lingering effects in the soil (Pezzarossa et al. 2012). The foliar approach uses the least amount of Se salts and is the most effective, safe, and economically justifiable method of increasing Se content in agricultural crops (Kápolna et al. 2009). Foliar treatment is more efficient since there is no root-to-shoot translocation, which is supported by the discovery that soils can function as significant Se sinks (Winkel et al. 2015). The type of applied selenium is the second most critical aspect for efficient agronomic biofortification (García Márquez et al. 2020), with selenate and selenite being the most widely employed. However, nanoselenium (nSe) is becoming increasingly popular (El-Ramady et al. 2020). The production and use of nSe as a nutrient and biofortifier have been demonstrated as an intriguing technique (El-Ramady et al. 2020), while reports of widespread use are still lacking. Experiments have revealed that selenate (Se(VI)) is the most efficient form of Se when applied to soil and is typically more effective than selenite (Se(IV)) when applied foliarly (Lyons 2018). Selenate is more efficient because of the faster absorption of Se(VI), translocation from root to stem and leaves, and faster transformation into its organic form, whereas selenite is more easily adsorbed to the soil, making root uptake more difficult (Izydorczyk et al. 2021). Selenite possesses phosphate-like properties, which explains its ease of adsorption to soil surfaces as compared to selenite.

Se and sulfur (S) rival for identical carriers, and high S levels often inhibit Se absorption (Malagoli et al. 2015). The high-affinity H^+/sulfate symporters, homologous to AtSULTR1;1 and AtSULTR1;2 of the model Brassicaceae plant *Arabidopsis thaliana* L., catalyze selenate absorption across the plasma membrane of root cells (White 2018). In contrast, selenite uptake occurs via passive diffusion. Furthermore, a metabolic inhibitor was shown to dramatically decrease the absorption of selenite, indicating that it is mediated by active transport (Gupta and Gupta 2017). Plants cannot directly absorb metallic selenide and elemental Se because these forms of Se are water-insoluble, but organo-Se compounds, such as seleno-amino

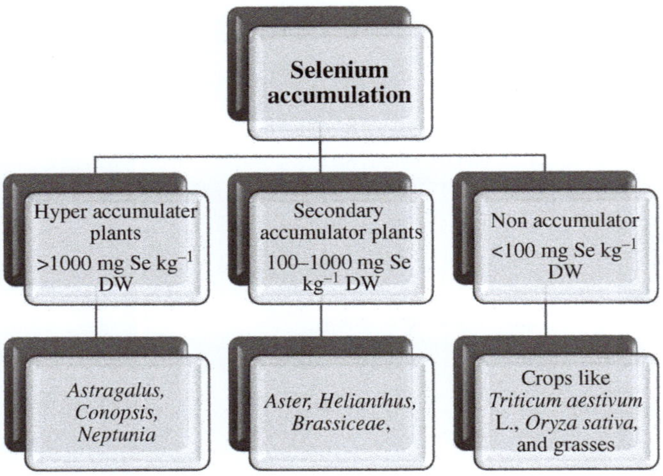

Figure 8.2 Classification of plants based on Se accumulation.

acids, have substantially better phytoavailability. The two Se-amino acids generated in the S assimilation route are SeCys and selenomethionine (SeMet), analogues of the S-amino acids cysteine (Cys) and methionine (Met; Schiavon et al. 2017). SeCys and SeMet are both absorbed at rates up to 20 times greater than selenate or selenite (Zhou et al. 2020). SeMet is one of the most efficiently stored Se species in several tissues (Pedrero and Madrid 2009). The distribution of Se in different regions of the plant varies depending on the species, stage of development, and physiological state (Terry et al. 2000). Higher plants have varying capacities to accumulate and tolerate Se; hence, they are classed as primary, secondary, and non-accumulator plants (Figure 8.2).

8.5 Se Biofortification in Wheat

Greek word "bios" means "life" and Latin word "fortificare" means "make strong," which means enrichment of food through the addition of micronutrients to crops. The progression of enhancing the bioavailable concentration of vital mineral elements in the seeds or edible portion of crops through agricultural intervention or selection through some breeding techniques is known as biofortification (Ríos et al. 2008). Se biofortification markedly enhances Se concentration in the edible portion of food crops, especially cereals including wheat can be a suitable option to cope with the global issue of mineral malnutrition, which is affecting approximately one billion people around the globe (Gupta and Gupta 2017). Different approaches are involved in improving the Se deficiency scenario worldwide which includes food fortification in the industries or milling factories, supplementation of required mineral element, and fertilization of required minerals to the specific crop (agronomic biofortification) or crop improvement or modification (genetic biofortification; Broadley, Alcock, and Alford 2010). Globally, cereals, and especially wheat crops, are grown on a large scale and consumed by most of the population, thus providing the main source of energy worldwide (Sarwar et al. 2013) and considered as the most appropriate food crop for agronomic biofortification of desired mineral element to incorporate that mineral and additionally to achieve better agronomic characters and yield (Bocchini et al. 2018). However, variations can be observed between various wheat varieties in terms of Se accumulation in grains due to their genetic makeup (Abenavoli et al. 2020; Ismail et al. 2024).

Cereals including wheat and rice are grown in larger areas, providing energy to the majority of population compared with any other crop; therefore, these are considered staple food crops (Sarwar et al. 2013). Cereals are the most important source of dietary Se, with major representatives being wheat, rice, and maize (Narwal et al. 2020), and Se-enriched wheat is considered as the best source of dietary Se intake for the population living in Se-deficit areas of the world (Finley 2005).

Many studies reported the positive impact of agronomic biofortification of Se in wheat, rice, barley, and maize grains (Bocchini et al. 2018; Abenavoli et al. 2020; Wang et al. 2020; Longchamp et al. 2015; Gupta, Winter, and Sanderson 1993). The average of Se in crop plants falls in 0.01–1.0 mg Se kg^{-1} (Läuchli 1993), while the majority of the agronomic crops have Se value of 0.001–30 mg Se kg^{-1} (Ros et al. 2016) and most of the wheat have the Se concentration of 0.020–0.600 mg Se kg^{-1} (Lyons et al. 2018). The recommended Se concentration in wheat grains either used as fodder or feed is 0.2–0.3 mg Se kg^{-1} and 01–0.2 mg Se kg^{-1}, respectively (Eich-Greatorex et al. 2007). The efficiency of cereal crops in terms of Se accumulation through agronomic biofortification is as wheat > rice > maize > barley > oats (Poblaciones et al. 2014); therefore, wheat crop agronomic biofortification with Se fertilizers is a better option to increase the Se intake through dietary intake in humans (Manojlovi´c et al. 2019). Adequate amounts of Se in wheat can also have a positive impact on productivity (Lara et al. 2019) and the yield of other crops, such as barley, rice, and maize (Bocchini et al. 2018). Overall, type of application, form of Se, and time of application play a critical role in achieving prosperous biofortification in cereal crops.

8.6 Factors Affecting Se Uptake and Absorption

Agronomic biofortification refers to the fertilizer application of desired mineral elements, which we need to incorporate in sufficient amounts in specific food crops like wheat through either foliar or soil application of that mineral (Ngigi et al. 2019). Se-containing fertilizer like sodium selenate application in both ways, i.e. soil and foliar application for agronomic biofortification in wheat, is the most appropriate short-term technique to increase Se concentration in wheat grains so that the Se can be added to daily die of humans (Wang et al. 2017). Compared to the foliar spray of Se, soil Se application results in multiple factors that involve less absorption by wheat including the colloidal formation with soil particles, thus lowering the availability of Se for wheat plants (Newman et al. 2019). Various factors are involved in the availability of Se to crops including Se fertilization method and type of Se source, irrigation scheduling, soil aeration, and soil physiochemical properties (Lyons et al.2003). Overall, the Se concentration in wheat grains is higher in regions of arid climatic conditions than those areas of humid environments (El-Ramady et al. 2015). Se uptake and assimilation in wheat are also affected by many factors like soil Se concentration, soil mineral status, soil type, and plant growth rate along with Se application methods, its quantity, and the presence of other anions, which can compete with Se for uptake and absorption especially sulfate ions (Stephen, Saville, and Watkinson 1989).

The mineral nutrition theory suggests that grain Se concentration varies due to differences in transport, absorption, and assimilation mechanisms, which are mainly the genotypic characteristics of various species (Zhang et al. 2019b). Therefore, plant breeding techniques for improved genetics in terms of Se assimilation in grains can be much effective strategy in this regard. Several findings revealed that the breeding of those varieties having the ability to accumulate more Se in their grains portion can be a suitable option for future Se biofortification strategies in food crops (Lyons et al., Stangoulis et al., and Graham 2003); however, the literature showed that genetic biofortification is good for enhancing the concentration of vitamin A, carotenoids, and iron (Fe) while in terms of iodine (I) and Se, agronomic biofortification is more appropriate (Sarwar et al. 2020). For agronomic biofortification, Se foliar spray and soil application were adopted in addition to Se seed-priming to enhance the Se contents in the grain parts (Sarwar et al. 2020). Various Se sources can be utilized for successful Se biofortification in wheat and other cereals, including selenate at top of the list, followed by selenite, slow-released Se fertilizer, amino acid–chelated Se, Se-enriched yeast, and nano-Se application (El-Ramady et al. 2016). Among these, nano-Se application is a considerably new methodology for enriching food crops with Se and is also considered to have many benefits due to its unique characteristics, including low toxicity, large surface area, high bioavailability, more biological activity, and high particle dispersion (Kumar and Prasad 2021). They represent a promising alternative to other forms of Se, where a reduction in application complexity may be achieved, and this leads to important results in the potentiation of antioxidant metabolism, the promotion of agronomic sustainability, and a reduction in waste (Table 8.1; García Márquez et al. 2020).

8.7 Conclusion and Future Prospects

Selenium is a vital mineral element for both plants and animals. In higher plants, the essentiality of Se is still unclear; however, in stress circumstances, positive impact of Se application is well reported. Se is a crucial mineral element for humans as it plays an important role in normal functioning of body, including gastrointestinal tract, thyroid glands, kidneys, and lung functioning and most importantly considered as anti-cancerous when taken up in optimum concentration. However, almost one billion people are facing Se deficiency due to insufficient intake of Se-rich foods, giving rise to the issue of nutritional security worldwide, especially in developing countries. Several strategies can overcome the Se deficiency in humans, including consumption of Se-rich food, supplementation, and using nutrient dense crops, but all these are costly and poor people of developing countries cannot afford it. Cereals including wheat is an important source of energy for most of the people around the globe as they consume it as a staple food crop. Therefore, enriching the wheat crop

Table 8.1 Summary of various studies on Se biofortification in wheat crop.

Type of experiment	Selenium source	Application method	Time of application	Dose	Se concentration in grains	Reference
Field trial	Sodium selenate (Na_2SeO_4)	Foliar spray	Tillering stage	0 g ha^{-1}	66.6 μg g^{-1}	Poblaciones et al. (2014)
				10 g ha^{-1}	153.6 μg g^{-1}	
				20 g ha^{-1}	254.8 μg g^{-1}	
				40 g ha^{-1}	430.4 μg g^{-1}	
	Sodium selenite (Na_2SeO_3)	Foliar spray	Tillering stage	0 g ha^{-1}	66.6 μg g^{-1}	
				10 g ha^{-1}	266.8 μg g^{-1}	
				20 g ha^{-1}	820 μg g^{-1}	
				40 g ha^{-1}	1383 μg g^{-1}	
Field trial	Sodium selenate (Na_2SeO_4)	Foliar spray	Second node on main stem stage	0 g ha^{-1}	32 μg g^{-1}	Ducsay et al. (2016)
				10 g ha^{-1}	51 μg g^{-1}	
				20 g ha^{-1}	72 μg g^{-1}	
	Sodium selenite (Na_2SeO_3)	Foliar spray	Second node on main stem stage	0 g ha^{-1}	32 μg g^{-1}	
				10 g ha^{-1}	190 μg g^{-1}	
				20 g ha^{-1}	350 μg g^{-1}	
Field trial	Sodium selenate (Na_2SeO_4)	Foliar spray	Pre-flowering	0 g ha^{-1}	120 μg g^{-1}	Wang et al. (2020)
				20 g ha^{-1}	1340 μg g^{-1}	
	Sodium selenite (Na_2SeO_3)			0 g ha^{-1}	120 μg g^{-1}	
				20 g ha^{-1}	610 μg g^{-1}	
Field trial	Sodium selenate (Na_2SeO_4)	Foliar spray	Pre-grain filling	0 g ha^{-1}	120 μg g^{-1}	
				20 g ha^{-1}	1590	
	Sodium selenite (Na_2SeO_3)			0 g ha^{-1}	120 μg g^{-1}	
				20 g ha^{-1}	970 μg g^{-1}	
Filed trial	Sodium selenate (Na_2SeO_4)	Foliar spray	Between growth stages GS31 and GS49	0 g ha^{-1}	150 μg g^{-1}	de Vita, Platani, and Fragasso (2017)
				5 g ha^{-1}	640 μg g^{-1}	
				25 g ha^{-1}	2390 μg g^{-1}	
				50 g ha^{-1}	2820 μg g^{-1}	
				80 g ha^{-1}	3930 μg g^{-1}	
Field trial	Sodium selenate (Na_2SeO_4)	Foliar spray	Mid-heading to the onset of flowering	0 g ha^{-1}	0.09 mg kg^{-1}	Manojlović et al. (2019)
				5 g ha^{-1}	0.2 mg kg^{-1}	
				10 g ha^{-1}	0.4 mg kg^{-1}	
		Fertigation	Mid-heading to the onset of flowering	10 g ha^{-1}	0.36 mg kg^{-1}	
Field trial	Sodium selenate (Na_2SeO_4)	Foliar spray	Tillering	0 g ha^{-1}	155 μg kg^{-1}	Radawiec, Szulc, and Rutkowska (2021)
				5 g ha^{-1}	253 μg kg^{-1}	
			Stem elongation	5 g ha^{-1}	294 μg kg^{-1}	
			Inflorescence emergence	5 g ha^{-1}	267 μg kg^{-1}	
			Ripening	5 g ha^{-1}	234 μg kg^{-1}	
		Soil application	Before sowing	0 g ha^{-1}	155 μg kg^{-1}	
				5 g ha^{-1}	368 μg kg^{-1}	

(*Continued*)

Table 8.1 Summary of various studies on Se biofortification in wheat crop. *(continued)*

Type of experiment	Selenium source	Application method	Time of application	Dose	Se concentration in grains	Reference
Pot trial	Sodium selenate (Na_2SeO_4)	Soil application	Before sowing	0 mg kg^{-1}	0.13 μg kg^{-1}	Wang et al. (2021)
				1 mg kg^{-1}	34.7 μg kg^{-1}	
				10 mg kg^{-1}	218.82 μg kg^{-1}	
	Sodium selenite (Na_2SeO_3)		Before sowing	0 mg kg^{-1}	0.13 μg kg^{-1}	
				1 mg kg^{-1}	1.94 μg kg^{-1}	
				10 mg kg^{-1}	25.11 μg kg^{-1}	
Pot trial	Sodium selenate (Na_2SeO_4)	Foliar spray	Grain filling	0 mg Se L^{-1}	8.4 μg kg^{-1}	Ismail et al. (2024)
				25 mg Se L^{-1}	36.6 μg kg^{-1}	
				50 mg Se L^{-1}	70.88 μg kg^{-1}	
				75 mg Se L^{-1}	112.6 μg kg^{-1}	
				100 mg Se L^{-1}	158 μg kg^{-1}	

with optimum levels of Se through agronomic techniques like application of Se-enriched fertilizer is necessary to cope with the Se deficiency problem in humans and global issue of nutritional security. Several literature has been published regarding the agronomic biofortification of Se on wheat; however, advanced genetic research to develop high-selenium wheat varieties utilizes CRISPR and other modern biotechnological tools. Additionally, combining agronomic and genetic strategies for more effective Se biofortification should be adopted worldwide.

References

Abenavoli, L., Milanovic, M., Procopio, A.C. et al. (2020). Ancient wheats: beneficial effects on insulin resistance. *Minerva Medica* 112 (5): 641–650.

Bocchini, M., D'Amato, R., Ciancaleoni, S. et al. (2018). Soil selenium (Se) biofortification changes the physiological, biochemical and epigenetic responses to water stress in *Zea mays* L. by inducing a higher drought tolerance. *Frontiers in Plant Science* 9: 329575.

Broadley, M.R., Alcock, J. and Alford, J. (2010). Selenium biofortification of high-yielding winter wheat (*Triticum aestivum* L.) by liquid or granular Se fertilisation. *Plant and Soil* 332: 5–18.

Chauhan, R., Awasthi, S. and Tripathi, P. (2017). Selenite modulates the level of phenolics and nutrient element to alleviate the toxicity of arsenite in rice (*Oryza sativa* L.). *Ecotoxicology and Environmental Safety* 138: 47–55.

D'Amato, R., Regni, L., Falcinelli, B. et al. (2020). Current knowledge on selenium biofortification to improve the nutraceutical profile of food: a comprehensive review. *Journal of Agricultural and Food Chemistry* 68 (14): 4075–4097.

Deng, X., Liu, K., Li, M. et al. (2017). Difference of selenium uptake and distribution in the plant and selenium form in the grains of rice with foliar spray of selenite or selenate at different stages. *Field Crops Research* 211: 165–171.

Ducsay, L.; Ložek, O.; Mar˘cek, M. et al. (2016). Possibility of selenium biofortification of winter wheat grain. *Plant, Soil and Environment* 62: 379–383.

Eich-Greatorex, S., Sogn, T.A., Øgaard, A.F. et al. (2007). Plant availability of inorganic and organic selenium fertiliser as influenced by soil organic matter content and pH. *Nutrient Cycling in Agroecosystems* 79: 221–231.

El-Ramady, H., Abdalla, N., Alshaal, T. et al. (2015). Selenium and its role in higher plants. Pollutants in buildings. *Water and Living Organisms* 235–296.

El-Ramady, H., Abdalla, N., Taha, H.S. et al. (2016). Selenium and nano-selenium in plant nutrition. *Environmental Chemistry Letters* 14: 123–147.

El-Ramady, H., Faizy, S.E.D., Abdalla, N. et al. (2020). Selenium and nano-selenium biofortification for human health: opportunities and challenges. *Soil Systems* 4 (3): 57.

Feng, R., Wei, C. and Tu, S. (2013). The roles of selenium in protecting plants against abiotic stresses. *Environmental and Experimental Botany* 87: 58–68.

Finley, J.W. (2005). Selenium accumulation in plant foods. *Nutrition Reviews* 63 (6): 196–202.

Galić, L., Vinković, T., Ravnjak, B. et al. (2021). Agronomic biofortification of significant cereal crops with selenium—a review. *Agronomy* 11 (5): 1015.

García Márquez, V., Morelos Moreno, Á., Benavides Mendoza, A. et al. (2020). Ionic selenium and nanoselenium as biofortifiers and stimulators of plant metabolism. *Agronomy* 10 (9): 1399.

Germ, M. and Stibilj, V. (2007). Selenium and plants. *Acta Agriculturae Slovenica* 89 (1): 65–71.

Gupta, M. and Gupta, S. (2017). An overview of selenium uptake, metabolism, and toxicity in plants. *Frontiers in plant science* 7: 234638.

Gupta, U.C., Winter, K.A. and Sanderson, J.B. (1993). Selenium content of barley as influenced by selenite and selenate-enriched fertilizers. *Communications in Soil Science and Plant Analysis* 24 (11–12): 1165–1170.

Hasanuzzaman, M., Bhuyan, M.B., Raza, A. et al. (2020). Selenium in plants: Boon or bane? *Environmental and Experimental Botany* 178: 104170.

Hemmati, M., Delkhosh, B., Rad, A.H. S. et al. (2019). Effect of the application of foliar selenium on canola cultivars as influenced by different irrigation regimes. *Journal of Agricultural Sciences* 25 (3): 309–318.

Ismail, M.S., Nawaz, F., Shehzad, M.A. et al. (2024). Selenium biofortification impacts nutritional composition and storage proteins in wheat grains. *Journal of Food Composition and Analysis* 127: 105961.

Izydorczyk, G., Ligas, B., Mikula, K. et al. (2021). Biofortification of edible plants with selenium and iodine—a systematic literature review. *Science of The Total Environment* 754: 141983.

Kápolna, E., Hillestrøm, P.R., Laursen, K.H. et al. (2009). Effect of foliar application of selenium on its uptake and speciation in carrot. *Food Chemistry* 115 (4): 1357–1363.

Kaur, N., Sharma, S., Kaur, S. et al. (2014). Selenium in agriculture: a nutrient or contaminant for crops?. *Archives of Agronomy and Soil Science* 60 (12): 1593–1624.

Khan, Z.I., Ashraf, M., Danish, M. et al. (2008). Assessment of selenium content in pasture and ewes in Punjab. *Pakistan Journal of Botany* 40 (3): 1159–1162.

Kumar, A. and Prasad, K.S. (2021). Role of nano-selenium in health and environment. *Journal of Biotechnology* 325: 152–163.

Lara, T.S, de Lima Lessa, J.H, De Souz, K.R.D. et al. (2019). Selenium biofortification of wheat grain via foliar application and its effect on plant metabolism. *Journal of Food Composition and Analysis* 81: 10–18.

Läuchli, A. (1993). Selenium in plants: uptake, functions, and environmental toxicity. *Botanica Acta* 106 (6): 455–468.

Longchamp, M., Castrec-Rouelle, M., Biron, P. et al. (2015). Variations in the accumulation, localization and rate of metabolization of selenium in mature *Zea mays* plants supplied with selenite or selenate. *Food Chemistry* 182: 128–135.

Lyons, G. (2018). Biofortification of cereals with foliar selenium and iodine could reduce hypothyroidism. *Frontiers in Plant Science* 9: 349532.

Lyons, G., Stangoulis, J. and Graham, R. (2003). High-selenium wheat: biofortification for better health. *Nutrition Research Reviews* 16 (1): 45–60.

Malagoli, M., Schiavon, M., Dall'Acqua, S. et al. (2015). Effects of selenium biofortification on crop nutritional quality. *Frontiers in Plant Science* 6: 280.

Manojlović, M.S., Lončarić, Z., Cabilovski, R.R. et al. (2019). Biofortification of wheat cultivars with selenium. *Acta Agriculturae Scandinavica, Section B—Soil & Plant Science* 69 (8): 715–724.

Narwal, S., Kumar, D., Kharub, A.S. et al. (2020). Barley biofortification: present status and future prospects. *Wheat and Barley Grain Biofortification* 275–294.

Nawaz, F., Ashraf, M.Y., Ahmad, R. et al. (2015). Supplemental selenium improves wheat grain yield and quality through alterations in biochemical processes under normal and water deficit conditions. *Food Chemistry*, 175: 350–357.

Newman, R., Waterland, N., Moon, Y. et al. (2019). Selenium biofortification of agricultural crops and effects on plant nutrients and bioactive compounds important for human health and disease prevention—a review. *Plant Foods for Human Nutrition* 74: 449–460.

Ngigi, P.B., Lachat, C., Masinde, P.W. et al. (2019). Agronomic biofortification of maize and beans in Kenya through selenium fertilization. *Environmental Geochemistry and Health* 41: 2577–2591.

Pedrero, Z. and Madrid, Y. (2009). Novel approaches for selenium speciation in foodstuffs and biological specimens: a review. *Analytica Chimica Acta* 634 (2): 135–152.

Pennanen, A., Xue TaiLin, X.T., and Hartikainen, H. (2002). Protective role of selenium in plant subjected to severe UV irradiation stress.

Pezzarossa, B., Remorini, D., Gentile, M.L. et al. (2012). Effects of foliar and fruit addition of sodium selenate on selenium accumulation and fruit quality. *Journal of the Science of Food and Agriculture* 92 (4): 781–786.

Pilon-Smits, E.A. (2019). On the ecology of selenium accumulation in plants. *Plants* 8 (7): 197.

Poblaciones, M.J., Rodrigo, S., Santamaría, O. et al. (2014). Agronomic selenium biofortification in *Triticum durum* under Mediterranean conditions: from grain to cooked pasta. *Food Chemistry* 146: 378–384.

Radawiec, A., Szulc, W. and Rutkowska, B. (2021). Selenium biofortification of wheat as a strategy to improve human nutrition. *Agriculture* 11 (2): 144.

Rayman, M.P. (2008). Food-chain selenium and human health: emphasis on intake. *British Journal of Nutrition* 100 (2): 254–268.

Ríos, J.J., Rosales, M.A. and Blasco, B. (2008). Biofortification of Se and induction of the antioxidant capacity in lettuce plants. *Scientia Horticulturae* 116 (3): 248–255.

Ros, G.H., Van Rotterdam, A.M.D., Bussink, D.W. et al. (2016). Selenium fertilization strategies for bio-fortification of food: an agro-ecosystem approach. *Plant and Soil* 404: 99–112.

Sarwar, M.H., Sarwar, M.F., Sarwar, M. et al. (2013). The importance of cereals (Poaceae: Gramineae) nutrition in human health: a review. *Journal of Cereals and Oilseeds* 4 (3): 32–35.

Sarwar, N., Akhtar, M., Kamran, M.A. et al. (2020). Selenium biofortification in food crops: key mechanisms and future perspectives. *Journal of Food Composition and Analysis* 93: 103615.

Schiavon, M., Berto, C., Malagoli, M. et al. (2016). Selenium biofortification in radish enhances nutritional quality via accumulation of methyl-selenocysteine and promotion of transcripts and metabolites related to glucosinolates, phenolics, and amino acids. *Frontiers in Plant Science* 7: 1371.

Schiavon, M., Lima, L.W., Jiang, Y. et al. (2017). Effects of selenium on plant metabolism and implications for crops and consumers. In: *Selenium in Plants: Molecular, Physiological, Ecological and Evolutionary Aspects*, 1e (ed. Elizabeth A.H. Pilon-Smits, Lenny H.E. Winkel, and Z.-Q. Lin), 257–275. Springer.

Schwarz, K. and Foltz, C.M. (1957). Selenium as an integral part of factor 3 against dietary necrotic liver degeneration. *Journal of the American Chemical Society* 79 (12): 3292–3293.

Shahid, M., Niazi, N.K., Khalid, S. et al. (2018). A critical review of selenium biogeochemical behavior in soil-plant system with an inference to human health. *Environmental pollution* 234: 915–934.

Shekari, L., Kamelmanesh, M.M., Mozafarian, M. et al. (2016). Beneficial effects of selenium on some morphological and physiological trait of hot pepper (*Capsicum annuum*). *Journal of Horticultural Science* 29 (4), 594–600.

Stephen, R.C., Saville, D.J. and Watkinson, J.H. (1989). The effects of sodium selenate applications on growth and selenium concentration in wheat. *New Zealand Journal of Crop and Horticultural Science* 17 (3): 229–237.

Terry, N., Zayed, A.M., De Souza, M.P. et al. (2000). Selenium in higher plants. *Annual Review of Plant Biology* 51 (1): 401–432.

de Vita, P., Platani, C. and Fragasso, M. (2017). Selenium-enriched durum wheat improves the nutritional profile of pasta without altering its organoleptic properties. *Food Chemistry* 214: 374–382.

Wang, M., Ali, F., Qi, M. et al. (2021). Insights into uptake, accumulation, and subcellular distribution of selenium among eight wheat (*Triticum aestivum* L.) cultivars supplied with selenite and selenate. *Ecotoxicology and Environmental Safety* 207: 111544.

Wang, M., Ali, F., Wang, M. et al. (2020). Understanding boosting selenium accumulation in Wheat (*Triticum aestivum* L.) following foliar selenium application at different stages, forms, and doses. *Environmental Science and Pollution Research* 27: 717–728.

Wang, Q., Yu, Y. and Li, J. (2017). Effects of different forms of selenium fertilizers on Se accumulation, distribution, and residual effect in winter wheat–summer maize rotation system. *Journal of Agricultural and Food Chemistry* 65 (6): 1116–1123.

White, P.J. (2018). Selenium in soils and crops. *Selenium*, 29–50.

Winkel, L.H., Vriens, B., Jones, G.D. et al. (2015). Selenium cycling across soil-plant-atmosphere interfaces: a critical review. *Nutrients* 7 (6): 4199–4239.

World Health Organization (WHO) (2004). *Vitamin and Mineral Requirements in Human Nutrition*. World Health Organization.

Yasin, M., El Mehdawi, A.F., Jahn, C.E. et al. (2015). Seleniferous soils as a source for production of selenium-enriched foods and potential of bacteria to enhance plant selenium uptake. *Plant and Soil* 386: 385–394.

Zhang, H., Zhao, Z., Zhang, X. et al. (2019a). Effects of foliar application of selenate and selenite at different growth stages on selenium accumulation and speciation in potato (*Solanum tuberosum* L.). *Food Chemistry* 286: 550–556.

Zhang, M., Xing, G., Tang, S. et al. (2019b). Improving soil selenium availability as a strategy to promote selenium uptake by high-Se rice cultivar. *Environmental and Experimental Botany* 163: 45–54.

Zhao, C., Ren, J., Xue, C. et al. (2005). Study on the relationship between soil selenium and plant selenium uptake. *Plant and Soil* 277: 197–206.

Zhou, F., Dinh, Q.T., Yang, W. et al. (2019). Assessment of speciation and in vitro bioaccessibility of selenium in Se-enriched *Pleurotus ostreatus* and potential health risks. *Ecotoxicology and Environmental Safety* 185: 109675.

Zhou, X., Yang, J., Kronzucker, H.J. et al. (2020). Selenium biofortification and interaction with other elements in plants: a review. *Frontiers in Plant Science* 11: 586421.

Zhu, Y.G., Pilon-Smits, E.A., Zhao, F.J. et al. (2009). Selenium in higher plants: understanding mechanisms for biofortification and phytoremediation. *Trends in Plant Science* 14 (8): 436–442.

9

Scope and Research Perspective of Lithium Biofortification in Crop Plants

Muaz Ameen[1], Athar Mahmood[2], Sajid Fiaz[3]*

[1] *Department of Botany, University of Agriculture Faisalabad, Faisalabad, Pakistan*
[2] *Department of Agronomy, University of Agriculture Faisalabad, Faisalabad, Pakistan*
[3] *Institute of Molecular Biology and Biotechnology, The University of Lahore, Lahore, Pakistan*

*Corresponding author: athar.mahmood@uaf.edu.pk

9.1 Introduction

It is estimated that the global population will reach 9.6 billion by the year 2050. At the same pace, food production is growing. However, as food production is expanding, over 2 billion individuals globally, particularly pregnant women, adolescents, and children, have been affected by issues like hidden hunger, which is a result of inadequacies in vital vitamins and minerals. The term "hidden hunger" refers to a starving state in which individuals consume enough food, mostly simple grains, but not enough vitamins or nutrients to maintain their general health and well-being. A portion of this may be attributed to plant breeding that prioritizes yield above food quality. Regarding human health, the most concerning deficiencies at the moment are those brought on by a lack of selenium (Se), iodine (I), zinc (Zn), iron (Fe), and vitamin A, particularly in developing nations. Bipolar disorder belongs to the type of mental disorders that are associated with lithium (Li) deficiency (Godswill et al. 2020). Furthermore, when Li is present in the body at ideal levels, it helps prevent conditions such as amyotrophic lateral sclerosis, Alzheimer's, schizophrenia, phobic anxiety, heart disease, and diabetes.

Among all the elements, lithium (Li), which was first identified in 1817, is the least dense in its solid state and is a naturally occurring metal that makes up 0.0017% of the earth's crust. The normothymic properties of carbonate (Li_2CO_3) have made it one of the most often utilized medications in psychiatric therapy. More and more people with bipolar disorder have been using it to treat their illness since 1949. For over 50 years, this medication has remained in use despite many changes in psychotherapy. It is particularly useful in managing manic episodes and acute mania (Plotnikov, Korotkova, and Voronova 2018). Lithium is not a significant nutrient for plants, but at relatively low quantities, it is stimulated to grow and accumulate in fascinating elements of the plant that are edible to humans. Although there is a lot of evidence that the population is lacking in this element, there is still little understanding regarding the use of lithium in plants. An adequate daily diet that includes sufficient amounts of lithium is necessary for the human body, and using biofortified crop products becomes an inexpensive way to improve food sustainability. In order to produce food that is enhanced with vital nutrients for human consumption, biofortification is the agricultural technique of incorporating one or more nutrients into the soil throughout the growing process (Franzaring et al. 2016).

An effective solution for preventing disorders caused by Li shortage is to increase the food intake of lithium, because its positive normothymic action may be achieved at lower dosages than those employed therapeutically. Therefore, it is not shocking that present psychiatry has advocated for the introduction of foods fortified with lithium to be given significant consideration. From this angle, it is worthwhile to take into account agronomic biofortification, shown to significantly improve nutritional food security by incorporating nutrients into plant edibles via fertilizer application or management. To enhance Li concentration in plant tissues, this method takes into account either foliar Li treatment or increasing the Li content of soils. Li biofortification in food crops is a novel assumption that has garnered significant attention, unlike other minerals (Naeem et al. 2021). The administration of lithium via the roots and leaves to edible portions of food crops and vegetables has been studied by researchers. Its potential impact on growth and other edible characteristics has also been

Crop Biofortification: Biotechnological Approaches for Achieving Nutritional Security under Changing Climate, First Edition. Edited by Adnan Noor Shah et al.

examined. This chapter seeks to assess the potential for Li accumulation and the agronomic traits of plants undergoing biofortification with this element. The potential of Li biofortification is investigated, with particular attention paid to food crops and vegetables, environmental variables, and integration with conventional biofortification techniques. In addition, the chapter discusses issues including hazards to human health and the environment and examines economic viability. It also looks forward to developing technologies, sustainable integration, and cooperative research prospects.

9.2 Historical Context of Lithium in Agriculture

9.2.1 Sources of Lithium in Environment

Lithium exists in its elemental form and exhibits a high level of reactivity in its pure form. Li interacts with other elements due to its very reactive nature, resulting in the formation of salts or minerals. A variety of sources (Table 9.1), including the disposal of Li batteries and Li brines, industrial wastewater, and other water sources in proximity to lithium-related organizations, contribute lithium to the environment. A significant source of Li introduction among these is Li brines. In contrast, other investigations revealed that lithium was present in greater quantities than those usually seen in naturally occurring surface waters and that these concentrations were comparable to those reported as a result of mineralization or enrichment connected to mining (Shahzad et al. 2016). Lithium is an alkali metal often present in the environment. Additionally, lithium constitutes around 0.006% comprising a certain proportion of the earth's crust; it is positioned as the 27th most-prevalent element. Combined research done by van Wyk, Dippenaar, and Ubomba-Jaswa (2024) found that lithium combines rapidly with water to produce hydrogen gas, posing a threat in certain situations. Compounds of lithium are very soluble and largely unreactive when dissolved in water, making them suitable for use as stable markers in different hydrological research.

9.2.2 Lithium Status of Soils and Waters

The lightest alkali metal, lithium, derives its name from the Greek word *lithos*, which translates to mean "stone," owing to its widespread occurrence in minerals. Redox and alkaline-acid reactions, as well as the kind and mineral makeup of the rocks that compose the soil, may all significantly influence the quantity of lithium in the soil. In igneous rocks, its average amount rarely goes above 40–45 mg kg^{-1}, and in sedimentary rocks, it rarely goes above 60–75 mg kg^{-1}. Lithium thus occurs in smaller amounts in soils formed from volcanic rock than in soils formed from sedimentary rocks. The concentration of Li in soils and its respective genetic layers ranges from 0.01 to 160 mg kg^{-1} (Economou-Eliopoulos and Kanellopoulos 2023). Most of the lithium is located in the clay fraction, with just a little amount identified in the organic soil fraction. The lack of correlation between the humus content and Li content of soils was verified by survey research. Lithium is usually found in higher amounts in dry and salty soils than in acidic soils where through weathering of sedimentary minerals, lithium is permeable and leaches downward through the soil profile. Consequently, compared to underlying layers, topsoil often contains less total and exchangeable Li. Between 4 and 8% of the total Li content in soil is made up of mobile forms. The content of Li in aquatic environments varies; it may be found in freshwater at concentrations between 0.07 and 40 μg L^{-1} and in saltwater at concentrations

Table 9.1 Different identified resources of lithium entry in the ecosystem.

Lithium source	Lithium concentration	Reference
Lithium brine	20 mg l^{-1} in dead sea 1500 mg l^{-1} in Salar de Atacama	(Alsabbagh, Aljarrah, and Almahasneh 2021)
Lithium pegmatite and/lithium brines	7.6×10^6 kg	(Christmann et al. 2015)
Authigenic clays	200–500 ppm in detrital clays	(Andrews et al. 2020)
Lithium shale rocks	20–100 mg kg^{-1}	(Gao et al. 2020)
Land disposal of Li-hydride and Li-deuteride material	—	—
Effluents from fabrics, ceramics, and cosmetics industry	—	(Khatoon et al. 2023)

between 170 and 190 μg L^{-1}. The amount of lithium in water is determined by types of minerals and rocks including both the chemical and physical components of the environment, which have an impact on the elements' movement and release into aquatic habitats. Groundwater typically has Li concentrations ranging from 0.05 to 150 μg l^{-1}, which is greater than concentrations in surface water. The amount of lithium in groundwater is lower in mild wet areas compared to hot dry areas (Shahzad et al. 2017).

9.2.3 Essential Nature of Li

Lithium's importance for plant development is uncertain since many research have indicated differing crucial functions of lithium in various species. Various plant species may uptake substantial quantities of lithium, despite it not being essential for their growth and development. Conflicting research suggests that lithium may play significant roles in halophytes metabolism. However, lithium is harmful to plants at high soil concentrations. Kastori, Maksimović, and Putnikdelić (2022) discovered that lithium affects the development of maize roots and their response to gravity when exposed to elevated Li levels. Research has shown that lithium inhibits calcium signaling in plants and the inositol cycle. *Brassica carinata* seedlings showed notable alterations in their metabolomes and transcriptomes when subjected to Li stress, suggesting the importance of lithium in plant biochemistry. Lower amounts of lithium have improved plant growth by increasing yield, speeding up maturity, and improving disease resistance in some situations. An amount of 5 mg dm^{-3} of Li notably boosted corn biomass by 15%. Currently, there is no definitive data to demonstrate that lithium is necessary or advantageous for plants. The need of lithium for plants is an ongoing subject of discussion.

Researchers have reservations about the vital role of lithium in animals, including humans. However, studies have shown that a shortage in lithium may lead to such issues as impaired reproduction and lactation in goats and rats. Low Li intakes lead to changes in negative behavior and aggression in both people and rats. According to a separate research, rats that were given a diet that was lacking in lithium (0.0066 g Li per gram) exhibited a decreased level of avoidance behavior when compared with rats that were given a diet that was enriched with lithium (0.11 g Li per gram). In addition, Shakoor et al. (2023b) demonstrated that lithium plays a part in several biological processes, including the activity of enzymes, the action of transporters and channels, the creation of energy, and the movement of living animals according to their circadian cycles. This highlights the significance of studying lithium in living organisms. Lithium is not bio-accumulative in animals; however, its compounds have hazardous effects that vary depending on the dosage of lithium and the type of animal. Inokuchi et al. (2015) discovered that a low dosage of 0.5 mg L^{-1} of lithium had a considerable impact on the reproduction of fathead minnows and the development of ceriodaphnia. According to the findings of Shakoor et al. (2023b), Li concentrations ranging from 526 to 840 mg kg^{-1} are responsible for the death of rats. Whenever the concentration of Li exceeds 20 mg L^{-1}, there is a potential for mortality as a result of Li poisoning. The tiny ionic radius of Li likely accounts for many of its biological roles. This substance has potent polarizing properties and has the potential to displace potassium (K^+), magnesium (Mg^{2+}), and calcium (Ca^{2+}) at various biological membranes and enzyme-binding sites. Li^+ might have a significant impact on inhibiting proteins that contain Mg^{2+} in both plant and human systems. The fundamental essence of Li has not been thoroughly investigated. Lithium may have analogous activities to Mg or Ca in plants and might potentially substitute for them in regular metabolic processes.

9.2.4 Importance of Lithium in Plants

Lithium-containing compounds are often present in nature, usually in small quantities. Li compounds are present in the earth's crust at an average concentration of around 30 mg kg^{-1}, in soil at 25 mg kg^{-1}, in drinking water at 2 mg kg^{-1}, in saltwater at 170–190 μg L^{-1}, and in the atmosphere at 2 ng m^{-3}. It is crucial to examine the presence of Li in plants because of the potential differences in Li concentrations in different kinds of water, including freshwater, lakes, seawater, and groundwater. Li levels in food vary greatly, with bananas having a concentration of 0.01 ppm and oats having a concentration of 55 ppm. Industries that use lithium include those engaged in manufacturing dyes, batteries, metal mixtures, and Li-based pool-cleaning agents. These organizations are significant contributors to Li pollution in water bodies and soil, causing environmental pollution and directly endangering plants due to toxicity (Otřísal and Sabol 2024). Concentrations of lithium over 1 mg L^{-1} have been seen in wetlands of Nevada, USA. Kavanagh, Keohane, and Cabellos (2018) found that citrus growth in southern California, USA, was diminished and stunted when heavily Li-contaminated water was used for irrigation. Their study showed that the concentration of lithium above 0.05 mg L^{-1} in over 100 out of 400 water samples resulted in hazardous and detrimental effects on citrus plants. Studies have shown that varying soil types with differing Li concentrations might lead to a decrease in food quality. The threshold levels of lithium in plants differ significantly, with citrus leaves displaying mild-to-severe toxicity effects when exposed to 4–40 mg of lithium per kilogram. Yu et al. (2022) conducted an analysis of 138 plants that were grown in eight different kinds of soil. The research found that most plants had an average Li concentration ranging from 0.15 to 0.3 mg kg^{-1}. The normal quantity of lithium in plants that grow in salty soils might be anywhere from two to ten times higher.

The average amount of lithium per kilogram of dried matter discovered by Yelatontsev and Mukhachev (2021) in various varieties of plants was 0.3–0.42 mg. Areas with higher concentrations of Li in minerals and sediments are anticipated to have a greater presence of water-soluble lithium in soil, which has reached approximately 5% of total soil content. Certain soil varieties appeared to be correlated with the elevated Li concentrations in surface soils. Levels of lithium in calcareous soils varied from 56 to 1.3 mg kg^{-1} in light organic soils. Regular consumption of Li from various food sources, which affects a number of metabolisms in plants and other organisms, including animals, prompted the investigation of the negative effects of lithium in food. To provide more insight into the mechanisms underlying Li toxicity in plants, this study compiled the available data on the subject.

9.3 Li Uptake, Translocation, and Accumulation in Plants

Oxidation, which occurs during the process of weathering, is responsible for the release of lithium from the initial component via clay soil. Organic materials or mineral oxides in the region are responsible for the absorption of lithium, which then progressively accumulates in phosphate rocks throughout this process. There are a number of factors that may influence the accumulation of lithium in soil, including the characteristics of the soil, the concentration of clay, and the initial Li content of the parent material. At the same time as Bolan et al. (2021) found an inverse link between clay content and water-soluble Li, they also found a direct relationship between the amount of clay in the soil and the total and exchangeable Li levels in the soil. Clay content and exchangeable lithium in Piedmont soils were shown to have a substantial relationship ($r = 0.623$), as indicated by the statistical analysis. Lithium is present in soil in trace quantities, with concentrations varying from 0.002 to 63 ppm in different soil types.

Bakhat et al. (2020) studied five ways in which minor elements might bond with minerals: as impurities mixed mechanically, in structural empty spaces or cavities, as inclusions and alteration products, and in adsorbed phase and solid solutions. Due to its extensive distribution, it is reasonable to assume that plants may readily absorb lithium. Within the realm of elements, lithium is regarded as a nonessential and nonphysiological component. Plants often possess the capacity to uptake lithium from their environment. Plant species vary in their ability to absorb and store lithium. Certain plants are adept at accumulating lithium, while others only absorb lithium up to a certain quantity before experiencing negative effects on growth. The precise process of Li absorption and movement inside plants remains uncertain. Some research has proposed that lithium has physicochemical characteristics with other alkali metals like sodium or potassium, suggesting it may be taken up using the same method as these components. Lithium may use the same pathways and transporters as potassium (K^+) for absorption and transportation inside leaves. During the process of root absorption, several plants demonstrated a competitive interaction between calcium and lithium. High quantities of lithium in soils may impact the absorption and circulation of calcium. Moreover, several plant species exhibited varying capacities for absorption and tolerance when exposed to Li stress.

Stolarz, Król, and Dziubińska (2015) found that higher levels of Li in the growth medium led to a proportional increase in Li absorption in maize and sunflower shoots. Sunflowers collected 3292 mg kg^{-1} of lithium in their dry weight, whereas maize accumulated 695 mg kg^{-1} when exposed to a nutritional solution containing 50 mg Li per dm^3. Arabidopsis plants displayed leaf bending and chlorosis in mature leaves when subjected to LiCl in hydroponic culture. Necrotic lesions induced by lithium were more evident in mature tobacco leaves. Variations in experimental circumstances, plant ages, sample date, and exposure may partly account for the differences in the absorption, accumulation, and translocation of lithium by various crop plants. Thus, precise details on the transport and uptake of lithium in plants could not be determined. Li absorption and translocation to the above-ground sections of crops increased when lithium was applied to the soil. Afzal et al. (2023) discovered that there was less Li accumulation in the shoots than in the roots. This difference may be due to how Li is dispersed between the underground and aboveground parts of plants when they are grown in a nutrient solution, which could vary from plants grown in soil.

da Silva et al. (2019) found meaningful alterations in the amount of Li that accumulated in several dicotyledonous plants cultivated hydroponically. Radish and watercress accumulated higher levels of lithium in their tissues than lettuce. Reich et al. (2017) used inductively coupled plasma mass spectrometry to analyze the pattern and behavior of various *B. carinata* seedling varieties in accumulating lithium. They saw a significant variation in response to dosage across these different kinds of seedlings. The yellow- and brown-seeded lines initially prevented the absorption of lithium, then absorbed the same quantities of lithium at low concentrations, and then showed reduced Li accumulation only in yellow-seeded lines when exposed to high concentrations (≤60 mM). Brown lines with seeds collected above 80 000 ppm of lithium per dry weight at a high concentration. Furthermore, many soil conditions, including pH and moisture levels, might affect the presence, absorption, and storage of lithium in plants. Plants belonging to the Solanaceae family may accumulate over 1000 mg kg^{-1} of lithium when grown in acidic environments. Plants cultivated in Solonetz and Solonchak soils or soils with elevated alkali metal levels had the highest uptake of lithium.

Behr and Zörb (2023) discovered that elevated levels of lithium are generally detrimental to all plants, with acidic soil allowing for more absorption of lithium compared to alkaline soil. The association between soil acidity and solubility of lithium likely influences the amounts of lithium in plants, because the concentration of these components in soil is linked to the amounts of lithium in plants. Lower soil pH reduces the negative surface charge, leading to a decrease in accessible cations-binding sites. Enhanced element solubility promotes greater absorption by plants, leading to increased uptake of lithium. Lithium does not transform into juvenile or other plant parts due to its larger concentration in older leaves and its static transport in the phloem tissue of wheat. Some research suggest that lithium is considered stationary and its hazardous effects are seen in older leaves. However, other investigations have shown that lithium exhibits movements comparable to potassium or calcium. Kalinowska, Hawrylak-Nowak, and Szymańska (2013) noted that potassium was not maintained in cotton leaves but was transported back to the roots, potentially slowing down the absorption of salt. Simultaneously, there was no retrograde movement of calcium and it was unable to translocate inside the phloem. This research implies that potassium may be removed from the plant without causing any negative consequences. Potassium may be obtained by new leaves by phloem transport or re-export from mature leaves. However, calcium cannot be obtained through re-export and must be received directly from roots. This requirement for calcium remains constant throughout plant growth. Thus, lithium is similar to calcium in that it is not re-exported from the leaf. This may result from a lack of mobilization in the leaf or Li's inability to migrate inside phloem.

Tanveer, Hasanuzzaman, and Wang (2019) found that the rate of Li transportation via the phloem was quite low. Wheat exhibited a comparable process on exposure to various Li treatments at varied sample intervals. Shakoor et al. (2023c) revealed that Li has a greater resemblance with calcium than other alkali metals. He noticed that immature wheat seedlings collected lithium in their roots after a few days of Li treatment. Older leaves had a notable quantity of lithium compared to roots, suggesting that Li is a stationary element. He observed a depletion of lithium from the roots into the soil during the plant's transfer from one container to another. Li transportation throughout the plant is primarily controlled by the Li gradient's direction. The movement of lithium inside the first wheat leaf impedes its redistribution. Increasing the concentration of Li to 12 ppm in the hydroponic experiment resulted in higher cucumber fruit output, sugar content, and ascorbic acid levels, without causing accumulation of Li in the fruit. Applying 1 kg of lithium per hectare in a field trial with soil having 1.4 to 2.4 ppm of lithium resulted in increased levels of citric and malic acids in potato tubers.

9.4 Conventional and Novel Biofortification Strategies

Hidden hunger is an indicator of malnutrition when individuals consume sufficient food, mostly staple crops, but these foods lack the necessary nutrients and vitamins for maintaining good health and well-being. This is partly because plant breeding has been primarily focused on increasing output without giving sufficient consideration to the nutritional content of food (Saltzman et al. 2017). The most concerning deficits for human health, particularly in poor nations like Brazil are iron (Fe), zinc (Zn), selenium (Se), iodine (I), and vitamin A. Li deficiency is linked to the development of mood disorders, such as bipolar disorder. Optimal levels of lithium in the body have a good function in preventing disorders including Alzheimer's, amyotrophic lateral sclerosis, schizophrenia, phobic anxiety, diabetes, heart disease, and alcohol dependency.

9.4.1 Foliar Biofortification

Despite the fact that lithium is not a nutrient that is required for plant development, it may stimulate plant growth and lead to the accumulation of this element in interesting portions that can be consumed by humans when it is present in relatively low quantities. With regard to the use of lithium in plants, there is still a lack of knowledge; nonetheless, there is a significant amount of evidence that the population is lacking in this element. Regularly consuming food rich in lithium is crucial for maintaining optimal amounts of this element in the human body. The use of biofortified agricultural products is an economical method to boost the cost-effectiveness of food sustainability. Biofortification is the process of adding important nutrients to food during production to enhance its nutritional value for humans. Biofortification should not change the fundamental traits of the goods or disrupt the current supply chain between producers and consumers. It also adds quality and value to the target product. Foliar fertilization complements soil fertilization by applying micronutrients directly to the leaves in experiments aiming for a rapid crop response, improved product distribution, and reduced production costs. Positive outcomes have been documented in several crops such as lettuce, cabbage, common bean, and soybean (Sheoran et al. 2022).

9.4.2 Breeding and Genetic Biofortification

Traditional plant breeding and genetic techniques are seen as long-term options for biofortification. Conventional plant breeding focuses on enhancing crop yield, disease resistance, and tolerance to environmental stresses, whereas biofortification methods emphasize increasing micronutrient and vitamin levels in the edible parts of the plant. This method makes use of the differences in absorption rates across various cultivars or even between male and female plants. Research on spinach showed different degrees of enrichment depending on the plant's gender. Male plants showed higher selenium accumulation rates than female plants. Moreover, male individuals exhibited elevated levels of nitrate, polyphenols, and ascorbic acid in biofortified spinach, indicating a possible role of phytohormones in the selenium accumulation process (Golubkina et al. 2017). Genetic engineering may be used to enhance mineral absorption of a desired micronutrient in plants by reconstructing accumulation routes from accumulator plants when the characteristic is not naturally present in the plant's genotypic variance. Progress in genetic engineering is comparable to that of traditional plant breeding; it is a one-time expense and may accelerate the regular plant breeding process. Conversely, the process of development might be lengthy. The main obstacle to this approach is the restricted acceptance of genetically modified crops by consumers and farmers; hence, organically grown foods are seen as healthier and more nutritious. This method's efficacy, like conventional breeding, is limited to minerals naturally found in soil (Jaiswal et al. 2022). To increase iodine content in *Arabidopsis thaliana* L., one may either decrease iodine volatilization by suppressing the *HOL-1* gene or boost iodine absorption by raising the expression of the human sodium-iodide symporter (Malik and Maqbool 2020).

9.4.3 Agronomic Biofortification

Agronomic biofortification is the act of increasing the percentage and availability of certain nutrients in crops by using mineral fertilizers on the soil or crops. Agronomic methods were first used to improve crop health and increase yield. Nutrition's significance has been emphasized throughout time, leading to the creation of agronomic techniques to enhance the nutritional characteristics of crops. To combat climate change and soil nutrient depletion, it is essential to develop and broaden agronomic techniques by focusing on improving the nutritional characteristics of crops. Agronomic biofortification aims to enhance the solubilization and mobilization of minerals (Bhardwaj et al. 2022).

The efficiency of agronomic interventions relies on factors such as mineral solubility and mobility, crop mineral absorption capacity, soil composition, and the safe buildup of bioavailable minerals in edible crop areas. Agronomic biofortification mostly emphasizes minerals over vitamins since vitamins are synthesized inside the crops. Agronomic biofortification is not enough to completely address micronutrient shortages and should be complemented with other strategies for effective biofortification. Appropriate use of fertilizers for agronomic biofortification is crucial to avoid unexpected and potentially serious impacts on the environment and crops. Implementing a well-rounded fertilization approach is both economical and ecologically responsible. Soil microorganisms are essential in soil ecology and are greatly impacted by fertilizer. Insufficient fertilization results in nutrient deficiency and alterations in the soil's microbial community. Unequal fertilizations might potentially damage soil biological health over time (Dhaliwal et al. 2022). The research focuses on agronomic biofortification using leaves to investigate the effectiveness of applying a trace element with restricted availability when used in soil. The Li content within soil decreases as weathering progresses. It is strongly absorbed by clay minerals and organic matter, as well as manganese oxides, which hinders its absorption by crops of agronomic importance.

9.5 Scope of Li Biofortification

9.5.1 Vegetables

Fortifying food with lithium is an approach worth examining to reduce aggressive behaviors and enhance mental therapy for the general population. Unfortunately, there is less research on Li biofortification of food crops owing to the absence of precise guidelines for recommended dietary allowance (RDA) by regulatory agencies like the World Health Organization (WHO) and regional health and nutrition institutes. The preceding sections indicate that a substantial amount of Li may be found in the edible portions of plants with the introduction of Li to the soil or other growth mediums, without affecting the yield. This section examines the levels of lithium in plant tissues when exposed to nontoxic levels of lithium in the rooting medium (Duborská et al. 2022). Research on Li biofortification mostly focuses on certain green crops. Studies on cereals have been conducted, with writers mostly focusing on issues such as Li toxicity, its positive effects, and its storage in plant tissues.

Xu et al. (2023) found a direct relationship between the Li levels in lettuce plants and the Li concentration in the nutrient solution, varying from 0 to 100 mg l^{-1}. The maximum shoot concentration that did not affect production was about 225 mg kg^{-1} dry weight, achieved using a nutrient solution with 2.5 mg L^{-1} of Li. Spinach grown in pots with alkaline calcareous soil collected 1490 mg kg^{-1} of lithium after three harvests, showing no decrease in yield even when up to 80 mg Li per kilogram of soil was added. The researchers noticed a rise in Li content to 165 mg Li per kilogram FW after applying 20 mg Li per kilogram soil throughout three harvests. The water content of fresh spinach is around 93%, which is equivalent to approximately 11.50 mg of lithium per kilogram of fresh biomass. There is a possibility that consuming 100 g of this spinach that has been biofortified with lithium might give 115% of the daily recommended intake for this element. In two distinct soils, the amount of lithium that was deposited by spinach and mustard leaves was 170 and 240 mg kg^{-1}, respectively. The soils were treated with 40 mg of Li per kilogram of soil as lithium sulfate in four separate applications at sowing, 32, 39, and 52 days after sowing. The study showed that incorporating Li (5.5 and 11 mg kg^{-1} soil) into sandy loam and silt loam soils led to a corresponding rise in Li levels in the leaves of spinach and mustard green plants, without adversely affecting plant development. Based on their first investigation, spinach acquired lower levels of Li in leaves compared to mustard greens at both administered Li rates. In a recent study, vegetables showed higher levels of Li when subjected to a lower application rate of 11 mg kg^{-1}, compared to a previous study that employed a higher rate of 40 mg kg^{-1}. The discrepancy is probably caused by differences in Li absorption from different types of soils. When exposed to a soil treatment rate of 50 mg kg^{-1}, the plants *Brassica napus, Brassica oleracea* (cv. Capitate), *Solanum lycopersicum, Helianthus annuus*, and *Cardamine hirsuta* absorbed different quantities of lithium. However, the treatments did not have any impact on the generation of dry matter. The amounts absorbed were 7.7, 971, 388, 16.9, and 1066 mg kg^{-1} of dry leaves, respectively. These significant accumulations suggest that optimizing soil application might help achieve the appropriate Li content in the edible sections of these crops. They also said that increased Li buildup might be achieved by adding EDTA and EDDS to the soil. When *Brassica juncea* is cultivated in a synthetic medium consisting of 25% dewatered municipal biosolids, 25% homogenized peat, and 25% bioavailable lithium at 389 mg kg^{-1}, and irrigated with an organic fertilizer containing 3 g of lithium chloride per liter, the plant has the capacity to accumulate 84 mg of lithium per kg of dry weight in the stem and 319 mg of lithium per kg of dry weight in the leaves. The experimental design failed to delineate the precise quantity of lithium introduced to the growth medium via irrigation, as well as the resulting effects on plant development. When vegetables, such as *Berberis vulgaris* (leaves), *Lactuca sativa*, and *Berberis nigra*, were exposed to a Li treatment rate of 5 mg kg^{-1}, they accumulated approximately 500 mg of Li per kilogram of dried matter.

Mezni et al. (2017) applied an organic Li–Se fertilizer to five different grape varieties. They observed an increase in the Li concentration in the liquid from 0.9 to 11.73 μg kg^{-1}, as well as in the dried skin from 62.2 to 311 μg kg^{-1}. The rise in one cultivar was accompanied by a little hindrance in vine development, whereas it was either enhanced or had no impact on other cultivars. The approximated potassium intake from 100 g of raw vegetables varies considerably, spanning from 84 to 7768 μg. In other instances, the ingestion values from lettuce (hydroponic study) and tomato exceed the RDA of 1000 mg^{-1}. This indicates that vegetable plants still require refinements in Li biofortification, a task that can be easily accomplished through hydroponics, particularly with regard to leafy greens such as spinach, lettuce, tomato, and cucumber. Additionally, further investigation is needed to enhance the circumstances and traits for Li biofortification in cereal crops. According to the results of the two-month hydroponic study, maize can assimilate 908 mg lithium per kilogram dried matter without experiencing any adverse effects on production or growth. Rzymski et al. (2017) produced a Li amount of 432 and 438 mg kg^{-1} DW in maize and sunflower, respectively, by using a nutrient solution with a Li content of 25 mg L^{-1}, without seeing any significant reduction in production. Furthermore, in agreement with the findings of earlier studies, the researchers found that the shoots of a dicotyledonous plant (sunflower) had a higher concentration of lithium than the shoots of a monocotyledonous plant (maize) when the experiment was conducted in a hydroponic environment. The research demonstrates that fodder crops such as maize may collect a substantial amount of lithium from lithium-enriched substrate. Enriching fodders with lithium is important and may provide Li intake via a secondary pathway. Feeding farm animals enhanced with Li fodder would increase the amounts of Li in their milk and meat, thus enhancing the Li intake of customers. Optimizing soil Li supplementation can achieve the necessary range of Li concentration in different fodders.

9.5.2 Crops

Although it does not contain a critical component for plant development, lithium does have an effect on the growth and metabolism of some halophytes. The amount of Li ions that are present in the soil has an effect on the amount of lithium that is accumulated in plants. To illustrate this point, consider the varied degrees of curling and necrotic lesions that were seen in tobacco (*Nicotiana tabacum*) that had been treated with 15–50 mM LiCl in a hydroponic solution. The toxicity of the solution increased as the concentration of LiCl increased above 50 mM. To investigate the accumulation of lithium, Solti et al. (2022) planted 34 different varieties of *Brassica* in soils that were contaminated with different quantities of lithium ranging from 20 to 1000 mg L^{-1}. The first screening revealed that nine different species exhibited germination rates over 90%. Among these species, *B. napus* and *B. oleracea* exhibited the greatest Li accumulation, with respective values of 2590 and 3091 mg kg^{-1}. The amount of lithium that could be produced from one ton of plant biomass was estimated to be 0.26 kg.

According to the findings of Shahzad et al. (2017), *B. carinata* seedlings demonstrated rapid development when exposed to concentrations of Li that were below 60 mM. However, when exposed to concentrations that were higher than 60 mM, the seedlings had a considerable loss (40–80%) in chlorophyll content and suffered growth inhibition. On lettuce (*L. sativa*) plants, Hawrylak-Nowak, Kalinowska, and Szymańska (2012) conducted a study to investigate the effects of various doses of LiCl or LiOH, ranging from 2.5 to 100 mg L^{-1}. When they used 2.5 mg L^{-1} LiOH, they found that the biomass of the shoots and roots grew by a considerable 62%. However, when they used 20–50 mg L^{-1} LiOH from 19 to 69% and LiCl from 13 to 58%, the biomass declined dramatically. At a concentration of 100 mg L^{-1}, growth was entirely halted. There is a growing need for natural Li remediation as a result of the increasing liquid contamination of soils.

According to Li et al. (2018), the presence of 10 and 15 mg kg^{-1} LiCl significantly increased the dry weight of leaves that ranged from 24 to 33%, stems that ranged from 20 to 26%, and roots that ranged from 29 to 44%. This indicates that *Apocynum venetum* is an excellent phytoremediation alternative for soils that have been contaminated with Li. However, once reaching dosages of 20 and 25 mg kg^{-1}, these increases were thereafter reduced. With a foliar treatment of 45.7 mg kg^{-1} Li_2SO_4, the output of soybean grain rose by 10.73%. However, the production of soybean grain dropped by 20% when treated with 120 mg kg^{-1} Li_2SO_4 and by 49% when treated with 120 mg kg^{-1} LiOH. Additionally, applying 10–40 mg kg^{-1} of LiOH and Li_2SO_4 by foliar spraying resulted in a 31% and 18% increase in lettuce root dry weight, respectively, at concentrations of 20 and 30 mg kg^{-1}. However, a notable decrease of 22% was seen at the 40 mg kg^{-1} dose (Table 9.1). Therefore, it may be inferred that high levels of Li have a distinct detrimental effect on plant physiological responses and growth, depending on the species and plant component. Further research is required to comprehend the mechanism of Li toxicity at elevated levels and its potential association with transmembrane transport. The modifications in plant physiology and nutritional content should be investigated over a longer period of time, especially in vegetable crops, *Brassica* species that accumulate Li, and leguminous bean crops.

Numerous studies have investigated the effects of lithium on soil and plant development. The application of lithium increased soil salinity and alkalinity while decreasing levels of essential nutrients such as sodium (Na), potassium (K), and nitrogen (N) (Hayyat et al. 2021). A number of plant species, including *B. carinata*, spinach, maize, and sunflower seedlings, exhibited reduced dried weight and growth when exposed to elevated Li concentrations in the soil. Large quantities of lithium may inhibit the development and growth of plant roots, according to scientific research. According to a study, phytotoxic symptoms were observed in plants when exposed to concentrations of 128.0 and 256.0 mg dm^{-3} of Li. Tissue mortality, leaf discoloration, leaf withering, and leaf yellowing were the symptoms that were noted (Elektorowicz and Keropian 2015). *Amaranthus viridis* responded well to Li, according to the findings of Gayathri, Sailesh, and Srinivas (2022). For dosages as low as 10 ppm, plant growth was stimulated, while higher concentrations led to an increase in biomass. The germination rates improved at lower levels (up to 10 ppm) on the stress tolerance index but dropped at higher concentrations. This demonstrated that lithium had a favorable influence on the plant. When the root and shoot lengths were measured in pot assessments, it was shown that they increased at low levels (up to 50 ppm), but decreased at higher dosages (75 and 100 ppm). There was a reduction in length but an increase in dry weights as a consequence of the plant's resistance to Li concentrations of up to 100 ppm. It is consistent with the results of earlier research on lithium and heavy metals, which indicate that these substances have an effect on the proliferation of cells and the plant's capacity to endure higher concentrations. Shakoor et al. (2023a) conducted a further study that showed that increased levels of lithium in soil led to a reduction in the synthesis of certain root metabolites. This, in turn, resulted in inadequate growth and development as well as decreased levels of calcium and magnesium in those plants. According to the findings of Sarma et al. (2024), increasing the dose of lithium from 40 to 400 ppm resulted in a decrease in the shoot biomass of Zea mays (maize) in contrast to the group that performed as the control. Some of the losses in biomass ranged from 10 to 87%. It has been shown that the accumulation of lithium in the plant, particularly at higher levels, is responsible for the loss of plant mass. This accumulation may be detrimental to the shoot mass.

9.6 Challenges and Limitations

Enhancing plants using lithium poses advantages and difficulties. Seidel et al. (2019) emphasize the potential advantages of this procedure, especially in dealing with inadequate Li consumption in diets. The assessment highlights the need for more research to enhance biofortification techniques. Robinson et al. (2018) investigated the impact of various Li sources and concentrations on plant development and lithium accumulation. Both researches conclude that excessive concentrations may adversely affect plants. Plant growth is affected by meteorological conditions, genetic qualities, terrain, soil fertility, and metal concentrations. Although some metals are necessary for plant development, an overabundance may be harmful and impede growth and productivity. Plant responses to Li toxicity differ across species, with some families showing tolerance and others, such as citrus plants, being sensitive. Lithium affects plant roots by changing their gravitropic growth in maize. Toxic effects of Li on roots include harm to root tips and the development of chlorotic patches. The concentration of lithium affects plant development, with low levels promoting growth and greater quantities hindering it. Symptoms generated by Li are difficult to detect; however, they may include necrosis, chlorosis, and leaf abscission. Lithium's influence on plants affects biochemical and metabolic processes, altering enzyme activity, inositol signaling pathways, and water content (Shen et al. 2020). Further quantification is needed to understand the dose–dependent interactions between the positive and harmful effects of lithium on plant development. Li's function as a cofactor in enzyme activation and its potential as a replacement component in metabolism need further research.

Lithium, a nonphysiological element, is harmful to photosynthetic pigments and carbon absorption in plants. Research has shown different reactions to Li supplementation, where sunflower chlorophyll levels remain unchanged at lower doses while there is a notable reduction in carotenoids at 25 mg dm^{-3} of Li. After being subjected to a greater concentration of lithium, namely, 50 mg dm^{-3}, maize exhibited a significant reduction in the amounts of chlorophyll and carotenoid, in addition to the presence of necrotic areas. Li has been reported to cause necrotic areas on the leaves of lettuce, sour orange, red kidney bean, and tomato, which might be associated with the breakdown of chlorophyll. The influence on chlorophyll levels in many plant species, such as *B. carinata* seedlings, indicates a wider impact on chloroplast organization and pigment–protein complexes. Although not required, lithium's impact on photosynthesis, sugar production, enzymatic activities, and nitrogen metabolism in plants is a cause for worry. Insufficient data regarding lithium's harmful effects on photosynthesis and pigments highlights the necessity for additional research. This emphasizes the importance of studying its influence on chloroplastic ultrastructure, electron chain reactions, stomatal conductance, Calvin cycle, and breakdown of plastoquinone and carotenoids to understand the mechanism responsible for lithium-induced decrease in photosynthesis (Fang et al. 2020).

The fundamental characteristics and roles of lithium are not completely known, even though it is categorized as an alkali metal in group IA of the periodic chart. Li has characteristics with other group IA and certain group IIA elements, showing a remarkable capacity to displace ions, for example, Mg^{2+}, K^+, Na^+, and Ca^{2+} because of its smaller size and strong polarizing ability. Lithium's complexation characteristics are stronger than K^+ and Na^+ but weaker than Mg^{2+}. It may replace half of the potassium in plants and greatly affect the function of enzymes that include magnesium ions. Lithium interacts with sodium, potassium, and magnesium ions, influencing enzyme function. The impact on nutrient absorption differs across plant species, with lettuce showing enhanced iron uptake and radish and watercress displaying altered potassium uptake (Alirezapour and Khanmohammadi 2021). Research indicates an intricate relationship between calcium and lithium, with findings showing that calcium hinders Li absorption. Further research is required for a complete understanding of the influence of lithium on plant mineral nutrition, including its connection with other minerals in soil and plant tissues.

High levels of lithium hinder plant development, whereas low levels may promote growth in some plant species. The specific processes responsible for these opposing effects are not fully understood; however, oxidative stress is suggested as a possible factor. Research indicates that lithium increases oxidative stress by promoting lipid peroxidation, leading to potential damage to cell membranes and changes in the composition of cellular organelles such as mitochondria, peroxisomes, and chloroplasts. Reactive oxygen species (ROS) produced during Li poisoning might hinder the synthesis of antioxidant enzymes including glutathione peroxidase (GPX), superoxide dismutase (SOD), and catalase (CAT). Lithium-induced ROS may potentially affect lipid membranes, causing deformation of the lipid bilayer in cell membranes. Disruption of membrane integrity, as shown by elevated lipid peroxidation levels, is linked to a reduction in plant biomass. Studies have shown equivocal findings about the accumulation of L-ascorbic acid (vitamin C) in plant tissues under Li stress, despite its putative significance as an antioxidant defense against lithium-induced ROS (Xu et al. 2024). Li toxicity in plants is associated with oxidative stress and disturbances in cellular functioning; however, the specific mechanisms need further research. These results emphasize the need for ongoing investigation and advancement in Li biofortification to fully use its prospective advantages.

9.7 Conclusion and Future Directions

In this review, an overview of the dispersion of lithium and its accumulation in the environment is provided. We have conducted an exhaustive analysis of the daily intake, bio-accessibility, mechanistic techniques, and possible risks associated with Li exposure throughout all environmental compartments (see Figures 9.1 and 9.2 for more information). For providing biologists and environmental researchers with a substantial and unified perspective on the growing relevance of Li in our ecosystems and society, this study has been conducted. High quantities of Li have a harmful impact on plant functions and chemical processes. Lithium disrupts the movement of other crucial elements including magnesium, calcium, and potassium. Lithium tends to concentrate in the aboveground parts of land plants, but it also impacts aquatic organisms

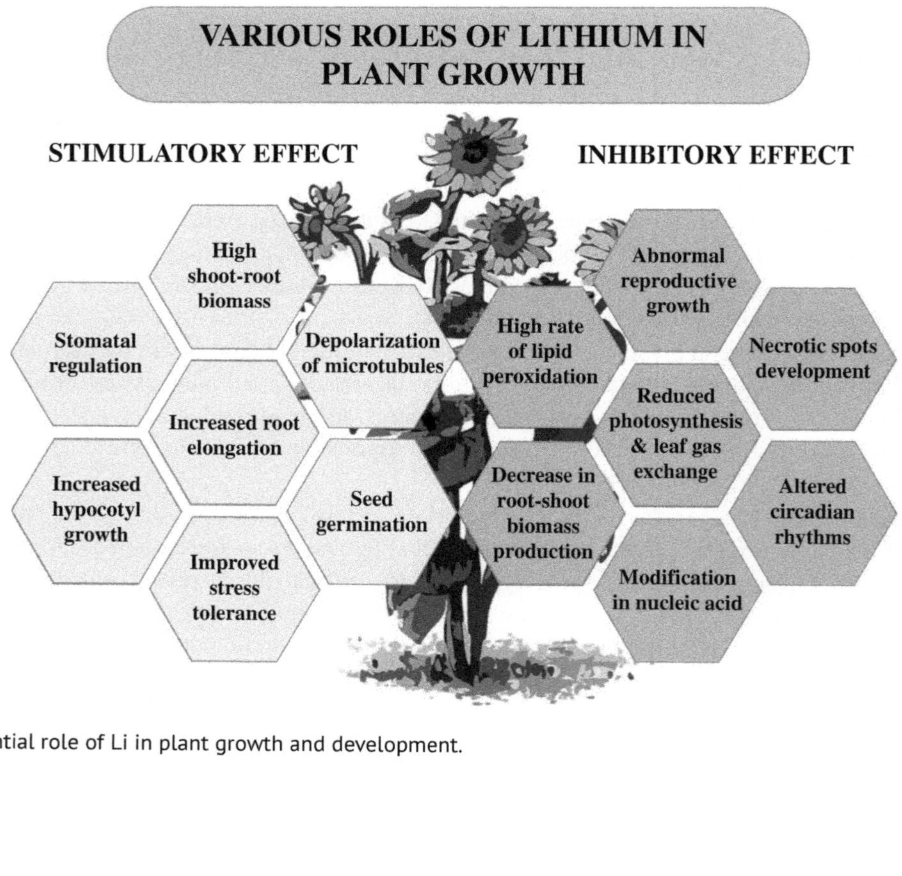

Figure 9.1 Differential role of Li in plant growth and development.

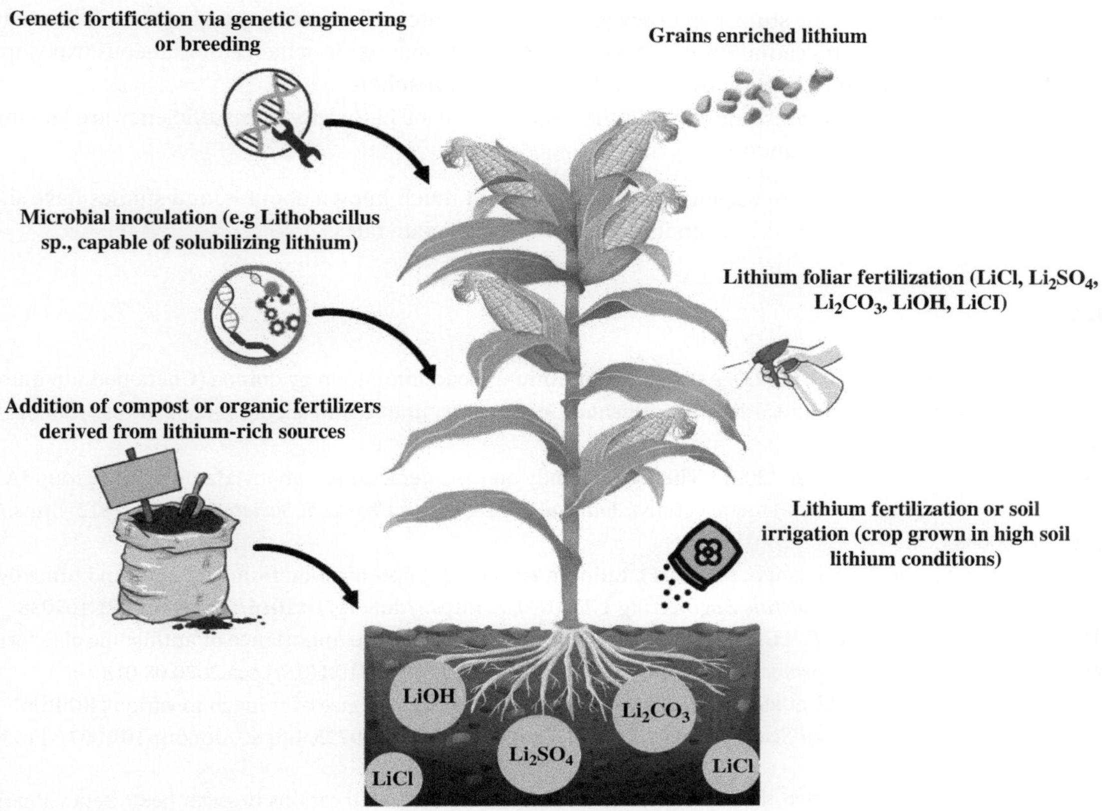

Figure 9.2 Examination of biofortification approaches for enhancing lithium levels in crops.

such as fish, amphibians, and crustaceans. Many research has examined the amounts of Li in various organs. In aquatic species, a significant release of domoic acid occurs when the concentration of Li is high enough to be 0.45 M L^{-1}. This release has a negative influence on embryonic development and may lead to microcephaly. It is via plants and drinking water that Li is introduced into the food chain. The amount of lithium that humans consume on a daily basis ranges from 0.65 to 3.1 mg. The range of Li concentrations that are considered appropriate in blood plasma is between 0.6 and 1.2 mmol L^{-1}. At concentrations of 1.5 mM L^{-1}, the onset of toxicity symptoms is seen, whereas life-threatening circumstances arise when the concentration reaches 2.0 mM L^{-1} or more. When it comes to the treatment of bipolar disorder and other manic diseases, Li is a useful therapeutic measure. Depression, inability to sleep, susceptibility to stress, and chronic pain have all been associated with low concentrations of Li in the plasma, which are less than 0.6 mmol L^{-1}. Kidney poisoning, neurological symptoms, coarse tremors, and muscle twitching are some of the harmful systemic outcomes that may occur when Li concentrations are more than 1.5 mM L^{-1}. On the basis of the facts that have been presented, the following concerns need further examination:

- Comprehensive data on Li content in the natural environment worldwide will provide insights into creating a secure environment and fostering sustainable development.
- Changes in the nutritional content of plants, notably in vegetables and Li-accumulating *Brassica* spp., as well as leguminous bean crops, should be investigated via research. This is especially important that research be carried out. In order to fully understand the core mechanism behind lithium's harmful effects on plants, it is necessary to do simultaneous research.
- Due to increased mobility, lithium may readily accumulate in plants, necessitating more research on the transfer of lithium to higher trophic levels.

- It is necessary to conduct specific studies in order to evaluate the interaction that occurs between lithium and heavy metals such as lead, copper, and cadmium in the soil, as well as the manner in which simultaneous exposure to these metals influences the amount of lithium that is available in the rhizosphere.
- Investigation into lithium daily consumption and the health effects of Li toxicity or insufficiency are lacking in locations where mining activities are underway.

Li biofortification of crops needs more research because there isn't much known about it, and studies have shown that this element is beneficial for human health when it is present at appropriate rates.

References

Afzal, S., Bakhat, H.F., Shahid, M. et al. (2023). Assessment of lithium bioaccumulation by quinoa (Chenopodium quinoa willd.) and its implication for human health. *Environmental Geochemistry and Health* 45: 6517–6532. https://doi.org/10.1007/s10653-023-01659-9.

Alirezapour, F. and Khanmohammadi, A. (2021). Theoretical study on the interaction of phenylalaninal with group IA (Li+, Na+, K+) and IIA (Be2+, Mg2+, Ca2+) metal cations. *Journal of the Chinese Chemical Society* 68: 1002–1012. https://doi.org/10.1002/jccs.202000383.

Alsabbagh, A., Aljarrah, S., and Almahasneh, M. (2021). Lithium enrichment optimization from Dead Sea end brine by chemical precipitation technique. *Minerals Engineering* 170: 107038. https://doi.org/10.1016/j.mineng.2021.107038.

Andrews, E., Pogge von Strandmann, P.A.E., Fantle, M.S. et al. (2020). Exploring the importance of authigenic clay formation in the global Li cycle. *Geochimica et Cosmochimica Acta* 289: 47–68. https://doi.org/10.1016/j.gca.2020.08.018.

Bakhat, H.F., Rasul, K., Farooq, A.B.U. et al. (2020). Growth and physiological response of spinach to various lithium concentrations in soil. *Environmental Science and Pollution Research* 27: 39717–39725. https://doi.org/10.1007/s11356-019-06877-2.

Behr, J.H. and Zörb, C. (2023). Interactive effect of lithium on concentration of alkali cations in sugar beet (Beta vulgaris L.) under saline conditions. *Journal of Plant Nutrition and Soil Science* 186: 38–49. https://doi.org/10.1002/jpln.202200079.

Bhardwaj, A.K., Chejara, S., Malik, K. et al. (2022). Agronomic biofortification of food crops: an emerging opportunity for global food and nutritional security. *Frontiers in Plant Science* 13: 1055278. https://doi.org/10.3389/fpls.2022.1055278.

Bolan, N., Hoang, S.A., Tanveer, M. et al. (2021). From mine to mind and mobiles – Lithium contamination and its risk management. *Environmental Pollution* 290: 118067. https://doi.org/10.1016/j.envpol.2021.118067.

Christmann, P., Gloaguen, E., Labbé, J.-F. et al. (2015). Global lithium resources and sustainability issues, (Chapter 1). In: *Lithium Process Chemistry* (eds. A. Chagnes and J. Światowska), 1–40. Amsterdam: Elsevier. https://doi.org/10.1016/B978-0-12-801417-2.00001-3.

da Silva, R.R., de Faria, A.J.G., Alexandrino, G.D.C. et al. (2019). Enrichment of lithium in lettuce plants through agronomic biofortification. *Journal of Plant Nutrition* 42: 2102–2113. https://doi.org/10.1080/01904167.2019.1648671.

Dhaliwal, S.S., Sharma, V., Shukla, A.K. et al. (2022). Biofortification – A frontier novel approach to enrich micronutrients in field crops to encounter the nutritional security. *Molecules 27* (4): 1340.

Duborská, E., Šebesta, M., Matulová, M. et al. (2022). Current strategies for selenium and iodine biofortification in crop plants. *Nutrients* 14: 4717. https://doi.org/10.3390/nu14224717.

Economou-Eliopoulos, M. and Kanellopoulos, C. (2023). Abundance and genetic significance of lithium in karst-type bauxite deposits: a comparative review. Minerals 13: 962. https://doi.org/10.3390/min13070962.

Elektorowicz, M. and Keropian, Z. (2015). Lithium, vanadium and chromium uptake ability of brassica juncea from lithium mine tailings. *International Journal of Phytoremediation* 17: 521–528. https://doi.org/10.1080/15226514.2013.876966.

Fang, T., Yu, X., Zhang, X. et al. (2020). A comparative investigation on lithium storage performance of carbon microsphere originated from agriculture bio-waste materials: sunflower stalk and walnut shell. *Waste and Biomass Valorization* 11: 6981–6992. https://doi.org/10.1007/s12649-019-00927-z.

Franzaring, J., Schlosser, S., Damsohn, W. et al. (2016). Regional differences in plant levels and investigations on the phytotoxicity of lithium. *Environmental Pollution* 216: 858–865. https://doi.org/10.1016/j.envpol.2016.06.059.

Gao, Y., Bagas, L., and Li, K. (2020). Newly discovered triassic lithium deposits in the Dahongliutan area, northwest China: a case study for the detection of lithium-bearing pegmatite deposits in rugged terrains using remote-sensing data and images. *Frontiers in Earth Science* 8. https://doi.org/10.3389/feart.2020.591966.

Gayathri, N., Sailesh, A.R., and Srinivas, N. (2022). Effect of lithium on seed germination and plant growth of Amaranthus viridis. *Journal of Applied and Natural Science* 14: 133–139. https://doi.org/10.31018/jans.v14i1.3165.

Godswill, A.G., Somtochukwu, I.V., Ikechukwu, A.O. et al. (2020). Health benefits of micronutrients (vitamins and minerals) and their associated deficiency diseases: a systematic review. *International Journal of Food Sciences* 3: 1–32. https://doi.org/10.47604/ijf.1024.

Golubkina, N.A., Kosheleva, O.V., Krivenkov, L.V. et al. (2017). Intersexual differences in plant growth, yield, mineral composition and antioxidants of spinach (Spinacia oleracea L.) as affected by selenium form. *Scientia Horticulturae* 225: 350–358. https://doi.org/10.1016/j.scienta.2017.07.001.

Hawrylak-Nowak, B., Kalinowska, M., Szymańska, M. (2012). A study on selected physiological parameters of plants grown under lithium supplementation. *Biological Trace Element Research* 149: 425–430. https://doi.org/10.1007/s12011-012-9435-4.

Hayyat, M.U., Nawaz, R., Siddiq, Z. et al. (2021). Investigation of lithium application and effect of organic matter on soil health. *Sustainability* 13: 1705. https://doi.org/10.3390/su13041705.

Inokuchi, A., Yamamoto, R., Morita, F. et al. (2015). Effects of lithium on growth, maturation, reproduction and gene expression in the nematode Caenorhabditis elegans. *Journal of Applied Toxicology* 35: 999–1006. https://doi.org/10.1002/jat.3058.

Jaiswal, D.K., Krishna, R., Chouhan, G.K. et al. (2022). Bio-fortification of minerals in crops: current scenario and future prospects for sustainable agriculture and human health. *Plant Growth Regulation* 98: 5–22. https://doi.org/10.1007/s10725-022-00847-4.

Kalinowska, M., Hawrylak-Nowak, B., and Szymańska, M. (2013). The influence of two lithium forms on the growth, l-ascorbic acid content and lithium accumulation in lettuce plants. *Biological Trace Element Research* 152: 251–257. https://doi.org/10.1007/s12011-013-9606-y.

Kastori, R., Maksimović, I., and Putnikdelić, M. (2022). lithium in the environment and its effects on higher plants. *Contemporary Agriculture* 71: 226–239. https://doi.org/10.2478/contagri-2022-0030.

Kavanagh, L., Keohane, J., and Cabellos, G.G. (2018). Induced plant accumulation of lithium. *Geosciences* 8: 56. https://doi.org/10.3390/geosciences8020056.

Khatoon, R., Raksasat, R., Ho, Y.C. et al. (2023). Reviewing advanced treatment of hydrocarbon-contaminated oilfield-produced water with recovery of lithium. *Sustainability* 15: 16016. https://doi.org/10.3390/su152216016.

Li, S., Jiang, M., Xie, Y. et al. (2018). Developing high-performance lithium metal anode in liquid electrolytes: challenges and progress. *Advanced Materials* 30: 1706375. https://doi.org/10.1002/adma.201706375.

Malik, K.A. and Maqbool, A. (2020). Transgenic crops for biofortification. *Frontiers in Sustainable Food Systems* 4. https://doi.org/10.3389/fsufs.2020.571402.

Mezni, A., Aoua, H., Khazri, O. et al. (2017). Lithium induced oxidative damage and inflammation in the rat's heart: protective effect of grape seed and skin extract. *Biomedicine & Pharmacotherapy* 95: 1103–1111. https://doi.org/10.1016/j.biopha.2017.09.027.

Naeem, A., Aslam, M., Saifullah. et al. (2021). Lithium: perspectives of nutritional beneficence, dietary intake, biogeochemistry, and biofortification of vegetables and mushrooms. *Science of The Total Environment* 798: 149249. https://doi.org/10.1016/j.scitotenv.2021.149249.

Otřísal, P. and Sabol, J. (2024). Distribution of lithium/nickel in soil and its uptake by plants. In: *Lithium and Nickel Contamination in Plants and the Environment*, 1–33. Singapore: World Scientific https://doi.org/10.1142/9789811283123_0001.

Plotnikov, E., Korotkova, E., and Voronova, O. (2018). Lithium salts of Krebs cycle substrates as potential normothymic antioxidant agents. *Journal of Pharmacy and Bioallied Sciences* 10: 240–245. https://doi.org/10.4103/jpbs.Jpbs_140_18.

Reich, M., Aghajanzadeh, T., Helm, J. et al. (2017). Chloride and sulfate salinity differently affect biomass, mineral nutrient composition and expression of sulfate transport and assimilation genes in Brassica rapa. *Plant and Soil* 411: 319–332. https://doi.org/10.1007/s11104-016-3026-7.

Robinson, B.H., Yalamanchali, R., Reiser, R. et al. (2018). Lithium as an emerging environmental contaminant: mobility in the soil-plant system. *Chemosphere* 197: 1–6. https://doi.org/10.1016/j.chemosphere.2018.01.012.

Rzymski, P., Niedzielski, P., Siwulski, M. et al. (2017). Lithium biofortification of medicinal mushrooms Agrocybe cylindracea and Hericium erinaceus. *Journal of Food Science and Technology* 54: 2387–2393. https://doi.org/10.1007/s13197-017-2679-4.

Saltzman, A., Birol, E., Oparinde, A. et al. (2017). Availability, production, and consumption of crops biofortified by plant breeding: current evidence and future potential. *Annals of the New York Academy of Sciences* 1390: 104–114. https://doi.org/10.1111/nyas.13314.

Sarma, H., Basumatary, T., Yousaf, B. et al. (2024). Nanoplastics and lithium accumulation in soil–plant systems: assessing uptake, toxicological effects, and potential synergistic interactions. *Current Research in Biotechnology* 7: 100170. https://doi.org/10.1016/j.crbiot.2023.100170.

Seidel, U., Baumhof, E., Hägele, F.A. et al. (2019). Lithium-rich mineral water is a highly bioavailable lithium source for human consumption. *Molecular Nutrition and Food Research* 63: 1900039. https://doi.org/10.1002/mnfr.201900039.

Shahzad, B., Mughal, M.N., Tanveer, M. et al. (2017). Is lithium biologically an important or toxic element to living organisms? An overview. *Environmental Science and Pollution Research* 24: 103–115. https://doi.org/10.1007/s11356-016-7898-0.

Shahzad, B., Tanveer, M., Hassan, W. et al. (2016). Lithium toxicity in plants: reasons, mechanisms and remediation possibilities – A review. *Plant Physiology and Biochemistry* 107: 104–115. https://doi.org/10.1016/j.plaphy.2016.05.034.

Shakoor, N., Adeel, M., Ahmad, M.A. et al. (2023a). Environment relevant concentrations of lithium influence soybean development via metabolic reprogramming. *Journal of Hazardous Materials* 441: 129898. https://doi.org/10.1016/j.jhazmat.2022.129898.

Shakoor, N., Adeel, M., Ahmad, M.A. et al. (2023b). Reimagining safe lithium applications in the living environment and its impacts on human, animal, and plant system. *Environmental Science and Ecotechnology* 15: 100252. https://doi.org/10.1016/j.ese.2023.100252.

Shakoor, N., Adeel, M., Azeem, I. et al. (2023c). Interplay of higher plants with lithium pollution: global trends, meta-analysis, and perspectives. *Chemosphere* 310: 136663. https://doi.org/10.1016/j.chemosphere.2022.136663.

Shen, J., Li, X., Shi, X. et al. (2020). The toxicity of lithium to human cardiomyocytes. *Environmental Sciences Europe* 32: 59. https://doi.org/10.1186/s12302-020-00333-6.

Sheoran, S., Kumar, S., Ramtekey, V. et al. (2022). Current Status and Potential of Biofortification to Enhance Crop Nutritional Quality: an Overview. *Sustainability* 14: 3301. https://doi.org/10.3390/su14063301.

Solti, I., Kolics, É., Keszthelyi, S. et al. (2022). Evaluation of the acaricidal activity of lithium chloride against tetranychus urticae (Acari: Tetranychidae). *Horticulturae* 8: 1127. https://doi.org/10.3390/horticulturae8121127.

Stolarz, M., Król, E., and Dziubińska, H. (2015). Lithium distinguishes between growth and circumnutation and augments glutamate-induced excitation of Helianthus annuus seedlings. *Acta Physiologiae Plantarum* 37: 69. https://doi.org/10.1007/s11738-015-1814-y.

Tanveer, M., Hasanuzzaman, M., and Wang, L. (2019). Lithium in environment and potential targets to reduce lithium toxicity in plants. *Journal of Plant Growth Regulation* 38: 1574–1586. https://doi.org/10.1007/s00344-019-09957-2.

van Wyk, Y., Dippenaar, M.A., and Ubomba-Jaswa, E. (2024). Enhancing hydrological analysis by incorporating environmental and artificial tracers of an altered vadose zone: a systematic review. *Journal of African Earth Sciences* 212: 105209. https://doi.org/10.1016/j.jafrearsci.2024.105209.

Xu, Y., Shui, X., Gao, M. et al. (2024). Toxicological effects and mechanisms of lithium on growth, photosynthesis and antioxidant system in the freshwater microalga chromochloris zofingiensis. *Journal of Hazardous Materials* 469: 133898. https://doi.org/10.1016/j.jhazmat.2024.133898.

Xu, Z., Zhang, Z., Peng, S. et al. (2023). Influences of lithium on soil properties and enzyme activities. *Chemosphere* 313: 137458. https://doi.org/10.1016/j.chemosphere.2022.137458.

Yelatontsev, D. and Mukhachev, A. (2021). Processing of lithium ores: industrial technologies and case studies – A review. *Hydrometallurgy* 201: 105578. https://doi.org/10.1016/j.hydromet.2021.105578.

Yu, J., Wu, Q., Bu, L. et al. (2022). Experimental study on improving lithium extraction efficiency of salinity-gradient solar pond through sodium carbonate addition and agitation. *Solar Energy* 242: 364–377. https://doi.org/10.1016/j.solener.2022.07.027.

10

Global Action Plan for Agricultural Diversification for Achieving Zero Hunger

Maryam Javid[1*], *Maria Javaid*[1], *Sajid Fiaz*[2]

[1] *Department of Botany, University of Peshawar, Peshawar, Khyber Pakhtunkhwa, Pakistan*
[2] *Institute of Molecular Biology and Biotechnology, The University of Lahore, Lahore, Pakistan*

*Corresponding author: maryamjavid@uop.edu.pk

10.1 Introduction

A highly specialized agri-food system known as agricultural diversification has helped increase food production globally by utilizing improved cultivars, advancing agricultural gear and technology, and bringing in outside inputs like pesticides, fertilizers, and irrigation (Francaviglia et al. 2022). However, through decreasing air and water quality, soil erosion, depletion of soil organic matter, resistance of weeds, pests, and pathogens to pesticides, and a decline in soil quality and agrobiodiversity, agricultural intensification and unsustainable soil management harm the environment (Di Bene et al. 2022). It is common knowledge that adjustments to agricultural systems are necessary to ensure their sustainability by striking a balance between environmental objectives and socioeconomic factors related to food production. Encouraging the redesign of intensive agricultural systems requires effective diversification methods and management techniques (DeBoe 2020). Many researchers believe that by providing various ecosystem services, crop diversity can increase crop output and resource use efficiency. Long-term increases in crop productivity and cropping system resilience can be achieved by combining low-input management techniques like agro-ecology, conservation agriculture, and organic farming with agricultural diversification techniques like crop rotation, cover crops, multiple cropping, or intercropping (Hawes, Iannetta, and Squire 2021).

The adoption of combined agricultural diversification strategies may require the use of certain financial instruments despite the broad scientific consensus regarding the potential agroecological and socioeconomic benefits of crop diversification. This is because the short-term economic costs of agricultural diversification may outweigh the long-term benefits to the environment and ecology (Alcon et al. 2024). In particular, there are transition costs associated with redesigning and diversifying agricultural intensification that farmers and consultants need to take into account in the short- and medium-term. These expenses are related to learning new technical skills and information necessary to manage the risks associated with "unknown" crops and their new market, particularly during the early stages of adoption (Mgale and Yunxian 2021).

Achieving zero hunger is an important component of the United Nations Sustainable Development Goals (SDGs), and requires novel technologies that address the intricate exchange of factors that affect food security (Pandey and Pandey 2023). One of the main methods for achieving zero hunger is agriculture diversification. This process involves the production of various varieties of livestock and crops to increase the agriculture systems' adaptability, which contributes to improved nutritional outcomes and food security (Singh, Sreenivasulu, and Prasad 2022).

In achieving zero hunger, agriculture diversification is of great importance. Researchers explore that agriculture diversification has consistently contributed to household food security, specifically in developing countries where farmers are the backbone of agriculture diversification (Sherifa 2021). Agriculture diversification was connected with household food security, as it provides a large yield and increases economic and nutrition security as studied in Malawi. Another research in Northern Ghana showed that crop diversification is the key factor in achieving zero hunger (Madsen et al. 2021).

For achieving zero hunger, agriculture diversification provides several benefits. The primary importance of agriculture diversification is the increase in crop resilience, which mitigates the risk associated with pests, climate change, and diseases. Second, it improves diversity, which is important for nutrient uptake and reduces the risk of malnutrition and micro-

Crop Biofortification: Biotechnological Approaches for Achieving Nutritional Security under Changing Climate, First Edition. Edited by Adnan Noor Shah et al.

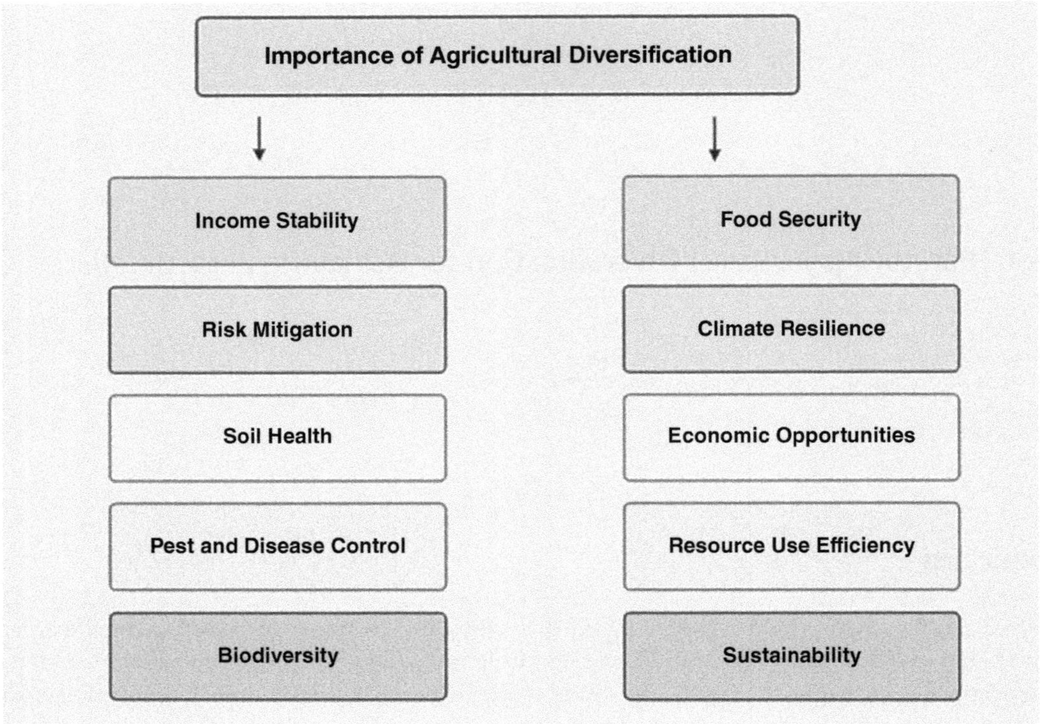

Figure 10.1 Importance of agricultural diversification.

nutrient deficiencies. Another importance of agriculture diversification is its improved income and household food security for farmers (Vernooy 2022). Moreover, the agriculture diversification strategies are hindered by many hurdles, such as limited inputs, credits, markets, and policies for research funding to support all these initiatives (Figure 10.1). However, the importance of crop diversification is not uniformly distributed, connected with some policymakers, experiencing more improvement in nutritional outcomes and food security (Zerssa et al. 2021).

Global action plan plays an important role in addressing the complicated challenges of food security. These plans identify the causes of malnutrition and hunger and also provide a framework for their coordination (Islam and Kieu 2020).

The G20 plan is one of the most important global action plans on food security and sustainable food systems, approved by 20 leaders in 2015. This plan is designed by 20 countries to reflect the comparative benefits and significant contributions to achieving G20 food security and malnutrition frameworks. These actions are planned to focus on developing and low-income countries for farmers, youth, and rural women (Zorbas et al. 2024).

The United Nations Updated Consolidated Framework for Action (UCFA) is another important global action plan; this global action plan is based on 10 key principles, twin track to nutrition and food security, smallholders especially women, the need for a detailed approach, focus of programs and policies, and step-by-step implementations of the right to adequate food, sustainability, participation, organizational development, coherence and coordination, good governance and accountability, and international cooperation (Sibanda and Mwamakamba 2021).

The World Food Security (WFS) developed a global strategic framework for food security and nutrition, providing a framework for nutrition policies and food security. The Global Action Plan aims to emphasize the need for a dual strategy that integrates short- and long-term sustainable agriculture, nutrition, food security, and rural development programs, with direct actions to combat hunger for the most vulnerable people (Muhie 2022) (Table 10.1).

10.2 Present Status of Global Food Security

Food security is a complicated concept that has attracted constant attention and gained economic significance since the 1974 World Food Conference, where issues of hunger, famine, and the food crisis were extensively discussed (Akbari et al. 2022). Though it has changed over time, the idea the definition of food security remains unchanged: "a state in which all

Table 10.1 The description of key role for global strategic framework.

Key role	Description	Reference
Common vision	Integrate the action of various groups like international organizations, governments, and civil societies.	Barrios et al. (2020)
Identify areas for intervention	Investment in agriculture, identifying food price fluctuation, and sustainable use of food.	Pawlak and Kołodziejczak (2020)
Inclusive plan	Tackling the main causes of hunger food security, and malnutrition including inequality, poverty, and limited access to resources.	Alaimo, Chilton, and Jones (2020)
Support for vulnerable population	Children, women, and farmers are unfairly affected by food insecurity.	Berry (2020)
Right adequate food	Fundamental human rights for ensuring food security for all.	Barrett (2021a)
International cooperation	Dealing with global challenges like global warming, economic shocks, and conflict.	Krishnamurthy, Choularton, and Kareiva (2020)

individuals, in all circumstances, have physical, social, and economic access to enough food that satisfies their dietary needs and food preferences for an active, healthy life." Four pillars support this complex definition: use, affordability, availability (i.e. sufficient food supply), stability (i.e. no food shortages or fluctuations in the seasons), and affordability (Murrell and Jones 2020).

People go without food not because it is hard to get in the market, but rather because they cannot afford to buy it. It is crucial to distinguish between the supply and demand of food. The situation of food supply is generally referred to as availability, and during the past few decades, availability has improved due to advancements in agricultural productivity (Hemathilake and Gunathilake 2022). In addition to the unequal distribution of food within households, the term "access" is frequently used to draw attention to the demand issue. The problem of rising food prices worldwide has the potential to undo decades of progress in reducing severe poverty and famine (Barrett 2021a).

Since 2017 more than 122 million people facing hunger in the year between 2021 and 2022 (Wudil et al. 2022).

About 2.4 billion individuals, especially residents of rural areas and women are faced with access to nutritious food. With 22.3% (148.1 million) stunted children, 6.8% (45 million) wasted children, and 5.6% (37 million) overweight children in 2021, child malnutrition is still a serious problem (Suri 2023). Reduced cognitive development and long-term health implications are associated with stunting, wasting, and overweight (Srivastava et al. 2024).

Approximately 1.9 billion people did not consistently have access to a healthy diet in 2017. Although the majority of people affected by food insecurity live in South Asia and sub-Saharan Africa's low-income nations, this problem affects people all over the world. Every country must be able to obtain and sustain a variety of nutrient-rich meals (Organization 2021). Two of the United Nations' 2030 Sustainable Development Goals are to eradicate hunger and achieve food security. Despite concerted international efforts to end world hunger, the number of undernourished individuals increased in 2015 after falling between 2003 and 2014.

The Food and Agriculture Organization (FAO) estimates that 2 billion people will have experienced moderate to severe food insecurity by 2020, and over 825 million people will have become famished (Gondwe 2021). Because of the extra 60 million victims of starvation since 2014, it has been predicted that there will be 840 million undernourished people by 2030, a substantial increase from the current projection of 380 million (Simon 2022). Approximately 830 million people globally experience hunger daily. The number of individuals experiencing severe food insecurity increased drastically from 135 million in 2019 to 345 million in 2022 as a direct result of COVID-19 and the ongoing conflict in Ukraine (Xιota\ acuteømegaτη 2023). It is projected that in 2030, 670 million people, or 8% of the world's population, will still be hungry despite the global economy's recovery. Similar goals were set in 2015 by the 2030 Agenda for Sustainable Development, which called for the eradication of world hunger, food insecurity, and starvation. Fifty million people in 45 countries are at risk of severe hunger, according to the World Food Programme. Droughts and other natural disasters have historically been the main contributors to food scarcity (Wudil et al. 2022).

10.2.1 Challenges in Achieving Zero Hunger

Achieving zero hunger is the most complicated and multifaceted process (Chichaibelu et al. 2021). Climate change, urbanization, and food prices are the main challenges in achieving zero hunger. Due to climate change, the crop yield is reduced, which impacts food security and leads to global food security. Africa has been identified as one of the continents most severely impacted by climate change for two reasons: first, due to its geographic makeup, which consists primarily of land spread across the warming tropics, and second, because African nations have limited resources to adapt socially, economically, and culturally to the effects of climate change.

Loss of biodiversity is another challenge in achieving zero hunger. To guarantee enough nutrition, biodiversity is essential. Without species diversity, ecological diversity, and genetic diversity in fish, plants, and animals, it is impossible to provide the micronutrient requirements for human health (Knez, Ranić, and Gurinović 2024). Many crops and fruits have enhanced nutrient contents as a result of pollination. But there's a strong argument to reconsider farming practices and dietary habits given the startling rate at which ecosystems are degrading and biodiversity is disappearing. Sustainable diets encourage the use of various foods, such as regional and traditional foods that draw on species that are high in nutrients, such as plants, animals, and wild, underappreciated, and neglected species (Figure 10.2). Reversing tendencies that cause ecological harm, deforestation, biodiversity loss, and general degradation of our natural resources depend on sustainable agriculture (Mondal and Palit 2022).

10.2.2 The Impact of Agriculture Diversification in Achieving a Sustainable Food System

Agricultural diversification is an important strategy for achieving a sustainable food system. Sustainable food system aims to ensure the food production, processing, and consumption of food in a way that is socially, economically, and environmentally sustainable. Agriculture diversification is important for achieving the goal of zero hunger by promoting biodiversity, enhancing ecosystem services, and improving the resilience of the agriculture system (Hoek et al. 2021).

Agriculture diversification systems strengthen biodiversity by incorporating various crops, livestock, and many other organisms. This also helps to maintain ecosystem services such as pest control, pollination and nutrient cycling, water quality, climate regulation, and soil health.

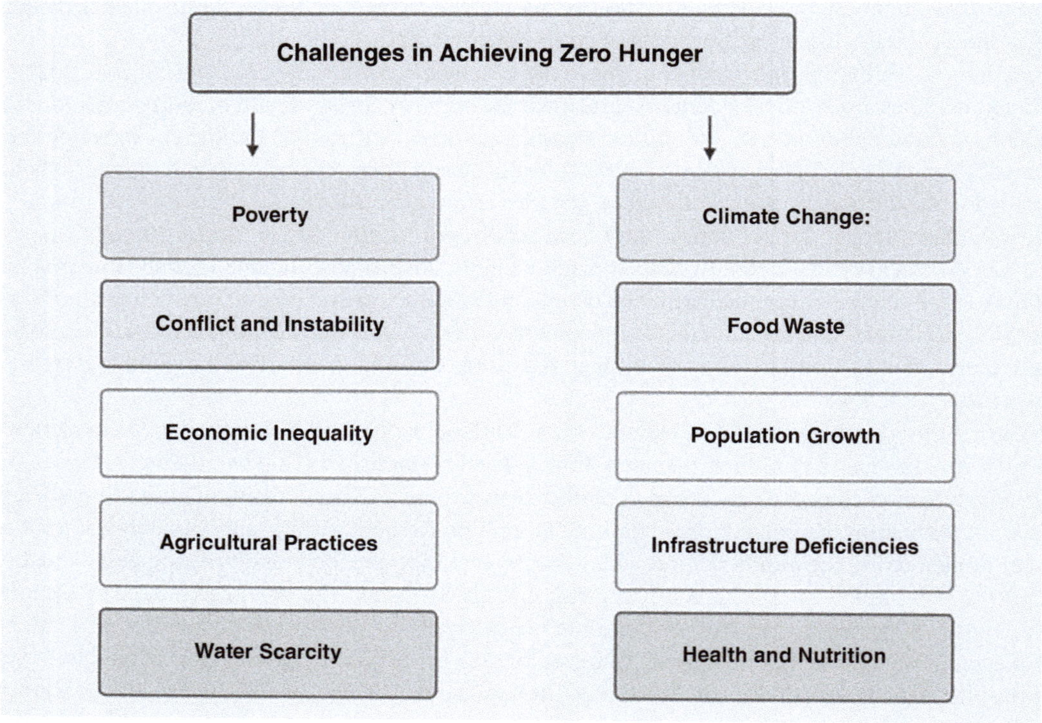

Figure 10.2 Challenges in achieving zero hunger.

Agricultural diversification systems also boost food security by providing a diverse range of food products. This is particularly essential for susceptible populations that have limited access to nutritious foods (Rehman et al. 2022).

10.3 Essential Principles of Agricultural Diversification

Crop diversity is important for maintaining sustainable ecosystem services and a sustainable agriculture system. It involves the improvement of healthy soil and pest management, as a result of which we can grow multiple crops from different varieties and families together. Livestock incorporation is another key factor of agriculture diversification. It involves the incorporation of live stocks into crops to decrease the need for synthetic inputs and improve soil health and biodiversity (Beillouin et al. 2021).

One of the key factors of agriculture diversification is agroecological practices which involve the agriculture management system in such a way that maintains the natural ecosystem, by using different types of techniques such as cover cropping, crop rotation, and intercropping to promote biodiversity and ecosystem services (Reiss and Drinkwater 2022).

Agricultural diversification has several significant social, economic, and environmental effects. Improved food security, job opportunities, and increased profitability are some of the economic importance. The environmental benefits include increased biodiversity, improved ecosystem services, and reduction of environmental hazards (Kurdyś-Kujawska, Strzelecka, and Zawadzka 2021).

The primary argument upheld was that growing food to feed the local population and achieve food autonomy. The production and consumption levels of the territorial food system are included from this perspective. This perspective holds that cattle grazing takes up too much land and prevents other forms of food production from occurring (Grandin 2022).

The second vision relied on the diversification of production and the creation of independent value chains. This vision pertains to the growth of diverse value chains and processing facilities inside the territory, and the expansion of farm activities, and involves both territorial and farm levels. This viewpoint, which criticizes CAP's intensification of dairy farms based mostly on per-hectare premiums, is defended by eight individuals. Degrowth and the creation of human-sized, diverse farms with more workers per hectare are called for (Pachoud, Dendoncker, and Girard 2024) (Figure 10.3).

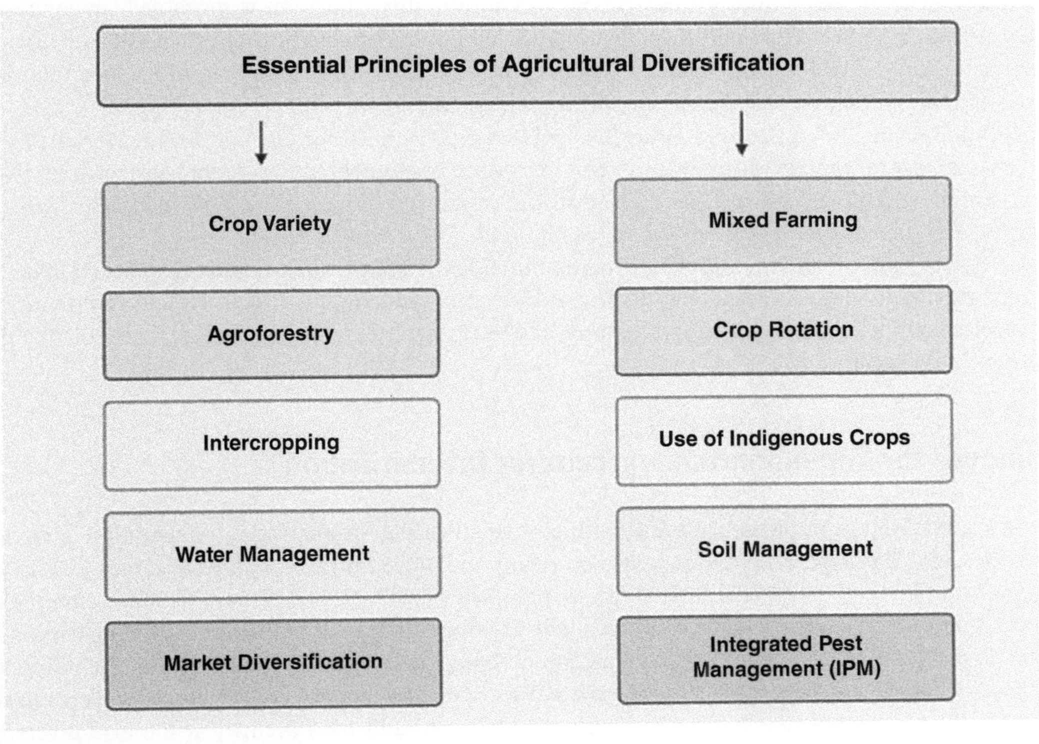

Figure 10.3 Essential principles of agricultural diversification.

10.4 Regulatory Framework for Expanding Agricultural Crop Diversity

Global communities have highly appreciated the critical role of agricultural diversification in food security and interrelated challenges such as climate change, environmental degradation, and loss of biodiversity. Various international agreements and commitments have highlighted the need for transitioning toward sustainable agriculture practices and more diversity (Raj et al. 2022).

The United Nations 2015 SDGs delivered a detailed policy for achieving a sustainable future. The second goal of this policy is to achieve zero hunger by promoting sustainable agriculture and implementing new strategies to increase biodiversity, improve crops, and maintain a sustainable ecosystem. Agriculture diversification is the key factor for achieving all these goals as it can increase food security, increase the resilience of the farming system, improve ecosystem services, and reduce the impacts of climate change (Saxena et al. 2021).

United Nations Framework Convention on Climate Change (UNFCCC) in 2015 offered the Paris Agreement; the aim of this agreement is to reduce global temperature rise to well below 2 °C above preindustrial level, with struggles to reduce it to 1.5 °C. Agricultural diversification also plays an important role in climate change mitigation and adaptation. This agreement encourages countries to adopt strategies and policies that promote environmental sustainability, including practices that improve soil carbon sequestration, enhance the resilience of agricultural systems to climate change impacts, and reduce the emission of greenhouse gases (Layachi 2021).

In 1992, an international agreement Convention on Biological Diversity (CBD) was adopted; this agreement aims to conserve biological diversity, promote sustainable use of ecological services, and fair sharing of resources arising from genetic resources. Biological diversity is closely related to agricultural diversification, as it involves the cultivation of various climate resilience crops and livestock, which can maintain the ecosystem and support the wild species Cond serration (Parks and Tsioumani 2023). Strategic plan for biodiversity 2011–2020 and after 2020 global biodiversity framework highlight the importance of biological diversity conservation across all sectors including agriculture and sustainable use and conservation of biological diversity (Perino et al. 2022).

10.4.1 National Policies for Promoting Agricultural Diversification

Different national policies also promote agricultural diversification in various geographical regions, such as Asia, the European Union, Africa, and India.

In Asia, through different policies and private sectors, agricultural diversification is promoted. In Japan and China, different policies are implemented to support the adaptation of agricultural diversification, such as using organic farming methods and the inclusion of livestock into farming systems. These policies are formed to improve food security, enhance rural livelihood, and reduce environmental impacts (Müller et al. 2021).

The European Union Common Agriculture Policy has pointed out the need for diversification. Measures such as crop rotation needs and support for organic farming and agroforestry are implemented to promote agricultural diversification practices. The common agriculture policy of the European Union aims to diminish the dependency on synthetic inputs, improve soil health, and increase ecological services (Rosati, Borek, and Canali 2021).

Comprehensive Africa Agriculture Development Programme (CAADP) targets to promote agricultural diversification in Africa. This program is focused on increasing agricultural outcomes, reducing food security, and improving rural livelihood. These strategies include crop rotation, intercropping, and the integration of legumes to improve soil health (Lulanga, Marinda, and Khayeka-Wandabwa 2022).

10.5 Techniques for Implementing Agricultural Diversification

Various practices are involved in implementing agricultural diversification, promoting the integration of multiple crops, agricultural activities, and livestock. Crop rotation, intercropping, and polycultural cropping are some of the important key strategies. Crop rotation is the practice of growing different crops on the same plot of land in succession (Sekaran et al. 2021). It can reduce pest and disease pressure, improve biodiversity, and improve soil health. Growing two or more crops at the same time on the same piece of land is known as intercropping, and it can have advantages, including better weed control, more effective use of resources, and greater ecosystem services. The practice of growing several crop species simultaneously, or polyculture, can increase agricultural systems' resilience and food security while also diversifying them further (Yang, Zhang, and Li 2021) (Table 10.2).

Table 10.2 Support Mechanisms and Incentives for Farmers to Diversify Agriculture.

Mechanism	Description	References
Financial incentives	Giving farmers that use varied farming techniques, such as crop rotation, intercropping, or agroforestry, direct financial subsidies.	Burgess, Cano, and Parkes (2022)
Tax credits	Providing tax breaks or other financial advantages to farmers who make investments in expanding their farming businesses.	Thilmany et al. (2022)
Ecosystem services	Putting in place programs that compensate farmers for the environmental advantages such as carbon sequestration, biodiversity preservation, or water management that come from their varied farming practices.	Jat et al. (2022)
Technical assistance	Bolstering agricultural extension services to give farmers technical advice, training, and demonstrations on how to apply varied farming methods.	Dabek (2022)
Research	Providing funding for the development of novel animal breeds, crop types, and agroecological technology that can be incorporated into diverse agricultural systems.	Kremsa (2021)
Capacity building	Funding initiatives that increase the expertise and understanding of farmers, extension agents, and other stakeholders so they can comprehend and implement varied farming methods.	Adamsone-Fiskovica and Grivins (2022)
Food system	Creating regional and local food value chains that open up markets for a variety of agricultural goods, including value-added goods, organic vegetables, and specialty crops.	Hermiatin et al. (2022)
Infrastructure development	Putting money into processing centers, transportation networks, and storage facilities to make it easier to distribute and sell a variety of agricultural products.	Nezamova and Olentsova (2021)
Income stabilization	Putting in place initiatives that lessen the risks connected to production shocks and price volatility in order to stabilize farm incomes.	Maurin et al. (2022)

To achieve agricultural diversity, it is also important to promote agroforestry and sustainable land management practices. Agroforestry, or the planting of trees and shrubs in agricultural settings, has the potential to improve soil fertility, increase carbon sequestration, and boost biodiversity, among other advantages (Marques et al. 2022). The overall sustainability and resilience of agricultural systems can also be enhanced by sustainable land management techniques like integrated pest management, conservation tillage, and the use of cover crops (Quintarelli et al. 2022).

Promoting native crops and conventional farming methods can also be a useful tactic for putting agricultural diversification into practice. Diverse and resilient agricultural practices that are well-suited to regional environmental and cultural conditions have been established by numerous traditional farming systems throughout the world. Agricultural diversification can be improved and cultural heritage and traditional ecological knowledge can be preserved by encouraging the cultivation and maintenance of these traditional crops and farming practices (Esquivel et al. 2021).

A multi-stakeholder strategy involving farmers, policymakers, academics, and community people is necessary for the successful implementation of these measures for agricultural diversification. This entails giving farmers the required financial and technical assistance in addition to establishing supportive legislative frameworks that encourage the use of diverse farming methods. Agricultural systems can become more resilient, sustainable, and equipped to handle future concerns related to food security and the environment by adopting these measures (Melaku and Minh 2021).

10.6 Evaluation for Assessing Progress in Agricultural Diversification

Assessing the development and effects of agricultural diversification programs requires efficient monitoring and assessment. This entails creating suitable metrics to assess the effectiveness of diversification initiatives (Adhikari et al. 2023). The variety of crops and livestock raised in agricultural systems, the rates at which diverse methods are adopted, the increase in food security and nutrition, the development of ecosystem services, and the financial and social gains made by farmers and communities are a few examples of key indicators. These metrics ought to be created with the complexity of agricultural diversification and its benefits to sustainable development in mind (Яценко et al. 2021).

To guarantee ongoing progress in the application of agricultural diversification, feedback loops are essential. To help stakeholders, learn from successes and setbacks and make evidence-based decisions, policies, programs, and interventions should be modified and improved in response to the results of monitoring and evaluation. In the end, this iterative process improves the long-term viability and impact of these projects by ensuring that diversification efforts remain sensitive to the changing needs and situations of farmers and communities (Stephenson et al. 2021).

Stakeholders may effectively follow the development and impact of agricultural diversification, make well-informed decisions, and constantly improve the strategies and interventions put into place by building thorough monitoring and evaluation frameworks. To fully utilize varied agricultural systems in addressing food security, environmental sustainability, and the resilience of rural communities, a comprehensive approach is necessary (Sekaran et al. 2021).

10.7 Importance of International Corporation

A worldwide strategy incorporating partnerships, multilateral organizations, and international cooperation is needed to promote agricultural diversification. The complexity of agricultural diversification, which crosses national boundaries and necessitates teamwork to address the interconnected problems of food security, climate change, and environmental sustainability, highlights the significance of international cooperation (Thomas, Barczak, and Zakhia-Rozis 2022). To assist agricultural diversification programs, multilateral organizations like the World Bank, the Food and Agriculture Organization (FAO), and the United Nations are essential because they offer a worldwide forum for collaboration, knowledge sharing, and resource mobilization (Canton 2021).

Promoting agricultural diversification also requires collaborations between the public and corporate sectors, as well as between civil society organizations. These collaborations can help to develop capacity and foster settings that are supportive of the adoption of diverse agricultural techniques by facilitating the exchange of best practices, knowledge, and resources. Promoting sustainable food systems and zero hunger objectives is essential as well because it increases understanding of the value of agricultural diversification and the teamwork required to meet these objectives (Kotykova et al. 2022).

By cooperating, nations and organizations can take advantage of their combined resources and strengths to advance agricultural diversity, deal with the underlying causes of poverty and malnutrition, and help achieve the SDGs. To create a more resilient, sustainable, and egalitarian food system that benefits everyone, there must be international cooperation and solidarity (Mensi and Udenigwe 2021).

10.8 Conclusion

Agricultural diversification is an important strategy for achieving zero hunger and addressing the complex challenge of food security. To achieve zero hunger, diversified farming practices, improving biodiversity, and enhancing the resilience of agricultural systems, played significant roles. Moreover, the successful implementation of agricultural diversification needs a comprehensive approach that includes national and international policies and corporations and targeted support mechanisms for farmers.

The chapters focused on the importance of global action plans and national and international policies in creating an environment for the adaptation of diversified agricultural practices. The techniques provide a clear vision, identify areas for intervention, and encourage stake-holders, including different organizations, and private sectors.

Overcoming these challenges will be important for farmers to realize the benefits and adopt agricultural diversification on a wider scale/tiny scale, given that lack of input access (seeds, fertilizers), market linkages and credit, as well as policy and research funding support have remained an issue with diversified farming. Developing effective monitoring and evaluation frameworks is vital to measure the implementation progress of diversification initiatives as well as assess their impact, which in turn will provide insights for enhancing these strategies.

With a commitment to agricultural diversification and tapping into the forces of global collaboration and solidarity, there is great potential for our world community to take steps closer to achieving zero hunger while building more sustainable food systems on which all depend.

References

Adamsone-Fiskovica, A. and Grivins, M. (2022). Knowledge production and communication in on-farm demonstrations: putting farmer participatory research and extension into practice. *The Journal of Agricultural Education and Extension* 28 (4): 479–502. https://doi.org/10.1080/1389224X.2021.1953551.

Adhikari, L., Komarek, A.M., de Voil, P. et al. (2023). A framework for the assessment of farm diversification options in broadacre agriculture. *Agricultural Systems* 210: 103724.

Akbari, M., Foroudi, P., Shahmoradi, M. et al. (2022). The evolution of food security: where are we now, where should we go next? *Sustainability* 14 (6): 3634.

Alaimo, K., Chilton, M., and Jones, S.J. (2020). Food insecurity, hunger, and malnutrition. In: *Present Knowledge in Nutrition*, 311–326. Amsterdam: Elsevier. https://www.sciencedirect.com/science/article/pii/B9780128184608000174.

Alcon, F., Albaladejo-García, J.A., Martínez-García, V. et al. (2024). Cost benefit analysis of diversified farming systems across Europe: incorporating non-market benefits of ecosystem services. *Science of the Total Environment* 912: 169272.

Barrett, C.B. (2021a). Overcoming global food security challenges through science and solidarity. *American Journal of Agricultural Economics* 103 (2): 422–447. https://doi.org/10.1111/ajae.12160.

Barrett, C.B. (2021b). Overcoming global food security challenges through science and solidarity. *American Journal of Agricultural Economics* 103 (2): 422–447. https://doi.org/10.1111/ajae.12160.

Barrios, E., Gemmill-Herren, B., Bicksler, A. et al. (2020). The 10 elements of agroecology: enabling transitions towards sustainable agriculture and food systems through visual narratives. *Ecosystems and People* 16 (1): 230–247. https://doi.org/10.1080/26395916.2020.1808705.

Beillouin, D., Ben-Ari, T., Malézieux, E. et al. (2021). Positive but variable effects of crop diversification on biodiversity and ecosystem services. *Global Change Biology* 27 (19): 4697–4710. https://doi.org/10.1111/gcb.15747.

Berry, E.M. (2020). Food insecurity, social inequity, and sustainability. *Hidden Hunger and the Transformation of Food Systems* 121: 95–104.

Burgess, A.J., Cano, M.E.C., and Parkes, B. (2022). The deployment of intercropping and agroforestry as adaptation to climate change. *Crop and Environment* 1 (2): 145–160.

Canton, H. (2021). Food and agriculture organization of the United Nations – FAO. In: *The Europa directory of international organizations* 2021: 297–305. Routledge. https://www.taylorfrancis.com/chapters/edit/10.4324/9781003179900-41/food-agriculture-organization-united-nations%E2%80%94fao-helen-canton.

Chichaibelu, B.B., Bekchanov, M., von Braun, J. et al. (2021). The global cost of reaching a world without hunger: Investment costs and policy action opportunities. *Food Policy* 104 (12): 102151.

Dabek, G. (2022). *Renewing International Extension to Equip Farmers for a Changing Climate*. Chicago: Chicago Council on Global Affairs. https://globalaffairs.org/sites/default/files/2021-09/FINALExtension%20Policy%20Brief%20CCGA_2021_v4.pdf.

DeBoe, G. (2020). Economic and environmental sustainability performance of environmental policies in agriculture. https://www.oecd-ilibrary.org/agriculture-and-food/economic-and-environmental-sustainability-performance-of-environmental-policies-in-agriculture_3d459f91-en.

Di Bene, C., Francaviglia, R., Farina, R. et al. (2022). Agricultural diversification. *Agriculture* 12 (3): 369. MDPI. https://www.mdpi.com/2077-0472/12/3/369.

Esquivel, K.E., Carlisle, L., Ke, A. et al. (2021). The "sweet spot" in the middle: why do mid-scale farms adopt diversification practices at higher rates? *Frontiers in Sustainable Food Systems* 5: 734088.

Francaviglia, R., Almagro, M., Lehtonen, H. et al. (2022). Agricultural diversification: benefits and barriers for sustainable soil management. *Frontiers in Environmental Science* 10: 1046354. https://www.frontiersin.org/articles/10.3389/fenvs.2022.1046354/full.

Gondwe, N.R. (2021). *UNSC Resolution 2417 and Food Insecurity in Yemen and South Sudan: International Humanitarian Law Obligations*. South Africa: University of Johannesburg. https://search.proquest.com/openview/3cdbed01064b5ac2639e8bde80a9d9ea/1?pq-origsite=gscholar&cbl=2026366&diss=y.

Grandin, T. (2022). Grazing cattle, sheep, and goats are important parts of a sustainable agricultural future. *Animals* 12 (16): 2092.

Hawes, C., Iannetta, P.P.M., and Squire, G.R. (2021). *Agroecological Practices for Whole-System Sustainability*. Wallingford: CABI Reviews. PAVSNNR202116005. https://doi.org/10.1079/PAVSNNR202116005.

Hemathilake, D. and Gunathilake, D. (2022). Agricultural productivity and food supply to meet increased demands. In: *Future Foods* 539–553. Amsterdam: Elsevier. https://www.sciencedirect.com/science/article/pii/B9780323910019000165.

Hermiatin, F.R., Handayati, Y., Perdana, T. et al. (2022). Creating food value chain transformations through regional food hubs: A review article. *Sustainability* 14 (13): 8196.

Hoek, A.C., Malekpour, S., Raven, R. et al. (2021). Towards environmentally sustainable food systems: decision-making factors in sustainable food production and consumption. *Sustainable Production and Consumption* 26: 610–626.

Islam, M.S. and Kieu, E. (2020). Tackling regional climate change impacts and food security issues: a critical analysis across ASEAN, PIF, and SAARC. *Sustainability* 12 (3): 883.

Jat, M.L., Chakraborty, D., Ladha, J.K. et al. (2022). Carbon sequestration potential, challenges, and strategies towards climate action in smallholder agricultural systems of South Asia. *Crop and Environment* 1 (1): 86–101.

Knez, M., Ranić, M., and Gurinović, M. (2024). Underutilized plants increase biodiversity, improve food and nutrition security, reduce malnutrition, and enhance human health and well-being. Let's put them back on the plate! *Nutrition Reviews* 82 (8): 1111–1124.

Kotykova, O., Babych, M., Iagodzinska, A. et al. (2022). Global initiatives on implementation of zero hunger policy. *Agricultural and Resource Economics: International Scientific E-Journal* 8 (2): 30–49.

Kremsa, V.Š. (2021). Sustainable management of agricultural resources (agricultural crops and animals). In: *Sustainable resource management* 99–145. Amsterdam: Elsevier. https://www.sciencedirect.com/science/article/pii/B9780128243428000109.

Krishnamurthy, P.K., Choularton, R.J., and Kareiva, P. (2020). Dealing with uncertainty in famine predictions: how complex events affect food security early warning skill in the Greater Horn of Africa. *Global Food Security* 26: 100374.

Kurdyś-Kujawska, A., Strzelecka, A., and Zawadzka, D. (2021). The impact of crop diversification on the economic efficiency of small farms in Poland. *Agriculture* 11 (3): 250.

Layachi, O.B. (2021). The impact of global warming on climate change and the role of the Paris agreement in maintaining the rate of temperature rise. *PalArch's Journal of Archaeology of Egypt/Egyptology* 18 (17): 317–344.

Lulanga, S., Marinda, P.A., and Khayeka-Wandabwa, C. (2022). Linkages between agricultural diversification, dietary diversity, and nutrition outcomes in sub-Saharan Africa: a systematic review. *Agricultural Sciences* 13 (7): 879–896.

Madsen, S., Bezner Kerr, R., Shumba, L. et al. (2021). Agroecological practices of legume residue management and crop diversification for improved smallholder food security, dietary diversity and sustainable land use in Malawi. *Agroecology and Sustainable Food Systems* 45 (2): 197–224. https://doi.org/10.1080/21683565.2020.1811828.

Marques, M.A., Anjos, L.H.C.D., and Sanchez Delgado, A.R. (2022). Land recovery and soil management with agroforestry systems. *Spanish Journal of Soil Science* 12: 10457.

Maurin, J., Calas, J., Godin, A. et al. (2022). Global biodiversity scenarios: what do they tell us for biodiversity-related financial risks? *AFD Research Papers* 266: 1–75.

Melaku, D. and Minh, T.T. (2021). *Multi-Stakeholder Platforms and Processes in Ethiopia: The Case From Agriculture and Water Management*. Research Program on Water, Land and Ecosystems. https://cgspace.cgiar.org/bitstream/handle/10568/115868/Multi_stakeholder%20Platforms%20in%20Ethiopia.pdf?sequence=1.

Mensi, A. and Udenigwe, C.C. (2021). Emerging and practical food innovations for achieving the Sustainable Development Goals (SDG) target 2.2. *Trends in Food Science & Technology* 111: 783–789.

Mgale, Y.J. and Yunxian, Y. (2021). Price risk perceptions and adoption of management strategies by smallholder rice farmers in Mbeya region, Tanzania. *Cogent Food & Agriculture* 7 (1): 1919370. https://doi.org/10.1080/23311932.2021.1919370.

Mondal, S. and Palit, D. (2022). Challenges in natural resource management for ecological sustainability. In: *Natural Resources Conservation and Advances for Sustainability*: 29–59. Amsterdam: Elsevier. https://www.sciencedirect.com/science/article/pii/B9780128229767000041.

Muhie, S.H. (2022). Novel approaches and practices to sustainable agriculture. *Journal of Agriculture and Food Research* 10: 100446.

Müller, M.F., Penny, G., Niles, M.T. et al. (2021). Impact of transnational land acquisitions on local food security and dietary diversity. *Proceedings of the National Academy of Sciences* 118 (4): e2020535118. https://doi.org/10.1073/pnas.2020535118.

Murrell, A. and Jones, R. (2020). Measuring food insecurity using the food abundance index: implications for economic, health and social well-being. *International Journal of Environmental Research and Public Health* 17 (7): 2434.

Nezamova, O.A. and Olentsova, J.A. (2021). Problems of improving the management of marketing channels in retail chains in the sale of agricultural products. *IOP Conference Series: Earth and Environmental Science* 839 (2):022060. https://doi.org/10.1088/1755-1315/839/2/022060/meta.

Organization, W.H. (2021). Asia and the Pacific Regional Overview of Food Security and Nutrition 2020: maternal and child diets at the heart of improving nutrition 2024. Rome: Food & Agriculture Organization. https://books.google.com/books?hl=en&lr=&id=pigXEAAAQBAJ&oi=fnd&pg=PR5&dq=Approximately+1.9+billion+people+did+not+consistently+have+access+to+a+healthy+diet+in+2017.+&ots=shy4MM9Uy9&sig=wO0sDf2Liy_xSnYO0SKLEDa3dpQ.

Pachoud, C., Dendoncker, N., and Girard, S. (2024). Identifying visions on agricultural diversification: a first step toward sustainable agrifood transformations in the Bauges. *Sustainability Science*. https://doi.org/10.1007/s11625-024-01513-6.

Pandey, P.C. and Pandey, M. (2023). Highlighting the role of agriculture and geospatial technology in food security and sustainable development goals. *Sustainable Development* 31 (5): 3175–3195. https://doi.org/10.1002/sd.2600.

Parks, L. and Tsioumani, E. (2023). Transforming biodiversity governance? Indigenous peoples' contributions to the convention on biological diversity. *Biological Conservation* 280: 109933.

Pawlak, K. and Kołodziejczak, M. (2020). The role of agriculture in ensuring food security in developing countries: considerations in the context of the problem of sustainable food production. *Sustainability* 12 (13): 5488.

Perino, A., Pereira, H.M., Felipe-Lucia, M. et al. (2022). Biodiversity post-2020: closing the gap between global targets and national-level implementation. *Conservation Letters* 15 (2): e12848. https://doi.org/10.1111/conl.12848.

Quintarelli, V., Radicetti, E., Allevato, E. et al. (2022). Cover crops for sustainable cropping systems: A review. *Agriculture* 12 (12): 2076.

Raj, S., Roodbar, S., Brinkley, C. et al. (2022). Food security and climate change: differences in impacts and adaptation strategies for rural communities in the global south and north. *Frontiers in Sustainable Food Systems* 5: 691191.

Rehman, A., Farooq, M., Lee, D.-J. et al. (2022). Sustainable agricultural practices for food security and ecosystem services. *Environmental Science and Pollution Research* 29 (56): 84076–84095.

Reiss, E.R. and Drinkwater, L.E. (2022). Promoting enhanced ecosystem services from cover crops using intra-and interspecific diversity. *Agriculture, Ecosystems & Environment* 323: 107586.

Rosati, A., Borek, R., and Canali, S. (2021). Agroforestry and organic agriculture. *Agroforestry Systems* 95: 805–821.

Saxena, A., Ramaswamy, M., Beale, J. et al. (2021). Striving for the United Nations (UN) Sustainable Development Goals (SDGs): what will it take? *Discover Sustainability* 2 (1): 20. https://doi.org/10.1007/s43621-021-00029-8.

Sekaran, U., Lai, L., Ussiri, D.A. et al. (2021). Role of integrated crop-livestock systems in improving agriculture production and addressing food security – a review. *Journal of Agriculture and Food Research* 5: 100190.

Sherifa, F.A. (2021). Livelihood diversification strategies and livelihood outcome of rural household in Ethiopia: A reviewed paper. *International Journal of African and Asian Studies* 74. https://www.academia.edu/download/73743569/third_article.pdf.

Sibanda, L.M. and Mwamakamba, S.N. (2021). Policy considerations for African food systems: towards the United Nations 2021 Food Systems Summit. *Sustainability* 13 (16): 9018.

Simon, M. (2022). *A Poison Like No Other: How Microplastics Corrupted our Planet and Our Bodies*. Washington: Island Press. https://books.google.com/books?hl=en&lr=&id=S_qZEAAAQBAJ&oi=fnd&pg=PP1&dq=Because+of+the+extra+60+million+victims+of+starvation+Since+2014,+it+has+been+predicted+that+there+will+be+840+million+undernourished+people+by+2030,+a+substantial+increase+from+the+current+projection+of+380+million&ots=7FI0B2lduu&sig=TegnxcC6MrNSf7PkNYxqUmQu2Tc.

Singh, R.K., Sreenivasulu, N., and Prasad, M. (2022). Potential of underutilized crops to introduce the nutritional diversity and achieve zero hunger. *Functional & Integrative Genomics* 22 (6): 1459–1465. https://doi.org/10.1007/s10142-022-00898-w.

Srivastava, S., Bisht, A., Laxmaiah, A. et al. (2024). *Global Health Challenges: Nutrition and Management*. Florida: CRC Press.

Stephenson, P.J., Bakarr, M., Bowles-Newark, N. et al. (2021). Conservation science in Africa: Mainstreaming biodiversity information into policy and decision-making. In: *Closing the Knowledge-Implementation Gap in Conservation Science* (eds. CC. Ferreira and CFC. Klütsch), vol. 4, 287–321. Berlin: Springer International Publishing. https://doi.org/10.1007/978-3-030-81085-6_11.

Suri, S. (2023). Navigating the nexus: challenges to global food security and the 2030 agenda – an overview. Food *Nutrition Chemistry* 1(3). https://www.researchgate.net/profile/Shoba-Suri/publication/378268959_Navigating_the_nexus_Challenges _to_global_food_security_and_the_2030_agenda-An_overview/links/65d08eab476dd15fb3436a60/Navigating-the-nexus-Challenges-to-global-food-security-and-the-2030-agenda-An-overview.pdf.

Thilmany, D., Bauman, A., Hadrich, J. et al. (2022). Unique financing strategies among beginning farmers and ranchers: differences among multigenerational and beginning operations. *Agricultural Finance Review* 82 (2): 285–309.

Thomas, A., Barczak, A., and Zakhia-Rozis, N. (2022). *Sustainable Food Systems for Food Security: Need for Combination of Local and Global Approaches*. France: Éditions Quae. https://library.oapen.org/handle/20.500.12657/57580.

Vernooy, R. (2022). Does crop diversification lead to climate-related resilience? Improving the theory through insights on practice. *Agroecology and Sustainable Food Systems* 46 (6): 877–901. https://doi.org/10.1080/21683565.2022.2076184.

Wudil, A.H., Usman, M., Rosak-Szyrocka, J. et al. (2022). Reversing years for global food security: a review of the food security situation in Sub-Saharan Africa (SSA). *International Journal of Environmental Research and Public Health* 19 (22): 14836.

Χιota\acuteømegaτη, E. (2023). *The Impact of the Russian-Ukrainian War on the Food Supply Chain: A Review of the Effects on the Global Economy and Malnutrition*. Thessaloniki: University of Macedonia. https://dspace.lib.uom.gr/handle/2159/29015.

Yang, H., Zhang, W., and Li, L. (2021). Intercropping: feed more people and build more sustainable agroecosystems. *Frontiers of Agricultural Science and Engineering* 8 (3): 373–386.

Zerssa, G., Feyssa, D., Kim, D.-G. et al. (2021). Challenges of smallholder farming in Ethiopia and opportunities by adopting climate-smart agriculture. *Agriculture* 11 (3): 192.

Zorbas, C., Resnick, D., Jones, E. et al. (2024). *From Promises to Action: Analyzing Global Commitments on Food Security and Diets since* 2015. International Food Policy Research Institute.

Яценко, О., Завадська, Ю., Христенко, О. et al. (2021). Innovative transformations of the agricultural complex in the context of global challenges of sustainable development. *Financial and Credit Activity Problems of Theory and Practice* 5 (40): 216–224.

11

Targeting Tissue-Specific Zinc Acquisition in Cereal

Azizullah Khalili[1, 2], *Yamin Bibi*[3], *and Abdul Qayyum*[1*]

[1] *Department of Agronomy, The University of Haripur, Haripur, Pakistan*
[2] *Department of Agronomy, Sayed Jamaluddin Afghani University (SJAU), Salar Bagh, Kunar, Afghanistan*
[3] *Department of Botany, Rawalpindi Women University, Rawalpindi, Pakistan*

*Corresponding author: aqayyum@uoh.edu.pk

11.1 Introduction to Tissue-Specific Zinc Acquisition

Food production and security stand as critical challenges for the twenty-first century, exacerbated by the continuous growth of the global population. Addressing these concerns aligns with the United Nations' Second Sustainable Development Goal (UN-SDG2), which aims to eradicate all forms of hunger by 2030. However, achieving this goal is complex due to various factors such as limited time, global health disparities, and socioeconomic barriers (Khalili et al. 2023). While agriculture has managed to meet the demand for food in the face of population growth, it faces constraints imposed by finite natural resources (UN 2012). Regrettably, agricultural systems have often prioritized objectives like farmers' profitability over broader goals such as human health and nutrition security (Mayer, Pfeiffer, and Beyer 2008). Cereals, including rice, wheat, maize, sorghum, and millet, serve as staple crops in many developing nations, forming the basis of diets and providing a significant portion of daily caloric intake (Serna Saldívar and Hernández 2020; Prasad, Shivay, and Kumar 2014). Despite the central role of cereals in global food security, there are concerns about the nutritional quality of these crops, particularly regarding micronutrients like zinc (Zn). Zn deficiency poses health risks and has been observed in various populations worldwide (Prasad et al. 1963; Sandstead et al. 1967; Caggiano et al. 1969; Hambidge et al. 1972). Efforts to address Zn deficiency in cereals have become increasingly important. Cereals are inherently rich in phytate compounds that inhibit Zn absorption, and certain processing methods further reduce their Zn content (Gowda et al. 2022; Aiqing et al. 2022). Consequently, populations reliant on these staple cereals are at risk of Zn deficiency; encouraging the consumption of whole grains, which retain essential nutrients like Zn in their bran and germ layers, presents a potential solution (Gallego-Castillo et al. 2021). Given that wheat, rice, and maize contribute significantly to daily energy intake globally, enhancing the Zn content and availability of these cereals is paramount (Poursarebani et al. 2014; Tilman et al. 2002). However, these cereals are considered poor sources of Zn and iron (Fe), particularly in terms of bioavailability (Timmer 2014; Cakmak 2008). Efforts to improve Zn uptake and utilization in cereals involve understanding the molecular, physiological, and genetic mechanisms underlying Zn acquisition (Bouain et al. 2018; Swamy et al. 2016). Optimizing Zn uptake activity and root system architecture can enhance Zn use efficiencies in crops (Khalili et al. 2023; Khalili et al. 2024). Furthermore, investigating the distribution of essential trace elements within cereal grains and exploring ways to improve their loading into the core endosperm are crucial steps in addressing nutrient deficiencies (Lombi et al. 2011; Kyriacou et al. 2014). Studies on Zn transporters, their expression patterns, and their response to environmental factors are essential for developing targeted strategies to enhance tissue-specific Zn uptake in cereals. Such endeavors hold promise for sustainable agriculture and improving both crop yields and nutritional quality. In this chapter, targeting tissue-specific Zn acquisition in cereal and its implications in biofortification is discussed.

Crop Biofortification: Biotechnological Approaches for Achieving Nutritional Security under Changing Climate, First Edition. Edited by Adnan Noor Shah et al.

11.2 Molecular Mechanisms of Zn Uptake in Cereal Roots

Zn is primarily absorbed by plant roots as Zn^{2+} ions through mechanisms involving mass flow and diffusion. These passive uptake processes depend on the movement of water molecules and the concentration gradient of Zn. The hyperpolarization of the root cell plasma membrane (PM), driven by the H^+-ATPase system's activity, plays a crucial role in the uptake of Zn^{2+}. This system uses energy from ATP hydrolysis to pump H^+ ions out of the cell, resulting in a hyperpolarized PM. This hyperpolarization enhances the uptake rates of cations and lowers the soil pH. Unlike water molecules, charged Zn^{2+} ions cannot freely pass through cell membranes (Alberts et al. 2007). Therefore, specific transporter proteins are required to facilitate the transport of these divalent cations across the cell membranes (Guerinot 2000; Clemens et al. 2002). The absence of ATP involvement in these transporter proteins indicates that Zn^{2+} uptake is passive rather than active. Additionally, Zn^{2+} absorption can occur through nonselective cation channels, which facilitate the passive transfer of various cations (Demidchik, Davenport, and Tester 2002). This supplementary mechanism likely arises due to the reduced cytoplasmic concentration of various metal cations, which are retained and interact with intracellular components such as Zn finger proteins, organic acids, and enzymes. Differences in Zn uptake, transport from roots to shoots, and sequestration depend on the specific Zn requirements and concentrations of different plant species. These divergences are rooted in the precise regulation of Zn absorption necessary for root cell survival and maintaining Zn homeostasis, which is essential for normal physiological processes such as transpiration and nutrient absorption. Plants primarily absorb Zn as divalent cations (Zn^{2+} ions), although organic ligand-Zn complexes can also be taken up. The uptake of Zn^{2+} relies on two physiological strategies determined by the ligands secreted by plant roots.

Strategy I involves the secretion of reductants, organic acids, and H^+ ions, which solubilize Zn complexes (such as Zn phosphates and hydroxides), releasing Zn^{2+} ions for absorption by root epidermal cells. These organic acids, such as citric acid, malic acid, oxalic acid, or tartaric acid, may be released in root exudates or directly by epidermal cells.

Strategy II involves the secretion of phytosiderophores (phytometallophores), which form stable complexes with Zn and are then transported into root epidermal cells. This mechanism is specific to cereal roots. Phytosiderophores are low molecular weight organic compounds, primarily nonprotein amino acids like nicotianamine, deoxymugineic acid, and avenic acid, which have a strong binding affinity for metals, facilitating their chelation and uptake. The synthesis of phytosiderophores involves the condensation of S-adenosylmethionine molecules, leading to the production of nicotianamine, a precursor to these compounds.

11.3 Enhancing Zn Transporters Expression in Specific Tissues of Cereal

Movement of Zn within cells and through various trafficking pathways in cereals requires crossing different biological membranes. This process is facilitated by numerous proteins responsible for the active transport of Zn and other divalent cations (Hanikenne et al. 2021). The role of these proteins in maintaining Zn homeostasis is closely linked to Fe homeostasis, highlighting the interplay between these essential nutrients. Proteins such as Zn-regulated transporter (ZRT) and iron-regulated transporter (IRT)-like protein, zinc-regulated, iron-regulated transporter-like protein (ZIP) are typically responsible for mediating the cellular uptake of Zn and other divalent metals, including manganese (Mn), Fe, and cadmium (Cd). This transport occurs either from outside the cell or from organelles into the cytoplasm (Guerinot 2000). In rice, several ZIP proteins are involved in Zn movement throughout the plant, particularly toward the grains (Figure 11.1). For instance, *OsZIP3* and *OsZIP1* are marker genes for Zn status in rice, with their expression levels correlating with grain Zn concentration (Swamy et al. 2016). *OsZIP1* is upregulated in roots and shoots under Zn deficiency and is linked to higher tolerance to Zn deficiency among rice cultivars (Ramesh et al. 2003). Additionally, *OsZIP1* and *ZmZIP5* are found in both the endoplasmic reticulum (ER) and PM, suggesting they play a role in detoxifying excess Zn from the cytosol and facilitating Zn translocation from roots to shoots (Li et al. 2013; Li et al. 2019; Liu et al. 2019; Mallikarjuna et al. 2020). Overexpression of *OsZIP8* in rice results in Zn accumulation in roots but a deficiency in shoots and seeds (Lee et al. 2010), a pattern also observed for *OsZIP5* (Lee et al. 2010). *OsZIP9* expression is regulated by Zn status in the shoots, indicating systemic regulation like *AtMTP2* and *AtHMA2* in Arabidopsis (Sinclair et al. 2018; Tan et al. 2020). *OsZIP3*, which is expressed in xylem transfer cells, contributes to Zn transportation (Sasaki et al. 2015). The barley ortholog *HvZIP7* is implicated in Zn translocation from roots to shoots (Tiong et al. 2014), while *TaZIP7* in wheat is highly expressed during leaf senescence, suggesting its role in Zn translocation to phloem and reproductive organs (Pearce et al. 2014). Similarly, *ZmZIP7*, like its counterparts in rice and wheat, enhances Zn concentration in seeds when expressed in Arabidopsis (Li,

Zeng and Liao 2016). In contrast to rice and wheat, *ZmZIP7* in maize is predominantly expressed in flag leaves rather than young kernels, and it is one of the less expressed ZIPs in maize (Figures 11.2 and 11.3) (Mondal et al. 2014). Homologs of these ZIP proteins in Arabidopsis, such as *AtIRT3* and *AtZIP4*, are believed to function in Zn xylem unloading and Zn uptake into root endodermis and pericycle cells, respectively (Lin et al. 2009, 2016; Tiong et al. 2015; Evens et al. 2017). This subgroup of plant ZIPs, which includes *AtIRT3, AtZIP4*, and *ZIP7* proteins from rice, barley, wheat, and maize, collectively plays a role in mediating root-to-shoot Zn translocation (Figure 11.4).

11.4 Strategies for Enhancing Zn Concentration in Plant Tissues

Numerous global research initiatives focused on Zn biofortification have been launched to address malnutrition issues. Cereal crops are commonly targeted in these projects due to their status as staple foods in many developing nations. These initiatives tackle the challenge of Zn deficiency through both genetic and agronomic biofortification methods. Biofortification methods include both genetic and agronomic approaches. Genetic biofortification aims to enhance micro-

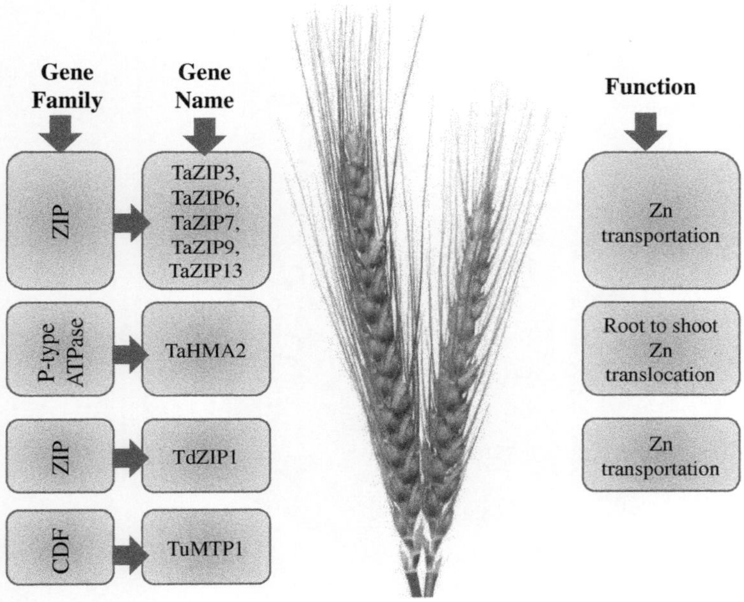

Figure 11.1 Genes involved in Zn transporters and function in wheat.

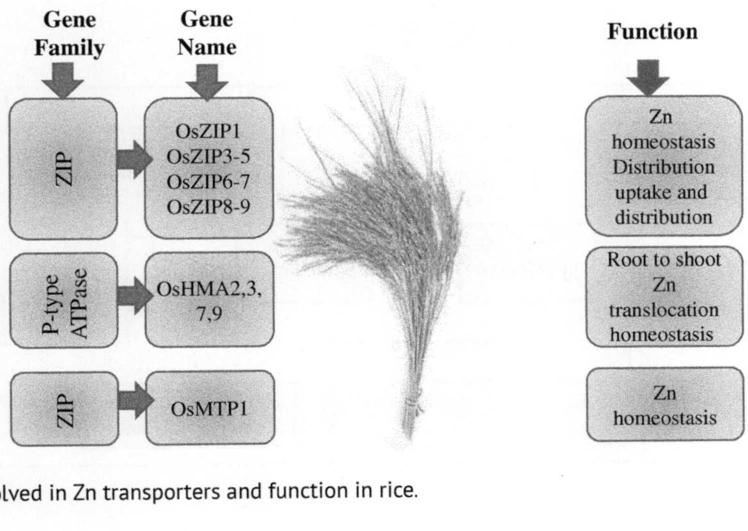

Figure 11.2 Genes involved in Zn transporters and function in rice.

nutrient levels in edible plant parts and can be achieved through conventional breeding or genetic engineering. Deficiencies in mineral elements in food crops may arise from various factors, including mineral-deficient soils and reduced mineral availability to plants due to environmental conditions (Figure 11.5).

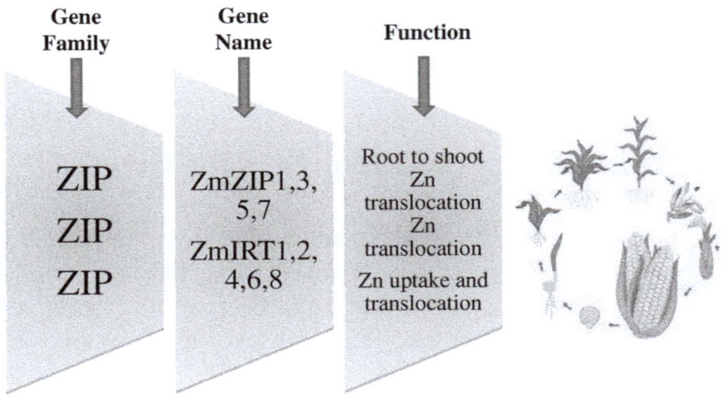

Figure 11.3 Genes involved in Zn transporters and function in maize.

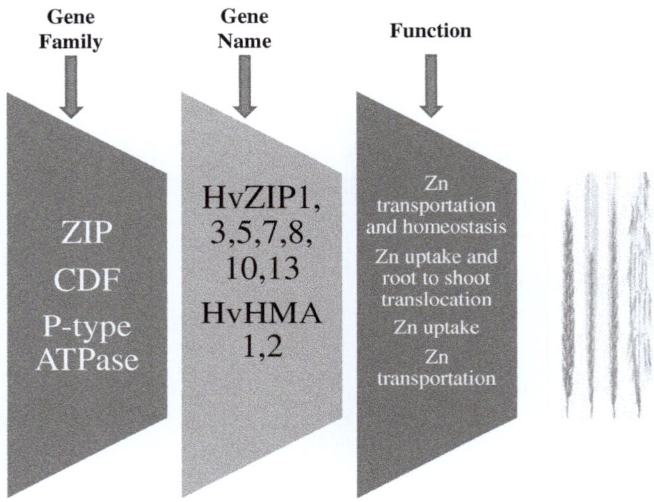

Figure 11.4 Genes involved in Zn transporters and function in barely.

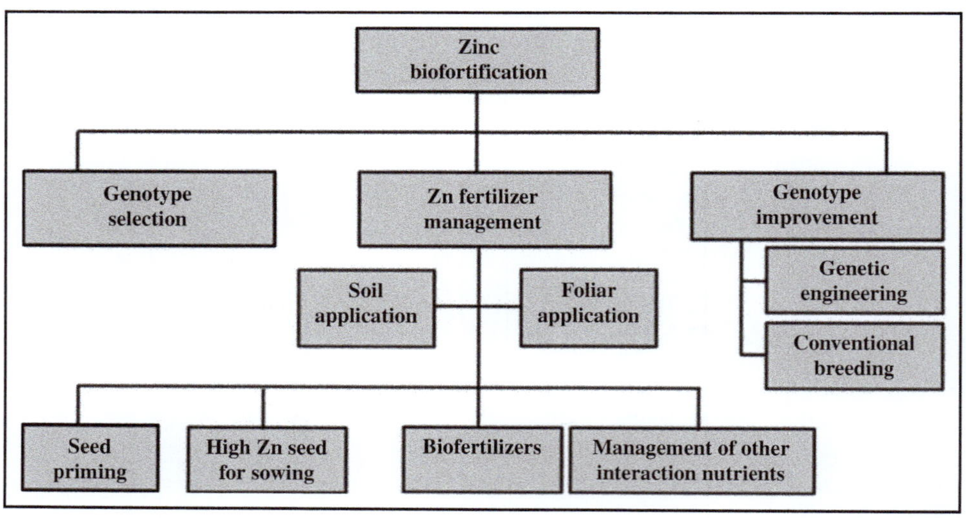

Figure 11.5 Zn biofortification approaches.

11.4.1 Genetic Approach

Biofortification, achieved through traditional and genetic plant-breeding methods, is employed to develop grains with high Zn content. Wang, Gong, and Ye (2020) noted that breeding approaches are more commonly utilized compared to genetically modified (GM) wheat due to the high costs and challenges associated with the generalization of GM wheat. Genetic biofortification techniques, including transgenic and breeding approaches, can be employed to create genetic biofortified varieties with elevated Zn content (Ali and Borrill 2020). Furthermore, genetic variation for nutrient acquisition, uptake, and translocation to shoot through the xylem may also be exploited (White and Broadley 2009). The HarvestPlus was part of a Consultative Group on International Agricultural Research (CGIAR) Challenge Program and has a vision of a world free of Zn, Fe, and Vitamin A deficiencies and has explored genetic variability, a broad range of major staple crops such as wheat, maize, rice, cassava, and beans for mineral concentrations and stability over different environments to enhance the mineral content and then to initiate a breeding approach of first assessing whether stable genetic variability exists for Zn and Fe with a crop, and this involved screening germplasm, varieties, and elite breeding lines (Andersson et al. 2017).

11.4.2 Agronomic Approach

Agronomic practices, in conjunction with breeding adjustments, can enhance crop nutrient content through the application of nutrients via soil or foliar methods. Several scientists have studied agronomic approaches to increase the bioavailable concentration of Zn in the edible parts of major crops, considering the perspectives of health and sustainable economic development. Agronomic biofortification is a rapid and effective strategy for increasing micronutrient amounts in crops and has been confirmed to address hidden hunger. Various reasons contribute to the success of agronomic biofortification, including soil nutrient bioavailability, mineral translocation within plants and food, and ultimately, bioavailability to humans (Khalili et al. 2024; Khalili et al. 2023; Ashraf et al. 2021). Agronomic biofortification offers advantages in terms of availability, effectiveness, application, and sustainability. Agronomic biofortification involves increasing the content of essential elements in the edible parts of crops through soil, foliar, and seed treatment techniques (Khalili et al. 2023; Stangoulis and Knez 2022).

11.4.2.1 Seed Treatments

Seed priming, which involves the treatment of hydrated seeds with a specific nutrient solution without actual germination, followed by dehydration to their original moisture content, allows for proper utilization. Seed nutripriming refers to soaking crop seeds with a specific concentration of nutrient solution for a specific duration before sowing, ensuring pre-germination metabolism occurs, followed by drying to obtain the original moisture content (McDonald 2000). Seed priming with micronutrients is an easily applicable method and cost-effective to enhance plant growth and improve micronutrient bioavailability (Parera and Cantliffe 2010). Primed seeds exhibit metabolic activities before germination, resulting in quicker sprouting and radical emergence compared to unprimed seeds (Farooq et al. 2010). This method also proves to be cost-effective in increasing productivity and net income Zulfiqar et al. (2020).

Kalyani et al. (2022) reported a positive impact on grain protein percentage and grain Zn concentration with Zn sulfate 0.6% seed priming, indicating its potential for improving grain protein quality, increasing Zn concentration, and reducing malnutrition caused by Zn deficiency. Nadeem et al. (2020) found that seed priming with Zn was the most effective approach to improving wheat grain yield. Hassan et al. (2019) observed that Zn deficiency adversely affected wheat grain yield, yield attributes, and quality, while Zn fertilization with seed priming enhanced these aspects. Rehman et al. (2018) concluded that a more effective way for enhancing productivity of wheat is by combining (Pseudomonas) with Zn sulfate seed priming and soil application of Zn. Several studies have demonstrated that seed priming stimulates plant emergence, improves competitiveness, and increases disease resistance (Singh et al. 2015); 0.5% Zn solution seed priming enhanced wheat seed germination, emergence, and growth of seedlings (Rehman et al. 2015).

11.4.2.2 Soil Applications

Zn management through soil application is a widely used method to enhance Zn concentration in plants. The efficacy of Zn fertilizers in terms of availability depends on the methods and timing of Zn and the chosen uses strategy. The predominant method for remedying Zn deficiency in soil is through soil application of Zn. Broadcasting, band placement, and fertigation are techniques employed for soil Zn fertilizer application. Fertilizing the soil with Zn increases Zn concentration in various parts of wheat grain such as embryo, aleurone, endosperm, and the whole grain. Cakmak (2008) discovered that the normal practice in wheat crops to address Zn deficiency and increase Zn bioavailability for human consumption is to apply Zn-containing fertilizers like Zn sulfate to the soil just before sowing. Kumar et al. (2022) conducted an investigation and found that soil applied of 10 kg Zn ha^{-1} resulted in the maximum yield and Zn uptake, which did not significantly differ from the soil applied of 7.5 kg Zn ha^{-1} in the cropping systems (maize-wheat). Gurjer et al. (2022) studied the critical Zn levels for soil and plants and found that the dry matter content of wheat significantly increased with all application rates of Zn ranging from 2.5 to 10 kg Zn ha^{-1} compared to the control.

The soil application of Zn is a favorable technique for enhancing Zn bioavailability in biofortified wheat crops, particularly in populations facing Zn deficiency. They also noted that the Zincol-2016 Zn-fortified wheat variety had significantly higher Zn content (53 ppm) in the grain compared to Jauhar-2016, thereby improving Zn bioavailability. Paramesh et al. (2020) conducted research, and the results indicated that applying 12.5 kg Zn ha^{-1} as a starter dose of soil, besides with a 0.5% Zn solution foliar spray after the anthesis stage was responsible for maximum wheat productivity and grain quality. Zulfiqar et al. (2020) verified that Zn application through various methods such as seed coating, priming, soil application, or foliar application increased the growth, yield, grain Zn content, Zn-use efficiency indices, and Zn bioavailability in common wheat. Studies have also indicated that soil application of Zn not only enhances crop growth and grain yield but also elevates seed Zn concentration. Similarly, Maqsood et al. (2009) demonstrated that soil-applied and foliar-sprayed Zn improved grain Zn concentration.

11.4.2.3 Foliar Applications

Fertilizer usage through foliar spraying is an important method to address nutrient deficiency in crop plants. When root absorption is hindered by unfavorable soil conditions or when nutrient requirements exceed root absorption capacity during peak growth stages, foliar application offers a higher possibility of nutrient uptake by the plant Swietlik and Faust (1984).

Numerous studies worldwide have shown that applying Zn fertilizers through foliar spraying significantly enhances grain Zn content in wheat. Zulfiqar et al. (2020) found that the highest grain Zn content and bioavailability were achieved when Zn fertilizers were applied through foliar spraying, followed by soil application. In a similar vein, Ramzan et al. (2020) illustrated that the application of Zn to wheat through both soil and foliar methods led to noteworthy enhancements in various growth parameters. These included plant height, number of tillers, spike length, spikelets per spike, spike grains, test weight, grain yield, biological yield, harvest index, protein content, and Zn content. They concluded that both Zn fertilization through soil and foliar applications improved the growth and wheat quality attributes, leading to enhanced grain micronutrient concentration and the ability to address hidden hunger. Nadeem et al. (2020) calculated that Zn-use efficiency indices were better improved with either foliar or soil application of Zn, and both methods increased grain Zn content and its bioavailability in rice and wheat.

However, limited research has been conducted to assess the efficiency of foliar application scheduling for Zn accumulation in cereal grains. Zn foliar application timing is a crucial factor in determining its efficacy in enhancing grain Zn content. Phattarakul et al. (2012) found that more effective application timing in terms of increasing grain Zn concentration was Zn applied through foliar spraying at either early milk or dough stages of rice. Similarly, Yang et al. (2020) reported that Zn application on the foliage of crop at a later phase of the wheat crop, i.e. grain development, increased grain Zn concentration, with the maximum concentration achieved when Zn was sprayed at the milking stage of grain development. Tuiwong et al. (2022) also found that foliar-applied Zn at the milky stage was a favorable approach to increase yield, yield components, and grain Zn content in rice.

In field research conducted by A. Jalal et al. (2020) to examine the foliar application of Zn and Fe, alone and in combination, on wheat, they observed significant enhancements in leaf area, number of tillers, yield attributes, grain and straw yield, grain Zn content, and grain Fe content with different foliar levels of Zn and Fe. They concluded that a foliar application of 0.4% Zn combined with 2% Fe was the most effective in enhancing grain quality and maximizing Zn and Fe content in wheat grain.

Hassan et al. (2019) indicated that foliar fertilization with Zn was more effective in improving grain quality compared to seed priming and soil application of Zn in wheat crops. Cakmak, Pfeiffer, and McClafferty (2010) also suggested that foliar-applied Zn was a better and highly active approach for wheat grain Zn concentration increasing and its bioavailability over to other approaches. Sher et al. (2020) recommended the 3 kg Zn ha^{-1} to achieve high grain Zn concentration and better grain quality in wheat.

11.4.2.4 Biofertilizers

Biofertilizers, including Zn-solubilizing microorganisms, owing to their utilization efficiency and environmentally friendly characteristics, must be widely adopted among farmers. However, the application of biofertilizers necessitates careful consideration of their storage and viable conditions for the microorganisms. Zn-solubilizing bacteria offer a potential alternative to Zn supplementation by converting applied inorganic Zn into forms readily accessible to plants. Some rhizobacteria species demonstrate the capability to mobilize Zn from soils into forms that plants can utilize. Bacteria able to solubilize Zn complex compounds achieve this due to the release of organic acids. Numerous investigations indicate that plant root zone bacteria, which exhibit environmental friendliness, play an essential role in the growing and development of plants for sustainable agriculture (Hayat et al. 2010). These bacteria could be found in diverse niches, such as the root zone on root surfaces or in association with roots. Root zone microbes facilitate nutrient accessibility for plants through various mechanisms, including fixation, release of nutrients in the soil, and the production of specific hormones. Instead of relying on Zn fertilization, these bacteria are employed for Zn solubilization by converting applied inorganic Zn into accessible forms through the synthesis and secretion of organic acids. Acidification, chelation, exchange reactions, and organic acid release are among the mechanisms employed by bacteria to solubilize essential micronutrients (Hafeez, Khanif, and Saleem, 2013). Bacteria not only enhance nutrient availability for plants but also improve nutrient utilization efficiencies (Khalid, Amanullah, and Ahmed 2022). The interactions among plants, soil, and microbes are intricate, and their symbiotic relationships can positively influence crop growth, yield, and quality. Numerous studies have demonstrated that the inoculation of cereal crops such as wheat, rice, barley, and maize with suitable Zn-solubilizing bacteria strains in the root zone can enhance crop yield. Naz et al. (2016) and Khalid et al. (2022) support the positive impact of biofertilizers, including ZSB, on Zn content in wheat plants.

Khalid et al. (2022) elucidated that the combination of Zn fertilizer application, Zn-solubilizing bacteria, and compost facilitates the effective uptake of Zn from the soil, leading to the production of high-quality, Zn-biofortified wheat crops, which represents a favorable option for farmers. Rezaeiniko, Enayatizamir, and Norouzi Masir (2022) conducted research that revealed the highest seed yield of 214% and Zn uptake of 46% when utilizing a mixed inoculum of bacteria (*Bacillus megaterium* + *Enterobacter cloacae*) at a rate of 10.2 mg kg^{-1}, over control. Nitu, Rajinder, and Sukhminderjit (2020) reported that the utilization of Zn-solubilizing bacteria alone or in combination with organic matter represents a cost-effective, sustainable, and innovative method to enhance Zn bioavailability, improve plant nutrient content, and facilitate nutrient absorption. Naz et al. (2016) demonstrated that the treatment of wheat seeds with different bacteria (*Pseudomonas, Rhizobium*, and *Azospirillum*) significantly increased Zn concentration in various plant parts at different growth stages, leading to the production of superior quality crops. Similarly, the use of Zn-solubilizing bacteria resulted in increased plant height, chlorophyll content, and grains per plant, along with a 31% increase in grain Zn content, thereby improving crop production and quality Joshi et al. (2013).

11.5 Conclusion

The chapter underscores the critical importance of addressing global food security and nutrition challenges to meet the United Nations' Second Sustainable Development Goal (UN-SDG2) of eradicating hunger by 2030. Despite agriculture's success in increasing food production to meet growing demands, the nutritional quality of staple cereals like rice, wheat, and maize remains a concern, especially regarding essential micronutrients such as Zn. Zn deficiency is a widespread issue, exacerbated by the presence of phytates in cereals that inhibit Zn absorption. Enhancing the Zn content and bioavailability in cereals is crucial for improving global health. This chapter explores various strategies for increasing Zn uptake and concentration in cereals, including understanding the molecular mechanisms of Zn absorption, optimizing Zn transporters, and employing genetic and agronomic biofortification methods.

Key strategies include:

1. **Molecular Mechanisms**: Plants absorb Zn primarily as Zn^{2+} ions, facilitated by specific transporter proteins. Strategies to enhance Zn uptake include manipulating the root system architecture and the expression of Zn transporters.
2. **Genetic Biofortification**: Traditional and genetic breeding methods aim to develop cereal varieties with higher Zn content. These methods exploit genetic variations in nutrient uptake and translocation.
3. **Agronomic Biofortification**: This approach involves soil and foliar applications of Zn fertilizers to increase Zn concentrations in crops. Techniques such as seed priming and the use of Zn-solubilizing biofertilizers are highlighted as cost-effective and sustainable methods to enhance Zn bioavailability.
4. **Zn Transporters**: Enhancing the expression of Zn transporters in specific tissues of cereals can improve Zn translocation from roots to shoots and grains, addressing Zn deficiency more effectively.

The chapter concludes that integrating these strategies can significantly improve the nutritional quality of staple cereals, contributing to better health outcomes and supporting global efforts to achieve food security and nutrition goals.

Targeting tissue-specific Zn acquisition in cereal crops represents a promising strategy to address both agricultural productivity and human nutritional deficiencies. By understanding and manipulating the molecular mechanisms involved in Zn uptake, transport, and accumulation, we can enhance Zn bioavailability in edible parts of cereal plants. This has significant implications for improving crop resilience to Zn-deficient soils and for addressing Zn malnutrition in human populations reliant on cereal-based diets. Recent advances in genomics, transcriptomics, and proteomics have identified key genes and pathways involved in Zn homeostasis. By employing biotechnological approaches, such as gene editing and transgenic techniques, it is possible to selectively enhance Zn accumulation in the grain while minimizing unwanted effects in other plant tissues. This targeted approach not only improves the nutritional quality of cereals but also contributes to sustainable agricultural practices by reducing the need for Zn fertilizers. Additionally, the integration of traditional breeding techniques with modern molecular tools can expedite the development of Zn-efficient cereal varieties. This requires a multidisciplinary effort involving plant scientists, nutritionists, agronomists, and policymakers to ensure the successful implementation of these strategies in diverse agro-ecological contexts.

In conclusion, targeting tissue-specific Zn acquisition in cereals is a viable and innovative approach to tackle the dual challenges of crop productivity and human nutrition. Continued research and collaborative efforts are essential to translate these scientific advancements into practical solutions that can benefit both farmers and consumers worldwide.

References

Aiqing, Z., Zhang, L., Ning, P. et al. (2022). Zinc in cereal grains: concentration, distribution, speciation, bioavailability, and barriers to transport from roots to grains in wheat. *Critical Reviews in Food Science and Nutrition* 62 (28): 7917–7928.

Alberts, B., Johnson, A., Lewis, J. et al. (2007). Molecular biology of the cell: reference edition. *Garland Science* 5.

Ali, M.W. and Borrill, P. (2020). Applying genomic resources to accelerate wheat biofortification. *Heredity* 125 (6): 386–395.

Andersson, M.S., Saltzman, A., Virk, P.S. et al. (2017). Progress update: crop development of biofortified staple food crops under HarvestPlus. *African Journal of Food, Agriculture, Nutrition and Development* 17 (2): 11905–11935.

Ashraf, S.A., Siddiqui, A.J., Abd Elmoneim, O.E. et al. (2021). Innovations in nanoscience for the sustainable development of food and agriculture with implications on health and environment. *Science of the Total Environment* 768: 144990.

Bouain, N., Satbhai, S.B., Korte, A. et al. (2018). Natural allelic variation of the AZI1 gene controls root growth under zinc-limiting condition. *PLoS Genetics* 14 (4): e1007304.

Caggiano, V., Schnitzler, R., Strauss, W. et al. (1969). Zinc deficiency in a patient with retarded growth, hypogonadism, hypogammaglobulinemia and chronic infection. *American Journal of Medical Sciences* 257 (5): 305–319.

Cakmak, I. (2008). Enrichment of cereal grains with zinc: agronomic or genetic biofortification? *Plant and Soil* 302 (1): 1–17.

Cakmak, I., Kalayci, M., Kaya, Y. et al. (2010a). Biofortification and localization of zinc in wheat grain. *Journal of Agricultural and Food Chemistry* 58 (16): 9092–9102.

Cakmak, I., Pfeiffer, W.H., and McClafferty, B. (2010). Biofortification of durum wheat with zinc and iron. *Cereal Chemistry* 87 (1): 10–20.

Clemens, S., Palmgren, M.G., and Krämer, U. (2002). A long way ahead: understanding and engineering plant metal accumulation. *Trends in Plant Science* 7 (7): 309–315.

Demidchik, V., Davenport, R.J., and Tester, M. (2002). Nonselective cation channels in plants. *Annual Review of Plant Biology* 53 (1): 67–107.

Evens, N.P., Buchner, P., Williams, L.E. et al. (2017). The role of ZIP transporters and group F bZIP transcription factors in the Zn-deficiency response of wheat (Triticum aestivum). *The Plant Journal* 92 (2): 291–304.

Farooq, M., Basra, S.M.A., Wahid, A. et al. (2010). Rice seed invigoration: a review. In: *Organic Farming, Pest Control and Remediation of Soil Pollutants: Organic Farming, Pest Control and Remediation of Soil Pollutants*, 137–175. Dordrecht: Springer.

Gallego-Castillo, S., Taleon, V., Talsma, E.F. et al. (2021). Effect of maize processing methods on the retention of minerals, phytic acid and amino acids when using high kernel-zinc maize. *Current Research in Food Science* 4: 279–286.

Gowda, N.N., Siliveru, K., Prasad, P.V. et al. (2022). Modern processing of Indian millets: a perspective on changes in nutritional properties. *Foods* 11 (4): 499.

Guerinot, M.L. (2000). The ZIP family of metal transporters. *Biochimica et Biophysica Acta (BBA)–Biomembranes* 1465 (1–2): 190–198.

Gurjer, K.L., Sharma, S.K., Yadav, K.K. et al. (2022). Establishment of critical limit of zinc for wheat crop. *The Pharma Innovation* 11 (2): 1457–1461.

Hafeez, B.M.K.Y., Khanif, Y.M., and Saleem, M. (2013). Role of zinc in plant nutrition – a review. *American Journal of Experimental Agriculture* 3 (2): 374–391.

Hambidge, K.M., Hambidge, C., Jacobs, M. et al. (1972). Low levels of zinc in hair, anorexia, poor growth, and hypogeusia in children. *Pediatric Research* 6 (12): 868–874.

Hanikenne, M., Esteves, S.M., Fanara, S. et al. (2021). Coordinated homeostasis of essential mineral nutrients: a focus on iron. *Journal of Experimental Botany* 72 (6): 2136–2153.

Hassan, M.U., Chattha, M.U., Ullah, A. et al. (2019). Agronomic biofortification to improve productivity and grain Zn concentration of bread wheat. *International Journal of Agriculture and Biology* 21 (3): 615–620.

Hayat, R., Ali, S., Amara, U. et al. (2010). Soil beneficial bacteria and their role in plant growth promotion: a review. *Annals of Microbiology* 60: 579–598.

Jalal, A., Shah, S., Teixeira Filho, M.C. et al. (2020). Agro-biofortification of zinc and iron in wheat grains. *Gesunde Pflanzen* 72 (3): 227–236.

Joshi, D., Negi, G., Vaid, S. et al. (2013). Enhancement of wheat growth and Zn content in grains by zinc solubilizing bacteria. *International Journal of Agriculture, Environment and Biotechnology* 6 (3): 363–370.

Kalyani, S.F., Siosemardeh, A., Hosseinpanahi, F. et al. (2022). Effect of seed priming on grain yield, protein content and concentration of some nutrients in dryland wheat grain. *Cereal Biotechnology and Biochemistry* 1 (1): 88–100.

Khalid, S., Amanullah, and Ahmed, I. (2022). Enhancing zinc biofortification of wheat through integration of zinc, compost, and zinc-solubilizing bacteria. *Agriculture* 12 (7): 968.

Khalili, A., Khalofah, A., Ramesh, A. et al. (2024). Temporal synchronization of nitrogen and sulfur fertilization: impacts on nutrient uptake, use efficiency, productivity, and relationships with other micronutrients in soybean. *Agronomy* 14 (3): 570.

Khalili, A., Qayyum, A., Khan, S.U. et al. (2023). Role of temporal Zn fertilization along with Zn solubilizing bacteria in enhancing zinc content, uptake, and zinc use efficiency in wheat genotypes and its implications for agronomic biofortification. *Agronomy* 13 (11): 2677.

Kumar, S., Verma, G., Dhaliwal, S.S. et al. (2022). Influence of zinc fertilization levels and frequencies on crop productivity, zinc uptake and buildup of soil zinc in maize–wheat system. *Journal of Plant Nutrition* 45 (12): 1774–1785.

Kyriacou, B., Moore, K.L., Paterson, D. et al. (2014). Localization of iron in rice grain using synchrotron X-ray fluorescence microscopy and high resolution secondary ion mass spectrometry. *Journal of Cereal Science* 59 (2): 173–180.

Lee, S., Jeong, H.J., Kim, S.A. et al. (2010). OsZIP5 is a plasma membrane zinc transporter in rice. *Plant Molecular Biology* 73 (4–5): 507–517.

Li, S., Liu, X., Zhou, X. et al. (2019). Improving zinc and iron accumulation in maize grains using the zinc and iron transporter ZmZIP5. *Plant and Cell Physiology* 60 (9): 2077–2085.

Li, S., Zhou, X., Huang, Y. et al. (2013). Identification and characterization of the zinc-regulated transporters, iron-regulated transporter-like protein (ZIP) gene family in maize. *BMC Plant Biology* 13: 1–14.

Li, X., Zeng, R., and Liao, H. (2016). Improving crop nutrient efficiency through root architecture modifications. *Journal of Integrative Plant Biology* 58 (3): 193–202.

Lin, Y.F., Hassan, Z., Talukdar, S. et al. (2016). Expression of the ZNT1 zinc transporter from the metal hyperaccumulator Noccaea caerulescens confers enhanced zinc and cadmium tolerance and accumulation to Arabidopsis thaliana. *PLoS One* 11 (3): e0149750.

Lin, Y.F., Liang, H.M., Yang, S.Y. et al. (2009). Arabidopsis IRT3 is a zinc-regulated and plasma membrane localized zinc/iron transporter. *New Phytologist* 182 (2): 392–404.

Liu, X.S., Feng, S.J., Zhang, B.Q. et al. (2019). OsZIP1 functions as a metal efflux transporter limiting excess zinc, copper and cadmium accumulation in rice. *BMC Plant Biology* 19: 1–16.

Lombi, E., Smith, E., Hansen, T.H. et al. (2011). Megapixel imaging of (micro) nutrients in mature barley grains. *Journal of Experimental Botany* 62 (1): 273–282.

Mallikarjuna, M.G., Thirunavukkarasu, N., Sharma, R. et al. (2020). Comparative transcriptome analysis of iron and zinc deficiency in maize (Zea mays L.). *Plants* 9 (12): 1812.

Maqsood, M.A., Rahmatullah, S.H.A.M.S.A., Kanwal, T. et al. (2009). Evaluation of Zn distribution among grain and straw of twelve indigenous wheat (Triticum aestivum L.) genotypes. *Pakistan Journal of Botany* 41 (1): 225–231.

Mayer, J.E., Pfeiffer, W.H., and Beyer, P. (2008). Biofortified crops to alleviate micronutrient malnutrition. *Current Opinion in Plant Biology* 11 (2): 166–170.

McDonald, M.B. (2000). Seed priming. *Seed Technology and Its Biological Basis* 287: 325.

Mondal, T.K., Ganie, S.A., Rana, M.K. et al. (2014). Genome-wide analysis of zinc transporter genes of maize (Zea mays). *Plant Molecular Biology Reporter* 32 (2): 605–616.

Nadeem, F., Farooq, M., Ullah, A. et al. (2020). Influence of Zn nutrition on the productivity, grain quality and grain biofortification of wheat under conventional and conservation rice–wheat cropping systems. *Archives of Agronomy and Soil Science* 66 (8): 1042–1057.

Naz, I., Ahmad, H., Khokhar, S.N. et al. (2016). Impact of zinc solubilizing bacteria on zinc contents of wheat. *American-Eurasian Journal Agriculture & Environmental Science* 16 (3): 449–454.

Nitu, R., Rajinder, K., and Sukhminderjit, K. (2020). Zinc solubilizing bacteria to augment soil fertility—a comprehensive review. *International Journal of Agricultural Sciences and Veterinary Medicine* 8 (1): 38–44.

Paramesh, V., Dhar, S., Dass, A. et al. (2020). Role of integrated nutrient management and agronomic fortification of zinc on yield, nutrient uptake and quality of wheat. *Sustainability* 12 (9): 3513.

Parera, C.A. and Cantliffe, D.J. (2010). Presowing seed priming. *Horticultural Reviews* 16: 109–141.

Pearce, S., Tabbita, F., Cantu, D. et al. (2014). Regulation of Zn and Fe transporters by the GPC1 gene during early wheat monocarpic senescence. *BMC Plant Biology* 14: 1–23.

Phattarakul, N., Rerkasem, B., Li, L.J. et al. (2012). Biofortification of rice grain with zinc through zinc fertilization in different countries. *Plant and Soil* 361: 131–141.

Poursarebani, N., Nussbaumer, T., Šimková, H. et al. (2014). Whole-genome profiling and shotgun sequencing delivers an anchored, gene-decorated, physical map assembly of bread wheat chromosome 6A. *The Plant Journal* 79 (2): 334–347.

Prasad, A.S., Miale Jr, A., Farid, Z. et al. (1963). Zinc metabolism in patients with the syndrome of iron deficiency anemia, hepatosplenomegaly, dwarfism, and hypogonadism. *Journal of Laboratory and Clinical Medicine* 61 (4): 537–49.

Prasad, R., Shivay, Y.S., and Kumar, D. (2014). Agronomic biofortification of cereal grains with iron and zinc. *Advances in Agronomy* 125: 55–91.

Ramesh, S.A., Shin, R., Eide, D.J. et al. (2003). Differential metal selectivity and gene expression of two zinc transporters from rice. *Plant Physiology* 133 (1): 126–134.

Ramzan, Y., Hafeez, M.B., Khan, S. et al. (2020). Biofortification with zinc and iron improves the grain quality and yield of wheat crop. *International Journal of Plant Production* 14: 501–510.

Rehman, A., Farooq, M., Ahmad, R. et al. (2015). Seed priming with zinc improves the germination and early seedling growth of wheat. *Seed Science and Technology* 43 (2): 262–268.

Rehman, A., Farooq, M., Naveed, M. et al. (2018). Seed priming of Zn with endophytic bacteria improves the productivity and grain biofortification of bread wheat. *European Journal of Agronomy* 94: 98–107.

Rezaeiniko, B., Enayatizamir, N., and Norouzi Masir, M. (2022). Changes in soil zinc chemical fractions and improvements in wheat grain quality in response to zinc solubilizing bacteria. *Communications in Soil Science and Plant Analysis* 53 (5): 622–635.

Sandstead, H.H., Prasad, A.S., Schulert, A.R. et al. (1967). Human zinc deficiency, endocrine manifestations and response to treatment. *The American Journal of Clinical Nutrition* 20 (5): 422–442.

Sasaki, A., Yamaji, N., Mitani-Ueno, N. et al. (2015). A node-localized transporter Os ZIP 3 is responsible for the preferential distribution of Zn to developing tissues in rice. *The Plant Journal* 84 (2): 374–384.

Serna Saldívar, S.O. and Hernández, D.S. (2020). Dietary fiber in cereals, legumes, pseudocereals and other seeds. In: *Science and Technology of Fibers in Food Systems*, 87–122. Cham: Springer.

Sher, A.L.I., Naveed, K.H.A.L.I.D., Ahmad, G.U.L.Z.A.R. et al. (2020). Grain zinc and iron enrichment through foliar application augments wheat yield under varying nitrogen regimes. *Pakistan Journal of Botany* 52 (1): 85–94.

Sinclair, S. A., Senger, T., Talke, I. N. et al. (2018). Systemic upregulation of MTP2-and HMA2-mediated Zn partitioning to the shoot supplements local Zn deficiency responses. *The Plant Cell* 30 (10): 2463–2479.

Singh, H., Jassal, R.K., Kang, J.S. et al. (2015). Seed priming techniques in field crops – A review. *Agricultural Reviews* 36 (4): 251–264.

Stangoulis, J.C. and Knez, M. (2022). Biofortification of major crop plants with iron and zinc – Achievements and future directions. *Plant and Soil* 474 (1): 57–76.

Swamy, B.M., Rahman, M.A., Inabangan-Asilo, M.A. et al. (2016). Advances in breeding for high grain zinc in rice. *Rice* 9: 1–16.

Swietlik, D. and Faust, M. (1984). Foliar nutrition of fruit crops. *Horticultural Reviews* 6: 287–355.

Tan, L., Qu, M., Zhu, Y. et al. (2020). Zinc transporter5 and Zinc transporter9 function synergistically in zinc/cadmium uptake. *Plant Physiology* 183 (3): 1235–1249.

Tilman, D., Cassman, K.G., Matson, P.A. et al. (2002). Agricultural sustainability and intensive production practices. *Nature* 418 (6898): 671–677.

Timmer, C.P. (2014). Food security in Asia and the Pacific: the rapidly changing role of rice. *Asia & The Pacific Policy Studies* 1 (1): 73–90.

Tiong, J., McDonald, G.K., Genc, Y. et al. (2014). H v ZIP 7 mediates zinc accumulation in barley (Hordeum vulgare) at moderately high zinc supply. *New Phytologist* 201 (1): 131–143.

Tiong, J., McDonald, G., Genc, Y. et al. (2015). Increased expression of six ZIP family genes by zinc (Zn) deficiency is associated with enhanced uptake and root-to-shoot translocation of Zn in barley (Hordeum vulgare). *New Phytologist* 207 (4): 1097–1109.

Tuiwong, P., Lordkaew, S., Veeradittakit, J. et al. (2022). Efficacy of nitrogen and zinc application at different growth stages on yield, grain zinc, and nitrogen concentration in rice. *Agronomy* 12 (9): 2093.

United Nations. (2012). *World Population Prospects: The 2010 Revision*. New York: United Nations Department of Economic and Social Affairs. http://esa.un.org/unpd/wpp/Documentation/publications.htm.

Wang, K., Gong, Q., and Ye, X. (2020). Recent developments and applications of genetic transformation and genome editing technologies in wheat. *Theoretical and Applied Genetics* 133 (5): 1603–1622.

White, P.J. and Broadley, M.R. (2009). Biofortification of crops with seven mineral elements often lacking in human diets – Iron, zinc, copper, calcium, magnesium, selenium and iodine. *New Phytologist* 182 (1): 49–84.

Yang, M., Li, Y., Liu, Z. et al. (2020). A high activity zinc transporter OsZIP9 mediates zinc uptake in rice. *The Plant Journal* 103 (5): 1695–1709.

Zulfiqar, U., Hussain, S., Ishfaq, M. et al. (2020). Zinc-induced effects on productivity, zinc use efficiency, and grain biofortification of bread wheat under different tillage permutations. *Agronomy* 10 (10): 1566.

12

Combating Mineral Malnutrition Through Iron Biofortification in Cereal Crops

Ambrin Rajput[1*], *Qurban Ali Panhwar*[2], *Umed Ali*[3], *Hafeezullah Babar*[1]

[1] *Soil Fertility Research Institute, Agriculture Research Center, Tandojam, Sindh Pakistan*
[2] *Soil and Environmental Sciences Division, Nuclear Institute of Agriculture (NIA), Tandojam, Sindh, Pakistan*
[3] *Department of Agriculture, Mir Chakar-Khan Rind University, Sibi, Balochistan, Pakistan*

*Corresponding author: ambrin2004@gmail.com

12.1 Introduction

The plant nutrition plays a pivotal role in sustainable agriculture, while enriching minerals in edible sources contributes to global human health. Therefore, bridging both strategies is essential for ensuring "One Health" initiatives. The rapid growth of the global populace over the last period has resulted in food insecurity. To address this challenge, high-yielding cereal varieties were introduced, coupled with increased nitrogen and nutrient applications to soils, resulting in the Green Revolution. However, this progress has exerted intense density on natural properties, prominent to severe climatic fluctuations that adversely affect crop production. Climate change has further exacerbated the issue by compromising the nutritional quality of plants, contributing to global malnutrition (FAO 2017a).

Plants need 16 vital mineral nutrients for plant growth promotion. While carbon, hydrogen, and oxygen are obtained from water and air, the other nutrients are derived from soil in the shape of inorganic forms. Macronutrients (N, P, K, Ca, Mg, and S) are requisite in greater quantities, whereas micronutrients (Fe, Cu, Mn, Zn, Mo, B, and Cl) are requisite in lower amounts (Yang et al. 2021; Cakmak et al. 2023). Despite their lower quantities, micronutrients are equally crucial. For instance, Fe plays a vital part in enhancing plant tissue and is elaborated in several mechanisms leading to Fe uptake, translocation, transport, and storing capacity. However, the world population is increasing with an effect on the atmospheric environment and climate change exaggerates, imposing Fe scarcity as a stimulating concern for agriculture (Tanin et al. 2024).

Acquiring sufficient nutrition and guaranteeing vigorous healthiness are intrinsic humanoid rights and essential constituents of active human resources development. There are many findings have been carried out on the microorganisms (rhizospheric and endophytic) that have been documented for their aptitude to make available nutrients in the soil, in that way enabling the relocation of microorganisms within plants, with the grains. Present studies have stated the useful effect of numerous microbes on enhancing the Fe and Zn amounts in plant grains (Ali et al. 2023).

In addition, a dire investigation of prevailing investigated data which indicates us to suggest which dominant efforts directed at extenuating worldwide starvation and malnourishment, predominantly relating to Fe insufficiencies, is inadequate. Wheat (*Triticum aestivum* L.) and rice (*Oryza sativa L.*) both are main grain crops and staple foods all over the globe. To emphasize the important nutritional role of these prominent cereal crops in the global humanoid diet, a scrupulous meta-analysis was carried out to assessing their impact on macronutrients and micronutrients. Contrariwise, biofortification has appeared as an effective, sustainable, and cost-effective manner to mend essential micronutrients directly related to health consequences, reducing death rates and illness among people in resource-poor and developing countries (Gupta et al. 2022).

Hence, Fe biofortification application in cereals keeps possibilities as a durable and justifiable resolution for enhancing public fitness and highlighting nutritious safety contests. Fe biofortification and dietary variation are essential interferences to lessening the occurrence of Fe deficiency (Da Silva Lopes et al. 2018). Fortifying staple food means, like wheat flour, is a profitable method to mend Fe without altering customers' food practices. Fe biofortified wheat crop is predominantly valuable for populaces that ingest momentous amounts of wheat-based products. Moreover, Fe biofortification is an active and dependable worth of giving Fe to every consumer at no supplementary charge (Field et al. 2021).

Crop Biofortification: Biotechnological Approaches for Achieving Nutritional Security under Changing Climate, First Edition. Edited by Adnan Noor Shah et al.

12.2 Mineral Nutrients

Nutrients serve as essential components in both soil and plant systems, playing a decisive part in regulating various physiological and evolving practices in plants. These processes include seed germination, root and shoot growth, photosynthesis, respiration, and flower development (Kirkby et al. 2023; Saleem et al. 2023).

Mineral nutrients contribute to improved crop yield by facilitating key physiological functions like photosynthesis, respiration, and transpiration. Conversely, the uptake of certain mineral nutrients (such as iron, zinc, manganese, and molybdenum) tends to decrease under abiotic stresses (D'Oria et al. 2022).

12.2.1 Role of Mineral Nutrients in Soil

Vital soil elements (N, P, and K) are needed in moderately higher quantities, serving as building blocks for fundamental biomolecules such as amino acids (Khan et al. 2019), nucleotides, and carbohydrates. These elements play a crucial role in maximizing grain yield (Rajput et al. 2014). Additionally, essential mineral nutrients like potassium (K) and magnesium (Mg) are indispensable for key plant physiological practices like photosynthesis and the translocation of photo-assimilates (Xie et al. 2021). K, functioning as a major osmolyte, regulates guard cell activities and stomatal conductance (Jákli et al. 2017), while Mg is essential for light absorption, chlorophyll synthesis, and the activation of Rubisco and Rubisco activase (Peng et al. 2015).

Consequently, maintaining optimum limits of these trace elements such as iron (Fe) and zinc (Zn) in the soil is fundamental for ensuring normal plant growth and achieving adequate crop yields (Shireen et al. 2018). Nutrients represent elemental essentials within soil crucial for the growth enhancement of plants and structural development. Deficiencies in these elements manifest in identifiable symptoms. Typically, these elements exist in the soil as salts corresponding to their composition. They are considered into two major categories micronutrients and macronutrients, related on plants' basic requirements. Soil and plant tissue investigates deliver insights into the presence of each element in plants and their permissible limits in soil. Maintaining optimal levels of each element benefits both plant growth and soil fertility, while deviations beyond permissible ranges can have adverse influences on soil health and plant development. Soil amendment techniques have proven effective in such scenarios.

Soil pH plays a significant role in regulating the availability of these elements, necessitating special attention. Nutrient acquisition, distribution, and availability within plants significantly influence growth and yield, especially under varying environmental conditions. Throughout their life cycle, plants encounter numerous environmental stresses that compromise cellular integrity and disrupt nutritional balance, leading to yield losses (Xie et al. 2021).

12.2.2 Role of Mineral Nutrients in Plants

There exist 16 crucial mineral nutrients that serve diverse roles in promoting plant growth and enhancing yields. Macroelements are indispensable for providing energy throughout the plant's body, requiring substantial intake quantities. However, micronutrients play vital roles in metabolic processes, cell growth, and immune responses, necessitating minimal amounts. As such, the term "malnutrition" encompasses both protein and deficiency of minor element deficiencies, posing a significant human health challenge (Graham et al. 2001; Kiran et al. 2022).

Micronutrient deficiencies, including Cu, Mn, Fe, B, Zn, and Mo, are mostly cited as "hidden hunger" a "silent epidemic" that could persist in regions with adequate food supplies, masking the affected individuals' apparent well-fed status (FAO 2021). While specific diseases may arise directly from micronutrient deficiencies, they frequently exacerbate infectious and chronic conditions, significantly influencing the transience rates and eminence of life (Graham et al. 2001; Kiran et al. 2022).

While micronutrient deficiencies affect populations worldwide, underprivileged communities in developing countries are particularly vulnerable due to their limited access to diverse food sources. These communities often allocate the majority of their limited income to staple foods, resulting in insufficient consumption of crucial micronutrients (FAO 2021). The multifaceted functions of the mineral elements perform a significant role in promoting the growth and development of several crops.

12.2.2.1 Nitrogen

Nitrogen (N) stands as a fundamental constituent of chlorophyll, a complex molecule crucial for the optimal functioning of photosynthesis (Simkin et al. 2022). Integral to proteins, N shows a pivotal function in diverse metabolic processes like photosynthesis and contributes to numerous structural and functional aspects within plant cells (Hawkesford et al. 2023).

As an indispensable nutrient for plant growth, the N accessibility profoundly impacts both the photosynthesis process and protein synthesis, which are indispensable for maximizing plant productivity (Anas et al. 2020).

12.2.2.2 Phosphorus

Phosphorus (P) stands as an indispensable element crucial for crop enhancing capabilities including root initiation systems, playing a pivotal function in various physiological procedures for photosynthesis, respiration, and the production of nucleic acids including proteins (Anas et al. 2020; Elhaissoufi et al. 2022; Tariq et al. 2023). However, P is quite immovable in the soil and often remains inaccessible to plants. Moreover, it has a tendency to bind to soil particles or become chemically bound to other elements, further reducing its availability to plants (Rajput et al. 2007, 2014). Optimizing P efficiency and uptake requires strategic application timing during the growing season (Bhatla et al. 2018).

12.2.2.3 Potassium

Potassium (K) is integral to the overall metabolic activity of plant enzymes, playing a vital role in enhancing photosynthesis and promoting direct growth. Additionally, K contributes to efficient water utilization (Mansour 2006). Its presence facilitates root development and enhances leaf efficiency in sugar and starch synthesis, essential for sugar translocation. Moreover, K acts as a balancer, influencing the effects of both nitrogen and phosphorus (Fernandes et al. 2011). Recognized as a quality-enhancing element, K improves various crop parameters such as color, glossiness, and dry matter accumulation, ultimately enhancing crop longevity.

12.2.2.4 Calcium

Calcium stands as an indispensable element crucial for better plant growing abilities, whether in non-stressed or stressful circumstances. Its role is dual-fold, serving as a constituent of cell walls and membranes and as an intracellular messenger. Consequently, the nutrient uptake, supply, and storage of calcium within the plant must be firmly controlled to fulfill these essential functions. In line with its structural significance, plants poor in calcium have demonstrated increased susceptibility to pathogens, while supplementation with exogenous calcium has been shown to enhance plant resistance.

12.2.2.5 Magnesium

Magnesium (Mg) plays a pivotal role as a crucial element in a myriad of ultimate physiological and biological functions within plants. Its functions encompass vital aspects such as chlorophyll development; the production, passage, and consumption of photoassimilates; enzyme initiation; and synthesis of proteins. Particularly crucial is Mg's role as cornerstone chlorophyll (pigments) with chloroplasts, essential for photosynthetic CO_2 assimilation. Approximately 15–35% of captivated Mg by the plants is integrated into chlorophyll, while residual about 65–85% fraction is either stored in or being used for the synthesis of proteins and associated biochemical functions (Ishfaq et al. 2022).

12.2.2.6 Sulfur

Sulfur (S) stands as an indispensable nutrient crucial for optimal plant growth promotion. It serves as a structural constituent of essential biomolecules such as protein disulfide bonds, amino acids, and vitamins. However, much of the sulfur present in soil exists within organic matter, rendering it inaccessible to plants. Deficiency in sulfur at any growth stage can lead to diminished yields. Sulfur performs a pivotal function in various plant growth–enhancing factors including nitrogen metabolism, enzyme activities, and the synthesis of proteins and oils. Typically, sulfur-deficient plants exhibit stunted growth with short and/or spindly stems, along with the yellowing of young (top) leaves (Narayan et al. 2023).

12.2.2.7 Copper

Copper (Cu) plays a pivotal role in numerous physiological procedures within plants, encompassing essential functions such as photosynthesis, respiration, growth and development, mitochondrial respiration, electron-transport chain, cell wall metabolism, lignin synthesis, oxidative stress, and hormonal indication (Kaur et al. 2022; Tariq et al. 2023). Symptoms of copper deficiency typically manifest with the yellowing of younger leaves, attributed to reduced chlorophyll content and impaired photosynthesis. As the deficiency progresses, leaves may exhibit symptoms such as wilting, curling, and mortality of tissues at the leaf tips (Mir et al. 2021).

Root uptake of copper is primarily facilitated by transportation proteins situated in the plasma membranes (root cells). These transporters possess the ability to selectively bind and transport copper ions through the membranes into root cells, where copper can then be distributed throughout the plant (Ashraf et al. 2021). Understanding the mechanisms of copper taken by plants can inform strategies for enhancing copper availability in soil and promoting plant growth.

12.2.2.8 Manganese

Manganese (Mn) serves as a significant element for many crops, playing a pivotal function in photosynthesis by catalyzing light-induced breakdown of water molecules in photosystem II (PSII) and facilitating the RuBP carboxylase reaction (Mondal and Bose 2019). Additionally, it contributes to chlorophyll synthesis and participates in numerous metabolic functions such as respiration, nitrogen metabolism, and lignin biosynthesis. However, akin to other micronutrients, excess Mn can be detrimental to plants, leading to signs such as leaf chlorosis, necrosis, and stunted growth (Cakmak et al. 2023).

Signs of manganese deficiency typically manifest as striping and light grey flaking at the base of newly emerged leaves, with the issue potentially spreading across the entire leaf (Mondal et al. 2018; Awino et al. 2020). Manganese is primarily captivated by roots in the form of Mn^{2+}, which represents the maximum predominant and accessible shape of manganese (Atajan et al. 2019; Hopkins and Hansen 2019).

12.2.2.9 Iron

Iron (Fe) stands as an indispensable micronutrient crucial for the plant growth, playing a pivotal role in innumerable physiological functions including photosynthesis, respiration, and DNA synthesis. Despite its significance, iron is classically existent in soil at levels of 3–5%, ranking it as the fourth greatest copious component of soil surface layer (Coelho et al. 2023). However, iron in soil often exists as insoluble iron oxides, posing a challenge for plants to access.

Moreover, Fe availability in soil is profoundly prejudiced by environmental aspects including soil pH, redox potential, and the occurrence of competing ions which can hinder iron uptake by plants (Johan et al. 2021). Certain soil microbes, with beneficial microbes (plant growth–promoting rhizobacteria [PGPR]) and fungi, play crucial roles in enhancing iron uptake by plants. Arbuscular mycorrhizal fungi (AMF), for instance, can boost iron uptake by increasing the root system's external area, producing organic acids, and releasing enzymes that aid in iron solubilization.

Additionally, the submission of organic modifications, i.e. compost and manure, can elevate iron availability in soil by stimulating soil microbial activities. Hence, enhanced activity in turn facilitates the release of iron from organic matter and mineral forms, further enhancing iron accessibility to plants (Thangavel et al. 2022).

12.2.2.10 Zinc

Zinc (Zn) stands as an indispensable micronutrient crucial for plants' growth and development, playing a critical part in a myriad of cellular, physiological, and biotic procedures including seed germination and flower development (Gondal et al. 2021). Numerous plant enzymes rely on zinc as a cofactor for optimal functioning, and zinc is intricately in diverse metabolic processes like photosynthesis, respiration, and protein synthesis (Rai et al. 2021).

Deficiency in zinc can result in stunted growth, chlorosis (yellowing of leaves), and diminished fruit and seed production (Al Jabri et al. 2022). Notably, Zn is not only crucial for plants but also for humans and animals. In physiology, Zn plays a crucial role in numerous physiological processes (Gondal et al. 2021; Brown et al. 2022).

Zinc insufficiency in soil may arise due to several factors, including the adsorption of soluble zinc in the soil by clay and limestone particles, as well as deficiencies in organic matter in eroded soils. Moreover, elevated concentrations of bicarbonate (HCO_3) can hinder the uptake of zinc by plant shoots (Kumar et al. 2022).

12.2.2.11 Boron

Boron (B) is not required by plants in large quantities, yet its absence or excess can lead to severe growth issues. Additionally, boron is essential for protein and enzymatic functions within cells, thereby enhancing membrane reliability (Shireen et al. 2018). Dissimilar to other micronutrients, B deficiency does not typically cause chlorosis; however, it can manifest similar toxicity symptoms. Ensuring adequate boron levels is essential for achieving high crop yields and qualities. Boron plays a major role, alongside calcium, in the structure of cell walls. It facilitates the movement of calcium into plants and contributes to normal nutrition in both plants and animals. Boron deficiencies can induce various anatomical, biochemical, and physiological alterations in plants (Brdar-Jokanović 2020).

12.2.2.12 Molybdenum

Molybdenum (Mo) serves as a vital constituent in different enzymes crucial for converting nitrate to nitrite, a noxious form of N, and subsequently into NH_4, essential for synthesizing amino acids within the plant. Moreover, symbiotic N_2-fixing bacteria in legumes rely on molybdenum to fix atmospheric N_2. Additionally, the plant utilizes molybdenum to transform inorganic P into organic forms. Given its close association with nitrogen, molybdenum deficiency symptoms can closely resemble those of nitrogen deficiency.

Notably, molybdenum is the sole micronutrient capable of mobility within the plant. Consequently, deficiency symptoms manifest primarily on older and middle leaves but gradually extend up the stem to affect newer leaves (Mondal and Bose 2019).

12.2.2.13 Chloride

Chloride (Cl) performs a main function in plants for growth promotion, despite being required in relatively small amounts. Here are some of the key roles of chloride in plants. It's a key component of chlorophyll and regulates water balance in plants. It is also involved in the structure and function of cell membranes and as a cofactor for various enzymes. Chloride promotes other major nutrients in plants. Its deficiency in plants can lead to various problems, including plant growth inhibition, and the yellow color of leaves resulting in yield losses (Xie et al. 2021).

12.3 Mineral Malnutrition

Mineral deficiency arises when a plant lacks an adequate supply of important nutrients crucial for its growth and development. Such plants often exhibit various symptoms indicating the deficiency, leading to ineffective growth. The Green Revolution played a pivotal role in enhancing crop yields across Asia, averting a potential 20% decrease in yields, mitigating a 19% increase in food prices, and preventing a 5% decline in calorie consumption, which would have resulted in a 2% rise in malnourished children (Evenson and Gollin 2003). While the Green Revolution significantly boosted food grain production, particularly in cereals, and substantially reduced global undernourishment, the persistence of malnutrition and the lack of access to quality food pose ongoing societal challenges (Pingali 2012; Garcia-Oliveira et al. 2018).

A significant portion of the populace suffers from "hidden hunger," lacking essential micronutrients with various vitamins in the diets. These elements are vital for human growth, development, and proper bodily functions. While short-term deficiencies may not pose immediate harm, prolonged shortages related to frequent diseases such as anemia (Fe deficiency), beriberi (vitamin "B" deficiency), pellagra (niacin deficiency), rickets (vitamin "D" deficiency), and scurvy (vitamin "C" deficiency) can be fatal (Ratajczak et al. 2021; Rempel et al. 2021). Iron deficiency, causing anemia, disturbs about 25% of the global population, with Fe deficiencies being the leading causes, moving 50–60% of those affected (Blicharz et al. 2020). These realities underscore the critical importance of biofortification of crops to provide nutritious diets worldwide and address these pressing issues. Cereals and pulses constitute a significant portion of the calorie and dietary needs, accounting for approximately 60% in the emerging world and about 30% in the developed world (Awika 2011).

12.4 Causes of Malnutrition in Plants

Mineral deficiency occurs when plants lack essential nutrients necessary for their growth and development (Agrios 2005). For instance, nitrogen, crucial for healthy plant proteins and chlorophyll formation, can result in yellowed, stunted growth if deficient. Phosphorus serves various plant functions and may be deficient due to nutrient imbalance or excess iron. Potassium regulates stomata and is often lacking in light, sandy soils, or imbalanced pH conditions (Bierman and Rosen 2005).

Calcium aids in cell formation, while magnesium and sulfur are vital for chlorophyll and protein synthesis, and essential for plant health (Bierman and Rosen 2005). Boron aids in cell expansion and carbohydrate transportation, with deficiency often linked to solubility issues in alkaline soils. Copper contributes to chlorophyll formation and is typically deficient in sandy soils with pH imbalances. Manganese performs diverse plant functions and is often deficient due to pH or nutrient imbalances (Shahane and Shivay 2022). Molybdenum is essential for nitrogen conversion, with deficiency commonly related to pH imbalances or nutrient ratios.

Zinc facilitates chlorophyll and enzyme production and is often scarce in sandy soils with excessive phosphorus. Iron, vital for chlorophyll formation and photosynthesis, may be deficient due to imbalanced pH or over/underwatering. Additionally, iron deficiency can occur alongside excessive manganese, copper, or zinc levels (Garcia-Oliveira et al. 2018).

Poor soil fertility, with deficiencies in nitrogen, phosphorus, and potassium, contributes to lower crop yields (Abbas et al. 2012). To address declining soil nutrient levels, proper application of organic and mineral nutrients is necessary. Microorganisms, in conjunction with fertilizers, enhance micronutrient uptake, leading to increased concentrations of vital nutrients like iron in plants (Rajput et al. 2015).

12.5 Biofortification

Micronutrient deficiencies in humans are a global concern, particularly impacting women and children. Despite being required in minute quantities, over three billion people suffer from these deficiencies, severely affecting human health (Cakmak and Kutman 2018; Kiran et al. 2022). To address issues like iron deficiency malnutrition, various approaches are proposed, with biofortification emerging as a promising solution. Cereals, as staple foods worldwide, offer a significant avenue for fortification, either through conventional breeding or agronomic interventions (FAO 2017b).

Traditionally, crop production prioritized yield and productivity over nutritional content, leading to widespread micronutrient malnutrition (Development initiatives 2017). Climate change, imbalanced fertilization, and soil degradation have further compromised the quality of crop-based foods, exacerbating malnutrition globally. While macronutrient fertilizers are extensively used, micronutrient fertilizers remain underutilized, contributing to both environmental issues and malnutrition.

Biofortification of cereals and staples presents a sustainable solution to address micronutrient deficiencies and improve human health. Compared to industrial fortification, biofortification offers a cost-effective and environmentally friendly method, aligning with broader food-based nutritious plans (FAO 2017b). International organizations like World Health Organization (WHO) and Consultative Group for International Agricultural Research (CGIAR) prioritize the development of nutritional improved biofortified crops to ensure food and nutritional sanctuary amid population growth and climate change.

Agronomic biofortification, although the older and extensively implemented method, faces challenges compared to newer biotechnological and breeding approaches. Advancements in fertilizer forms and techniques have improved agronomic biofortification efficiency, enhancing micronutrient uptake by crop plants. This review highlights the progress, challenges, and future prospects of biofortification in addressing human nutrition and health concerns.

12.5.1 Approaches of Biofortification

Numerous kinds of biofortification approaches and mostly three tactics can be characterized (Figure 12.1).

12.5.1.1 Conventional Breeding

The conventional breeding technique stands out as a widely embraced and cost-effective approach, deemed feasible and widely accepted associated with the transgenic approach, for fortifying microelements (Van Der Straeten et al. 2020). Its efficacy relies on the existence of sample genotypic dissimilarity in the targeted attribute. Through successive generations, higher parent appearances boasting high-nutrient content are hybridized with receiver lines bearing further desired agronomic traits, yielding progeny with the desired nutrient levels and agricultural characteristics.

Genetic approaches (with conventional and molecular breeding practices) are extensively observed as active ways for improving biofortification of iron and zinc. Conventional breeding comprises a prevailing inherited variety (primary, secondary, and tertiary) genic pool to supplement the iron and zinc amount in grains of wheat. Nevertheless, the absence of genetic variety within the genetic pool poses a significant challenge to this strategy. While crossing with distant relatives

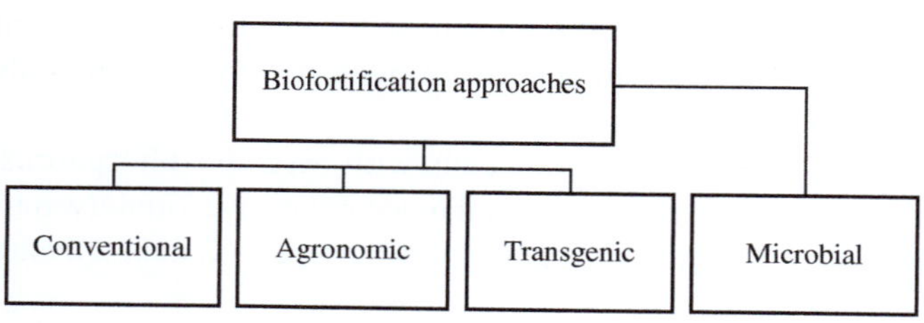

Figure 12.1 Approaches of biofortification.

may address this limitation, it inevitably reduces the integration of promising qualities into the envisioned marketable cultivars. Recognizing breeding as a rapid means to enhance plants, numerous organizations worldwide have initiated breeding programs aimed at augmenting the micronutrient content of crops (Van Der Straeten et al. 2020).

12.5.1.2 Biofortification Through Agronomic Methods

Agronomic biofortification, achieved by micronutrient-enriched fertilizer applications, stands as a straightforward and swift method to enhance crop nutritional value, subsequently enhancing human nutrition (Cakmak and Kutman 2018; Bhardwaj et al. 2022). This approach predominantly relies on fertilizer application techniques, aiming to solubilize and mobilize nutrients from the source to sink the usable portions of plants. For the duration of 1960s, the intensified use of macronutrient fertilizers spearheaded the Green Revolution, bolstering crop productivity and alleviating starvation, especially in developing nations. However, in today's context, with the imperative to feed a burgeoning global population of around seven billion, the focus extends beyond mere yield augmentation. The goal now encompasses enriching plant consumables with micronutrients to promote good health (Bhardwaj et al. 2022).

Although micronutrients are naturally present in varying concentrations across diverse plant portions and are primarily fascinated from the soil, fertilizer applications can rectify soil deficiencies and enhance plant and human nutrition. Nevertheless, microelements given to the soil often become rapidly secure and fail to efficiently translocate to consumable plant parts. In such cases, alternative application methods like foliar sprays of soluble forms are recommended. With due attention to fertilizer form, application technique, and timing, agronomic biofortification emerges as a cost-effective and straightforward tool. While agronomic biofortification has proven effective across various crops, nutrient uptake efficiency can fluctuate by differing growing situations, such as various kinds of soil, crop managing practices, and fertilizer kind. Given the escalating challenges posed by climate change and global pandemics like COVID-19, the significance of agronomic biofortification has become even more pronounced (Bhardwaj et al. 2022).

The advent of different fertilizers boasting enhanced plant uptake efficacy and translocations has further augmented the efficacy of agronomic biofortification. Additionally, novel techniques such as nutri-priming, foliar use, soilless stimulation, and mechanical methods have expanded the scope of agronomic biofortification. The improvements, attached with an increased indulgence of trace elements nutrition, underline the pivotal role of agronomic biofortification in ensuring global food and nutritional safety (Bhardwaj et al. 2022).

Maize biofortification with critical trace elements (iron and zinc) could be achieved by a combination of common applies, breeding for effective accumulators, and chemical fortification of crop grains (Table 12.1). Among these approaches, agronomic biofortification stands out as the most efficient and widely accepted method. Previous findings have exposed that the plant micronutrient contents (Fe and Zn) can be expressively improved by adding inorganic forms of these elements into organically bound and naturally chelated forms, thereby enhancing their availability and uptake (Arabhanvi and Hulihalli 2018).

12.5.1.3 Biofortification Through Transgenic/Genetic Means

In settings where genetic variation is limited or absent, the transgenic method emerges as a substitute route for developing biofortified crops (Brinch-Pedersen et al. 2007). Efficient biofortified genotypes of various crops are being developed and cultivated worldwide (Table 12.2). This method taps into a vast genetic reservoir, allowing for the transformation of beneficial genes from diverse plant types, irrespective of their progress. The crop plants' natural capacity to fascinate an element is insufficient, and transgenic techniques provide a very feasible solution for fortifying the crop (Perez-Massot et al. 2013). The important factor for growth in transgenic crops is identifying the important functions of genes and utilizing them to

Table 12.1 Influence of Fe and Zn application on maize grains.

Treatments	Fe contents (mg kg^{-1})	Increase (%)	Zn contents (mg kg^{-1})	Increase (%)
NPK	74.10	—	14.30	—
NPK + Fe (10 kg)	91.60	23.60	18.30	28.00
NPK + Fe (10 kg)	107.60	45.20	23.20	62.20
NPK + Fe (10 kg)	122.70	65.60	25.10	75.50
NPK + foliar spray of Fe (0.1%)	153.60	107.30	31.80	122.40

Table 12.2 Efficient biofortified genotypes used in the world.

Crop	Variety	Location	References
Rice	CR Dhan-310 (protein-rich)	India	Yadava et al. (2020)
	Zhongguangxiang (iron-rich)	China	Saltzman et al. (2013)
	GR2 (provitamin A–rich transgenic biofortified)	USA	Saltzman et al. (2013)
	BRRI Dhan-62 (zinc-rich)	Bangladesh	Andersson et al. (2017)
Wheat	WB-02, HPBW-01 (zinc- and iron-rich)	India	Yadava et al. (2020)
	BHU-3 and BHU-6 (zinc-rich)	India	Andersson et al. (2017)
Maize	Pusa Vivek-QPM9 (provitamin "A", lysine- and tryptophan rich hybrid), Pusa-HQPM-5/7 (provitamin "A", tryptophan, lysine rich)	India	Yadava et al. (2020)
Cowpea	Pant (Lobia-1) and Pant (Lobia-2) (iron-rich cowpea)	India	Saltzman et al. (2013)
Sweet potato	Nanshu-0101 (provitamin "A" rich)	China	Saltzman et al. (2013)
Lentil	Pusa-Ageti Masoor (iron-rich)	India	Yadava et al. (2020)

modify plant metabolism. The paths from various microorganisms (bacteria) could be familiarized into crop plants to explore novel metabolic engineering routes (Newell-McGloughlin 2008). Concurrently, gene incorporation can enhance micronutrient concentrations and bioavailability while minimizing nutrient fixation that impedes their absorption in the physiological system of the plants.

Genetic biofortification contains usual breeding methods for identification and leverages genetic dissimilarities for mineral content, alongside novel tactics such as gene finding and marker-assisted breeding. Researchers have employed genome-wide relationship studies (GWAS) to pinpoint specific genomic areas in maize conducive to kernel Zn and Fe biofortification and have proposed genomic selection (GS) as a potential breeding practice for Fe and Zn biofortification in wheat (Hindu et al. 2018). From a sustainability perspective, nutrition-focused crop breeding offers several advantages. Current improvements allow for the modulation of indicating pathways, although breeding mainly relies on prolonged cycles of hybridization and collection that are time-consuming and labor-intensive.

However, in recent years, modern molecular tools like DNA markers and marker-assisted selection (MAS) technologies have accelerated the progress of nutrient-rich varieties. Researchers have identified quantitative trait loci (QTLs) for iron and zinc biofortification in pearl millet using diversity array technology (DArT) and simple sequence repeat (SSR) markers. Numerous iron- and zinc-biofortified rice, wheat, and maize varieties have been globally being used to combat malnutrition (Kumar et al. 2018).

Additionally, genetic or plant breeding approaches and numerous transgenic mediations have been deployed successfully for biofortifying food crops. Transgenic practices facilitate gene exchange between unrelated species or the introduction of novel genes into several crops (Ramesh et al. 2014). Researchers have introduced innovative methods to enhance seed Zn and Fe amounts by overexpressing specific genes in barley and achieved substantial increases in Fe supplements in rice grains by genetic modifications. Other studies have created transgenic rice plants with elevated iron content, enriched with specific enzymes, thereby facilitating enhanced iron uptake and bioavailability.

12.5.2 Microbial-based Approaches for Fe Biofortification

Beneficial microbes like PGPR are key contributors to enhancing the biofortification of micronutrients in cereal grains. They achieve this through processes like nutrient solubilization, siderophore production, and exopolysaccharide synthesis (Jalal et al. 2021; Shahane and Shivay 2022). The use of PGPR for biofortification has gained momentum, especially for improving iron (Fe) and other microelements levels in crops. For instance, Gopalakrishnan et al. (2016) identified species from genera such as *Pseudomonas, Brevibacterium, Bacillus, Enterobacter*, and Acinetobacter as promising candidates for biofortification and plant biocontrol. These microorganisms can increase the solubility of micronutrients from otherwise insoluble sources in the soil, facilitating their incorporation into edible plant tissues (Khalid et al. 2015; Gopalakrishnan et al. 2016).

Using PGPR for biofortification not only addresses malnutrition issues but also enhances cereal yield, soil fertility, and biodiversity (Ditta et al. 2022; Jalal et al. 2022; Ahmed et al. 2023). The soil microbial communities show an important function in enriching the soil and regulating supplement availability to crop plants. Changes in this community can affect supplement uptake by crops (Nuttall et al. 2017; Panhwar et al. 2021). PGPR also plays a significant role in fortifying both macronutrients and micronutrients in food crops involving processes such as siderophore production, N_2 fixation, and P-mobilization (Khan et al. 2019; Singh et al. 2017). While both rhizospheric and endophytic microorganisms influence micronutrient availability to plants, endophytic microorganisms are considered more effective in enhancing iron (Fe) and zinc (Zn) uptake and translocation due to their indirect regulation of metal transporters.

To biofortify Zn and Fe, formulations of that microbes (bacteria or fungi) could be discovered as seed primers or dressing in soil methods. Contemporary implements like transcriptomics, metaproteomics, and genomics can identify the genes and proteins intricate in their translocation within major crop plants, thereby enhancing the efficacy of plant–microbe interactions (Singh and Prasanna 2020). Additionally, the inoculation or co-inoculation of PGPR isolates significantly improved plant height and shoot fresh biomass compared to uninoculated controls (Table 12.3; Ahmad et al. 2023; Panhwar et al. 2020).

Figure 12.2 illustrates the effectiveness of novel bio-inoculants in enhancing Fe uptake in maize shoots. The application of microbial inoculants notably increases both zinc and iron levels in maize shoots compared to the non-applied treatments. Specifically, the consortium inoculation (ZM27 + ZM63 + S10) leads to a significant 18% increase in maize shoot Fe amounts associated to the non-inoculation. Similarly, the bacterial consortium inoculation results in a 15% boost in plant shoot Zn quantity, demonstrating the efficacy of microbial inoculation in enhancing nutrient uptake in maize (Ahmad et al. 2023; Othman et al. 2022).

The various plant beneficial microbes are crucial for iron (Fe) biofortification as they aid in the production and release of siderophores into the environment. Siderophore-producing bacteria such as *Salmonella, Klebsiella, Aerobacter, Yersinia* sp., *Mycobacterium* sp., *Aspergillus nidulans, Mucor*, and *Rhizopus* significantly enhance Fe accessibility in plants (Kannahi and Senbagam 2014). Furthermore, microorganisms can produce signals that regulate gene expression linked to nutrient

Table 12.3 Impact of innovative bio-inoculants on plant growth parameters of maize.

Treatments	Shoot length height (cm)	Fresh plant biomass (g plant^{-1})	Dry shoot biomass (g plant^{-1})	Fresh root biomass (g plant^{-1})	Dry root biomass (g plant^{-1})
Control	73.33h	37.67j	11.72e	20.97e	5.43g
ZM27	74.33h	42.15fg	12.71ce	24.27cd	6.00ef
ZM31	82.0df	42.9eg	13.02ce	22.83d	6.13df
ZM63	88.0b	39.81hi	13.56ad	24.10d	6.00ef
S10	89.33b	38.06ij	13.93ac	26.93ab	5.83fg
ZM27 + ZM31	84.33cd	44.60ce	12.40de	24.07d	6.37ef
ZM27 + ZM63	78.00g	41.33gh	12.49ce	23.13d	6.13df
ZM27 + S10	94.00a	47.03ab	14.80ab	26.9ab	7.00ab
ZM31 + ZM63	88.7b	46.44ac	12.31de	24.07d	5.93eg
ZM31 + S10	86.7bc	44.50ce	13.37bd	25.73bc	6.40ce
ZM63 + S10	83.0de	43.91df	12.51ce	23.60d	6.25df
ZM27 + ZM31 + ZM63	82.0df	43.80df	12.92ce	23.30d	6.55bd
ZM27 + ZM31 + S10	79fg	42.17fg	12.31de	23.90d	6.59bd
ZM27 + ZM63 + S10	96a	48.03a	14.94a	27.8a	7.28a
ZM31 + ZM63 + S10	79.7eg	44.58ce	13.06ce	23.17d	6.63bd
ZM27 + ZM31 + ZM63 + S10	81.33dg	45.2bd	13.21cd	23.26d	6.90ac
LSD ($p < 0.05$)	3.5199	2.091	1.4840	1.6179	0.5420

Note: The mean of the same letters does not differ significantly at $p \leq 0.05$.
Source: Ahmad et al., 2023 / Frontiers Media S.A. / CC BY 4.0

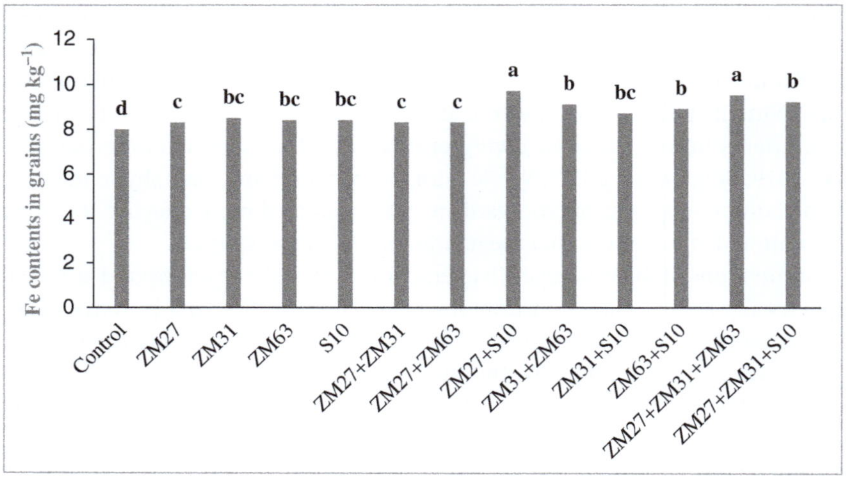

Figure 12.2 Influence of PGPR-inoculation on shoot Fe content in maize. *Source*: Adapted from Ahmad et al. (2023).

uptake and transport in plants. Some microorganisms have been employed in biofortification programs for wheat, resulting in enhanced nutrient levels (Table 12.4). Additionally, various microorganisms, with endophytes and mycorrhizal fungi, have been studied for influencing the uptake of both macronutrients and micronutrients (Zhang et al. 2009).

12.6 Iron Biofortification for Combating Malnutrition in Cereal Crops

About two billion people worldwide are affected by micronutrient deficiencies, which significantly impact their health and socioeconomic well-being (Sheoran et al. 2022). This deficiency is largely attributed to the ingesting of cereal-based foods that deliver ample calories but lack essential phytochemicals like minerals, vitamins, antioxidants, and fiber. Biofortified crops including numerous important microelements are emerging as an advanced and encouraging tactic to combat these extensive deficits and mend overall humanoid nourishment (Straeten et al. 2020). To address micronutrient shortages sustainably, fortification – adding essential micronutrients to staple foods – is considered a viable approach (Kiran et al. 2022; FAO 2021).

Fe is an acute minor element essential for plant, human, and animal health; however, mostly Fe in the soil exists in an inaccessible form to plants (ferric form, Fe^{3+}), leading to low bioavailability (Singh and Prasanna 2020). Although various methods have been settled to combat Fe shortage in humans, biofortification-breeding nutrients into food crops is recognized as a sustainable policy, especially for people with inadequate admittance to unlike diets. The practice of iron (Fe) fertilization has demonstrated an ability to enhance iron concentrations in plant grains when administered as a foliar fertilizer. Research

Table 12.4 Microorganisms with potential for wheat biofortification with micronutrients.

S. No.	Species	Microbes	Improved micronutrient contents (%)	References
1	*T. aestivum*	*Providencia* sp.	106.0	Rana et al. (2012)
2	*T. turgidum*	*E. cloacae* sp.	21.11	Ramesh et al. (2014)
3	*T. turgidum*	*Bacillus aryabhattai*	45.00	Ramesh et al. (2014)
4	*T. aestivum*	*P. fluorescens* stain	85.00	Sirohi et al. (2015)
5	*T. aestivum*	*B. subtilis, Arthrobacter* sp.	100.0	Singh et al. (2017)
6	*T. aestivum*	*P. fragi*	Twofolds	Kamran et al. (2017)
7	*T. aestivum*	*Exiguobacterium aurantiacum*	Sixfold	Shaikh and Saraf (2017)
8	*T. aestivum*	*Pseudomonas* sp.	Twofold	Rehman et al. (2018)
9	*T. aestivum*	*B. subtilis*	14–20	Singh et al. (2017)

Source: Adapted from Zhang et al. (2009).

suggests that the inclusion of iron sulfate during the germination of rice seeds resulted in a 15.6-fold increase in iron concentration in germinated brown rice related to scenarios where $FeSO_4$ was not used. Furthermore, combining iron with urea fertilizer in foliar sprays has been associated with elevated iron accretion in wheat grain.

12.6.1 Iron Biofortification in Wheat

Iron is a critical element for humans, necessary for suitable bodily function, yet the body cannot synthesize it and must obtain it from dietary sources. Introducing iron-rich mineral fertilizers into the soil and practicing effective soil management, including pH balancing and maintaining optimal nutrient levels, can improve iron absorption by crops. However, the efficacy of these approaches may vary depending on the crop type and prevailing growing conditions. For instance, individuals following a vegan diet typically require 1.8 times more than the indorsed daily intake (Recommended Dietary Allowance [RDA]) of iron compared to those who consume meat.

Wheat grain plays a proper role in the human diet, providing numerous elements and mineral deposits. Therefore, the production of wheat grains needs to be doubled by 2050 to ensure global food safety (Hughes and Ortiz 2018). Extensive screening of wheat germplasm has been conducted to assess its metal contents, including iron (Fe), zinc (Zn), selenium (Se), manganese (Mn), magnesium (Mg), and proteins (Hakki et al. 2012; Cakmak et al. 2000; Gupta et al. 2020; Oury et al. 2006; Esfandiari et al. 2016), as well as vitamins (Kumar et al. 2018). Screening also considers phytic acid due to its role in limiting nutrient bioavailability (Sakai et al. 2015). In recent periods, efforts in wheat biofortification have explored breeding, agronomic, and genetic solutions (Jha and Warkentin 2020; Liu et al. 2007; Sakai et al. 2015). Initiatives like the "International Maize and Wheat Improvement Center" (CIMMYT) gene bank and the "Harvest Plus project" have laid the groundwork by releasing economical cultivar of wheat, such as cv. "Zincol-2016 in Pakistan, cv. Bari Gom-33 in Bangladesh, and cv. Zinc-Shakti (Chitra), WB-02, and HPBW-01 in India." High-yielding genotypes have reported Fe concentrations ranging from 20 to 60 mg kg^{-1} (Gupta et al. 2020). Additionally, up to a threefold enhancement of Fe-concentrations in wheat grains has been achieved over soil and foliar use (Hakki et al. 2012; Lockyer et al. 2018). Promoting stable fertilization among the farmers in the developing world can promote to encounter micronutrient content goals to combat hidden starvation. Although increasing Fe contents in the most edible part, the endosperm, through agronomic practices is challenging, some findings have stated an upsurge in Fe concentration through soil application (Cakmak et al. 2010; Kiran et al. 2021; Ozturk et al. 2006).

12.6.2 Iron Biofortification in Rice

Rice holds significant importance for enhancing micronutrients due to its global role as a primary staple food crop, making rice biofortification a powerful tool for addressing malnutrition worldwide. In 2013, the high-zinc (Zn) rice variety, with 20–22 mg kg^{-1} Zn in brown rice, was distributed by Harvest-Plus and the Bangladesh Rice Research Institute. Studies have recorded increases of 17.4 mg kg^{-1} for Zn, 0.123 mg kg^{-1} for selenium (Se), and 14.2 mg kg^{-1} for iron (Fe) in rice (Fang et al. 2008). The provitamin "A" biofortified "Golden Rice" has emerged as a cost-effective solution in regions where rice is a staple crop (Dubock 2017). Lately, Pradhan et al. (2020) selected 484 rice varieties and identified co-localized quantitative trait locus (QTL) regions for Fe alongside high-yield traits. Understanding the rice grain composition, including the localization of Fe chelators, transporters, promoters, and inhibitors, is crucial for improving the bioavailability of microelements in rice and thereby enhancing the nourishment and people's health (Pradhan et al. 2020).

12.6.3 Iron Biofortification in Maize

Rice holds significant importance for enhancing micronutrients due to its global role as a primary staple food crop, making rice biofortification a powerful tool for addressing malnutrition worldwide. In 2013, the first high-zinc (Zn) rice cultivar, containing 20–22 mg kg^{-1} Zn in brown rice, was disseminated through Harvest Plus and the Bangladesh Rice Research Institute. Increases of 17.4, 0.123, and 14.2 mg kg^{-1} for Zn, selenium (Se), and iron (Fe), respectively, have been documented in rice (Fang et al. 2008). The provitamin A biofortified "Golden Rice" has emerged as a cost-effective intervention in regions where rice is the staple crop (Dubock 2017). Recently, Pradhan et al. (2020) screened 484 rice lines and identified co-localized QTL regions for Fe alongside high-yield attributes. Understanding the composition of rice grains, including the localization of Fe chelators, transporters, promoters, and inhibitors, is essential for enhancing the bioavailability of micronutrients in rice and subsequently improving the nutrition and health of consumers (Pradhan et al. 2020).

Iron biofortification in maize is also crucial, considering maize's role as both a cash crop and a primary food in most countries, providing sustenance for people and animals worldwide. The exogenous use of Zn, whether through seed priming, foliar spray, or soil incorporation, enhances seed germination, potency in plants, and reception against various anxieties (Imran and Rehim 2017; Liu et al. 2017). Maqbool and Beshir (2018) documented high accretion of Zn, reaching 36 mg kg^{-1}, in maize grains with the use of ZnO nanoparticles. Previous studies have highlighted substantial genome-wide associations between micronutrient absorption in maize kernels and yield, suggesting the feasibility of maize biofortification by consuming specialized phenotyping utensils and conservative plant breeding practices (Subbaiah et al. 2016; Prasanna et al. 2020; Jin et al. 2013). Among maize lines (1000-CIMMYT), kinds of Zn, Fe, and "provitamin A" concentrations have been described. Specifically, maize lines with Zn concentrations ranging from 15 to 35 mg kg^{-1}, an average of 20 mg kg^{-1} Fe, and approximately 0 to 15 mg kg^{-1} total "provitamin A" contents have been recognized (Hindu et al. 2018).

12.7 Conclusion

The word "hidden hunger" refers to the inadequacy of micronutrients, often overlooked compared to the more visible issue of starvation. Malnutrition is primarily associated with foods lacking sufficient micronutrients, falling below the optimal concentrations required for health. Mineral nutrient deficiencies, particularly in iron (Fe), remain a significant concern in human nutrition, affecting approximately two billion people globally and posing a serious healthcare challenge. Biofortification, an emerging, economical, and viable technique, holds promise for supplying micronutrients with partial access to varied regimes and other mediations. Conventional, agronomic, and transgenic approaches are likely to be effective in enriching food crops with these essential micronutrients. Promising avenues include exploring transgenic and genomic techniques for enhancing Fe levels in cereals, besides integrating microbial and genetic strategies to further enhance Fe bioavailability, thereby impacting uptake and accumulation in plants.

12.8 Future Prospects, Challenges, and Recommendations

The world population is increasing and nutrient-rich food requirement is also increasing, making biofortification an essential area of study. Researchers are striving to develop new crop varieties that are rich in crucial "vitamins and minerals," i.e. Fe, Zn, and vitamin "A." This ongoing investigation aims to cultivate a diverse array of nutrient-rich crops, which could play a crucial role in combatting global malnutrition and enhancing public health. Biofortification research advancements are also concentrating on improving plants' absorption of vital nutrients from the soil, exploring the practice of chemical minerals and agricultural methods to proliferate nutrient accessibility. While staple crops like rice, wheat, and maize have been the primary targets of biofortification, there is potential to extend these efforts to other foods like fruits, vegetables, and legumes, given their widespread consumption. Enhancing the efficiency of biofortification by increasing nutrient levels with fewer resources and in a more cost-effective and sustainable manner is imperative. Moreover, improving the distribution and accessibility of biofortified crops requires the development of new storage and transportation facilities. Equally significant is raising awareness and understanding of biofortification to address any concerns or misconceptions about the safety and nutritional benefits of biofortified foods. By addressing malnutrition and improving human health, biofortification also contributes to nutrient safety by cumulative the accessibility of healthy food. Prioritizing these endeavors is crucial for enhancing the nutritional status of people worldwide.

12.9 Recommendations and Challenges

Several tasks must be addressed to effectively implement biofortification programs aimed at combating malnutrition:

1. Inadequate obtainability of biofortified varieties: Biofortified crop varieties are not widely accessible for cultivation, which hinders their impact on addressing nutritional deficiencies.
2. Lack of awareness and understanding: Many people are unaware of the benefits of biofortified crops, which affects their acceptance and adoption.
3. Infrastructure and availability issues: The availability of biofortified varieties is hindered by challenges, i.e. insufficient infrastructure, insufficient storing amenities, and high budgets associated with distribution.

4. Political and regulatory barriers: Political and regulatory hurdles can delay the development and dissemination of biofortified crops, impacting their availability to those who need them.
5. Challenges with genetically modified (GM) crops: GM crops, which can be important for biofortification efforts, face complex and sometimes controversial regulations that impede their development and commercialization.
6. Yield variability: Biofortified various crops may not consistently yield as highly as nonbiofortified cultivars in specific rising circumstances like dearth or pest outbreaks, which can deter farmer adoption.
7. Farmer awareness and prioritization: Farmers may not prioritize the adoption of biofortified varieties due to limited knowledge about their yield potential and other considerations.
8. Funding requirements: Implementing biofortification techniques requires substantial funding to assist in investigation, development, and employment factors.
9. Coordination challenges: There is often a lack of effective coordination among biofortification programs involving various stakeholders such as governments, NGOs, farmers, and the private sector, which can hinder program success.

Addressing these challenges through increased research, awareness campaigns, policy advocacy, infrastructure development, and stakeholder collaboration is necessary for the effective implementation of biofortification programs to alleviate malnutrition.

References

Abbas, G., Khattak, J.Z.K., Mir, A.M. et al. (2012). Effect of organic manures with recommended dose of NPK on the performance of wheat (Triticum aestivum L.). *The Journal of Animal and Plant Sciences* 22 (3): 683–687.

Agrios, G.N. (2005). Nutrient deficiencies in plants. In: *Plant Pathology* (ed. KD Sonnack), 372–373. London, UK: Elsevier Academic Press.

Ahmad, M., Hussain, A., Dar, A. et al. (2023). Combating iron and zinc malnutrition through mineral biofortification in maize through plant growth promoting *Bacillus* and *Paenibacillus* species. *Frontiers in Plant Science* 13: 1094551. doi: 10.3389/fpls.2022.1094551.

Ali, M, Ahmed, I., Tariq, H. et al. (2023). Growth improvement of wheat (*Triticum aestivum* L.) and zinc biofortification using potent zinc-solubilizing bacteria. *Frontiers in Plant Science* 14: 1140454. doi: 10.3389/fpls.2023.1140454.

Anas, M., Liao, F., Verma, K.K. et al. (2020). Fate of nitrogen in agriculture and environment: agronomic, eco-physiological and molecular approaches to improve nitrogen use efficiency. *Biological Research* 53 (1): 1–20.

Andersson, M., Turesson, H., Nicolia, A., et al. (2017). Efficient targeted multiallelic mutagenesis in tetraploid potato (*Solanum tuberosum*) by transient CRISPR-Cas9 expression in protoplasts. *Plant Cell Reports* 36(1): 117–128.

Arabhanvi, F. and Hulihalli, U.K. (2018). Agronomic fortification with zinc and iron to enhancing micronutrient concentration in sweet corn grain to ameliorate the deficiency symptoms in human beings. *International Journal of Current Microbiology and Applied Sciences* 2018:7 (2): 333–340.

Ashraf, M.A., Akihiro, T., Ito, K. et al. (2021). ATP binding cassette proteins ABCG37 and ABCG33 function as potassium-independent cesium uptake carriers in Arabidopsis roots. *Molecular Plant* 14 (4): 664–678.

Atajan, F.A., Mozafari, V., Abbaszadeh-Dahaji, P. et al. (2019). Fractionation and speciation of manganese in rhizosphere soils of Pseudomonas sp. rhizobacteria inoculated Pistachio (Pistacia vera L.) seedlings under salinity stress. *Communications in Soil Science and Plant Analysis* 50 (7): 894–908.

Awika, J.M. (2011). Major cereal grains production and use around the world. In: *Advances in Cereal Science: Implications to Food Processing and Health Promotion*, (eds. J.M. Awika, V. Piironen and S. Bean), 1–13. Washington, DC: ACS Publications.

Awino, F.B., Maher, W.A., Krikowa, F. et al. (2020). Occurrence of trace metals in food crops grown on the Mbale dumpsite, Uganda, and human health risks. *Integrated Environmental Assessment and Management* 16 (3): 362–377.

Bhardwaj, A.K., Chejara, S., Malik, K. et al. (2022) Agronomic biofortification of food crops: An emerging opportunity for global food and nutritional security. *Frontiers In Plant Science* 10: e5278. 13: 1055278. doi: 10.3389/fpls.2022.1055278.

Bhatla, S.C.A., Lal, M., Kathpalia, R. et al. (2018). Plant mineral nutrition. In: *Plant Physiology, Development and Metabolism* 37–81. Singapore: Springer.

Bierman, M.P. and Rosen, C.J. (2005). Diagnosing nutrient disorders in fruit and vegetable crops. www.extension.umn.edu/distribution/horticulture/M1190.html.

Blicharz, A., Marzeda, M., Drozd, M. et al. (2020). Beriberi disease–a picture of thiamine deficiency in under developed and highly developed countries. *Journal of Education Health and Sport* 10 (9): 407–414. doi: 10.12775/JEHS.2020.10.09.048.

Brinch-Pedersen, H., Borg, S., Tauris, B. et al. (2007). Molecular genetic approaches to increasing mineral availability and vitamin content of cereals. *Journal of Cereal Science* 46 (3): 308–326. doi: 10.1016/j.jcs.2007.02.004.

Brown, P.H., Zhao, F.J. and Dobermann, A. (2022). What is a plant nutrient Changing definitions to advance science and innovation in plant nutrition. *Plant and Soil* 476 (1–2): 11–23.

Cakmak, I., Brown, P., Colmenero-Flores, J.M. et al. (2023). Micronutrients in marschner's mineral nutrition of plants. In: *Plant and soil*, 283–385. Cambridge: Academic Press.

Cakmak, I., Kutman, U.B. (2018). Agronomic biofortification of cereals with zinc: A review. *European Journal of Soil Science* 69 (1 Suppl 2): 172–180.

Cakmak, I., Ozkan, H., Braun, H.J. et al. (2000) Zinc and iron concentrations in seeds of wild, primitive, and modern wheats. *Food and Nutrition Bulletin* 21 (4): 401–403.

Cakmak, I., Pfeiffer, W.H. and McClafferty, B. (2010). Biofortification of durum wheat with Zinc and Iron. *Cereal Chemistry* 87 (1): 10–20.

Coelho, D.G., Da Silva, V.M., Gomes Filho, A.A.P. et al. (2023). Bioaccumulation and physiological traits qualify Pistia stratiotes as a suitable species for phytoremediation and bioindication of iron-contaminated water. *Journal of Hazardous Materials* 446: 130701.

Da Silva Lopes, K., Takemoto, Y., Garcia-Casal, M.N. et al. (2018). Nutrition-specific interventions for preventing and controlling anaemia throughout the life cycle: an overview of systematic reviews. *Cochrane Database of Systematic Reviews* 8 (9154): 013092. doi: 10.1002/14651858.

D'Oria, A., Courbet, G., Billiot, B. et al. (2022). Drought specifically downregulates mineral nutrition: Plant ionomic content and associated gene expression. *Plant Direct* 6 (8): e402. doi: 10.1002/pld3.402.

Development initiatives. (2017). *Global Nutrition Report: Nourishing the SDGs*. Bristol, UK: Development Initiatives. (accessed 20 August 2022).

Singh, D. and Prasanna, R. (2020). Potential of microbes in the biofortification of Zn and Fe in dietary food grains. A review. *Agronomy for Sustainable Development*. 40 (2): 15.

Ditta, A., Ullah, N., Imtiaz, M. et al (2022). Zn biofortification in crops through Zn-solubilizing plant growth promoting rhizobacteria. In: Sustainable plant nutrition under contaminated environments (ed. Q. Mahmood). Switzerland: Springer Nature Switzerland AG.

Dubock, A. (2017). An overview of agriculture, nutrition and fortification, supplementation and biofortification: Golden Rice as an example for enhancing micronutrient intake. *Agriculture & Food Security* 6 (1): 59.

Elhaissoufi, W., Ghoulam, C., Barakat, A. et al. (2022). Phosphate bacterial solubilization: A key rhizosphere driving force enabling higher P use efficiency and crop productivity. *Journal of Advanced Research* 38 (9): 13–28.

Esfandiari, E., Abdoli, M., Mousavi, S.-B. et al. (2016). Impact of foliar zinc application on agronomic traits and grain quality parameters of wheat grown in zinc deficient soil. *Indian Journal of Plant Physiology* 21 (3): 263–270.

Evenson, R.E. and Gollin, D. (2003). Assessing the impact of the Green Revolution, 1960 to 2000. *Science* 300 (5620): 758–762. doi: 10.1126/science.1078710.

Fang, Y., Wang, L., Xin, Z. et al. (2008). Effect of Foliar Application of Zinc, Selenium, and Iron Fertilizers on Nutrients Concentration and Yield of Rice Grain in China. *Journal of Agricultural and Food Chemistry* 56 (6): 2079–2084.

FAO. (2017a). Food *and Agriculture Organization of the United Nations; The Future of Food and Agriculture– Trends and Challenges*. Rome: FAO.

FAO. (2017b). Nutrient-sensitive agriculture and food systems in practice. In: *Food and Agriculture Organization of the United Nations*. Rome: FAO. https://ww.fao.org/3/i7848e/i7848e.pdf (accessed 10 May 2022).

FAO. (2021). *The State of Food Security and Nutrition in the World. In Building Climate Resilience for Food Security and Nutrition*. Rome, Italy: Food and Agriculture Organization.

Fernandes, L.J.C., Villas Boas, R.L., Backes, C. et al. (2011). Contribution of nitrogen concentrations in garlic cloves treated with doses of N in side dressing. *Pakistan Journal of Botany* 49 (1): 101–108.

Field, M.S., Mithra, P. and Pena-Rosas, J.P. (2021). Wheat flour fortification with iron and other micronutrients for reducing anaemia and improving iron status in populations, *Cochrane Database of Systematic Reviews* 1 (3): 11302. doi: 10.1002/14651858.

Garcia-Oliveira, A.L., Chander, S., Ortiz, R. et al. (2018). Genetic Basis and Breeding Perspectives of Grain Iron and Zinc Enrichment in Cereals. *Frontiers in Plant Science* 9: 937. doi: 10.3389/fpls.2018.00937.

Gondal, A.H., Zafar, A., Zainab, D. et al. (2021). A detailed review study of zinc involvement in animal, plant and human nutrition. *Indian Journal of Pure and Applied Bioscience* 9 (2): 262–271.

Gopalakrishnan, S., Vadlamudi, S., Samineni, S. et al. (2016). Plant growth-promotion and biofortification of chickpea and pigeonpea through inoculation of biocontrol potential bacteria, isolated from organic soils. *Springerplus* 5 (1): 1882. doi: 10.1186/s40064-016-3590-6.

Graham, R.D., Welch, R.M., Bouis, H.E. (2001). Addressing micronutrient malnutrition through enhancing the nutritional quality of staple foods: Principles, perspectives and knowledge gaps. *Advances in Agronomy* 70: 77–142.

Gupta, O.P., Singh, A.K., Singh, A. et al. (2022). Wheat biofortification: utilizing natural genetic diversity, genome-wide association mapping, genomic selection, and genome editing technologies. *Frontiers Nutrition* 9: 826131. doi: 10.3389/fnut.2022.826131.

Gupta, P.K., Balyan, H.S., Sharma, S. et al. (2020). Biofortification and bioavailability of Zn, Fe and Se in wheat: Present status and future prospects. *Theoretical and Applied Genetics* 134 (1): 1–35.

Hakki, E., Hamurcu, M., Khan, M.K. et al. (2012). Wheat biofortification—A potential key to human malnutrition. *Journal of Elementology* 22 (3): 937–944.

Hawkesford, M.J., Cakmak, I., Coskun, D. et al. (2023). Functions of macronutrients. *In Marschner's Mineral Nutrition of Plants* 201–281. Academic press.

Hindu, V., Palacios-Rojas, N., Raman, B. et al. (2018). Identification and validation of genomic regions influencing kernel zinc and iron in maize. *Theoretical and Applied Genetics* 131 (2): 1443–1457.

Hopkins, B.G. and Hansen, N.C. (2019). Phosphorus management in high-yield systems. *Journal of Environmental Quality* 48 (5): 1265–1280.

Hughes, J. and Ortiz, O. (2018). Biofortification strategy. In: *Proceedings of the CGIAR System Management Board Meeting*, 1–48. SMB10-BP1f, Nairobi, Kenya, 26–27 September: Nairobi, Kenya: Business Plan Companion.

Imran, M. and Rehim, A. (2017). Zinc fertilization approaches for agronomic biofortification and estimated human bioavailability of zinc in maize grain. *Archives of Agronomy and Soil Science* 63 (1): 106–116.

Ishfaq, M., Wang, Y., Yan, M. et al. (2022). Physiological essence of magnesium in plants and its widespread deficiency in the farming system of china. *Frontiers Plant Science* 25:13: 802274. doi: 10.3389/fpls.2022.802274. PMID: 35548291; PMCID: PMC9085447.

Jabri, A. l., Saleem, H., Rizwan, M.H. et al. (2022). Zinc oxide nanoparticles and their biosynthesis: overview. *Life* 12:4: 594.

Jákli, B., Tavakol, E., Tränkner, M. et al. (2017). Quantitative limitations to photosynthesis in K deficient sunflower and their implications on water-use efficiency. *Journal of Plant Physiology* 209: 20–30. doi: 10.1016/j.jplph.2016.11.010.

Jalal, A., da Silva Oliveira, C.E., Freitas, L.A. et al. (2022). Agronomic biofortification and productivity of wheat with soil zinc and diazotrophic bacteria in tropical savannah. *Crop and Pasture Science* 11: 959. doi: 10.1071/CP21457.

Jalal, A., Galindo, F.S., Boleta, E.H.M. et al. (2021). Common bean yield and zinc use efficiency in association with diazotrophic bacteria co-inoculations. *Agronomy* 11(5): 959. doi: 10.3390/agronomy11050959.

Jha, A.B. and Warkentin, T.D. (2020). Biofortification of pulse crops: Status and future perspectives. *Plants* 9 (1): 73. doi: 10.3390/plants9010073.

Jin, T., Zhou, J., Chen, J. et al. (2013). The genetic architecture of zinc and iron content in maize grains as revealed by QTL mapping and meta-analysis. *Breeding Science* 63 (3): 317–324.

Johan, P.D., Ahmed, O.H., Omar, L. et al. (2021). Phosphorus transformation in soils following co-application of charcoal and wood ash. *Agronomy* 11: 10.

Kamran, S., Shahid, I., Baig, D.N. et al. (2017). Contribution of zinc solubilizing bacteria in growth promotion and zinc content of wheat. *Frontiers in Microbiology* 8: 2593. doi: 10.3389/fmicb.2017.02593.

Kannahi, M. and Senbagam, N. (2014). Studies on siderophore production by microbial isolates obtained from rhizosphere soil and its antibacterial activity. *Journal of Chemical and Pharmaceutical Research* 6 (4): 1142–1145.

Kaur, H., Kaur, H., Kaur, H. et al. (2022). The beneficial roles of trace and ultratrace elements in plants. *Plant Growth Regulation* 100: 1–18.

Khalid, S., Asghar, H.N., Akhtar, M.J. et al. (2015). Biofortification of iron in chickpea by plant growth promoting rhizobacteria. *Pakistan Journal of Botany* 47 (3): 1191–1194.

Khan, S., Yu, H., Li, Q. et al. (2019). Exogenous application of amino acids improves the growth and yield of lettuce by enhancing photosynthetic assimilation and nutrient availability. *Agronomy* 9 (5): 266. doi: 10.3390/ agronomy9050266.

Kiran, A., Wakeel, A., Mahmood, K. et al. (2022). Biofortification of staple crops to alleviate human malnutrition: Contributions and potential in developing countries. *Agronomy* 12 (2): 452. doi: 10.3390/agronomy12020452.

Kiran, A., Wakeel, A., Sultana, R. et al. (2021). Concentration and localization of Fe and Zn in wheat grain as affected by its application to soil and foliage. *Bulletin of Environmental Contamination and Toxicology* 106 (5): 852–858.

Kirkby, E.A., Nikolic, M., White, P.J. et al. (2023). Mineral nutrition, yield, and source–sink relationships. In: *Marschner's Mineral Nutrition of Plants*, 131–200. Cambridge: Academic Press.

Kumar, J., Saripalli, G., Gahlaut, V. et al. (2018). Genetics of Fe, Zn,-carotene, GPC and yield traits in bread wheat (Triticum aestivum L.) using multi-locus and multi-traits GWAS. *Euphytica* 214 (11): 219.

Kumar, S., Wani, J. A., Mehraj, K. et al. (2022). Fortification of micronutrients for sustainable development in field crops: A review. *The Pharma Innovation Journal* 11: 363–373.

Liu, D.Y., Zhang, W., Pang, L.L. et al. (2017). Effects of Zinc application rate and zinc distribution relative to root distribution on grain yield and grain Zn concentration in wheat. *Plant and Soil* 411 (1): 167–178.

Liu, Z., Wang, H., Wang, X. et al. (2007). Phytase activity, phytate, iron, and zinc contents in wheat pearling fractions and their variation across production locations. *Journal of Cereal Science* 45 (3): 319–326.

Lockyer, S., White, A., Walton, J. et al. (2018). Proceedings of the 'Working together to consider the role of biofortification in the global food chain' workshop. *Nutrition Bulletin* 43 (4): 416–427.

Mansour, F.Y.O. (2006). *Physiological Studies on Garlic (Allium Sativum L.). M.Sc. Thesis, Faculty of Agriculture*. Egypt: Minufiya University.

Maqbool, M.A. and Beshir, A. (2018) Zinc biofortification of maize (Zea mays L.): Status and challenges. *Plant Breeding* 138 (1–13): 1–28.

Mir, A.R., Pichtel, J. and Hayat, S. (2021). Copper: uptake, toxicity and tolerance in plants and management of Cu-contaminated soil. *Biometals* 34 (4): 737–759.

Mondal, G., Dasgupta, B. and Sharma, R. (2018). Diseases of medicinal and aromatic plants and their management. *Recent Approaches for Management of Plant Diseases* 251: 283.

Mondal, S. and Bose, B. (2019). Impact of micronutrient seed priming on germination, growth, development, nutritional status and yield aspects of plants. *Journal of Plant Nutrition* 42 19: 2577–2599.

Narayan, O.P., Kumar, P., Yadav, B. et al. (2023). Sulfur nutrition and its role in plant growth and development. *Plant Signal Behavior* 18 (1): 2030082. doi: 10.1080/15592324.2022.2030082.

Newell-McGloughlin, M. (2008). Nutritionally improved agricultural crops. *Plant Physiology* 147: 939–953. doi: 10.1104/pp.108.1 21947.

Nuttall, J.G., O'leary, G.J., Panozzo, J.F. et al. (2017). Models of grain quality in wheat–a review. *Field Crops Research* 202: 136–145. doi: 10.1016/j.fcr.2015.12.011.

Othman, N.M.I., Othman, R., Zuan, A.T.K. et al. (2022). Isolation, characterization, and identification of zinc-solubilizing bacteria (ZSB) from wetland rice fields in Peninsular Malaysia. *Agriculture* 12: 1823.

Oury, F.-X., Leenhardt, F., Rémésy, C. et al. (2006). Genetic variability and stability of grain magnesium, zinc and iron concentrations in bread wheat. *European Journal of Agronomy* 25 (2): 177–185.

Ozturk, L., Yazici, M.A., Yucel, C. et al. (2006). Concentration and localization of zinc during seed development and germination in wheat. *Physiologia Plantarum* 128 (1): 144–152.

Panhwar, Q.A., Ali, A., Depar, N. (2021). Influence of phosphatic fertilizer on growth, yield and bacterial abundance in rice (*Oryza sativa* L.) genotypes. *Pakistan Journal of Botany* 53 (1): 293–300. doi: 10.30848/PJB2021-1(3).

Panhwar, Q.A., Ali, A., Depar, N. et al. (2020). Screening of plant growth promoting rhizobacteria for sustainable wheat (*Triticum aestivum* L.) crop production. *Pakistan Journal of Botany* 52 (1). doi: 10.30848/PJB2020-1(20).

Peng, H.Y., Qi, Y.P., Lee, J. et al. (2015). Proteomic analysis of Citrus sinensis roots and leaves in response to long-term magnesium-deficiency. *BMC Genomics [Electronic Resource]* 16 (1): 1–24. doi: 10.1186/s12864-0151462-z.

Perez-Massot, E., Banakar, R., Gomez-Galera, S. et al. (2013). The contribution of transgenic plants to better health through improved nutrition: opportunities and constraints. *Genes & Nutrition* 8 (1): 29–41. doi: 10.1007/s12263-012-0315-5.

Pingali, P.L. (2012). Green Revolution: Impacts, limits, and the path ahead. *Proceedings of the National Academy of Sciences* 109: 12302–12308. U.S.A. doi: 10.1073/pnas.0912953109.

Pradhan, S.K., Pandit, E., Pawar, S. et al. (2020). Linkage disequilibrium mapping for grain Fe and Zn enhancing QTLs useful for nutrient dense rice breeding. *BMC Plant Biology* 20 (1): 1–24.

Prasanna, B.M., Palacios-Rojas, N., Hossain, F. et al. (2020). Molecular breeding for nutritionally enriched maize: Status and prospects. *Frontiers in Genetics* 10: 1392.

Rai, S., Singh, P.K., Mankotia, S. et al. (2021). *Iron Homeostasis in Plants and its Crosstalk with Siderophores in the Biogeochemical Process Controlling the Mobility of Copper in Vineyard Soils and its Transfer to the Plant.* Nantes: Doctoral dissertation.

Rajput, A., Memon, M., Memon, K.S. et al. (2015). Integrated nutrient management for better growth and yield of banana under Southern Sindh climate of Pakistan. *Soil Environment* 34 (2): 126–135.

Rajput, A., Panhwar, Q.A., Naher, U.A. et al. (2014). Influence of incubation period, temperature and different phosphate levels on phosphate adsorption in soil. *American Journal of Agricultural and Biological Sciences* 9 (2): 251–260.

Rajput, A., Waggan, M.R., Khan P. et al. (2007). Role of calcium carbonate on phosphate sorption in soil. *International Journal of Applied Biology and Biotechnology* 4 (4): 357–359.

Rajput. A., Solangi, M.A., Irshad, M. et al. (2014). Evaluation of the fertility status of the rice tract area of Jafarabad district, Baluchistan Pakistan. *E3 Journal of Agriculture Research and Development* 5 (5): 85–96.

Ramesh, A., Sharma, S.K., Sharma, M.P. et al. (2014). Plant growth-promoting traits in Enterobacter cloacae sub sp. dissolvens MDSR9 isolated from soybean rhizosphere and its impact on growth and nutrition of soybean and wheat upon inoculation. *Agricultural Research* 3 (1): 53–66. doi: 10.1007/ s40003-014-0100-3.

Rana, A., Joshi, M., Prasann, R. et al. (2012). Biofortification of wheat through inoculation of plant growth promoting rhizobacteria and cyanobacteria. *European Journal of Soil Biology* 50: 118–126. doi: 10.1016/j.ejsobi.2012.01.005.

Ratajczak, A.E., Rychter, A.M., Zawada, A. et al. (2021). Do only calcium and vitamin D matter? Micronutrients in the diet of inflammatory bowel diseases patients and the risk of osteoporosis. *Nutrients* 13 (2): 525. doi: 10.3390/nu13020525.

Rehman, A, Farooq, M., Naveed, M. et al. (2018). Seed priming of Zn with endophytic bacteria improves the productivity and grain biofortification of bread wheat. *European Journal of Agronomy* 94: 98–107. doi: 10.106/j. eja.2018.01.017.

Rempel, J., Grover, K. and El-Matary, W. (2021). Micronutrient Deficiencies and Anemia in Children with Inflammatory Bowel Disease. *Nutrients* 13: 236. doi: 10.3390/nu13010236.

Sakai, H., Iwai, T., Matsubara, C. et al. (2015). A decrease in phytic acid content substantially affects the distribution of mineral elements within rice seeds. *Plant Science* 238: 170–177.

Saleem, S., Mushtaq, N.U., Rasool, A. et al. (2023). Plant nutrition and soil fertility: Physiological and molecular avenues for crop improvement. In: *Sustainable Plant Nutrition*, 23–49. Cambridge: Academic Press.

Saltzman, A., Birol, E., Bouis, H. E., et al. (2013). Biofortification: Progress toward a more nourishing future. *Global Food Security* 2(1): 9–17.

Shahane, A.A. and Shivay, Y.S. (2022). Agronomic biofortification of crops: Current research status and future needs. *Indian Journal of Fertilisers* 18 2: 164–179.

Shaikh, S. and Saraf, M. (2017). Biofortification of *Triticum aestivum* through the inoculation of zinc solubilizing plant growth promoting rhizobacteria in field experiment. *Biocatalysis and Agricultural Biotechnology* 9: 120–126. doi: 10.1016/j. bcab.2016.12.008.

Sheoran, S., Kumar, S., Ramtekey, V. et al. (2022). Current status and potential of biofortification to enhance crop nutritional quality: An overview. *Sustainability* 14: 3301. doi: 10.3390/su14063301.

Shireen, F., Nawaz, M.A., Chen, C. et al. (2018). Boron: Functions and approaches to enhance its availability in plants for sustainable agriculture. *International Journal of Molecular Sciences* 19 (7): 1856. doi: 10.3390/ ijms19071856.

Simkin, A.J., Kapoor, L., Doss, C.G.P. et al. (2022). The role of photosynthesis related pigments in light harvesting, photoprotection and enhancement of photosynthetic yield in planta. *Photosynthesis Research* 152 (1): 23–42.

Singh, D., Rajawat, M.V.S., Kaushik, R. et al. (2017). Beneficial role of endophytes in biofortification of Zn in wheat genotypes varying in nutrient use efficiency grown in soils sufficient and deficient in Zn. *Plant and Soil* 416: 107–116. doi: 10.1007/s1110 4-017-3189-x.

Sirohi, G., Upadhyay, A., Srivastava, P.S. et al. (2015). PGPR mediated zinc biofertilization of soil and its impact on growth and productivity of wheat. *Journal of Soil Science and Plant Nutrition* 15: 202–216. doi: 10.4067/S0718- 95162015005000017.

Straeten, V.D., Bhullar, D., De Steur, N.K. (2020). Multiplying the efficiency and impact of biofortification through metabolic engineering. *Nature Communications* 11: 5203. doi: 10.1038/s41467-020-19020-4.

Subbaiah, L.V., Prasad, T.N.V.K.V., Krishna, T.G. et al. (2016). Novel effects of nanoparticulate delivery of zinc on growth, productivity, and zinc biofortification in maize (Zea mays L.). *Journal of Agricultural and Food Chemistry* 64 (19): 3778–3788.

Tanin, M.J., Saini, D.K., Kumar, P. et al. (2024). Iron biofortification in wheat: Past, present, and future. *Current Plant Biology* 38; 100328.

Tariq, A., Zeng, F., Graciano, C. et al. (2023). Regulation of metabolites by nutrients in plants. In: *Plant Ionomics: Sensing, Signaling, and Regulation*, 1–18. Hoboken: Wiley.

Thangavel, P., Anjum, N.A., Muthukumar, T. et al. (2022). Arbuscular mycorrhizae: natural modulators of plant–nutrient relation and growth in stressful environments. *Archives of Microbiology* 204 (5): 264.

Van Der Straeten, D., Bhullar, N.K., De Steur, H. et al. (2020). Multiplying the efficiency and impact of biofortification through metabolic engineering. *Nature Communication* 11 (1): 1–10. doi: 10.1038/541467020-19020-4.

Xie, K., Cakmak, I., Wang, S., Zhang, F. et al. (2021). Synergistic and antagonistic interactions between potassium and magnesium in higher plants. *The Crop Journal* 9 (2): 249–256. doi: 10. 1016/j.cj.2020.10.005.

Yadava, D.V., Choudhary, P.R, Hossain F. et al. (2020). Biofortified varieties: Sustainable way to alleviate malnutrition, 3e. New Delhi: Indian Council of Agricultural Research.

Yang, F., Tang, C. and Antonietti, M. (2021). Natural and artificial humic substances to manage minerals, ions, water, and soil microorganisms. *Chemical Society Reviews* 50 (10): 6221–6239.

Zhang, H., Sun, Y., Xie, X. et al. (2009). A soil bacterium regulates plant acquisition of iron via deficiency inducible mechanisms. *The Plant Journal* 58 (4): 568–577. doi: 10.1111/j.1365-313X.2009.

13

Selenium Biofortification in Horticultural Crops

Ghulam Murtaza[1*], *Zeeshan Ahmed*[2,3,4*], *Muhammad Rizwan*[5], *Muhammad Usman*[6,7]

[1] *Faculty of Environmental Science and Engineering, Kunming University of Science and Technology, Kunming, China*
[2] *Xinjiang Institute of Ecology and Geography, Chinese Academy of Sciences, Urumqi, China*
[3] *Cele National Station of Observation and Research for Desert-Grassland Ecosystems, Chinese Academy of Sciences, Urumqi, China*
[4] *College of Life Science, Shenyang Normal University, Shenyang, China*
[5] *School of Energy Science and Engineering, Central South University, Changsha, China*
[6] *School of Agriculture and Biology, Shanghai Jiao Tong University, China*
[7] *Department of Botany, Government College University, Katcheri Road, Lahore, Pakistan*

*Corresponding authors: zeeshanagronomist@yahoo.com; murtazabotanist@gmail.com

13.1 Introduction

While selenocysteine and selenium's functions in signaling translational termination, both of which are encoded by the UGA codon, were found in the 1990s, selenium was first discovered by Berzelius in 1817 (Wen 2021). Selenium, a trace element, is essential for human health and has a function in preventing COVID-19 infections (Guo et al. 2023). Selenium is essential for plant growth, developmental control, and signal transduction and is abundant in horticultural crops (Syed et al. 2021). Although selenium is usually thought to be good for plants and necessary for people, consuming more than 300 $\mu g\ day^{-1}$ of it through diet and supplements for five years increased the risk of death from all causes ten years later. It is predicted that up to one billion people globally suffer from either excessive (more than 400 $\mu g\ day^{-1}$) or insufficient (less than 40 $\mu g\ day^{-1}$) selenium intake (Abishek et al. 2023). Selenium deficiencies are associated with a number of illnesses in humans, such as immune system, thyroid, and reproductive issues, as well as cancer and even mortality (Hegedusova et al. 2021). Both plants and people may become hazardous from too much selenium. Excessive amount of selenium can cause oxidative stress and improper protein integration, both of which can be harmful. The amount of selenium in the soil and its bioavailability to crops, which are the primary sources of selenium for humans, determine how much selenium is consumed through diet (Ryant et al. 2020). Thus, research on selenium in horticultural plants is important.

Selenium comes in two primary forms: organic and inorganic. SeO_4^{2-}, H_2SeO^3, $HSeO^{3-}$, and SeO_3^{2-} are the primary forms of inorganic selenium, while selenocysteine and selenomethionine are the primary forms of organic selenium (Shekari et al. 2016). Moreover, selenium in other forms, methyl selenocysteine, and selenoglucosinolates are present (Syed et al. 2021). It appears to be safer to produce organic selenium from plants than to absorb inorganic selenium directly (Mezeyova et al. 2020). The methylation of selenocysteine and selenomethionine to methyl selenocysteine and methyl selenomethionine, respectively, detoxifies selenium in hyperaccumulators and results in a safe and nontoxic accumulation (White 2018). Therefore, methylation is one method of detoxifying too much selenium in plants. Different settings include different types of selenium. While selenite is the most common form of accessible selenium in agricultural soils, it also exists in acidic soils (El-Ramady et al. 2022). For horticultural vegetable soils that produce selenium slowly, selenite offers a longer lasting selenium fertilizer in comparison to the comparatively mobile selenium (Ali et al. 2023).

Due to its narrow safety threshold and requirement as a trace element, selenium has garnered significant attention in plant and agricultural sciences globally, making it a popular research topic. Plant physiology, cell biology theory, molecular biology, sequencing techniques, selenium uptake, transportation, valence state alterations, roles, and modifications have all been extensively investigated due to the rapid advancement of material separation and identification. Studies on selenium in horticulture crops have expanded as a result of the mineral's many uses and necessity for human health. This chapter reviews selenium and its functions in horticultural plants/crops to deliver a theoretical basis for practical applications.

Crop Biofortification: Biotechnological Approaches for Achieving Nutritional Security under Changing Climate, First Edition. Edited by Adnan Noor Shah et al.

13.2 Selenium in Horticultural Plants/Crops

Regarding the question of whether selenium is necessary for vascular plants, there is insufficient data. Nonetheless, it has been suggested that it might serve advantageous biological purposes in animals that can store large levels of selenium and that require selenium for normal growth (Jalali et al. 2022). Plant selenium accumulation differs depending on the type of plant and is influenced by soil selenium content, soil characteristics, and selenium's chemical form (Rahim et al. 2020). Most plants that are classified as nonaccumulators often have leaf selenium concentrations less than 100 mg selenium kg^{-1} dry weight. Selenium accumulation in leaves is limited to a few plant species that are cultivated in selenium-rich environments (Francini et al. 2023), *Stanleya pinnata* and *Astragalus* are considered hyperaccumulators when they accumulate more than 1000 mg kg^{-1} dry weight, while *Atriplex, Melilotus*, and *Brassica juncea* are considered indicators or secondary accumulators when they acquire 100 to 1000 mg kg^{-1} dry weight (Hussain et al. 2023).

Due to their high sulfur compound content, Brassicaceae family plants, garlic, and onions have a high selenium concentration, but most other agricultural crops have a much lower tolerance (less than 50 mg selenium kg^{-1}). Selenium can perform the same function as sulfur in biological systems and shares many of the same characteristics as sulfur. Because selenium's uptake, translocation, and metabolism resemble those of sulfur, substituting selenium for sulfur produces molecules that are higher in selenium, known as selenium analogues (Shalaby et al. 2021). Legumes, particularly lentils, are a good source of selenium; however, the highest selenium level has been discovered in nuts with high protein content, such as Brazil nuts, walnuts, and pistachios. Because they are abundant in water and poor in protein, fruits often have low selenium contents (Chao et al. 2022).

Selenium has a beneficial effect on plant growth and seed germination at suitable levels (Saffan et al. 2022). Plants that are exposed to selenium are protected from several biotic stressors such as diseases and herbivores, as well as abiotic stressors such as heavy metals, ultraviolet (UV) radiation, and arsenic. By preventing lipid peroxidation and increasing the activity of glutathione peroxidase, selenium mitigates the effects of oxidative stress (Garousi 2017). Increased glutathione peroxidase activity and improved antioxidation may prevent postharvest losses and postpone plant senescence.

Fruit and vegetable quality can be impacted by selenium. When *Camellia oleifera* are treated with selenium, there is a decrease in the amount of oleic acid and an increase in the amount of linoleic acid and sterols in the cells (Song et al. 2015). By lowering the ethylene production and respiratory intensity in broccoli (McKenzie et al. 2015), phenylalanine ammonia-lyase activity and ethylene production in chicory and lettuce (Ghafouri et al. 2022), and production of ethylene in tomatoes, selenium treatment had a positive impact on preserving the sensory and postharvest quality (Marfetan et al. 2023). Selenium enhanced the amount of vitamin C content, total amino acids, and plant yield in green tea (Puccinelli et al. 2017). Selenium spraying of the canopy extended the shelf life of pears and peaches by slowing down the pace of fruit softening (Pezzarossa et al. 2012). *Botrytis cinerea*–caused postharvest gray mold disease in *Solanum lycopersicum* fruits may be effectively controlled by using selenium (Wu et al. 2016). Selenium has been observed to postpone fruit ripening and plant senescence in several horticultural crops, which may reduce postharvest loss because of its capacity to boost the antioxidant defense system in plants (Puccinelli et al. 2020).

13.3 Enrichment of Horticultural Crops with Selenium

A varied diet, dietary supplements, food fortification, and biofortification of crops are some methods to raise a deficient selenium status (Shiriaev et al. 2022). A varied diet can offer an adequate amount of vitamins, minerals, and proteins; nevertheless, access to different foods is often unattainable in many socioeconomic circumstances worldwide (Motesharezadeh et al. 2021). Supplements containing selenium comprise selenium-contained yeast, selenocysteine, and selenomethionine as well as sodium selenite and selenate. However, as of 2002, only sodium selenite and selenate are allowed as dietary supplements in the European Union (Abdalla et al. 2021). Although selenium-enriched yeast seems to vary greatly in terms of both speciation and selenium concentration, it is nevertheless a useful method for increasing human selenium consumption (Meucci et al. 2021).

Fortification is the practice of adding nutrients, like vitamins and minerals, to processed food to improve its nutritional value. It is used to replenish nutrients lost during food processing or in regions where malnutrition issues are prevalent, with the goal of addressing dietary deficiencies in one or more nutrients (Reich and Hondal 2016). Biofortification is the technique of increasing the concentration of micronutrients in plants during plant growth rather than during agricultural processing in order to increase the nutritional content of plant-derived food (Diaz-Gomez et al. 2016). Some crops have the capacity to collect selenium, which is essential for human health and nutrition. The food chain starts with plants and finishes with humans. Elevating the selenium levels in plant products, such as cereals, fruit crops, and fruity and leafy vegetables, without going over the hazardous threshold, is therefore a good strategy to raise the selenium intake of animals and humans and may benefit long-term health. The two main methods for raising the amounts of minerals and their bioavailability in the edible portion of staple crops are plant breeding and agronomic intervention.

Applying mineral elements to the soil and plants that have good mobility, such as iron, zinc, and selenium, is known as agronomic biofortification (Mimmo et al. 2017). Due to the extremely low selenium consumption of the Finnish people, this technique has primarily been implemented in Central and Northern Europe. It was implemented nationally in Finland in 1984 when selenium was added to all agricultural fertilizers. The program was successful in raising the amount of selenium in food and in human consumption (Liu et al. 2021). Legumes, which provide a significant portion of the world's population with their primary source of dietary proteins and have greater selenium content in grains than cereals, may be included in selenium biofortification projects under Mediterranean circumstances. A small amount of sodium selenate applied to peas increased the concentration of selenium in the grain significantly (Jakabova et al. 2020).

According to other reviews, biofortifying cereals, vegetables, and fruits with selenium is an effective method of increasing human selenium supplementation (Conversa et al. 2019). Most selenium (Se) in many plants that are enriched with selenium is found in its main organic form (methylselenocysteine, selenocysteine, and selenomethionine). Compared to inorganic forms, it is more readily available to people and more effective at raising selenium levels, particularly in the blood. Thus, it seems good to eat vegetables that are high in selenium (Yue et al. 2019). For instance, selenium is typically found in broccoli, garlic, and onions as γ-glutamyl-selenium-methylselenocysteine, though there may be minor variations depending on the kind of plant and dosage. The γ-glutamyl-selenium-methylselenocysteine, which makes up around 63% of the total selenium in selenium-enriched onions, is the primary form of selenium, followed by selenomethionine (5%) and selenate (10%; Hurst et al. 2010). The most common form of selenium (73%) is γ-glutamyl-selenium-methylselenocysteine, which is also present in enriched *Allium sativum*. Lower amounts of γ-glutamyl-selenomethionine (4%), selenomethionine (13%), selenate (2%), and selenium-methyl-selenocysteine (3%) are among the other chemical species present (Saeedi et al. 2021). Selenium-methylselenocysteine (45%) makes up most of the selenium in selenium-enriched broccoli, with trace levels of selenate and selenomethionine (Izydorczyk et al. 2021). Table 13.1 lists the amount of selenium found in horticulture crops.

Since selenium is mixed into organic substances that can function as selenium analogues (selenomethionine and selenocysteine) and of vital sulfur compounds (methionine and cysteine), intervention with sulfur metabolism is primarily responsible for any potential negative impacts of selenium improvement on plants (Wang et al. 2023). According to a review by Huang et al. (2019), the primary foods that provide selenium in the diet are milk/dairy, eggs, fish, meat, cereals, bread, and nuts-based products. The quantity and chemical type of selenium contained in natural products are also well recognized. The biofortification-induced selenium accumulation in staple crops needs to meet the necessary Recommended Dietary Allowance for selenium application in an appropriate way, without causing phytotoxic effects, decreasing agricultural productivity, or inducing selenium intoxication.

13.4 Biofortification of Horticultural Crops and Plants with Selenium

Selenium can be added to plants in four major ways: (1) through soil enrichment, (2) by soaking seeds in a solution containing selenium before planting, (3) by fruit or foliar spraying, and (4) by hydroponically cultivating plants in a selenium-rich nutrient solution.

One suitable approach to biofortify large volumes of food and raise the selenium level in soils with low selenium concentration is to add selenium fertilizers to the soil. To get selenium plant concentrations comparable to those obtained with other fertilization techniques, substantial amounts of selenium must be added to the soil (Perez et al. 2019). By adding selenium-hyperaccumulator plants to the soil or fertilizing it with salt (selenite or selenate), one can raise the amount of selenium in the soil. By enriching the soil with selenium, *Stanleya pinnata* plants have proven to be an effective biofortifier for broccoli and carrots (Kathi et al. 2023). *Stanleya pinnata* is a selenium-hyperaccumulator that can be utilized to phytoremediate high-selenium soil, the plant material that has been enhanced with selenium can then be put into the soil, combining biofortification and phytoremediation (Bideshki et al. 2019). The concentration of selenium in grapes, pears,

Table 13.1 The amount of selenium found in horticulture crops (Wen 2021).

Crops	Selenium (Se) form	Organ	Amount of selenium content
Tea	Total	Leaves	0.02–0.1 mg kg^{-1}
Tomato	Total	Fruits	0–80.4 mg kg^{-1} dry weight
Tomato	Total	Leaves	0–35 µg g^{-1} dry weight
Potato	Total	Stolon	0–42 µg g^{-1} dry weight
Potato	Total	Roots	0–35 µg g^{-1} dry weight
Potato	Total	Leaves	0–90 µg g^{-1} dry weight
Strawberry	Total	Fruit	0–46 µg g^{-1} dry weight
Strawberry	Total	Root	0–174 µg g^{-1} dry weight
Strawberry	Total	Shoot	0–125 µg g^{-1} dry weight
Cowpea	Selenate	Lateral root cap	0–44 µg g^{-1}
Cowpea	Selenate	Columella of root cap	0–147 µg g^{-1}
Cowpea	Selenate	Rhizodermis	0–249 µg g^{-1}
Cowpea	Selenate	Root cortex	0–359 µg g^{-1}
Cowpea	Selenate	Endodermis	0–447 µg g^{-1}
Cowpea	Selenate	Pericycle	0–183 µg g^{-1}
Cowpea	Selenate	Root stele	0–144 µg g^{-1}
Cowpea	Selenate	Lateral root cap	0–30 µg g^{-1}
Cowpea	Selenate	Columella of root cap	0–60 µg g^{-1}
Cowpea	Selenate	Rhizodermis	0–44 µg g^{-1}
Cowpea	Selenate	Root cortex	0–132 µg g^{-1}
Cowpea	Selenate	Endodermis	0–145 µg g^{-1}
Cowpea	Selenate	Pericycle	0–73 µg g^{-1}
Cowpea	Selenate	Root stele	0–136 µg g^{-1}
Basil	Total	Leaves	0–58 µg g^{-1} dry weight
Maize	Total	Roots	0–420 mg kg^{-1}
Maize	Total	Shoots	0–150 mg kg^{-1}
Pumpkin	Total	seeds	0–19 µg g^{-1}
Cucumber	Total	Fruits	0–237 mg kg^{-1} dry weight
Cucumber	Total	Roots	0–0.35 µg g^{-1} dry weight
Cucumber	Total	Leaves	0–0.09 µg g^{-1} dry weight
Broccoli	Total	Plant	0.38–45 µg g^{-1} dry weight
Arabidopsis thaliana	Total	Roots	7.70 µmol g^{-1} dry weight
Arabidopsis thaliana	Total	Shoots	0.50 µmol g^{-1} dry weight
Mustard	Total	Roots	0.58 µmol g^{-1} dry weight
Chicory	Total	Leaves	0–29 µg g^{-1} dry weight
Lettuce	Total	Leaves	0–41 mg kg^{-1} dry weight

peaches, tomatoes, basil, radish, cabbage, broccoli, carrots, garlic, pumpkins, buckwheat, soybeans, rice, and potatoes has been increased through the use of selenium spraying (De Vita et al. 2017).

The properties of the fruit and leaf surface, including the presence of hairs, the traits of the epicarp, the makeup of the epicuticular wax, or the deposition of wax platelets, all affect how spraying works (Yan et al. 2021). The biofortification of *Cichorium intybus* with foliar application of selenium has shown to be beneficial; plants sprayed with selenium had higher concentrations of the mineral than plants grown in a selenium-containing nutrient solution (Li et al. 2019). Additionally, because no residual impacts have been found and because foliar spraying uses less selenium than soil application, it is preferred. Applying foliar fertilizer to crops is a safe, efficient, and cost-effective method of increasing their selenium content while requiring a minimal amount of selenium salt intake (Pan et al. 2023).

Although positive findings have been attained in grains, the effects of biofortifying plants by soaking seeds in a solution containing selenium are still unknown. While Nawaz et al. (2013) discovered improved drought tolerance in wheat, Ryant et al. (2020) showed increased selenium concentration in buckwheat plants without any drop in yield. In comparison with the other techniques, nevertheless, the selenium levels found in these biofortified plants were lower. Vegetables enhanced with selenium can be obtained by hydroponic culture. Research has been done on sweet basil, lettuce, tomatoes, chard, spinach, and chicory (Wenjiang et al. 2021). By using a floating system, salt loss may be prevented by conveniently adjusting the supply of selenium to the growth phase of plants and controlling the concentration of selenium in the growth medium (Kieliszek 2019). Selenium content in plant tissues may generally be increased by foliar application or selenium supplementation in the growth medium without sacrificing quality or productivity.

13.5 Beneficial Effects of Selenium Supplementation on the Production, Quality, and Senescence of Leafy Vegetables

The biofortification of leafy vegetables with selenium has been the subject of extensive research. Plants have been cultivated on substrates that are high in selenium, in a nutrient solution that contains selenium, or in situations where they have been treated with foliar applications of selenium (Golubkina et al. 2023). The numerous researches that have been conducted on the selenium enrichment of leafy greens are outlined in Table 13.2.

It has been discovered that adding selenium to plants can raise their selenium level considerably without having a detrimental effect on their biomass or leaf quality, as in the instance of chicory, basil, lettuce, and spinach (Golubkina et al. 2023). It was found that chicory plants did not exhibit any harmful effects when small amounts of selenium, about 1.5 mg L^{-1} in the nutritional solution or 1 mg L^{-1} through foliar fertilization, were added (Trod et al. 2023). Similarly, sweet basil plants' biomass was unaffected by foliar spray with 50 mg selenium L^{-1}. Additionally beneficial to plant growth, selenium may boost lettuce, chicory, and basil yields (Mezeyova et al. 2016). Puccinelli et al. (2017) found that selenium had a favorable impact on plant development when applied at rates between 5 and 20 μmol L^{-1}, had no effect when applied at 40 μmol L^{-1}, and decreased biomass output when applied at a rate of 120 μmol L^{-1}. The toxicity of selenium in plants is dependent on its chemical state. Selenate is generally less hazardous than selenite, and when selenium is provided, biomass output increases. Furthermore, compared to selenate, selenite is harmful to plants at lower doses (Liao et al. 2017).

Because the antioxidant enzymes glutathione peroxidase, lipoxigenase, catalase, superoxide dismutase, and ascorbate peroxidase are more active in selenium-enriched leafy greens, there is an apparent antioxidant impact from selenium (Meng et al. 2019). Measuring the levels of various oxidative indicators and chemicals, including flavonoids, phenols, hydrogen peroxide, 2,2-diphenyl-1-picrylhydrazyl, and malondialdehyde as well as activity of antioxidants using the Ferric Reducing Antioxidant Power assay, also revealed a better oxidative status (Placzek et al. 2019). Selenium raises the amount of this molecule, and the decreased amount of glutathione is another indication of the oxidative state of plant tissues (Zhang et al. 2019). Nevertheless, selenium can occasionally have a beneficial impact on the level of anthocyanin while having no impact on the content of phenolic compounds (Qin et al. 2019). Selenium may be beneficial at low doses by promoting plant growth because of its antioxidant function. However, excessive amounts of selenium, particularly as selenite, can increase lipid peroxidation, hydrogen peroxide, and malondialdehyde, which can all lead to oxidative stress in green vegetables (Buturi et al. 2021).

It is unclear how selenium affects pigments that are photosynthetic. The photosynthetic pigment found in basil (used with 50 mg selenium L^{-1}), chicory (treated with 1 mg selenium L^{-1}), and lettuce (used with 0.31 mg selenium L^{-1} selenite and 2.37 mg selenium L^{-1} selenate) has generally not been shown to be affected by low amounts of selenium (Huang et al. 2023). Conversely, it was discovered that higher selenium levels reduced the amount of pigments used for photosynthesis in basil (treated with 30 mg selenium L^{-1}); lettuce (applied 0.47 mg selenium L^{-1} selenite, 3.1 mg selenium L^{-1} selenate as well as treated with 40 μg selenium plant^{-1} selenite (Schiavon et al. 2020). It has also been observed that there is an increase in the level of carotenoids in basil (applied dosages ranging from 30 to 120 mg selenium L^{-1}) and chlorophyll contents in lettuce (applied dosage 1 mg selenium L^{-1}; Puccinelli et al. 2017).

Table 13.2 The accumulation of selenium in the edible sections of leafy vegetables in connection to the concentration, the chemical form of selenium that is supplemented to plants, and the method that is used to supplement selenium.

Crops	Applied dose of selenium	Chemical form of selenium	Supplementation techniques	Selenium (in edible portion)	References
Beetroot	10 and 20 mg selenium L^{-1}	Selenate	Nutrient solution enrichment	Around 1390 μg of selenium per shoot	Puccinelli et al. (2017)
Spinach	1–10 mg selenium L^{-1}	Selenate	Nutrient solution enrichment	3.90 mg g^{-1} dry weight	Saffaryazdi et al. (2012)
Spinach	0.2–0.4 mg selenium L^{-1}	Selenate	Nutrient solution enrichment	15 mg kg^{-1} dry weight	Ferrarese et al. (2012)
Spinach	0.8 and 1.6 mg selenium L^{-1}	Selenate	Foliar method	41 mg kg^{-1} dry weight	Kopsell et al. (2009)
Basil	25 and 50 mg m^{-2}	Selenate	Foliar method	7.80 mg kg^{-1} dry weight	Mezeyova et al. (2016)
Basil	1–50 mg selenium L^{-1}	Selenate	Foliar method	150 mg kg^{-1} dry weight	Puccinelli et al. (2017)
Chicory	0.5 and 1 mg selenium L^{-1}	Selenate	Nutrient solution enrichment	29 mg kg^{-1} dry weight	Puccinelli et al. (2017)
Chicory	1 mg selenium L^{-1}	Selenate	Foliar method	45 mg kg^{-1} dry weight	Ghafouri et al. (2022)
Chicory	0.5 and 1 mg selenium L^{-1}	Selenate	Nutrient solution enrichment	30 mg kg^{-1} dry weight	Saffan et al. (2022)
Lettuce	40 μg per plant	Selenium urea and Selenate	Vermiculite–peat–sand application	413 μg kg^{-1} dry weight	Goicoechea et al. (2015)
Lettuce	1–1000 mg selenium kg^{-1}	Selenate	Peat fertilization	219 mg kg^{-1} dry weight	Businelli et al. (2015)
Lettuce	0.16–2.4 mg selenium L^{-1}	Selenate	Nutrient solution enrichment	30 mg kg^{-1} dry weight	Hawrylak-Nowak (2013)
Lettuce	0.16–5.12 mg selenium L^{-1}	Selenate	Nutrient solution enrichment	30 mg kg^{-1} dry weight	Hawrylak-Nowak (2013)
Lettuce	0.5–1 mg selenium L^{-1}	Selenate	Nutrient solution enrichment	26 mg kg^{-1} dry weight	Puccinelli et al. (2017)
Lettuce	5 mg selenium kg^{-1}	Selenate	Soil fertilization	170 mg kg^{-1} dry weight	Puccinelli et al. (2017)

It is unclear how selenium affects the metabolism of nitrogen. Huang et al. (2023) discovered that basil and lettuce plants treated with selenium had lower nitrate contents but chard, spinach, and chicory showed no changes. In chicory and lettuce, decreased rates of ethylene production have been linked to a longer shelf life and retained quality (Newman et al. 2019). Furthermore, it has been discovered that selenium improves stress tolerance in lettuce plants by boosting superoxide dismutase activity and reducing the decline in tocopherol levels (Sun et al. 2022). The key effects of selenium in leafy greens are outlined in Table 13.3.

13.6 Impact of Selenium Fortification on Fruit Crops: Its Influence on Crop Yield, Fruit Quality, and Senescence

Various experiments have been carried out to investigate the impact of selenium biofortification on the quality of fruit and postharvest, owing to selenium's antioxidant capability and its influence on ethylene production. Sodium selenate, namely in the form of tomato selenium, has been included in nutritional solutions according to Huang et al. (2023). There have been limited investigations undertaken on the enrichment of fruit plants with selenium. Pears and peaches were treated with a solution containing selenate, table grapes were treated with a solution containing organic selenium, and peaches

Table 13.3 Major impacts of selenium supplementation on green leafy vegetables.

Crops	Biomass	Antioxidant enzymes	Oxidative markers	Antioxidant compounds	Photosynthetic content	Ethylene production	Nitrate content	References
Spinach	None	–	–	–	–	–	–	Puccinelli et al. (2017)
Spinach	None	–	–	Increase	Enhance	–	–	Puccinelli et al. (2017)
Lettuce	None	–	Reduced	–	Reduced	–	–	Goicoechea et al. (2015)
Lettuce	Reduced at higher selenium	–	–	–	–	–	Reduced	Rios et al. (2010)
Lettuce	Reduced at higher selenium	Increase	–	–	–	–	–	Puccinelli et al. (2017)
Lettuce	Reduced at higher selenium	Enhance	–	Increase	–	–	–	Rios et al. (2008)
Lettuce	None	Enhance	–	–	–	–	–	Ramos et al. (2010)
Basil	None	–	–	Increase	Reduced at higher selenium	–	–	Oraghi Ardebili et al. (2015)
Lettuce	None	–	–	–	No effect	Reduced	No effect	Puccinelli et al. (2017)
Chicory	None	–	–	–	–	Reduced	–	Puccinelli et al. (2017)
Beetroot	–	–	–	–	–	–	No effect	Hernandez-Castro et al. (2015)
Lettuce	None	–	Improve	–	Reduced at higher selenium	–	–	Hawrylak-Nowak (2013)
Basil	None	-	Increase	Enhance	No effect	–	–	Puccinelli et al. (2017)

and grapes were treated with a solution containing amino acid–chelated selenium (Newman et al. 2019). The numerous studies on the selenium enrichment of fruit crops are compiled in Table 13.4.

Selenium supplementation has been observed to elevate the selenium levels in the leaves and fruits of both fruit trees and tomato plants while having no impact on crop yield (Puccinelli et al. 2017). Studies have shown that selenium has a beneficial impact on various quality characteristics of tomato fruit, including firmness, total sugar, fructose, glucose, titratable acidity, and soluble solid content (Zhu et al. 2017). Selenium has a beneficial impact on the concentration of soluble solids in peaches and pears (Andrejiova et al. 2016), as well as on the levels of fructose, glucose, proteins, and organic acids in grapes (Marfetan et al. 2020). Furthermore, selenium protects tomato fruits from losing weight and from losing titratable acidity and firmness during storage (Saffan et al. 2022). Selenium in tomato fruits can stimulate the production of pigments and antioxidant molecules (Zhu et al. 2017).

Selenium treatments showed a beneficial impact on enhancing the oxidative condition of fruits and delaying the ripening process of tomatoes. Consequently, this significantly influenced the duration of fruit preservation after harvest, without negatively impacting its quality (Consentino et al. 2023). The study conducted by Zhu et al. (2017) found that Selenium has the potential to enhance the activity of antioxidant enzymes and reduce the presence of reactive oxygen species during storage. The study conducted by Chao et al. (2022) revealed a delay in the initiation of the ripening process in tomato plants enhanced with selenium. These plants also exhibited a decreased rate of change in color and were harvested earlier compared

Table 13.4 Relationship between the concentration and chemical type of selenium given to plants, as well as the technique of selenium supplementation, and its subsequent accumulation in fruit.

Crops	Used form of selenium	Applied dose of selenium	Application method	Selenium in fruits	References
Grape	Organic	120 mg L^{-1}	Foliar	25 μg kg^{-1} fresh weight	Puccinelli et al. (2017)
Pear	Selenate	0–1 mg selenium L^{-1}	Foliar	200 μg kg^{-1} dry weight	Pezzarossa et al. (2012)
Peach	Selenate	0–1 mg selenium L^{-1}	Foliar	75 μg kg^{-1} dry weight	Pezzarossa et al. (2012)
Tomato	Selenate	150–300 gha^{-1}	Foliar	Not measured	Andrejiova et al. (2016)
Tomato	Selenate	1 mg selenium L^{-1}	Foliar	Not measured	Zhu et al. (2017)
Tomato	Selenate	1 mg selenium L^{-1}	Foliar	500 μg kg^{-1} dry weight	Zhu et al. (2017)
Tomato	Selenate	1–30 mg selenium kg^{-1} dry weight	Peat supplementation	201 μg kg^{-1} fresh weight	Puccinelli et al. (2017)
Tomato	Selenate	1 mg selenium L^{-1}	Nutrient solution enrichment	201 mg Kg^{-1} dry weight	Puccinelli et al. (2017)

to the plants that were not treated with selenium. The delay in the ripening of fruit may depend on a lower production of ethylene. The decrease in ethylene production may be attributed to the greater abundance of selenomethionine compared to methionine, which serves as a precursor for ethylene in the biosynthesis of the ethylene process (Chao et al. 2022). Another possibility relates to the genes responsible for the enzymes that regulate the ethylene production process. The enzyme 1-aminocyclopropane-1-carboxylic acid synthase catalyzes the conversion of sulfur-adenosylmethionine to 1-aminocyclopropane-1-carboxylic acid in this route. 1-Aminocyclopropane-1-carboxylic acids are transformed into ethylene through the action of 1-aminocyclopropane-1-carboxylic acid oxidase (Tripathy et al. 2020).

Selenium might potentially improve the preservation of vegetables after harvesting by inhibiting the germination and proliferation of a few hazardous funguses, such as *Penicillium expansum* and *Botrytis cinerea* (Wu et al. 2014). *Botrytis cinerea* is responsible for gray mold decay, a significant disease that affects fruits and vegetables both before and after harvest. This disease results in financial losses (Cabot et al. 2013). Selenium has the ability to combat infections caused by *Fusarium* and *Alternaria brassicicola*. It also increases the resistance against fungal diseases, such as decreasing the negative effects caused by *Fusarium* wilt disease in tomatoes (Companioni et al. 2012). Furthermore, elevated levels of selenium in plant tissues can potentially mitigate the negative impact of invertebrate grazing damage (Puccinelli et al. 2017). Table 13.5 provides an overview of the major impacts caused by selenium in fruit crops.

Table 13.5 Major impacts of selenium applications on different fruit crops.

Crops	Antioxidants enzyme	Reactive oxygen species	Qualitative parameters	Ethylene production	Postharvest quality	References
Grapes	–	–	Increased	–	–	Zhu et al. (2017)
Tomato	–	–	Enhanced	–	–	Andrejiova et al. (2016)
Tomato	Enhanced	Reduced	–	Reduced	–	Zhu et al. (2017)
Tomato			Enhanced		Enhanced	Puccinelli et al. (2017)
Pear	–	–	Increased	–	Increased	Puccinelli et al. (2017)
Peach	–	–	Increased	–	Increased	Pezzarossa et al. (2012)
Tomato	–	–	–	Reduced	–	Puccinelli et al. (2017)
Tomato	–	–	Increased	–	–	Businelli et al. (2015)
Tomato	Increased	–	–	–	–	Businelli et al. (2015)

13.7 Selenium Metabolism

The process of selenium metabolism in plants is categorized into five major stages. The initial process involves the absorption of selenium. Plant roots mostly absorb selenium in the form of selenate, although organic selenium and selenite are also readily absorbed (Wen 2021). The sulfate carrier Sultr1; 2, which plays a crucial role in facilitating the transportation of sulfate into the roots, is also responsible for the absorption of selenate in *Arabidopsis thaliana* (Silva et al. 2023). The absorption of selenite in rice is facilitated by the silicon influx carrier NIP2; 1 and the phosphate carrier PT2 (Cheng et al. 2022). The absorption rates of several selenium forms were ranked as follows: Selenomethionine > selenocysteine > selenite > selenate, following exposure to a concentration of 5 μM selenium (Guo et al. 2023). While it is known that organic forms of selenium are absorbed more quickly than inorganic forms and that plants may take up intact amino acids, there is limited understanding of the relationship between the uptake of organic forms and any protein or specific ingestion process.

The second phase involves the transportation of selenium. Selenite and selenate are carried inside tomato roots for a duration of up to four hours after the plant tops have been removed, as evidenced by the use of ^{75}selenium as a tracer. The concentration of ^{75}selenium in the xylem discharge for selenate was found to be 6–13 times more than in the exterior solution. Chromatographic investigations revealed that the selenium moved as inorganic selenate (Zhu et al. 2022). The research indicated that selenite and selenate employ different absorption and transport pathways in tomatoes. Selenite circulates via a phosphate transport pathway, while selenate is delivered via sulfate channels and transporters (Guo et al. 2023). Selenite undergoes quick conversion into organic selenium molecules in roots, while selenate is promptly transported to xylem (Wang et al. 2020). Therefore, selenite is comparatively less hazardous than selenate when present in consumable portions such as leaves, fruits, and non-root parts. Nevertheless, the process of converting selenium into its organic form is not entirely devoid of risk, as organic selenium can potentially have harmful effects, especially when selenocysteine is integrated into proteins in a nonspecific way (Ding et al. 2020).

The third stage, the conversion into organic selenium, involves two diminutions. Both adenosine phosphoselenate reductase and adenosine triphosphate (ATP) sulfurylase facilitate the transformation of selenate into selenite (Wen 2021). Additionally, they have crucial functions in regulating the process of selenate uptake (Zhu et al. 2017). Subsequently, the conversion of selenite to selenide occurs through the activation of sulfite reductase or through an interaction between glutathione and selenite (Silva et al. 2023). Organic selenocysteine is produced through the reaction of *O*-acetyl-L-serine and selenide. Selenomethionine, a significant organic form of selenium, is derived from selenocysteine (Yan et al. 2021). The fourth stage involves the retention of selenium. Plant cells typically store selenate within their vacuoles. SULTR4; 1 and 4; 2 are believed to facilitate the removal of selenate from vacuoles by catalysis (Pan et al. 2023). Selenium hyperaccumulators have the ability to accumulate significant amounts of selenium and transform it into methyl selenocysteine. This compound is then stored in the outer layer of young leaves as well as in reproductive organs such as seeds, ovules, and pollen (Ali et al. 2023). The fifth stage involves the evaporation of selenium and its subsequent reutilization. Both selenium-accumulating and non-accumulating plants have the ability to produce precursor chemicals that can be released as volatile substances (Hussain et al. 2023). There are further processes that are not well understood, including the absorption of organic selenium-dependent proteins, the movement of selenite into plastids, and the conversion of organic selenium into inorganic forms. It is crucial to thoroughly examine these processes, particularly the transformation and control of inorganic selenium into organic selenium, as well as other stages involved in the accumulation of beneficial components in horticulture crops. This research aims to enhance the nutritional value of food.

13.8 Conclusion

Biofortified food containing selenium can effectively combat selenium deficits in people, thereby augmenting the dietary selenium intake and mitigating the potential hazards associated with excessive selenium consumption that can be induced by mineral supplements. Selenium has the ability to improve the shelf life and storage of horticulture crops. Selenium in fruit can influence the process of ripening by utilizing its antioxidant and anti-senescence characteristics. This can have positive impacts on the fruit's postharvest shelf life and provide additional advantages for human health. Additional research is required to provide a comprehensive understanding of the specific biochemical and molecular processes that are directly or indirectly influenced by selenium in fruit tissues during maturing and postharvest.

References

Abdalla, M.A., Wick, J.E., Famuyide, I.M. et al. (2021). Selenium enrichment of green and red lettuce and the induction of radical scavenging potential. *Horticulturae* 7 (11): 488.

Abishek, R., Yazhini, G., Poovarasan, T. et al. (2023). *Selenium and Its Role in Horticulture Crops*. In: *Archives of agriculture and allied sciences.*

Ali, J., Jan, I., Ullah, H. et al. (2023). Biochemical response of okra (*Abelmoschus esculentus* L.) to Selenium (Se) under drought stress. *Sustainability* 15 (7): 5694.

Andrejiova A., Hegedűsová A. and Mezeyová I. (2016). Effect of genotype and selenium biofortification on content of important bioactive substances in tomato (Lycopersicon esculentum Mill.) fruits. *Journal of International Scientific Publications* 4: 8–18.

Bideshki, A., Arvin, M.J., Aien, A. et al. (2019). Interactive effects of Foliar 24-Epibrassinolide and selenium applications on yield, reduce nitrate accumulation and selenium enrichment in potato tuber in field. *Cogent Food & Agriculture* 5 (1): 1690315.

Businelli D., D'Amato R., Onofri A. et al. (2015). Se-enrichment of cucumber (*Cucumis sativus* L.), lettuce (*Lactuca sativa* L.) and tomato (*Solanum lycopersicum* L. Karst) through fortification in pre-transplanting. *Scientia Horticulturae* 197: 697–704.

Buturi, C.V., Mauro, R.P., Fogliano, V. et al. (2021). Mineral biofortification of vegetables as a tool to improve human diet. *Foods* 10 (2): 223.

Cabot C., Gallego B., Martos S. et al. (2013). Signal cross talk in Arabidopsis exposed to cadmium, silicon, and *Botrytis cinerea*. *Planta* 237 (1): 337–349.

Chao, W., Rao, S., Chen, Q. et al. (2022). Advances in research on the involvement of selenium in regulating plant ecosystems. *Plants* 11 (20): 2712.

Cheng, B., Wang, C., Chen, F. et al. (2022). Multiomics understanding of improved quality in cherry radish (*Raphanus sativus* L. var. *radculus pers*) after foliar application of selenium nanomaterials. *Science of the Total Environment* 824 (11): 153712.

Companioni B., Medrano J., Torres J.A. et al. (2012). Protective action of sodium selenite against Fusarium wilt in tomato: total protein contents, levels of phenolic compounds and changes in antioxidant potential. *International Symposium on Soilless Culture and Hydroponics* 947 (1): 321–327.

Consentino, B.B., Ciriello, M., Sabatino, L. et al. (2023). Current acquaintance on agronomic biofortification to modulate the yield and functional value of vegetable crops: a review. *Horticulturae* 9 (2): 219.

Conversa, G., Lazzizera, C., Chiaravalle, A.E. et al. (2019). Selenium fern application and arbuscular mycorrhizal fungi soil inoculation enhance Se content and antioxidant properties of green asparagus (Asparagus officinalis L.) spears. *Scientia Horticulturae* 252: 176–191.

De Vita, P., Platani, C., Fragasso, M., Ficco, D.B.M. et al. (2017). Selenium-enriched durum wheat improves the nutritional profile of pasta without altering its organoleptic properties. *Food Chemistry* 214 (7): 374–382.

Diaz-Gomez J., Twyman R.M., Zhu C. et al. (2016). Biofortification of crops with nutrients: factors affecting utilization and storage. *Current Opinion Biotechnology* 44: 115–123.

Ding, Y., Di, X., Norton, G.J., Beesley, L. et al. (2020). Selenite foliar application alleviates arsenic uptake, accumulation, migration and increases photosynthesis of different upland rice varieties. *International Journal of Environmental Research and Public Health (IJERPH)* 17 (10): 3621.

El-Ramady, H., El-Sakhawy, T., Omara, A.E.D. et al. (2022). Selenium and nano-selenium for plant nutrition and crop quality. In: *Selenium and Nano-Selenium in Environmental Stress Management and Crop Quality Improvement*, 55–78. Cham: Springer International Publishing.

Ferrarese, M., Mahmoodi Sourestani, M., Quattrini, E. et al. (2012). Ferrante A. Biofortification of spinach plants applying selenium in the nutrient solution of floating system. *Vegetable Crops Research Bulletin* 76 (1): 127–136.

Francini, A., Quattrini, E., Giuffrida, F. et al. (2023). Biofortification of baby leafy vegetables using nutrient solution containing selenium. *Journal of the Science of Food and Agriculture* 103 (11): 5472–5480.

Garcia Marquez, V., Morelos Moreno, Á., Benavides Mendoza, A. et al. (2020). Ionic selenium and nanoselenium as biofortifiers and stimulators of plant metabolism. *Agronomy* 10 (9): 1399.

Garousi, F. (2017). Toxicity of selenium, application of selenium in fertilizers, selenium treatment of seeds, and selenium in edible parts of plants. *Acta Universitatis Sapientiae Alimentaria* 10 (1): 61–74.

Ghafouri, M., Razavi, F., Arghavani, M. et al. (2022). Delaying postharvest senescence and improving antioxidant capacity of kiwifruits cv. Hayward by preharvest selenium application. *Journal Food Processing and Preservation* 46 (12): e17185.

Goicoechea N., Garmendia I., Fabbrin E.G. et al. (2015). Selenium fertilization and mycorrhizal technology may interfere in enhancing bioactive compounds in edible tissues of lettuces. *Scientia Horticulturae* 195: 163–172.

Golubkina, N., Antoshkina, M., Bondareva, L. et al. (2023). Effect of foliar application of sodium selenate on mineral relationships in Brassicaceae crops. *Horticulturae* 9 (5): 535.

Guo, Q., Ye, J., Zeng, J. et al. (2023). Selenium species transforming along soil–plant continuum and their beneficial roles for horticultural crops. *Horticulture Research* 10 (2): uhac270.

Hawrylak-Nowak, B. (2013). Comparative effects of selenite and selenate on growth and selenium accumulation in lettuce plants under hydroponic conditions. *Plant Growth Regulation* 70 (2): 149–157.

Hegedusova, A., Hegedűs, O., Jakabová, S. et al. (2021). *Selenium Supplementation in Horticultural Crops*. Springer International Publishing.

Hernandez-Castro E., Trejo-Téllez L., Gómez-Merino F. et al. (2015). Bioaccumulation of iron, selenium, nitrate, and proteins in chard shoots. *Journal of Soil Science and Plant Nutrition* 15 (3): 694–710.

Huang, A., Huang, K., Peng, J. et al. (2019). Effects of foliar spraying of selenium fertilizer on selenium-enriched content, heavy metal content and yield of sweet corn grain. *Journal of Southern Agriculture* 50: 40–44.

Huang, J.Q., Wang, Z.H., Sun, L.H. et al. (2023). Selenium in modern agriculture. *Modern Agriculture* 1 (1): 34–42.

Hurst R., Armah C.N., Dainty J.R. et al. (2010). Establishing optimal selenium status: results of a randomized, double-blind, placebo-controlled trial. *American Journal of Clinical Nutrition* 91 (4): 923–931.

Hussain, S., Ahmed, S., Yasin, N.A. et al. (2023). In vitro and in silico study of salt stress resilience in *Brassica rapa* through selenium seed priming. *South African Journal of Botany* 160 (4): 504–515.

Izydorczyk, G., Ligas, B., Mikula, K. et al. (2021). Biofortification of edible plants with selenium and iodine – a systematic literature review. *The Science of the Total Environment* 754 (3): 141983.

Jakabova, S., Čurlej, J., Fikselová, M. et al. (2020). The use of biofortification for production of selenium enriched garden pea. *Slovak Journal Food Science* 14: 437–442.

Jalali, P., Roosta, H.R., Khodadadi, M. et al. (2022). Effects of brown seaweed extract, silicon, and selenium on fruit quality and yield of tomato under different substrates. *PLoS One* 17 (12): e0277923.

Kathi, S., Laza, H., Singh, S. et al. (2023). A decade of improving nutritional quality of horticultural crops agronomically (2012–2022): a systematic literature review. *The Science of the Total Environment* 911 (1): 168665.

Kieliszek, M. (2019). Selenium–fascinating microelement, properties and sources in food. *Molecules* 24 (7): 1298.

Kopsell D.A., Sams C.E., Barickman T.C. et al. (2009). Selenization of basil and cilantro through foliar applications of selenate-selenium and selenite-selenium. *HortScience* 44 (2): 438–442.

Li, F., Huang, M., Li, L., Geng, C. et al. (2019). Research progress of influence of selenium-enriched fertilizer on tea plant and new selenium fertilizer. *Journal of Agricultural Science* 48: 1–7.

Liao, Q., Xing, Y., Liang, P. et al. (2017). Natural selenium-enriched ability of sweet potatoes and effect of ways of applying selenium on selenium content of sweet potato. *Southwest China Journal Agricultural Sciences* 30: 1838–1842.

Liu, Y., Huang, S., Jiang, Z. et al. (2021). Selenium biofortification modulates plant growth, microelement and heavy metal concentrations, selenium uptake, and accumulation in black-grained wheat. *Frontiers in Plant Science* 12: 748523.

Marfetan, J.A., Gallo, A.L., Farias, M.E. et al. (2023). *Exiguobacterium sp.* as a bioinoculant for plant-growth promotion and Selenium biofortification strategies in horticultural plants. *World Journal of Microbiology and Biotechnology* 39 (5): 134.

McKenzie, M., Lill, R., Trolove, S. et al. (2015). Biofortification of fruit and vegetables with selenium. *Selenium: Chemistry, Analysis, Function and Effects* 304–323.

Meng, J., Dong, M., Zhang, J. et al. (2019). Study on cultivation of selenium-rich bean sprouts. *Food Research Development* 40: 57–61.

Meucci, A., Shiriaev, A., Rosellini, I. et al. (2021). Se-enrichment pattern, composition, and aroma profile of ripe tomatoes after sodium selenate foliar spraying performed at different plant developmental stages. *Plants* 10 (6): 1050.

Mezeyova I., Hegedűsová A., Andrejiová A. et al. (2016). Phytomass and content of essential oils in Ocimum basilicum after foliar treatment with selenium. *Journal of International Scientific Publication* 4 (1): 19–27.

Mezeyova, I., Hegedűsová, A., Hegedűs, O. et al. (2020). Basil seeds as a source of antioxidants affected by fortification with selenium. *Folia Horticulturae* 32: 11–20.

Mimmo, T., Tiziani, R., Valentinuzzi, F. et al. (2017). Selenium biofortification in Fragaria×ananassa: implications on strawberry fruits quality, content of bioactive health beneficial compounds and metabolomic profile. *Frontiers in Plant Science* 8: 1887.

Motesharezadeh, B., Ghorbani, S., Alikhani, H.A. et al. (2021). Investigation of different selenium sources and supplying methods for selenium enrichment of basil vegetable (a case study under calcareous and non-calcareous soil systems). *Recent Patents on Food Nutrition & Agriculture* 12 (1): 73–82.

Nawaz F., Ashraf M.Y., Ahmad R. et al. (2013). Selenium (Se) seed priming induced growth and biochemical changes in wheat under water deficit conditions. *Biological Trace Element Research* 151 (2): 284–293.

Newman, R., Waterland, N., Moon, Y. et al. (2019). Selenium biofortification of agricultural crops and effects on plant nutrients and bioactive compounds important for human health and disease prevention – a review. *Plant Foods Human Nutrition* 74 (4): 449–460.

Oraghi Ardebili Z., Oraghi Ardebili N., Jalili S. et al. (2015). The modified qualities of basil plants by selenium and/or ascorbic acid. *Turkish Journal of Botany* 39 (3): 401–407.

Pan, Z., Chen, J., Wang, M. et al. (2023). Selenium speciation and characteristics of selenium-enriched crops in Guiyang seleniferous soil, southwestern China. *Environmental Chemistry* 20 (3): 114–123.

Perez, M.B., Lipinski, V.M., Filippini, M.F. et al. (2019). Selenium biofortification on garlic growth and other nutrients accumulation. *Horticultura Brasileira* 37 (3): 294–301.

Pezzarossa B., Remorini D., Gentile M.L. et al. (2012) Effects of foliar and fruit addition of sodium selenate on selenium accumulation and fruit quality. *Journal of The Science of Food and Agriculture* 92 (4): 781–786.

Placzek, A., Stepien, P., Zarczynski, P. et al. (2019). Methods for enrichment of animal diets with selenium. *Journal of Elementology* 24 (3).

Puccinelli, M., Malorgio, F. and Pezzarossa, B. (2017). Selenium enrichment of horticultural crops. *Molecules* 22 (6): 933.

Puccinelli, M., Pezzarossa, B., Rosellini, I. et al. (2020). Selenium enrichment enhances the quality and shelf life of basil leaves. *Plants* 9 (6): 801.

Qin, C., Shi, C., Wan, Q. et al. (2019). Speciation analysis of selenium in selenium-enriched wheat by HPLC-ICP-MS. *Food Research Development* 40: 140–144.

Rahim, F.P., Rocio, C.G., Adalberto, B.M. et al. (2020). Agronomic biofortification with selenium in tomato crops (*Solanum lycopersicon* L. Mill). *Agriculture* 10 (10): 486.

Ramos S.J., Faquin V., Guilherme L.R.G. et al. (2010). Selenium biofortification and antioxidant activity in lettuce plants fed with selenate and selenite. *Plant Soil Environment* 56 (12): 584–588.

Reich H.J. and Hondal, R.J. (2016). Why nature chose selenium? *ACS Chemical Biology* 11 (4): 821–841.

Rios J.J., Blasco B., Rosales M.A. et al. (2010). Response of nitrogen metabolism in lettuce plants subjected to different doses and forms of selenium. *Journal of The Science of Food and Agriculture* 90 (11): 1914–1919.

Rios J.J., Rosales M.A., Blasco B. et al. (2008). Biofortification of Se and induction of the antioxidant capacity in lettuce plants. *Scientia Horticulturae* 116 (3): 248–255.

Ryant, P., Antošovský, J., Adam, V. et al. (2020). The importance of selenium in fruit nutrition. *In Fruit crops* 241–254.

Saeedi, M., Soltani, F., Babalar, M. et al. (2021). Selenium fortification alters the growth, antioxidant characteristics and secondary metabolite profiles of cauliflower (Brassica oleracea var. botrytis) cultivars in hydroponic culture. *Plants* 10 (8): 1537.

Saffan, M.M., Koriem, M.A., El-Henawy, A. et al. (2022). Sustainable production of tomato plants (*Solanum lycopersicum* L.) under low-quality irrigation water as affected by bio-nanofertilizers of selenium and copper. *Sustainability* 14 (6): 3236.

Saffaryazdi A., Lahouti M., Ganjeali A. et al. (2012). Impact of selenium supplementation on growth and selenium accumulation on spinach (*Spinacia oleracea* L.) plants. *Notulae Scientia Biologicae* 4 (4): 95–100.

Schiavon, M., Nardi, S., Dalla Vecchia, F. et al. (2020). Selenium biofortification in the 21st century: status and challenges for healthy human nutrition. *Plant and Soil* 453 (3): 245–270.

Shalaby, T.A., Abd-Alkarim, E., El-Aidy, F. et al. (2021). Nano-selenium, silicon and H_2O_2 boost growth and productivity of cucumber under combined salinity and heat stress. *Ecotoxicology and Environmental Safety* 212: 111962.

Shekari, L., Kamelmanesh, M.M., Mozafarian, M. et al. (2016). Beneficial effects of selenium on some morphological and physiological trait of hot pepper (*Capsicum annuum*). *Journal Of Horticultural Science* 29 (4): 594–600.

Shiriaev, A., Pezzarossa, B., Rosellini, I. et al. (2022). Efficacy and comparison of different strategies for selenium biofortification of tomatoes. *Horticulturae* 8 (9): 800.

Silva, M.A., Sousa, G.F.D., Van Opbergen, G.A.Z. et al. (2023). Foliar application of selenium associated with a multi-nutrient fertilizer in soybean: yield, grain quality, and critical Se threshold. *Plants* 12 (10): 2028.

Song Y.R., Jiang X.G., Peng S.F. et al. (2015). Effect of selenium content on the quality and functional components of selenium-riched *Camellia oleifera* oil. *Journal of Chinese Institute of Food Science* and *Technology* 15 (1): 142–149.

Sun, M.F., Wang, J.J., Wei, L.I.U. et al. (2022). Effect and mechanism of exogenous selenium on selenium content and quality of fresh tea leaves. *Notulae Botanicae Horti Agrobotanici Cluj-Napoca* 50 (3): 12814–12814.

Syed, F., Zahid, N.Y., Hayat, R. et al. (2021). Biofortification of horticultural crops with selenium. *Agrobiological Records* 6: 36–42.

Tripathy, B., Tripathy, P., Sahu, P. et al. (2020). Biofortification of vegetable crops – a new tool to alleviate micronutrient malnutrition. In: *Agriculture and Rural Development: Spatial Issues, Challenges and Approaches*, 14: 83–97. Pune: Jyotikiran Publication.

Trod, B.S., Buttarelli, M.S., Stoffel, M.M. et al. (2023). Postharvest commercial quality improvement of broccoli (Brassica oleracea L.) after foliar biofortification with selenium. *Crop Science* 63 (2): 784–800.

Wang, C., Rong, H., Zhang, X. et al. (2020). Effects and mechanisms of foliar application of silicon and selenium composite sols on diminishing cadmium and lead translocation and affiliated physiological and biochemical responses in hybrid rice (*Oryza sativa* L.) exposed to cadmium and lead. *Chemosphere* 251 (4): 126347.

Wang, Z., Li, D., Gruda, N.S., Zhu, C. et al. (2023). How to efficiently produce the selenium-enriched cucumber fruit with high yield and qualities via hydroponic cultivation? The balance between selenium supply and CO_2 fertilization. *Agronomy* 13 (3): 922.

Wen D. (2021). Selenium in horticultural crops. *Scientia Horticulturae* 289 (3): 110441.

Wenjiang, W.U., Lei, W.A.N.G., Haijun, M.E.N.G. et al. (2021). Current situation of selenium-enriched crops research in China based on bibliometric. *Journal of Agriculture* 11 (12): 109.

White P.J. (2018). Selenium in soils and crops. *Selenium. Molecular and Integrative Toxicology*, 29–50. Cham: Springer.

Wu Z., Yin X., Bañuelos G.S. et al. (2016). Effect of selenium on control of postharvest gray mold of tomato fruit and the possible mechanisms involved. *Frontiers in Microbiology* 6: 1441.

Wu Z., Yin X., Lin Z. et al. (2014). Inhibitory effect of selenium against Penicillium expansum and its possible mechanisms of action. *Current Microbiology* 69: 192–201.

Yan, Y., Wang, Z., Wang, R. et al. (2021). Effects of leaves and trunk selenium application on selenium enrichment and quality of persimmon fruits. *VII International Symposium on Persimmon* 1338: 255–262.

Yue, S., Zhang, H., Zhen, H. et al. (2019). Selenium accumulation, speciation and bioaccessibility in selenium-enriched earthworm (*Eisenia fetida*). *Microchemical Journal* 145: 1–8.

Zhang, H., Xu, G., Pang, Z. et al. (2019). Effects of slow-released sand selenium on selenium enrichment and growth of green cabbage. *Guangxi Zhiwu/Guihaia* 39: 238–247.

Zhu Z., Chen Y., Shi G. et al. (2017). Selenium delays tomato fruit ripening by inhibiting ethylene biosynthesis and enhancing the antioxidant defense system. *Food Chemistry* 219: 179–184.

Zhu, Y., Dong, Y., Zhu, N. et al. (2022). Foliar application of biosynthetic nano-selenium alleviates the toxicity of Cd, Pb, and Hg in Brassica chinensis by inhibiting heavy metal adsorption and improving antioxidant system in plant. *Ecotoxicology and Environmental Safety* 240: 113681.

14

Magnesium Mysteries Unveiled: Insights into Its Impact on Plants and Human Health and Biofortification Strategies to Enhance Magnesium Content in Cereal Crops

Umed Ali[1], Maqsood Ahmed Khaskheli[2], Ambrin Rajput[3], Tahmina Shar[4], Mahpara Khatoon Bhutto[5], Ghulam Hussain Jatoi[6]*

[1] Department of Agriculture, Mir Chakar Khan Rind University, Sibi, Balochistan, Pakistan
[2] Department of Plant Pathology, College of Agriculture, Guizhou University, Guiyang, China
[3] Soil Fertility Research Institute, Agriculture Research Center, Tandojam, Sindh Pakistan
[4] Agriculture Research Center, Tandojam, Sindh, Pakistan
[5] Department of Agronomy, Sindh Agriculture University, Tandojam, Sindh, Pakistan
[6] Department of Plant Pathology, Faculty of Crop Protection, Sindh Agriculture University, Tandojam, Sindh, Pakistan

*Corresponding author: ualilaghari@hotmail.com

14.1 Introduction

Originally the word "cereals" came from the Latin language and it is honored a high rank in the world of grains. These have been staple foods around the world for a very long time and have historical relevance in human culture. Edible grains from the Gramineae family are cultivated all over the world and are known as cereals. Basically, cereal crops are cultivated for grain purposes. Those crops include rice, wheat, sorghum, millet, oat, barley, triticale, fonio, millet, and quinoa. As a source of nutrients and dietary energy, the grains of these crops are referred to as staple food in both developed and developing nations. They account for 75% carbohydrates, 25% starch, 13% fiber, 6–15% protein, and very low about 2% fat in their products and serves as a major source of energy for humans and domestic animals (Ram and Mishra 2010). These nutrients can differ depending on their species and account for about half of the world's energy terms. The fact that 2800 million tons of cereals are produced annually (FAO 2024) indicates how important cereals are, as their availability affects global food security to a greater extent. Global cereal production increased by approximately 30.4 million tons (1.1%) between 2022 and 2023, reaching 2839.5 million tons in 2023 – a higher amount than in both 2022 and 2021 (Table 14.1). This information is based on a report released by the FAO in March of 2024.

All living things, including humans, need mineral nutrients for proper metabolism and homeostasis. Each key mineral nutrient serves a different purpose in the body and is just as crucial to a person's diet. The occurrence of numerous diseases and metabolic problems is heightened when these nutrients are inadequate or absent from the diet. To alleviate malnutrition, it is imperative to identify the various sources of these minerals, quantify their content, and determine their bioavailability. Additionally, the biofortification of edible crops is necessary to improve the population's nutritional status and optimize mineral intake through a balanced and enriched diet (Stein 2010). By understanding the nature of mineral elements, the quality of edible crops can significantly be improved in terms of biofortification and bioavailability. According to Martínez-Ballesta et al. (2010), over 20 distinct dietary minerals are probably necessary for the human body to function properly and to be healthy.

Numerous macro-minerals, trace elements, vitamins such as E and B complex, bioactive compounds, and nonstarch polysaccharides found in cereal grains may have positive effects on health. Due to its involvement in hundreds of enzymatic activities, magnesium (Mg) is one of the macrominerals that is essential to humans, animals, and plants. Mg is the second most prevalent cation in plants. As a result of the actions of several transporters, Mg is taken up from the soil by the roots and moved to the sprouts and leaves, where it is accumulated in chloroplasts (Sun et al. 2017). This vital mineral influences human post-receptor insulin signaling and vascular smooth muscle tone by acting as a calcium antagonist. Furthermore, according to Bo and Pisu (2008), it has an impact on endothelial cell activities, energy metabolism, and neurotransmitter release. As a cofactor for about 300 enzymes, Mg also plays a role in controlling the excitability of muscles

Crop Biofortification: Biotechnological Approaches for Achieving Nutritional Security under Changing Climate, First Edition. Edited by Adnan Noor Shah et al.

Table 14.1 World cereal production (in million tons).

Continents	2021	2022	2023	% Change (2023 over 2022)
Asia	1236.9	1260.9	1277.0	1.3
Africa	203.3	201.2	197.4	−1.9
Central America and the Caribbean	43.0	42.5	41.1	−3.0
South America	228.6	246.6	246.1	−0.2
North America	495.2	471.7	518.7	10.0
Europe	548.8	525.0	517.1	−1.5
Oceania	55.9	61.1	42.0	−31.2
Total	2811.7	2809.1	2839.5	1.1

Source: FAO report published on 1 March 2024 about cereal supply and demand brief.

and nerves (Huskisson, Maggini, and Ruf 2007). An adult's body has between 20 and 28 g of Mg. Apatite-bound bone contains approximately 55% of it, followed by muscle (25%) and soft tissues (19.3%) (Jahnen-Dechent and Ketteler 2012). According to Witkowski, Hubert, and Mazur (2012), less than 1% is present in serum, mostly as the bioactive form known as free cation, which is also conjugated to proteins. Malnutrition is a common cause of Mg deficiency, which is linked to aging and age-related disorders (Durlach et al. 1998; Killilea and Maier 2008). According to research, consuming more Mg may help prevent chronic diseases like diabetes, metabolic syndrome, hypertension, and a number of cardiovascular disorders (Bo and Pisu 2008). When coupled with a high-fructose diet, a low-Mg diet may exacerbate insulin resistance. Furthermore, consumption of less Mg is associated with an inflammatory response due to changes in the intracellular-Ca concentration (Ahokas et al. 2005; Rayssiguier et al. 2006).

Crop biofortification has considerable potential as a cost-effective and sustainable way to supply essential nutrients to communities that do not have the availability of a variety of food sources. To maintain the quality of diets, biofortification involves increasing the density of vital components such as amino acids, mineral elements, and various vitamins in staple foods (Bouis and Saltzman 2017). It guarantees the supply of food crops with high nutrient content produced sustainably, particularly for the impoverished populations of developing countries. Improving the necessary minerals' bioavailability in biofortified crops provides a practical way to increase the accessibility of foods through conventional agricultural methods and food business initiatives. They may assist in supplying a wide range of foods in a practical manner to undernourished and low-income populations. From the economic point of view, producing crops through biofortification strategies is an investment done for a single time and only offers a financially viable way toward the battle against mineral deficiencies. Following their development, biofortified crops require no further investment in supplementation since, when they are processed, they naturally become integrated into food supplies required by the human population (Pfeiffer and McClafferty 2007). Thus, biofortification entails raising the vitamin and mineral content of crops' edible parts using agronomic practices, genetic engineering, conventional breeding, and other comparable approaches (Bouis 2018). Generally, two main approaches are used to increase the intake of Mg^{2+} by humans and livestock by improving the absorption of Mg^{2+} in plants (Hermans et al. 2013; Marschner 2012). The first is agronomic biofortification, which is widely used to improve Mg^{2+} fertilization of plants by soil and foliar layer. Genetic biofortification, which involves breeding crops with higher Mg^{2+} levels, genetically modifying certain genes to increase plant tolerance to Mg stressors, or improving Mg^{2+} uptake pathways, is the second and more widely used technique.

Agronomic techniques can improve the transmission of bioavailable minerals to humans and livestock (Broadley et al. 2009). Still, to maintain economic sustainability and reduce environmental contamination, mineral fertilizer use must be optimized for the sake of sustainable agriculture (Fageria, Baligar, and Li 2008). Thus, genetic biofortification may become a desirable alternative in the future. Genetic biofortification can be applied to horticultural and field species using forward genetics (mutant screening for altered Mg profile) and reverse genetics (knockout mutants of known or unknown genes), screening of natural varieties, and biparental crosses progeny (recombinant inbred lines [RILs]) if a gene is found to be involved in Mg^{2+} homeostasis in some plants (Hermans et al. 2013). One example is the earliest attempt to modify the expression of known Mg^{2+} transporters in plants by Deng et al. (2006). By introducing *AtMGT1* in tobacco and regulating it with the constitutive *CaMV 35S* promoter, they were able to significantly increase the Mg^{2+} content in transgenic lines by 30% in tobacco plants. But in normal growing conditions, this improvement was accompanied by a decrease in plant biomass. This problem might be resolved by using different regulatory promoters, such as stress-, development-, or cell-specific promoters (Hermans et al. 2013). Increasing Mg^{2+} transport and/or accumulation could be both a feasible and beneficial strategy for developing plants resistant to Al^{3+} (Bose, Babourina, and Rengel 2011; 2013). Further, components of phytohormone production pathways and the genes that code for these compounds may be sources of improved absorption and utilization of Mg^{2+} in crops (Guo et al. 2015; Hermans et al. 2013; Werner et al. 2010). Quantitative trait loci (QTL) mapping is a useful technique for locating gene clusters in plants that respond to Mg stressors (Broadley et al. 2008).

Importantly, other environmental conditions such as transport system antagonistic effects, heat, drought, high radiation, low pH, and metal toxicity must be taken into account when employing fertilizers and crop-improvement biofortification techniques (Gransee and Führs 2013). Additionally, the methods of plant breeding and genetics may be useful in shedding light on the interaction mechanisms and identifying novel components related to the utilization of Mg^{2+}. In light of this, the chapter explores a number of important topics, such as the roles that Mg plays in plants, the symptoms of Mg toxicity and deficiency, the mechanisms that control the transport of Mg within plants, the impact of Mg on cereal quality, the nutritional value of cereals, the current state of Mg in cereal crops, the critical role that Mg plays in human physiology, the indicators of Mg deficiency in humans, and methods for biofortifying cereal crops to increase Mg levels.

14.2 Functions of Magnesium in Plant Systems

14.2.1 The Role of Magnesium in Regulating Photosynthesis

Mg is especially crucial to photosynthesis. Up to 35% of Mg is devoted to chloroplasts alone in plant cells. According to Chen et al. (2018), plants have developed a gradient of Mg distribution, with the thylakoid > stroma > cytosol, in order to maximize photosynthesis. Mg ions found in chloroplasts are crucial components of chlorophyll because they facilitate light absorption and function as cofactors for several enzymes involved in photosynthetic processes (Guo et al. 2016).

According to Ye et al. (2019), low Mg levels frequently lead to damaged photosystems, reduced amounts of chlorophyll in leaves, and a slowing in photosynthetic electron transport. These effects are mostly brought on by regulatory effects of Mg on the production as well as decay of chlorophyll (Chen et al. 2018; Peng et al. 2019), alongside the feedback mechanism triggered by the accumulation of sugar in source leaves (Cakmak and Kirkby 2008). Earlier theories state that Mg deficit reduces the uptake of CO_2 by hindering electron transport in the dark reaction. However, according to recent studies, Mg insufficiency affects CO_2 absorption even before light-dependent reaction takes place (Li et al. 2020b; Jaghdani, Jahns, and Trankner 2021). The maximum activity related to ribulose-1,5-bisphosphate carboxylase (Rubisco) and the net assimilation rate both significantly decrease during the initial phase of Mg deficiency in the leaves of rice crop, while other parameters such as the electron transport rate, chlorophyll concentrations, and the optimal efficiency of PSII photochemical reactions (Fv/Fm) stay constant (Li et al. 2020b). Similar results in barley are reported where an Mg deficit mostly hampered CO_2 absorption but had no effect on the utilization of photosynthetic light (Jaghdani, Jahns, and Trankner 2021). Photosynthetic enzymes, including Rubisco and phosphoenolpyruvate (PEP) carboxylase, are known to be strongly dependent on Mg concentrations, particularly on the free Mg^{2+} in the stroma (Andreo, Gonzalez, and Iglesias 1987; Kai, Matsumura, and Izui 2003). This phenomenon may be explained by the fact that, in contrast to photosynthetic light reactions, photosynthetic enzyme activities related to carbon fixation are more susceptible to decrease in tissue Mg concentrations.

It is known that diurnally regulated CO_2 fixation promotes the growth of plants (Dodd et al. 2014), the precise regulatory mechanisms accountable for the rhythmic fixation of CO_2 remain largely unidentified. A study recently conducted on rice by Li et al. (2020b) revealed that Rubisco activity is directly influenced by rhythmic fluctuation in free Mg^{2+} levels inside the chloroplast stroma, which in turn affects the daily pattern of CO_2 fixation (Figure 14.1). Moreover, subcellular pattern of Mg exchange sustains the efficacy of Mg utilization intracellularly which allows an increased amount of Mg throughout the night for the breakdown of energy and production of proteins while ensuring adequate Mg during the day for optimal photosynthetic efficiency. Mg concentrations inside the stroma of chloroplasts facilitates carboxylation activity of rubisco, an enzyme essential for CO_2 fixation. By carbamylating the active site *Lys 210*, Mg ions (Mg^{2+}) are essential in preventing Rubisco from collapsing down and binding to oxygen (Stec 2012). Given this, the variations in the activity of rubisco enzyme influenced via Mg fluctuations within chloroplasts could potentially be a key regulator governing the rate of photosynthetic activity. The Mg transporter *OsMGT3*, on the other hand, has been effectively identified as being positioned in the chloroplast envelope of rice. According to Li et al. (2020b), this transporter is accurately controlled through the circa-

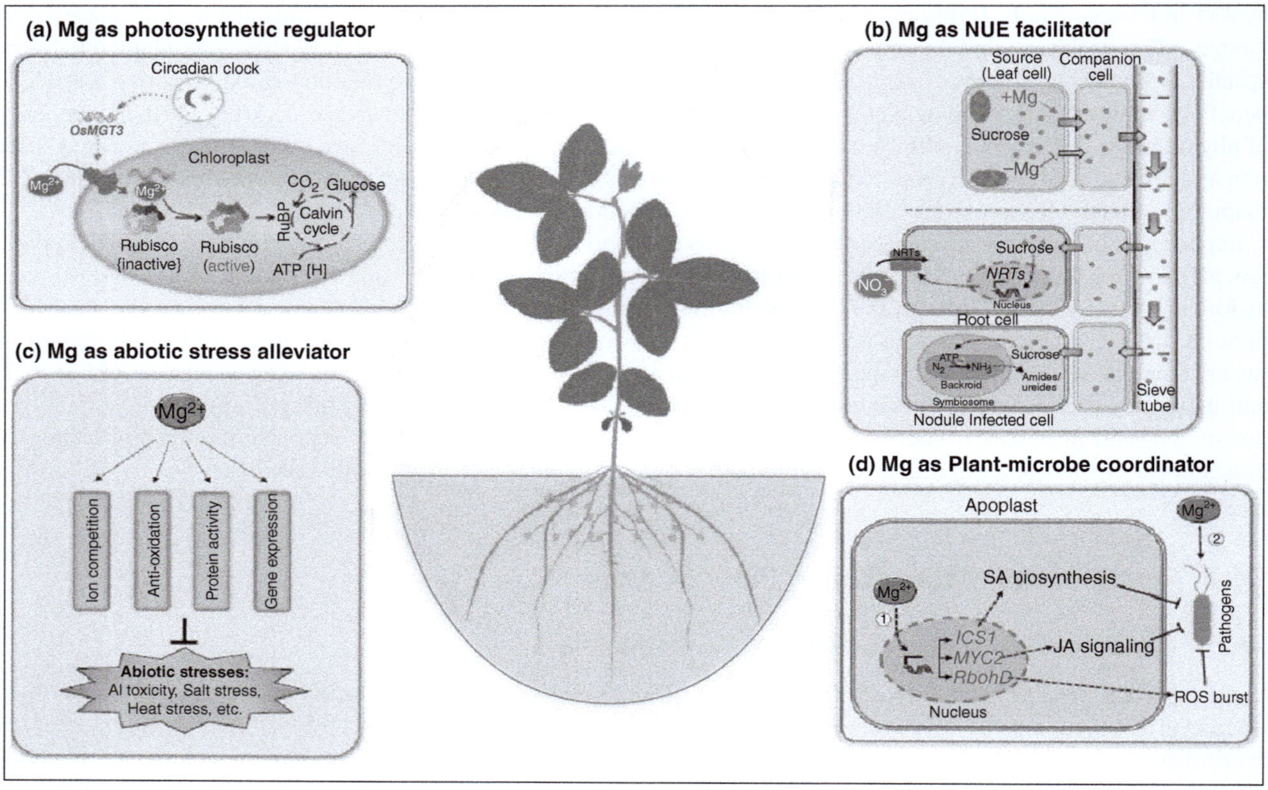

Figure 14.1 Insights into the functions of magnesium (Mg) in plants. (a) The function of magnesium in the regulation of photosynthetic processes: A magnesium influx transporter *OsMGT3* is located at the chloroplast membrane of rice mesophyll cells. The circadian rhythm precisely controls the expression of *OsMGT3*, and is responsible for magnesium uptake and the development of diel fluctuations in magnesium concentrations in the stroma, enabling efficient CO_2 fixation through photosynthetic processes during the cycle of day and night; (b) Mg as a promoter of nitrogen-use efficiency (NUE): The transfer of sucrose from shoots to roots is facilitated by an adequate magnesium supply. Aside from serving as a signaling molecule to activate the genes encoding the nitrate transporter (NRT) at the root level, the allocated sucrose also provides energy for nodule growth and N fixation, which in turn enhances NUE; (c) Mg functions as a mitigator of abiotic stresses in plants: In plants, magnesium (Mg) engages in competitive interactions with cation binding sites in both the apoplast and symplast pathways to successfully mitigate abiotic stresses such as heat stress, salt stress, and aluminum toxicity. Additionally, Mg boosts antioxidant systems and regulates protein activity and gene expression to combat these stressors; (d) Mg functions as a mediator between plants and microbes: First, an increase in Mg levels in plant cytoplasm stimulates ROS production and defense-related hormone signaling pathways, preventing the spread of pathogens. Second, when pathogens invade plant cells, their apoplastic magnesium content rises, thereby enhancing the pathogen's survival and pathogenicity (Tian et al. 2021 / Springer Nature).

dian clock which is in charge of rhythmic Mg loading into rice chloroplasts (Figure 14.1). The expression profiles of *OsMGT3* homologs similarly shows a diurnal rhythm in a variety of plant species, including *Arabidopsis*, maize, and sugarcane demonstrating potential for functional commonalities (Drummond et al. 2006; Wang et al. 2019). The precise functional differences between *OsMGT3* homologs are yet unknown, although it is thought that plants have a common mechanism for controlling photosynthesis through changes in Mg levels in chloroplasts during the day. It is noteworthy to mention that artificially altering the expressions of *OsMGT3* gene in mesophyll cell of rice may enhance the biomass yield of plants by upregulating plant's ability to photosynthesis (Li et al. 2020b). This control of Mg transport within chloroplasts provides a strong tool to enhance the ability of plants for photosynthesis and in turn increases the yield of crops.

14.2.2 Magnesium's Contribution to Improving Nitrogen-Use Efficiency

In living organisms, nitrogen (N) is necessary for the synthesis of nucleotides and proteins. N application in agriculture, mostly through fertilizers, is always the highest because of plants' high requirement for it (Heffer and Prud'homme 2016). Over the years, research has consistently demonstrated Mg's beneficial impacts on absorption and use of N, even though Mg is present in plant cells at much lower levels than N (Grzebisz 2013; Szulc 2010; Szulc et al. 2013). The studies conducted in field conditions reflected that Mg supplements can improve crop yield, protein content, and N-use-efficiency

(NUE) in a variety of crops (Grzebisz 2013). Furthermore, this stimulating effect can be stronger in situations where the N is deficient in soil. Interestingly, it has been observed that tuber crops such as *Beta vulgaris* generally show greater increases in yield and N levels following the supplementation of Mg (Grzebisz et al. 2010). This is most likely because of their higher needs for N as well as carbohydrates. Similar results have also been reported for hydroponic experiments. Increased N concentration and total N uptake were seen in soybeans when Mg supply was increased; this effect was mainly ascribed to its stimulation of nitrate rather than ammonium root absorption (Peng et al. 2020). Additionally, Mg plays a critical role in the symbiotic relationship found in legumes. According to Peng et al. (2018), external Mg supplementation has been demonstrated to increase N fixation and nodule formation by facilitating the allocation of carbohydrates in plant roots and so improving N utilization. Still, the root N absorption is not the only indicator of Mg's promotion of N. Several studies show that Mg application to foliage parts of cereal crops can significantly raise their N levels and yields (Grzebisz 2013). Especially interesting is the fact that applying Mg increased N content more significantly than Mg content in plants (Grzebisz 2013). This suggests that N enhancement, rather than direct Mg contributions, is more likely to be responsible for the yield improvements facilitated by foliar Mg.

Until now, the most widely accepted explanation of Mg's role in facilitating N uptake has been that it helps photosynthates transfer from shoots to roots. As a result, the deposited carbon increases the root's ability to absorb N by acting as an energy source (Verbruggen and Hermans 2013). But, according to recent findings, the carbon that is transported within plant roots functions as a signaling compound which significantly affects the expression patterns of transporters involved in the uptake of N (Peng et al. 2020). The reduced expression patterns of *NRT* family transporters which regulate nitrate N were observed in the conditions of Mg deficiency, thereby decreasing the concentration of N in the roots of soybean crop. Without observable changes in root biomass, these inhibitions were overcome by a brief application of exogenous sucrose (Peng et al. 2020). These findings imply that the positive regulation of NRTs expression by sucrose is most likely the cause of Mg-facilitated N uptake. Mg controls not only the uptake of N by the roots but also the assimilation process. Because Mg directly contributes to ribosome aggregation for protein production, improved N assimilation by Mg supplementation may also be linked to improved conversion of the absorbed N into proteins (Chen et al. 2018). However, further research is still required to fully understand the deeper mechanisms underlying the synergism between Mg and N.

14.2.3 Role of Magnesium in Alleviating Various Abiotic Stresses

14.2.3.1 Aluminum Toxicity

In acidic soils, aluminum (Al) toxicity is a major barrier to plant growth. Mg can lessen the toxicity of Al in both extracellular and intracellular environments because of the comparable hydrated radii of Al^{3+} and Mg^{2+} (Bose, Babourina, and Rengel 2011). The main mechanism by which Mg can reduce external accumulation of Al in roots in a variety of plant species, including rice (Watanabe and Okada 2005), wheat (Ryan, Kinraide, and Kochian 1994), sorghum (Tan, Keltjens, and Findenegg 1992), and rice bean (Yang et al. 2007), is its capacity to either directly engage in competition with Al for binding sites on the plasma membrane and root cell wall or to displace it. Moreover, Mg competes with Al for protein synthesis, energy metabolism, and photosynthesis inside the cell on binding sites of molecules that provide oxygen. This helps to reduce subcellular stress brought on by Al (Hasni et al. 2015). Mg and Al in roots and shoots, respectively, may interact indirectly in addition to direct ionic antagonism. According to Chen et al. (2015), Mg improves the ability of the roots of some legume crops to withstand Al stress by inducing the release of organic acids (OAs) from the roots, which interact with Al^{3+} ions and facilitate external detoxification of Al. One explanation for how Mg promotes OA secretion is that it enhances the transfer of carbon from shoots to roots, which provides a source of carbon for OA secretion. Mg may offer an alternative possible mechanism via regulating the protein activity and gene expression of efflux transporters for OAs and H^+-ATPase on the cell's outer membrane (Chen et al. 2015; Neuhaus et al. 2013). According to research, nitric oxide (NO) production was stimulated by Mg deficiency and Al toxicity which reduced the growth of primary root in *Arabidopsis* plants (Li et al. 2020a). This suggests that NO, a molecule involved in cellular signaling, may play a role in the relationship between Mg and Al in influencing the growth of the root. According to research, Mg improved plasma membrane H^+-ATPase activity and proton motive force in alfalfa (*Medicago sativa* L.) shoots, which mitigated the damage caused by Al on photosystems. Su et al. (2020) suggest that Mg may be responsible for this behavior via controlling auxin metabolism, which is involved in regulating photosynthetic electron transport. In response to Al stress, a defense mechanism in increasing Mg absorption has been developed by plants. Mg absorption capacity positively correlates with genotype resistance to Al toxicity in both maize and *Arabidopsis thaliana* (Bose et al. 2013; Li et al. 2016). According to Chen et al. (2012), when external Al is applied to rice, the roots actively absorb more Mg in order to reduce Al stress. It has been discovered that Al-induced Mg uptake is caused by the overexpression of the root Mg transporter gene *OsMGT1* (Chen et al. 2012). Results from the above studies tell that plant Mg levels and tolerance against Al toxicity can be increased by following some viable strategies related to plant breeding and genetics. Studies provide evidence that this method has been successfully applied against tobacco (Chen and Ma 2013; Deng et al. 2006).

14.2.3.2 Heat Stress

Heat stress causes metabolic disturbances that severely limit crop growth and development. Mg serves as a beneficial element for plant tolerance against these kinds of environmental stresses. In crops such as chickpeas, maize, and wheat, necrotic symptoms caused by high temperatures can be significantly reduced by an adequate supply of Mg (Kaushal et al. 2013; Mengutay et al. 2013). As per Hazra et al. (2015), Mg not only aids in maintaining the rubisco activation state but also improves catalyst efficacy and heat resilience of rubisco activase. This lessens the generation of reactive oxygen species (ROS), which are caused by disturbances in photoelectron transport brought on by heat stress. Additionally, sufficient Mg helps keep antioxidants and ROS in equilibrium within plant cells (Silva et al. 2017). Siddiqui et al. (2016) and Boaretto et al. (2020) have reported that Mg supplementation increases the activities of osmoprotectants like proline and glycine betaine, as well as antioxidant enzymes like superoxide dismutase (SOD), ascorbate peroxidase (APX), catalase (CAT), and glutathione reductase (GR), in preventing photooxidation brought on by heat stress. In addition to the boosted antioxidant systems, there's another potential mechanism that insufficient Mg causes a decrease in CO_2 fixation and the closure of stomata, which in turn lowers transpiration rate and heat dissipation indirectly. However, the availability of Mg can counteract the effects of heat stress on root growth, as evidenced by the rising root-to-shoot ratio in wheat as well as maize (Mengutay et al. 2013). Because Mg has a unique role in the distribution of carbohydrates, sufficient Mg may aid roots in obtaining enough carbon sources to maintain strong respiration, hence improving growth status under heat stress.

14.2.3.3 Salt Stress

Soil salinity is a major environmental problem that reduces crop yield and productivity globally. It stunts the growth of plants because of osmotic stress and too many sodium ions (Na^+) in the soil (Horie, Karahara, and Katsuhara 2012). Plant salt stress is known to be lessened by Mg under a variety of circumstances. For instance, it was discovered that applying Mg foliarly to the leaves of strawberries grown in a saline environment increased their biomass and chlorophyll content (Yildirim, Karlidag, and Turan 2009). The suppression of reproductive growth of *Plantago crassifolia* was lessened by the use of Mg fertilizer in highly salinized soils (Grigore et al. 2012). Furthermore, Mg supplementation mitigated the detrimental impact on pepper seedlings caused by salt stress in a hydroponic experiment (Zirek and Ozlem 2020). Since Na^+ and Mg^{2+} are both cations, Mg^{2+} has the ability to compete with Na^+ for the attachment sites on the surface of roots. Though ion antagonism is a widely known phenomenon, it is still difficult to describe how Mg specifically helps to mitigate salt stress. Interestingly, a functional investigation conducted in rice on the Mg transporter *OsMGT1* has provided new molecular evidence supporting Mg's ability to mitigate salinity stress (Chen et al. 2017). Ren et al. (2005) stated that *OsHKT1;5* functions as a transporter permeable to Na^+, playing a vital role in preventing the accumulation of Na^+ in the shoots by removing it from the xylem. The function of *OsHKT1;5* to transport Na^+ is enhanced by the expression of *OsMGT1* that transports Mg^{2+} within the mature zones of root stellar tissues. This reduces sodium accumulation in the shoots and ultimately contributes to rice's ability to tolerate higher salt conditions (Chen et al. 2017). While the exact activation mechanism remains unidentified, Mg regulation of *OsHKT1;5* activities may be explained by its potential impact on protein stereo structure. It's crucial to remember that Mg itself, especially when given in excessive amounts, can also function as a salt stressor under certain conditions. So, careful consideration of the right dosage of Mg should be kept in mind while applying it to mitigate salt stress.

14.2.4 The Significance of Magnesium in Plant–Microbe Relationships

Fungi, bacteria, viruses, and nematodes are examples of plant pathogenic microorganisms that feed on their hosts and drastically reduce agricultural yields globally (Khater, de la Escosura-Muñiz, and Merkoçi 2017). Mg is thought to have a role in plant–pathogen interaction since it is a necessary mineral element for both microorganisms and plants. But up until now, there have only been a few findings regarding Mg's roles in plant pathology. The primary explanation for this is that Mg has complex impacts on the incidence of disease that can be either direct due to its unique physiological and metabolic functions or indirect due to its general effects on plant health (Elmer and Datnoff 2014). An excessive amount of Mg might antagonistically prevent cations like Ca and K from being absorbed by plant hosts (Tang and Luan 2017). Calcium and potassium are both crucial for triggering immunity in plants (Cheval et al. 2013). The Ca influx of plant cells is a notable plant pathogen-associated molecular pattern (PAMPS) triggered by immune reactions during the early stages of pathogen infection (Aldon et al. 2018). Because of Mg's involvement, the antagonistic effects of minerals may thus play a major role in regulating the plant's resistance to a range of examined plant diseases (Jones and Huber 2007).

A defense mechanism against necrotrophic pathogens and herbivorous insects is mediated by ethylene (ET) and jasmonic acid (JA) in dicotyledonous plants, whereas salicylic acid (SA) among defense-related hormones contributes to the enhancement of resistance against (hemo) biotrophic pathogens (Pieterse et al. 2012). Soil-applied supplementation of Mg oxide (MgO) to healthy plants of *Arabidopsis* significantly increased the endogenous levels of SA and also enhanced the expression of *isochorismate synthase 1* (*ICS1*), an enzyme that is responsible for SA production and conversion of chorismate to isochorismate. Moreover, *AtRbohD* transcription was stimulated, which resulted in the accumulation of H_2O_2 *Arabidopsis* roots (Ota et al. 2019). MgO application has been shown to improve *Arabidopsis* resistance to *Ralstonia solanacearum* by promoting the biosynthesis of SA and the generation of ROS mediated by *AtRbohD*, without affecting the JA and ET pathways (Ota et al. 2019). On the other hand, MgO treatment in soil improves tomatoes' resistance to Fusarium wilt mostly via the JA pathway, as shown by its inefficacy within the *def-1* mutant lacking functional JA signaling (Fujikawa et al. 2021). Therefore, it seems that Mg activates innate immunity through distinct hormonal mechanisms (Figure 14.1). A number of studies have demonstrated the effectiveness of MgO nanoparticles (ranging from 1 to 100 nm) as effective antimicrobial agents for the control of plant diseases caused by bacteria and fungi (Huang et al. 2005; Imada et al. 2016; Wani and Shah 2012). However, it is unclear whether these positive effects were due to the properties of the nanoparticles or MgO itself.

However, there is a competitive interaction between plants and their pathogens regarding the uptake of Mg. For instance, when common beans were inoculated with strains of *Pseudomonas syringae Pv. Phaseolicola*, there was an increase in the level of Mg in the apoplast. This may be an attempt employed by pathogenic bacteria to enter the plant because Mg maintains the integrity of the outer membrane of bacteria (O'Leary et al. 2016). According to Reza et al. (2016), the growth of *Magnaporthe oryzae* increased by Mg, a fungus responsible for the blast disease of rice. *Gaeumannomyces graminis var tritici*, the cause of wheat take-all disease, can be lessened by decreasing the amount of soil Mg (Huber and Jones 2013). The following studies suggest that Mg has a significant role in the pathogenicity of plant infections caused by bacteria and fungi (Figure 14.1). Given the intricate nature of Mg in plant physiology and pathogenesis, further research is required to elucidate the fluctuations in Mg levels in the cytoplasm of different organelles. Moreover, understanding how cellular Mg transporters are upregulated during interactions between plants and pathogens will shed light on the complex mechanisms governing Mg regulation in the competition between pathogens and plants.

14.3 Detecting Magnesium Deficiency and Toxicity in Plants, Symptoms, and Threshold Values

A Mg deficit in agriculture can impede plant growth, leading to notable reductions in agricultural yield and quality traits (Verbruggen and Hermans 2013). Mg is essential for the physiological processes involved in photosynthesis, including the synthesis and distribution of photosynthates. Thus, a Mg deficit slows down the rate of photosynthesis and disturbs the transport of carbohydrates to sink organs. As a result, sink organ growth is ultimately inhibited, which lowers productivity in a variety of plant species (Yang et al. 2012; Fischer 1997). A subtle Mg deficit can have an effect on plant metabolism, which can therefore have an impact on several quality characteristics of ornamental plants, fruits, vegetables, and other agricultural crops (Gerendás and Führs 2013). For example, adding Mg sulfate to soybean crops can raise oil and protein levels (Deliboran et al. 2011; Vrataric et al. 2006). Also, in black tea, it can raise the concentrations of amino acids, polyphenols, and catechins (Jayaganesh, Venkatesan, and Senthurpandian 2011). Therefore, a small Mg deficiency may put crop quality and biomass production at risk.

The major symptom of Mg deficiency in plants is usually interveinal chlorosis in the leaves. However, there can be variations in the degree of Mg deficiency and the site of leave chlorosis. For instance, when exposed to low Mg levels, younger mature leaves of soybean plants show more pronounced chlorosis as compared to new developing or older leaves (Chen et al. 2018). According to Chen and Fan (2018), banana plants that were cultivated in sand culture to induce Mg deficit showed symptoms of chlorosis on their second latest leaf after 35 days. The highest level of leaf chlorosis resulting from Mg deficiency in hydroponically grown rice plants was observed in middle-aged leaves, which occurred at the time when Mg had maximum mobility (Peng et al. 2019). Based on this result, it appears that young mature leaves that have large Mg pool and high photosynthetic capacities are more susceptible to photooxidative stress caused by Mg deficit, in results, increasing the levels of ROS and rapid chlorophyll-mediated degradation by STAY-Green (SGR) proteins (Peng et al. 2019). This theory also explains why, in low-Mg environments, increased light intensity accelerates the development of chlorosis (Cakmak and Yazici 2010). Furthermore, oxidative stress is thought to be triggered by low Mg levels, which interfere with the iron's ability to sequester into vacuoles in young mature leaves and consequently induce Fe toxicity in plants (Kobayashi et al. 2018).

Mg deficits in plants grown in open field conditions show more complicated symptoms than in carefully regulated greenhouse experiments. In experiments conducted on open field conditions, the initial symptoms of Mg deficiency are usually seen on older leaves (Castro et al. 2021; Wang 2020). This discrepancy could be explained by the fact that exchangeable and slowly accessible Mg is present in soils and may serve as a buffer against the quick Mg deficiency which typically occurs in highly regulated environments, that is, medium culture experiments or hydroponics. Tian et al. (2021) conducted an experiment on rice to check the effects of different Mg in hydroponics culture. The symptoms of Mg deficiency were first observed in young mature leaves when Mg was applied in very low (5 μM) concentrations. However, leaf chlorosis was observed on older leaves when Mg supplementation was in medium (20 μM) concentrations. Furthermore, they observed that a Mg deficit may cause metabolic problems in plants and make them more vulnerable to abiotic stresses, even if it is not noticed at the early growth stages. Hence, adverse effects resulting from latent Mg deficiency must be taken into consideration to fully understand the significant values of Mg in agriculture systems.

Mg nutrition and plant growth in different species have been documented in numerous research studies. Mg deficiency is considered a more interesting topic than its toxicity. Symptoms of Mg toxicity were very hard to observe even when tobacco plants were exposed to high (60 mM Mg^{2+}) concentrations (Shaul et al. 1999). The plant's large vacuole capacity for storing Mg may be the reason for few reports of Mg toxicity (Hawkesford et al. 2012). However, some authors have connected the formation of necrotic spots (Messenger and Hruby 1992) or copper-colored spots along the margins of veins on leaves (Venkatesan and Jayaganesh 2010) to Mg toxicity. It is commonly accepted that the optimal Mg concentration for growth occurs in vegetative tissues when the dry weight falls between 1.5 and 3.5 mg g^{-1} (Grzebisz 2011; Römheld 2012). Ding, Luo, and Xu (2006) reported that the onset of leaf chlorosis is often linked to Mg concentrations less than 1–2 mg g^{-1} leaf dry weight. A decrease in plant growth and a change in the distribution of biomass among the organs are frequently observed below that threshold level.

14.4 Transport Mechanism of Magnesium in Plant Systems

Two primary mechanisms are responsible for the uptake of Mg by plants from soil solutions: (1) mass flow refers to the passive transport of Mg ion driven by transpiration stream, and (2) diffusion, which is the movement of Mg^{2+} from areas with greater concentration to areas with lesser concentration. Therefore, Mg concentration in soil solution and the ability of soil to replenish it with Mg determine the amounts of Mg that plants absorb.

A highly efficient mechanism for the absorption, storage, and transportation of Mg had been developed by plants, which enables them to maintain high Mg concentrations in their tissues, usually in the millimolar range. The majority of Mg enters within the cells as Mg chlorophyll, Mg enzymes, and Mg ATP for utilizing itself in a variety of biological activities. Only 5–10% Mg binds to pectate to stabilize the cell wall (Figure 14.2). In plant cells, chloroplasts and mitochondria accumulate (0.5–2 mM and 2–4 mM Mg^{2+}, respectively). It is significantly more than the cytosol (0.2–0.4 mM) (Gout et al. 2014; Igamberdiev and Kleczkowski 2001; Ishijima et al. 2003). This requires the help of Mg transporters in the corresponding envelopes. Moreover, transporters are also necessary to enable the exchange of Mg between the cytosol and the vacuole, ensuring a consistent supply of Mg to cytosol.

14.4.1 Gene Families Responsible for Magnesium Transport

MGT or *MRS2* family of Mg transporters is well-known for the transport of Mg in plants. Their members are found in several species of plants (Chen et al. 2012; Li et al. 2016; Saito et al. 2013). Although functional complementation experiments in bacteria and yeast have shown Mg transport activity; however, members of this family function independently, displaying distinct subcellular localization and spatial and temporal patterns of transcript expression (Gebert et al. 2009; Saito et al. 2013). The *MGT* family members play a wide variety of functions throughout the life of plants from contributing to

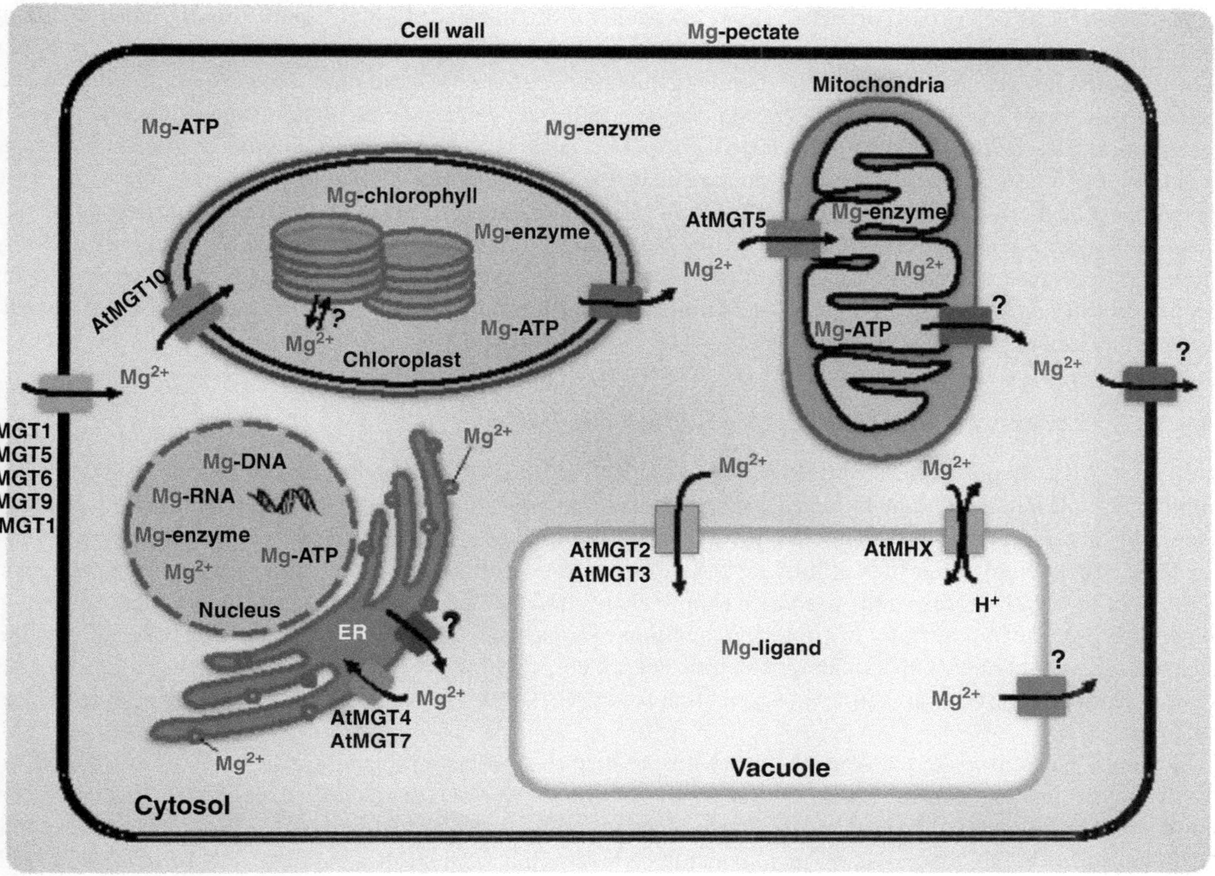

Figure 14.2 A schematic illustration from Chen et al. 2018 / Frontiers Media S.A showing how magnesium is transported and distributed inside the cells of plants. Following the entry into plant cell, intracellular magnesium exhibits the capability to associate with a wide range of substrates such as DNA, RNA, ATP, enzymes, and chlorophyll. This interaction helps with a number of metabolic processes, including the production of proteins and nucleic acids, photosynthesis, and enzyme activation. There are multiple known transporters involved in the entry of magnesium into organelles. Nevertheless, little is known about the transporters responsible for the magnesium efflux from cells and organelles.

the growth of roots to contributing to photosynthesis and production of pollen grains and responses to environmental stresses (Chen et al. 2012; Chen et al. 2017; Xu et al. 2015). The *MGT* family members were first identified through the cobalt resistance screening technique and are homologous to CorA proteins originating in bacteria. According to Klein, Moore, and Steitz (2004), nickel tracer tests demonstrated that members of the MGT family might also be involved in the transportation of other divalent cations, such as copper (Cu^{2+}), manganese (Mn^{2+}), iron (Fe^{2+}), nickel (Ni^{2+}), and cobalt (Co^{2+}) (Li et al. 2001). However, owing to the presence of the conserved tripeptide motif Gly-Met-Asn (GMN) at the end of the first domain of transmembrane, the affinity of Mg is notably higher compared to other cations (Knoop et al. 2005). This shows that the GMN motif is specifically designed for Mg because changing it prevents Mg transport (Lunin et al. 2006). Interestingly, other divalent cations like zinc and cadmium are preferred for transport by GVN and GIN, two variations of the GMN motif (Knoop et al. 2005). In conclusion, it appears that the GxN motif is conserved across a wide range of metal transporters in plants, providing scientists with a useful avenue for further investigation.

Researchers discovered that CorA's crystal structure was unique and did not resemble any other channel or transporter. It forms a funnel-shaped homopentamer, permitting only the passage of divalent cations, and has two transmembrane helices per monomer (Lunin et al. 2006). Furthermore, CorA's divalent cation sensor relies heavily on the interaction pair *Asp253–Asp89*, which allows it to detect the amount of Mg in the cytosol and start basic conformational changes that open and close channels (Dalmas et al. 2014). That is why, the transportation of Mg through CorA is a self-regulatory mechanism that is highly dependent on intracellular Mg concentrations.

There are a lot of other transporter families responsible for the transport of Mg^{2+} apart from the *CorA* homologous of *MGT* family. For example, *MgtA, MgtB*, and *MgtE* in bacteria; *ATMHX* in *Arabidopsis*; and *TRPM6/TRPM7* in human cells bear no resemblance to members of *MGT* family (Chubanov et al. 2004; Gaash et al. 2013; Monteilh-Zoller et al. 2003). According to Hmiel et al. (1986 and 1989), Mg does not affect the expression of *CorA*, whereas low Mg concentrations strongly stimulated *MgtA* and *MgtB* in bacteria. Transporters such as MgtA and MgtB function as P-type ATPases, actively facilitating Mg inflow in contrast to CorA (Soncini et al. 1996; Tao et al. 1995). Furthermore, *MgtA* and *MgtB* expressions are regulated by the system of PhoP-PhoQ (Groisman 2001). When Mg level is low, PhoQ acts as a sensor kinase triggering the phosphorylation and activation of PhoP which is a transcription factor for regulating several genes specifically *MgtA* and *MgtB*. In general, various Mg transporters found so far indicate that many unanswered questions remain regarding the involvement of additional Mg-inducible gene families in the transportation of Mg, with the exception of constitutively expressed members of the *MGT* family (Waters 2011).

14.4.2 Mechanisms for Magnesium Uptake, Storage, and Transport

Chen et al. (2012) identified that *OsMGT1* is responsible for the uptake of Mg in rice roots (Figure 14.2). *AtMGT6* has been found to be a plasma-membrane localized transporter in *Arabidopsis* that mediates low-Mg tolerance and root Mg uptake (Figure 14.2; Mao et al. 2014). The reported Km values for Mg uptake via *AtMGT6* and *OsMGT1* were 19.54 μM and 30 μM, respectively, confirming their high affinity for Mg. However, the significant accumulation of Mg in roots was not stopped by the loss of *OsMGT1* in rice or the decreased expression of *AtMGT6* in *Arabidopsis*, suggesting that some other Mg transporters are also active in roots for Mg uptake in addition to these two genes. *AtCNGC10*, a protein belonging to the cyclic nucleotide-gated channel (CNGC) family has been identified as an Mg mediator in the meristem and distal elongation zones of plant roots (Guo et al. 2010). However, because *AtCNGC10* also transports K^+, Ca^{2+}, and H^+, it is less selective for Mg^{2+} than *MGTs*.

Once roots have taken up Mg from the soil with the help of the above transporters, it is then translocated to shoots through xylem loading. Unfortunately, there has not been much research done on the transporters that cause the Mg to transport from the root to the shoot. According to Shaul et al. (1999), an *Arabidopsis* gene *MHX*, which codes for a Mg^{2+}/H^+ exchanger, is highly expressed in vascular tissues. This finding suggests that the gene may be involved in either the loading or the retrieval of Mg from the xylem. *MHX* has been identified as being located at the tonoplast (Figure 14.2), but its function related to the physiology of plants remains unidentified. Consequently, to ascertain whether the *MHX* gene plays any involvement in Mg transportation and its homeostasis, the functional studies of *MHX* knockout mutants might be given priority. Additionally, two *CorA*-like homologs that might be involved in the loading of Mg into the xylem are *OsMGT2* and *OsMGT6* in rice and *OsMGT9* in *Arabidopsis* (Gebert et al. 2009). These homologs are largely expressed within the root maturation zones as well as root vascular tissues. When Mg is translocated from the roots to the shoots, it is spread across the various tissues giving preference to the growing parts of plants (Yamaji and Ma 2014). Nevertheless, it is unknown which transporter distributes Mg in shoots. Because being a mobile element in plants, Mg deficiency also causes Mg to redistribute across leaves necessitating the requirement of phloem-targeted Mg transporters, which are still unknown. As Mg is a necessary requirement for photosynthesis, plants preferentially transfer leaf Mg into chloroplasts for chlorophyll production where it also facilitates the fixation of CO_2. *AtMGT10* is abundant in leaf chloroplast thus locating at chloroplast envelope membrane (Figure 14.2; Drummond et al. 2006). The expression patterns of *AtMGt10* exhibit a circadian rhythm, indicating its role in regulating the activity of photosynthetic enzymes within chloroplast stroma, a site where diurnal Mg homeostasis occurs. *OsMGT3* and *OsMGT9* have been found to be localized in the chloroplast of rice (Saito et al. 2013); however, no roles for these genes have been observed in plants to date.

Following their entry into plant cells, the majority of minerals are allocated into vacuoles for subcellular compartmentalization (Ward, Mäser, and Schroeder 2009). It has been suggested that the movement of magnesium into vacuoles stores extra magnesium and maintains osmotic potential (Shaul et al. 1999). Mesophyll cells have been found to exhibit the greatest vacuolar Mg contents, up to 80 mM (Conn et al. 2011a). In *Arabidopsis*, elevated vacuolar Mg accumulation is facilitated by two transporters *AtMGT2* and *AtMGT3* located at the mesophyll and tonoplast of the plant cell, respectively (Figure 14.2). Mesophyll-specific vacuolar Mg accumulation can be reduced by the knockout of anyone *AtMGT2* or *AtMGT3* from both. In low Mg conditions, a significant reduction in the growth and development of plants was observed probably due to the deficiency of Mg because *MGT1, MGT2*, and *MGT3* were all knocked out from those lines (Lenz et al. 2013). Conn et al. (2011a) found a correlation between Mg levels and expressions of these genes within the leaves of several *Arabidopsis* ecotypes, indicating the significance of these genes in preserving the homeostasis of Mg mesophyll cells of the leaf. Apart from *MGTs, MHX*, which is also restricted to the tonoplast, could be an additional route by which Mg can enter into vacuoles of plant cells (Shaul et al. 1999). As *MHX* transports Mg ion (Mg^{2+}), zinc ion (Zn^{2+}), and hydrogen (H^+), it is more likely that it contributes to general ion balance and pH equilibrium than that it stores Mg specifically.

Male fertility and pollen formation have been shown to be two physiological functions of Mg in recent years. Pollen formation in *Arabidopsis* was significantly inhibited after the disruption of Mg transport by *ATMGT4, ATMGT5*, and *ATMGT9* (Chen et al. 2009; Li et al. 2015; Xu et al. 2015). Microspores and bicellular pollen exhibit significant expression levels for both *AtMGT5* and *AtMGT9*. When these genes' expression was disrupted, there was not enough Mg produced for pollen mitosis and intine production, which reduced the quantity of mature pollen grains (Chen et al. 2009; Xu et al. 2015). The *AtMGT5* and *AtMGT9* genes may work in coordination to transfer Mg from the tapetum to the microspore because they are both plasma membrane-localized Mg transporters. *AtMGT4* is another member of the MGT family primarily expressed in pollen grains from bicellular to the stage of mature pollens and is found in the endoplasmic reticulum. The precise processes connecting this gene for the production of pollen grains are yet unknown, despite the fact that *AtMGT4*-RNAi transgenic lines produce pollens with roughly 50% sterility (Li et al. 2015).

14.5 Magnesium's Functions in the Human Body and the Health Issues Caused by Low Magnesium Intake

A lack of Mg in the body can usually be caused by inadequate consumption, insufficient absorption, or excessive excretion. Mg deficiency has been related to a broad range of human disorders, such as metabolic and cardiovascular diseases, skeletal disorders, respiratory illnesses, and brain abnormalities (stress, sadness, and anxiety) (Figure 14.3). Mg aids in the mineralization of bones and is essential for bone health. Additionally, it helps to synthesize active forms of vitamin D (Risco and Traba 1994), which helps the intestines absorb calcium and phosphate (Brown et al. 2015). It has been shown that consuming Mg can lower the mortality risk linked with hypovitaminosis D (Swaminathan 2003). Results from the National Health and Nutrition Examination Survey (NHANES) showed that increased Mg consumption lowered the incidence of vitamin D deficiency general population (Deng et al. 2013). Consequently, consuming enough Mg or Mg-rich foods could help minimize the need for vitamin D supplements since Mg activates and stimulates the body's own endogenous vitamin D, which is

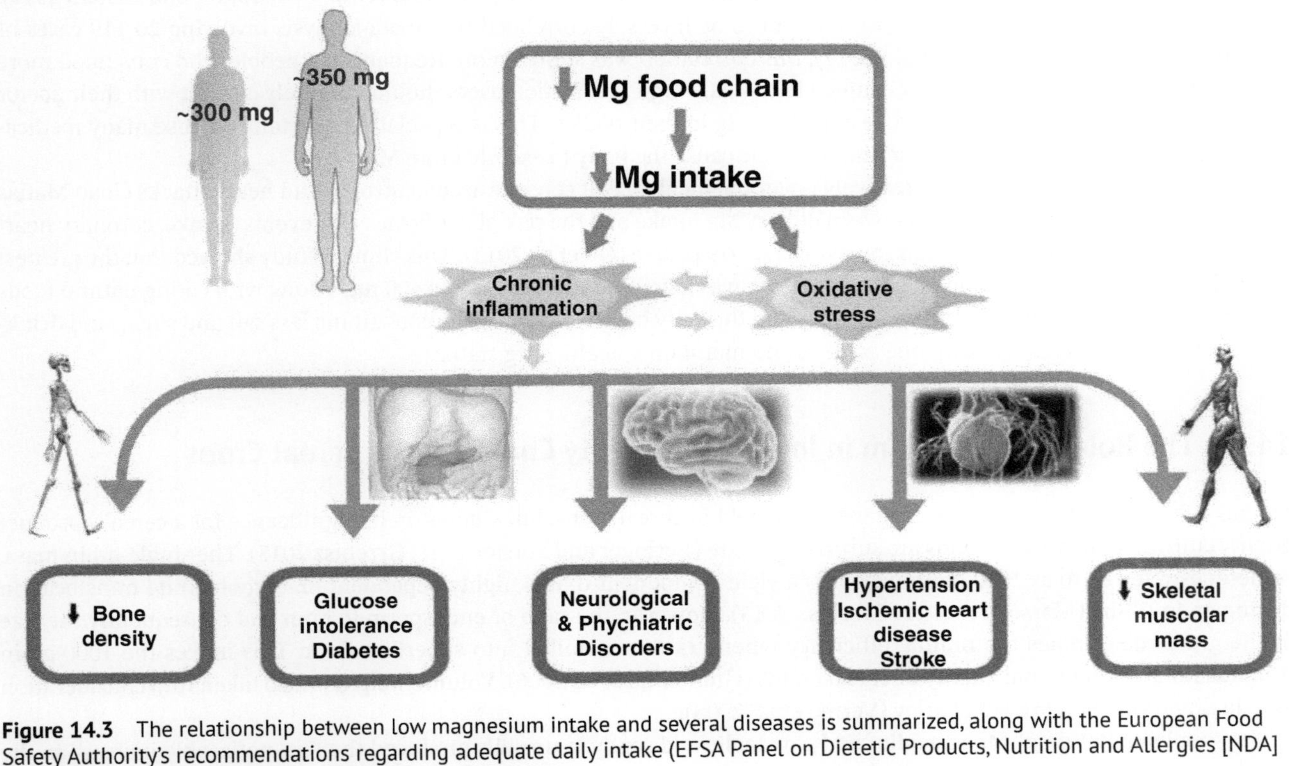

Figure 14.3 The relationship between low magnesium intake and several diseases is summarized, along with the European Food Safety Authority's recommendations regarding adequate daily intake (EFSA Panel on Dietetic Products, Nutrition and Allergies [NDA] 2015) (Cazzola et al. 2020 / with permission of ELSEVIER).

already present and helps the body perform its activities (Uwitonze and Razzaque 2018). This is advantageous for people who are dependent on taking vitamin D pills because those supplements are not always risk-free, especially if one uses them frequently or at high dosages (Razzaque 2018). From a clinical perspective, the symptoms linked to high vitamin D intake are similar to those seen in situations of Mg deficiency. Muscle cramps and spasms are typically caused by elevated calcium levels in comparison to Mg levels; in these cases, the dosage of the vitamin D supplement may need to be readjusted.

Mg, which is irregularly distributed throughout the adult human body, normally weighs 20–28 g. Of the total Mg, it is shared by bones (60–65%), muscles' internal cells (27%) other cells (6–7%), and extracellular space (< 1%, serum and red blood cells) (Grzebisz 2011). The amount of Mg that is available, or its capacity to be stored in the human body, depends on a number of factors, including the concentration of Mg in the food that is ingested. Mg content in body is reduced when foods that have high fiber, fat, sugar and/or sodium content are consumed in the diets (Grzebisz 2011). Urine excretes high concentrations of Mg that the body cannot not retain (Vormann 2003). About 90% of Mg's total content is found in bound structures within human cells. The rest is present in ionic forms, mainly as Mg^{2+}. These bonds are particularly important because they form their bonds with proteins, negatively charged phospholipids, ATP, and nucleic acids. The divalent Mg cation is involved in thousands of processes within the human body and directly activates about 350 enzymes. Mg plays a key role in the biophysical activities that require energy in the body by producing, storing, and using ATP molecules. Furthermore, Mg controls two additional important biological processes: (i) protein synthesis; and (ii) transmembrane transfer of calcium, sodium, and potassium (Fawcett, Haxby, and Male 1999).

Research indicates that dietary Mg intake and bone mineral density are positively correlated with each other (Orchard et al. 2014). Aydin et al. (2010) reported that biochemical characteristics of osteoporotic postmenopausal women who took oral Mg citrate for 30 days showed a decrease in bone turnover. It is believed that Mg can interact with fluoride, a common environmental element, to generate magnesium fluoride, or sellaite. This substance may take the place of Mg in bones and cartilage, which could make the skeleton more prone to fractures.

Furthermore, a healthy Mg level stabilizes the heart's rhythm, which is necessary for regular cardiac activities. The heart has been shown to contain significantly more Mg than other muscles. Nonetheless, the heart's content of Mg^{2+} falls and its functions are somewhat disrupted when there are slight but unfavorable variations in the concentration of this mineral in the extracellular and/or intracellular spaces. A lack of Mg causes a number of heart problems, including arrhythmias, hypertension, arteriosclerosis, blood clot formation in blood arteries, and ultimately myocardial infarction (Figure 14.3; Delva 2003; Touyz 2003). It is also thought that Mg helps to maintain blood pressure (Grober, Schmidt, and Kisters 2015). Dietary Mg uptake and hypertension were found to be inversely correlated in a meta-analysis involving 20 119 cases of hypertension. According to Han et al. (2017), this correlation was shown more frequently in people who consumed more Mg (> 300 mg) than in people who consumed less (< 200 mg). Medication users should definitely consult with their doctor to find out whether their drugs could be depleting Mg in their bodies. This is especially important because many medications that treat high blood pressure or diuretics may cause the body to excrete more Mg.

Research indicates that Mg can effectively lower the incidence of cerebrovascular strokes and heart attacks (Joao Matias et al. 2014). A negative relationship between dietary Mg intake and the risk of cardiovascular events (stroke, coronary heart disease) was found in a meta-analysis involving 532 979 people (Qu et al. 2013). This clinical study showed that the greatest risk reduction was observed when the dosage of Mg was increased from 150 mg to 400 mg. Along with eating natural foods high in Mg, one should also minimize Mg loss from the body by avoiding alcohol, consuming less salt and sugar, and drinking coffee and soda. This will help one achieve and maintain a healthy Mg status.

14.6 The Role of Magnesium in Improving Quality Characters of Cereal Crops

Mg has a favorable effect for increasing the grain yield of cereal crops, but studies on its significance for a cereal's product quality cultivated in farmers' fields are surprisingly rare (Beringer and Forster 1981; Grzebisz 2013). The single-grain mass, sometimes referred to as 1000-grain weight is a yield component that is highly dependent on carbohydrate translocation during grain filling (Marschner 2012; Grzebisz 2013). However, the ratio of endosperm to bran and consequently the size of the grains determines the milling efficiency when grains are milled into superfine flour. This makes the 1000-grain weight a notable functional quality of cereal grains (Greffeuille et al. 2006). Volume weight is also taken into consideration as a quality trait when brewing barley (Verma et al. 2008).

The supplementation of Mg to acidic soddy–podzolic and sandy loam soils produced the first noticeable increase in the starch concentration of rye grains (Magnitskii, Sokolova, and Zhukov 1970). Furthermore, the 1000-grain weight was increased when winter wheat (*Triticum aestivum* L.) was fertilized with $MgSo_4$ on Mg-deficient soil (Al'shevskii and

Derebon 1982). Moreover, when the Mg concentration of the seed was increased up to 0.1% in the grain, the weight of 1000-grains was also enhanced in a pot-grown barley (Beringer and Forster 1981). A greater accumulation of phytates was also observed in conjunction with this increase in grain weight. This accumulation of phytate was especially apparent at Mg concentrations higher than those required for maximum yield. Moreover, the detrimental effects of phytates on the intestinal absorption of trace minerals and Mg have been documented (Bohn et al. 2004; Gibson et al. 2010). A different study indicates that dietary phytates may have advantages, including decreasing the levels of cholesterol and blood sugar besides also having antioxidative and anticancer qualities (Schlemmer et al. 2009).

Mg may also improve the quality of cereal crops simply by enhancing the nutritional status of crop plants. The importance of Mg in N nutrition was observed (Ding, Luo, and Xu 2006). It is possible to speculate that better N metabolism leads to better assimilation of N to a greater status of Mg (Grzebisz 2013). The content of raw gluten and crude protein in grains was enhanced by the application of MgSO4 (Al'shevskii and Derebon 1982). Magnitskii et al. (1970) also showed that adding Mg to acidic soddy-podzolic, sandy-loamy soils increased the protein concentration of rye grain. The results are in agreement with Fecenko and Francakova, (1980), who studied barley plants in a pot culture experiment. In these studies, Mg fertilization decreased the amount of soluble N within the total N concentration in grains, especially when given in the form of Mg sulfate, but it had no pronounced effect on the total N concentration. Additionally, further research has focused on investigating the effects of Mg combined with N supplementation for the yield as well as quality attributes of winter wheat grown in soil that is Mg-deficient but rich in potassium (Chwil 2001 2009). Mg sulfate was responsible for the increased amount of crude protein and gluten in the grain when the total N rates were taken into account. While increased N supplies positively affected grain yield and protein concentration, achieving higher yields required Mg to be applied at the highest rates consistent with the corresponding N levels.

14.6.1 Significance of Cereal Grains to Living Beings and Their Nutritional Composition

Despite regional and socioeconomic differences, cereal crops with their grain products have consistently remained fundamental in the human diet since the beginning of agriculture. This significance is underscored by the widespread cultivation of cereal crops, which occupy approximately half of the world's harvested land area (FAO 2023). Given their yearly production of more than 2800 million tons (FAO 2024), they significantly contribute to global food security, which highlights the vital significance of cereals and cereal products. A significant supply of macronutrients, mostly in the form of carbs (about 75%), proteins (ranging from 5% to 15%), and fats (1% to 5%), as well as a variety of micronutrients like vitamins and minerals, cereals and cereal-based products supply about 75% of the world's population's daily energy and nutritional needs (WHO 2003). The major grain crops and their nutritional significance are listed in Table 14.2. The energy content of cereals is 10 000–15 000 kJ kg^{-1}, or around 15–20 times greater than vegetables and fruits. Additionally, cereals have a major contribution of bioactive compounds and eatable fibers, which have been shown to improve health when consumed as whole grains (Hall, Hillen, and Garden-Robinson 2017).

Table 14.2 Nutritive content of major cereals.

Cereal	Protein (g)	Fat (g)	Carbohydrates (g)	Minerals (g)	Calcium (g)	Fiber (g)	Energy (kcal)
Wheat (whole)	11.9	1.5	71.2	1.5	41	1.2	346
Rice (raw)	6.8	0.5	78.2	0.6	10	0.2	345
Finger millet	12.3	1.3	72	2.7	344	3.6	328
Foxtail millet	12.3	2.7	66	2.1	31	8	365
Pearl millet	11.6	5.0	67.5	2.3	42	1.2	361
Barley	13.6	1.2	74	1.5	26	2.4	352
Sorghum	10.4	1.9	72.6	1.6	25	1.6	349
Amaranth	14	7.0	65	1.5	37	2.7	371
Maize	9.2	4.7	72	1.9	1.2	7.3	365
Oat	16.9	6.9	66.3	1.8	4.0	10.4	389

Source: Yadav et al., 2023 / Springer Nature.

In order to improve human health, standard dietary guidelines have been developed with an emphasis on increasing the percentage of whole grains within their diets (US Department of Agriculture and US Department of Health and Human Services 2015; Dietitians Association of Australia). The crucial element for the well-being of the host is the prebiotic activity of whole grains for gut microflora identified for a variety of cereals containing minerals, vitamins, essential oils, carbohydrates, lipids, proteins, and fibers when consumed in their natural or unprocessed state. However, value-added and commercial post-harvest processing operations, like dehusking, dehulling, and polishing, frequently result in the loss of vital minerals and oils. Cereals are frequently offered for human consumption as processed food additives or in their natural state. Cereals are mostly consumed by poultry and animals as animal feed, and subsequently, poultry products, dairy products, and meat are consumed by humans to get health benefits from both plant and animal sources. Additionally, cereals have multifaceted commercial goals in the paper manufacturing industry, adhesives, oils, textiles, laundry preparations, and cosmetics (Rosentrater and Evers 2018).

14.6.2 The Present Status of Mg in Cereal Grain Composition

The type of grain, the state of the soil, and agricultural techniques are some of the variables that affect the Mg concentration of grains in cereal crops. The reason whole grains usually have higher Mg content than processed grains is that they retain the mineral-rich outer bran and germ layers. For instance, data from the USDA National Nutrient Database reveals that cooked brown rice typically yields around 21 mg of Mg per 100 g, while cooked white rice contains approximately 9 mg per 100 g. Similarly, whole wheat flour boasts about 138 mg of Mg per 100 g, whereas refined wheat flour registers at approximately 29 mg per 100 g. Raw oats contain 138 mg of Mg per 100 g, while oatmeal cooked from these oats provides 29 mg of Mg per 100 g. Raw barley stands out with the highest Mg content at 158 mg per 100 g, but this value decreases to only 33 mg in cooked barley. Corn, whether raw or cooked, consistently maintains a Mg content of 37 mg per 100 g (Figure 14.4).

Depending on certain cultivars, growing circumstances, and processing techniques, these values may differ slightly. Still, whole grains are a great source of Mg in the diet, helping the body meet its overall nutritional needs and facilitating a number of physiological processes.

14.7 Biofortification Techniques for Enhancing Mineral Nutrition in Crop Plants

Biofortification is a viable method for supplementing crops with essential nutrients to communities without having access to a wide range of food options. It is an economical, sustainable, and environmentally friendly technique. Biofortification is a technique used to improve the nutritional value of vital food crops by increasing the density of essential elements such as vitamins, minerals, and amino acids (Bouis and Saltzman 2017). It guarantees the supply of food crops rich in nutrients and promotes their sustainable production, particularly for the impoverished populations of developing countries. Because biofortified crops have higher bioavailability of essential minerals, they are a powerful tool to increase food accessibility using conventional farming and food trade methods. It can be a practical means of supplying a diverse range of foods to

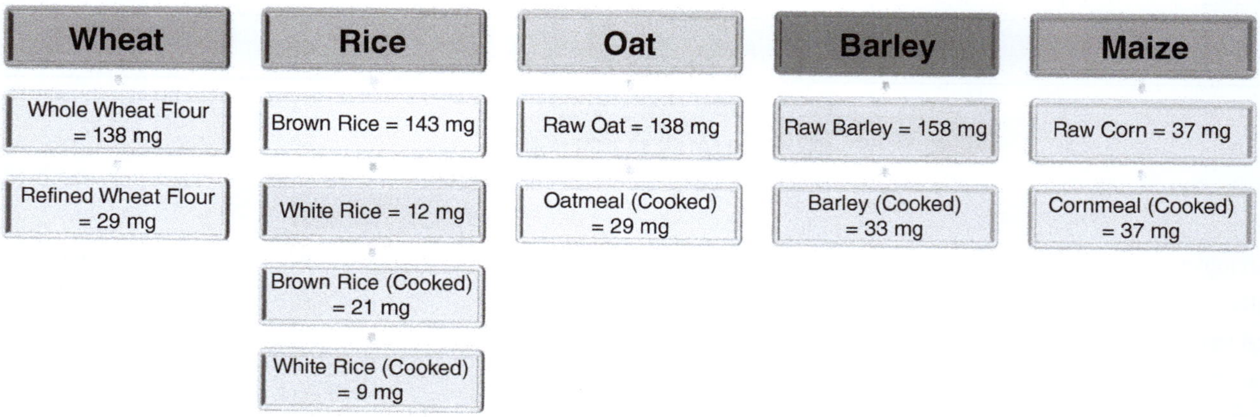

Figure 14.4 Current status of magnesium in cereal grains (Haytowitz et al., 2019 / US Department of Agriculture).

undernourished and low-income populations. According to an economic point of view, developing biofortified crops is an investment that pays off because once established, these crops do not need to be supplemented again and can be easily incorporated into food resources during processing (Pfeiffer and McClafferty 2007).

The major goals of biofortification are to enhance food quantities of mineral nutrients such as iron, zinc, copper, calcium, Mg, and Iodine and to lower the amounts of antinutrients that hinder human absorption of these nutrients. The bioavailability and solubility of minerals are influenced by the presence of specific phenolic compounds and toxins, as well as by the oxidation state of all dietary matrices (Etcheverry, Grusak, and Fleige 2012). As the principle suppliers of minerals, food crops play a vital role in maintaining optimal levels of important vitamins and mineral plant based diets which in turn improves the health of humans (Welch and Graham 2004). In recent decades, problems with mineral deficiencies have been resolved through biofortification of food crops that supply enormous amount of these elements to human population (Tan et al. 2017). Since cereal crops are staple food sources, biofortification is a feasible and safe way to increase their nutritional value, guaranteeing that people from whole world have access to a healthy and balanced diet. In the current global scenario, reducing malnutrition among children and people from all over the world is one of the major issues of today's world. By lowering hunger and malnutrition, biofortification of cereal crops has the potential to solve global food insecurity problems. Its primary goal is to lower the number of diseases and death rates linked to micronutrient deficiencies in staple diets. This strategy raises food productivity while also improving life quality for vulnerable communities.

In order to combat malnutrition, a variety of direct and indirect methods are employed to raise the concentration of vital nutrients in grains and other plant parts (Garg et al. 2018). Malnutrition is combated via taking a diversity of diets, exogenous supplementation, and biofortification. However, biofortification is a practical, affordable, and long-term way to lessen nutrient deficiencies. All of the important vitamins, minerals, and fatty acids needed by an individual to meet their nutritional needs are present in biofortified crops (Sanjeeva et al. 2020; Singh, Govindan, and Andersson 2017; Tan et al. 2018). Additionally, biofortified crops are sustainable and eco-friendly since they increase soil health and nutrient uptake (Waters and Sankaran 2011; Singla and Grover 2017). Enhancing the nutritional composition of food grains by means of agronomic techniques, conventional breeding procedures, and modern breeding technology is known as biofortification (Figure 14.5).

14.7.1 Agronomic Biofortification Strategies to Enhance Magnesium Content in Crop Plants

Two common agronomic techniques to raise the level of minerals in eatable parts are either by supplementing crops with exogenous fertilizers or optimizing solubility as well as mobility of minerals present in the soil. Soluble inorganic fertilizers are applied specifically to the roots or leaves of crops growing in areas where the soil lacks essential mineral elements. Foliar treatments of liquid fertilizers are used in cases when the plants are unable to uptake and transport minerals to their edible parts. It has been noted that low-input agriculture has become unable to feed the increasing global population. So, factory-made exogenous fertilizers have become obligatory for achieving higher yield outputs of crops to avoid famine (Graham et al. 2007). Therefore, the recommendations are that any forthcoming strategy for food security should integrate the utilization of synthetic fertilizers in the future. Before inorganic fertilizers are distributed, it may be feasible to add mineral elements that are necessary for human nutrition if their broad usage is encouraged.

The availability of Mg to crops in agricultural production systems is impacted by various factors such as type and texture of soil, cation exchange capacity, site-specific meteorological and anthropogenic conditions, agronomic management techniques, and the crop species itself (Mikkelsen 2010). The bioavailability of Mg in sufficient amounts within the soils promotes vigorous crop growth and improves yields. Mg-deficient soils significantly hinder the uptake of Mg by crop roots. This is primarily due to the small amount of Mg in source rocks, loss of Mg by leaching and mobilization, depletion of Mg as a result of intensive farming, and competition between different cations (Wang et al. 2020). Soil acidity is another significant component that affects agricultural productivity (Mohebbi and Mahler 1989). It is directly linked to nutrient deficiencies, including K, Ca, Mn, P, and Zn, while the availability of Mg is hampered by the toxicity of Al and Mn (Nguyen et al. 2018; Zhu, Jiang, and Ji 2004). Furthermore, because Mg^{2+} is very mobile, excessive rains may cause it to be released from the root zone (Gransee and Führs 2013), especially in acidic soils. This would lower crop output and nutrient utilization efficiency.

14.7.1.1 Inorganic Mg Fertilizers and Foliar Applications

Agronomic biofortification applies Mg fertilizers, which are often delivered as oxide, carbonate, or sulfate salts, in order to maintain high concentrations of Mg while promoting plant growth (Draycott and Allison 1998). Recently, a Mg fertilizer

Figure 14.5 Various techniques for crop biofortification. These can be broadly categorized into three categories: (1) agronomic practices; (2) conventional breeding methods, and (3) modern breeding technologies. (1) Agronomic practices: Nutrient-enriched seed priming for crop biofortification. Nutrient-rich seed priming increases crop nutritional quality by giving vital nutrients at the germination stage, ensuring strong growth and better nutrient uptake throughout the life of plants. This method increases crop resilience and helps produce foods that are higher in nutrients. Foliar application of nutrients to the crop. Consumers will ultimately receive healthier foods when liquid nutrients are applied foliarly to the aerial organs of plants, where they are absorbed through the stomata and the epidermis in the leaves. This improves nutrient absorption in reproductive organs and integrates them into the food chain. Enhancing crop biofortification by flooding-based mineral fertilization. Minerals such as selenium, zinc, iron, calcium, magnesium, and others are applied through irrigation. So, they mix with water and are easily absorbed by plant roots, leading to a rise in the accumulation of these minerals in plant parts that are edible. Biofortification of crops through soil-applied mineral fertilization. By applying mineral fertilizers – typically NPK – to the soil either prior to or during sowing, with the help of different seed and fertilizer drills can improve the biofortification of crops. This process helps plant roots absorb nutrients and incorporate them into the food chain. Using microbe-mediated mechanisms to increase mineral uptake for crop biofortification: increased nutrient absorption via microbial-mediated mechanism is achieved when various microbial species such as mycorrhizal fungi and rhizobium bacteria form the mutual beneficial interactions with plants to promote nutrient acquisition. (2) Conventional breeding: It involves crop biofortification by crossing two parents with dissimilar phenotypes and making a selection in future segregating generations according to the trait of interest (mainly high in mineral uptake) can increase the mineral quantity in staple crops. (3) New breeding techniques: Knocking out of genes accountable for producing antinutrient substances. The knocking out of the genes that produce antinutrient elements in plants is a novel breeding technique. The bioavailability of essential elements is limited by a variety of antinutrients, including tannins, lectins, phytic acid, saponins, lathyrogens, protease inhibitors, and alpha-amylase (a-amylase) inhibitors. These factors contribute to the production of malnourished crops. Reduced accumulation of these chemicals can be achieved by suppressing genes linked to antinutrient production through the use of RNA interference (RNAi). Overexpression of genes promotes plant nutrient accumulation for biofortification. Increasing the expression of genes that cause plants to accumulate macro- and micronutrients facilitates mineral biofortification, which causes higher nutrient accumulation and increases its deposition in plant portions that are edible. Crop biofortification improved via gene transfer for enhanced nutritional quality. Crop biofortification through the use of genes from other species improves nutritional quality and reduces malnutrition by allowing different genes to cross species boundaries, encouraging the production of balanced diets, and eventually helping to prevent diseases associated with malnutrition. This technique does have certain drawbacks, though, such as being costly, time-consuming, and labor-intensive. In addition, because of regulatory concerns, it is not as widely accepted by the population (Shahzad et al., 2021 / Frontiers Media S.A).

MgNH4(PO4) ·6H2O has gained much attention because it also provides phosphorus (P) for use in agriculture (Parsons and Smith 2008). Moreover, $MgSO_4$ offers easily accessible Mg^{2+}, while MgO is a fertilizer that slowly releases Mg ions (White and Broadley 2009). Fertilizers containing Mg are either applied on the soil's surface or incorporated into the subsoil if they have low solubility (Draycott and Allison 1998). Applying $MgSO_4$ is frequent in some crops, and it has been shown to enhance Mg content and correct its deficiency when applied as foliar treatment (Jezek et al. 2015). Tomato yield (7.7–17.9 t ha^{-1}) (Kashinath et al. 2013), barley grain yield (8.6%) (Babaeian et al. 2012), hazelnut fruit yield (51%), and total oil content (4.8%) were all been observed to increase with Mg fertilization (Özenç and Özenç 2015). Furthermore, optimum Mg levels were found to increase the yield of a variety of crops in a meta-analysis (Wang et al. 2020). These Mg levels with specific crops include 94.1 kg ha^{-1} for fruit, 46.9 kg ha^{-1} for grass, 54.1 kg ha^{-1} for tobacco, 58.3 kg ha^{-1} for tuber, 43.5 kg ha^{-1} for vegetable, 27.8 kg ha^{-1} for cereal, 47.2 kg ha^{-1} for oil crop, 34.1 kg ha^{-1} for tea, and 76.8 kg ha^{-1} for other crops. MgO was used as a source. Oury et al. (2006) stated supplementing Mg by fertilizers raised the amounts of Mg in plant tissues thereby creating a strong positive correlation between Mg^{2+} in the soil solution and Mg concentrations in produce. Therefore, it may be possible to manage and remedy Mg deficiency by supplying Mg^{2+} through soil-based fertilizers and foliar sprays.

14.7.1.2 Enhancing Soil Magnesium Uptake via Agronomic Approaches

Increased tissue concentration and crop production can be achieved by crop rotations and intercropping crops which possess superior abilities to accumulate and mobilize Mg, especially when dealing with soil conditions where Mg availability and mobility are constrained (Graham et al. 2007; Inal et al. 2007). According to Welch and Graham, (2005), these plants may include native food crops that are high in nutrients and have the ability to effectively absorb minerals from unfertilized soils. Hence, increasing the diversity of crop rotations to incorporate species that naturally accumulate higher levels of Mg^{2+} in their edible parts could serve as a distinct approach to enhancing Mg intake in the human diet (Graham et al. 2007).

The microorganisms within soils have the ability to maximize soil volume and improve the availability of nutrients required by plants (Lynch 2007; Kirkby and Johnston 2008). Mycorrhizal fungi, which make beneficial interactions with a variety of crops, have the ability to enhance the volume of soil that is used to acquire immobile mineral elements. Moreover,

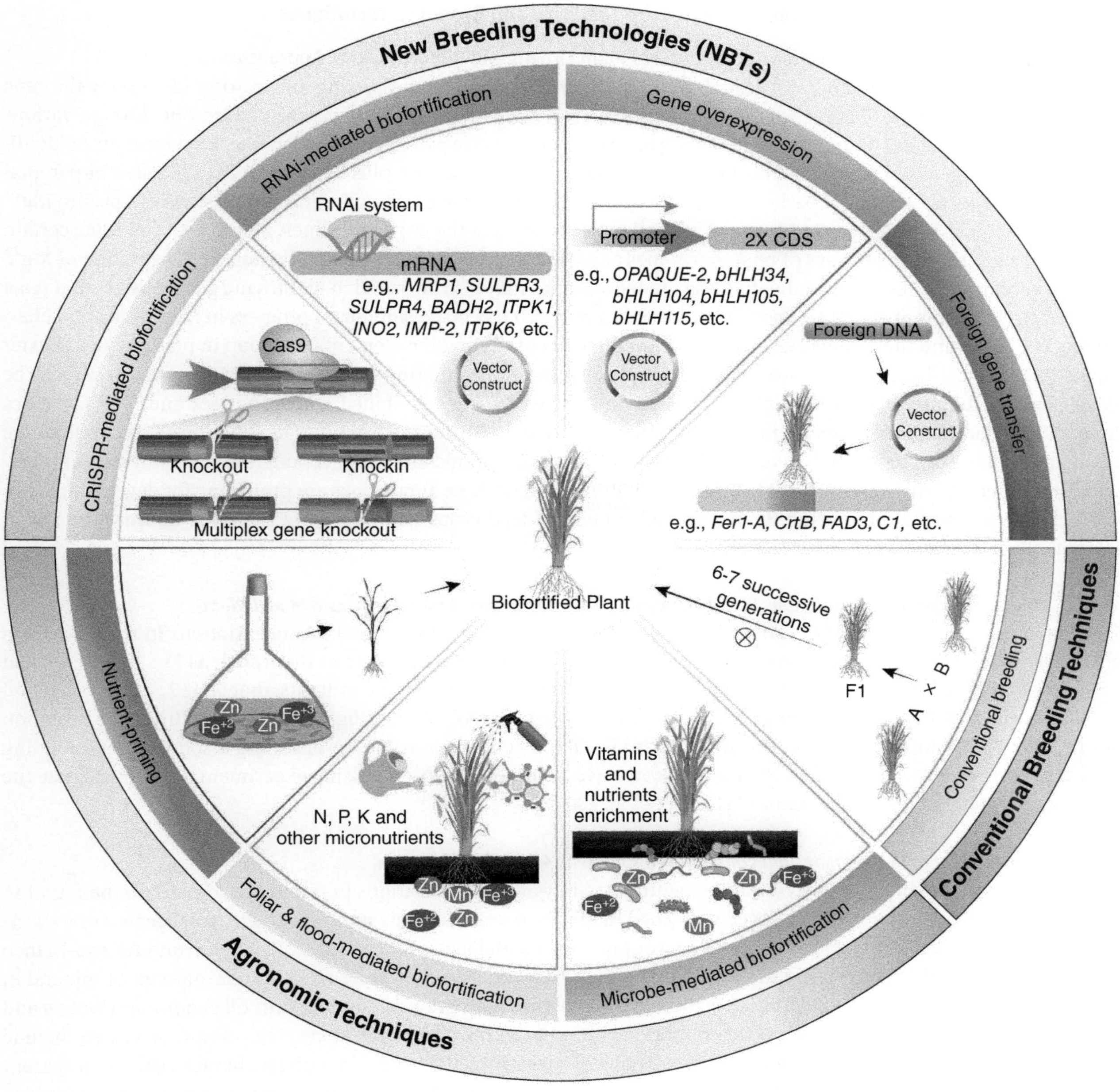

they excrete enzymes, OAs, and siderophores, facilitating the decomposition of organic compounds (Morgan, Bending, and White 2005; Smith and Read 2007). The concentrations of these minerals can be increased in edible produce by managing mycorrhizal fungi in agriculture systems (He and Nara 2007). Also, numerous studies have reported that mycorrhizal associations raise the amounts of Se, Zn, Fe, Mg, and Cu in various crops (Cavagnaro 2008; Harrier and Watson 2003; Larsen et al. 2006). However, such relationships can lower yields in well-fertilized soils since photosynthate from plants fuels the mutual interactions of mycorrhizal fungi and plants (Morgan, Bending, and White 2005; Lynch 2007). In situations where N is limited, symbiotic or associative partnerships with N_2-fixing bacteria have particular significance (Hardarson and Broughton 2003). Lastly, some soil microorganisms that influence the phyto-availability of mineral elements can receive carbon from the exudates of plant roots and mycorrhizal fungi. Therefore, growth-promoting bacterial inoculants can enhance plant growth and yield, tissue mineral concentrations, and the uptake of iron, zinc, copper, and Mg via the roots of plants (Whiting, de Souza, and Terry 2001).

14.7.2 Enhancing Magnesium Levels Through Crop Genetics and Breeding Techniques

14.7.2.1 Genetic Diversity in Magnesium Content Across Plant Organs and Biofortification Approaches

Genetic biofortification may emerge as a desirable and attractive technique in the coming future. After identifying the gene or genes responsible for Mg^{2+} homeostasis in certain plants, genetic biofortification can be carried out through various techniques such as screening mutants for altered Mg profiles, using reverse genetics to develop knockout mutants of identified or unidentified genes, screening plants for natural variations in Mg content, plus developing RILs through bi-parental crosses (Hermans et al. 2013). An early effort by Deng et al. (2006) involved changing the expressions of specific Mg^{2+} transporters in plants, leading to a 30% increase in Mg^{2+} concentration in the transgenic lines. Moreover, by altering certain aspects of the biosynthesis pathways of phytohormones and their responsive genes, the utilization and absorption of Mg^{2+} can also be enhanced. QTLs associated with these phytohormones may also be useful in identifying gene clusters that react to Mg stresses (Hermans et al. 2013). The field of ionomics research has made significant progress in identifying and characterizing the genes and their networks responsible for directly controlling the elemental makeup in plant tissues (Baxter 2010; Salt, Baxter, and Lahner 2008). Identification of genes related to regulating the amount of Mg in plants may also be accomplished by the use of high-throughput mineral analysis technology combined with genomics and bioinformatics techniques. The largest plant ionomics project was initially started at the Purdue Ionomics facility (PiiMS) (www.ionomicshub.org) and was associated with *A. thaliana*. Currently, studies on ionomics are being conducted on several crops, including rice, soybeans, and maize (Simic′ et al. 2012). Findings from all of these approaches are promising for developing additional techniques aimed at enhancing the nutritional content of food crops with Mg. Here is a brief summary of recent developments in the field.

14.7.2.1.1 Mutagenesis-induced Variations in the Magnesium Concentrations (Arabidopsis as a Model Plant)

A list of more than 50 mutants of the *Arabidopsis* plant that were modified for mineral concentrations in the leaves was published by Lahner et al. (2003). When these mutants were grown in the field, some of them such as 71 : 13; 79 : 48; and 89 : 54 showed reduced or increased up to 50% Mg concentrations, together with the mutants, that is, 117 : 20; 132 : 01; and 136 : 31 that elevated the concentrations of lithium (Li) and arsenic (As). According to Tian et al. (2010), the mutation found in 136 : 31 mutant was mapped to *AtNPCC6/AtNaKR1*, a gene responsible for encoding a soluble metal-binding protein exclusively expressed in companion cells. Nevertheless, subsequent backcrossing experiments did not validate the high Mg content phenotype for that mutant (Lahner et al. 2003).

14.7.2.1.2 Natural Variation of Magnesium Content in Arabidopsis Accession

A substantial source of genetic variety for investigating possibly adaptive variations in tissue Mg content can be found in natural populations of *A. thaliana*. The species is a good model for researching adaptation and selection because of its wide geographic spread and resulting exposure to a variety of environmental factors (Koornneef, Reymond, and Alonso-Blanco 2011). It is anticipated that genetic variation will be observed for a variety of attributes related to the amount of mineral in plant tissue and the ability of mineral uptake as those natural populations grow under different soil conditions (Waters and Grusak 2008). Many studies have used different approaches to examine accessions at varying resolutions: these include mapping QTLS related to offspring originating from pairs of accessions that differ in a given characteristic, multiparent mapping techniques (Huang et al. 2011; Kover et al. 2009), and analyses of larger populations under various conditions (Atwell et al. 2010; Conn and Gilliham 2010).

QTL or linkage mapping: RIL populations are used in linkage mapping. They are produced through controlled crosses between two parental lines with different phenotypes. Waters and Grusak, (2008) discovered that three *MRS2* family members (*AtMRS2-2, 2-3,* and *2-11*) were found to co-localize with a QTL associated with seed Mg content in crosses involving *Cvi* × *Ler* and *Col* × *Ler*. Contradict to this finding, the cross between *Cvi* × *Ler* did not report the QTLs associated with the *MRS2* family were not reported. Instead, another study identified *AtCNGC3, 11, 12,* or *ATMHX1* as contributing factors (Vreugdenhil et al. 2004). The former report is noteworthy because it suggests that the three transporters in the plant may work together in root Mg loading to the leaves, redistributing it from the chloroplasts, and eventually storing it in the seeds. This happens because of their nonoverlapping localization. To be more precise, *AtMRS2-2, AtMRS2-3,* and *AtMRS2-11* are located in the vasculature, the seed, and the chloroplast, respectively. These transporters are in charge of providing sufficient Mg for photosynthesis, enriching the Mg present in the leaf, and loading the generated photosynthates within phloem to deliver it to seeds.

Genome-wide association mapping (GWAS): Alleles or loci responsible for the innate variability of particular traits can now be identified using powerful mapping techniques such as genome-wide association or linkage disequilibrium (Atwell et al. 2010). More research into the genetic bases of variations in Mg content is possible now because the 1001 genomes project (www.1001genomes.org) has been launched. This project aims to sequence the genomes of 1001 *Arabidopsis* accessions. Additionally, the mineral profile data for almost half of these accessions is accessible through iHUB (www.ionomicshub.org) and some other public databases.

The assessments of ionomic variation are expanded to include several tissues in other reports. One of the studies revealed a significant difference in Mg content amongst the most contrasted *Arabidopsis* accessions grown hydroponically with roots showing over an 80% variance and shoots showing a 50% variance (Baxter et al. 2012). This work provides modern support for the long-held belief that the control of the plant ionome by genes and gene × environment is extremely important.

Upon further dissecting the 96 accessions Nordborg panel utilized in this study into seeds, roots, and leaves, it becomes evident that there is no consistent correlation in synteny between different elements in one tissue or across different tissues for a particular element. This finding highlights the flexibility in ionome, thus emphasizing the necessity for paying closer consideration to ionomics at higher resolutions. According to research, cell localized ion transporters for Mg and calcium are responsible for the nonuniform accumulation of elements among adjacent cells within the same tissue (Conn et al. 2011a; 2011b; Conn and Gilliham 2010). A framework for correlation was developed based on the transcriptome and ionome of more than 30 *Arabidopsis* accessions. This allowed for the identification of several potential genes that contribute to the accumulation of calcium, Mg, and zinc across all ontologies (Conn et al. 2012). It was demonstrated that more than one-third of potential transporters were disrupted for their targeted element, and a large number of them were expressed only in those cells where the elements were accumulated mostly. *AtMRS2-5*, a vacuolar transporter specifically expressed in mesophyll cells was implicated in Mg homeostasis in leaves and it was one gene that was found to be linked with shoot Mg accumulation (Conn et al. 2011b). The transporter transcript most strongly associated with the accumulation of Mg is *Metal Tolerance Protein 5* (*AtMTP5/AtMTPc2*). Notably, its homolog is more abundant in the zinc hyperaccumulator *Noccaea caerulescens* than in *Thlaspi arvense* (Tian et al. 2010). When the ionome study was conducted for *AtMTPc2* T-DNA insertion line, which was modified especially for Mg only, further supported the observation that despite being Zn hyperaccumulator, *N. caerulescens* showed no response to zinc accumulation. There is a significant opportunity to discover novel targets associated with plant Mg homeostasis, which could be utilized in biofortification strategies. This opportunity arises from integrating findings obtained thus far across various resolutions and environmental conditions.

Polymorphisms: The development of high throughput sequencing has accelerated research into the genomic foundation of plant adaptability to diverse soil types. Turner et al. (2010) studied innate polymorphisms in *Arabidopsis lyrata* species cultivated in granitic soils across two continents, as well as in serpentine soils characterized by calcium and Mg ratio typically exceeding 1 : 10. It was discovered that several cation transporters including those (*MRS2-2, MRS2-6, MRS2-7*) responsible for Mg transport showed polymorphisms and were considered as good contributor of Mg to ecology of serpentine regions. Nevertheless, a large number of additional loci were discovered to have polymorphisms driven by soil type. Those include proteins that bind calcium, kinases, and transcriptional factors that may have a role in post-transitional modifications. Bradshaw (2005) further noted that *A. thaliana*, a low serpentine tolerant species, has an advantage in growing in serpentine conditions due to a mutation in Cation exchanger 1 (*AtCAX1*), which codes for a vacuolar Ca^{2+}/H^+ antiporter. However, given that this is just one of several implicated gene loci reported by Turner et al. (2010). It was observed that neither the *cax1-1* mutant nor *mrs2-1* and nor *mrs2-5* were capable of restoring plant biomass to the level observed in untreated Col-0 wild-type plants (Conn et al. 2011b). It is possible to increase the amount of Mg in cereal crops by investigating the transport of Mg, as demonstrated by Turner et al. (2010), who also provide evidence for the multi-locus contribution to plant adaptation to serpentine soils.

14.7.2.2 Genetic Variations Impacting Mineral Element's Content in Crop's Edible Parts

Using mineral fertilizers to raise the concentrations of important minerals in products may be complemented by breeding efforts aimed at developing crops with improved abilities to absorb as well as retain those minerals in their eatable parts. This is important to note that genotypic improvement is the only strategy to increase a crop's ability to acquire, utilize, or accumulate accessible mineral elements. In the eatable parts of most crop species, there appears to be significant genetic diversity in the concentrations of the mineral nutrients that humans typically lack in their diets.

The majority of studies investigating genetic variation with regard to mineral concentration have focused on carbohydrate-rich portions of staple food crops. They include rice (*Oryza sativa*), wheat (*Triticum spp.*), and maize (*Zea mays*) grains; in addition, they include the seeds of common beans (*Phaseolus vulgaris*), sweet potato tubers (*Ipomoea batatas*), and the underground parts of yam (*Dioscorea alata*) and cassava (*Manihot esculenta*) (Grusak and Cakmak 2005; Pfeiffer and McClafferty 2007; White and Broadley 2005). Research has also been done on genetic diversity with respect to nutrient levels in the eatable parts of a variety of commonly consumed crops that make a substantial contribution to the human diet. These crops include peas, soybean, lentils, pigeon pea, cowpea, groundnut, chickpea and peanut (legume crops), barley, millet sorghum (cereal), tomato and banana (fruits), and cabbage, cauliflower, broccoli, spinach, carrot, potato (vegetable) (Grusak and Cakmak 2005; Johns and Eyzaguirre 2007; Pfeiffer and McClafferty 2007; White and Broadley 2005). These findings imply that domesticated species possess high genetic potential to be used in breeding initiatives to control the tissue concentrations of all the mineral components that are deficient in human diets.

Most cereal species have different mineral element concentrations in their grains due to genetic variation. Many cereals' seeds are frequently eaten after being ground or polished, it is important to think about how genetic diversity in mineral element distribution within seeds might be used to biofortify crops. Elevated concentrations of most mineral elements are found in the husk and/or aleurone layers of the seed; these concentrations are not evenly distributed throughout the seed. Cereal grains that have been ground or polished may lose a significant amount of these mineral components from the diet; the amount lost will vary on the crop and its varieties (Lyons et al. 2005; Prom-u-thai et al. 2007). Grain morphological traits, that is, the size of grains, the size of the embryo, and the quantity and density of tissue layers affect the distribution of mineral elements including iron, zinc, and Mg inside cereal grains (Cakmak et al. 2004; Oikeh et al. 2003; Welch et al. 1993). For instance, there is a cultivar-dependent relationship between the amounts of iron, zinc, cu, Mg, and calcium and how many aleurone cell layers in maize, rice, and barley are present (Prom-u-thai et al. 2007; Welch et al. 1993).

14.7.2.3 Genetic Variation in Promoter and Antinutrient Levels in Edible Parts of Cultivated Crops

The availability of essential mineral components is largely dependent on the presence of substances in a meal that promote or inhibit their absorption in the digestive system (Bohn, Meyer, and Rasmussen 2008). The primary compounds that are known to impede the absorption of Fe, Zn, Ca, and Mg are polyphenolics found in tea, coffee, beans, and sorghum, as well as phytate found within seeds of both cereal and legume plants. According to Roberfroid, (2007), inulin, a fructose polysaccharide that frequently has a terminal glucose unit and is found in considerable quantities in a variety of edible crops, facilitates the absorption of calcium and Mg. Different crops, such as cereal grains, legumes, and vegetables, have different levels of phytate (Hotz and Brown 2004; Reddy 2002; Vreugdenhil, Aarts, and Koornneef 2005). Research has indicated significant genetic variation within species with respect to the amount of phytate present in grains. This variation has been noted in wheat (Welch et al. 2005), barley (Dai et al. 2007), pearl millet (Abdalla et al. 1998), oats (Lolas, Palamidis, and Markakis 1976), triticale (Feil and Fossati 1997), sorghum (Slingerland et al. 2006), and rice (Glahn, Cheng, and Welch 2002; Welch et al. 2000).

Furthermore, low phytic acid (lpa) mutants have been reported in wheat, barley, maize, rice, and soybean (Bohn, Meyer, and Rasmussen 2008; Raboy 2007). They have been shown to reduce seed phytate concentrations, but their total P concentrations are comparable to those of normal types. Their genetic variability regarding phytic acid concentration in these mutants is either naturally occurring or artificially induced. It is common to observe two different types of lpa mutants. One kind of seed shows no discernible rise in the accumulation of inositol phosphates (IPS). However, the seeds of the other type show higher levels of IP3, IP4, and IP5 accumulation (Raboy 2007). Replacing conventional varieties with lpa mutants in diets improves the mineral nutrition of monogastric animals, including humans (Adams et al. 2002; Hambidge et al. 2005; Mazariegos et al. 2006). However, the concentrations and distributions of mineral elements within the seed appear to be less affected by the lpa mutations (Bryant et al. 2005; Guttieri, Peterson, and Souza 2006; Liu, Petersonn and Raboy 2007). This advantageous impact is especially seen in persons with inadequate dietary mineral intake. A number of QTLs that influence the amount of phytate in rice seeds have been identified and can be used as molecular markers in breeding initiatives to biofortify crops (Stangoulis et al. 2007). Weak correlations have been found between phytate and the concentrations of iron, zinc, calcium, or Mg in sorghum grain (Reddy, Ramesh, and Longvah 2005), soybean seeds (Raboy, Dickinson and Below 1984), and common bean seeds (Cichy et al. 2005). Additionally, there has been no overlap between quantitative trait loci (QTLs) influencing seed phytate and those influencing concentrations of iron, zinc, or Mg (Cichy et al. 2005; Stangoulis et al. 2007; Waters and Grusak 2008; Vreugdenhil et al. 2004). These findings suggest that different genes are responsible for these traits. This implies the possibility of producing edible portions with low phytate levels while being rich in iron, zinc, calcium, and/or Mg through breeding efforts.

14.7.2.4 New Breeding Technologies to Modifying Levels of Antinutritive Compounds and Promotor Substances

In order to enhance the bioavailability of Fe, Zn, Ca, and Mg in food, transgenic methods have concentrated on decreasing the levels of antinutrients, like phytate, oxalate, and polyphenolics, and raising the levels of promoter substances, like ascorbate, β-carotene, and cysteine-rich polypeptides, in staple crops (Davies 2007; Kopsell and Kopsell 2006; Lucca, Poletti, and Sautter 2006; Zhu et al. 2007).

To lower the levels of phytate in staple crops, two transgenic approaches have been used (Lönnerdal 2003; Raboy 2007). The initial approach which has been effective in lowering the levels of phytate in rice, maize, and soybean seeds, is to downregulate the expression of genes encoding enzymes involved in the synthesis or sequestration of IP6 (Nunes et al. 2006; Kuwano et al. 2006; Shi et al. 2007). Increasing the expression of phytases in edible tissues is the second strategy. This method has been effective in lowering endogenous phytate concentrations in seeds and/or raising phytase activity in feed derived from alfalfa (Flachowski, Chesson, and Aulrich 2005), canola (Ponstein et al. 2002), rice (Hong et al. 2004), wheat (Brinch-Pedersen et al. 2006), maize (Chen et al. 2008), and soybean (Bilyeu et al. 2008).

A number of crop plants have undergone genetic modification (GM) to increase the amounts of the vitamin A precursor β-carotene in their edible parts. Among these are maize (Aluru et al. 2008) and golden rice (Paine et al. 2005). The two approaches that have been used are either overexpressing carotenogenic transgenes from nuclear or plastid genomes or changing the expression of genes that regulate plastid development. Likewise, methods including GM have been employed to raise ascorbate levels within *Zea mays* kernels (Chen et al. 2003). These types of new breeding technologies can be employed in breeding programs to biofortify cereal crops in order to increase the levels of Mg in edible grains and reduce the concentrations of antinutritive compounds.

14.8 Conclusion

Mg is an important component of plant physiology that regulates photosynthesis and NUE and reduces abiotic stress. Its function in photosynthesis includes the manufacture of chlorophyll and enzyme co-factorship, both of which are essential for maintaining the integrity of the photosystem and electron transport. It also facilitates the uptake and assimilation of N, improving protein content and yield in a range of crops. Mg also reduces Al toxicity, heat stress, and salt stress through a variety of pathways. Mg's complex role in coordinating plant defensive responses is further highlighted by its involvement in plant–pathogen interactions, which may provide new insights into boosting crop resilience against pathogenic attacks.

A Mg deficit impairs the growth and quality of crops, reducing agricultural productivity by impeding photosynthesis and the distribution of carbohydrates to sink organs, which affects biomass and quality. A prevalent symptom of Mg deficiency that varies depending on the plant type and leaf position is leaf interveinal chlorosis. Although Mg toxicity is rare, healthy growth depends on maintaining appropriate concentrations of Mg which are normally within 1.5–3.5 mg per gram of dry weight. A comprehensive understanding of Mg's metabolic role is necessary to improve farming practices and sustain crop yields under different environmental conditions. Various processes are used by plants to regulate the amount of Mg in response to changes in the environment. In times of deficit, *MGT/MRS2* family members promote fast absorption of Mg through the roots, whereas in times of excess, transporters such as *MGT2/MRS2-1* sequester Mg into vacuoles. Internal signaling molecules and interactions with other ions also play crucial roles in Mg homeostasis and tolerance in plants.

Mg plays a vital role in numerous biological activities. And field crops use complex transport systems to obtain, store, and translocate them to plant organs. Special transporter genes such as *AtMGT6* in *Arabidopsis* and *OsMGT1* in rice, together with other transporters like *AtCNGC10* help to enhance the uptake of Mg into the roots. After that, Mg is translocated to aerial parts of plants preferably to growing tissues and organelles like chloroplasts and vacuoles with the help of several transporter families such as *MHX* and *MGT/MRS2*. These transporters are crucial for Mg uptake as well as stress responses, pollen production, and photosynthesis, underscoring the complexity of plant Mg transport systems and their importance for plant growth and environmental change adaptability.

The quality of cereals cultivated in fields benefits greatly from the addition of Mg, an element that is often ignored. Although most studies have concentrated on its beneficial effects on grain yield, studies also indicate that Mg levels have an impact on important quality attributes including single grain mass and volume weight. It has been discovered that adding Mg to soil increases the 1000-grain weight in winter wheat and barley as well as the starch concentration in rye. However, adding too much Mg may cause phytate formation which could hinder the absorption of nutrients. Furthermore, Mg increases the concentrations of gluten and crude protein in grains, boosts the assimilation of N into crops, and improves crop nutrition in an indirect manner. These results demonstrate how Mg plays a variety of roles in improving grain yield and quality across a range of nutrient- and soil-rich environments.

Numerous health problems, including metabolic disorders, heart diseases, and neurological abnormalities like depression and stress, are associated with Mg deficiency in the human body. Its critical function in the mineralization of bone and its effect on the metabolism of vitamin D highlight its importance in maintaining bone health and lowering the risk of hypovitaminosis D-related mortality. Research suggests that consuming more Mg may reduce the likelihood of developing a vitamin D deficiency, offering a possible substitute for taking excessive amounts of vitamin D supplements. Moreover, Mg is involved in blood pressure regulation, heart function regulation, and other cellular activities. Studies have shown that Mg can effectively lower the risk of heart attacks and strokes. Hypomagnesaemia, which is defined by low blood Mg levels, poses major health hazards, such as arrhythmia and type-2 diabetes. In order to prevent these chronic diseases, it is crucial to consume an adequate amount of Mg through diet. Improving the amount of Mg in the diet, particularly in stable foods (cereal crops), can help lower the risk of the diseases listed above that are brought on by Mg deficiency in the body by using biofortification techniques.

Biofortification is a viable strategy to combat malnutrition since it increases essential nutrients in food crops and offers long-term solutions to nutrient shortages worldwide. Agronomic biofortification strategies involve applying Mg fertilizers by soil or foliar sprays play a crucial role in getting sustainable yield with improved nutrition status of crops. Crop growth and productivity are affected by a variety of factors that affect Mg accessibility, including soil quality, climate, and management practices. Crop rotations can also increase the uptake of Mg and the use of soil microbes also encourage soil exploration and plant growth, which in turn increases crop yields and nutritional quality. Genetic biofortification strategies, which identify target genes for Mg^{2+} homeostasis using methods like mutant screens and natural variability screening, show potential for increasing the amount of Mg in plants. Genetic diversity in the Mg content of tissues has been found in studies conducted on model plants such as *Arabidopsis*, where mutant plants display different mineral profiles. Furthermore, natural variations in *Arabidopsis* populations provide insights into adaptive variations in Mg concentration under various environmental conditions. Important genes involved in Mg translocation and accumulation have been found by genome-wide association studies and linkage mapping, offering prospective targets for biofortification initiatives into basic food crops such as wheat, rice, and maize. Moreover, transgenic techniques for manipulating antinutrient concentrations and promoter substances provide ways to correct nutrient deficits in food crops and boost Mg bioavailability.

References

Abdalla, A.A., El Tinay, A.H., Mohamed, B.E. et al. (1998). Proximate composition, starch, phytate and mineral contents of 10 pearl millet genotypes. *Food Chemistry* 63: 243–246.

Adams, C.L., Hambidge, M., Raboy, V. et al. (2002). Zinc absorption from a low-phytic acid maize. *The American Journal of Clinical Nutrition* 76 (3): 556–559.

Ahokas, R.A., Sun, Y., Bhattacharya, S.K. et al. (2005). Aldosteronism and a proinflammatory vascular phenotype: Role of Mg^{2+}, Ca^{2+}, and H_2O_2 in peripheral blood mononuclear cells. *Circulation* 111: 51–57.

Al'shevskii, N.G. and Derebon, Y.G. (1982). Effect of magnesium fertilizers on yield and quality of winter wheat. *Khimiya v Sel'skom Khozyaistve* 4: 17–20.

Aldon, D., Mbengue, M., Mazars, C. et al. (2018). Calcium signalling in plant biotic interactions. *International Journal of Molecular Sciences* 19 (3): 665.

Aluru, M., Xu, Y., Guo, R. et al. (2008). Generation of transgenic maize with enhanced provitamin A content. *Journal of Experimental Botany* 59 (13): 3551–3562.

Andreo, C.S., Gonzalez, D.H. and Iglesias, A.A. (1987). Higher plant phosphoenolpyruvate carboxylase: Structure and regulation. *FEBS Letters* 213 (1): 1–8.

Atwell, S., Huang, Y.S., Vilhja´lmsson, B.J. et al. (2010). Genome-wide association study of 107 phenotypes in Arabidopsis thaliana inbred lines. *Nature* 465 (7298): 627–631.

Aydin, H., Deyneli, O., Yavuz, D. et al. (2010). Short-term oral magnesium supplementation suppresses bone turnover in postmenopausal osteoporotic women. *Biological Trace Element Research* 133 (2): 136–143.

Babaeian, M., Esmaeilian, Y., Tavassoli, A. et al. (2012). Efficacy of different iron, zinc and magnesium fertilizers on yield and yield components of barley. *African Journal of Microbiology Research* 6 (25): 5754–5756.

Baxter, I. (2010). Ionomics: The functional genomics of elements. *Briefings in Functional Genomics* 9 (2): 149–156.

Baxter, I., Hermans, C., Lahner, B. et al. (2012). Mineral accumulation of Arabidopsis natural populations. *PLoS One* 7 (3): e35121.

Beringer, H. and Forster, H. (1981). Einfluß variierter Mg-Ernährung auf Tausendkorngewicht und P-Fraktionen des Gerstenkorns. *Z. Pflanzenernähr Bodenkd* 144: 8–15.

Bilyeu, K.D., Zeng, P., Coello, P. et al. (2008). Quantitative conversion of phytate to inorganic phosphorus in soybean seeds expressing a bacterial phytase. *Plant Physiology* 146 (2): 468–477.

Bo, S. and Pisu, E. (2008). Role of dietary magnesium in cardiovascular disease prevention, insulin sensitivity and diabetes. *Current Opinion in Lipidology* 19 (1): 50–56.

Boaretto, R.M., Hippler, F.W.R., Ferreira, G.A. et al. (2020). The possible role of extra magnesium and nitrogen supply to alleviate stress caused by high irradiation and temperature in lemon trees. *Plant Soil* 457 (3): 57–70.

Bohn, T., Davidsson, L., Walczyk, T. et al. (2004). Phytic acid added to white-wheat bread inhibits fractional apparent magnesium absorption in humans. *The American Journal of Clinical Nutrition* 79: 418–423.

Bohn, L., Meyer, A.S. and Rasmussen, S.K. (2008). Phytate: Impact on environment and human nutrition. A challenge for molecular breeding. *Journal of Zhejiang University Science B* 9 (3): 165–191.

Bose, J., Babourina, O. and Rengel, Z. (2011). Role of magnesium in alleviation of aluminium toxicity in plants. *Journal of Experimental Botany* 62 (7): 2251–2264.

Bose, J., Babourina, O., Shabala, S. et al. (2013). Low-pH and aluminum resistance in Arabidopsis correlates with high cytosolic magnesium content and increased magnesium uptake by plant roots. *Plant Cell Physiology* 54 (7): 1093–1104.

Bouis, H.E. (2018). Biofortification: An agricultural tool to address mineral and vitamin deficiencies. In: *Food Fortification in a Globalized World*, 69–81. Amsterdam: Elsevier.

Bouis, H.E. and Saltzman, A. (2017). Improving nutrition through biofortification: A review of evidence from HarvestPlus 2003 through 2016. *Global Food Security* 12: 49–58.

Bradshaw, H.D. Jr. (2005). Mutations in *CAX1* produce phenotypes characteristic of plants tolerant to serpentine soils. *New Phytologist* 167 (1): 81–88.

Brinch-Pedersen, H., Hatzack, F., Stöger, E. et al. (2006). Heat-stable phytases in transgenic wheat (*Triticum aestivum* L.): Deposition pattern, thermostability, and phytate hydrolysis. *Journal of Agricultural and Food Chemistry* 54 (13): 4624–4632.

Broadley, M.R., Hammond, J.P., King, G.J. et al. (2008). Shoot calcium and magnesium concentrations differ between subtaxa, are highly heritable, and associate with potentially pleiotropic loci in *Brassica oleracea*. *Plant Physiology* 146 (4): 1707–1720.

Brown, R.B., Haq, A., Stanford, C.F. et al. (2015). Vitamin D, phosphate, and vasculotoxicity. *Canadian Journal of Physiology and Pharmacology* 93 (12): 1077–1082.

Bryant, R.J., Dorsch, J.A., Peterson, K.L. et al. (2005). Phosphorus and mineral concentrations in whole grain and milled low phytic acid (lpa) 1-1 rice. *Cereal Chemistry* 82 (5): 517–522.

Cakmak, I. and Yazici, A.M. (2010). Magnesium: A forgotten element in crop production. *Better Crops* 94: 23–25.

Cakmak, I. and Kirkby, E.A. (2008). Role of magnesium in carbon partitioning and alleviating photooxidative damage. *Physiologia plantarum* 133 (4): 692–704.

Cakmak, I., Torun, A., Millet, E. et al. (2004). *Triticum dicoccoides*: an important genetic resource for increasing zinc and iron concentration in modern cultivated wheat. *Soil Science and Plant Nutrition* 50: 1047–1054.

Castro, C.D., Oliveira Junior, A.D., Oliveira, F.A.D. et al. (2021). Magnesium: Management for the nutritional balance of soybean. *Documentos - Embrapa Soja* 430: 54. https://www.infoteca.cnptia.embrapa.br/infoteca/handle/doc/1125328.

Cavagnaro, T.R. (2008). The role of arbuscular mycorrhizas in improving plant zinc nutrition under low soil zinc concentrations: A review. *Plant Soil* 304 (1): 315–325.

Cazzola, R., Della Porta, M., Manoni, M. et al. (2020). Going to the roots of reduced magnesium dietary intake: A tradeoff between climate changes and sources. *Heliyon* 6 (11): e05390.

Chen, H.B. and Fan, X.L. (2018). Effects of magnesium remobilization and allocation on banana plant growth. *Journal of Plant Nutrition* 41: 1312–1320.

Chen, Z.C., Peng, W.T., Li, J. et al. (2018). Functional dissection and transport mechanism of magnesium in plants. *Seminars in Cell & Developmental Biology* 74: 142–152.

Chen, J., Li, L.G., Liu, Z.H. et al. (2009). Magnesium transporter *AtMGT9* is essential for pollen development in Arabidopsis. *Cell Research* 19 (7): 887–898.

Chen, Q., Kan, Q., Wang, P. et al. (2015). Phosphorylation and interaction with the 14-3-3 protein of the plasma membrane H^+-ATPase are involved in the regulation of magnesium-mediated increases in aluminum-induced citrate exudation in broad bean (*Vicia faba* L.). *Plant and Cell Physiology* 56 (6): 1144–1153.

Chen, R., Xue, G., Chen, P. et al. (2008). Transgenic maize plants expressing a fungal phytase gene. *Transgenic Research* 17 (4): 633–643.

Chen, Z.C., Yamaji, N., Motoyama, R. et al. (2012). Up-regulation of a magnesium transporter gene *OsMGT1* is required for conferring aluminum tolerance in rice. *Plant Physiology* 159 (4): 1624-1633.

Chen, Z., Young, T.E., Ling, J. et al. (2003). Increasing vitamin C content of plants through enhanced ascorbate recycling. *Proceedings of the National Academy of Sciences* 100 (6): 3525–3530.

Chen, Z.C. and Ma, J.F. (2013). Magnesium transporters and their role in Al tolerance in plants. *Plant Soil* 368: 51–56.

Chen, Z.C., Yamaji, N., Horie, T. et al. (2017). A magnesium transporter *OsMGT1* plays a critical role in salt tolerance in rice. *Plant Physiology* 174 (3): 1837–1849.

Cheval, C., Aldon, D., Galaud, J.P. et al. (2013). Calcium/calmodulin-mediated regulation of plant immunity. *Biochimica et Biophysica Acta* 1833 (7): 1766–1771.

Chubanov, V., Waldegger, S., Schnitzler, M. et al. (2004). Disruption of *TRPM6/TRPM7* complex formation by a mutation in the *TRPM6* gene causes hypomagnesemia with secondary hypocalcemia. *Proceedings of the National Academy of Sciences* s101 (9): 2894–2899.

Chwil, S. (2001). Wpływ dolistnego i doglebowego stosowania magnezu na wielkość i strukturę plonu pszenicy ozimej. *Biul Magnezol* 6: 118–124.

Chwil, S. (2009) Oddziaływanie magnezu i azotu na cechy jakościowe plonu pszenicy ozimej. *Ann. Universitat Maria Curie Skłodowska Lublin-Polonia* 64: 61–66.

Cichy, K.A., Forster, S., Grafton, K.F. et al. (2005). Inheritance of seed zinc accumulation in navy bean. *Crop Science* 45: 864–870.

Conn, S.J., Berninger, P., Broadley, M.R. et al. (2012). Exploiting natural variation to uncover candidate genes that control element accumulation in Arabidopsis thaliana. *New Phytologist* 193 (4): 859–866.

Conn, S.J., Conn, V., Tyerman, S.D. et al. (2011b). Magnesium transporters, *MGT2/MRS2-1* and *MGT3/MRS2-5*, are important for magnesium partitioning within Arabidopsis thaliana mesophyll vacuoles. *New Phytologist* 190 (3): 583–594.

Conn, S.J., Gilliham, M., Athman, A. et al. (2011a). Cell-specific vacuolar calcium storage mediated by *CAX1* regulates apoplastic calcium concentration, gas exchange, and plant productivity in Arabidopsis. *Plant Cell* 23 (1): 240–257.

Conn, S. and Gilliham, M. (2010). Comparative physiology of elemental distributions in plants. *Annals of Botany* 105 (7): 1081–1102.

Dai, F., Wang, J., Zhang, S. et al. (2007). Genotypic and environmental variation in phytic acid content and its relation to protein content and malt quality in barley. *Food Chemistry* 105 (2): 606–611.

Dalmas, O., Sompornpisut, P., Bezanilla, F. et al. (2014). Molecular mechanism of Mg^{2+}-dependent gating in CorA. *Nature Communications* 5: 3590.

Davies, K.M. (2007). Genetic modification of plant metabolism for human health benefits. *Mutation Research* 622 (1–2): 122–137.

Deliboran, A., Sakin, E., Aslan, H. et al. (2011). Effects of different water, phosphorus and magnesium doses on the quality and yield factors of soybean (*Glycine max* L.) in Harran plain conditions. *International Journal of. Physical Sciences* 6 (6): 1484–1495.

Delva, P. (2003). Magnesium and coronary heart disease. *Molecular Aspects of Medicine* 24 (1–3): 63–78.

Deng, W., Luo, K., Li, D. et al. (2006). Overexpression of an Arabidopsis magnesium transport gene, *AtMGT1*, in *Nicotiana benthamiana* confers Al tolerance. *Journal of Experimental Botany* 57 (15): 4235–4243.

Deng, X., Song, Y., Manson, J.E. et al. (2013). Magnesium, vitamin D status and mortality: Results from US National Health and Nutrition Examination Survey (NHANES) 2001 to 2006 and NHANES III. *BMC Medicine* 11 (1): 187.

Dietitians Association of Australia (DAA). (n.d.). *The Australian Dietary Guidelines.* https://daa.asn.au/smart-eating-for-you/smart-eating-fast-facts/healthy-eating/the-australian-dietary-guidelines.

Ding, Y., Luo, W. and Xu, G. (2006). Characterization of magnesium nutrition and interaction of magnesium and potassium in rice. *Annals of Applied Biology* 149 (2): 111–123.

Dodd, A.N., Kusakina, J., Hall, A. et al. (2014). The circadian regulation of photosynthesis. *Photosynthesis Research* 119 (1–2): 181–190.

Draycott, P. and Allison, M. (1998). Magnesium fertilizers in soil and plants: Comparisons and usage. In: *Proceedings of the International Fertilizer Society*, 412. New York, UK: International Fertiliser Society.

Drummond, R.S.M., Tutone, A., Li, Y.C. et al. (2006). A putative magnesium transporter *AtMRS2-11* is localized to the plant chloroplast envelope membrane system. *Plant Science* 170 (1): 78–89.

Durlach, J., Bac, P., Durlach, V. et al. (1998). Magnesium status and ageing: An update. *Magnesium Research* 11 (1): 25–42.

EFSA Panel on Dietetic Products, Nutrition and Allergies (NDA). (2015). Scientific opinion on dietary reference values for magnesium. *EFSA Journal* 13 (7): 4186.

Elmer, W. and Datnoff, L. (2014). Mineral nutrition and suppression of plant disease. *Encyclopedia of agriculture and food systems* 4: 231–244.

Etcheverry, P., Grusak, M.A. and Fleige, L.E. (2012). Application of in vitro bioaccessibility and bioavailability methods for calcium, carotenoids, folate, iron, magnesium, polyphenols, zinc, and vitamins B6, B12, D, and E. *Frontiers in Physiology* 3: 317.

Fageria, N.K., Baligar, V.C. and Li, Y.C. (2008). The role of nutrient efficient plants in improving crop yields in the twenty first century. *Journal of Plant Nutrition* 31 (6): 1121–57.

FAO. (2023). Food Outlook: Biannual Report on Global Food Markets. chrome-extension://efaidnbmnnnibpcajpcglclefindmkaj/https://openknowledge.fao.org/server/api/core/bitstreams/6e04f2b4-82fc-4740-8cd5-9b66f5335239/content.

FAO. (2024). Crop Prospects and Food Situation. Published on 1 March 2024. chrome-extension://efaidnbmnnnibpcajpcglclefindmkaj/https://www.fao.org/3/cd0022en/cd0022en.pdf.

Fawcett, W., Haxby, E. and Male, D. (1999). Magnesium: physiology and pharmacology. *British Journal of Anaesthesia* 88 (2): 302–320.

Fecenko, J. and Francakova, H. (1980). Effect of magnesium fertilization on the nitrogen compounds of spring barley. *Pol'nohospodarstvo* 26: 700–706.

Feil, B. and Fossati, D. (1997). Phytic acid in triticale grains as affected by cultivar and environment. *Crop Science* 37 (3): 916–921.

Fischer, E.S. (1997). Photosynthetic irradiance response curves of Phaseolus vulgaris under moderate or severe magnesium deficiency. *Photosynthetica* 33: 385–390.

Flachowski, G., Chesson, A. and Aulrich, K. (2005). Animal nutrition with feeds from genetically modified plants. *Archives of Animal Nutrition* 59 (1): 1–40.

Fujikawa, I., Takehara, Y., Ota, M. et al. (2021). Magnesium oxide induces immunity against Fusarium wilt by triggering the jasmonic acid signaling pathway in tomato. *Journal of Biotechnology* 325: 100–108.

Gaash, R., Elazar, M., Mizrahi, K. et al. (2013). Phylogeny and a structural model of plant *MHX* transporters. *BMC Plant* Biology 13 (1): 75.

Garg, M., Sharma, N., Sharma, S. et al. (2018). Biofortified crops generated by breeding, agronomy, and transgenic approaches are improving lives of millions of people around the world. *Frontiers in Nutrition* 5: 12. doi: 10.3389/fnut.2018.00012.

Gebert, M., Meschenmoser, K., Svidova, S. et al. (2009). A root-expressed magnesium transporter of the *MRS2/MGT* gene family in *Arabidopsis thaliana* allows for growth in low-Mg^{2+} environments. *Plant Cell* 21 (12): 4018–4030.

Gerendás, J. and Führs, H. (2013). The significance of magnesium for crop quality. *Plant Soil* 368: 101–128.

Gibson, R.S., Bailey, K.B., Gibbs, M. et al. (2010). A review of phytate, iron, zinc, and calcium concentrations in plant based complementary foods used in low-income countries and implications for bioavailability. *Food and Nutrition Bulletin* 31 (2 Suppl) (Supplement): S134–S146.

Glahn, R.P., Cheng, Z. and Welch, R.M. (2002). Comparison of iron bioavailability from 15 rice genotypes: studies using an in vitro digestion/Caco-2 cell culture model. *Journal of Agricultural and Food Chemistry* 50 (12): 3586–3591.

Gout, E., Rébeillé, F., Douce, R. et al. (2014). Interplay of Mg^{2+}, ADP, and ATP in the cytosol and mitochondria: unravelling the role of Mg^{2+} in cell respiration. *Proceedings of the National Academy of Sciences* 111: 4560–4567.

Graham, R.D., Welch, R.M., Saunders, D.A. et al. (2007). Nutritious subsistence food systems. *Advances in agronomy* 92: 1–74.

Gransee, A. and Führs, H. (2013). Magnesium mobility in soils as a challenge for soil and plant analysis, magnesium fertilization and root uptake under adverse growth conditions. *Plant Soil* 368: 5–2.

Greffeuille, V., Abecassis, J., Lapierre, C. et al. (2006). Bran size distribution at milling and mechanical and biochemical characterization of common wheat grain outer layers: A relationship assessment. *Cereal Chemistry* 83: 641–646.

Grigore, M.N., Boscaiu, M., Llinares, J. et al. (2012). Mitigation of salt stress-induced inhibition of *Plantago crassifolia* reproductive development by supplemental calcium or magnesium. *Notulae Botanic ae Horti Agrobotanici Cluj-Napoca* 40: 58–66.

Grober, U., Schmidt, J. and Kisters, K. (2015). Magnesium in prevention and therapy. *Nutrients* 7 (9): 8199–8226.

Groisman, E.A. (2001). The pleiotropic two-component regulatory system PhoP-PhoQ. *Journal of Bacteriology* 183 (6): 1835–1842.

Grusak, M.A. and Cakmak, I. (2005). Methods to improve the crop-delivery of minerals to humans and livestock. In:, *Plant nutritional genomics* (eds. M.R. Blackwell and P.J. White), 265–286. Oxford, UK: Blackwell.

Grzebisz, W. (2011). Magnesium–food and human health. *Journal of Elementology* 16 (2): 299–323.

Grzebisz, W. (2013). Crop response to magnesium fertilization as affected by nitrogen supply. *Plant Soil* 368: 23–39.

Grzebisz, W., Przygocka-Cyna, K., Szczepaniak, W. et al. (2010). Magnesium as a nutritional tool of nitrogen efficient management-plant production and environment. *Journal of Elementology* 15: 771–788.

Guo, W., Chen, S., Hussain, N. et al. (2015). Magnesium stress signaling in plant: Just a beginning. *Plant Signaling & Behavior* 10: e992287.

Guo, K.M., Babourina, O., Christopher, D.A. et al (2010). The cyclic nucleotide-gated channel *AtCNGC10* transports Ca^{2+} and Mg^{2+} in Arabidopsis. *Physiologial Plantarum* 139: 303–312.

Guo, W., Nazim, H., Liang, Z. et al. (2016). Magnesium deficiency in plants: An urgent problem. *The Crop Journal* 4: 83–91.

Guttieri, M.J., Peterson, K.M. and Souza, E.J. (2006). Mineral distributions in milling fractions of low phytic acid wheat. *Crop Science* 46: 2692–2698.

Hall, C., Hillen, C. and Garden-Robinson, J. (2017). Composition, nutritional value and health benefits of pulses. *Cereal Chemistry* 94: 11–31.

Hambidge, K.M., Krebs, N.F., Westcott, J.L. et al. (2005). Absorption of calcium from tortilla meals prepared from low-phytate maize. *The American Journal of Clinical Nutrition* 82: 84–87.

Han, H., Fang, X., Wei, X. et al. (2017). Dose-response relationship between dietary magnesium intake, serum magnesium concentration and risk of hypertension: A systematic review and meta-analysis of prospective cohort studies. *Nutrition Journal* 16 (1): 26.

Hardarson, G. and Broughton, W.J. (2003). Maximising the use of biological nitrogen fixation in agriculture. *Springer Science and Business Media* 99.

Harrier, L.A. and Watson, C.A. (2003). The role of arbuscular mycorrhizal fungi in sustainable cropping systems. *Advances in Agronomy* 20: 185–225.

Hasni, I., Msilini, N., Hamdani, S. et al. (2015). Characterization of the structural changes and photochemical activity of photosystem I under Al^{3+} effect. *Journal of Photochemistry and Photobiology B: Biology* 149: 292–299.

Hawkesford, M., Horst, W., Kichey, T. et al. (2012). Functions of macronutrients. In: *Mineral Nutrition of Higher Plants*, 3e (ed. P. Marschner), 135–189. Amsterdam: Elsevier.

Haytowitz, D.B., Ahuja, J.K.C., Wu, X. et al. (2019). *USDA National Nutrient Database for Standard Reference, Legacy Release [Dataset]*. Elsevier: Nutrient Data Laboratory, Beltsville Human Nutrition Research Center, ARS, USDA. doi: 10.15482/USDA.ADC/1529216.

Hazra, S., Henderson, J.N., Liles, K. et al. (2015). Regulation of Ribulose-1,5-bisphosphate Carboxylase/Oxygenase (Rubisco) Activase. *Journal of Biological Chemistry* 290: 24222–24236.

He, X. and Nara, K. (2007). Element biofortification: Can mycorrhizas potentially offer a more effective and sustainable pathway to curb human malnutrition? *Trends in Plant Science* 12: 331–333.

Heffer, P. and Prud'homme, M. (2016). Global nitrogen fertilizer demand and supply: Trend, current level and outlook. Melbourne, Australia: International Nitrogen Initiative Conference.

Hermans, C., Simon, J.C., Jiugeng, C. et al. (2013). An update on magnesium homeostasis mechanisms in plants. *Metallomics* 5 (8): 1170–1183.

Hmiel, S.P., Snavely, M.D., Miller, C.G. et al. (1986). Magnesium transport in *Salmonella typhimurium*: Characterization of magnesium influx and cloning of a transport gene. *Journal of Bacteriology* 168: 1444–1450.

Hong, C.-Y., Cheng, K.-J., Tseng, T.-H. et al. (2004). Production of two highly active bacterial phytases with broad pH optima in germinating transgenic rice seeds. *Transgenic Research* 13: 29–39.

Horie, T., Karahara, I. and Katsuhara, M. (2012). Salinity tolerance mechanisms in glycophytes: An overview with the central focus on rice plants. *Rice* 5 1–18.

Hotz, C. and Brown, K.H. (2004). Assessment of the risk of zinc deficiency in populations and options for its control. *Food and Nutrition Bulletin* 25: 94–204.

Huang, L., Li, D.Q., Lin, Y.J. et al. (2005). Controllable preparation of Nano-MgO and investigation of its bactericidal properties. *Journal of Inorganic Biochemistry* 99: 986–993.

Huang, X., Paulo, M.J., Boer, M. et al. (2011). Analysis of natural allelic variation in Arabidopsis using a multiparent recombinant inbred line population. *Proceedings of the National Academy of Sciences* 108 (10): 4488–4493.

Huber, D.M. and Jones, J.B. (2013). The role of magnesium in plant disease. *Plant Soil* 368: 73–85.

Huskisson, E., Maggini, S. and Ruf, M. (2007). The role of vitamins and minerals in energy metabolism and well-being. *Journal of International Medical Research* 35: 277–289.

Igamberdiev, A.U. and Kleczkowski, L.A. (2001). Implications of adenylate kinase-governed equilibrium of adenylates on contents of free magnesium in plant cells and compartments. *Biochemical Journal* 360: 225–231.

Imada, K., Sakai, S., Kajihara, H. et al. (2016). Magnesium oxide nanoparticles induce systemic resistance in tomato against bacterial wilt disease. *Plant Pathology* 65: 551–560.

Inal, A., Gunes, A., Zhang, F. et al. (2007). Peanut/maize intercropping induced changes in rhizosphere and nutrient concentrations in shoots. *Plant Physiology and Biochemistry* 45: 350–356.

Ishijima, S., Uchibori, A., Takagi, H. et al. (2003). Light-induced increase in free Mg^{2+} concentration in spinach chloroplasts: Measurement of free Mg^{2+} by using a fluorescent probe and necessity of stromal alkalinization. *Archives of Biochemistry and Biophysics* 412: 126–132.

Jaghdani, S.J., Jahns, P. and Trankner, M. (2021). Mg deficiency induces photo-oxidative stress primarily by limiting CO_2 assimilation and not by limiting photosynthetic light utilization. *Plant Science* 302: 110751.

Jahnen-Dechent, W. and Ketteler, M. (2012). Magnesium basics. *Clinical Kidney Journal* 5 (Suppl_1): i3–i14.

Jayaganesh, S., Venkatesan, S. and Senthurpandian, V.K. (2011). Impact of different sources and doses of magnesium fertilizer on biochemical constituents and quality parameters of black tea. *Asian Journal of Biochemistry* 6: 273–281.

Jezek, M., Geilfus, C.M., Bayer, A. et al. (2015). Photosynthetic capacity, nutrient status, and growth of maize (*Zea mays* L.) upon $MgSO_4$ leaf-application. *Frontiers in Plant Science* 5: 781.

Joao Matias, P., Azevedo, A., Laranjinha, I. et al. (2014). Lower serum magnesium is associated with cardiovascular risk factors and mortality in haemodialysis patients. *Blood Purification* 38 (3–4): 244–252.

Johns, T. and Eyzaguirre, P.B. (2007). Biofortification, biodiversity and diet: A search for complementary applications against poverty and malnutrition. *Food Policy* 32: 1–24.

Jones, J.B. and Huber, D.M. (2007). Magnesium and plant disease. In: *Mineral Nutrition and Plant Disease* (eds. L.E. Datnoff, W.H. Elmer and D.M. Huber). Saint Paul, Minnesota, America: Am. Phytopatholgical Society.

Kai, Y., Matsumura, H. and Izui, K. (2003). Phosphoenolpyruvate carboxylase: Three-dimensional structure and molecular mechanisms. *Archives of Biochemistry and Biophysics* 414: 170–179.

Kashinath, B.L., Murthy, A.G., Senthivel, T. et al. (2013). Effect of applied magnesium on yield and quality of tomato in Alfisols of Karnataka. *Journal of Horticultural Sciences* 8 (1): 55–59.

Kaushal, N., Awasthi, R., Gupta, K. et al. (2013). Heat-stress-induced reproductive failures in chickpea (*Cicer arietinum*) are associated with impaired sucrose metabolism in leaves and anthers. *Functional Plant Biology* 40: 1334.

Khater, M., de la Escosura-Muñiz, A. and Merkoçi, A. (2017). Biosensors for plant pathogen detection. *Biosensors and Bioelectronics* 93: 72–86.

Killilea, D.W. and Maier, J.A.M. (2008). A connection between magnesium deficiency and aging: New insights from cellular studies. *Magnesium Research* 21: 77–82.

Kirkby, E.A. and Johnston, A.E. (2008). Soil and fertilizer phosphorus in relation to crop nutrition. In: *The Ecophysiology of Plant-Phosphorus Interactions*. (eds. J.P. Hammond and P.J. White), 177–223. Dordrecht, the Netherlands: Springer.

Klein, D.J., Moore, P.B. and Steitz, T.A. (2004). The contribution of metal ions to the structural stability of the large ribosomal subunit. *RNA* 10: 1366–1379.

Knoop, V., Groth-Malonek, M., Gebert, M. et al. (2005). Transport of magnesium and other divalent cations: Evolution of the *2-TM-GxN* proteins in the *MIT* superfamily. *Molecular Genetics and Genomics* 274: 205–216.

Kobayashi, N.I., Ogura, T., Takagi, K. et al. (2018). Magnesium deficiency damages the youngest mature leaf in rice through tissue-specific iron toxicity. *Plant Soil* 428: 137–152.

Koornneef, M., Reymond, M. and Alonso-Blanco, C. (2011). Natural variation in *Arabidopsis thaliana*. In: *Genetics and Genomics of Brassicaceae* (eds. R. Schmidt and I. Bancroft), 123–151. Singapore: Springer.

Kopsell, D.A. and Kopsell, D.E. (2006). Accumulation and bioavailability of dietary carotenoids in vegetable crops. *Trends in Plant Science* 11: 499–507.

Kover, P.X., Valdar, W., Trakalo, J. et al. (2009). A multiparent advanced generation inter-cross to fine-map quantitative traits in *Arabidopsis thaliana*. *PLoS Genetics* 5 (7): e1000551.

Kuwano, M., Ohyama, A., Tanaka, Y. et al. (2006). Molecular breeding for transgenic rice with low phytic-acid phenotype through manipulating myo-inositol 3-phosphate synthase gene. *Molecular Breeding* 18: 263–272.

Lahner, B., Gong, J., Mahmoudian, M. et al. (2003). Genomic scale profiling of nutrient and trace elements in Arabidopsis thaliana. *Nature Biotechnology* 21: 1215–1221.

Larsen, E.H., Lobinski, R., Burger-Meyer, K. et al. (2006). Uptake and speciation of selenium in garlic cultivated in soil amended with symbiotic fungi (mycorrhiza) and selenate. *Analytical and Bioanalytical Chemistry* 385: 1098–1108.

Lenz, H.N., Dombinov, V., Dreistein, J. et al. (2013). Magnesium deficiency phenotypes upon multiple knockout of *Arabidopsis thaliana MRS2* clade B genes can be ameliorated by concomitantly reduced calcium supply. *Plant and Cell Physiology* 54: 1118–1131.

Li, D., Ma, W., Wei, J. et al. (2020a). Magnesium promotes root growth and increases aluminum tolerance via modulation of nitric oxide production in Arabidopsis. *Plant Soil* 457: 83–95.

Li, J., Huang, Y., Tan, H. et al. (2015). An endoplasmic reticulum magnesium transporter is essential for pollen development in Arabidopsis. *Plant Science* 231: 212-220.

Li, J., Yokosho, K., Liu, S. et al. (2020b). Diel magnesium fluctuations in chloroplasts contribute to photosynthesis in rice. *Nature Plants* 6: 848–859.

Li, L., Tutone, A.F., Drummond, R.S. et al. (2001). A novel family of magnesium transport genes in Arabidopsis. *Plant Cell* 13: 2761–2775.

Liu, K., Peterson, K.L. and Raboy, V. (2007). Comparison of the phosphorus and mineral concentrations in bran and abraded kernel fractions of a normal barley (*Hordeum vulgare*) cultivar versus four low phytic acid isolines. *Journal of Agricultural and Food Chemistry* 55: 4453–4460.

Lolas, G.M., Palamidis, N. and Markakis, P. (1976). The phytic acid–total phosphorus relationship in barley, oats, soybeans, and wheat. *Cereal Chemistry* 53: 867–871.

Lönnerdal, B. (2003). Genetically modified plants for improved trace element nutrition. *The Journal of Nutrition* 133: 1490–1493.

Lucca, P., Poletti, S. and Sautter, C. (2006). Genetic engineering approaches to enrich rice with iron and vitamin A. *Physiologia Plantarum* 126: 291–303.

Lunin, V.V., Dobrovetsky, E., Khutoreskaya, G. et al. (2006). Crystal structure of the CorA Mg^{2+} transporter. *Nature* 440: 833–837.

Lynch, J.P. (2007). Roots of the second green revolution. *Australian Journal of Botany* 55: 493–512.

Lyons, G.H., Judson, G.J., Ortiz-Monasterio, I. et al. (2005). Selenium in Australia: Selenium status and biofortification of wheat for better health. *Journal of Trace Elements in Medicine and Biology* 19: 75–82.

Magnitskii, K.P., Sokolova, V.A. and Zhukov, S.N. (1970). Effect of Mg fertilizers on yield and quality of winter rye. *Khimiya v Sel'skom Khozyaistve* 8: 565–568.

Mao, D., Chen, J., Tian, L. et al. (2014). Arabidopsis transporter *MGT6* mediates magnesium uptake and is required for growth under magnesium limitation. *Plant Cell* 26: 2234–2248.

Marschner, P. (ed.) (2012). *Marschner's Mineral Nutrition of Higher Plants*, 3e. Amsterdam: Elsevier.

Martínez-Ballesta, M.C., Dominguez-Perles, R., Moreno, D.A. et al. (2010). Minerals in plant food: Effect of agricultural practices and role in human health. A review. *Agronomy for Sustainable Development* 30: 295–309.

Mazariegos, M., Hambidge, K.M., Krebs, N.F. et al. (2006). Zinc absorption in Guatemalan schoolchildren fed normal or low-phytate maize. *The American Journal of Clinical Nutrition* 83: 59–64.

Mengutay, M., Ceylan, Y., Kutman, U.B. et al. (2013). Adequate magnesium nutrition mitigates adverse effects of heat stress on maize and wheat. *Plant Soil* 368: 57–72.

Messenger, A.S. and Hruby, B.A. (1992). Excess calcium and magnesium associated with lower crown mortality of pin oak. *Journal of Arboriculture* 18: 241–248.

Mikkelsen, R. (2010). Soil and fertilizer magnesium. *Better Crops* 94 (4): 26–28.

Mohebbi, S. and Mahler, R. (1989). The effect of soil pH on wheat and lentils grown on an agriculturally acidified northern Idaho soil under greenhouse conditions. *Communications in Soil Science and Plant Analysis* 20 (1–2): 359–381.

Morgan, J.A.W., Bending, G.D. and White, P.J. (2005). Biological costs and benefits to plant-microbe interactions in the rhizosphere. *Journal of Experimental Botany* 56: 1729–1739.

Neuhaus, C., Geilfus, C.M., Zorb, C. et al. (2013). Transcript expression of Mg-chelatase and H^{+}-ATPase isogenes in *Vicia faba* leaves as influenced by root and foliar magnesium supply. *Plant Soil* 368: 41–5.

Nguyen, B.T., Thanh, K.D., Thanh, V.T. et al. (2018). High soil Mn and Al, as well as low leaf P concentration, may explain for low natural rubber productivity on a tropical acid soil in Vietnam. *Journal of Plant Nutrition* 41 (7): 903–914.

Nunes, A.C.S., Vianna, G.R., Cuneo, F. et al. (2006). RNAi-mediated silencing of the myo-inositol-1-phosphate synthase gene (*GmMIPS1*) in transgenic soybean inhibited seed development and reduced phytate content. *Planta* 224: 125–132.

O'Leary, B.M., Neale, H.C., Geilfus, C.M. et al. (2016). Early changes in apoplast composition associated with defence and disease in interactions between *Phaseolus vulgaris* and the halo blight pathogen *Pseudomonas syringae Pv. phaseolicola*. *Plant, cell & environment* 39: 2172–2184.

Oikeh, S.O., Menkir, A., Maziya Dixon, B. et al. (2003). Genotypic differences in concentration and bioavailability of kernel-iron in tropical maize varieties grown under field conditions. *Journal of Plant Nutrition* 26: 2307–2319.

Orchard, T.S., Larson, J.C., Alghothani, N. et al. (2014). Magnesium intake, bone mineral density, and fractures: Results from the Women's Health Initiative Observational Study. *The American Journal of Clinical Nutrition* 99 (4): 926–933.

Ota, M., Imada, K., Sasaki, K. et al. (2019). MgO-induced defence against bacterial wilt disease in *Arabidopsis thaliana*. *Plant Pathology* 68: 323–333.

Oury, F.-X., Leenhardt, F., Rémésy, C. et al. (2006). Genetic variability and stability of grain magnesium, zinc and iron concentration in bread wheat. *European Journal of Agronomy* 25: 177–185.

Özenç, N. and Özenç, D.B. (2015). Effect of magnesium fertilization on some plant nutrient interactions and nut quality properties in Turkish hazelnut (*Corylus avellana* L.). *Scientific Research and Essays* 10 (4): 465–470.

Paine, J.A., Shipton, C.A., Chaggar, S. et al. (2005). Improving the nutritional value of Golden Rice through increased pro-vitamin A content. *Nature Biotechnology* 23: 482–487.

Parsons, S.A. and Smith, J.A. (2008). Phosphorus removal and recovery from municipal wastewaters. *Elements* 4: 109–112.

Peng, W.T., Qi, W.L., Nie, M.M. et al. (2020). Magnesium supports nitrogen uptake through regulating *NRT2.1/2.2* in soybean. *Plant Soil* 457: 97–111.

Peng, W.T., Zhang, L.D., Zhou, Z. et al. (2018). Magnesium promotes root nodulation through facilitation of carbohydrate allocation in soybean. *Physiologia Plantarum* 163: 372–385.

Peng, Y.Y., Liao, L.L., Liu, S. et al. (2019). Magnesium deficiency triggers SGR-mediated chlorophyll degradation for magnesium remobilization. *Plant Physiology* 181: 262–275.

Pfeiffer, W.H. and McClafferty, B. (2007). HarvestPlus: Breeding crops for better nutrition. *Crop Science* 47: S88–S105.

Pieterse, C.M., Van der Does, D., Zamioudis, C. et al. (2012). Hormonal modulation of plant immunity. *Annual Review of Cell and Developmental Biology* 28: 489–521.

Ponstein, A.S., Bade, J.B., Verwoerd, T.C. et al. (2002). Stable expression of phytase (*phyA*) in canola (*Brassica napus*) seeds: Towards a commercial product. *Molecular Breeding* 10: 31–44.

Prom-u-thai, C., Fukai, S., Godwin, I.D. et al. (2007). Genotypic variation of iron partitioning in rice grain. *Journal of the Science of Food and Agriculture* 87: 2049–2054.

Qu, X., Jin, F., Hao, Y. et al. (2013). Magnesium and the risk of cardiovascular events: A meta-analysis of prospective cohort studies. *PLoS One* 8 (3): e57720.

Raboy, V. (2007). Seed phosphorus and the development of low-phytate crops. In: *Inositol Phosphates: Linking Agriculture and the Environment* (eds. B.L. Turner, A.E. Richardson and E.J. Mullaney), 111–132. Wallingford, UK: CABI.

Raboy, V., Dickinson, D.B. and Below, F.E. (1984). Variation in seed total phosphorus, phytic acid, zinc, calcium, magnesium and protein among lines of *Glycine max* and *G. soja*. *Crop Science* 24: 431–434.

Ram, S. and Mishra, B. (2010). *Cereals Processing and Nutritional Quality*, 188. New Delhi: New India Publishing Agency.

Rayssiguier, Y., Gueux, E., Nowacki, W. et al. (2006). High fructose consumption combined with low dietary magnesium intake may increase the incidence of the metabolic syndrome by inducing inflammation. *Magnesium Research* 19: 237–243.

Razzaque, M.S. (2018). Can adverse effects of excessive vitamin D supplementation occur without developing hypervitaminosis D? *The Journal of Steroid Biochemistry and Molecular Biology* 180 (1): 81–86.

Reddy, B.V.S., Ramesh, S. and Longvah, T. (2005). Prospects of breeding for micronutrients and β-carotene-dense sorghums. *International Sorghum and Millets Newsletter* 46: 10–14.

Reddy, N.R. (2002). Occurrence, distribution, content, and dietary intake of phytate. In: *Food Phytates* (eds. N.R. Reddy and S.K. Sathe), 25–51. Boca Raton, FL, USA: CRC Press.

Ren, Z.H., Gao, J.P., Li, L.G. et al. (2005). A rice quantitative trait locus for salt tolerance encodes a sodium transporter. *Nature Genetics* 37: 1141–1146.

Reza, M.H., Shah, H., Manjrekar, J. et al. (2016). Magnesium uptake by CorA transporters is essential for growth, development and infection in the rice blast fungus *Magnaporthe oryzae*. *PLoS One* 11: e0159244.

Risco, F. and Traba, M.L. (1994). Possible involvement of a magnesium dependent mitochondrial alkaline phosphatase in the regulation of the 25-hydroxyvitamin D3-1 alpha-and 25-hydroxyvitamin D3-24R-hydroxylases in LLC-PK1 cells. *Magnesium Research* 7 (3-4): 169–178.

Roberfroid, M.B. (2007). Inulin-type fructans: Functional food ingredients. *Journal of Nutrition* 137: S2493–S2502.

Römheld, V. (2012). Diagnosis of deficiency and toxicity of nutrients. In: *Mineral Nutrition of Higher Plants*, 3e (ed. P. Marschner), 299–311. Amsterdam: Elsevier.

Rosentrater, K.A. and Evers, A.D. (2018). Feed and industrial uses for cereals. In: *Kent's Technology of Cereals*, 785–837. Abington: Woodhead Publishing.

Ryan, P., Kinraide, T. and Kochian, L. (1994). Al^{3+}-Ca^{2+} interactions in aluminum rhizotoxicity: I. Inhibition of root growth is not caused by reduction of calcium uptake. *Planta* 192: 98–103.

Saito, T., Kobayashi, N.I., Tanoi, K. et al. (2013). Expression and functional analysis of the *CorA-MRS2-ALR*-type magnesium transporter family in rice. *Plant and Cell Physiology* 54: 1673–1683.

Salt, D.E., Baxter, I. and Lahner, B. (2008). Ionomics and the study of the plant ionome. *Annual Review of Plant Biology* 59: 709–733.

Sanjeeva, D., Neeraja, C., Madhu, P. et al. (2020). Zinc biofortified rice varieties: challenges, possibilities, and progress in India. *Frontiers in Nutrition* 7: 26. doi: 10.3389/fnut.2020.00026.

Schlemmer, U., Frolich, W., Prieto, R.M. et al. (2009). Phytate in foods and significance for humans: food sources, intake, processing, bioavailability, protective role and analysis. *Molecular Nutrition & Food Research* 53: S330–S375.

Shahzad, R., Jamil, S., Ahmad, S. et al. (2021). Biofortification of cereals and pulses using new breeding techniques: current and future perspectives. *Frontiers in Nutrition* 8: 721728.

Shaul, O., Hilgemann, D.W., de-Almeida-Engler, J. et al. (1999). Cloning and characterization of a novel Mg^{2+}/H^{+} exchanger. *The EMBO Journal* 18: 3973–3980.

Shi, J., Wang, H., Schellin, K. et al. (2007). Embryo-specific silencing of a transporter reduces phytic acid content of maize and soybean seeds. *Nature Biotechnology* 25: 930–937.

Siddiqui, M.H., Alamri, S.A., Al-Khaishany, M.Y.Y. et al. (2016). Mitigation of adverse effects of heat stress on *Vicia faba* by exogenous application of magnesium. *Saudi Journal of Biological Sciences* 25: 1393–2140.

Silva, D.M., Souza, K.R., Boas, L.V. et al. (2017). The effect of magnesium nutrition on the antioxidant response of coffee seedlings under heat stress. *Scientia Horticulturae* 224: 115–125.

Simic′, D., Drinic′, S.M., Zdunic′, Z. et al. (2012). Quantitative trait loci for biofortification traits in maize grain. *Journal of Heredity* 103: 47–54.

Singh, R., Govindan, V. and Andersson, M.S. (2017). Zinc-biofortified wheat: harnessing genetic diversity for improved nutritional quality. Available online at: https://www.harvestplus.org/content/zinc-biofortified-wheat-harnessinggenetic-diversity-improved-nutritional-quality.

Singla, P. and Grover, K. (2017). Biofortified cereal crops: A sustainable approach for food and nutritional security. *Current Journal of Applied Science and Technology* 24: 1–13. doi: 10.9734/CJAST/2017/37865.

Slingerland, M.A., Traore, K., Kayodé, A.P.P. et al. (2006). Fighting Fe deficiency malnutrition in West Africa: An interdisciplinary program on a food chain approach. *Netherlands Journal of Agricultural Science* 53: 253–279.

Smith, S.E. and Read, D.J. (2007). Mycorrhizal symbiosis, 3e. Elsevier. London, UK: Academic Press.

Soncini, F.C., Véscovi, G.E., Solomon, F. et al. (1996). Molecular basis of the magnesium deprivation response in *Salmonella typhimurium*: Identification of PhoP-regulated genes. *Journal of Bacteriology* 178: 5092–5099.

Stangoulis, J.C.R., Huynh, B.L., Welch, R.M. et al. (2007). Quantitative trait loci for phytate in rice grain and their relationship with grain micronutrient content. *Euphytica* 154: 289–294.

Stec, B. (2012). Structural mechanism of RuBisCO activation by carbamylation of the active site lysine. *Proceedings of the National Academy of Sciences* 109: 18785–18790.

Stein, J.A. (2010). Global impacts of human mineral malnutrition. *Plant Soil* 335: 133–154.

Su, L., Lv, A., Wen, W. et al. (2020). Auxin Is Involved in Magnesium-Mediated Photoprotection in Photosystems of Alfalfa Seedlings Under Aluminum Stress. *Frontiers in Plant Science* 11: 746.

Sun, Y., Yang, R., Li, L. et al. (2017). The magnesium transporter *MGT10* is essential for chloroplast development and photosynthesis in *Arabidopsis thaliana*. *Molecular Plant* 10 (12): 1584–1587.

Swaminathan, R. (2003). Magnesium metabolism and its disorders. *The Clinical Biochemist Reviews* 24 (2): 47–66.

Szulc, P. (2010). Effects of differentiated levels of nitrogen fertilization and the method of magnesium application on the utilization of nitrogen by two different maize cultivars for grain. *Polish Journal of Environmental Studies* 19: 407–412.

Szulc, P., Bocianowski, J., Kruczek, A. et al. (2013). Response of two cultivar types of maize (*Zea mays* L.) expressed in protein content and its yield to varied soil resources of N and Mg and a form of nitrogen fertilizer. *Polish Journal of Environmental Studies* 22: 1845–1853.

Tan, G.Z., Das Bhowmik, S.S., Hoang, T.M. et al. (2017). Finger on the pulse: Pumping iron into chickpea. *Frontiers in Plant Science* 8: 1755.

Tan, G.Z., Das Bhowmik, S.S., Hoang, T.M. et al. (2018). Investigation of baseline iron levels in Australian chickpea and evaluation of a transgenic biofortification approach. *Frontiers in Plant Science* 9: 788. doi: 10.3389/fpls.2018.00788.

Tan, K., Keltjens, W.G. and Findenegg, G.R. (1992). Aluminium toxicity with sorghum genotypes in nutrient solutions and its amelioration by magnesium. *Zeitschrift Für Pflanzenernährung Und Bodenkunde* 155 (2): 81–86.

Tang, R.J. and Luan, S. (2017). Regulation of calcium and magnesium homeostasis in plants: From transporters to signaling network. *Current Opinion in Plant Biology* 39: 97–105.

Tao, T., Snavely, M.D., Farr, S.G. et al. (1995). Magnesium transport in *Salmonella typhimurium*: *Mgta* encodes a P-type ATPase and is regulated by Mg^{2+} in a manner similar to that of the *mgtB* P-type ATPase. *Journal of bacteriology* 177: 2654–2662.

Tian, H., Baxter, I.R., Lahner, B. et al. (2010). Arabidopsis *NPCC6/NaKR1* is a phloem mobile metal binding protein necessary for phloem function and root meristem maintenance. *Plant Cell* 22 (12): 3963–3979.

Tian, X.-Y., He, D.-D., Bai, S. et al. (2021). Physiological and molecular advances in magnesium nutrition of plants. *Plant Soil* 468 (1): 1–17. doi: 10.1007/s11104-021-05139-w.

Touyz, R. (2003). Role of magnesium in the pathogenesis of hypertension. *Molecular Aspects of Medicine* 24: 107–136.

Turner, T.L., Bourne, E.C., Von Wettberg, E.J. et al. (2010). Population resequencing reveals local adaptation of *Arabidopsis lyrata* to serpentine soils. *Nature Genetics* 42 (3): 260–264.

US Department of Health and Human Services and US Department of Agriculture. (2015). 2015–2020 Dietary Guidelines for Americans, 8e. https://health.gov/sites/default/files/2019-09/2015-2020_Dietary_Guidelines.pdf.

Uwitonze, A.M. and Razzaque, M.S. (2018). Role of magnesium in vitamin D activation and function. *Journal of Osteopathic Medicine* 118 (3): 181–189.

Venkatesan, S. and Jayaganesh, S. (2010). Characterization of magnesium toxicity, its influence on amino acid synthesis pathway and biochemical parameters of tea. *Research Journal of Phytochemistry* 4: 67–77.

Verbruggen, N. and Hermans, C. (2013). Physiological and molecular responses to magnesium nutritional imbalance in plants. *Plant Soil* 368: 87–99.

Verma, R.P.S., Sarkar, B., Gupta, R. et al. (2008). Breeding barley for malting quality improvement in India. *Cereal Research Communications* 36: 135–145.

Vormann, J. (2003). Magnesium: nutrition and metabolism. *Molecular aspects of medicine* 24: 27–37.

Vrataric, M., Sudaric, A., Kovacevic, V. et al. (2006). Response of soybean to foliar fertilization with magnesium sulfate (Epsom salt). *Cereal Research Communications* 34: 709–712.

Vreugdenhil, D., Aarts, M.G.M. and Koornneef, M. (2005). Exploring natural genetic variation to improve plant nutrient content. In: *Plant Nutritional Genomics* (eds. M.R. Broadley and P.J. White), 201–219. Oxford, UK: Blackwell Publishing.

Vreugdenhil, D., Aarts, M.G.M., Koornneef, M. et al. (2004). Natural variation and QTL analysis for cationic mineral content in seeds of *Arabidopsis thaliana*. *Plant, Cell & Environment* 27: 828–839.

Wang, Z. (2020). *Evaluation of Magnesium Fertilization Effects on Improving Yield and Quality of Horticultural Crops in China*. China, Plant Nutr: China Agricultural University.

Wang, Y., Hua, X., Xu, J. et al. (2019). Comparative genomics revealed the gene evolution and functional divergence of magnesium transporter families in Saccharum. *BMC Genomics* 20: 1–18.

Wang, Z., Mahmood, H.U., Faisal, N. et al. (2020). Magnesium fertilization improves crop yield in most production systems: A meta-analysis. *Frontiers in plant science* 10: 1727.

Wani, A. and Shah, M. (2012). A unique and profound effect of MgO and ZnO nanoparticles on some plant pathogenic fungi. *Journal of Applied Pharmaceutical Science* 2: 40–44.

Ward, J.M., Mäser, P. and Schroeder, J.I. (2009). Plant ion channels: Gene families, physiology, and functional genomics analyses. *Annual Review of Physiology* 71: 59–82.

Watanabe, T. and Okada, K. (2005). Interactive effects of Al, Ca and other cations on root elongation of rice cultivars under low pH. *Annals of Botany* 95: 379–385.

Waters, B.M. (2011). Moving magnesium in plant cells. *New Phytologist* 190: 510–513.

Waters, B.M. and Grusak, M.A. (2008). Quantitative trait locus mapping for seed mineral concentrations in two *Arabidopsis thaliana* recombinant inbred populations. *New Phytologist* 179: 1033–1047.

Waters, B.M. and Sankaran, R.P. (2011). Moving micronutrients from the soil to the seeds: genes and physiological processes from a biofortification perspective. *Plant Science* 180: 562–574. doi: 10.1016/j.plantsci.2010.12.003.

Welch, R.M. and Graham, R.D. (2004). Breeding for micronutrients in staple food crops from a human nutrition perspective. *Journal of Experimental Botany* 55: 353–364.

Welch, R.M. and Graham, R.D. (2005). Agriculture: The real nexus for enhancing bioavailable micronutrients in food crops. *Journal of Trace Elements in Medicine and Biology* 18 (1): 299–307.

Welch, R.M., House, W.A., Beebe, S. et al. (2000). Testing iron and zinc bioavailability in genetically enriched beans (*Phaseolus vulgaris* L.) and rice (*Oryza sativa* L.) in a rat model. *Food and Nutrition Bulletin* 21: 428–433.

Welch, R.M., House, W.A., Ortiz-Monasterio, I. et al. (2005). Potential for improving bioavailable zinc in wheat grain (*Triticum species*) through plant breeding. *Journal of Agricultural and Food Chemistry* 53: 2176–2180.

Welch, R.M., Smith, M.E., Van Campen. et al. (1993). Improving the mineral reserves and protein quality of maize (*Zea mays* L.) kernels using unique genes. *Plant Soil* 156: 215–218.

Werner, T., Nehnevajova, E., Kollmer, I. et al. (2010). Root-specific reduction of cytokinin causes enhanced root growth, drought tolerance, and leaf mineral enrichment in Arabidopsis and tobacco. *Plant Cell* 22: 3905–3920.

White, P.J. and Broadley, M.R. (2005). Biofortifying crops with essential mineral elements. *Trends in Plant Science* 10: 586–593.

White, P.J. and Broadley, M.R. (2009). Biofortification of crops with seven mineral elements often lacking in human diets – iron, zinc, copper, calcium, magnesium, selenium, and iodine. *New Phytologist* 182 (1): 49–84.

Whiting, S.N., de Souza, M.P. and Terry, N. (2001). Rhizosphere bacteria mobilize Zn for hyperaccumulation by *Thlaspi caerulescens*. *Environmental Science & Technology* 35: 3144–3150.

Witkowski, M., Hubert, J. and Mazur, A. (2012). Methods of assessment of magnesium status in humans: A systematic review. *Magnesium Research* 24 (4): 163–180.

World Health Organization. (2003). *Diet, nutrition, and the prevention of chronic diseases: Report of a joint WHO/FAO expert consultation*, (Vol. 916). WHO: Geneva.

Xu, X.F., Wang, B., Lou, Y. et al. (2015). Magnesium transporter 5 plays an important role in Mg transport for male gametophyte development in Arabidopsis. *The Plant Journal* 84: 925–936.

Yadav, S.K., Yadav, P. and Chinnusamy, V. (2023). Nutrigenomics in cereals. In: *Biofortification in Cereals; Progress and Prospects* (eds. R. Deshmukh, A. Nadaf, WA. Ansari. et al.), 311–345. Singapore: Springer Nature.

Yamaji, N. and Ma, J.F. (2014). The node, a hub for mineral nutrient distribution in graminaceous plants. *Trends in Plant Science* 19: 556–563.

Yang, G.H., Yang, L.T., Jiang, H.X. et al. (2012). Physiological impacts of magnesium-deficiency in Citrus seedlings: Photosynthesis, antioxidant system and carbohydrates. *Trees* 26: 1237–1250.

Yang, J.L., You, J.F., Li, Y.Y. et al. (2007). Magnesium enhances aluminum-induced citrate secretion in rice bean roots (*Vigna umbellata*) by restoring plasma membrane H^+-ATPase activity. *Plant and Cell Physiology* 48: 66–73.

Ye, X., Chen, X.F., Deng, C.L. et al. (2019). Magnesium-deficiency effects on pigments, photosynthesis and photosynthetic electron transport of leaves, and nutrients of leaf blades and veins in *Citrus sinensis* seedlings. *Plants* 8: 389.

Yildirim, E., Karlidag, H. and Turan, M. (2009). Mitigation of salt stress in strawberry by foliar K, Ca and Mg nutrient supply. *Plant, Soil and Environment* 55: 213–221.

Zhu, C., Naqvi, S., Gomez-Galera, S. et al. (2007). Transgenic strategies for the nutritional enhancement of plants. *Trends in Plant Science* 12: 548–555.

Zhu, M., Jiang, X. and Ji, G. (2004). Experimental investigation on aluminum release from haplic acrisols in southeastern China. *Applied Geochemistry* 19 (7): 981–990.

Zirek, N.S. and Ozlem, U. (2020). The developmental and metabolic effects of different magnesium doses in pepper plants under salt stress. *Notulae Botanicae Horti Agrobotanici Cluj-Napoca* 48: 967–977.

15

Combating Fe Biofortification Under Heavy Metal Pollution

Ijaz Rasool Noorka[1], Muhammad Tamoor Qureshi[1], Mohamed Behnassi[2,3], Kafeel Ahmad[4], Muhammad Amjed Nawaz[5], Muhammad Salman Hameed[6], FNU Abdullah[7], Bilal Ahmad Khan[8]*

[1] *Department of Plant Breeding and Genetics, University of Sargodha, Pakistan*
[2] *Economic Social & Environmental Council (ESEC-Rabat)*
[3] *Ministry of Higher Education, Scientific Research and Innovation, Morocco*
[4] *Department of Botany, University of Sargodha, Pakistan*
[5] *National Research Tomsk State University, Tomsk, Russia*
[6] *National Key Laboratory of Green Pesticide, International Joint Research Center for Intelligent Biosensor Technology and Health, Central China Normal University Wuhan, P.R. China*
[7] *Department of Soil and Crop Sciences, Texas A&M University, College Station, TX, USA*
[8] *College of Earth and Environmental Sciences, University of the Punjab, Lahore*

*Corresponding author: ijazphd@yahoo.com

15.1 Introduction

The global human population is increasing rapidly and it will be 9.6 billion by 2050 as per the United Nations Population Fund. To feed this increased population would require a 70% increase in current production (Grafton, Daugbjerg, and Qureshi 2015). Uncontrolled increase in population is creating serious threats for most of the nations to fulfill their basic need of food, feed, fuel, fiber, and shelter (McNeely and Scherr 2012). In developing countries, the situation is worse because resources are limited and the growth rate of the population is uncontrolled (De Janvry and Sadoulet 2010). To overcome this condition, the need of the hour is to improve crop productivity by increasing the cultivated areas and/or improving yield per unit area. Among these two suggestions, improvement yield per unit area in existing farmland is preferred because of its feasibility (Fan et al. 2012). It is investigated that a 1% increase in yield decreases 0.6–1% in poverty (Mellor and Malik 2017). Continuous changes in climate and deterioration of the environment due to human activities are becoming key threats to food security (Spiertz 2013).

Plants experience several environmental variations but fail to prevent them externally because of their sessile nature (van Loon 2016); however, they respond endogenously to mitigate these unfavorable conditions (Fahad et al. 2015). Agricultural productivity is highly dependent upon multiple factors including growth conditions such as abiotic and biotic stresses (Mareri, Parrotta, and Cai 2022). Abiotic stresses may include climatic stress such as drought, heat, cold, temperature, rainfall, humidity, light, salinity, radiation, and heavy metals (Ahmad, Raina, and Khan 2019). Soil characteristics may include pH, texture, composition, structure of soil, and nutrient availability (Lerner 2018). These stresses induce physical, morphological, chemical, and biological changes in plants (Ashraf et al. 2018). Biotic factors include living organisms such as nematodes, insects, fungi, viruses, bacteria, and weeds (Suzuki et al. 2014). These factors cause serious preharvest and postharvest losses (Bray 2000). Nutritional imbalance, extreme high and low temperatures, water deficit, and salt stress are considered significant limitations for the crop's productivity (Cramer et al. 2011; Gull, Lone, and Wani 2019).

Among abiotic factors, heavy metals are gaining more and more attention because heavy metals are not only toxic to plant life but are serious threats to human health as well (Tuteja, Gill, and Tuteja 2011). Globally, toxicity caused by heavy metals is one of the major constraints and sources of pollution, which contaminates the productivity of crops leading to abiotic stress. So, exposure, absorption, and intake of toxic metals are serious concerns for life. Scientifically, heavy metals are those elements having a density ≥ 5 g cm^{-3} and an atomic number greater than 20 (Morkunas et al. 2018). From this definition, 51 elements (Weast 1984) from periodic tables are categorized into heavy metals and metalloids (Ali and Khan 2018). In trace amounts, some of these heavy metals are essentially required for plant upkeep, yield, and growth but they are toxic in excessive concentration for plants so their intake can be dangerous (Mourato, Reis, and Martins 2012). So heavy

Crop Biofortification: Biotechnological Approaches for Achieving Nutritional Security under Changing Climate, First Edition. Edited by Adnan Noor Shah et al.

Table 15.1 Anthropogenic activities to accelerate heavy metals in soil and environment.

Agricultural sources	**Pesticides** **Fertilizers (organic and inorganic)**	**Soil pollution**
Industrial sources	Burning of metals Metals recycling Chemical manufacturing industries Oil refineries Petrochemical plant, combustion from coal burning plant Power generation Smelling and mining of metals	Soil pollution + air pollution
Domestic source	Vehicle exhaust Smoking Urban runoff Construction	Soil pollution + air pollution

metals can be divided into three types; primary heavy metals are essential (iron [Fe], Mn, and Mo). Secondary heavy metals are required in low quantities but become toxic at high concentrations (Zn, Ni, Cr, Co, and Cu), while tertiary heavy metals are toxic at any concentration (Ag, Pb, cadmium [Cd], As, and Hg) (Nies 1984). Heavy metals, toxic chemicals, and metalloids accumulate in soil and environment through natural processes and by human activities or anthropogenic activities and affect the soil–plant environmental system (Schutzendubel and Polle 2002). Anthropogenic activities induce climate change, soil, environmental pollution, overexploitation, acid precipitation, and soil degradation.

In soil and environment, major sources of heavy metals are from the agricultural sector, which may include the application of pesticides, insecticides, fungicides, organic and inorganic fertilizers, irrigation water mixed with industrial effluent, municipal waste, and sewage sludge (Alloway 2013). Phosphorus inorganic fertilizers are one of the significant sources of deposition of toxic metals in soil their impact depends upon the source of phosphorus used (Pacyna et al. 2007). Industrial sources are burning of fuels, oil refineries, industrial discharge, petrochemical plants, metals recycling, chemical industries, combustion from coal-burning plants, power generation, urban runoff, vehicle exhaust, smelting and mining of metals and usages of metals in industries (Mohammadi et al. 2015) (Figure 15.1; Table 15.1). Moreover, metals are also used for multiple industrialization and mechanization processes in the fields of dry batteries, aviation (Azimi, et al. 2017), automobile fuels, and manufacturing of steel and photographic film (Buccolieri et al. 2006). Natural sources are soil parental material, continental dust, burning of fossil fuels, atmosphere, weathering, volcanic eruption, and erosion from enriched metallic soils (Sörme and Lagerkvist 2002).

All these sources may contribute to increasing the undesirable concentration of heavy metals such as Cd, nickel, lead, copper, arsenic, zinc, mercury, and chromium in soil, air, and the environment (Mahurpawar 2015). Major routes of exposure to humans, plants, and animals are through ingestion, inhalation, dermal, soil, sedimentation, and intake of contaminated water (Mudgal et al. 2010). Heavy metal accumulates in the human body through daily diet which may include cereals and leafy vegetables and fruits. It has been reported that cereal-based diets are the major contributor of heavy metals in human intake. Carcinogenic and neurotoxic action is well documented in humans due to exposure, absorption, and intake of heavy metals. Tobacco smoking is one of the major contributors to Cd intake in the human body.

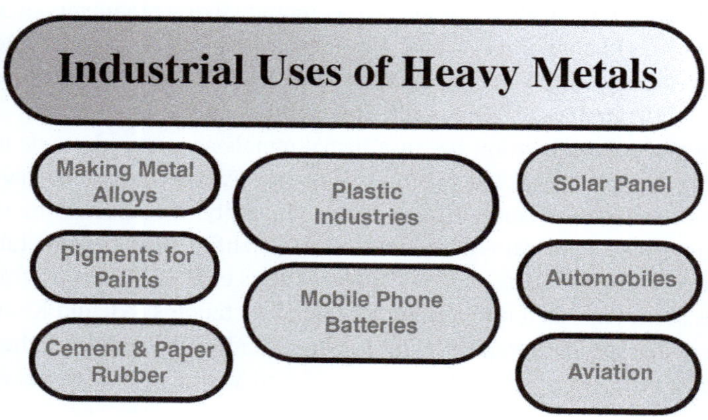

Figure 15.1 Possible uses of heavy metals in industries.

Epidemiologic studies revealed that chronic exposure to Cd is linked with cardiovascular effects, respiratory disease, osteoporosis, renal dysfunction, kidney disease, and cancer and may cause death (Rahimzadeh et al. 2017). Lethal effects on the central nervous system, reproductive system, renal, and hematopoietic are reported due to Pb exposure (Papanikolaou et al. 2015). Diseases such as skin lesions, hyperkeratosis, and cancer are observed because of chronic exposure to arsenic (Fatoki and Badmus 2022; Mazumder 2008). Chromium bioaccumulation and its derivative cause pathophysiological defects (Hossini et al. 2022) such as allergic reactions, affect the male reproductive system, damage small intestine and stomach, anemia, and alterations in DNA epigenetic results in modification in gene expression (Pavesi and Moreira 2020). Mercury exposure has been associated with numerous types of disorders, diseases, and risks such as hemolysis, atherosclerosis, hypertension, nephrological, coronary dysfunction, and myocardial infarction (Zulaikhah, Wahyuwibowo, and Pratama 2020). The insurgency of nickel in the human body causes allergy, cardio disease, and cancer of multiple types depending upon exposure (Genchi et al. 2020). Zinc toxification is rare but a high concentration of Zn accumulation is linked with the deficiency of copper. Cardiovascular effects and clinical syndromes such as visual impairment, hearing, and endocrine conditions are because by carbon monoxide toxicity in humans (Leyssens et al. 2017).

15.2 Effect of Heavy Metals on Plants

Heavy metals can enter the plant system through roots or be absorbed by leaves during respiration and transpiration (Mench and Martin 1991); however, accumulation of toxic metals depends upon several factors such as soil pH, humidity, temperature, fertilizers applied, cation exchange capacity, and air index (Sarwar et al. 2010). These are in the form of soluble components, which are available for plant uptake and others are solubilized by root exudates (Göthberg et al. 2004). Plants' ability to absorb essential nutrients attain other unnecessary elements and in response plant metabolism cannot break down these nonessential elements with undesirable concentrations resulting in drastic outcomes (Pantazis, Kalavrouziotis, and Deligiannakis 2007). Visible changes in plant morpho-physiological and biochemical traits can indicate toxicity symptoms (Polle and Schützendübel 2003). Phenotypical changes are anatomical and morphological changes including restricted plant growth, reduced seed germination, poor seed viability, deprived growth of shoot and root (Ueno et al. 2009), root blackening, changes in pigmenting, compromised pollen germination, chlorosis, severe decline in yield, and, in some, causes death of the plant.

Chemical changes include damage to the cellular membrane and membrane permeability (Obata, Inoue, and Umebayashi 1996), damage to plant tissues, disturbance to the ATP synthesis (Van Assche and Clijsters 1990), restriction mitosis stages, inhibition of structural protein, and reactive oxygen species ROS production. Biochemical changes include inhabitation of photosynthesis by affecting the photochemical process and chlorophyll biosynthesis (Khan, Samiullah Singh and Nazar 2007), peroxidation of lipids, limiting the rate of respiration and transpiration, restricting metabolic activities in various cell compartments, particularly in chloroplasts. Mineralization, accumulation of toxic metals in edible parts, hormonal imbalance (Kumari et al. 2022), altered bioavailability of other essential nutrients, and water stress are also caused by heavy metal toxicity in plants (Göthberg et al. 2004). (Table 15.2) Naturally, plants have strong antioxidant defense mechanisms

Table 15.2 Impact of heavy metals on crop plants.

Morphological traits	Low seed germination, poor seed viability, stunt growth, short length of roots and shoots, poor pollen viability	Ueno et al. (2009) Polle and Schützendübel (2003) Barman and Kundu (2023)
Metabolic and enzymatic	Restrict metabolic activities, high production of ROS, high concentration of ethylene, malondialdehyde carbonylated, peroxidation of lipid damaging cell membrane	Asada (1999) Mourón et al. (2004) Noctor and Foyer (1998) Van Assche and Clijsters (1990)
Biochemical traits	Damage plant tissues, disturb the synthesis of ATP, restrict mitosis stages, inhibition of structural protein	Azooz, Abou-Elhamd, and Al-Fredan (2012) Ghori et al. (2019)
Photosynthesis parameters	Low rate of photosynthesis, low rate of transpiration, poor stomata conductance, chlorophyll biosynthesis	Hafeez et al. (2023) Bharwana et al. (2013) Ci et al. (2010)
Yield-related parameters	High concentration of HM in edible part, loss in biomass	Qaswar, Hussain, and Rengel (2017)
Other traits	Hormonal imbalance, mineralization, hinder the accumulation of other essential nutrient, death of plant	Kumari et al. (2022)

to prevent them from oxidative damage. It is composed of enzymatic and nonenzymatic metabolites. Ascorbic acid is good example of nonenzymatic antioxidant (Noctor and Foyer 1998) while superoxide dismutase, catalase, peroxidase, glutathione reductase and ascorbate peroxidase are enzymatic antioxidant (Polle and Rennenberg 1993). Under stress conditions, reactive oxygen species generation is increased. This increase causes peroxidation of membrane phospholipids and oxidation of cellular proteins. Moreover, ROS oxidize amino acid and nucleic acids (Mourón et al. 2004). High concentrations of ethylene, malondialdehyde, and carbonylated proteins are indicators of increased ROS and damaged plant tissues (Asada 1999).

15.3 Remediation Techniques

In situ and ex situ remediation techniques can be used to rectify heavy metal toxicity including soil washing and flushing, surface capping, solidification, encapsulation, electrokinetic extraction, and excavation (Edelstein and Ben-Hur 2018). These techniques employ immobilization and extraction of toxic metals from the polluted soil. However, these remediation techniques are less productive, costly, and feasibly not viable. So, the primary concern is to use appropriate methods to rectify toxicity that should be cost-effective, feasible, easily accessible, long-lasting, and environment friendly (Liu et al. 2018). Heavy metals concentration in soil can also be alleviated using proper management practices and using proper technologies for better crop productivity. The one significant technique to mitigate these metals is the use of chemical approaches, specifically with the application of micronutrients. Around the globe, the use of micronutrients, such as Zn, Fe, and Se, is an effective, cost-effective, and practical approach to remove the metallic toxicity in crop plants (Szerement et al. 2022). Micronutrients are required in trace amounts for humans and plants which is essentially important for growth, development, and improvement and crucial for many biological and physiological functions.

15.4 Micronutrients to Reduce Heavy Metals Toxicity

Micronutrients are required in trace amounts for humans and plants which is essentially important for growth, development, and improvement and crucial for many biological and physiological functions but around the world billions of people are facing deficiency of micronutrients or they are undernourished and the situation is worse in developing counties. In developing countries, resources are limited and the population growth rate is uncontrolled so these deficiencies prevail and affect growing children and pregnant women, labor, and sportsmen. The dynamic role of micronutrients is well documented for human health and plant survival and development. For humans, plants are the most significant and economical source to stave off this micronutrient malnutrition. However, no single plants provide all nutrients essential for human nourishment and most of the crop plants lack micronutrients even required for their survival and growth. The importance of micronutrients in crop sciences is inevitable because plants primarily depend upon nutrients and these nutrients are involved in numerous ranges of plant activities. In plants, micronutrients are required for cellular and metabolic activities such as germination, growth, photosynthesis, chlorophyll synthesis, respiration, transpiration, reproductive growth, metabolic activities, activation of defense mechanisms against biotic and abiotic stresses, and essential for better yield. In plants, the acceleration of reactive oxygen species is stimulated because of abiotic stresses such as heavy metals, salinity, drought, temperature extreme high/low, humidity, and radiation. High ROS production is the stimulus of oxidative stress in crop plants. One of the main features of micronutrients is to control the production of ROS homeostasis, and oxidative stress and activate the defense mechanism of plants during heavy metals stress. Kandhol et al. (2024) reported that heavy metals toxicity restricts the accumulation of essential nutrients such as calcium, nitrogen, and iron.

15.5 Strategies to Improve Micronutrient Deficiency in Plants

Biofortification is one of the best adaptive techniques to improve the desirable concentration of micronutrients in the edible part or to improve the overall nutritive status of crop plants without compromising any other essential characteristics such as quality, taste, and yield (Taylor, Taylor, and Kini 2012), through using plant-breeding approaches, agronomic approaches, biotechnology, and genetic engineering (Saltzman et al. 2013). Biofortification is a novel method, eco-friendly, cost-effective, and a one-time investment (Bouis et al. 2020). This strategy not only improves the quality of crop but also enhances the fertility of the soil, improves protein, oil contents, amino acids, and carotenoids, improves grain yield, a key role in enzymatic and metabolic activities in plant, and develops resistance against stresses (Qaim, Stein, and Meenakshi 2007).

15.6 Role of Biofortification Approaches to Mitigate Heavy Metals Toxicity

Application of micronutrients such as Fe, Zn, and Se is reported to produce enzymes that scavenge ROS production (Bayanati et al. 2022), including superoxide dismutase, catalase, peroxidase, glutathione S-transferase, glutathione reductase, glutathione peroxidase, and ascorbate peroxidase. Nonenzymatic metabolites such as ascorbic acid, carotenoids, and proline are nonenzymatic antioxidants. These nonenzymatic antioxidants are also effective in controlling ROS development. Barman and Kundu (2023) revealed that the damaging effect of Cd in the reproductive parts of rice plants including germination, viability of pollens, and Cd translocation in various parts of the plant, concentration in grain was efficiently alleviated by the foliar application of Se at various stages. Application of Se improves Se concentration in grain and reduces ROS production. Combined foliar and basal application of selenium and silicon in rice crops alleviates the toxicity of Cd by improving yield, reducing the accumulation of Cd in rice grain enhancing the activity of antioxidant enzymes, improving the rate of photosynthesis and transpiration and stomata conductance (Gao et al. 2018; Li et al. 2022). Zhou et al. (2021) also reported similar results in wheat crops. Results showed that basal and foliar applications of Se and Si reduce the malondialdehyde (MDA) value and improve wheat grain yield and antioxidant enzyme activities by regulating the Cd transport gene. They also recommend that foliar application showed better results than basal application and due to foliar application of both nutrients, Cd concentration in vegetative parts is reduced. While against arsenic toxicity, combined fertigation of selenium and silicon also displays better results by improving photosynthetic parameters such as CO_2 concentration transpiration rate, and stomata opening, and modulating pigment levels (Das and Biswas 2022). In wheat crops, silicon improves the growth of wheat seedlings under Cd toxicity by enhancing the antioxidant defense mechanism of wheat crops specifically the AsA-GSH cycle and reducing Cd accumulation in vegetative parts of wheat seedlings (Singh et al. 2022). In corn crops, the phytotoxicity of Cd is ameliorated by the application of silicon and zinc. Both micronutrients application reduces the accumulation of Cd, improves biomass, and increases Zn and Si concentration in roots and leaves (Da-Cunha and do Nascimento 2019; Vaculík et al. 2012). Applications of zinc reduce the harmful effect of Cd in mung bean by restricting Cd translocation in several parts of the plant (Rashid et al. 2022).

15.7 Fe Alleviates the Toxicity of Heavy Metals

Fe is one of the most significant essential nutrients for humans, animals, and plants. It plays a crucial role in electron transfer reactions and as a cofactor in photosynthesis, maintenance of mitochondria, and chloroplasts integration of N and S (Abdel-Ghany et al. 2005) in the form of Fe-sulfur and hemi proteins (Atamna, Walter, and Ames 2002). Deficiency of Fe is the most common nutritional disorder in the world. In crops, Fe deficiency, particularly in alkaline soils or calcareous soils, is a major cause of chlorosis, compromised quality, decreased yield, and vegetative growth (Abadía et al. 2011). Fe deficiency induces chlorosis, because the production of light-harvesting pigments, carotenoids, and chlorophylls decreases due to which there is a marked decline in electron transport, induced formation of the thylakoid membrane, and slow photosynthesis rate (Abadía 1992). There is wide variation in absorption and transport mechanisms due to the bioavailability of Fe in plant tissues. Fe can play an imperative role in mitigating the toxicity of heavy metals by restricting the accumulation and translocation of metals and enhancing the growth of plants through the accumulation of photosynthetic pigment (Nada et al. 2007). Biofortification of Fe is the best-suited way to stave off Fe deficiency in humans and in plants as well. There is a need to develop crops that proficiently mobilize Fe with limited Fe bioavailability and can accumulate more Fe in edible parts of plants. In the world, different countries are releasing Fe-biofortified varieties with the help of Harvest Plus and local agriculture researchers and researchers have made tremendous achievements by evolving a number of Fe-fortified varieties (Figure 15.2).

Cd toxicity, decrease in chlorophyll content, and damage to chloroplast-thylakoids were observed with an increase in oxidative stress along with Fe-deficient conditions (Qureshi et al. 2010). But, impressively, Fe application under Cd toxicity

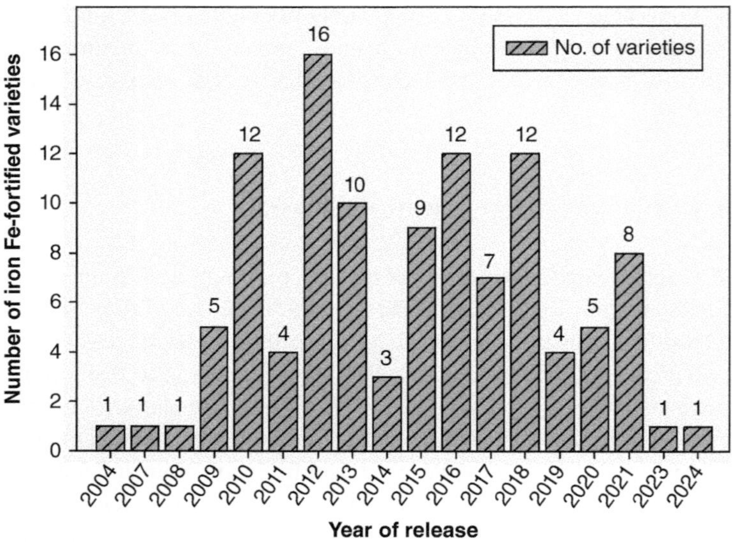

Figure 15.2 Fe-fortified varieties year by year.

mitigates the toxicity of Cd by improving chlorophyll content, chloroplast-thylakoids, and suppression of oxidative stress. Zaheer et al. (2020) recommended that chromium toxicity in spinach can be alleviated by the application of Fe along with amino acid lysine. Application of Fe improves Fe concentration in leaves, hinders chromium uptake, and mediates the toxic effect of chromium by reducing electrolyte leakage, decreasing hydrogen peroxide contents and malondialdehyde which means it inhibits oxidative damage in vegetative parts of spinach. Danish et al. (2019) stated that agronomic biofortification of Fe along with plant growth-promoting rhizobacteria is a suitable approach to mitigate the toxic effect of chromium. Fortification of Fe improves fresh and dry weight and length of shoot and root, the height of the plant, number of leaves, and improves biochemical and metabolic activities of corn crops and in rice crops similar findings were observed (Bashir et al. 2018). Application of Fe by foliar and lysine in various growth and development stages of rice crops exposed and improved the yield and quality of rice crops under Cd toxicity.

15.8 Conclusion

An uncontrolled increase in population is creating serious threats for most of the nations to fulfill their basic need of food, feed, fuel, fiber, and shelter. On the other hand, continuous changes in climate and deterioration of the environment due to anthropogenic activities are becoming a key threat to food security. Among abiotic factors, heavy metals are not only toxic to plant life but are serious threats to human health as well and contaminate the productivity of agriculture because plants are more vulnerable to this abiotic stress. The use of chemical approaches to mitigate the heavy metals effects in plants is one of the significant techniques, specifically with the application of micronutrients. Around the globe, the use of micronutrients, such as Zn, Fe, and Se, is a cost-effective and practical approach to alleviate the metallic toxicity in plants. Biofortification is one of the best adaptive techniques to improve the concentration of micronutrients in edible parts of crop plants. Biofortification not only improves micronutrient deficiency but also develops resistance against biotic and abiotic stress. Fe plays a crucial role in electron transfer reactions and as a cofactor in photosynthesis and maintenance of mitochondria and development of resistance against heavy metals toxicity.

References

Abadía, J. (1992). Leaf responses to Fe deficiency: a review. *Journal of Plant Nutrition* 15 (10): 1699–1713.

Abadía, J., Rellán-Álvarez, R., Vázquez, S. et al. (2011). Towards a knowledge-based correction of iron chlorosis. *Plant Physiology and Biochemistry* 49 (5): 471–482.

Abdel-Ghany, S.E., Ye, H., Garifullina, G.F. et al. (2005). Iron-sulfur cluster biogenesis in chloroplasts. Involvement of the scaffold protein CpIscA. *Plant Physiology* 138 (1): 161–172.

Ali, H. and Khan, E. (2018). What are heavy metals? Long-standing controversy over the scientific use of the term "heavy metals"–proposal of a comprehensive definition. *Toxicological & Environmental Chemistry* 100 (1): 6–19.

Alloway, B.J. (2013). Sources of heavy metals and metalloids in soils. In: *Heavy Metals in Soils: Trace Metals and Metalloids in Soils and Their Bioavailability*, 11–50. Dordrecht: Springer.

Asada, K. (1999). The water-water cycle in chloroplasts: scavenging of active oxygens and dissipation of excess photons. *Annual Review of Plant Biology* 50 (1): 601–639.

Ashraf, M.A., Iqbal, M., Rasheed, R., et al. (2018). Environmental stress and secondary metabolites in plants: an overview. *Plant Metabolites and Regulation Under Environmental Stress* 153–167.

Atamna, H., Walter, P.B., and Ames, B.N. (2002). The role of heme and iron-sulfur clusters in mitochondrial biogenesis, maintenance, and decay with age. *Archives of Biochemistry and Biophysics* 397 (2): 345–353.

Azimi, A., Azari, A., Rezakazemi, M. et al. (2017). Removal of heavy metals from industrial wastewaters: a review. *Chemical and Biochemical Engineering Reviews* 4 (1): 37–59.

Azooz, M.M., Abou-Elhamd, M.F. and M.A. Al-Fredan. (2012). Biphasic effect of copper on growth, proline, lipid peroxidation and antioxidant enzyme activities of wheat (Triticum aestivum'cv. Hasaawi) at early growing stage. *Australian Journal of Crop Science* 6 (4): 688–694.

Barman, F. and Kundu, R. (2023). Foliar application of selenium affecting pollen viability, grain chalkiness, and transporter genes in cadmium accumulating rice cultivar: a pot study. *Chemosphere* 313: 137538.

Bayanati, M., Al-Tawaha, A.R., Al-Taey, D. et al. (2022). Interaction between zinc and selenium bio-fortification and toxic metals (loid) accumulation in food crops. *Frontiers in Plant Science* 13: 1001992.

Bharwana, S.A., Ali, S., Farooq, M.A. et al. (2013). Alleviation of lead toxicity by silicon is related to elevated photosynthesis, antioxidant enzymes suppressed lead uptake and oxidative stress in cotton. *Journal of Bioremediation Biodegradation* 4 (4): 10–4172.

Bouis, H., Birol, E., Boy, E. et al. (2020). Food biofortification: reaping the benefits of science to overcome hidden hunger. In: *October Webinar on the Need for Agricultural Innovation to Sustainably Feed the World by 2050 (No. 69)*. Council for Agricultural Science and Technology (CAST).

Bray, E.A. (2000). Responses to abiotic stresses. *Biochemistry and Molecular Biology of Plants* 1158–1203.

Buccolieri, A., Buccolieri, G., Dell'Atti, A. et al. (2006). Natural sources and heavy metals. *Annali Di Chimica: Journal of Analytical, Environmental and Cultural Heritage Chemistry* 96 (3-4): 167–181.

Ci, D., Jiang, D., Wollenweber, B. et al. (2010). Cadmium stress in wheat seedlings: growth, cadmium accumulation and photosynthesis. *Acta Physiologiae Plantarum* 32(2): 365–373.

Cramer, G.R., Urano, K., Delrot, S. et al. (2011). Effects of abiotic stress on plants: a systems biology perspective. *BMC Plant Biology* 11: 1–14.

Da-Cunha, K.P.V. and do Nascimento, C.W.A., (2009). Silicon effects on metal tolerance and structural changes in maize (Zea mays L.) grown on a cadmium and zinc enriched soil. *Water Air Soil Pollution* 197 (1): 323–330.

Danish, S., Kiran, S., Fahad, S., et al. (2019). Alleviation of chromium toxicity in maize by Fe fortification and chromium tolerant ACC deaminase producing plant growth promoting rhizobacteria. *Ecotoxicology and Environmental Safety* 185 (18): 109706.

Das, S. and Biswas, A.K. (2022). Comparative study of silicon and selenium to modulate chloroplast pigments levels, Hill activity, photosynthetic parameters and carbohydrate metabolism under arsenic stress in rice seedlings. *Environmental Science Pollution Research International* 29 (13): 19508–19529.

De Janvry, A. and Sadoulet, E. (2010). Agricultural growth and poverty reduction: additional evidence. *The World Bank Research Observer* 25 (1): 1–20.

Edelstein, M. and Ben-Hur, M. (2018). Heavy metals and metalloids: sources, risks and strategies to reduce their accumulation in horticultural crops. *Scientia Horticulturae* 234: 431–444.

Fahad, S., Nie, L., Chen, Y. et al. (2015). Crop plant hormones and environmental stress. Sustainable. *Agriculture Reviews: Volume* 15: 371–400.

Fan, M., Shen, J., Yuan, L. et al. (2012). Improving crop productivity and resource use efficiency to ensure food security and environmental quality in China. *Journal of Experimental Botany* 63 (1): 13–24.

Fatoki, J.O. and Badmus, J.A. (2022). Arsenic as an environmental and human health antagonist: a review of its toxicity and disease initiation. *Journal of Hazardous Materials Advances* 5: 100052.

Gao, M., Zhou, J., Liu, H. et al. (2018). Foliar spraying with silicon and selenium reduces cadmium uptake and mitigates cadmium toxicity in rice. *Science of The Total Environment* 631 (1): 1100–1108.

Genchi, G., Carocci, A., Lauria, G. et al. (2020). Nickel: human health and environmental toxicology. *International Journal of Environmental Research and Public Health* 17 (3): 679.

Ghori, N.H., Ghori, T., Hayat, M.Q. et al. (2019). Heavy metal stress and responses in plants. *International Journal of Environmental Science and Technology* 16: 1807–1828.

Göthberg, A., Greger, M., Holm, K. et al. (2004). Influence of nutrient levels on uptake and effects of mercury, cadmium, and lead in water spinach. *Journal of Environmental Quality* 33 (4): 1247–1255.

Grafton, R.Q., Daugbjerg, C., and Qureshi, M.E. (2015). Towards food security by 2050. *Food Security* 7: 179–183.

Gull, A., Lone, A.A., and Wani, N.U.I. (2019). Biotic and abiotic stresses in plants. *Abiotic and Biotic Stress in Plants* 7: 1–19.

Hafeez, A., Rasheed, R., Ashraf, M.A. et al. (2023). Effect of heavy metals on growth, physiological and biochemical responses of plants. In: *Plants and Their Interaction to Environmental Pollution*, 139–159. Amsterdam: Elsevier.

Hossini, H., Shafie, B., Niri, A.D. et al. (2022). A comprehensive review on human health effects of chromium: insights on induced toxicity. *Environmental Science and Pollution Research* 29 (47): 70686–70705.

Kandhol, N., Srivastava, A., Rai, P. et al. (2024). Cytokinin and indole-3-acetic acid crosstalk is indispensable for silicon mediated chromium stress tolerance in roots of wheat seedlings. *Journal of Hazardous Materials* 468: 133134.

Khan, N.A., Samiullah Singh, S. and Nazar, R. (2007). Activities of antioxidative enzymes, sulphur assimilation, photosynthetic activity and growth of wheat (Triticum aestivum) cultivars differing in yield potential under cadmium stress. *Journal of Agronomy and Crop Science* 193 (6): 435–444.

Kumari, V.V., Banerjee, P., Verma, V.C. et al. (2022). Plant nutrition: an effective way to alleviate abiotic stress in agricultural crops. *International Journal of Molecular Sciences* 23 (15): 8519.

Lerner, H.R. (2018). Introduction to the response of plants to environmental stresses. In: *Plant Responses to Environmental Stresses*, 1–26. United Kingdom: Routledge.

Leyssens, L., Vinck, B., Van Der Straeten, C. et al. (2017). Cobalt toxicity in humans—a review of the potential sources and systemic health effects. *Toxicology* 387: 43–56.

Li, D., Hongyan, L.I.U., Gao, M. et al. (2022). Effects of soil amendments, foliar sprayings of silicon and selenium and their combinations on the reduction of cadmium accumulation in rice. *Environmental Science and Pollution Research* 32 (4): 649–659.

Liu, L., Li, W., Song, W. et al. (2018). Remediation techniques for heavy metal-contaminated soils: principles and applicability. *Science of the Total Environment* 633: 206–219.

Mahurpawar, M. (2015). Effects of heavy metals on human health. *International Journal of Research Granthaalayah* 530 (516): 1–7.

Mareri, L., Parrotta, L., and Cai, G. (2022). Environmental stress and plants. *International Journal of Molecular Sciences* 23 (10):5416.

Mazumder, D.G. (2008). Chronic arsenic toxicity and human health. *Indian Journal of Medical Research* 128 (4): 436–447.

McNeely, J.A. and Scherr, S.J. (2012). *Ecoagriculture: Strategies to Feed the World and Save Wild Biodiversity*. Washington: Island Press.

Mellor, J.W. and Malik, S.J. (2017). The impact of growth in small commercial farm productivity on rural poverty reduction. *World Development* 91 (C): 1–10.

Mench, M. and Martin, E. (1991). Mobilization of cadmium and other metals from two soils by root exudates of Zea mays L., Nicotiana tabacum L. and Nicotiana rustica L. *Plant and soil* 132: 187–196.

Mohammadi, R.M., Sobhan, A.S., Karimi, H. et al. (2015). Natural and anthropogenic source of heavy metals pollution in the soil samples of an industrial complex; a case study. *Iranian Journal of Toxicology*.

Morkunas, I., Woźniak, A., Mai, V.C. et al. (2018). The role of heavy metals in plant response to biotic stress. *Molecules* 23 (9): 2320.

Mourato, M., Reis, R. and Martins, L.L. (2012). Characterization of plant antioxidative system in response to abiotic stresses: a focus on heavy metal toxicity. *Advances in Selected Plant Physiology Aspects* 12: 1–17.

Mourón, S.A., Grillo, C.A., Dulout, F.N. et al. (2004) A comparative investigation of DNA strand breaks, sister chromatid exchanges and K-ras gene mutations induced by cadmium salts in cultured human cells. *Mutation Research/Fundamental and Molecular Mechanisms of Mutagenesis* 568 (2): 221–231.

Mudgal, V., Madaan, N., Mudgal, A. et al. (2010). Effect of toxic metals on human health. *The Open Nutraceuticals Journal* 3 (1): 94–99.

Nada, E., Ferjani, B.A., Ali, R. et al. (2007). Cadmium-induced growth inhibition and alteration of biochemical parameters in almond seedlings grown in solution culture. *Acta Physiologiae Plantarum* 29 (1): 57–62.

Nies, D.H. (1984). Microbial heavy-metal resistance. *Applied Microbiology Biotechnology* 51: 730–750.

Noctor, G. and Foyer, C.H. (1998). Ascorbate and glutathione: keeping active oxygen under control. *Annual Review of Plant Biology* 49 (1): 249–279.

Obata, H., Inoue, N. and Umebayashi, M. (1996). Effect of Cd on plasma membrane ATPase from plant roots differing in tolerance to Cd. *Soil Science and Plant Nutrition* 42 (2): 361–366.

Pacyna, E.G., Pacyna, J.M., Fudala, J. et al. (2007). Current and future emissions of selected heavy metals to the atmosphere from anthropogenic sources in Europe. *Atmospheric Environment* 41 (38): 8557–8566.

Pantazis, V., Kalavrouziotis, I. and Deligiannakis, I. (2007). *Cu-Zn Accumulation on Soil Plant System Irrigated with Wastewater*. Proceedings in IWA Facing Sludge Diversities: Challenges, Risks and Opportunities. Antalya.

Pavesi, T. and Moreira, J.C. (2020). Mechanisms and individuality in chromium toxicity in humans. *Journal of Applied Toxicology* 40 (9): 1183–1197.

Polle, A. and Rennenberg, H. (1993). Significance of antioxidants in plant adaptation to environmental stress. In: Mansfield T., Fowden L., Stoddard, F. *Plant Adaptation to Environmental Stress*. London: Chapman & Hall, 263–273.

Polle, A. and Schützendübel, A. (2003). Heavy metal signalling in plants: linking cellular and organismic responses. In: *Plant Responses to Abiotic Stress*, 187–215. Berlin, Heidelberg: Springer.

Qaim, M., Stein, A.J. and Meenakshi, J.V. (2007). Economics of biofortification. *Agricultural Economics* 37 (s1): 119–133.

Qureshi, M.I., D'Amici, G.M., Fagioni, M. et al. (2010). Iron stabilizes thylakoid protein–pigment complexes in Indian mustard during Cd-phytoremediation as revealed by BN-SDS-PAGE and ESI-MS/MS. *Journal of Plant Physiology* 167 (10): 761–770.

Rahimzadeh, M.R., Rahimzadeh, M.R., Kazemi, S. et al. (2017). Cadmium toxicity and treatment: an update. *Caspian Journal of Internal Medicine* 8 (3): 135.

Rashid, M.H., Rahman, M.M., Halim, M.A. et al. (2022). Growth, metal partitioning and antioxidant enzyme activities of mung beans as influenced by zinc oxide nanoparticles under cadmium stress. Zed Rengel: Crop and Pasture Science.

Saltzman, A., Birol, E., Bouis, H.E. et al. (2013). Biofortification: progress toward a more nourishing future. *Global Food Security* 2(1): 9–17.

Sarwar, N., Saifullah Malhi, S.S., Zia, M.H. et al. (2010). Role of mineral nutrition in minimizing cadmium accumulation by plants. *Journal of the Science of Food and Agriculture* 90 (6): 925–937.

Schutzendubel, A. and Polle, A. (2002). Plant responses to abiotic stresses: heavy metal-induced oxidative stress and protection by mycorrhization. *Journal of Experimental Botany* 53 (372): 1351–1365.

Singh, S., Prasad, S.M., Sharma, S. et al. (2022). Silicon and nitric oxide-mediated mechanisms of cadmium toxicity alleviation in wheat seedlings. *Physiologia Plantarum* 174 (5): p.e13065.

Sörme, L. and Lagerkvist, R.J.S.O.T T.E. (2002). Sources of heavy metals in urban wastewater in Stockholm. *Science of the Total Environment* 298 (1–3): 131–145.

Spiertz, H. (2013). Challenges for crop production research in improving land use, productivity and sustainability. *Sustainability* 5 (4): 1632–1644.

Suzuki, N., Rivero, R.M., Shulaev, V. et al. (2014). Abiotic and biotic stress combinations. *New Phytologist* 203 (1): 32–43.

Szerement, J., Szatanik-Kloc, A., Mokrzycki, J. et al. (2022). Agronomic biofortification with Se, Zn, and Fe: an effective strategy to enhance crop nutritional quality and stress defense – a review. *Journal of Soil Science and Plant Nutrition* 22 (1): 1129–1159.

Taylor, J., Taylor, J.R.N., and Kini, F. (2012). Cereal biofortification: strategies, challenges and benefits. *Cereal Foods World* 57 (4): 165–169.

Tuteja, N., Gill, S.S., and Tuteja, R. (2011). Plant responses to abiotic stresses: shedding light on salt, drought, cold and heavy metal stress. *Omics and Plant Abiotic Stress Tolerance* 1: 39–64.

Ueno, D., Koyama, E., Kono, I. et al. (2009). Identification of a novel major quantitative trait locus controlling distribution of Cd between roots and shoots in rice. *Plant and Cell Physiology* 50 (12): 2223–2233.

Vaculík, M., Landberg, T., Greger, M. et al. (2012). Silicon modifies root anatomy, and uptake and subcellular distribution of cadmium in young maize plants. *Annals of Botany* 110 (2): 433–443.

Van Assche, F. and Clijsters, H. (1990). Effects of metals on enzyme activity in plants. *Plant, Cell and Environment* 13 (3): 195–206.

van Loon, L. C. (2016). The intelligent behavior of plants. *Trends in Plant Science* 21 (4): 286–294.

Weast, R.C. (1984). CRC *Handbook of Chemistry and Physics*, 64e. Boca Raton: CRC Press.

Zaheer, I.E., Ali, S., Saleem, M.H. et al. (2020). Iron–lysine mediated alleviation of chromium toxicity in spinach (Spinacia oleracea L.) plants in relation to morpho-physiological traits and iron uptake when irrigated with tannery wastewater. *Sustainability* 12 (16): 6690.

Zhou, J., Zhang, C., Du, B. et al. (2021). Soil and foliar applications of silicon and selenium effects on cadmium accumulation and plant growth by modulation of antioxidant system and Cd translocation: comparison of soft vs. durum wheat varieties. *Journal of Hazardous Materials* 402: 123456.

Zulaikhah, S.T., Wahyuwibowo, J., and Pratama, A. A. (2020). Mercury and its effect on human health: a review of the literature. *International Journal of Public Health Science* 9 (2): 103–114.

16

Biofortification in Vegetables: Enhancing Nutritional Value for Improved Human Health

Waseem Hassan[1], Sundas Riaz[1], Muhammad Asif Shehzad[2], Muhammad Nawaz[4], Ahmad Mahmood[1], Umair Riaz[1], Shakeel Ahmad[1], Muhammad Imran[1], Muqarrab Ali[3], Adnan Noor Shah[4]*

[1] *Department of Soil and Environmental Sciences, Muhammad Nawaz Shareef University of Agriculture, Multan*
[2] *Institute of Plant Breeding and Biotechnology, Muhammad Nawaz Shareef University of Agriculture, Multan*
[3] *Department of Climate Change, Muhammad Nawaz Shareef University of Agriculture, Multan*
[4] *Department of Agricultural Engineering, Khwaja Fareed University of Engineering and Information Technology*

*Corresponding author: adnan.shah@kfueit.edu.pk

16.1 Introduction

16.1.1 Definition and Concept of Biofortification

Biofortification is a process aimed at increasing the nutritional value of food crops by enhancing their levels of essential vitamins, minerals, and other nutrients through either conventional breeding methods or biotechnology (Ofori et al. 2022). The concept revolves around increasing the concentration of essential vitamins, minerals, and other nutrients such as iron, zinc, vitamin A, and vitamin D in staple crops to address malnutrition, particularly in populations that heavily rely on these crops for sustenance.

16.1.2 Importance of Biofortification in Addressing Global Malnutrition

Every human has the fundamental right to access nutritious food. A large section of the global population suffers from vitamin and mineral deficiencies, especially in underdeveloped nations. Severe health issues such as stunted growth, compromised immune systems, and elevated mortality rates might result from them. According to Stein (2010), biofortification aids in boosting these vital nutrients in staple crops, increasing their accessibility for people at risk of malnutrition. Approximately 3 billion people suffer from micronutrient deficiencies, while over 1 billion people, primarily in poor nations, go hungry every day. Over 1.5 million children worldwide pass away each year as a result of nutritional deficiencies, especially those involving vitamin A and iron and zinc, with the majority of those fatalities taking place in sub-Saharan Africa and South Asia.

As a result, undernutrition in protein and calories as well as deficiencies in micronutrients are included in the general term "malnutrition," which is a serious public health issue, especially in southern Asia and sub-Saharan Africa (Graham et al. 2001). Plant-breeding programs can increase the amounts of minerals and vitamins in crops since plants frequently exhibit genetic diversity in the content of critical micronutrients (Hirschi 2009). The HarvestPlus program's primary approach is conventional breeding, which aims to target the mainstay grains across Asia and Africa, including wheat, rice, corn, cassava, pearl millet, beans, and sweet potatoes (Zhao and Shewry 2011). Genetic engineering can be used to increase the micronutrient content of grains. Through Agrobacterium-mediated transformation, the soybean iron storage protein ferritin gene was introduced into the rice variety Kitaake (Goto et al. 1999). The expression of the ferritin gene was selectively localized in the endosperm using the promoter for the rice seed storage protein glutelin *GluB-1–*. The transgenic seeds had up to three times more iron in them than the control group. Using molecular engineering to increase the concentration of zinc in edible sections of crops may result in the build-up of some poisonous

Crop Biofortification: Biotechnological Approaches for Achieving Nutritional Security under Changing Climate, First Edition. Edited by Adnan Noor Shah et al.

metals, such as cadmium (Cd), and a toxic analogue of zinc. The creation of crops that are biofortified with zinc hence requires methods to reduce the buildup of any hazardous substances that share chemical characteristics. Enhancing Cd sequestration in root vacuoles or expressing Cd exporters in roots are two potential strategies (Lee et al. 2003). Cassava has the lowest protein-to-carbohydrate ratio among the 10 major crops grown worldwide, with 1–2% protein found in its roots. Usually cereals supply 50% of the protein in human diets and can make up as much as 70% of protein intakes for those living in developing nations (Gibbon and Larkins 2005). Vitamin A deficiency (VAD) is extremely common in the poor, whose diets are mainly composed of cereal grains and tubers. Thus, cereal and tuber crops are the primary targets of crop development to increase vitamin A levels. The main target for provitamin A biofortification is β-carotene, and greater carotenoid accumulation can be achieved by directing and boosting flow into the carotenoid biosynthetic pathway or by decreasing β-carotene turnover (Sayre et al. 2011; Ye et al. 2000; Yan et al. 2010). The best example of the former technique is the development of so-called Golden Rice (Ye et al. 2000).

CIAT (International Centre for Tropical Agriculture) research found that certain common bean types contained up to 60–80% more zinc than commercially grown variants. Breeding efforts can be used to add more zinc into improved cultivars (Beebe et al. 2000). Rapeseed, maize, and soybean lysine content are being increased using biotechnology techniques. The insertion of bacterial genes for dihydrodipicolinic acid (DHPHS) and aspartokinase (AK) enzymes represented by the *dapA* gene from *Corynebacterium* and the *lysC* gene from *Escherichia coli* resulted in a fivefold rise in lysine in canola, maize, and soybean (Falco et al. 1995). Similarly, the amino acid profile and total protein content of potatoes were improved by introducing the *AMAI1* gene from *Amaranthus hypochondriacus*. The biofortified potatoes have been tested in field experiments and are currently awaiting a decision from authorities on approval or the necessity for additional testing (Ramaswamy 2007).

16.1.3 Scope and Objectives of Biofortification

Vegetable biofortification has a wide range of applications and tremendous promise to alleviate nutritional inadequacies globally. Biofortification is a sustainable and economical method of enhancing the quality of food by breeding or genetically modifying crops to contain higher levels of vital nutrients such as vitamins, minerals, and antioxidants. Vegetables that have been biofortified can also be customized to target certain vitamin deficits that are common in various places, which makes them very useful for meeting local nutritional demands. Furthermore, communities in both rural and urban locations can get biofortified vegetables due to their cultural acceptance and integration into current farming systems. The major goal of biofortification in vegetables is to improve their nutritional value by boosting the levels of vital vitamins, minerals, and other beneficial chemicals. This tries to address micronutrient deficiencies such as vitamin A, iron, zinc, and iodine, which are common worldwide and contribute to malnutrition and other health problems. Specifically, the aims of biofortification in vegetables are (i) improving public health; (ii) addressing malnutrition; (iii) addressing food security; and (iv) promoting sustainable agriculture.

16.2 Nutritional Challenges and Opportunities in Vegetables

16.2.1 Overview of the Nutritional Composition of Vegetables

Vegetables are a varied collection of crops that contain a wide range of important nutrients, including:

1. Vitamins: Vegetables contain vitamins A, C, K, and folate, among others (Ebabhi and Adebayo 2022).
2. Minerals: Vegetables are rich in minerals such as potassium, calcium, magnesium, and iron (Marles 2017).
3. Fiber: Vegetables include dietary fiber, which promotes digestive health and fullness.
4. Antioxidants: Vegetables contain antioxidants such as polyphenols, carotenoids, and anthocyanins that prevent oxidative stress and inflammation (Jaiswal 2020).
5. Phytochemicals: Vegetables contain phytochemicals such as isothiocyanates, indoles, and lignans that have been associated with health benefits (Williams et al. 2013).
6. Protein: Vegetables, particularly legumes such as beans and lentils, include protein, though not as much as other food groups (Natesh et al. 2017).

16.2.1.1 Factors Affecting Nutritional Makeup of Vegetables

The nutritional makeup of vegetables can vary based on factors such as considerations for vegetable preparation including kind, growing circumstances, ripeness, and cooking method.

In general, eating a variety of veggies can supply critical nutrients, improve overall health, and lower the risk of chronic diseases such as heart disease, diabetes, and some malignancies.

16.2.2 Common Nutrient Deficiencies Addressed by Biofortification

Biofortification targets the following common nutrient deficiencies:

Iron (Fe) deficiency: Iron deficiency is a major food-related issue worldwide, causing anemia, impaired physical and mental development, illness, and mortality (Okwuonu et al. 2021).

Iodine (I) deficiency: Iodine is essential for thyroid function, and a lack of it can cause goiters, cretinism, and other health issues (Avnee et al. 2023).

Selenium (Se) deficiency: Selenium is a vital antioxidant whose lack can impair immune function and raise the risk of certain diseases (Avnee et al. 2023).

Protein deficiency: Protein is required for growth and development, and a lack can result in malnutrition, decreased immune function, and other health issues (Avnee et al. 2023).

Vitamin D deficiency: Vitamin D is required for bone health, and lack can result in rickets, osteomalacia, and an increased risk of osteoporosis (Neill et al. 2023).

Zinc (Zn) deficiency: Zinc is an important micronutrient whose lack can lead to a variety of health problems, including decreased immunological function and development retardation (Okwuonu et al. 2021).

In conclusion, biofortification is a cost-effective and sustainable approach to addressing common nutrient deficiencies, particularly in developing nations where access to diverse diets, supplements, and fortified foods is restricted. Biofortification can benefit public health by raising the levels of key vitamins and minerals in staple crops, lowering the risk of micronutrient deficiencies (Table 16.1).

Table 16.1 Effects of biofortification of iron, iodine, selenium, protein, vitamin D, and zinc and their effects on different vegetables.

Vegetable names	Biofortification method	Before biofortification	After biofortification	Reference
Leafy greens	Iron	- Iron content: 20–30 mg/kg - Yield: 8–10 tons/ha - Leaf size: 5–7 cm - Leaf color: Pale green - Chlorophyll content: 0.5–1.0 mg/g - Antioxidant activity: 20–30% - Consumer acceptance: 40–50% willing to try	- Iron content: 40–50 mg/kg - Yield: 12–15 tons/ha - Leaf size: 7–9 cm - Leaf color: Deep green - Chlorophyll content: 1.5–2.5 mg/g - Antioxidant activity: 50–60% - Consumer acceptance: 80–90% willing to try	Giordano et al. (2019)
Tomato	Iron	- Iron content: 5–7 mg/kg - Yield: 20–25 tons/ha - Fruit size: 60–70 g - Fruit color: Red (chromameter value: 20–25) - Lycopene content: 50–60 mg/kg - Antioxidant activity: 30–40% - Consumer acceptance: 50–60% willing to try	- Iron content: 15–20 mg/kg - Yield: 30–35 tons/ha - Fruit size: 80–90 g - Fruit color: Deep red (chromameter value: 30–35) - Lycopene content: 80–100 mg/kg - Antioxidant activity: 60–70% - Consumer acceptance: 90–95% willing to try	Hass et al. (2016)

(Continued)

Table 16.1 Effects of biofortification of iron, iodine, selenium, protein, vitamin D, and zinc and their effects on different vegetables. *(continued)*

Vegetable names	Biofortification method	Before biofortification	After biofortification	Reference
Onion	Iodine	- Iodine content: 5–10 μg/kg - Yield: 20–25 tons/ha - Bulb size: 5–6 cm - Bulb color: Yellow (chromameter value: 20–25) - Antioxidant activity: 30–40% - Consumer acceptance: 50–60% willing to try	- Iodine content: 20–30 μg/kg - Yield: 25–30 tons/ha - Bulb size: 6–7 cm - Bulb color: Deep yellow (chromameter value: 30–35) - Antioxidant activity: 50–60% - Consumer acceptance: 80–90% willing to try	Broadley et al. (2006)
Garlic	Iodine	- Iodine content: 10–15 μ/kg - Yield: 15–20 tons/ha - Clove size: 2–3 cm - Clove color: White (chromameter value: 20–25) - Antioxidant activity: 40–50% - Consumer acceptance: 60–70% willing to try	- Iodine content: 30–40 μg/kg - Yield: 20–25 tons/ha - Clove size: 3–4 cm - Clove color: Deep white (chromameter value: 30–35) - Antioxidant activity: 60–70% - Consumer acceptance: 90–95% willing to try	Barbosa et al. (2020)
Beans	Selenium	- Selenium content: 10–20 μg/kg - Yield: 15–20 tons/ha - Seed size: 4–5 cm - Seed color: White (chromameter value: 20–25) - Antioxidant activity: 30–40% - Consumer acceptance: 50–60% willing to try	- Selenium content: 50–70 μg/kg - Yield: 20–25 tons/ha - Seed size: 5–6 cm - Seed color: Deep white (chromameter value: 30–35) - Antioxidant activity: 60–70% - Consumer acceptance: 90–95% willing to try	Ligowe et al. (2020)
Cabbage	Selenium	- Selenium content: 5–10 μg/kg - Yield: 18–22 tons/ha - Head size: 1.5–2 kg - Head color: Green (chromameter value: 25–30) - Antioxidant activity: 30–40% - Consumer acceptance: 50–60% willing to try	- Selenium content: 30–40 μg/kg - Yield: 22–25 tons/ha - Head size: 2–2.5 kg - Head color: Deep green (chromameter value: 35–40) - Antioxidant activity: 60–70% - Consumer acceptance: 90–95% willing to try	Kápolna et al. (2012)
Chickpeas	Protein	- Protein content: 18–20% - Yield: 15–18 tons/ha - Seed size: 4–5 cm - Seed color: Beige (chromameter value: 20–25) - Antioxidant activity: 30–40% - Consumer acceptance: 50–60% willing to try	- Protein content: 25–28% - Yield: 18–22 tons/ha - Seed size: 5–6 cm - Seed color: Light brown (chromameter value: 25–30) - Antioxidant activity: 50–60% - Consumer acceptance: 90–95% willing to try	Thavarajah (2012)
Sweet corn	Protein	- β-carotene content: 0.5–1.0 μg/g - Yield: 18–22 tons/ha - Ear size: 15–18 cm - Kernel color: Yellow (chromameter value: 20–25) - Antioxidant activity: 30–40% - Consumer acceptance: 50–60% willing to try	- β-carotene content: 2.0–3.0 μg/g - Yield: 22–25 tons/ha - Ear size: 18–20 cm - Kernel color: Orange-yellow (chromameter value: 25–30) - Antioxidant activity: 50–60% - Consumer acceptance: 90–95% willing to t	Dhakal et al. (2022)

(*Continued*)

Table 16.1 Effects of biofortification of iron, iodine, selenium, protein, vitamin D, and zinc and their effects on different vegetables. *(continued)*

Vegetable names	Biofortification method	Before biofortification	After biofortification	Reference
Mushrooms	Vitamin D	- Vitamin D content: 0–5 μg/100g - Yield: 10–12 tons/ha - Cap size: 3–5 cm - Cap color: White (chromameter value: 20–25) - Antioxidant activity: 30–40% - Consumer acceptance: 50–60% willing to try	- Vitamin D content: 10–15 μg/100g - Yield: 12–15 tons/ha - Cap size: 5–6 cm - Cap color: Light brown (chromameter value: 25–30) - Antioxidant activity: 50–60% - Consumer acceptance: 90–95% willing to try	Pinto et al. (2020)
Bamboo shoots	Vitamin D	- Vitamin B12 content: 0–0.5 μg/100g - Yield: 8–10 tons/ha - Shoot size: 10–12 cm - Shoot color: Green (chromameter value: 20–25) - Antioxidant activity: 30–40% - Consumer acceptance: 50–60% willing to try	- Vitamin B12 content: 2–2 μg/100g - Yield: 10–12 tons/ha - Shoot size: 12–14 cm - Shoot color: Dark green (chromameter value: 25–30) - Antioxidant activity: 50–60% - Consumer acceptance: 90–95% willing to try	Das et al. (2014)
Olives	Zinc	- Zinc content: 0.5 mg/100g - Yield: 5 tons/acre - Size: Medium - Color: Green - Flavor: Bitter - Zinc absorption rate: 10%	- Zinc content: 2.5 mg/100g (500% increase) - Yield: 6 tons/acre (20% increase) - Size: Large - Color: Dark green - Flavor: Mild - Zinc absorption rate: 25% (150% increase)	Signorell et al. (2023)
Turnips	Zinc	- Zinc content: 0.3 mg/100g - Yield: 4 tons/acre - Size: Small to medium - Color: White to purple - Flavor: Sweet and peppery - Zinc absorption rate: 8% - Moisture content: 90% - Shelf life: 2 wks	- Zinc content: 1.2 mg/100 g (300% increase) - Yield: 5.5 tons/acre (38% increase) - Size: Medium to large - Color: Deep purple - Flavor: Sweet and slightly sweet - Zinc absorption rate: 18% (125% increase) - Moisture content: 92% - Shelf life: 4 wks (100% increase)	Mangueze et al. (2018)

16.2.3 Challenges in Achieving Optimal Nutrient Levels in Vegetables

Achieving appropriate nutritional levels in vegetables can be problematic for numerous reasons, including the following:

1. Soil quality: Soil degradation, low fertility, and insufficient nutrient availability can all impede plant nutrition uptake (Kalkhajeh et al. 2021).
2. Climate and meteorological conditions: Extreme temperatures, droughts, and heavy rains can all have an impact on nutrient availability and plant development (Steenson et al. 2020).
3. Crop variety and breeding: Certain crop types may have lower nutritional content or a decreased ability to absorb nutrients from the soil (Van Delden et al. 2021).

4. Postharvest processing and storage: Improper handling, storage, and transportation can result in nutritional loss and deterioration (Van Delden et al. 2021).
5. Genetic constraints: Some crops may have innate genetic restrictions that limit their ability to collect particular nutrients (Van Delden et al. 2021).
6. Pest and disease pressure: Pests and illnesses can deplete crop yields and nutritional value (Mahapatra et al. 2022).
7. Lack of micronutrient fertilization: Failure to use micronutrient fertilizers might lead to nutrient deficits (Mahapatra et al. 2022).

16.2.4 Opportunities Presented by Biofortification Techniques

Following are the opportunities presented by the biofortification techniques:

1. Nutrient-dense food crops: By using biotechnology, agronomic techniques, and conventional plant breeding, biofortification can raise the nutrient density of food crops (Athar et al. 2020).
2. Reducing micronutrient deficiencies: In impoverished nations with limited access to a variety of foods and supplements, biofortification can help minimize micronutrient deficiencies (Ofori et al. 2022).
3. Increasing nutritional accessibility: People who live in rural and urban areas who primarily rely on staple crops for their daily caloric intake may benefit most from biofortification (Athar et al. 2020).
4. Fighting hidden hunger: By boosting the micronutrient content of staple crops – which are frequently deficient in important vitamins and minerals – biofortification can help fight hidden hunger (Ofori et al. 2022).
5. Sustainable agriculture: By lowering the need for synthetic fertilizers and enhancing soil health, biofortification can support sustainable agriculture (Athar et al. 2020).

In conclusion, biofortification presents a number of options for addressing deficits in certain micronutrients, enhancing nutritional accessibility, and advancing sustainable agriculture. Biofortification can improve public health and well-being by making food crops more nutrient-dense, especially in impoverished nations.

16.3 Biofortification Techniques in Vegetable Crops

16.3.1 Traditional Breeding Methods for Biofortification

Billions of people worldwide suffer from micronutrient deficiencies, also known as "hidden hunger," especially in underdeveloped nations. A series of health issues, including stunted growth, weakened immune systems, and elevated death rates, can result from these deficiencies (Kennedy et al. 2003). The technique of biofortification, which involves enhancing crops with vital vitamins and minerals, provides a potent weapon to address this issue. Conventional breeding techniques continue to be essential to biofortification initiatives, especially in the case of vegetables.

16.3.1.1 Utilizing Natural Variation

Conventional breeding techniques take advantage of the innate genetic diversity seen in plant populations to determine the micronutrient content. Veggies with higher concentrations of specific micronutrients are carefully chosen by plant breeders, who then use these veggies as parents for their offspring. When this process is repeated over several generations, completely new vegetable kinds with far higher vitamin and mineral content emerge. It is similar to animal selective breeding, but it uses the inherent genetic diversity found in populations of plants.

16.3.1.2 Success Stories in Vegetable Biofortification

16.3.1.2.1 Vitamin A–Rich Orange Sweet Potato

Through traditional breeding programs, scientists have identified and propagated sweet potato varieties naturally rich in β-carotene, the precursor to vitamin A. This biofortified orange sweet potato offers a potent weapon against VAD (Burri 2011).

16.3.1.2.2 Iron-biofortified Amaranth

Amaranth, a grain-like vegetable, is a valuable source of protein and micronutrients. Breeding programs have successfully developed amaranth varieties with significantly higher iron content, addressing deficiencies prevalent in certain populations (Chaudhari et al. 2021).

16.3.1.3 Collaboration for Enhanced Impact

While traditional breeding offers a powerful approach, it's not without limitations. Breeding programs often prioritize goals such as disease resistance and yield improvement. Integrating nutritional quality as an additional objective necessitates collaboration between plant breeders and nutrition scientists. This collaborative approach ensures that newly developed vegetable varieties meet not only yield targets but also offer enhanced nutritional value.

16.3.1.4 Future

Traditional breeding methods have played and will continue to play a vital role in biofortification efforts. Continued investment in these programs, alongside collaboration with other biofortification techniques such as genetic engineering, holds immense promise for a healthier and more food-secure future. As we strive to create a world free from hidden hunger, traditional breeding methods will undoubtedly remain a cornerstone in this ongoing battle.

16.3.2 Genetic Engineering Approaches for Enhancing Nutrient Content

Although conventional breeding techniques hold considerable importance, genetic engineering offers a more focused alternative for augmenting the nutritional composition of vegetables.

16.3.2.1 Promise of Genetic Modification

Through genetic engineering, scientists can precisely alter a plant's genetic composition to promote a certain micronutrient. Compared to conventional breeding techniques, this strategy has the following benefits:

1. Targeted manipulation: Through genetic engineering, desirable nutrient biosynthetic genes can be introduced, resulting in a more targeted increase in nutrient content (Colombo and, Osmond 2019).
2. Greater control: By controlling gene expression, this method makes sure that nutrient production takes place in the plant at the ideal time and place (Blumwald and Mittler 2010).
3. Overcoming limitations: Genetic engineering provides a method to get over natural genetic variety limitations for a given micronutrient by introducing genes from other organisms (Old and Primrose 1981).

16.3.2.2 Engineering for Enhanced Nutrition

Several successful examples showcase the potential of genetic engineering in biofortification.

1. Iron-fortified rice: Scientists have successfully transferred the iron storage protein ferritin gene from soybeans into rice. This transgenic rice variety boasts an iron content multiple times higher than traditional varieties (Goto et al. 1999).
2. Vitamin A–rich vegetables: Scientists are investigating the possibility of introducing genes related to the carotenoid biosynthetic pathway – which produces precursors of vitamin A – into plants such as sweet potatoes and tomatoes. This strategy has the potential to raise the vitamin A content of these important food sources (Stanley and Yuan 2019).

16.3.2.3 Challenges and Considerations

Despite its potential, genetic engineering for biofortification faces certain challenges.

1. Consumer acceptance: Adoption of biofortified crops may be hampered by public worries about the safety of genetically modified organisms (GMOs). To solve these issues, open discussion and solid scientific evidence are essential.
2. Regulatory hurdles: The distribution of biofortified cultivars to consumers can be delayed by the protracted and costly regulatory approval process for genetically modified agriculture.
3. Unintended consequences: Comprehensive safety evaluations are necessary due to the possibility of unforeseen repercussions resulting from genetic alterations.

16.3.2.4 Future

A potent biofortification technique made possible by genetic engineering has the potential to greatly increase the nutritious content of plants. Sufficient research and cooperation among scientists, authorities, and the general public, together with an emphasis on consumer education, are all necessary to guarantee the conscientious advancement and implementation of this technology. Through the application of genetic engineering in conjunction with other biofortification methods, we can pave the way for a time when everyone has easy access to nutrient-dense vegetables, thereby promoting global health and increased food security.

16.3.3 Agronomic Practices Influencing Nutrient Levels in Vegetables

The technique of enhancing crops with vital vitamins and minerals, or biofortification, is a potent weapon against world-wide starvation. The nutrient content of vegetables is greatly influenced by agronomic practices, even if genetic engineering and conventional breeding techniques also play important roles. Farmers may grow veggies full of essential micronutrients by knowing these techniques.

16.3.3.1 Optimizing the Soil Environment

1. Soil fertility: Well-nourished soil is the starting point for wholesome plants. Regular soil testing to detect deficiencies and subsequent additions of mineral fertilizers or organic matter can greatly increase the number of micronutrients found in plants (Ganeshamurthy et al. 2015).
2. Microbial activity: A healthy soil population of microorganisms is essential to the cycling of nutrients and their availability to plants. Crop rotation, cover crops, and the use of compost all help to maintain a healthy soil microbiome, which in turn increases the amount of nutrients that vegetables absorb (Krishnakumar and Thomas 2024).

16.3.3.2 Precise Nutrient Management

1. Nutrient timing and application methods: A healthy soil population of microorganisms is essential to the cycling of nutrients and their availability to plants. Crop rotation, cover crops, and the use of compost all help to maintain a healthy soil microbiome, which in turn increases the amount of nutrients that vegetables absorb (Shah and Wu 2019).
2. Balanced fertilization: It is possible to overlook important micronutrients while concentrating just on main nutrients such as potassium, phosphorus, and nitrogen. Vegetables obtain a full nutritional profile when fertilizers made with a balanced blend of macro- and micronutrients are used (Yadav et al. 2022).

16.3.3.3 Cultivation Techniques

1. Spacing and planting density: Plant spacing should be such that roots can grow and absorb nutrients to the fullest. Vegetables may find it difficult to acquire nutrients if their roots are overcrowded (Casper and Jackson 1997).
2. Irrigation management: Nutrient intake may be adversely affected by water stress. Vegetables with the highest nutrient content are those that are grown with effective irrigation techniques that supply enough moisture during the growing season (Rouphael et al. 2012).

Agronomic approaches like these can let farmers actively participate in biofortification initiatives. Vegetables that have higher concentrations of vital vitamins and minerals are produced through a combination of nutrient-rich soil, intelligent nutrient management, and appropriate production practices. This enhances the nutritious content of our food and makes everyone's future healthier.

16.3.4 Emerging Technologies and Their Potential for Biofortification

Novel technologies have intriguing opportunities to enhance the nutritional value of vegetables by adding vital micronutrients (Figure 16.1).

16.3.4.1 Gene Editing for Precise Modifications

- CRISPR-Case: With the use of this ground-breaking gene editing technology, scientists can precisely alter a plant's DNA in specific areas. By modifying the genes involved in micronutrient production or uptake, it has enormous potential for biofortification and could result in the creation of crop types with higher nutritional contents (Chen et al. 2019).

16.3.4.2 Precision Fermentation for Micronutrient Production

- Microbial engineering: It is possible to modify microorganisms, such as yeast, to create particular vitamins and minerals. These can then be added to fertilizers, sprayed directly on crops during growth, or utilized as biofortificants to directly provide vegetables with vital micronutrients (Branduardi and Porro 2012).

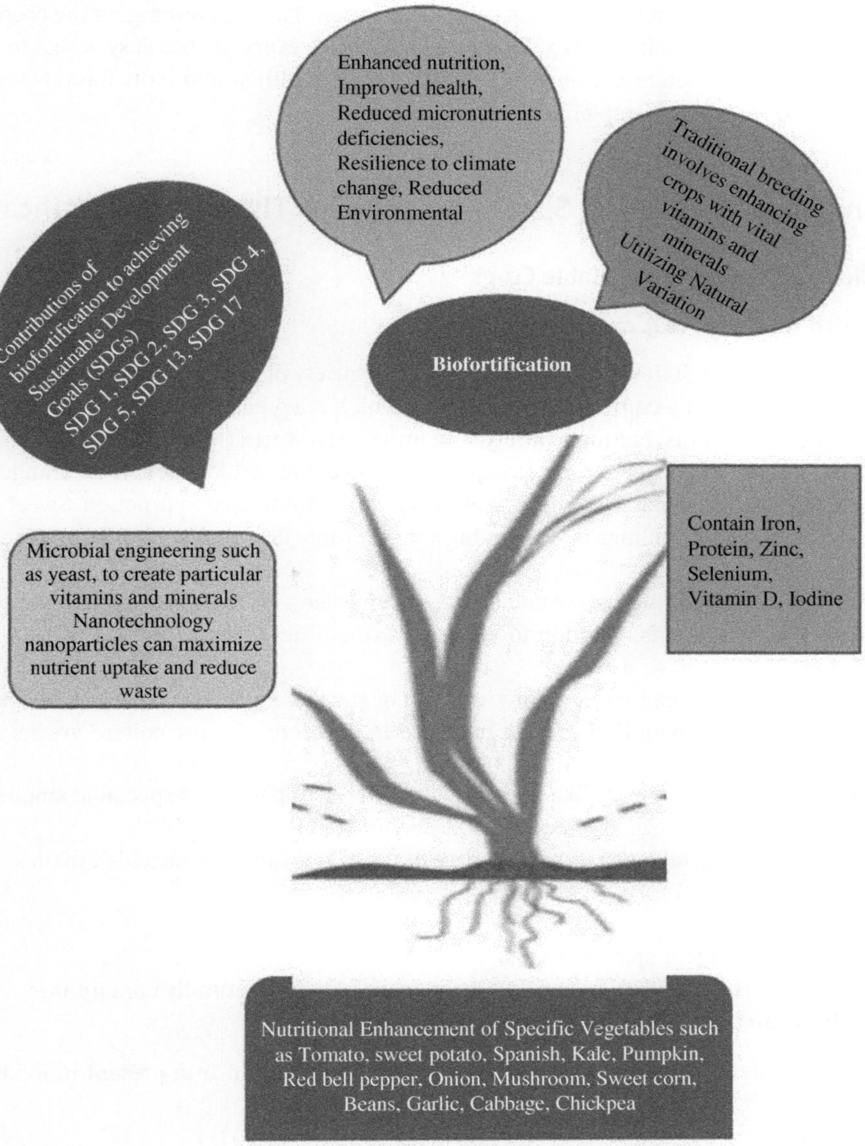

Figure 16.1 Multidimensional effects of bioremediation in vegetables and its positive effects.

16.3.4.3 Nanotechnology for Targeted Delivery

- Nanoparticles: Nanoparticle-encapsulated micronutrients provide a targeted delivery mechanism. By precisely releasing their payload inside plant cells, these nanoparticles can maximize nutrient uptake and reduce waste (Mittal et al. 2020).

16.3.4.4 Future

Despite the enormous potential of these new technologies, there are still many obstacles to overcome. To guarantee the security and effectiveness of these methods, a great deal of study is required. Laws and regulations need to change to make room for these new methods. In order to address possible issues regarding GMOs, public engagement and education are essential.

One effective tool in the battle against malnutrition is biofortification. Through leveraging the promise of both emerging technologies and conventional approaches, we can build a future where everyone has easy access to nutrient-rich veggies. This cooperative endeavor has the potential to support a world that is healthier and more food secure since it is motivated by scientific innovation and responsible implementation.

16.4 Nutritional Enhancement of Specific Vegetables Through Biofortification

16.4.1 Case Studies of Biofortified Vegetable Crops

16.4.1.1 Vitamin A–rich Vegetables (e.g. carrots, sweet potatoes)

1. Carrot: A popular vegetable high in vitamin A, carrots offer a variety of health benefits and are an integral element of a balanced diet due to their high β-carotene concentration, which is a precursor to vitamin A. This vivid orange vegetable is prized for its health benefits, culinary variety, and nutritional worth (Nankumbi et al. 2023).
2. Sweet Potato: Rich in β-carotene, a precursor to vitamin A, and renowned for their vivid orange color, sweet potatoes are a highly nutritious root vegetable. Vitamin A is essential for immune system function. Sweet potatoes include antioxidants that may help lower the chance of developing some chronic illnesses, such as heart disease and some types of cancer (Nankumbi et al. 2023).
3. Spinach: This leafy green is rich in β-carotene and other essential elements such as calcium and iron (Begum et al. 2021).
4. Kale: Another leafy green that adds nutrition to salads and smoothies, kale is high in β-carotene and other vitamins (Adebayo and Ebabhi 2022).
5. Red bell peppers: These peppers can be eaten raw, roasted, or sautéed and contain β-carotene (Begum et al. 2021).
6. Collard greens: Another leafy green that may be prepared in a variety of ways, collard greens are rich in β-carotene (Nankumbi et al. 2023).
7. Broccoli: This food offers extra nutrients like fiber and vitamin C along with a respectable amount of β-carotene, however, not as much as some other veggies (Adebayo and Ebabhi 2022).
8. Pumpkins: These versatile vegetables are a great source of β-carotene and can be added to pies, soups, and other recipes (Begum et al. 2021).

16.4.1.2 Iron-fortified Vegetables (e.g. spinach, beans)

Certain vegetables, especially those in the legume and leafy green families, naturally contain iron. When following a vegetarian or vegan diet, they can be an excellent source of iron.

1. Spinach: A leafy green that is high in nonheme iron, which is the kind of iron present in foods derived from plants (Taneja et al. 2020).
2. Lentils: Iron and protein abound in these legumes (Firdose and Firdose 2021).
3. Beans: High in iron are varieties such as kidney beans, chickpeas, and black beans (Taneja et al. 2020).
4. Swiss chard: A vegetable that is lush green and high in iron (Dias 2012).
5. Broccoli: This vegetable has some iron in it as well as other vital elements such as vitamin C, which aids in the absorption of iron (Firdose and Firdose 2021).
6. Tofu: Tofu, which is made from soybeans, has a high iron content (Dias 2012).

16.4.1.3 Zinc-enriched Vegetables (e.g. wheat, rice)

Although zinc is more frequently found in animal products, it is naturally present in several vegetables, among them are:

1. Legumes: Chickpeas, lentils, and beans are excellent providers of zinc (Shukla et al. 2016).
2. Seeds and nuts: Although not technically considered vegetables, zinc can be found in cashews, sesame seeds, and pumpkin seeds (Shukla et al. 2016).
3. Whole grains: Although they aren't quite considered vegetables, whole grains such as brown rice, quinoa, and oats do contain zinc (Chadare et al. 2019).

16.4.2 Comparative Analysis of Nutritional Profiles Before and After Biofortification

Biofortification is the process of enhancing the nutritional content of crops using traditional plant breeding, genetic engineering, or agronomic approaches. This technique tries to alleviate micronutrient deficiencies, particularly in places with limited dietary diversity. A comparison of nutritional profiles before and after biofortification in vegetables shows considerable improvements in micronutrient content.

1. **Carrot:**
   Before biofortification: Regular carrots are high in β-carotene, a precursor to vitamin A, but poor in iron and zinc.
   After biofortification: Carrots have higher levels of β-carotene and may also have more iron and zinc (Bouis et al. 2017).
2. **Sweet potato:**
   Before biofortification: Standard sweet potatoes are high in carbs and β-carotene, but deficient in other minerals.
   After biofortification: Orange-fleshed sweet potatoes contain increased quantities of β-carotene (White and Broadley 2005).
3. **Tomato:**
   Before biofortification: Regular tomatoes were rich in vitamin C, potassium, and lycopene.
   After biofortification: Tomatoes may contain higher levels of β-carotene, folate, and iron (Mayer et al. 2008).
4. **Spinach:**
   Before biofortification: Spinach was a rich source of iron and folate, but had antinutritional elements such as oxalates that reduced mineral absorption.
   After biofortification: Efforts are directed at boosting bioavailable iron and zinc while lowering antinutritional components (Saltzman et al. 2013).
5. **Cassava:**
   Before biofortification: Standard cassava is deficient in micronutrients yet high in carbs.
   After biofortification: Cassava varieties have been produced with increased levels of β-carotene, iron, and zinc (Bouis et al. 2017).

16.4.3 Consumer Acceptance and Market Potential of Biofortified Vegetables

To better understand the consumer acceptability of food items made from biofortified crops and the factors influencing this acceptance, interdisciplinary research methodologies are employed. The food science literature primarily uses sensory evaluation and hedonic testing, whereas the economics literature usually uses preference elicitation or willingness to pay approaches. Vegetables that have been biofortified may find favor with consumers and have a large market depending on a number of variables, such as availability, cost, taste, availability, perceived benefits, and cultural preferences. Here are some important things to think about:

1. Benefits to nutrition: Vegetables that are biofortified are cultivated to include higher concentrations of vital elements such as vitamins and minerals.
2. Awareness and education: Knowledge of the advantages of vegetables enhanced with biofortification is typically a prerequisite for consumer acceptance. Successful marketing initiatives emphasizing the benefits of nutrition can change consumers' perceptions for the better.
3. Taste and quality: Although biofortified veggies are nutritious, consumer adoption may be restricted if they fall short of expectations in terms of flavor, texture, and general quality.
4. Pricing and affordability: The cost of a product has a big impact on what consumers decide to buy. When veggies that have been biofortified are priced similarly to normal selections, buyers could be more inclined to buy them.
5. Support from the government: Government initiatives that encourage the use of biofortified crops, including food producer incentives or farmer subsidies, can boost market expansion and consumer acceptability.

Overall, even though biofortified veggies have a lot of promise to combat hunger and enhance public health, their successful adoption hinges on a number of variables, such as consumer awareness, taste preferences, price, and infrastructure and policies that support them.

16.5 Impact of Biofortified Vegetables on Human Health

16.5.1 Evaluation of Nutritional Benefits in Reducing Micronutrient Deficiencies

Analyzing the contribution of nutrition to the reduction of micronutrient deficits, food fortification, and supplementation are the most common and frequently utilized approaches to treat micronutrient deficiency (Gibson 2011). These measures may not be the most acceptable, accessible, or appropriate strategies for rural and/or ultra-poor households. They also may not address the root cause of inadequate nutritional intake due to food insecurity. In these situations, complementary approaches are required to promote community- and agriculture-based interventions, culturally acceptable dietary modifications, and concurrent attempts to increase the effectiveness and reach of fortification and supplementation (Girard et al. 2012).

Typical measures to treat deficiencies in micronutrients:

1. Fortification: Fortification is the process of incorporating necessary vitamins and minerals-such as folic acid in flour, vitamin D in milk, or iodine in salt – into frequently eaten foods. Approximately 80% of the entire programmed costs are related to fortification (Imdad and Bhutta 2012).
2. Supplementation: Giving concentrated quantities of vitamins and minerals, usually in the form of pills, capsules, or liquids.
3. Convenience: For people with dietary limitations or limited access to a varied diet, supplements provide a handy alternative to maintaining appropriate nutritional intake.
4. Preventive health: Certain supplements, including calcium and vitamin D for bone health, might lower the chance of contracting specific illnesses or ailments.
5. Assisting special populations: Targeted supplementation may be beneficial for specific populations, such as vegetarians, the elderly, and expectant mothers.
6. Correcting deficiencies: By giving vital nutrients to prevent negative health effects, supplements can quickly cure deficiencies.

16.5.2 Health Outcomes Associated with Increased Consumption of Biofortified Vegetables

Increased consumption of biofortified vegetables has been linked to a number of good health outcomes due to their enhanced nutrient content when compared to regular kinds. The following are some of the potential health benefits:

1. Improved micronutrient intake: Biofortified veggies have higher quantities of critical micronutrients such as iron, zinc, and vitamin A and C. Increased intake can help treat micronutrient deficiencies, which are frequent in communities with limited access to diverse meals (Pasricha et al. 2013).
2. Improved immune function: Proper vitamin and mineral intake supports a strong immune system. Biofortified veggies, particularly those high in vitamin C and zinc, can boost immune function and assist the body in fighting infections (Marcos 2021).
3. Improved cognitive performance: Iron and folate, two micronutrients included in biofortified veggies, are critical for brain growth and cognitive performance. Frequent ingestion may enhance overall cognitive function, including memory and focus (DellaPenna 1999).
4. Reduced risk of chronic diseases: Antioxidant components found in biofortified vegetables, such as flavonoids and carotenoids, have been associated with a lower risk of chronic diseases, such as cancer, age-related macular degeneration, and cardiovascular disease (Black et al. 2008).
5. Better maternal and child health: Micronutrient deficiencies can have long-term health effects, and pregnant women and young children are especially susceptible to them. During pregnancy and the early years of life, consuming more biofortified veggies can benefit the mother's health, fetal development, and child growth (Black et al. 2008).

16.5.3 Contributions of Biofortification to Achieving Sustainable Development Goals (SDGs)

By tackling major issues with nutrition, health, poverty, and sustainability, biofortification helps to achieve a number of Sustainable Development Goals (SDGs). In order to achieve the SDGs, biofortification helps in the following ways:

SDG 1: No Poverty: By boosting their earnings and enhancing their standard of living, biofortified crops and improved nutritional quality can assist in bringing smallholder farmers out of poverty (Saddique et al. 2024). Through the provision of biofortified seeds and culture skills, biofortification aids in the reduction of poverty.

SDG 2: Zero Hunger: In order to combat hunger and malnutrition, biofortified crops
Enhance the nutritional value of basic meals. Biofortification helps to achieve food security and reduce hunger by expanding the availability of nutrient-rich crops, especially in areas where micronutrient shortages are common (Burlingame et al. 2019).

SDG 3: Good Health and Well-Being: Foods enriched with important vitamins and minerals boost general health outcomes by lowering the risk of diseases associated with malnutrition (Fernandez et al. 2020). By promoting access to nutrient-rich diets, biofortification contributes to better maternal health, child development, and overall well-being (Sheoran et al. 2022).

SDG 4: Quality Education: Better nutrition from biofortified foods can improve academic achievement and cognitive development, particularly in young children. Biofortification helps to achieve the objective of guaranteeing inclusive and equitable high-quality education for everyone by promoting improved nutrition (Unterhalter 2019).

SDG 5: Gender Equality: Women are disproportionately affected by food insecurity and malnutrition, especially during pregnancy and lactation, therefore, biofortification programs frequently target them. Biofortification promotes gender equality by enabling women to have healthier lives and take part more fully in social and economic activities by enhancing the nutritional status of women and children (Mutuku et al. 2020).

SDG 13: Climate Action: By increasing crop yield and decreasing the demand for chemical pesticides and fertilizers, biofortified crops can help build climate resilience. Biofortification helps lessen the effects of climate change on food security and agricultural livelihoods by fostering biodiversity and sustainable farming practices (Doni et al. 2020).

SDG 17: Collaboration to Achieve the Goals: Collaboration between governments, private industry, civil society, and other stakeholders is necessary to achieve the SDGs. In order to create and promote biofortified crops, agricultural research institutions, governments, non-governmental organizations, and the corporate sector frequently collaborate on biofortification projects. Biofortification facilitates multi-stakeholder relationships, which in turn aids in the collective endeavor to accomplish the SDGs (Maltais et al. 2018).

In general, biofortification presents a comprehensive strategy for concurrently tackling several development obstacles, rendering it an invaluable instrument for promoting sustainable development and enhancing the well-being of millions of individuals worldwide.

16.6 Challenges and Future Directions

16.6.1 Regulatory and Policy Considerations for Biofortified Vegetable Crops

Depending on the nation, the particular crop, and the planned application of the biofortified variants, there are different regulatory and policy implications for biofortified vegetable crops. Nonetheless, a number of recurring ideas and tenets direct the regulatory structures overseeing biofortification:

1. Safety evaluations: Biofortified vegetable crops are put through extensive safety testing to make sure they are safe for both the environment and human consumption before being made available for commercial cultivation and consumption. This usually entails assessing the possible hazards connected to the altered features, such as toxicity, allergenicity, and unintentional environmental consequences (Glenn et al. 2008).
2. Regulatory approval: Biofortified vegetable crops cannot typically be grown, sold, or eaten without regulatory approval. Prior to awarding permission, these organizations evaluate the biofortified crops for safety, effectiveness, and environmental impact (Mejia and associates 2017).
3. Labeling and traceability: To guarantee consumer awareness and choice, biofortified vegetable crops must have clear labeling and traceability procedures. Effective monitoring and control are made possible by traceability systems, which follow the movement of biofortified crops from farm to fork (Kápolna et al. 2012).
4. Intellectual property rights (IPR): The development and marketing of vegetable crops fortified with biofortification depend heavily on intellectual property rights. In order to secure their discoveries and guarantee reasonable returns on their investment, breeders, and developers of biofortified varieties may apply for patents or plant variety protection. Promoting fair access to biofortified crops requires striking a balance between IPR and access and benefit-sharing considerations (Dias et al. 2011).

5. International trade regulations: Vegetable crops that have been biofortified may be governed by international trade laws, especially if they are traded internationally. Trade must be facilitated while maintaining safety and environmental protection, and this requires adherence to international accords such as the Cartagena Protocol on
   a. Biosafety and the World Trade Organization's (WTO) Sanitary and Phytosanitary
   b. Agreement on the Application of Sanitary and Phytosanitary Measures (SPS Agreement) (Mejia et al. 2017)
6. Public consultation and engagement: Developing and adopting regulatory frameworks for biofortified crops depends on involving stakeholders, such as farmers, consumers, civil society organizations, and legislators (Garcia et al. 2017).
7. Monitoring and post-market surveillance: Following the commercialization of biofortified vegetable crops, mechanisms for monitoring and post-market surveillance are put in place to keep an eye on the products' efficacy, safety, and effects on the environment and public health. Ongoing observation makes it possible to identify any negative consequences and, if necessary, to take prompt corrective action (Vlaicu et al. 2023).

Equitable access, sustainability, and socioeconomic growth should all be supported by efficient legislative and policy frameworks for biofortified vegetable crops, which strike a balance between innovation and safety. Governments, academic institutions, business players, and civil society must work together to create and execute strong regulatory frameworks that optimize the potential advantages of biofortification while reducing hazards (Sadikiel et al. 2024).

16.6.2 Socioeconomic Implications of Widespread Adoption of Biofortification

The act of increasing a crop's nutritional content by conventional plant breeding, agronomy, or biotechnology is known as biofortification. Its widespread use has the potential to have substantial socioeconomic effects. Here are a few crucial points to think about:

1. Better public health: Biofortified foods can combat pervasive malnutrition, especially in areas with restricted access to a variety of meals (Kumar et al. 2023).
2. Improved food security: By raising the nutritional value of basic crops, biofortification can help to increase food security (Bhardwa et al. 2022).
3. Lower healthcare expenses: Treating diseases and ailments linked to nutrient deficiencies can result in lower healthcare expenses when malnutrition is addressed by biofortification.
4. Greater productivity and economic growth: Especially in youngsters, improved nutrition can result in enhanced cognitive development and productivity (Mayer et al. 2008).
5. Empowerment of smallholder farmers: By raising the market value of their produce and lowering their susceptibility to crop failures, biofortified crops can provide smallholder farmers greater leverage (Bashaasha et al. 2022).
6. Environmental sustainability: Increasing crop resilience to environmental challenges like drought or soil degradation is the main goal of several biofortification strategies (Sheoran et al. 2022).
7. Ethical points to remember: Biofortification, like any technology, presents ethical questions about access to seeds, intellectual property rights, and possible unintended effects on biodiversity and conventional farming methods (Sheoran et al. 2022).

It has become clear that the broad implementation of biofortification has promise for resolving a number of socioeconomic issues, ranging from promoting economic growth and environmental sustainability to enhancing public health and food security.

16.6.3 Future Prospects and Research Priorities for Advancing Biofortification Technologies

While research and development efforts are being made to improve the nutritional content, agronomic performance, and uptake of biofortified crops, the future prospects for biofortification technologies seem bright. In order to advance biofortification technology, the following are some important research areas and priorities:

1. Improving nutritional traits: Scientists are always looking for ways to find and add more nutrients to biofortified crops than only the iron, zinc, and vitamin A that are usually the focus (Yadava et al. 2018).

2. **Bioavailability and absorption:** To guarantee biofortified crops' efficacy in preventing malnutrition, it is imperative to enhance the nutrients' bioavailability and absorption (Wakeel et al. 2018).
3. **Agronomic performance and yield stability:** In order for farmers to widely embrace biofortified crop varieties, they must demonstrate agronomic performance and yield stability that are comparable to those of conventional varieties (Srivastav et al. 2022).
4. **Breeding methods and genomics:** The production of biofortified crop varieties with specific nutritional properties is being sped up by advancements in breeding methods, such as gene editing, genomic selection, and marker-assisted selection.
5. **Crop diversification and multinutrient approaches:** In order to address the complexity of malnutrition and encourage dietary diversity, there is a growing interest in crop diversification and multinutrient biofortification techniques in addition to single-nutrient biofortification (Yadava et al. 2018).
6. **Scaling up and delivery mechanisms:** Research on efficient delivery mechanisms and approaches for reaching target populations, particularly in low-resource contexts, is necessary for efforts to scale up biofortification programs (Birol et al. 2023).
7. **Consumer acceptance and behaviors change:** In order to promote uptake and impact, it is imperative that consumers become more aware of, accepting of, and demanding of biofortified foods (Bouis et al. 2017).

The development of biofortification technologies has significant promise for enhancing nutrition, health, and livelihoods by tackling these research goals and working across disciplines and sectors, especially in areas where malnutrition is a problem.

16.7 Conclusion

16.7.1 Call to Action for Promoting Biofortification Initiatives

Become a part of the campaign for a healthier future! Encourage biofortification programs now. By working together, we can fight malnutrition, provide crops with vital nutrients, and guarantee everyone has access to food. Act immediately to advance sustainable agriculture and enhance world health. Contribute, advocate, share – every action matter in sustaining our planet. We can address shortages in essential micronutrients like zinc, iron, and vitamin A by adding more nutrients to regularly consumed crops. This strategy gives people access to more nutrient-dense diets while also providing a long-term way to improve food security. We are getting closer to realizing a society where everyone is healthier and more equal as long as we keep pushing biofortification research and implementation efforts forward.

References

Arscott, S.A. and Tanumihardjo, S.A. (2010). Carrots of many colors provide basic nutrition and bioavailable phytochemicals acting as a functional food. *Comprehensive Reviews in Food Science and Food Safety* 9: 223–239.

Athar, T., Khan, M.K., Pandey, A. et al. (2020). Biofortification and the involved modern approaches. *Journal of Elementology* 25 (2): 717–731.

Avnee, Sood, S., Chaudhary, D.R., Jhorar, P. et al. (2023). Biofortification: an approach to eradicate micronutrient deficiency. *Frontiers in Nutrition* 10: p.1233070.

Baldassano, S., Di Gaudio, F., Sabatino, L. et al. (2022). Biofortification: effect of iodine fortified food in the healthy population, double-arm nutritional study. *Frontiers in Nutrition* 9: 871638.

Barbosa, Vera, Ana Luísa Maulvault, Patrícia Anacleto et al. (2020). Enriched feeds with iodine and selenium from natural and sustainable sources to modulate farmed gilthead seabream (Sparus aurata) and common carp (Cyprinus carpio) fillets elemental nutritional value. *Food and Chemical Toxicology* 140: 111330.

Bashaasha, B., Emegu, R.I., Webb, P. et al. (2022). Determinants of uptake of biofortified crop varieties by smallholder farmers in Uganda. *African Journal of Food Agriculture, Nutrition and Development* 22 (10): 21724–21742.

Beebe, S., Gonzalez, A.V., and Rengifo, J. (2000). Research on trace minerals in common bean. *The Food and Nutrition Bulletin* 21 (4): 387–391.

Begum, R., Rahman, M.N., Islam, R. et al. (2021). Vitamin A knowledge and household consumption frequency of vitamin A–rich foods in Tangail, Bangladesh. *Malaysian Journal of Nutrition* 27 (1): 001–013.

Bhardwaj, A.K., Chejara, S., Malik, K. et al. (2022). Agronomic biofortification of food crops: an emerging opportunity for global food and nutritional security. *Frontiers in Plant Science* 13: 1055278.

Birol, E. and Bouis, H.E. (2023). Role of socio-economic research in developing, delivering and scaling new crop varieties: the case of staple crop biofortification. *Frontiers in Plant Science* 14: 1099496.

Black, R.E., Allen, L.H., Bhutta, Z.A. et al. (2008). Maternal and child under nutrition: global and regional exposures and health consequences. *Lancet* 371 (9608): 243–260.

Bouis, H. E., and Saltzman, A. (2017). Improving nutrition through biofortification: a review of evidence from Harvest Plus, 2003 through 2016. *Global Food Security* 12: 49–58.

Bouis, H.E. (2018). Biofortification: an agricultural tool to address mineral and vitamin deficiencies. In: *Food Fortification in a Globalized World* (69–81). Academic Press.

Branduardi, P. and Porro, D. (2012). Yeasts in biotechnology. In: H. Feldmann (ed.) *Yeast*, 347–370. Wiley.

Burlingame, B., Vogliano, C., and Eme, P.E. (2019). Leveraging agricultural biodiversity for sustainable diets, highlighting Pacific Small Island Developing States. *Advances in Food Security and Sustainability* 4: 133–173.

Burri, B.J. (2011). Evaluating sweet potato as an intervention food to prevent vitamin A deficiency. *Comprehensive Reviews in Food Science and Food Safety* 10 (2): 118–130.

Cakmak, I. (2009). Enrichment of fertilizers with zinc: an excellent investment for humanity and crop production in India. *Journal of Trace Elements in Medicine and Biology* 29 (4): 281–289. doi: 10.1016/j.jtemb.2009.05.002.

Casper, B.B. and Jackson, R.B. (1997). Plant competition underground. *Annual Review of Ecology, Evolution, and Systematics* 28 (1): 545–570.

Chadare, F.J., Idohou, R., Nago, E. et al. (2019). Conventional and food-to-food fortification: an appraisal of past practices and lessons learned. *Food Science & Nutrition* 7 (9): 2781–2795.

Chakraborty, S., Chakraborty, N., and Datta, A. (2000). Increased nutritive value of transgenic potato by expressing a nonallergenic seed albumin gene from Amranthus hypochondriacus. *The Proceeding of the National Academy of Science* USA 97 (7):3724–3729.

Chaudhari, D., Mehta, P., Chaudhary, N. et al. (2021). Iron ferti-fortification in grain Amaranthus (Amaranthus hypochondriacus L.). *The Pharma Innovation Journal* 10 (3): 635–638.

Chen, K., Y. Wang, R. Zhang, H. et al. (2019). CRISPR/Cas genome editing and precision plant breeding in agriculture. *Annual Review of Plant Biology* 70 (1): 667–697.

Chow, Z., Cuadra, S., Brenes, D. et al. (2011). Aceptacion de tres lineas de frijol (Phaseolus vulgaris L.) mejorado nutricionalmente por los consumidores nicaraguenses. *Perspectivas En Nutricion Humana* 13 (2): 179–189.

Das, P.K., Bhavani, R.V., and Swaminathan, M.S. (2014). A farming system model to leverage agriculture for nutritional outcomes. *Agricultural Research* 3: 193–203.

DellaPenna, D. (1999). Nutritional genomics: manipulating plant micronutrients to improve human health. *Science* 285 (5426): 375–379.

Dhakal, K., Magar, A.B.P., Pokhrel, K.R. et al. (2022). Zinc and provitamin A biofortified maize genotypes exhibited potent to reduce hidden – hunger in Nepal. *Plants* 11 (21): 2898.

Dias, J.C.S. (2011). Biodiversity and vegetable breeding in the light of developments in intellectual property rights. *Ecosystems Biodiversity* 389–428.

Dias, J.S. (2012). Major classes of phytonutriceuticals in vegetables and health benefits: a review. *Journal of Nutritional Therapeutics* 1 (1): 31–62.

Doni, F., Gasperini, A., and Soares, J.T. (2020). What is the SDG 13. In: *SDG13–Climate Action: Combating Climate Change and Its Impacts*, 21–30. Emerald Publishing Limited.

Ebabhi, A. and Adebayo, R. (2022). Nutritional values of vegetables. *Vegetable Crops-Health Benefits and Cultivation.*

Falco, S.C., Guida, T., Locke, M. et al. (1995). Transgenic canola and soybean seeds with increased lysine. *Biotechnology* 13 (6): 577–582.

Fanzo, J. and Glass, S. (2018). Ethical and sociocultural considerations of biofortified crops: ensuring value and sustainability for public health. In: *Advances in Food Security and Sustainability*, vol. 3, 93–133. Elsevier.

Fernandez, R.M. (2020). SDG3 good health and well-being: integration and connection with other SDGs. *Good Health and Well-Being*, 629–636.

Firdose, K. and Firdose, N. (2021). Dietary iron. In: *Iron Metabolism-A Double-Edged Sword*. Intech Open.

Ganeshamurthy, A.N., Kalaivanan, D., Selvakumar, G. et al. (2015). Nutrient management in horticultural crops. *Indian Journal of Fertilisers* 11 (12): 30–42.

Garcia-Casal, M.N., Peña-Rosas, J.P., Giyose, B., and Consultation Working Groups (2017). Staple crops biofortified with increased vitamins and minerals: considerations for a public health strategy. *Annals of the New York Academy of Sciences* 1390 (1), 3–3.

Gibbon, B.C. and Larkins, B.A. (2005). Molecular genetic approaches to developing quality protein maize. *Trends Genet* 21 (4): 227–233.

Gibson, R.S. (2011). Strategies for preventing multi-micronutrient deficiencies: a review of experiences with food-based approaches in developing countries. In: *Combating Micronutrient Deficiencies: Food-Based Approaches* (Thompson, B., Amoroso, L., eds.). Rome, Italy: CAB International and FAO: 7–27.

Giordano, M., El-Nakhel, C., Pannico, A. et al. (2019). Iron biofortification of red and green pigmented lettuce in closed soilless cultivation impacts crop performance and modulates mineral and bioactive composition. *Agronomy* 9 (6): 290.

Girard, A.W., Self, J.L., McAuliffe, C. et al. (2012). The effects of household food production strategies on the health and nutrition outcomes of women and young children: a systematic review. *Paediatric and Perinatal Epidemiology* 26: 205–222.

Glenn, K. (2008). Nutritional and safety assessment of foods and feeds improved through biotechnology – case studies by the International Food Biotechnology Committee of ILSI. *Asia Pacific Journal of Clinical Nutrition* 17 (1): 229–232.

Goto, F., Yoshimura, T., Shigemoto, N. et al. (1999). Iron fortification of rice seed by the soybean ferritin gene. *Nature Biotechnology* 17 (3): 282–286.

Grace, M.H., Yousef, G.G., Esposito, D. et al. (2014). Bioactive capacity, sensory properties, and nutritional analysis of a shelf stable protein-rich functional ingredient with concentrated fruit and vegetable phytoactives. *Plant Foods for Human Nutrition* 69: 372–378.

Graham, R.D., Welch, R.M., Bouis, H.E. et al. (2001). Addressing micronutrient malnutrition through enhancing the nutritional quality of staple foods: principles, perspectives and knowledge gaps. *Advances in Agronomy* 70: 77–142.

Haas, J.D., Luna, S.V., Lung'aho. et al. (2016). Consuming iron biofortified beans increases iron status in Rwandan women after 128 days in a randomized controlled feeding trial. *The Journal of nutrition* 146 (8): 1586–1592.

Hess, S. and King, J. (2009). Effects of maternal zinc supplementation on pregnancy and lactation outcomes. *Food and Nutrition Bulletin* 30 (1):S60–S78. doi: 10.1177/15648265090301S10.

Hirschi, K.D. (2009). Nutrient biofortification of food crops. *Annual Review of Nutrition* 29 (1): 401–421.

Imdad, A. and Bhutta, Z.A. (2012). Routine iron/folate supplementation during pregnancy: effect on maternal anemia and birth outcomes. *Paediatric and Perinatal Epidemiology* 26 (1): 168–177.

J., Bae H., Jeong, J., Lee, J.Y. et al. (2003). Functional expression of a bacterial heavy metal transporter in Arabidopsis enhances resistance to and decreases uptake of heavy metals. *Plant Physiology* 133 (2): 589–596.

Jaiswal, A.K. ed. (2020). *Nutritional Composition and Antioxidant Properties of Fruits and Vegetables*. Academic Press.

Jha, A.B. and Warkentin, T.D. (2020). Biofortification of pulse crops: status and future perspectives. *Plants* 9 (1): 73. doi: 10.3390/plants9010073.

Kabir, A.H., Paltridge, N., and Stangoulis, J. (2016). Chlorosis correction and agronomic biofortification in field peas through foliar application of iron fertilizers under Fe deficiency. *Journal of Plant Interactions* 11 (1): 1–4.

Kalkhajeh, Y.K., Huang, B., Hu, W. et al. (2021). Environmental soil quality and vegetable safety under current greenhouse vegetable production management in China. *Agriculture, Ecosystems Environment* 307: 107230.

Kápolna, E., Laursen, K.H., Husted, S. et al. (2012). Bio-fortification and isotopic labelling of Se metabolites in onions and carrots following foliar application of Se and 77Se. *Food Chemistry* 133 (3): 650–657.

Kennedy, G., Nantel, G., and Shetty, P. (2003). The scourge of "hidden hunger": global dimensions of micronutrient deficiencies. *Food Nutrition-Sensitive Agriculture* 32: 8–16.

Khattak, S.G., Dominy, P.J., Ahmad, W. et al. (2015). Effect of Zn as soil addition and foliar Application on yield and protein content of wheat in alkaline soil. *Journal of Natural Science Found* 43 (4): 303–312.

Kumar, A., Reddy, B.V., Ramaiah, B. et al. (2012) Genetic variability and character association for grain iron and zinc contents in sorghum germplasm accessions and commercial cultivars. *The European Journal of Plant Science and Biotechnology* 6 (1): 1–5.

Kumar, S., DePauw, R.M., Kumar, S. et al. (2023). Breeding and adoption of biofortified crops and their nutritional impact on human health. *Annals of the New York Academy of Sciences* 1520 (1): 5–19.

Laurie, S.M. and Van Heerden, S.M. (2012). Consumer acceptability of four products made from beta-carotene-rich sweet potato. *African Journal of Food Science* 6 (4): 96–103.

Ligowe, I.S., Young, S.D., Ander, E.L. et al. (2020). Selenium biofortification of crops on a Malawi Alfisol under conservation agriculture. *Geoderma* 369: 114315.

Mahapatra, D.M., Satapathy, K.C., and Panda, B., (2022). Biofertilizers and nanofertilizers for sustainable agriculture: phyco prospects and challenges. *Science of the Total Environment* 803: 149990.

Maltais, A., Weitz, N., and Persson, Å. (2018). SDG 17: partnerships for the goals. *A Review of Research Needs*.

Mangueze, A.V.D.J., Pessoa, M.F., Silva, M.J. et al. (2018). Simultaneous zinc and selenium biofortification in rice. Accumulation, localization and implications on the overall mineral content of the flour. *Journal of Cereal Science* 82: 34–41.

Marcos, A. (2021). Editorial: a review of micronutrients and the immune system – working in harmony to reduce the risk of infection. *Nutrients* 13 (11):259. doi:10.3390/nu13114180.

Marles, R.J. (2017). Mineral nutrient composition of vegetables, fruits and grains: the context of reports of apparent historical declines. *Journal of Food Composition and Analysis* 56: 93–103.

Mayer, J.E., Pfeiffer, W.H., and Beyer, P. (2008). Biofortified crops to alleviate micronutrient malnutrition. *Current Opinion in Plant Biology* 11 (2): 166–170.

Meilgaard, M. C., Civile, G. V., and Carr, B. T. (2006). *Sensoryevaluation techniques*, 4e. Florida: CRC Press.

Mejia, L.A., Dary, O., and Boukerdenna, H. (2017). Global regulatory framework for production and marketing of crops biofortified with vitamins and minerals. *Annals of the New York Academy of Sciences* 1390 (1): 47–58.

Mills, J.P., Simon, P.W., and Tanumihardjo, S.A. (2008). Biofortified carrot intake enhances liver antioxidant capacity and vitamin A status in mongolian gerbils1. *The Journal of Nutrition* 138 (9): 1692–1698.

Mittal, D., G. Kaur, P. Singh, K. et al. (2020). Nanoparticle-based sustainable agriculture and food science: recent advances and future outlook. *Frontiers in Nanotechnology* 2: 579954.

Mittler, R. and E. Blumwald. (2010). Genetic engineering for modern agriculture: challenges and perspectives. *Annual Review of Plant Biology* 61 (1): 443–462.

Mutuku, J.M., Mwaniki, M.W., Onjong, H.A. et al. (2020). The biofortification continuum: implications for food and nutrition security in developing countries. *African Journal of Food, Agriculture, Nutrition and Development* 20 (1): 15317–15330.

Muzinghi, T., Langyintuo, A.S., Malaba, L.C. et al. (2008). Consumer acceptability of yellow maize products in Zimbabwe. *Food Policy* 33, 352–361.

Nankumbi, J., Grant, F.K., Sibeko, L. et al. (2023). Predictors of vitamin A rich food consumption among women living in households growing orange fleshed sweetpotatoes in selected regions in Uganda. *Frontiers in Public Health* 10: 880166.

Natesh, H.N., Abbey, L., and Asiedu, S.K., (2017). An overview of nutritional and antinutritional factors in green leafy vegetables. *Horticulture International Journal* 1 (2): 58–65.

Neill, H.R., Gill, C.I., McDonald, E.J. et al. (2023). The future is bright: biofortification of common foods can improve vitamin D status. *Critical Reviews in Food Science and Nutrition* 63 (4): 505–521.

Netto Rangel, C., Moreira da Silva, E.M., Salvador, L. et al. (2011). Evaluación sensorial de pasteles con harina de camote (Ipomoea batatas L.). *Perspectivas En Nutricion Humana* 13 (2): 203–211.

Nkhata, S.G., Chilungo, S., Memba, A. et al. (2020). Biofortification of maize and sweetpotatoes with provitamin A carotenoids and implication on eradicating vitamin A deficiency in developing countries. *Journal of Agriculture and Food Research* 2: 100068.

Ofori, K.F., Antoniello, S., English, M.M. et al. (2022) Improving nutrition through biofortification–A systematic review. *Frontiers in Nutrition* 9: 1043655.

Okwuonu, I.C., Narayanan, N.N., Egesi, C.N. et al. (2021). Opportunities and challenges for biofortification of cassava to address iron and zinc deficiency in Nigeria. *Global Food Security* 28: 100478.

Old, R.W. and S.B. Primrose. (1981). *Principles of Gene Manipulation: An Introduction to Genetic Engineering*. University of California Press.

Olson, R., Gavin-Smith, B., Ferraboschi, C. et al. (2021) Food fortification: the advantages, disadvantages and lessons from sight and life programs. *Nutrients* 13: 1118. doi: 10.3390/nu13041118.

Olson, R, Gavin-Smith, B, Ferraboschi, C. et al. (2021) Food fortification: the advantages, disadvantages and lessons from sight and life programs. *Nutrients* 13 (4): 1118. doi: 10.3390/nu13041118.

Osmond, A.T.Y. and S.M. Colombo. (2019). The future of genetic engineering to provide essential dietary nutrients and improve growth performance in aquaculture: advantages and challenges. *Journal of the World Aquaculture Society* 50 (3): 490–509.

Pillay, K., Derera, J., Siwela, M. et al. (2011). Consumer acceptance of yellow, provitamin A-biofortified maize in KwaZuluNatal. *South African Journal of Clinical Nutrition* 24 (4): 186–191.

Pinto, J.M., Merzbach, V., Willmott, A.G. et al. (2020). Assessing the impact of a mushroom-derived food ingredient on vitamin D levels in healthy volunteers. *Journal of the International Society of Sports Nutrition* 17: 1–10.

Radcliffe, C. and Singh, J., (2021). Achieving food and nutrition security and climate change: clash of the titans or alignment of the stars? Exploring synergies and trade-off between climate change and the sustainable development goals 1–36.

Ram, H., Sohu, V.S., Cakmak, I. et al. (2015). Agronomic fortification of rice and wheat grains with zinc for nutritional security. *Current Science* 109 (6): 1171–1176.

Ramaswamy, B. (2007). Biofortified crops and biotechnology: a political economy landscape for India. *The Journal of AgrobiotechnologyForum* 10: 170–177.

Rizwan, M., Zhu, Y., Qing, P. et al. (2021). Factors determining consumer acceptance of biofortified food: case of zinc-fortified wheat in Pakistan's Punjab Province. *Frontiers in nutrition* 8: 647823.

Rouphael, Y., M. Cardarelli, D. Schwarz. et al. (2012). Effects of drought on nutrient uptake and assimilation in vegetable crops. In: R. Aroca (ed.), *Plant Responses to Drought Stress*. Springer Berlin Heidelberg, Berlin, Heidelberg 171–195.

Saddique, M.A.B., Ali, Z., Sher, M.A. et al. (2024). Biofortified rice for zero hunger: current status, challenges, and prospects. *Biofortification of Grain and Vegetable Crops* 59–80.

Sadikiel Mmbando, G., (2024). The adoption of genetically modified crops in Africa: the public's current perception, the regulatory obstacles, and ethical challenges. *GM Crops & Food* 15 (1): 1–15.

Saltzman, A., et al. (2013). Biofortification: progress toward a more nourishing future. *Global Food Security* 2 (1): 9–17.

Sayre R, Beeching JR, Cahoon EB. et al. (2011) The BioCassava Plus program: biofortification of cassava for sub-Saharan Africa. *Annual Review of Plant Biology* 62: 251–272.

Serrano Romero, A.D.C., Vílchez Morales, E.Y., Sandino Silva, C.M. et al. (2011). Evaluación sensorial de tortas de camote (Ipomoea batatas), elaboradas con o sin hojas de camote, con niños en edad escolar en Nicaragua. *Perspectiv as EnNutricion Humana* 13 (2): 191–202.

Shah, F. and W.Wu. (2019). Soil and crop management strategies to ensure higher crop productivity within sustainable environments. *Sustainability* 11: 1485.

Shenkin, A. (2006). Micronutrients in health and disease. *Postgraduate Medical Journal* 82: 559–567. doi: 10.1136/pgmj.2006.047670.

Sheoran, S., Kumar, S., Ramtekey, V. et al. (2022). Current status and potential of biofortification to enhance crop nutritional quality: an overview. *Sustainability*, 14 (6): 3301.

Shikuku, K.M., Okello, J.J., Wambugu, S. et al. (2019). Nutrition and food security impacts of quality seeds of biofortified orange-fleshed sweetpotato: quasi-experimental evidence from Tanzania. *World Development* 124: 104646.

Shukla, A.K., Tiwari, P.K., Pakhare, A. et al. (2016). Zinc and iron in soil, plant, animal and human health. *Indian Journal Fertilisers* 12 (11): 133–149.

Signorell, C., Kurpad, A.V., Pauline, M. et al. (2023). The effect of zinc biofortified wheat produced via foliar application on zinc status: a randomized, controlled trial in Indian children. *The Journal of Nutrition* 153 (10): 3092–3100.

Srivastav, P., Vutukuru, M., Ravindran, G. et al. (2022). Biofortification – present scenario, possibilities and challenges: a scientometric approach. *Sustainability* 14 (18): 11632.

Stanley, L. and Y.-W. Yuan. (2019). Transcriptional regulation of carotenoid biosynthesis in plants: so many regulators, so little consensus. *Frontiers of Plant Science* 10: 1017.

Steenson, S. and Buttriss, J.L., (2020). The challenges of defining a healthy and "sustainable" diet. *Nutrition Bulletin* 45 (2): 206–222.

Stein, A.J. (2010). Global impact of human mineral malnutrition. *Plant Soil* 335: 133–154.

Taneja, D.K., Rai, S.K., and Yadav, K. (2020). Evaluation of promotion of iron-rich foods for the prevention of nutritional anemia in India. *Indian Journal of Public Health* 64 (3): 236–241.

Thavarajah, P. (2012). Evaluation of chickpea (Cicer arietinum L.) micronutrient composition: biofortification opportunities to combat global micronutrient malnutrition. *Food Research International* 49 (1): 99–104.

Thomas, G.V. and V. Krishnakumar. (2024). Plantation crops and soil health management: an overview. In: G.V. Thomas and V. Krishnakumar (eds.), *Soil Health Management for Plantation Crops*. Springer Nature Singapore, Singapore, 1–36.

Titcomb, T.J., Sheftel, J., Sowa, M. et al. (2018). β-Cryptoxanthin and zeaxanthin are highly bioavailable from whole-grain and refined biofortified orange maize in humans with optimal vitamin A status: a randomized, crossover, placebo-controlled trial. *The American Journal of Clinical Nutrition* 108 (4) 793–802.

Tomlins, K., Ndunguru, G., Stambul, K. et al. (2007). Sensory evaluation and consumer acceptability of pale-fleshed and orange-fleshed sweetpotato by school children and mothers with preschool children. *Journal of the Science of Food and Agriculture* 87 (13):2436–2446.

Unnevehr, L., Pray, C., Paarlberg, R. et al. (2007). Addressing micronutrient deficiencies: alternative interventions and technologies. *AgBioForum* 10: 124–134.

Unterhalter, E. (2019). The many meanings of quality education: politics of targets and indicators in SDG 4. *Global Policy* 10: 39–51.

Van Delden, S.H., Sharath Kumar, M., Butturini, M. et al. (2021). Current status and future challenges in implementing and upscaling vertical farming systems. *Nature Food* 2(12): 944–956.

Vlaicu, P.A., Untea, A.E., Varzaru, I. et al. (2023). Designing nutrition for health – incorporating dietary by-products into poultry feeds to create functional foods with insights into health benefits, risks, bioactive compounds, food component functionality and safety regulations. *Foods* 12 (21): 4001.

Wakeel, A., Farooq, M., Bashir, K., Ozturk, L., et al. (2018). Micronutrient malnutrition and biofortification: recent advances and future perspectives. *Plant Micro Nutrient Use Efficiency*. 225–243.

White, P.J., Brown, P.H. (2010). Plant nutrition for sustainable development and global health. *Annals of Botany* 105: 1073–1080. doi: 10.1007/s11104-10-677-7.

White, P.J. and Broadley, M.R. (2005). Biofortifying crops with essential mineral elements. *Trends in Plant Science* 10 (12): 586–593.

Williams, D.J., Edwards, D., Hamernig, I. et al. (2013). Vegetables containing phytochemicals with potential anti-obesity properties: a review. *Food Research International* 52 (1): 323–333.

Yadav, A.K., G.G. Gurnule, N.I. Gour, U. et al. (2022). Micronutrients and fertilizers for improving and maintaining crop value: a review. *International Journal of Environment Agriculture Biotechnology* 7.

Yadava, D.K., Hossain, F., and Mohapatra, T. (2018). Nutritional security through crop biofortification in India: status and future prospects. *Indian Journal of Medical Research* 148 (5): 621–631.

Yan J, Kandianis CB, Harjes CE. et al. (2010). Rare genetic variation at Zea mays crtRB1 increases beta-carotene in maize grain *Nature genetics* 42 (4): 322–327.

Ye X, Al-Babili S, Kloti A, Zhang J. et al. (2000). Engineering the provitamin A (b-carotene) biosynthetic pathway into (carotenoid free) rice endosperm. *Science* 287 (5451): 303–305.

Zhao, F.J., Shewry, P.R. (2011). Recent developments in modifying crops and agronomic practice to improve human health. *Food Policy* 36: S94–S101.

17

Genetic Engineering for Crop Biofortification

*Taufiq Nawaz[1], Dillon Nelson[1], Anam Saleem[2], Marya Bibi[3], Nitish Joshi[3], Shah Fahad[4], Shah Saud[5], Muhammad Adnan[6], Muhammad Nasir Rasheed Khan[4], Shah Hassan[7], Tanzeel Ur Rahman[8], Muhammad Aaqil[9], Maqzia Mushtaq[10], Imran Khan[11]**

[1] *Department of Biology/Microbiology, South Dakota State University, Brookings, South Dakota, USA*
[2] *Department of Biosciences, COMSATS University Islamabad, Pakistan*
[3] *Department of Agronomy, Horticulture & Plant Science, South Dakota State University, Brookings, South Dakota, USA*
[4] *Department of Agronomy, Abdul Wali Khan University Mardan, Khyber Pakhtunkhwa, Pakistan*
[5] *College of Life Science, Linyi University, Linyi, Shandong, China*
[6] *Department of Agriculture, University of Swabi, Pakistan*
[7] *Department of Agricultural Extension Education & Communication, The University of Agriculture, Peshawar, Pakistan*
[8] *Department of Hotel Management & Tourism, University of Swabi, Pakistan*
[9] *College of Food Science and Technology, Yunnan Agricultural University, Kunming, China*
[10] *Principal Pharmacist, Shifa International Hospital Ltd, Islamabad, Pakistan*
[11] *Center for Life Science Ventures, Cornell University, Ithaca, New York, USA*

*Corresponding author: Ik298@cornell.edu

17.1 Introduction to Genetic Engineering in Agriculture

17.1.1 Historical Perspective

The advent of genetic engineering in agriculture stands as a paradigm shift in the realm of crop improvement, ushering in a new era of precision and targeted enhancement (Singha and Singha 2024). The historical trajectory of genetic engineering traces its roots to the mid-twentieth century, an epoch marked by the unraveling of the intricacies encoded within the DNA molecule (Randhawa and Sengar 2021). Pivotal breakthroughs, such as the elucidation of the double helix structure by James Watson and Francis Crick in 1953, served as foundational keystones in deciphering the genetic codes governing life (Davies 2002; Markel 2021). The seminal period for the integration of genetic engineering into agriculture emerged during the 1970s, propelled by the inception of recombinant DNA technology (Herdt 2006). This groundbreaking development granted scientists the unprecedented ability to manipulate and transfer genetic material across organisms, transcending the natural barriers that once constrained the breeding process (Selinger 2002). Notably, the successful insertion of foreign genes into bacteria during this era marked a watershed moment, culminating in the production of insulin through genetic modification (Rasmussen 2014). This monumental achievement not only revolutionized medical science but also laid the critical groundwork for extending genetic engineering to the realm of crops (Council 1984; Pisano 2006). The transition from manipulating bacterial DNA to the modification of plant genomes was a logical progression in the evolutionary narrative of genetic engineering (Nicholl 2023). As scientists gained mastery over the intricate techniques involved in altering genetic codes, the prospect of enhancing crops' nutritional content and resilience became an increasingly attainable objective (Kumar et al. 2015). The marriage of genetic engineering with agriculture held the promise of addressing not only crop yield challenges but also addressing the pervasive issue of nutrient deficiencies in the global food supply (Conway 2012; Lederman et al. 2014). This historical perspective underscores the transformative journey from the elucidation of the DNA structure to the application of genetic engineering in agriculture. The evolution of genetic engineering techniques, accelerated by breakthroughs in recombinant DNA technology, has set the stage for a pivotal juncture where the manipulation of plant genomes seeks to augment the nutritional quality and robustness of crops, heralding a new era in sustainable and fortified agriculture (Chen et al. 2022; Lappé and Bailey 2014). An overview of different approaches for production of biofortified crops are shown in Figure 17.1.

Crop Biofortification: Biotechnological Approaches for Achieving Nutritional Security under Changing Climate, First Edition. Edited by Adnan Noor Shah et al.

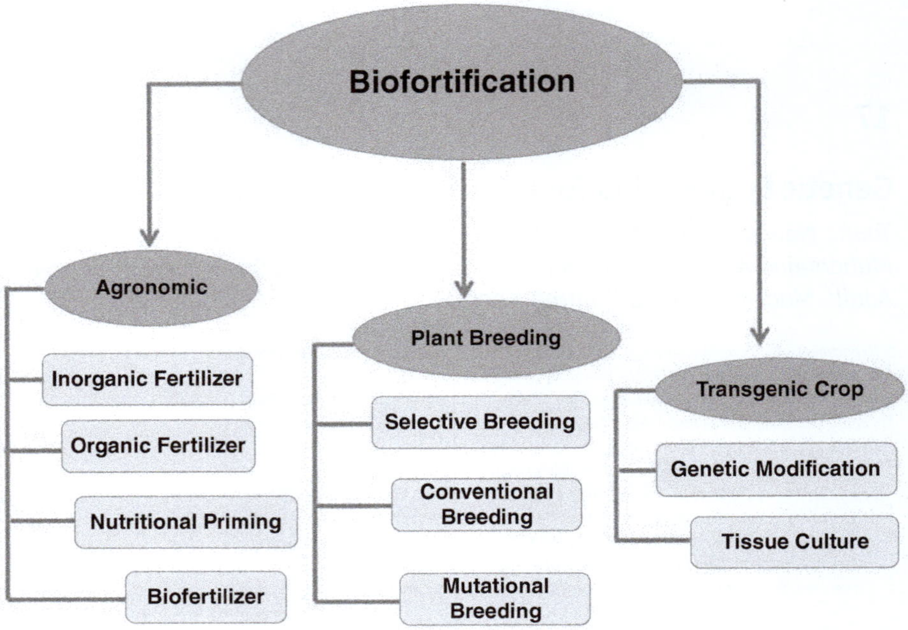

Figure 17.1 Different approaches for production of biofortified crops.

17.1.2 Overview of Genetic Engineering Techniques in Crops

Genetic engineering techniques in crops have evolved significantly over the years, enabling scientists to precisely modify the genetic makeup of plants (Lawlor 2013). One of the pioneering methods is agrobacterium-mediated gene transfer, discovered in the 1980s. This technique involves using a soil bacterium, *Agrobacterium tumefaciens*, to transfer desired genes into plant cells. Although revolutionary at the time, this method has been supplemented and, in some cases, replaced by more advanced technologies (Imani and Kogel 2020; Nester 2015; Özyiğit 2012). The introduction of transgenic crops in the 1990s marked a major milestone. Transgenic crops, also known as genetically modified organisms (GMOs), involve the insertion of specific genes into the plant's DNA, imparting traits such as resistance to pests, diseases, or tolerance to herbicides. This era saw the commercialization of genetically modified (GM) soybeans, corn, and cotton (Coupe and Capel 2016; James 1998; Kumar et al. 2020). In recent years, the CRISPR-Cas9 system has emerged as a game-changer in genetic engineering (Roy et al. 2022). This revolutionary technique allows for precise and targeted modifications to the plant genome by editing specific DNA sequences (Chen and Gao 2014). Unlike traditional genetic engineering, CRISPR-Cas9 offers greater accuracy, efficiency, and versatility, making it a powerful tool for crop improvement (Zhang et al. 2021). The overview encompasses a spectrum of techniques, from classical breeding to modern gene-editing technologies. The chapter will delve into the nuances of these methods, providing a comprehensive understanding of the evolution of genetic engineering in agriculture and setting the stage for exploring its applications in biofortification.

17.2 Need for Biofortification

Malnutrition remains a pressing global health challenge, affecting millions of individuals across diverse socioeconomic landscapes. The consequences of malnutrition are profound, encompassing a range of health issues, developmental impediments, and compromised immune systems. Deficiencies in essential micronutrients such as vitamins and minerals contribute significantly to this global health crisis (Morales et al. 2023; Mahwish et al. 2022; Tulchinsky 2010).

17.2.1 Global Malnutrition and its Impact

The staggering scale of global malnutrition, as highlighted by the World Health Organization (WHO), paints a dire picture of the state of public health worldwide. WHO estimates reveal that over two billion people grapple with deficiencies in critical micronutrients such as vitamin A, iron, zinc, and folate. These deficiencies, pervasive across both developing and developed nations, cast a wide net of consequences that extend beyond mere nutritional insufficiency (Bamji and Nair 2016; Graham et al. 2007; Sanders 2023). Among the myriad repercussions, stunted growth stands out as a prominent outcome of micronutrient deficiencies. Children, especially those in their formative years, bear the brunt of malnutrition, experiencing physical development setbacks that can have enduring impacts on their overall health and well-being. The insidious effects extend to impairments in cognitive development, hindering intellectual growth and potentially limiting future opportunities for affected individuals (Perera 2018; Verschuren 2002). Compromised immune function is another critical facet of the health implications arising from malnutrition. Insufficient intake of key micronutrients weakens the body's defense mechanisms, rendering individuals more susceptible to a host of infectious diseases. The interplay between malnutrition and increased vulnerability to diseases creates a vicious cycle that further exacerbates health disparities, particularly in regions where access to medical interventions is limited (Chandra and Newberne 2012; Hughes and Kelly 2006l Siddiqui et al. 2020). Moreover, the global reach of malnutrition underscores its complexity. While traditionally associated with developing nations grappling with food insecurity, malnutrition is not confined to geographical boundaries (Hassan 2016). Developed nations also face challenges related to inadequate nutrient intake, shedding light on the multifaceted nature of this pervasive global issue (Greig et al. 2007).

In essence, the impact of global malnutrition extends far beyond the statistics, infiltrating various aspects of human health and development. The interconnectedness of stunted growth, impaired cognition, weakened immune systems, and susceptibility to diseases paints a comprehensive portrait of the urgent need for innovative solutions such as biofortification to break the cycle of malnutrition and uplift the health and well-being of populations worldwide. The prevalence of undernourishment (in percentage) in the Global South has been shown in Figure 17.2.

17.2.2 Role of Biofortification in Addressing Nutrient Deficiencies

Biofortification emerges as a promising and sustainable solution to combat nutrient deficiencies on a global scale. Unlike traditional supplementation methods, biofortification integrates nutrient enhancement directly into the crops

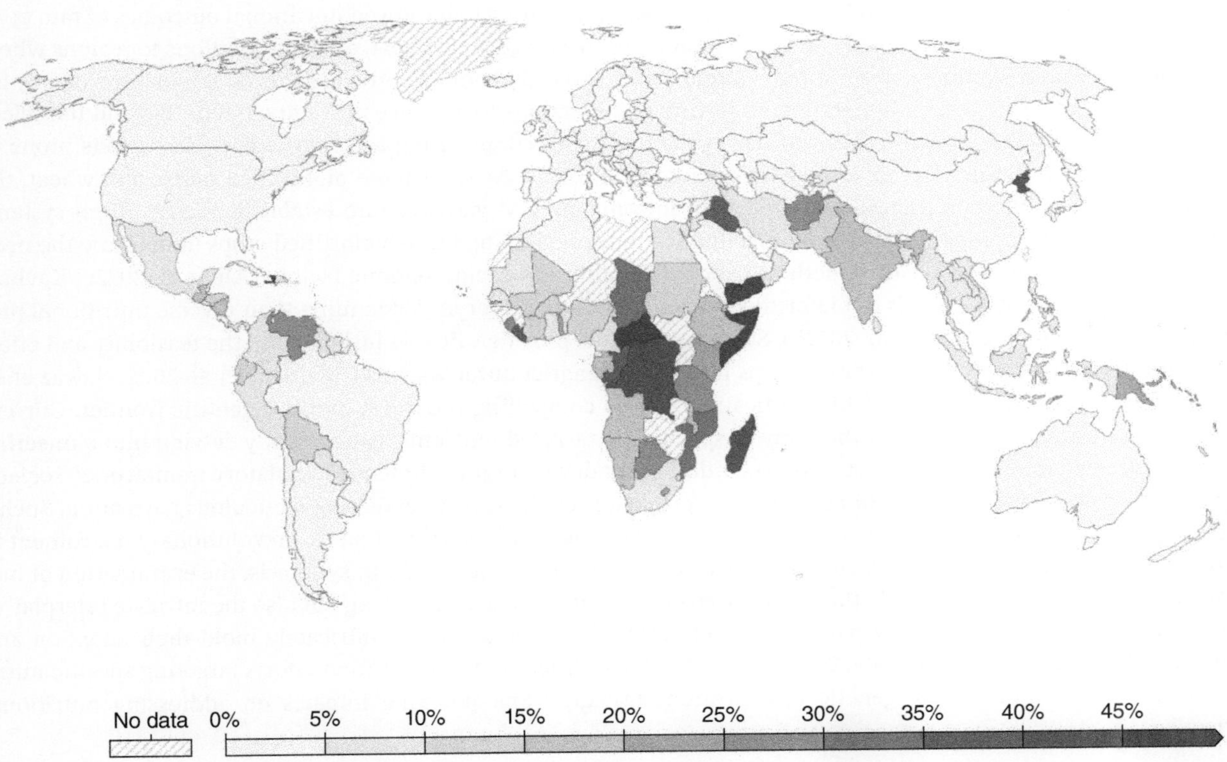

Figure 17.2 The prevalence of undernourishment (in percentage) in Global South. *Source:* Asia 2021; https://ourworldindata.org/hunger-and undernourishment#long-term-decline-of-undernourishment. Adapted from Sangam et al. (2023).

themselves, ensuring a continuous and natural source of essential vitamins and minerals through the regular diet of the population (Athar et al. 2020; Naz et al. 2023; Sandhu et al. 2023). This approach leverages genetic engineering techniques to enhance the nutrient content of staple crops, making them more nutritionally dense without altering their fundamental characteristics. By fortifying crops with targeted micronutrients, biofortification addresses the root cause of malnutrition, offering a holistic and long-term solution (Bouis et al. 2003; Gu et al. 2023; Malo 2017). Moreover, biofortified crops provide a cost-effective and accessible means of delivering essential nutrients to communities that may lack access to diverse diets or nutritional supplements (Bouis et al. 2020). This is particularly relevant in regions where staple crops form a substantial portion of daily food intake. Biofortification thus aligns with the principles of food security and sustainable agriculture, promoting health at the intersection of nutrition and crop production (Akbar et al. 2023; Mrabet 2023). Conclusively, the need for biofortification is underscored by the pervasive impact of malnutrition on global health (Bakar Ijaz et al. 2021). Biofortified crops present a transformative strategy to alleviate nutrient deficiencies, offering a sustainable and accessible solution that can positively impact the well-being of communities around the world.

17.3 Biofortified Crops: An Overview

Genetic engineering has revolutionized agriculture by offering innovative solutions to address global malnutrition. In the realm of biofortification, crops are tailored to enhance specific nutrients crucial for human health. This approach holds immense promise in combatting deficiencies, with key focuses on vitamin A, iron, and zinc (Kadapa et al. 2023; Singhal et al. 2023). Vitamin A biofortification is exemplified by Golden Rice, a groundbreaking development engineered to produce beta-carotene, a precursor to vitamin A (Stone 2010). This variety holds the potential to alleviate vitamin A deficiency, especially in regions heavily reliant on rice as a dietary staple. Despite its potential, the acceptance and implementation of Golden Rice face challenges, shedding light on the complex interplay between science, public perception, and policy (Dawe et al. 2002; Stone and Glover 2017). Iron biofortification targets staple crops like beans and rice, aiming to address iron deficiency prevalent in vulnerable populations (Khan et al. 2023c). Genetic modification strategies are employed to elevate iron levels in these crops, contributing to improved nutritional outcomes (Amin et al. 2023). Similarly, efforts in zinc biofortification focus on crops like maize and wheat, offering solutions to regions where zinc deficiency poses a significant health risk (Kumar et al. 2019; Stangoulis and Knez 2022). Biofortified beans, for instance, showcase the successful integration of genetic modifications to enhance nutrient content. Beyond the scientific aspects, the adoption of such crops can have profound socioeconomic impacts, particularly in regions prone to nutrient deficiencies (Romman et al. 2023; Storozhenko et al. 2007). As we explore biofortified maize and wheat, the discussion extends to the challenges associated with introducing GM varieties into established agricultural systems (Ullah et al. 2022). The agronomic and economic implications of adopting these biofortified crops underscore the need for a nuanced approach that considers both nutritional benefits and socioeconomic factors (Ali et al. 2022c; Kavhiza et al. 2022). A comparative analysis of biofortified varieties is essential for understanding their diverse nutritional profiles and potential synergies (Ali et al. 2022b). Such assessments provide valuable insights into the feasibility and effectiveness of integrating multiple biofortified crops into regional agricultural systems (Bhardwaj et al. 2022; Nawaz et al. 2023a). Anticipating the trajectory of biofortification reveals compelling prospects on the scientific frontier. Current research endeavors aspire to broaden the array of specifically targeted nutrients, concurrently delving into pioneering genetic engineering methodologies. Yet, this expedition is not devoid of impediments. Regulatory frameworks, societal acceptance, and ecological considerations emerge as formidable challenges necessitating meticulous navigation. Such a nuanced approach is imperative to ensure the judicious progression of biofortification as a revolutionary instrument in global agriculture (Almeida and Ranisch 2022; Black et al. 2006; Dixit et al. 2023). In synthesis, the examination of biofortified crops encapsulates not solely the scientific strides within genetic engineering but also the intricate interplay of socioeconomic dynamics, regulatory paradigms, and ethical considerations that intricately mold their adoption and consequential influence (Ahmad et al. 2022). Table 17.1 shows different biofortification efforts targeting specific nutrients, along with example crops, genetic modification strategies, and potential impacts on addressing nutritional deficiencies.

Table 17.1 Overview of different biofortification efforts targeting specific nutrients, along with example crops, genetic modification strategies, and potential impacts on addressing nutritional deficiencies.

Nutrient	Crop	Genetic modification strategy	Potential impact
Vitamin A	Golden Rice	Engineered to produce beta-carotene	Alleviates vitamin A deficiency
Iron	Beans, Rice	Elevating iron levels in staple crops	Addresses iron deficiency in vulnerable populations
Zinc	Maize, Wheat	Genetic modifications to enhance zinc content	Mitigates zinc deficiency health risks
Biofortified beans	Beans	Successful integration of genetic modifications	Improves nutrient content

17.4 Genetic Modification Techniques

Genetic engineering techniques have evolved significantly in the pursuit of enhancing crop nutritional content, with CRISPR-Cas9 standing out as a transformative tool. The CRISPR-Cas9 system, based on the bacterial adaptive immune system, enables precise genomic modifications in crops (Bao et al. 2019; Sami et al. 2021). Through the design of guide RNA sequences, scientists can target specific genes associated with nutrient biosynthesis, resulting in crops with enhanced nutritional profiles. This targeted approach reduces unintended genetic alterations and accelerates the development of biofortified crops (Marx 2020). Beyond CRISPR-Cas9, other sophisticated gene-editing tools like TALENs and zinc finger nucleases contribute additional precision to genetic modifications (Gupta and Musunuru 2014). TALENs, relying on specific DNA-binding proteins, offer a modular and programmable approach, while zinc finger nucleases utilize engineered zinc finger proteins for targeted DNA cleavage (Carlson et al. 2012). The diversity of available gene-editing tools allows researchers to choose the most suitable method for achieving desired trait modifications in crops (Hua et al. 2019). Traditional breeding, a time-tested method for trait enhancement, involves the selection and cross-breeding of plants with desirable characteristics. However, this process is often time-consuming and lacks the precision offered by genetic engineering (Holmes 2008). Genetic modification techniques, including CRISPR-Cas9, overcome these limitations by enabling rapid and specific trait alterations, expediting the development of crops with improved nutritional attributes (Arora and Narula 2017). The integration of traditional breeding and genetic engineering is gaining attention as a synergistic approach. By combining the genetic diversity harnessed through traditional breeding with the precision of gene-editing tools, researchers aim to optimize the speed and efficacy of biofortification efforts. This integrated strategy holds promise for addressing complex challenges associated with nutrient enhancement in crops (Huang et al. 2022; Narain 2018). Regulatory considerations play a pivotal role in shaping the landscape of genetic modification in agriculture (Nawaz et al. 2024b). Policies governing CRISPR-Cas9, and other gene-editing tools vary globally, and discussions on appropriate regulatory frameworks continue (Turnbull et al. 2021). Striking a balance between fostering innovation and ensuring the safety of GM crops is paramount in regulatory decision-making (Rapela 2019). The environmental and ecological impacts of genetic modification techniques are critical considerations (Aaqil et al. 2023). Assessing the unintended effects on ecosystems and biodiversity is essential. Comparisons between the environmental footprints of CRISPR-Cas9 and traditional breeding methods provide insights into the ecological implications of these technologies, aiding in the development of sustainable agricultural practices (Ferry and Gatehouse 2009; Ullah et al. 2019). Figure 17.3 indicates the comparison of classical breeding with transformative technologies involved in production of biofortified crops.

Conducting a rigorous risk–benefit analysis is imperative for informed decision-making. Evaluating potential risks, such as unintended ecological consequences and genetic off-target effects, alongside the benefits, including improved crop yields and enhanced nutritional content, guides researchers, policymakers, and stakeholders in navigating the complexities of adopting genetic modification techniques for biofortification (Ullah et al. 2023). Table 17.2 shows different aspects of genetic engineering techniques in crop biofortification, their advantages, and challenges.

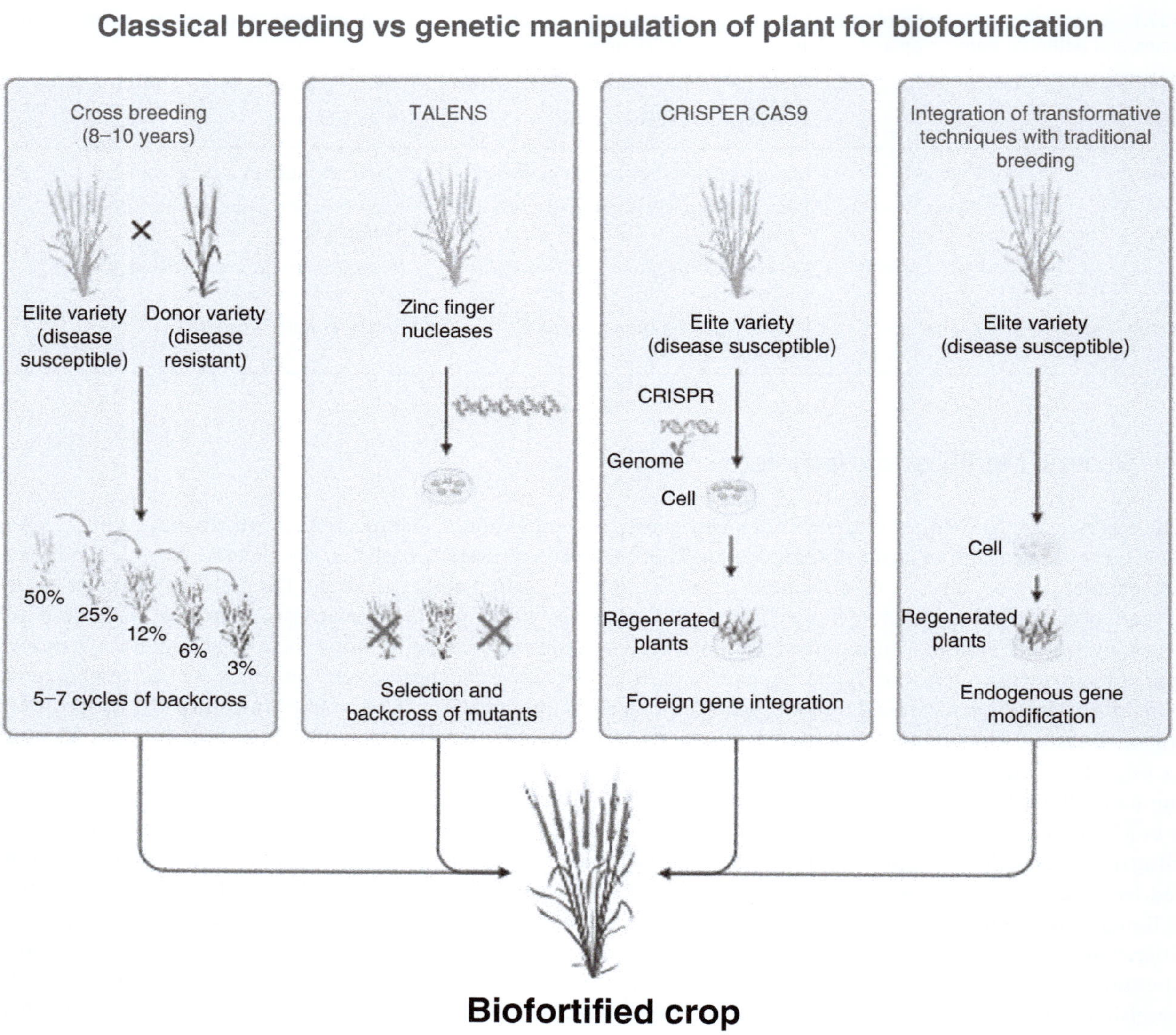

Figure 17.3 Comparison of classical breeding with transformative technologies involved in production of biofortified crops.

Table 17.2 Overview of the different aspects of genetic engineering techniques in crop biofortification.

Genetic engineering techniques	Description	Advantages	Challenges
CRISPR-Cas9	Enables precise genomic modifications in crops	Rapid and specific trait alterations	Regulatory uncertainties
TALENs	Modular and programmable approach utilizing DNA-binding proteins	Additional precision in genetic modifications	Technical complexity
Zinc Finger Nucleases	Utilizes engineered zinc finger proteins for targeted DNA cleavage	Targeted DNA cleavage	Efficiency and off-target effects
Integration with Traditional Breeding	Synergistic approach combining genetic diversity with precision	Optimizes speed and efficacy of biofortification efforts	Balancing traditional and modern methods

17.5 Regulatory and Ethical Considerations

17.5.1 Regulatory Landscape for Genetically Modified Crops

Understanding the global regulatory frameworks governing GM crops is vital for the nuanced discussion on biofortification (Kameri-Mbote 2005). International agreements such as the Cartagena Protocol on Biosafety and the Codex Alimentarius Commission provide the foundational guidelines (Burgiel 2002). An in-depth exploration of risk assessment protocols applied by regulatory bodies is imperative. This involves a thorough analysis of methodologies employed to evaluate the safety and environmental impact of GM crops, encompassing both known and potential risks (König et al. 2004; van der Heijden 2021). Additionally, scrutinizing variations in regulatory approaches among different regions is crucial, shedding light on their implications for the cross-border movement of biofortified crops (Kalaitzandonakes et al. 2023). The intricacies of labeling regulations for GMOs necessitate a comprehensive examination. This includes an in-depth analysis of challenges and benefits associated with the implementation of traceability systems throughout the supply chain for biofortified crops. A robust understanding of these regulatory aspects is essential for navigating the complex landscape of GM crop dissemination (Irfan et al. 2022).

17.5.2 Ethical Considerations in Genetic Engineering for Agriculture

Informed consent and stakeholder engagement represent fundamental ethical considerations in genetic engineering. This involves a detailed exploration of ethical principles related to obtaining informed consent from diverse stakeholders, including farmers, consumers, and local communities (Nawaz et al. 2024a). The strategies employed for effective engagement and decision-making processes in the development and deployment of GM crops must be meticulously examined (Eastwood et al. 2019; National Academies of Sciences and Medicine 2016; Resnik 2018). Ethical considerations pertaining to the potential environmental impact of GM crops are of paramount importance. This involves a nuanced discussion on issues related to biodiversity, ecosystem services, and unintended consequences (Abushal et al. 2021). The exploration of ethical frameworks for assessing and mitigating environmental risks associated with biofortification efforts should encompass both theoretical considerations and practical applications (Mampan et al. 2011). Ensuring equitable access to biotechnological innovations demands a comprehensive analysis of ethical concerns related to access and benefit-sharing. This includes an examination of initiatives promoting inclusivity and fairness in the adoption of GM crops, with a focus on addressing socioeconomic disparities and ensuring a just distribution of benefits (Devos et al. 2008). Transparency in research methodologies, data sharing, and the publication of results is a foundational ethical principle. This encompasses a detailed discussion on conflicts of interest and the critical role of scientific integrity in shaping public trust within the dynamic field of genetic engineering for agriculture (Harfouche et al. 2021; Prager et al. 2019). Ethical considerations regarding the potential long-term health effects of consuming biofortified crops require a thorough investigation. This involves scrutinizing the methodologies and frameworks used for monitoring and assessing health impacts over extended periods (Nawaz et al. 2023b). Additionally, exploring societal implications, including socioeconomic factors and potential unintended consequences on cultural practices and food systems, is crucial for a holistic understanding of the ethical dimensions associated with genetic engineering for crops biofortification (Shahab et al. 2023).

17.6 Environmental Impact Assessment of GM Biofortified Crops

GM biofortified crops have emerged as a promising solution to address global nutritional deficiencies, offering enhanced nutrient content to combat malnutrition. However, the cultivation and widespread adoption of GM crops raise concerns about potential environmental consequences. A thorough evaluation of the environmental impact is crucial to ensure the sustainable development and responsible deployment of these biofortified crops (Bhardwaj et al. 2022; Rehman et al. 2022).

17.6.1 Environmental Consequences of GM Biofortified Crops

17.6.1.1 Gene Flow and Hybridization

Genetic modification of crops for biofortification introduces the potential for gene flow, a process by which genetic material, including transgenes, moves from GM crops to wild relatives or non-GM crops (Senjam et al. 2021). The consequence of this gene flow is hybridization, a phenomenon wherein the genetic material from the GM crop integrates into the genome of the recipient organism (Messer). This aspect is of paramount concern due to its potential to induce unintended genetic modifications in natural ecosystems, thereby instigating uncertain and complex ecological consequences (Prakash et al. 2011). The ecological implications of gene flow and hybridization are intricate and multifaceted. The integration of transgenic traits into nontarget organisms can alter their genetic makeup and, consequently, influence their ecological interactions (Quist 2010). Such modifications may lead to changes in the phenotype, behavior, or reproductive capabilities of the recipient organisms, potentially affecting their fitness and survival in the ecosystem (Candolin and Wong 2012). The uncertainty surrounding the ecological consequences of gene flow is heightened by the dynamic nature of natural ecosystems. The intricate web of interactions among different species, environmental factors, and ecological processes makes it challenging to predict the long-term effects accurately (Smith et al. 2020). Additionally, the extent and pattern of gene flow may vary depending on factors such as the specific crop, its reproductive biology, and the surrounding environment (Andersson and De Vicente 2010). Careful and systematic monitoring is imperative to comprehend the full scope and implications of gene flow from biofortified crops. This monitoring should encompass a range of methodologies, including molecular techniques to detect the presence of transgenes in non-GM populations, ecological surveys to assess changes in population dynamics, and long-term studies to capture the cumulative effects over successive generations (Chambers et al. 2014; Kurkela and Brown 2009).

Furthermore, it is essential to consider the potential for gene flow not only to wild relatives but also to traditional, non-GM crops (Andersson and De Vicente 2010). Hybridization with non-GM crops may have implications for agricultural biodiversity, impacting the genetic integrity of locally adapted varieties. This emphasizes the need for comprehensive assessments that account for the broader genetic landscape and potential interactions between GM and non-GM crops (Tsatsakis et al. 2017). To address the challenges posed by gene flow and hybridization, sophisticated modeling approaches can be employed to simulate and predict the spread of transgenes in different scenarios (Ellstrand 2003). These models consider factors such as dispersal mechanisms, reproductive biology, and landscape characteristics, aiding in the development of strategies to mitigate and manage gene flow effectively (Epps et al. 2007).

In conclusion, the potential for gene flow and hybridization from GM biofortified crops necessitates meticulous investigation and ongoing scrutiny. The uncertainty surrounding the ecological consequences underscores the importance of a precautionary approach and the implementation of comprehensive monitoring and modeling strategies (Ali et al. 2022a). Only through a thorough understanding of the intricacies involved can we develop effective measures to minimize the unintended environmental impacts associated with gene flow in the context of GM biofortified crops.

17.7 Ecological Disruption in the Context of GM Biofortified Crops

The introduction of GM biofortified crops into natural ecosystems carries the inherent risk of ecological disruption (Mannion and Morse 2013). The alteration of characteristics in the targeted crops, particularly changes in nutritional content, has the potential to disturb existing ecological balances and provoke cascading effects throughout the ecosystem (Brühl and Zaller 2021). Understanding the intricate ecological dynamics is imperative for predicting, assessing, and effectively mitigating environmental disruptions associated with the cultivation of GM biofortified crops (Chávez-Dulanto et al. 2021; Shah et al. 2023). The ecological disruption resulting from the cultivation of GM crops primarily stems from the modified traits and their influence on interactions within the ecosystem (Snow et al. 2005). In the context of biofortification, where crops are engineered to possess enhanced nutritional profiles, the shifts in plant physiology and biochemistry can have repercussions on various trophic levels (Sati et al. 2023). Changes in the nutritional content of biofortified crops may alter the dynamics of herbivore–plant interactions (Bouis and Saltzman 2017). For instance, increased nutrient levels could impact the feeding preferences and performance of herbivores, potentially leading to changes in herbivore abundance and distribution (Fahad et al. 2023). The repercussions of these alterations extend to higher trophic levels, influencing the predators and parasitoids that rely on herbivores as a food source (Nawaz et al. 2022; Tylianakis and Binzer 2014; Vermeulen et al. 2020).

Furthermore, the modification of crop traits may influence the composition and activity of soil microbial communities (Khan et al. 2023b). Root exudates, altered by genetic modifications, can affect the rhizosphere microbiome, potentially disrupting symbiotic relationships between plants and beneficial microbes. This disruption may have cascading effects on nutrient cycling, soil structure, and overall soil health (Dawar et al. 2023; Jing et al. 2023; Vives-Peris et al.

2020). The potential for unintended ecological consequences is compounded by the complex web of interactions within ecosystems (Mehmood et al. 2024). Keystone species, which play crucial roles in maintaining ecological balance, may be indirectly affected by changes in the abundance or behavior of nontarget organisms. Such indirect effects can propagate throughout the ecosystem, leading to unforeseen ecological consequences (Fleeger 2020; Lurgi et al. 2018). Understanding the ecological dynamics necessitates comprehensive ecological studies that consider the intricate interplay of biotic and abiotic factors (Adnan et al. 2023). Long-term field studies, coupled with controlled experiments, can provide insights into the responses of various components of the ecosystem to the presence of GM biofortified crops. Analyzing community structure, population dynamics, and ecosystem functions is essential for identifying potential disruptions and their underlying mechanisms (Khan et al. 2022). Mitigating ecological disruptions associated with GM biofortified crops requires a proactive and adaptive approach. This involves the development of strategies that minimize unintended consequences while optimizing the benefits of biofortification (Krishnappa et al. 2022). Integrating ecological considerations into the design and assessment of GM crops, along with the implementation of sustainable agricultural practices, can contribute to the preservation of ecological balance (Conner et al. 2003). The expression of novel traits in GM biofortified crops has the potential to exert inadvertent impacts on nontarget organisms within agroecosystems (Khan et al. 2023a). Nontarget organisms encompass a diverse array of life forms, ranging from beneficial insects like pollinators and predators to soil microorganisms and various components of wildlife. The intricate interactions and dependencies within these ecosystems necessitate a nuanced examination of the direct and indirect consequences associated with the cultivation of GM crops (Müller et al. 2020). Figure 17.4 indicates the ecological disruption governed by GM biofortified crops.

In conclusion, the cultivation of GM biofortified crops has the potential to disrupt existing ecological balances, necessitating a nuanced understanding of the ecological dynamics at play. Through rigorous ecological research and a holistic approach to risk assessment, it becomes possible to predict, prevent, and mitigate the environmental disruptions associated with the deployment of GM crops in biofortification initiatives.

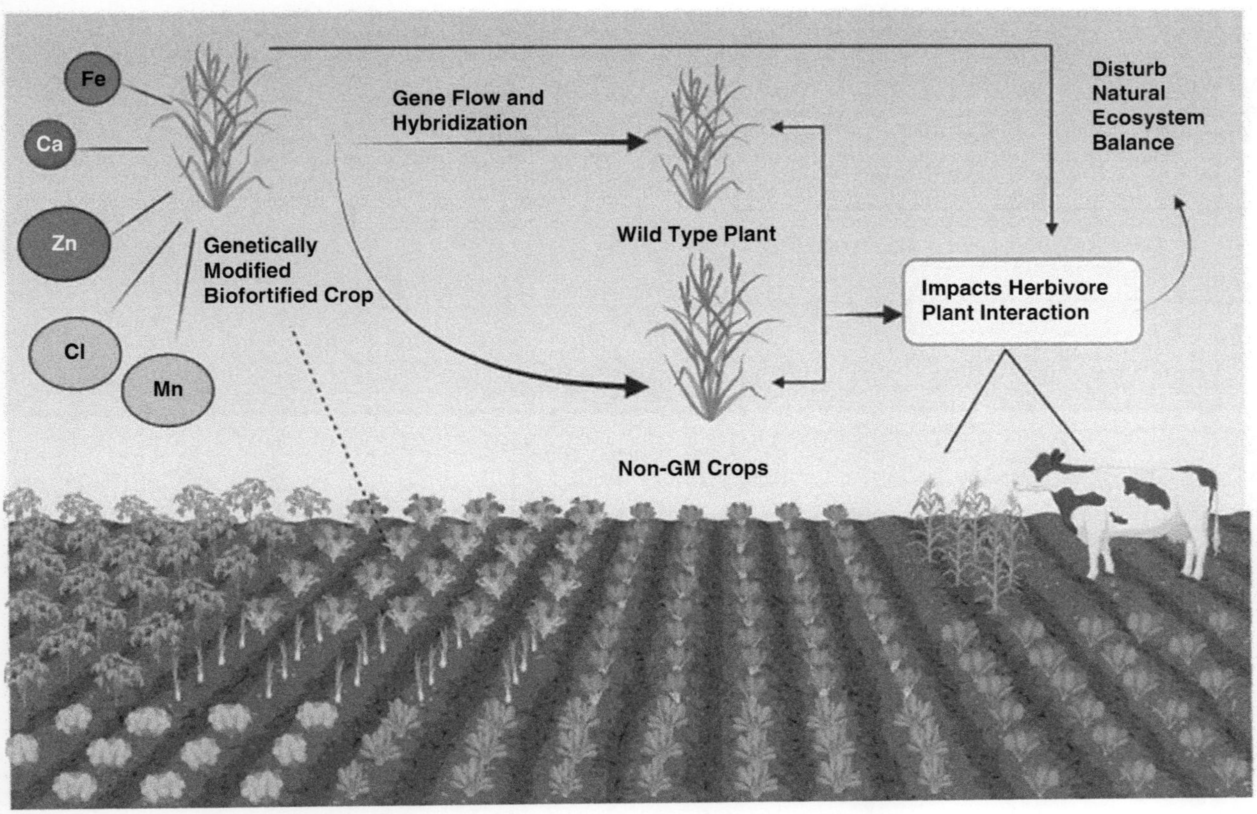

Figure 17.4 Ecological disruption governed by genetically modified biofortified crop.

17.8 Impact on Nontarget Organisms

In the context of GM biofortified crops, the inadvertent impacts on nontarget organisms are multifaceted. Beneficial insects, crucial for pollination and natural pest control, may be affected directly through the altered traits of the biofortified crops or indirectly by changes in the availability of resources. For example, modifications in floral traits or nectar composition may influence the foraging behavior and reproductive success of pollinators (Brookes 2022; Zimmerman 1988). Soil microorganisms, playing pivotal roles in nutrient cycling, organic matter decomposition, and disease suppression, are also susceptible to the introduction of GM crops. Changes in root exudates or plant residues resulting from genetic modifications can influence the composition and function of soil microbial communities. Understanding these alterations is imperative for maintaining soil health and fertility (Gayan et al. 2023; Irfan et al. 2021). Wildlife, which encompasses various organisms ranging from mammals to birds and amphibians, may experience both direct and cascading effects due to the cultivation of GM biofortified crops (Holt-Giménez 2019). Changes in crop characteristics, including nutritional content or resistance traits, can influence the availability and palatability of crops to herbivores. These alterations may, in turn, affect the abundance and distribution of higher trophic levels in the ecosystem (Hamann et al. 2021). Assessing the impact on nontarget organisms is not solely confined to evaluating immediate effects. The broader consideration involves understanding the repercussions on biodiversity and ecosystem services (Tallis et al. 2015). Biodiversity, as a measure of the variety of life forms and their interactions within an ecosystem, may be influenced by changes in species composition, abundance, and trophic dynamics resulting from GM crop cultivation (Brose and Hillebrand 2016). Ecosystem services, encompassing the benefits that ecosystems provide to human societies, are intricately linked to biodiversity and functional ecological processes. Evaluating the impact of GM biofortified crops on ecosystem services involves considering alterations in pollination services, soil fertility, and pest regulation, among others (Daily 1997). Balancing the enhancement of nutritional traits in crops with the preservation of these critical ecosystem functions is pivotal for ensuring the overall ecological sustainability of GM biofortified crops (Bhardwaj et al. 2022). Figure 17.5 depicts the possible ecological disruptions of soil microflora induced by GM biofortified crops.

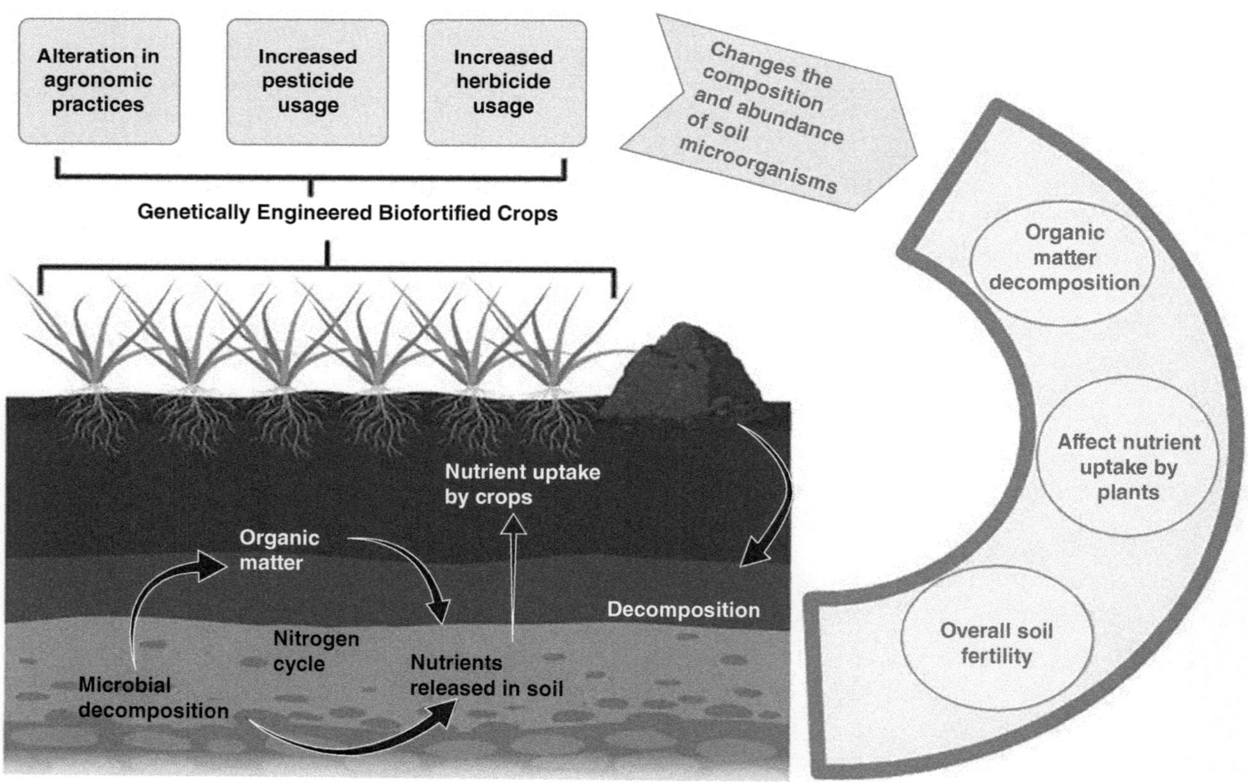

Figure 17.5 Ecological disruption of soil microflora induced by genetically modified biofortified crops.

In summary, the expression of novel traits in GM biofortified crops necessitates a comprehensive evaluation of its impact on nontarget organisms, encompassing beneficial insects, soil microorganisms, and wildlife (Olembo et al. 2010). The complexity of ecological interactions and dependencies within agroecosystems underscores the importance of assessing both direct and indirect effects on biodiversity and ecosystem services (Garbach et al. 2014). Such evaluations are paramount for making informed decisions regarding the ecological sustainability of GM crops and their integration into global agricultural systems (Organisms 2011).

17.9 Soil and Water Quality

The assessment of the environmental impact associated with the cultivation of GM biofortified crops requires a thorough examination of potential repercussions on soil and water quality (Mathur et al. 2017). Cultivation practices linked to GM crops often involve alterations in agronomic practices, such as heightened use of herbicides or pesticides (Carpenter 2011; Saljoqi et al. 2022). Understanding the enduring consequences of these practices on soil microbial communities, nutrient cycling processes, and water ecosystems is vital for preserving environmental health and preventing unintended pollution (Bawazeer et al. 2021; Jain et al. 2023).

The introduction of GM crops, particularly those engineered for herbicide tolerance or pest resistance, may exert both direct and indirect effects on soil microbial communities (Dunfield and Germida 2004). Changes in the composition and abundance of soil microorganisms have implications for nutrient cycling, organic matter decomposition, and overall soil fertility (Condron et al. 2010). Comprehensive assessments are necessary to unravel the complex interactions between GM crops and soil microbiota, considering factors such as microbial diversity, functional redundancy, and ecosystem resilience (Martin et al. 2019). Cultivation practices associated with GM crops, particularly those involving the use of specific herbicides or pesticides, may impact nutrient cycling processes in the soil. The potential alteration of nutrient availability and cycling dynamics can have cascading effects on plant growth, microbial activity, and nutrient uptake by crops (Briat et al. 2020; Mandal et al. 2020). A thorough understanding of intricate nutrient fluxes within the soil ecosystem is essential for predicting and mitigating disruptions arising from the cultivation of GM biofortified crops (Das et al. 2022). The influence of GM crops on water quality extends beyond agricultural fields to adjacent water ecosystems. Runoff from fields treated with herbicides or pesticides may carry these substances into nearby water bodies, posing risks to aquatic organisms and overall ecosystem health (Agarwal and Singh 2021). Evaluating the long-term effects on water quality, including the persistence and transport of GM-related substances, is crucial for comprehensively assessing the environmental impact (Glencross et al. 2020). Mitigating potential negative impacts on soil and water quality necessitates the implementation of preventive measures (Schoumans et al. 2014). These may include optimized application techniques for herbicides or pesticides, the development of environmentally friendly formulations, and the incorporation of soil conservation practices (Singh et al. 2020). Additionally, the adoption of precision agriculture technologies can help minimize the overuse of chemical inputs, reducing the environmental footprint associated with GM crop cultivation (Shah and Wu 2019). Figure 17.5 depicts the possible ecological disruptions of soil microflora induced by GM biofortified crops.

Therefore, the environmental assessment of GM biofortified crops demands a nuanced understanding of their impact on soil and water quality. Rigorous investigations into the long-term effects on soil microbial communities, nutrient cycling, and water ecosystems are essential for ensuring sustainable agricultural practices. By identifying potential risks and implementing targeted preventive measures, the cultivation of GM biofortified crops can contribute to improved nutritional outcomes without compromising environmental integrity.

17.10 Strategies for Minimizing Negative Environmental Effects

17.10.1 Containment Measures

In the context of cultivating GM biofortified crops, the implementation of containment measures is imperative to curtail the inadvertent spread of GM traits to non-GM crops or wild relatives (Mall et al. 2018). Containment strategies are designed to minimize gene flow, thereby preventing the unintended dissemination of GMOs into the broader environment (Council 2004). These measures encompass a range of physical, agronomic, and spatial practices aimed at achieving effective isolation and confinement of GM crops (Prabhu 2009).

Physical barriers represent a primary means of containment, often involving the use of structures or enclosures to restrict the movement of pollen, seeds, or vegetative propagules carrying modified genetic material (Hokanson 1996). This may include the establishment of isolation zones, such as physical barriers like greenhouses or netting, to confine the cultivation of GM crops within controlled environments. The goal is to create a physical barrier that limits the interactions between GM crops and non-GM crops, reducing the likelihood of gene transfer (Beckman 1980). Isolation distances play a critical role in containment strategies, delineating the spatial separation between GM and non-GM crops. Determining appropriate isolation distances involves considering factors such as the crop's reproductive biology, pollination mechanisms, and the potential for gene flow. By establishing adequate separation, the probability of cross-pollination and gene exchange is minimized, contributing to the containment of modified genetic material (Beckie and Hall 2008; Eastham and Sweet 2002; Gupta 2021). Alternative cultivation practices also contribute to containment efforts by introducing agricultural techniques that further restrict the potential for gene flow (Lal et al. 2020). These practices may involve altering planting densities, modifying planting times, or implementing specific agronomic practices tailored to the unique characteristics of GM crops (Norsworthy et al. 2012). Through these alternative cultivation methods, the emphasis is on mitigating the opportunities for unintended gene transfer while maintaining the nutritional benefits associated with biofortified crops (Unnevehr et al. 2007). The effectiveness of containment measures is contingent upon a comprehensive understanding of the biology and ecology of both the GM and non-GM crops involved (Rüdelsheim 2015). Besides, ongoing monitoring and assessment are integral components of containment strategies, allowing for the identification of any breaches or unintended gene flow (Craig et al. 2008). Continuous refinement of containment measures based on empirical observations and scientific data ensures that cultivation practices remain robust and adaptive to evolving environmental conditions (Weiss et al. 2020).

In brief, containment measures in the cultivation of GM biofortified crops involve a multifaceted approach encompassing physical barriers, isolation distances, and alternative cultivation practices. These strategies aim to limit the spread of GM traits, safeguarding against unintended gene flow and preserving the integrity of both GM and non-GM crop populations.

17.10.2 Monitoring and Surveillance

Monitoring and surveillance are integral components of a comprehensive risk assessment framework for the cultivation of GM biofortified crops. Rigorous programs must be established to systematically and continuously evaluate the environmental dynamics associated with these crops (Marx 2010). The primary objective of such initiatives is the early detection of any unforeseen or unintended effects on the environment, facilitating prompt identification and mitigation of potential risks (Magnan et al. 2016). Continuous monitoring involves the systematic collection and analysis of data related to the cultivation and interaction of GM biofortified crops within their respective ecosystems. This process encompasses various environmental parameters, including but not limited to soil composition, water quality, biodiversity, and the presence of unintended genetic material in non-GM crops or wild relatives (Mathur et al. 2017; Tsatsakis et al. 2017). The establishment of robust monitoring and surveillance programs requires the collaboration of interdisciplinary teams comprising ecologists, agronomists, geneticists, and environmental scientists. These teams utilize advanced monitoring technologies, such as remote sensing, molecular assays, and ecological surveys, to gather quantitative and qualitative data (Danovaro et al. 2016; Eigenbrode et al. 2018). By implementing continuous surveillance, scientists and regulatory bodies can promptly identify deviations from expected environmental conditions or any unanticipated consequences associated with the cultivation of GM crops (Meckawy et al. 2022). Early detection is crucial for devising effective intervention strategies and minimizing the potential harm that may arise from unintended effects (Pashayan et al. 2020). The data obtained through monitoring and surveillance programs are subjected to rigorous analysis, employing statistical models and ecological assessments. This analytical phase aims to discern patterns, trends, and correlations within the collected data, providing insights into the potential environmental impacts of GM biofortified crops (Bhatia et al. 2021; Chinedu et al. 2018).

Likewise, the surveillance efforts extend to assessing the persistence and dispersal patterns of GM traits in the environment. This involves monitoring the spread of genetic material to non-GM crops or wild populations, ensuring a thorough understanding of the potential ecological consequences associated with gene flow (Djordjevic et al. 2023; Kjellson and Strandberg 2001). The information garnered from monitoring and surveillance activities serves as a basis for evidence-based decision-making. Regulatory agencies utilize this data to make informed assessments of the environmental safety of

GM crops and to develop adaptive management strategies when necessary (Craig et al. 2008). In summary, the establishment of robust monitoring and surveillance programs is imperative for the responsible cultivation of GM biofortified crops. Through continuous data collection, interdisciplinary analysis, and early intervention, these programs contribute to minimizing potential environmental harm and fostering the sustainable integration of GM crops into agricultural systems.

17.10.3 Ecological Modeling

The application of ecological modeling in the context of cultivating GM biofortified crops is instrumental in gaining profound insights into the potential long-term ecological consequences. Ecological modeling encompasses a range of quantitative techniques that simulate complex ecological systems, enabling researchers to predict and evaluate the dynamic interactions between GM crops and their surrounding ecosystems (Hoffman 2016; Tixier et al. 2013). These predictive models serve as invaluable tools for systematically assessing various scenarios related to the cultivation of GM biofortified crops. By employing mathematical algorithms and computational simulations, researchers can project the ecological impact of introducing GMOs into diverse environments. This process involves considering multiple variables, including genetic traits of the biofortified crops, environmental conditions, and interactions with other organisms within the ecosystem (Mulualem 2015; Tan et al. 2023; Zomorrodi and Segrè 2016). The complexity of ecological systems necessitates a comprehensive understanding of how changes in one component may propagate through the entire ecosystem. Ecological models facilitate the exploration of potential ecological consequences by capturing intricate relationships and feedback loops within these systems. This includes studying the potential impacts on biodiversity, nutrient cycling, and ecosystem services (Anand et al. 2010; Riva et al. 2023). Moreover, ecological modeling allows for the assessment of the spatial and temporal dynamics of GM biofortified crops. Researchers can simulate the dispersal of genetic material, such as pollen, and analyze its potential consequences on neighboring non-GM crops or wild relatives (Appugol et al. 2022). This spatial modeling aids in predicting the extent of gene flow and potential hybridization events, providing critical information for the design of containment measures (Bohling 2016). The refinement of cultivation strategies is a direct outcome of the insights gained from ecological modeling. Researchers can iteratively adjust parameters in the models to simulate alternative scenarios and evaluate the potential outcomes (Jakeman et al. 2006). This iterative process enables the identification of strategies that mitigate negative ecological impacts while optimizing the desired nutritional traits in biofortified crops (Venkataram and Hefferon 2023).

In conclusion, ecological modeling stands as a sophisticated and indispensable approach in the scientific toolkit for assessing the environmental implications of cultivating GM biofortified crops. By providing a platform for simulating diverse ecological scenarios, these models contribute to evidence-based decision-making, guiding the development and implementation of biofortification strategies that are both nutritionally beneficial and environmentally sustainable.

17.10.4 Integrated Pest Management Practices

Implementing integrated pest management (IPM) practices can help reduce the reliance on chemical inputs associated with some GM crops (Anderson et al. 2019). By incorporating natural predators and adopting sustainable farming practices, the environmental footprint of biofortified crops can be minimized (Pakeerathan and Vaishnavi 2022).

17.10.5 Adaptive Management Strategies

Recognizing the dynamic nature of ecosystems and adopting adaptive management strategies are essential. This involves a flexible approach that allows for adjustments in cultivation practices based on ongoing environmental monitoring and research findings (Duru et al. 2015; Rist et al. 2013). In essence, the cultivation of GM biofortified crops necessitates a comprehensive and ongoing environmental impact assessment. By understanding the potential consequences and implementing effective strategies to minimize negative effects, scientists and policymakers can pave the way for the responsible and sustainable integration of GM crops into global agriculture, ensuring both improved nutrition and environmental conservation (Huesing et al. 2016; Lawler 2009). Figure 17.6 shows steps to minimize negative influence of environment posed by biofortified crops.

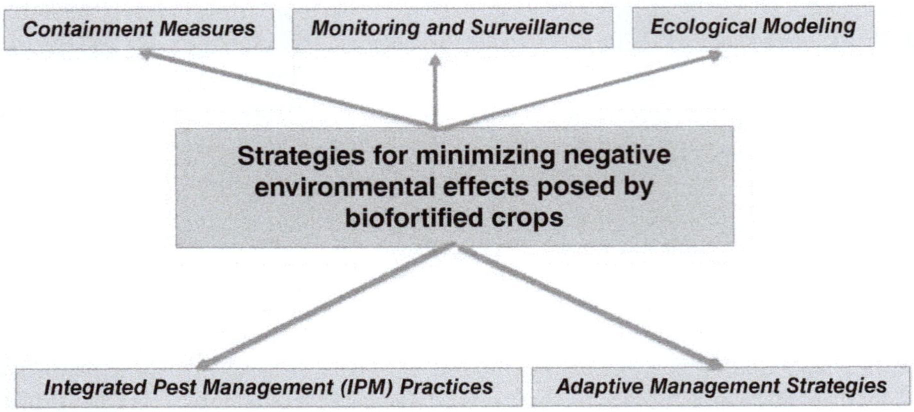

Figure 17.6 Steps to minimize negative influence of environment posed by biofortified crops.

17.11 Genetic Modifications and Agronomic Performance in Biofortified Crops

Genetic engineering has become a potent tool for improving crop characteristics, not only to address nutritional deficiencies but also to enhance overall agronomic performance. This discussion explores the intricate relationship between genetic modifications, crop yield, and key agronomic traits, emphasizing the need for a delicate balance in successful biofortification (Dhaliwal et al. 2022; Newell-McGloughlin 2008). The impact of genetic modifications on crop yield is a primary concern in biofortification. While the objective is to improve nutritional content, it is essential to ensure that these modifications do not compromise fundamental crop productivity (Khaipho-Burch et al. 2023). Studies have reported varying effects, with some genetic modifications positively influencing yield through enhanced resilience to environmental stressors or improved photosynthetic efficiency (Morales et al. 2020). However, unintended consequences such as altered plant architecture or disrupted hormonal balances may lead to reduced yields in other cases, highlighting the complexity of genetic engineering's impact on crop productivity (Zalabák et al. 2013). Balancing nutrient enhancement with other crucial agronomic traits, such as pest resistance and drought tolerance, presents a significant challenge in biofortification efforts. Crops engineered for nutrient enhancement often incorporate genes for pest resistance, aiming to protect against yield-limiting pests while improving nutritional profiles. However, the intricate interplay between pest resistance mechanisms and nutrient-related pathways necessitates careful consideration to avoid unintended consequences. Similarly, drought tolerance, a vital trait in water-scarce regions, must be maintained alongside nutrient enhancement. Identifying and engineering genes that contribute to both nutrient accumulation and drought tolerance requires a deep understanding of plant physiology (Afridi et al. 2023; Rosero et al. 2020; Tripathy and Rath). Successful biofortification programs integrate multiple approaches, considering the synergies and trade-offs between nutrient enhancement and other agronomic traits. This involves advanced breeding techniques, such as marker-assisted selection, to introgressive multiple traits without sacrificing overall crop performance (Peprah 2020; Sheoran et al. 2022a).

In summary, achieving a balance between nutrient enhancement and other desirable agronomic traits is critical for the success of biofortification efforts. Ongoing advancements in genetic engineering techniques, coupled with a profound understanding of plant biology, offer promising avenues for developing biofortified crops that not only address nutritional deficiencies but also ensure sustainable and resilient agriculture. Continued research and careful field trials are imperative to validate the efficacy and safety of GM crops in diverse agroecological settings.

17.12 Multi-Biofortification Approaches

Multi-biofortification approaches represent an innovative and promising avenue in the realm of genetic engineering for crop improvement. This strategy involves the simultaneous enhancement of multiple nutritional attributes in a single crop, aiming to address complex dietary deficiencies and improve overall nutritional quality. The exploration of such approaches brings to light a myriad of opportunities, synergies, and challenges that merit in-depth scientific scrutiny (Dhaliwal et al. 2022). The integration of multiple biofortification strategies holds the potential to create crops with enhanced nutritional profiles that surpass the capabilities of single-biofortified varieties. By targeting multiple essential nutrients concurrently, these crops can offer a more comprehensive solution to malnutrition, especially in regions where diverse nutrient deficiencies prevail. For instance, a crop engineered to accumulate increased levels of both iron and zinc could be particularly beneficial in areas where deficiencies in both minerals contribute significantly to public health concerns (Gómez-Galera et al. 2010; Ofori et al. 2022; Wang et al. 2023). One of the primary advantages of multi-biofortification approaches is the potential for synergistic interactions among the enhanced traits. The co-accumulation of different nutrients in a crop has the capacity to amplify the nutritional benefits, offering a cumulative impact that goes beyond the sum of individual enhancements. Synergies may arise through metabolic pathways that intersect or mutually reinforce each other, leading to a more robust and effective biofortification outcome (Jiang et al. 2021). Additionally, multi-biofortified crops have the potential to provide a more balanced nutrient profile, minimizing the risk of unintended consequences associated with excessive accumulation of a single nutrient. This balance is crucial to prevent potential antagonistic interactions that could compromise the overall nutritional quality or agronomic performance of the engineered crop. Achieving this delicate balance requires a nuanced understanding of the metabolic pathways involved and careful fine-tuning of genetic modifications (Graham et al. 2000; Sheoran et al. 2022b). However, the pursuit of multi-biofortification is not without its challenges. One significant hurdle lies in the intricate regulatory frameworks governing GMOs (Bindraban et al. 2015). The approval processes for crops engineered with multiple biofortification traits are often more complex and time-consuming than those for single-trait biofortified crops. Regulatory bodies may require comprehensive assessments of potential interactions and unintended effects, demanding a thorough understanding of the molecular and biochemical aspects of each modification (Cellini et al. 2004; Kiran 2020). Another challenge is the potential for trade-offs between different biofortification traits. Genetic modifications intended to enhance one nutrient may inadvertently interfere with the absorption or utilization of another, leading to unintended consequences. Striking the right balance between nutrient enhancements while avoiding negative interactions demands a meticulous and sophisticated approach to the design and evaluation of multi-biofortified crops (Fanzo and Glass 2018; Rietra et al. 2017). Moreover, the implementation of multi-biofortification approaches necessitates a high level of technical expertise and precision in genetic engineering. Coordinating the expression of multiple genes and ensuring their simultaneous activation in relevant tissues without compromising the overall fitness of the crop is a complex task. Advances in genome editing technologies, such as CRISPR-Cas9, offer powerful tools for more precise control over gene expression, but challenges in the delivery and stability of genetic modifications persist (Bouis et al. 2020; Wang et al. 2017). In conclusion, the exploration of multi-biofortification approaches represents a pivotal advancement in the quest for nutritionally enhanced crops. The potential benefits, including synergies among nutrients and a more balanced nutrient profile, hold promise for addressing complex nutritional challenges. However, the realization of this potential is accompanied by significant scientific, regulatory, and technical challenges that demand careful consideration and continuous innovation. The journey toward crops with optimal nutritional profiles requires interdisciplinary collaboration, ethical scrutiny, and ongoing research to navigate the complexities inherent in the simultaneous enhancement of multiple biofortification traits.

17.13 Targeted Nutrient Delivery

Precision in nutrient delivery to specific plant tissues is a crucial aspect of genetic engineering for crop biofortification. This precision allows scientists to enhance the nutrient content of crops efficiently, addressing specific nutritional deficiencies without compromising the overall health and development of the plant. Various technologies and methods have been developed to achieve targeted nutrient distribution within the plant, each with its advantages and challenges (Kumar et al. 2019; Santana et al. 2020). One prominent approach in targeted nutrient delivery involves the use of nanotechnology. Nanoparticles, with dimensions on the nanometer scale, can be engineered to encapsulate specific nutrients (Jafari and McClements 2017). These nanoparticles serve as carriers, protecting the nutrients from degradation and ensuring their controlled release. The surface properties of these nanoparticles can be modified to enhance their interaction with specific plant tissues, facilitating efficient nutrient uptake (Mishra et al. 2014). Another innovative technique is the use of precision gene editing tools like CRISPR-Cas9. With these tools, scientists can modify the plant's genome to express genes responsible for targeted nutrient accumulation in specific tissues (Razzaq et al. 2019). By precisely editing the plant's genetic code, researchers can influence the expression of transporters, receptors, and enzymes involved in nutrient uptake and assimilation. This approach enables the development of crops with enhanced nutrient content in edible parts without affecting other plant tissues (Carvalho and Vasconcelos 2013). Furthermore, advancements in hydroponics and aeroponics systems contribute to targeted nutrient delivery. These soilless cultivation techniques provide a controlled environment where the composition and concentration of nutrients in the growing medium can be precisely regulated (Banerjee et al. 2022). By tailoring nutrient solutions to the specific needs of the plant at different growth stages, scientists can optimize the bioavailability of essential nutrients (Briat et al. 2020). In addition to nanotechnology and gene editing, researchers are exploring the potential of mycorrhizal associations to enhance targeted nutrient delivery (Joel et al. 2023). Mycorrhizal fungi form symbiotic relationships with plant roots, improving nutrient absorption by extending the root's reach and facilitating nutrient transfer. By understanding and manipulating these interactions, scientists aim to develop crops with increased nutrient uptake efficiency (Abobaker 2017). The use of precision agriculture practices complements these technologies, allowing for the fine-tuning of nutrient delivery based on real-time data. Sensor technologies, including drones and satellite imaging, provide valuable information on plant health, nutrient status, and growth patterns. This data-driven approach enables farmers to adjust nutrient application with precision, ensuring that crops receive the right amount of nutrients at the right time and in the right location (Paul et al. 2022; Singh 2022). Despite the promising strides in targeted nutrient delivery, challenges remain. Ensuring the safety of nanomaterials used in agriculture, addressing ethical concerns related to gene editing, and understanding the long-term effects of these technologies on ecosystems are areas that require careful consideration. Additionally, the scalability and cost-effectiveness of these approaches need to be optimized to facilitate their widespread adoption in agricultural systems worldwide (Ijaz et al. 2023; Mishra et al. 2023). In close, achieving precision in nutrient delivery to specific plant tissues represents a paradigm shift in crop biofortification. The integration of nanotechnology, gene editing, mycorrhizal associations, and precision agriculture holds tremendous potential for developing crops with tailored nutrient profiles. As researchers continue to refine these technologies and address associated challenges, the future of targeted nutrient delivery in agriculture appears promising, offering sustainable solutions to global malnutrition challenges.

17.14 Integration with Traditional Breeding Programs

Collaboration between genetic engineering and traditional breeding methods is a pivotal aspect of enhancing biofortification through a holistic approach to crop improvement. This integration seeks to leverage the strengths of both approaches, harnessing the precision and rapidity of genetic engineering alongside the robustness and adaptability inherent in traditional breeding. By combining these methodologies, scientists aim to accelerate the development of crops with improved nutritional content while maintaining desirable agronomic traits (Bakala et al. 2020; Mulualem 2015). Traditional breeding has been a cornerstone of agriculture for centuries, relying on natural processes to bring about genetic variation and select desired traits (Singer et al. 2021). This method involves controlled crossbreeding and selection based on the phenotype of the plants. It is a time-consuming process, often requiring multiple generations to achieve the desired outcome. However, traditional breeding is crucial for preserving genetic diversity and adapting crops to diverse environmental conditions (Begna 2021a, Begna 2021b). Genetic engineering, on the other hand, introduces targeted genetic modifications to achieve specific traits in a more controlled and rapid manner. Technologies such as CRISPR-Cas9 have revolutionized the precision with which scientists can edit the genome of a plant, allowing for the direct manipulation of genes responsible for nutrient accumulation. This technique provides a powerful tool for enhancing biofortification by directly addressing specific nutritional deficiencies (Li et al. 2023; Tiedje et al. 1989). The integration of genetic engineering with traditional breeding begins by identifying target genes associated with both nutritional content and desirable agronomic traits (Gascuel et al. 2017). Traditional breeding helps introduce these genes into existing crop varieties, facilitating the incorporation of beneficial traits without sacrificing the genetic diversity crucial for long-term crop resilience. This collaborative approach enables the development of biofortified crops that not only meet nutritional needs but also thrive in various agroecological contexts (Bharadwaj 2016). Besides, the integration of genetic engineering and traditional breeding contributes to the sustainability and acceptance of biofortified crops (Yadav et al. 2020). The genetic diversity introduced through traditional breeding helps buffer crops against evolving pests, diseases, and environmental stresses. This resilience is essential for the long-term success of biofortified crops, ensuring their continued viability in dynamic agricultural systems. The collaborative efforts also extend to addressing consumer preferences and cultural acceptance of biofortified crops. Traditional breeding allows for the incorporation of traits related to taste, texture, and overall culinary suitability, aligning the biofortified varieties with local food preferences. This consideration is vital for fostering acceptance and uptake of GM crops, as consumer perception plays a crucial role in the successful adoption of new agricultural technologies (Custodio et al. 2023; Luh et al. 2014; Samuel et al. 2023). Additionally, the integration of genetic engineering and traditional breeding enhances the potential for developing crop varieties with multiple beneficial traits. For instance, a biofortified crop could simultaneously exhibit increased nutrient content, resistance to pests, and tolerance to abiotic stresses such as drought or salinity (National Academies of Sciences and Medicine 2016). This multifaceted approach addresses not only nutritional needs, but also broader challenges faced by agricultural systems, promoting sustainable and resilient food production (Horton et al. 2017). Overall, the collaboration between genetic engineering and traditional breeding represents a synergistic approach to biofortification and crop improvement. By integrating the precision of genetic engineering with the adaptability of traditional breeding, scientists can develop biofortified crops that are not only nutritionally enhanced but also well-suited to diverse agroecological conditions. This holistic strategy ensures the continued relevance and sustainability of biofortified crops in contributing to global food security and addressing malnutrition.

17.15 Global Perspectives on Biofortification

In examining global perspectives on biofortification, it is imperative to recognize the diverse nutritional needs and priorities across different regions (Horton et al. 2017). Nutritional deficiencies vary significantly worldwide, and understanding these regional disparities is crucial for tailoring biofortification strategies to address specific challenges (Bhardwaj et al. 2022). This discussion delves into the complex interplay between regional nutritional requirements, biofortification priorities, and the adaptation of strategies to meet the unique needs of diverse geographic areas (Meybeck et al. 2017). Nutritional needs are shaped by a myriad of factors, including cultural practices, dietary habits, and prevalent health issues. For instance, certain regions may grapple with a high prevalence of vitamin A deficiency, leading to an increased risk of blindness and compromised immune function (Khanna 2021). In contrast, other areas may face challenges related to iron or zinc deficiencies, impacting cognitive development and overall health. Recognizing these differences is fundamental to the success of biofortification initiatives (Ashoka et al. 2023). In Sub-Saharan Africa, where vitamin A deficiency is prevalent, biofortification efforts have focused on enhancing the vitamin A content in staple crops like maize and sweet potatoes (Low et al. 2017). This targeted approach aligns with the specific nutritional needs of the population, addressing a deficiency that poses severe health risks. Similarly, in regions where iron deficiency anemia is a significant health concern, crops such as rice and wheat have been biofortified to increase their iron content, contributing to improved nutritional outcomes (Stover et al. 2020).

Cultural and dietary practices also play a pivotal role in shaping nutritional needs (FOOD 2018). In Asian countries, where rice is a dietary staple, biofortification efforts have concentrated on developing rice varieties with increased zinc content. This approach aligns with the dietary habits of the population and aims to combat the widespread zinc deficiency prevalent in these regions (Junaid-ur-Rahman et al. 2022; Sohail et al. 2022). In Latin America, where beans are a dietary cornerstone, biofortification strategies have targeted this crop to enhance iron content, directly addressing prevalent nutritional challenges (Gautam et al. 2021). Geographic factors further complicate the nutritional landscape. Mountainous regions, for example, may face unique challenges in agriculture and nutrient absorption due to the nature of their terrain. Biofortification strategies in such areas need to consider not only the nutritional needs of the population but also the specific agroecological conditions that influence crop growth and nutrient uptake (Gay et al. 2022; Ros et al. 2016). Tailoring biofortification strategies involves a nuanced understanding of the socioeconomic landscape of each region. Affordability and accessibility of biofortified crops are critical factors in ensuring the success of these initiatives. In poor regions, biofortification efforts must prioritize crops that are not only nutrient-rich but also cost-effective and viable for local farmers (Gupte and Longhurst 2019; Jaiswal et al. 2022).

The success of global biofortification initiatives hinges on collaboration between scientists, policymakers, and local communities. Implementing strategies that resonate with the cultural, dietary, and economic contexts of each region fosters acceptance and long-term sustainability (Jakhar and Kuri 2022). Furthermore, ongoing research and monitoring are essential to adapt biofortification programs to evolving nutritional needs, ensuring that these initiatives remain dynamic and responsive to the ever-changing global landscape of health and nutrition (Baima 2005).

17.16 Regulatory Framework and Policy Implications

GMOs represent a subject of significant regulatory scrutiny in the context of genetic engineering for crop biofortification (Kameri-Mbote 2005). The current regulatory landscape governing GMOs encompasses a multifaceted evaluation of their environmental, human health, and socioeconomic impacts (Mannion and Morse 2013). Regulatory agencies worldwide, such as the US Environmental Protection Agency (EPA), the European Food Safety Authority (EFSA), and others, have established guidelines and assessment protocols to ensure the safety and ethical considerations associated with the development and release of GM crops.

These regulations typically involve rigorous testing procedures to evaluate the potential environmental risks, allergenicity, toxicity, and unintended effects associated with GMOs (Freese and Schubert 2004). Furthermore, assessments often consider the potential for gene flow to wild relatives, the development of resistance in target pests, and the overall impact on nontarget organisms within the ecosystem (Raybould and Wilkinson 2005). Policy recommendations for the responsible development and deployment of biofortified crops are integral to navigating the intricate regulatory landscape. Policymakers must strike a delicate balance between promoting technological advancements for addressing malnutrition and ensuring the safety of the environment and human consumers (Baral and Birol 2020; Vora et al. 2023). Recommendations may include the establishment of comprehensive labeling systems to inform consumers about the presence of GM traits in biofortified crops. Additionally, policies may emphasize the importance of monitoring and surveillance programs to track the long-term effects of these crops on both human health and the environment (O'Fallon et al. 2007; Yadav et al. 2020).

Internationally, harmonizing regulatory standards and fostering collaboration among countries can facilitate the global acceptance and trade of biofortified crops. Policymakers may consider adopting science-based decision-making frameworks, considering the latest research findings and technological advancements, to inform regulatory policies effectively (Kim 2017; Singer 2004). Moreover, there is a growing need for adaptive regulatory frameworks that can accommodate emerging genetic engineering technologies. Policymakers may need to engage in continuous dialogue with scientific experts, industry stakeholders, and the public to ensure that regulations remain robust, transparent, and reflective of the evolving landscape of genetic engineering in agriculture (Lescrauwaet et al. 2022; Mandel 2020).

In summary, the regulatory framework and policy implications surrounding genetic engineering for crop biofortification involve a meticulous examination of safety and ethical considerations. Policy recommendations must align with scientific advancements and international standards, aiming to strike a balance between technological progress and the safeguarding of environmental and human well-being.

17.17 Technological Advancements and Innovations

The field of genetic engineering for crop biofortification is subject to continual evolution, propelled by ongoing technological advancements. Vigilance in staying abreast of the latest innovations is imperative for sustained progress. Researchers and scientists actively engage in exploring cutting-edge approaches to enhance the nutrient content of crops, pushing the boundaries of genetic engineering methodologies (Andrade et al. 2021; National Academies of Sciences and Medicine 2019). Technological advancements in genetic engineering encompass a spectrum of techniques, with CRISPR-Cas9 standing out as a particularly influential tool. Continuous improvements and refinements in gene-editing technologies allow for more precise and targeted modifications, minimizing unintended effects and optimizing the success of biofortification efforts (Damian and Porteus 2013; Singh and Dhar 2020).

In the pursuit of enhancing nutrient content in crops, researchers delve into innovative strategies. This may involve the identification and manipulation of specific genes responsible for nutrient accumulation, fine-tuning metabolic pathways, or introducing novel genes from other organisms that contribute to heightened nutritional profiles (Shahid et al. 2020; Varshney and Bohra 2023). Synthetic biology approaches, such as gene synthesis and pathway engineering, also play a role in the creation of biofortified crops with enhanced nutritional attributes (Roy 2022). The exploration of innovative delivery mechanisms for nutrients within plant tissues represents another frontier. Scientists investigate methods to ensure the efficient and targeted distribution of enhanced nutrients throughout the plant, optimizing their bioavailability for both plant growth and eventual consumption by humans or animals (Elemike et al. 2019; Trivedi et al. 2020). Moreover, interdisciplinary collaborations between genetic engineers, biotechnologists, and agronomists contribute to the development of holistic solutions (Vanloqueren and Baret 2017). These collaborative efforts aim to integrate genetic engineering advancements with traditional breeding programs, ensuring a comprehensive approach to crop improvement that considers not only nutrient content but also other crucial agronomic traits such as yield, pest resistance, and adaptability to varying environmental conditions (Varshney et al. 2021).

As the field advances, it is essential to remain attuned to ethical considerations, regulatory frameworks, and societal attitudes surrounding GMOs (Kjeldaas et al. 2023). Embracing responsible research practices and fostering transparent communication about the benefits and potential risks of innovative genetic engineering technologies will be pivotal in shaping a sustainable and widely accepted future for biofortified crops (Juma 2015). In essence, the chapter explores the dynamic interplay between technological progress and scientific creativity in the realm of genetic engineering, illuminating the path toward more resilient, nutritious, and sustainable crops.

17.18 Future Prospects and Research Directions

The future of genetic engineering for crop biofortification holds tremendous promise as researchers continue to explore and develop innovative technologies to address global nutritional challenges (Shahzad et al. 2021). Emerging technologies in this field are poised to revolutionize the way we enhance the nutritional content of crops, offering more precise and efficient methods for nutrient enrichment (Elemike et al. 2019). Moreover, the ongoing research has identified several critical research gaps and highlighted potential areas for future exploration, guiding the scientific community toward a more comprehensive understanding and application of genetic engineering in agriculture (Mishra et al. 2017). One of the prominent emerging technologies is the continued refinement and widespread adoption of the CRISPR-Cas9 system. This revolutionary gene-editing tool enables precise modifications to the plant genome, allowing scientists to target specific genes responsible for nutrient synthesis. The application of CRISPR-Cas9 facilitates the development of biofortified crops with enhanced nutrient profiles, offering an unparalleled level of precision in genetic modifications (Erdoğan et al. 2023).

Beyond CRISPR-Cas9, other advanced gene-editing techniques, such as CRISPR-Cas12 and CRISPR-Cas13, are under exploration (Bigini et al. 2023). These systems present unique capabilities, including multiplex gene editing and RNA targeting, which can be harnessed to simultaneously enhance multiple nutrients in crops or regulate gene expression for optimal nutrient accumulation (Gogolev et al. 2021). The exploration of these technologies opens new avenues for the development of biofortified crops with a more diverse and comprehensive nutritional composition (Gautam et al. 2021).

Nanotechnology is another frontier that holds promise in the future of genetic engineering for crop biofortification. Nanoparticles can be engineered to deliver specific nutrients directly to plant cells, improving bioavailability and absorption (Ul din et al. 2023). This approach minimizes nutrient losses in the soil and ensures efficient utilization by the plant, addressing challenges related to nutrient transport and uptake (Cherry et al. 2008).

Furthermore, the integration of omics technologies, such as genomics, transcriptomics, and metabolomics, provides a holistic understanding of the complex interactions within the plant system (Yang et al. 2021). This systems biology approach enables scientists to identify key regulatory pathways and target genes for biofortification more effectively. Harnessing big data and computational modeling enhances the predictive capabilities of these omics technologies, facilitating the design of crops with optimal nutrient profiles (Singh and Prabha 2019; Wanjek 2011). Despite the rapid progress in genetic engineering for crop biofortification, several research gaps remain to be addressed. The long-term effects of GM crops on soil health, ecosystem dynamics, and nontarget organisms require comprehensive investigation (Kumar et al. 2019; Kumar et al. 2020). Understanding the potential unintended consequences and ecological impacts is crucial for the responsible deployment of biofortified crops on a large scale (Sharma et al. 2017). Additionally, research should focus on optimizing the expression of nutrient-related genes to ensure sustained and stable enrichment in crops. Improving the efficiency of nutrient uptake, transport, and storage mechanisms within plants is essential for maximizing the impact of biofortification on human nutrition (Bhardwaj et al. 2022; Soares et al. 2019). The development of biofortified crops tailored to specific agro-ecological zones and climates represents an important area for future exploration (Bouis et al. 2020). Customizing genetic engineering strategies for diverse environments ensures that biofortification interventions are locally relevant and adapted to the unique challenges faced by different regions (Singh, Paroda, and Dadlani 2022). Moreover, the socioeconomic aspects of biofortification demand attention (De Steur et al. 2017). Research should explore the economic feasibility, market dynamics, and social acceptance of GM biofortified crops. Understanding the perspectives of farmers, consumers, and policymakers is crucial for the successful integration of these crops into agricultural systems (Chinedu et al. 2018).

In summary, the prospects of genetic engineering for crop biofortification are shaped by the convergence of cutting-edge technologies and the identification of key research priorities. The ongoing exploration of emerging technologies, coupled with a targeted focus on research gaps, will pave the way for the development of sustainable, nutritionally enhanced crops that contribute to global food security and address malnutrition challenges. As the scientific community continues to push the boundaries of knowledge in this field, the potential for transformative impact on agriculture and human health remains both exciting and promising.

17.19 Conclusion

In conclusion, the scientific exploration of genetic engineering for crop biofortification underscores its pivotal role in addressing global malnutrition and nutrient deficiencies. The historical overview illuminates the evolution of genetic engineering techniques, with a particular focus on CRISPR-Cas9 and its contemporary counterparts. The imperative need for targeted nutrient enhancement in crops is emphasized, acknowledging the specific benefits derived from biofortified crops such as Golden Rice and biofortified beans.

Regulatory and ethical considerations remain paramount in the implementation of genetic engineering for biofortification, necessitating ongoing dialogues and adherence to international guidelines. Success stories exemplify the tangible impact of these innovations on communities, reinforcing the potential to mitigate malnutrition-related health challenges.

Despite these successes, challenges persist, spanning technological hurdles and socio-environmental concerns. Strategies to enhance consumer acceptance and mitigate apprehensions are crucial for the widespread adoption of biofortified crops. Economic and agricultural impacts are dynamic, with implications for both farmers and broader sustainability efforts.

Looking ahead, this chapter anticipates ongoing innovation in genetic engineering, potentially revealing new methodologies for improved biofortification. Emphasizing the importance of collaborations and partnerships, both at the international and public–private levels, is recognized as pivotal for advancing research and ensuring successful implementation globally.

The imperative to incorporate biofortification into the framework of sustainable agriculture is underscored, recognizing the necessity for nuanced, context-specific strategies. As this scientific discussion draws to a close, it advocates for a comprehensive perspective that harmonizes technological progress with ethical considerations, economic feasibility, and environmental sustainability. This chapter establishes itself as a foundational contribution to the continuous scientific discourse, providing guidance for future research and the execution of initiatives in the field of genetic engineering for crop biofortification.

References

Aaqil, M., Peng, C., Kamal, A. et al. (2023). Tea harvesting and processing techniques and its effect on phytochemical profile and final quality of black tea: A review. *Foods* 12 (24): 4467.

Abobaker, A.M. (2017). *The Influence of Fertiliser Type on Mycorrhizal Colonisation and Plant Growth in Horticultural Systems.* Tasmania: University of Tasmania.

Abushal, L.T., Salama, M., Essa, M.M. et al. (2021). Agricultural biotechnology: Revealing insights about ethical concerns. *Journal of Biosciences* 46 (3): 81.

Adnan, M., Khan, M.A., Basir, A. et al. (2023). Biochar as soil amendment for mitigating nutrients stress in crops. In: *Sustainable Agriculture Reviews 61: Biochar to Improve Crop Production and Decrease Plant Stress under a Changing Climate.* Cham: Springer.

Afridi, G.M., Ullah, N., Ullah, S. et al. (2023). Modulation of salt stress through application of citrate capped silver nanoparticles and indole acetic acid in maize. *Plant Physiology and Biochemistry* 201: 107914.

Agarwal, A.V. and Singh, R.P. (2021). Assessment of the environmental and health impacts of genetically modified crops. In: *Policy Issues in Genetically Modified Crops.* Amsterdam, Netherlands: Elsevier.

Ahmad, K., Din, Z.U., Ullah, H. et al. (2022). Preparation and characterization of bio-based nanocomposites packaging films reinforced with cellulose nanofibers from unripe banana peels. *Starch-Stärke* 74 (5–6): 2100283.

Akbar, J., Gul, M., Jahangir, M. et al. (2023). Global trends in halal food standards: A review. *Foods* 12 (23): 4200.

Ali, J., Jan, I., Ullah, H. et al. (2022a). Influence of ascophyllum nodosum extract foliar spray on the physiological and biochemical attributes of okra under drought stress. *Plants* 11 (6): 790.

Ali, S., Hassan Bin Asad, M.H., Javed, M.A. et al. (2022b). A novel synthetic derivative of biaryl guanidine as a potential bace1 inhibitor, to treat Alzheimer's disease: In-silico, in-vitro and in-vivo evaluation. *Pakistan Journal of Pharmaceutical Sciences* 35 (5): 1339–1345.

Ali, Y., Nawaz, T., Ahmed, N. et al. (2022c). Maize (Zea mays) response to abiotic stress. In: *Maize Genetic Resources-Breeding Strategies and Recent Advances.* England: Intechopen.

Almeida, M. and Ranisch, R. (2022). Beyond safety: Mapping the ethical debate on heritable genome editing interventions. *Humanities and Social Sciences Communications* 9 (1): 1–14.

Amin, F., Ullah, S., Saud, S. et al. (2023). Hydrothermal time analysis of mung bean (vigna radiata (L.) Wilczek) seed germination at different water potential and temperatures. *South African Journal of Botany* 157 (2): 447–456.

Anand, M., Gonzalez, A., Guichard, F. et al. (2010). Ecological systems as complex systems: Challenges for an emerging science. *Diversity* 2 (3): 395–410.

Andrade, J.E., Ali, A.N.M.A., Chowdhury, R. et al. (2021). *Rice Fortification in Bangladesh: Technical Feasibility and Regulatory Requirement for Introducing Rice Fortification in Public Modern Storage/Distribution of Fortified Rice through pfds Channels.* Washington: International Food Policy Research Institute.

Appugol, K.A., Mangang, I.B., and Loganathan, M. (2022). Rising concerns in entomophagy. *Agriculture & Food E-Newsletter* 10.

Arora, l. And Narula, A. (2017). Gene editing and crop improvement using crispr-cas9 system. *Frontiers in Plant Science* 8: 1932.

Ashoka, P., Spandana, B., Saikanth, D. et al. (2023). Bio-fortification and its impact on global health. *Journal of Experimental Agriculture International* 45 (10): 106–115.

Athar, T., Khan, M.K., Pandey, A. et al. (2020). Biofortification and the involved modern approaches. *Journal of Elementology* 25.

Baima, S. (2005). Plant genomics and plant breeding: At the root of human nutrition and health. *Current Topics in Nutraceutical Research* 3 (2): 95–112.

Bakar Ijaz, S.S., Bisma Imran, C., Taufiq Nawaz, S. et al. (2021). Alleviation of zinc deficiency from humans through plants by organic sources: A powerful tonic. *International Journal of Advanced Research* 7: 240–243.

Bamji, M.S. and Nair, K.M. (2016). Food-based approach to combat micronutrient deficiencies. *Proceedings of the National Academy of Sciences India a Physical Sciences* 82: 1529–1540.

Banerjee, A., Paul, K., Varshney, A. et al. (2022). Soilless indoor smart agriculture as an emerging enabler technology for food and nutrition security amidst climate change. In: *Plant Nutrition and Food Security in the Era of Climate Change*. Amsterdam, Netherlands: Elsevier.

Bao, A., Burritt, D.J., Chen, H. et al. (2019). The CRISPR/cas9 system and its applications in crop genome editing. *Critical Reviews in Biotechnology* 39 (3): 321–336.

Baral, A. and Birol, E. (2020). *Catalyzing the Scale-Up of Crop Biofortification*. Washington, US: International Food Policy Research Institute.

Bawazeer, S., Rauf, A., Nawaz, T. et al. (2021). Punica granatum peel extracts mediated the green synthesis of gold nanoparticles and their detailed in vivo biological activities. *Green Processing and Synthesis* 10: 882–892.

Beckie, H.J. and Hall, L.M. (2008). Simple to complex: Modelling crop pollen-mediated gene flow. *Plant Science* 175: 615–628.

Beckman, C. (1980). Defenses triggered by the invader: Physical defenses. *Plant Disease and Advanced Treatise* 224–245.

Begna, T. (2021a). Conventional breeding methods widely used to improve self-pollinated crops. *International Journal of Research* 7 (1): 1–16.

Begna, T. (2021b). Role and economic importance of crop genetic diversity in food security. *International Journal of Agricultural Science and Food Technology* 7: 164–169.

Bharadwaj, D.N. (2016). Sustainable agriculture and plant breeding. *Advances in Plant Breeding Strategies: Agronomic, Abiotic and Biotic Stress Traits* 3–34.

Bhardwaj, A.K., Chejara, S., Malik, K. et al. (2022). Agronomic biofortification of food crops: An emerging opportunity for global food and nutritional security. *Frontiers in Plant Science* 13: 1055278.

Bhatia, S., Lassmann, B., Cohn, E. et al. (2021). Using digital surveillance tools for near real-time mapping of the risk of infectious disease spread. *Nature Partner Journals Digital Medicine* 4 (1): 73.

Bigini, F., Lee, S.H., Sun, Y.J. et al. (2023). Unleashing the potential of crispr multiplexing: Harnessing cas12 and cas13 for precise gene modulation in eye diseases. *Vision Research* 213 (7675): 108317.

Bindraban, P.S., Dimkpa, C., Nagarajan, L. et al. (2015). Revisiting fertilisers and fertilisation strategies for improved nutrient uptake by plants. *Biology and Fertility of Soils* 51 (8): 897–911.

Black, J., Lodge, M., and Thatcher, M. (2006). *Regulatory Innovation: A Comparative Analysis* Massachusetts: Edward Elgar Publishing.

Bohling, J.H. (2016). Strategies to address the conservation threats posed by hybridization and genetic introgression. *Biological Conservation* 203: 321–327.

Bouis, H., Birol, E., Boy, E. et al. (2020). Food biofortification: Reaping the benefits of science to overcome hidden hunger. In: *October Webinar on the Need for agricultural innovation to sustainably feed the world by 2050*. Council for Agricultural Science and Technology (cast).

Bouis, H.E. and Saltzman, A. (2017). Improving nutrition through biofortification: A review of evidence from HarvestPlus, 2003 through 2016. *Global Food Security* 12: 49–58.

Bouis, H.E., Chassy, B.M. and Ochanda, J.O. (2003). 2. Genetically modified food crops and their contribution to human nutrition and food quality. *Trends in Food Science & Technology* 14: 191–209.

Briat, J.F., Gojon, A., Plassard, C. et al. (2020). Reappraisal of the central role of soil nutrient availability in nutrient management in light of recent advances in plant nutrition at crop and molecular levels. *European Journal of Agronomy* 116: 126069.

Brookes, G. (2022). Farm income and production impacts from the use of genetically modified (gm) crop technology 1996–2020. *GM Crops & Food* 13 (1): 171–195.

Brose, u. and Hillebrand, H. (2016). *Biodiversity and Ecosystem Functioning in Dynamic Landscapes*. London, UK: The Royal Society.

Brühl, C.A. and Zaller, J.G. (2021). *Indirect Herbicide Effects on Biodiversity, Ecosystem Functions, and Interactions with Global Changes. Herbicides*. Amsterdam, Netherlands: Elsevier.

Burgiel, S.W. (2002). The Cartagena Protocol on Biosafety: Taking the steps from negotiation to implementation. *Review European Comparative and International Environmental Law* 11: 53.

Candolin, U. and Wong, B.B. (2012). *Behavioural Responses to a Changing World: Mechanisms and Consequences*. Oxford: Oxford University Press.

Carlson, D.F., Fahrenkrug, S.C., and Hackett, P.B. (2012). Targeting DNA with fingers and talens. *Molecular Therapy-Nucleic Acids* 1 (1): e3.

Carpenter, J.E. (2011). Impact of GM crops on biodiversity. *Gm Crops* 2 (1): 7–23.

Carvalho, S.M. and Vasconcelos, M.W. (2013). Producing more with less: Strategies and novel technologies for plant-based food biofortification. *Food Research International* 54: 961–971.

Cellini, F., Chesson, A., Colquhoun, I. et al. (2004). Unintended effects and their detection in genetically modified crops. *Food and Chemical Toxicology* 42 (7): 1089–1125.

Chambers, J.A., Zambrano, P., Falck-Zepeda, J.B. et al. (2014). *Gm Agricultural Technologies for Africa: A State of Affairs* Washington, US: International food policy research institute.

Chandra, R.K. and Newberne, P. M. (2012). *Nutrition, Immunity, and Infection: Mechanisms of Interactions*. Berlin, Germany: Springer Science & Business Media.

Chávez-dulanto, P.N., Thiry, A.A., Glorio-Paulet, P. et al. (2021). Increasing the impact of science and technology to provide more people with healthier and safer food. *Food and Energy Security* 10: e259.

Chen, K. and Gao, C. (2014). Targeted genome modification technologies and their applications in crop improvements. *Plant Cell Reports* 33 (4): 575–583.

Chen, Z., Debernardi, J.M., Dubcovsky, J. et al. (2022). Recent advances in crop transformation technologies. *Nature Plants* 8 (12): 1343–1351.

Cherry, K., Shepherd, M., Withers, P. et al. (2008). Assessing the effectiveness of actions to mitigate nutrient loss from agriculture: A review of methods. *Science of the Total Environment* 406 (1–2): 1–23.

Chinedu, O., Sanou, E., Tur-Cardona, J. et al. (2018). Farmers' valuation of transgenic biofortified sorghum for nutritional improvement in Burkina Faso: A latent class approach. *Food Policy* 79 (1): 132–140.

Condron, L., Stark, C., O'callaghan, M. et al. (2010). The role of microbial communities in the formation and decomposition of soil organic matter. In: *Soil Microbiology and Sustainable Crop Production*, 81–118. Dordrecht: Springer.

Conner, A.J., Glare, T.R., and Nap, J.P. (2003). The release of genetically modified crops into the environment: Part ii. Overview of ecological risk assessment. *The Plant Journal* 33 (1): 19–46.

Conway, G. (2012). *One Billion Hungry: Can We Feed the World*? New York, US: Cornell University Press.

Council, N.R. (1984). *Genetic Engineering of Plants: Agricultural Research Opportunities and Policy Concerns*. Washington, DC: National Academies Press.

Council, N.R. (2004). *Biological Confinement of Genetically Engineered Organisms*. Washington, DC: National Academies Press.

Coupe, R.H. and Capel, P.D. (2016). Trends in pesticide use on soybean, corn and cotton since the introduction of major genetically modified crops in the United States. *Pest Management Science* 72 (5): 1013–1022.

Craig, W., Tepfer, M., Degrassi, G. et al. (2008). An overview of general features of risk assessments of genetically modified crops. *Euphytica* 164: 853–880.

Custodio, M.C., Demont, M. and De steur, H. (2023). Market intelligence for guiding crop improvement: A systematic review of stakeholder preference studies in the rice sector in the global south and beyond. *Comprehensive Reviews in Food Science and Food Safety* 22 (6): 4404–4432.

Daily, G.C. (1997). Introduction: What are ecosystem services. *Nature's Services: Societal Dependence on Natural Ecosystems* 1: 1–10.

Damian, M. and Porteus, M.H. (2013). A crisper look at genome editing: RNA-guided genome modification. *Molecular Therapy* 21 (4): 720–722.

Danovaro, R., Carugati, L., Berzano, M. et al. (2016). Implementing and innovating marine monitoring approaches for assessing marine environmental status. *Frontiers in Marine Science* 3: 213.

Das, B.S., Wani, S.P., Benbi, D.K. et al. (2022). Soil health and its relationship with food security and human health to meet the sustainable development goals in India. *Soil Security* 8: 100071.

Davies, K. 2002. *Cracking the Genome: Inside the Race to Unlock Human DNA*. Maryland, US: JHU press.

Dawar, K., Mian, I., Khan, S. et al. (2023). Alleviation of cadmium toxicity and fortification of zinc in wheat cultivars cultivated in cd contaminated soil. *South African Journal of Botany* 162: 611–621.

Dawe, D., Robertson, R., and Unnevehr, L. (2002). Golden rice: What role could it play in alleviation of vitamin a deficiency? *Food Policy* 27: 541–560.

De steur, H., Demont, M., Gellynck, X. et al. (2017). The social and economic impact of biofortification through genetic modification. *Current Opinion in Biotechnology* 44: 161–168.

Devos, Y., Maeseele, P., Reheul, D. et al. (2008). Ethics in the societal debate on genetically modified organisms: A (re) quest for sense and sensibility. *Journal of Agricultural and Environmental Ethics* 21 (1): 29–61.

Dhaliwal, S.S., Sharma, V., Shukla, A.K. et al. (2022). Biofortification – A frontier novel approach to enrich micronutrients in field crops to encounter the nutritional security. *Molecules* 27 (4): 1340.

Dixit, Y., Yadav, P., Sharma, A.K. et al. (2023). Multiplex genome editing to construct cellulase engineered saccharomyces cerevisiae for ethanol production from cellulosic biomass. *Renewable and Sustainable Energy Reviews* 187: 113772.

Djordjevic, S.P., Jarocki, V.M., Seemann, T. et al. (2023). Genomic surveillance for antimicrobial resistance – A one health perspective. *Nature Reviews Genetics* 1–16.

Dunfield, K.E. and Germida, J.J. (2004). Impact of genetically modified crops on soil-and plant-associated microbial communities. *Journal of Environmental Quality* 33 (3): 806–815.

Duru, M., Therond, O., Martin, G. et al. (2015). How to implement biodiversity-based agriculture to enhance ecosystem services: A review. *Agronomy for Sustainable Development* 35: 1259–1281.

Eastham, K. and Sweet, J. (2002). *Genetically Modified Organisms (GMOS): The significance of Gene flow through pollen transfer*. Copenhagen: European Environment Agency Copenhagen.

Eastwood, C., Klerkx, L., Ayre, M. et al. (2019). Managing socio-ethical challenges in the development of smart farming: From a fragmented to a comprehensive approach for responsible research and innovation. *Journal of Agricultural and Environmental Ethics* 32: 741–768.

Eigenbrode, S.D., Binns, W.P., and Huggins, D.R. (2018). Confronting climate change challenges to dryland cereal production: a call for collaborative, transdisciplinary research, and producer engagement. *Frontiers in Ecology and Evolution* 5: 164.

Elemike, E.E., Uzoh, I.M., Onwudiwe, D.C. et al. (2019). The role of nanotechnology in the fortification of plant nutrients and improvement of crop production. *Applied Sciences* 9 (3): 499.

Ellstrand, N.C. (2003). Current knowledge of gene flow in plants: implications for transgene flow. *Philosophical Transactions of the Royal Society of London. Series B: Biological Sciences* 358 (1434): 1163–1170.

Epps, C.W., Wehausen, J.D., Bleich, V.C. et al. (2007). Optimizing dispersal and corridor models using landscape genetics. *Journal of Applied Ecology* 44 (4): 714–724.

Erdoğan, I., Cevher-Keskin, B., Bilir, Ö. et al. (2023). Recent developments in crispr/cas9 genome-editing technology related to plant disease resistance and abiotic stress tolerance. *Biology* 12 (7): 1037.

Fahad, S., Ali, I., Hussain, I. et al. (2023). Modulation of maize growth, yield and soil enzymes activities by introducing wheat straw mulching and tillage practices. *Plant and Soil* 1–21.

Fanzo, J. And Glass, S. (2018). Ethical and sociocultural considerations of biofortified crops: ensuring value and sustainability for public health. *Advances in Food Security and Sustainability*. Amsterdam: Elsevier.

Ferry, N. And Gatehouse, A.M. (2009). *Environmental Impact of Genetically Modified Crops* cabi.

Fleeger, J.W. (2020). How do indirect effects of contaminants inform ecotoxicology? A review. *Processes* 8 (12): 1659.

Food, T.S. (2018). The 10 elements of agroecology.

Freese, W. and Schubert, D. (2004). Safety testing and regulation of genetically engineered foods. *Biotechnology and Genetic Engineering Reviews* 21 (1): 299–324.

Garbach, K., Milder, J., Montenegro, M. et al. (2014). Biodiversity and ecosystem services in agroecosystems. *Encyclopedia of Agriculture and Food Systems* 2: 21–40.

Gascuel, Q., Diretto, G., Monforte, A.J. et al. (2017). Use of natural diversity and biotechnology to increase the quality and nutritional content of tomato and grape. *Frontiers in Plant Science* 8: 652.

Gautam, D., Shrestha, B., and Subedi, B. (2021). Bio-fortified maize: Cornerstone in plant breeding to combat hidden hunger in developing countries. *Archives of Agriculture and Environmental Science* 6: 100–107.

Gay, J.D., Currey, B. And Brookshire, E. (2022). Global distribution and climate sensitivity of the tropical montane forest nitrogen cycle. *Nature Communications* 13: 7364.

Gayan, A., Borah, P., Nath, D. et al. (2023). Soil microbial diversity, soil health and agricultural sustainability. In: *Sustainable Agriculture and the Environment*. Amsterdam, Netherlands: Elsevier.

Glencross, B.D., Baily, J., Berntssen, M.H. et al. (2020). Risk assessment of the use of alternative animal and plant raw material resources in aquaculture feeds. *Reviews in Aquaculture* 12: 703–758.

Gogolev, Y.V., Ahmar, S., Akpinar, B.A. et al. (2021). Omics, epigenetics, and genome editing techniques for food and nutritional security. *Plant* 10 (7): 1423.

Gómez-Galera, S., Rojas, E., Sudhakar, D. et al. (2010). Critical evaluation of strategies for mineral fortification of staple food crops. *Transgenic Research* 19 (2): 165–180.

Graham, R.D., Humphries, J.M., and Kitchen, J.L. (2000). Nutritionally enhanced cereals: A sustainable foundation for a balanced diet. *Asia Pacific Journal of Clinical Nutrition* 9: S91–S96.

Graham, R.D., Welch, R.M., Saunders, D.A. et al. (2007). Nutritious subsistence food systems. *Advances in Agronomy* 92: 1–74.

Greig, A., Hulme, D., and Turner, M. (2007). *Challenging Global Inequality: Development Theory and Practice in the 21st Century*. London, UK: Bloomsbury Publishing.

Gupta, P. (2021). *Strategies for Non-GM Segregation in the US Grain and Feed Supply Chain*. United States: Iowa State University.

Gupta, R.M. and Musunuru, K. (2014). Expanding the genetic editing tool kit: ZFNs, TALENs, and CRISPR-Cas9. *The Journal of Clinical Investigation* 124 (10): 4154–4161.

Gupte, J. and Longhurst, R. (2019). How do the state's organisational capacities at the micro- and macro-levels influence agriculture-nutrition linkages in fragile contexts? *Food Policy* 82: 74–83.

Hamann, E., Blevins, C., Franks, S. J. et al. (2021). Climate change alters plant–herbivore interactions. *New Phytologist* 229 (4): 1894–1910.

Harfouche, A.L., Petousi, V., Meilan, R. et al. (2021). Promoting ethically responsible use of agricultural biotechnology. *Trends in Plant Science* 26 (6): 546–559.

Hassan, A. (2016). *Food Security and Child Malnutrition: The Impact on Health, Growth, and Well-Being*. Florida: CRC Press.

Herdt, R.W. (2006). Biotechnology in agriculture. *Annual Review of Environment Resources* 31 (1): 265–295.

Hoffman, W. (2016). Ecosystems, food crops, and bioscience: A symbiosis for the anthropocene. *Asian Biotechnology & Development Review*. 18 (1): 39–68.

Hokanson, S.C. (1996). *Risk Assessment of Transgenic Plants: Evaluation of Border Rows as a Containment Strategy for Transgenic Pollen and A Comparison of Pollen Dispersal Patterns for Native and Transgenes*. Michigan: Michigan State University.

Holmes, C. (2008). *Seeds, Scientists & Genetically Modified Organisms: Genetic Engineering Practices and Global Connections*. Halifax, Nova Scotia: Dalhousie University.

Holt-Giménez, E. (2019). *Can We Feed the World Without Destroying It?* Hoboken, New Jersey: John Wiley & Sons.

Horton, P., Banwart, S.A., Brockington, D. et al. (2017). An agenda for integrated system-wide interdisciplinary agri-food research. *Food Security* 9: 195–210.

Hua, K., Zhang, J., Botella, J.R. et al. (2019). Perspectives on the application of genome-editing technologies in crop breeding. *Molecular Plant* 12 (8): 1047–1059.

Huang, X., Huang, S., Han, B. et al. (2022). The integrated genomics of crop domestication and breeding. *Cell* 185 (15): 2828–2839.

Huesing, J.E., Andres, D., Braverman, M.P. et al. (2016). Global adoption of genetically modified (GM) crops: Challenges for the public sector. *Journal of Agricultural and Food Chemistry* 64 (2): 394–402.

Hughes, S. and Kelly, P. (2006). Interactions of malnutrition and Immune impairment, with specific reference to immunity against parasites. *Parasite immunology* 28 (11): 577.

Ijaz, M., Khan, F., Ahmed, T. et al. (2023). Nanobiotechnology to advance stress resilience in plants: Current opportunities and challenges. *Materials Today Bio* 100759.

Irfan, M., Dawar, K., Fahad, S. et al. (2022). Exploring the potential effect of Achnatherum splendens L.–derived biochar treated with phosphoric acid on bioavailability of cadmium and wheat growth in contaminated soil. *Environmental Science and Pollution Research* 29 (25): 37676–37684.

Irfan, M., Mudassir, M., Khan, M.J. et al. (2021). Heavy metals immobilization and improvement in maize (Zea mays L.) growth amended with biochar and compost. *Scientific Reports* 11 (1): 18416.

Jafari, S.M. and Mcclements, D.J. (2017). Nanotechnology approaches for increasing nutrient bioavailability. *Advances in Food and Nutrition Research* 81: 1–30.

Jaiswal, D.K., Krishna, R., Chouhan, G.K. et al. (2022). Bio-fortification of minerals in crops: Current scenario and future prospects for sustainable agriculture and human health. *Plant Growth Regulation* 98 (2): 5–22.

Jakeman, A.J., Letcher, R.A. and Norton, J.P. (2006). Ten iterative steps in development and evaluation of environmental models. *Environmental Modelling & Software* 21: 602–614.

James, C. 1998. *Global Review of Commercialized Transgenic Crops: 1998*. Ithaca, NY: ISAAA.

Jiang, L., Strobbe, S., Van Der Straeten, D. et al. (2021). Regulation of plant vitamin metabolism: Backbone of biofortification for the alleviation of hidden hunger. *Molecular Plant* 14 (1): 40–60.

Jing, H., Wang, H., Wang, G. et al. (2023). The mechanism effects of root exudate on microbial community of rhizosphere soil of tree, shrub, and grass in forest ecosystem under N deposition. *ISME Communications* 3 (1): 120.

Joel, J.M., Johnson, R. and Puthur, J.T. (2023). Co-application of arbuscular mycorrhizal fungi and engineered nanomaterials: A promising strategy for crop resilience against abiotic stresses. *South African Journal of Botany* 162 (1): 314–323.

Juma, C. 2015. *The New Harvest: Agricultural Innovation in AFRICA*. Oxford: Oxford University Press.

Junaid-Ur-Rahman, S., Chughtai, M.F.J., Khaliq, A. et al. (2022). Rice: A potential vehicle for micronutrient fortification. *Clinical Phytoscience* 8: 1–14.

Kadapa, S., Gunturi, A., Gundreddy, R. et al. (2023). Agronomic biofortification of millets: New way to alleviate malnutrition. In: *Millets - Rediscover Ancient Grains*. London: IntechOpen.

Kalaitzandonakes, N., Willig, C. and Zahringer, K. (2023). The economics and policy of genome editing in crop improvement. *The Plant Genome* 16 (2): e20248.

Kameri-Mbote, P. (2005). *Regulation of GMO Crops and Foods*. Kenya: Case study of Kenya.

Kavhiza, N.J., Zargar, M., Prikhodko, S.I. et al. (2022). Improving crop productivity and ensuring food security through the adoption of genetically modified crops in sub-Saharan Africa. *Agronomy* 12 (2): 439.

Khaipho-Burch, M., Cooper, M., Crossa, J. et al. (2023). Genetic modification can improve crop yields – But stop overselling it. *Nature* 621(7979): 470–473.

Khan, F.U., Khan, A.A., Li, K. et al. (2022). Influences of long-term crop cultivation and fertilizer management on soil aggregates stability and fertility in the Loess Plateau, Northern China. *Journal of Soil Science and Plant Nutrition* 22 (2): 1446–1457.

Khan, M.A., Basir, A., Adnan, M. et al. (2023a). Biochar to improve crops yield and quality under a changing climate. In: *Sustainable Agriculture Reviews 61: Biochar to Improve Crop Production and Decrease Plant Stress under a Changing Climate*. Cham: Springer.

Khan, M.S., Ali, S., Ali, N. et al. (2023b). Response of okra (Abelmoschus esculentus L.) F5: 6 population of for earliness and yield traits. *Pakistan Journal of Botany* 55: 689–695.

Khan, S.A., Sohni, S., Akhtar, K. et al. (2023c). Lignocellulose biomatrix zero-valent cobalt nanoparticles: A dip-catalyst for organic pollutants degradation. *Industrial Crops and Products* 198: 116694.

Khanna, S.K. (2021). *Cultural Influences on Food: Dietary and Health Implications*. UK: Taylor & Francis.

Kim, Y. (2017). *The Role of Science in Nanotechnology Decision-Making: Toward Evidence-Based Policy Making*. Arizona: Arizona State University.

Kiran, K. (2020). Advanced approaches for biofortification. In: *Advances in Agri-Food Biotechnology*, 29–55. Singapore: Springer.

Kjeldaas, S., Dassler, T., Antonsen, T. et al. (2023). With great power comes great responsibility: Why 'safe enough'is not good enough in debates on new gene technologies. *Agriculture and Human Values* 40: 533–545.

Kjellson, G. and Strandberg, M. (2001). *Monitoring and Surveillance of Genetically Modified Higher Plants: Guidelines for Procedures and Analysis of Environmental Effects*. Berlin/Heidelberg: Springer Science & Business Media.

König, A., Cockburn, A., Crevel, R. et al. (2004). Assessment of the safety of foods derived from genetically modified (GM) crops. *Food and Chemical Toxicology* 42 (7): 1047–1088.

Krishnappa, G., Tyagi, B.S., Gupta, V. et al. P. (2022). Wheat breeding. In: *Fundamentals of Field Crop Breeding*. Singapore: Springer.

Kumar, A., Pathak, R.K., Gupta, S.M. et al. (2015). Systems biology for smart crops and agricultural innovation: Filling the gaps between genotype and phenotype for complex traits linked with robust agricultural productivity and sustainability. *Omics: A Journal of Integrative Biology* 19 (1): 581–601.

Kumar, K., Gambhir, G., Dass, A. et al. (2020). Genetically modified crops: Current status and future prospects. *Planta* 251 (91): 1–27.

Kumar, S., Palve, A., Joshi, C. et al. (2019). Crop biofortification for iron (Fe), zinc (Zn) and vitamin A with transgenic approaches. *Heliyon* 5 (6): e01914.

Kurkela, S. and Brown, D.W. (2009). Molecular diagnostic techniques. *Medicine* 37 (10): 535–540.

Lal, M., Bhardwaj, E., Chahar, N. et al. (2020). (Trans) gene flow: Mechanisms, biosafety concerns and mitigation for containment. *Reproductive Ecology of Flowering Plants: Patterns and Processes* 335–394.

Lappé, M. and Bailey, B. (2014). *Against the Grain: Genetic Transformation of Global Agriculture*. England: Routledge.

Lawler, J.J. (2009). Climate change adaptation strategies for resource management and conservation planning. *Annals of the New York Academy of Sciences* 1162: 79–98.

Lawlor, D.W. (2013). Genetic engineering to improve plant performance under drought: Physiological evaluation of achievements, limitations, and possibilities. *Journal of Experimental Botany* 64 (1): 83–108.

Lederman, N.G., Antink, A. and Bartos, S. (2014). Nature of science, scientific inquiry, and socio-scientific issues arising from genetics: A pathway to developing a scientifically literate citizenry. *Science & Education* 23 (2): 285–302.

Lescrauwaet, L., Wagner, H., Yoon, C. et al. (2022). Adaptive legal frameworks and economic dynamics in emerging technologies: Navigating the intersection for responsible innovation. *Law and Economics* 16 (3): 202–220.

Li, T., Yang, Y., Qi, H. et al. (2023). CRISPR/Cas9 therapeutics: Progress and prospects. *Signal Transduction and Targeted Therapy* 8 (1): 36.

Low, J.W., Mwanga, R.O., Andrade, M. et al. (2017). Tackling vitamin A deficiency with biofortified sweetpotato in sub-Saharan Africa. *Global Food Security* 14: 23–30.

Luh, Y.-H., Jiang, W.-J., and Chien, Y.-N. (2014). Adoption of genetically-modified seeds in Taiwan: The role of information acquisition and knowledge accumulation. *China Agricultural Economic Review* 6: 669–697.

Lurgi, M., Ritchie, E.G., and Fordham, D.A. (2018). Eradicating abundant invasive prey could cause unexpected and varied biodiversity outcomes: The importance of multispecies interactions. *Journal of Applied Ecology* 55: 2396–2407.

Magnan, A. K., Schipper, E.L. F., Burkett, M. et al. (2016). Addressing the risk of maladaptation to climate change. *Wiley Interdisciplinary Reviews: Climate Change* 7: 646–665.

Mahwish, Saeed, F., Afzaal, M., Hussain, M. et al. (2022). Dietary guidelines to boost immunity during pre and post covid-19 conditions. *International Journal of Food Properties* 25: 1246–1265.

Mall, T., Han, L., Tagliani, L. et al (2018). Transgenic crops: status, potential, and challenges. In: *Biotechnologies of Crop Improvement, Volume 2: Transgenic Approaches* 451–485. Cham: Springer.

Mampan, K.P., Hill, J. and Saleem, M. (2011). Natural resources management and food security in the context of sustainable development. *Sains Malaysiana* 40: 1331–1340.

Mandal, A., Sarkar, B., Owens, G. et al. (2020). Impact of genetically modified crops on rhizosphere microorganisms and processes: A review focusing on Bt cotton. *Applied Soil Ecology* 148: 103492.

Mandel, G.N. (2020). *Regulating Emerging Technologies. Emerging Technologies*. London: Routledge.

Mannion, A. and Morse, S. (2013). GM crops 1996–2012: A review of agronomic, environmental and socio-economic impacts. *University of Surrey, Centre for Environmental Strategy Working Paper* 4: 1–40.

Markel, H. (2021). *The Secret of life: Rosalind Franklin, James Watson, Francis Crick, and the discovery of DNA's double helix*. New York: WW Norton & Company.

Martin, E.A., Feit, B., Requier, F. et al. (2019). Assessing the resilience of biodiversity-driven functions in agroecosystems under environmental change. *Advances in Ecological Research* Amsterdam, Netherlands: Elsevier.

Marx, G.M. (2010). *Monitoring of Genetically Modified Food Products in South Africa*. Bloemfontein: University of the Free State.

Marx, V. (2020). Guide RNAs: It's good to be choosy. *Nature Methods* 17: 1179–1182.

Mathur, V., Javid, L., Kulshrestha, S. et al. (2017). World cultivation of genetically modified crops: Opportunities and risks. In: *Sustainable Agriculture Reviews*, 45–87. Cham: Springer.

Meckawy, R., Stuckler, D., Mehta, A. et al. (2022). Effectiveness of early warning systems in the detection of infectious diseases outbreaks: A systematic review. *BMC Public Health* 22: 1–62.

Mehmood, A., Naveed, K., Liu, K. et al. (2024). Exogenous application of ascorbic acid improves physiological and productive traits of Nigella sativa. *Heliyon* 10 (7): e28766.

Meybeck, A., Laval, E., Lévesque, R. et al. (2017). Food security and nutrition in the age of climate change. In: *Proceedings of the International Symposium organized by the Government of Québec in collaboration with FAO*, 132. Québec City: Food and Agriculture Organization of the United Nations.

Mishra, S., Keswani, C., Abhilash, P. et al. (2017). Integrated approach of agri-nanotechnology: Challenges and future trends. *Frontiers in Plant Science* 8: 471.

Mishra, V., Mishra, R.K., Dikshit, A. et al. (2014). Interactions of nanoparticles with plants: An emerging prospective in the agriculture industry. *Emerging Technologies and Management of Crop Stress Tolerance*. Amsterdam: Elsevier.

Mishra, V., Mudgal, N., Rawat, D. et al. (2023). Integrating microalgae into textile wastewater treatment processes: Advancements and opportunities. *Journal of Water Process Engineering* 55: 104128.

Morales, F., Ancín, M., Fakhet, D. et al. (2020). Photosynthetic metabolism under stressful growth conditions as a bases for crop breeding and yield improvement. *Plants* 9: 88.

Morales, F., Montserrat-De la paz, S., Leon, M.J. et al. (2023). Effects of malnutrition on the immune system and infection and the role of nutritional strategies regarding improvements in children's health status: A literature review. *Nutrients* 16: 1.

Mrabet, R. (2023). Sustainable agriculture for food and nutritional security. In: *Sustainable Agriculture and the Environment*. Amsterdam: Elsevier.

Müller, B., Hoffmann, F., Heckelei, T. et al. (2020). Modelling food security: Bridging the gap between the micro and the macro scale. *Global Environmental Change* 63: 102085.

Mulualem, T. (2015). Application of bio-fortification through plant breeding to improve the value of staple crops. *Biomedicine and Biotechnology* 3: 11–19.

Narain, P. (2018). Scientific and Technological Interventions for Attaining Precision in Plant Genetics and Breeding. *Journal of Agronomy and Research* 1: 5–21.

National Academies of Sciences and Medicine (2016). *Genetically Engineered Crops: Experiences and Prospects*. Washington, US: National Academies Press.

National Academies of Sciences and Medicine (2019). *Science Breakthroughs to Advance Food and Agricultural Research by 2030*. Washington, US: National Academies Press.

Nawaz, T., Gu, L., Fahad, S. et al. (2023a). Unveiling the antioxidant capacity of fermented foods and food microorganisms: A focus on cyanobacteria. *Journal of Umm Al-Qura University for Applied Sciences* 10: 232–243.

Nawaz, T., Gu, L., Fahad, S. et al. (2023b). A comprehensive review of the therapeutic potential of cyanobacterial marine bioactives: Unveiling the hidden treasures of the sea. *Food and Energy Security* 12: e495.

Nawaz, T., Gu, L., Fahad, S. et al. (2024a). Sustainable protein production through genetic engineering of cyanobacteria and use of atmospheric N2 gas. *Food and Energy Security* 40: 100399.

Nawaz, T., Rehman, Z.U., Ullah, R. et al. (2022). Physicochemical and adulteration study of fresh milk collected from different locations in Pakistan. *Saudi Journal of Biological Sciences* 29 (12): 103449.

Nawaz, T., Saud, S., Gu, L. et al. (2024b). Cyanobacteria: Harnessing the power of microorganisms for plant growth promotion, stress alleviation, and phytoremediation in the era of sustainable agriculture. *Plant Stress* 100399.

Naz, T., Iqbal, M.M., Fahad, S. et al. (2023). Bio-fortification of two wheat cultivars with iron and zinc through their soil and foliar application in salt-factored soil: Growth, ionic, physiological, and biochemical modifications. *Journal of Plant Growth Regulation* 42 (3): 1–19.

Nester, E.W. (2015). Agrobacterium: Nature's genetic engineer. *Frontiers in Plant Science* 5: 730.

Newell-Mcgloughlin, M. (2008). Nutritionally improved agricultural crops. *Plant Physiology* 147 (3): 939–953.

Nicholl, D.S. (2023). *An Introduction to Genetic Engineering*. Cambridge, United Kingdom: Cambridge University Press.

Norsworthy, J.K., Ward, S.M., Shaw, D.R. et al. (2012). Reducing the risks of herbicide resistance: Best management practices and recommendations. *Weed Science* 60: 31–62.

O'fallon, M.J., Gursoy, D. and Swanger, N. (2007). To buy or not to buy: Impact of labeling on purchasing intentions of genetically modified foods. *International Journal of Hospitality Management* 26 (1): 117–130.

Ofori, K.F., Antoniello, S., English, M.M. et al. (2022). Improving nutrition through biofortification – A systematic review. *Frontiers in Nutrition* 9: 1043655.

Olembo, N.K., M'mboyi, F., Nyende, B. et al. (2010). *Status of Crop Biotechnology in Sub-Saharan Africa: A Cross-Country Analysis*. Nairobi Kenya: African Biotechnology Stakeholders Forum (ABSF).

Organisms, E.P.O.G.M. (2011). Guidance on the post-market environmental monitoring (PMEM) of genetically modified plants. *EFSA Journal* 9: 2316.

Özyiğit, İ.İ. (2012). Agrobacterium tumefaciens and its use in plant biotechnology. In: *Crop Production for Agricultural Improvement*, 317–361. Dordrecht: Springer.

Pakeerathan, K. and Vaishnavi, K. (2022). SMART crop protection in organic agriculture and policy recommendations for the Sri Lankan context. *Environment Sustenance and Food Safety: Need for More Vibrant Policy Initiatives for Sri Lanka*, 132: 132. Battramulla: National Science and Technology Commission of Sri Lanka.

Pashayan, N., Antoniou, A.C., Ivanus, U. et al. (2020). Personalized early detection and prevention of breast cancer: ENVISION consensus statement. *Nature Reviews Clinical Oncology* 17 (11): 687–705.

Paul, K., Chatterjee, S.S., Pai, P. et al. (2022). Viable smart sensors and their application in data driven agriculture. *Computers and Electronics in Agriculture* 198: 107096.

Peeters, M. and Deketelaere, K. (2006). *EU Climate Change Policy: The Challenge of New Regulatory Initiatives*. Hardback: Edward Elgar Publishing.

Peprah, B.B. (2020). *Genetic Improvement of Beta Carotene in Cassava (Manihot esculenta Crantz) Landraces*. Bloemfontein: University of the Free State.

Perera, F. (2018). Pollution from fossil-fuel combustion is the leading environmental threat to global pediatric health and equity: Solutions exist. *International Journal of Environmental Research and Public Health* 15: 16.

Pisano, G.P. (2006). *Science Business: The Promise, the Reality, and the Future of Biotech*, Massachusetts: Harvard Business Press.

Prabhu, K. (2009). Use of GMOs under containment, confined and limited field trials and post-release monitoring of GMOs. In: *Biosafety of Genetically Modified Organisms: Basic Concepts, Methods and Issues* (eds. M.K.A. Chowdhury, M.I. Hoque and A. Sonnino), 157–220. Rome: Food and Agriculture Organization of the United Nations.

Prager, E.M., Chambers, K.E., Plotkin, J.L. et al. (2019). *Improving Transparency and Scientific Rigor in Academic Publishing*. Hoboken: Wiley Online Library.

Prakash, D., Verma, S., Bhatia, R. et al. (2011). Risks and precautions of genetically modified organisms. *ISRN Ecology* 2011: 1–13.

Quist, D. (2010). *Vertical (Trans) Gene Flow: Implications for Crop Diversity and Wild Relatives*. Biotechnology & Biosafety Series *11*. Penang, Malaysia: Third World Network.

Randhawa, S. and SENGAR, S. (2021). The evolution and history of gene editing technologies. *Progress in Molecular Biology and Translational Science* 178: 1–62.

Rapela, M.A. (2019). *Fostering Innovation for Agriculture 4.0*. Berlin, Germany: Springer.

Rasmussen, N. (2014). *Gene Jockeys: Life Science and the Rise of Biotech Enterprise*. Baltimore: Johns Hopkins University Press.

Raybould, A. and Wilkinson, M.J. (2005). Assessing the environmental risks of gene flow from GM crops to wild relatives. In: *Gene Flow From GM Plants*, 169–185.

Razzaq, A., Saleem, F., Kanwal, M. et al. (2019). Modern trends in plant genome editing: An inclusive review of the CRISPR/Cas9 toolbox. *International Journal of Molecular Sciences* 20 (16): 4045.

Rehman, G., Sardar, S., Alkhateeb, M.A. et al. (2022). Evaluation of functional and bioactive properties of crude gill extract of Tor putitora using different assays. *International Journal of Food Science & Technology* 57: 6459–6470.

Resnik, D.B. (2018). Ethics of community engagement in field trials of genetically modified mosquitoes. *Developing World Bioethics* 18: 135–143.

Rietra, R.P., Heinen, M., Dimkpa, C.O. et al. (2017). Effects of nutrient antagonism and Synergism on yield and fertilizer use efficiency. *Communications in Soil Science and Plant Analysis* 48 (16): 1895–1920.

Riva, F., Graco-Roza, C., Daskalova. et al. (2023). Toward a cohesive understanding of ecological complexity. *Science Advances* 9 (25): eabq4207.

Romman, M., Parvez, R., Adnan, M. et al. (2023). Nitrogen-fixing biofertilizers. In: *Biofertilizers for Sustainable Soil Management*. Florida: CRC Press.

Ros, G., Van Rotterdam, A., Bussink, D. et al. (2016). Selenium fertilization strategies for bio-fortification of food: An agro-ecosystem approach. *Plant and Soil* 404 (1): 99–112.

Rosero, A., Granda, L., Berdugo-Cely, J.A. et al. (2020). A dual strategy of breeding for drought tolerance and introducing drought-tolerant, underutilized crops into production systems to enhance their resilience to water deficiency. *Plants* 9 (10): 1263.

Roy, B. (2022). Bio-fortification; Future challenges toward nutritional security: Mini review. *Plant Archives* 22 (1): *09725210*.

Roy, S., Kumar, V., Behera, B.K. et al. (2022). CRISPR/Cas genome editing – Can it become a game changer in future fisheries sector? *Frontiers in Marine Science* 9: 924475.

Rüdelsheim, P. (2015). Containment management for research and development of GM crops. *Collection of Biosafety Reviews* 9: 40–71.

Saljoqi, A.-U.-R., Amin, S., Salim, M. et al. (2022). Pesticide residue analysis of three different pesticides used against Helicoverpa Armigera (Hubner) in tomato crop. *Sarhad Journal of Agriculture* 38: 448.

Sami, A., Xue, Z., Tazein, S. et al. (2021). CRISPR–Cas9-based genetic engineering for crop improvement under drought stress. *Bioengineered* 12 (1): 5814–5829.

Samuel, L., De barcellos, M.D., Watabaji, M.D. et al. (2023). Factors affecting farmers' acceptance and adoption of biofortified crops: A systematic review. *Outlook on Agriculture* 53 (1): 15–29.

Sanders, D. (2023). *The Struggle for Health: Medicine and the Politics of Underdevelopment* Oxford, UK: Oxford University Press.

Sandhu, R., Chaudhary, N., Shams, R. et al. (2023). A critical review on integrating bio fortification in crops for sustainable agricultural development and nutritional security. *Journal of Agriculture and Food Research* 14 (8): 100830.

Santana, I., Wu, H., Hu, P. et al. (2020). Targeted delivery of nanomaterials with chemical cargoes in plants enabled by a biorecognition motif. *Nature Communications* 11 (1): 2045.

Sati, D., Pande, V., Pandey, S.C. et al. (2023). Recent advances in PGPR and molecular mechanisms involved in drought stress resistance. *Journal of Soil Science and Plant Nutrition* 23 (9): 106–124.

Schoumans, O., Chardon, W., Bechmann, M. et al. (2014). Mitigation options to reduce phosphorus losses from the agricultural sector and improve surface water quality: A review. *Science of the Total Environment* 468: 1255–1266.

Selinger, E. (2002). The Invention of Modern Science. *JSTOR* 77 (2): 182–184.

Senjam, B.D., Chamling, N., Parameshwar, P. et al. (2021). *Chapter-9 Biofortification in Fruit Crops* 55: 141.

Shah, F. and Wu, W. (2019). Soil and crop management strategies to ensure higher crop productivity within sustainable environments. *Sustainability* 11 (5): 1485.

Shah, S., Chen, C., Sun, Y. et al. (2023). Mechanisms of nitric oxide involvement in plant-microbe interaction and its enhancement of stress resistance. *Plant Stress* 10: 100191.

Shahab, M., Roberto, S.R., Adnan, M. et al. (2023). Phenolic compounds as a quality determinant of grapes: A critical review. *Journal of Plant Growth Regulation* 42: 5325–5331.

Shahid, A., Rehman, A.U., Usman, M. et al. (2020). Engineering the metabolic pathways of lipid biosynthesis to develop robust microalgal strains for biodiesel production. *Biotechnology and Applied Biochemistry* 67 (1): 41–51.

Shahzad, R., Jamil, S., Ahmad, S. et al. (2021). Biofortification of cereals and pulses using new breeding techniques: Current and future perspectives. *Frontiers in Nutrition* 8: 721728.

Sharma, D., Jamra, G., Singh, U.M. et al. (2017). Calcium biofortification: Three pronged molecular approaches for dissecting complex trait of calcium nutrition in finger millet (Eleusine coracana) for devising strategies of enrichment of food crops. *Frontiers in Plant Science* 7: 2028.

Sheoran, S., Kumar, S., Ramtekey, V. et al. (2022). Current status and potential of biofortification to enhance crop nutritional quality: An overview. *Sustainability* 14: 3301.

Siddiqui, F., Salam, R. A., Lassi, Z.S. et al. (2020). The intertwined relationship between malnutrition and poverty. *Frontiers in Public Health* 8: 453.

Singer, D. (2004). Capital rules: The domestic politics of international regulatory harmonization. *International Organization* 58: 531–565.

Singer, S.D., Laurie, J.D., Bilichak, A. et al. (2021). Genetic variation and unintended risk in the context of old and new breeding techniques. *Critical Reviews in Plant Sciences* 40: 68–108.

Singh, A., Dhiman, N., Kar, A.K. et al. (2020). Advances in controlled release pesticide formulations: Prospects to safer integrated pest management and sustainable agriculture. *Journal of Hazardous Materials* 385: 121525.

Singh, A.K. (2022). Precision agriculture in India – Opportunities and challenges. *Indian Journal of Fertilisers* 18: 308–331.

Singh, D.P. and Prabha, R. (2019). Omics data integration in microbial research for agricultural and environmental applications. *Microbial Interventions in Agriculture and Environment: Soil and Crop Health Management* 3: 461–491.

Singh, R., Paroda, R., and Dadlani, M. (2022). Science, technology and innovation. In: *Indian Agriculture Towards 2030: Pathways for Enhancing Farmers' Income, Nutritional Security and Sustainable Food and Farm Systems*. Singapore: Springer Nature.

Singh, V. and Dhar, P.K. (2020). *Genome Engineering via CRISPR-Cas9 System*. Massachusetts: Academic Press.

Singha, S. and Singha, R. (2024). Crop improvement strategies and principles of selective breeding. In: *Water-Soil-Plant-Animal Nexus in the Era of Climate Change*. Pennsylvania: IGI Global.

Singhal, R.K., Fahad, S., Kumar, P. et al. (2023). Beneficial elements: New players in improving nutrient use efficiency and abiotic stress tolerance. *Plant Growth Regulation* 100 (2): 237–265.

Smith, A.L., Hodkinson, T.R., Villellas, J. et al. (2020). Global gene flow releases invasive plants from environmental constraints on genetic diversity. *Proceedings of the National Academy of Sciences* 117 (8): 4218–4227.

Snow, A.A., Andow, D.A., Gepts, P. et al. (2005). Genetically engineered organisms and the environment: Current status and recommendations. *Ecological Applications* 15: 377–404.

Soares, J.C., Santos, C.S., Carvalho, S.M. et al. (2019). Preserving the nutritional quality of crop plants under a changing climate: Importance and strategies. *Plant and Soil* 443: 1–26.

Sohail, M.I., Ur Rehman, M.Z., Aziz, T. et al. (2022). Iron bio-fortification and heavy metal/(loid) s contamination in cereals: Successes, issues, and challenges. *Crop and Pasture Science* 73 (7–8): 877–895.

Stangoulis, J.C. and Knez, M. (2022). Biofortification of major crop plants with iron and zinc-achievements and future directions. *Plant and Soil* 474: 57–76.

Steinwand, M.A. and Ronald, P.C. (2020). Crop biotechnology and the future of food. *Nature Food* 1 (5): 273–283.

Stone, G. (2010). The anthropology of genetically modified crops. *Annual Review of Anthropology* 39: 381–400.

Stone, G.D. and Glover, D. (2017). Disembedding grain: Golden Rice, the Green Revolution, and heirloom seeds in the Philippines. *Agriculture and Human Values* 34: 87–102.

Storozhenko, S., De Brouwer, V., Volckaert, M. et al. (2007). Folate fortification of rice by metabolic engineering. *Nature Biotechnology* 25: 1277–1279.

Stover, P.J., Garza, C., Durga, J. et al. (2020). Emerging concepts in nutrient needs. *The Journal of Nutrition* 150: 2593S–2601S.

Tallis, H., Kennedy, C.M., Ruckelshaus, M. et al. (2015). Mitigation for one & all: An integrated framework for mitigation of development impacts on biodiversity and ecosystem services. *Environmental Impact Assessment Review* 55: 21–34.

Tan, F., Wang, J., Guo, Y. et al. (2023). Cost-effectiveness of zinc interventions in China: A cohort-based Markov model. *Agribusiness* 39: 1437–1457.

Tiedje, J.M., Colwell, R.K., Grossman, Y. L. et al. (1989). The planned introduction of genetically engineered organisms: Ecological considerations and recommendations. *Ecology* 70: 298–315.

Tixier, P., Peyrard, N., Aubertot, J.-N. et al. (2013). Modelling interaction networks for enhanced ecosystem services in agroecosystems. *Advances in Ecological Research*. Amsterdam, Netherlands: Elsevier.

Trivedi, P., Leach, J.E., Tringe, S.G. et al. (2020). Plant–microbiome interactions: From community assembly to plant health. *Nature Reviews Microbiology* 18: 607–621.

Tsatsakis, A.M., Nawaz, M.A., Kouretas, D. et al. (2017). Environmental impacts of genetically modified plants: A review. *Environmental Research* 156: 818–833.

Tulchinsky, T.H. (2010). Micronutrient deficiency conditions: Global health issues. *Public Health Reviews* 32: 243–255.

Turnbull, C., Lillemo, M., and Hvoslef-eide, T. A. 2021. Global regulation of genetically modified crops amid the gene-edited crop boom – A review. *Frontiers in Plant Science* 12: 630396.

Tylianakis, J.M. and Binzer, A. (2014). Effects of global environmental changes on parasitoid–host food webs and biological control. *Biological Control* 75: 77–86.

Ul din, K., Naeem, M.S., Zulifqar, U. et al. (2023). Nanoparticles based biofortification in food crops: Overview, implications, and prospects. *Mineral Biofortification in Crop Plants for Ensuring Food Security* 173–201.

Ullah, I., Dawar, K., Tariq, M. et al. (2022). Gibberellic acid and urease inhibitor optimize nitrogen uptake and yield of maize at varying nitrogen levels under changing climate. *Environmental Science and Pollution Research* 29: 6568–6577.

Ullah, R., Junaid, M., Kanwal, M. et al. (2019) Status of research and applications of bio-fertilizers: Global scenario. In: *Biofertilizers for Sustainable Soil Management*. Florida: CRC Press.

Ullah, S., Saud, S., Liu, K. et al. (2023). Germination response of Oat (Avena sativa L.) to temperature and salinity using halothermal time model. *Plant Stress* 10: 100263.

Unnevehr, L., Paarlberg, R.L. and Pray, C.E. (2007). Addressing micronutrient deficiencies: Alternative interventions and technologies. *AgBioForum* 10 (3).

Van der heijden, J. (2021). Risk as an approach to regulatory governance: An evidence synthesis and research agenda. *Sage Open* 11: 21582440211032202.

Vanloqueren, G. and Baret, P.V. (2017). How agricultural research systems shape a technological regime that develops genetic engineering but locks out agroecological innovations. In: *Food Sovereignty, Agroecology and Biocultural Diversity*. London, UK: Routledge.

Varshney, R.K. and Bohra, A. (2023). Boosting resilience of global crop production through sustainable stress management. In: *Global Climate Change and Plant Stress Management*, 1–5. Chichester, UK: John Wiley & Sons, Ltd.

Varshney, R.K., Bohra, A., YU, J. et al. (2021). Designing future crops: Genomics-assisted breeding comes of age. *Trends in Plant Science* 26: 631–649.

Venkataram, V. and Hefferon, K. (2023). *Agricultural Biotechnology: Genetic Engineering for a Food Cause*. Amsterdam, Netherlands: Elsevier.

Vermeulen, S.J., Park, T., Khoury, C.K. et al. (2020). Changing diets and the transformation of the global food system. *Annals of the New York Academy of Sciences* 1478: 3–17.

Verschuren, P.M. 2002. Functional foods: Scientific and global perspectives. *British Journal of Nutrition* 88: S126–S130.

Vives-Peris, V., De Ollas, C., Gómez-Cadenas, A. et al. (2020). Root exudates: From plant to rhizosphere and beyond. *Plant Cell Reports* 39: 3–17.

Vora, Z., Pandya, J., Sangh, C. et al. (2023). The evolving landscape of global regulations on genome-edited crops. *Journal of Plant Biochemistry and Biotechnology* 32(4): 1–15.

Wang, H.-X., Li, M., Lee, C.M. et al. (2017). CRISPR/Cas9-based genome editing for disease modeling and therapy: Challenges and opportunities for nonviral delivery. *Chemical Reviews* 117: 9874–9906.

Wang, X., Dou, Z., Feng, S. et al. (2023). Global food nutrients analysis reveals alarming gaps and daunting challenges. *Nature Food* 4: 1007–1017.

Wanjek, C. 2011. Systems biology as defined by NIH: An intellectual resource for integrative biology. *The NIH Catalyst* 19: 1–12.

Weiss, M., Jacob, F. and Duveiller, G. (2020). Remote sensing for agricultural applications: A meta-review. *Remote Sensing of Environment* 236: 111402.

Yadav, D.N., Bansal, S., Tushir, S. et al. (2020). Advantage of biofortification over fortification technologies. *Wheat and Barley Grain Biofortification*. Amsterdam, Netherlands: Elsevier.

Yang, Y., Saand, M.A., Huang, L. et al. (2021). Applications of multi-omics technologies for crop improvement. *Frontiers in Plant Science* 12: 563953.

Zalabák, D., Pospíšilová, H., Šmehilová, M. et al. (2013). Genetic engineering of cytokinin metabolism: Prospective way to improve agricultural traits of crop plants. *Biotechnology Advances* 31: 97–117.

Zhang, D., Zhang, Z., Unver, T. et al. (2021). CRISPR/Cas: A powerful tool for gene function study and crop improvement. *Journal of Advanced Research* 29: 207–221.

Zimmerman, M. (1988). Nectar production, flowering phenology, and strategies for pollination. *Plant Reproductive Ecology: Patterns and Strategies* 41: 157–178.

Zomorrodi, A.R. and Segrè, D. (2016). Synthetic ecology of microbes: Mathematical models and applications. *Journal of Molecular Biology* 428: 837–861.

18

Development of Biofortified Crops through Marker-Assisted Selection

Muhammad Adeel Ghafar[1], *Muhammad Majeed*[2]*, *Konain Zahra Maqbool*[3], *Waseem Ahmed Khattak*[4], *Usama Ahmad Khan*[5], *Arslan Rafique*[2]

[1] *Department of Botany, Government College University, Faisalabad, Pakistan*
[2] *Department of Botany, University of Gujrat, Pakistan*
[3] *Department of Botany, University of Baltistan Skardu, Gilgit Baltistan, Pakistan*
[4] *Department of Plant Sciences, Quaid-I-Azam University, Islamabad, Pakistan*
[5] *Department of Botany, Ghazi University, Dera Ghazi Khan, Pakistan*

*Corresponding author: m.majeed@uog.edu.pk

18.1 Introduction

Biofortification is a potential method in the battle against malnutrition since it aims to increase the necessary nutritional content of staple crops. This strategy has enormous potential to alleviate the pervasive deficits in vital vitamins and minerals, which are especially common among people who are vulnerable in developing nations. Through biofortification, crops are fortified with increased concentrations of essential elements, such as iron, zinc, vitamin A, and others, in an effort to enhance the general nutritional status and health outcomes of populations that depend on these staple foods (Hassan et al. 2022b; Jamil et al. 2022; Majeed, Bhatti, and Amjad 2021a; Majeed et al. 2021b). Crop nutritional value has traditionally been improved by conventional breeding techniques, which can involve drawn-out and labor-intensive procedures. However, with the introduction of marker-assisted selection (MAS), agricultural biotechnology has entered a new chapter. Breeders may more quickly and efficiently generate biofortified crop varieties thanks to MAS, which enables them to identify desirable features at the molecular level. Humans worldwide frequently lack certain minerals and micronutrients, particularly women and children. Although humans only require trace amounts of micronutrients in their diet, over three billion individuals are thought to be deficient in these essential elements, which has a negative influence on their health (Allen et al. 2006; Cakmak and Kutman 2018; Graham, Welch, and Bouis 2001; Tulchinsky 2010). Micronutrient deficiencies are more common and have severe effects when they occur in places where a diet high in cereal grains, low in dietary diversity, and/or without adequate supplementation or fortification programs are the norm (Allen et al. 2006). Deficits in micronutrients affect immunological response, growth, and cognitive development, with frequently permanent effects (White and Broadley 2009). As a result, this "hidden hunger" hinders attempts to reduce poverty and places a heavy weight on economic growth and human health. The need for food is rising sharply due to a constantly expanding population and shifting dietary habits, yet there is not enough land for agriculture, which results in a heavy reliance on natural resources and frequently low-quality agricultural output. Human nutrition may be improved by agricultural interventions to increase the nutritional content of food and fodder crops fed to animals, particularly in underdeveloped nations.

Biofortification is a viable strategy to mitigate micronutrient shortages. It involves enhancing the nutritional content of consumable crops by sophisticated transgenic techniques, mineral fertilization, or selective traditional breeding. Along with improving food taste, cooking time, yield, post-harvest losses, and insect resistance, biofortification can also increase the quality of the produce. The primary focus of biofortification initiatives is on rural populations that mostly rely on their staple crops to meet their nutritional needs. Biofortified crops may find their way into neighborhood stores, which would provide easy access to individuals, especially those living in remote regions. Their diets are mostly composed of basic foods,

Crop Biofortification: Biotechnological Approaches for Achieving Nutritional Security under Changing Climate, First Edition. Edited by Adnan Noor Shah et al.

making it difficult to boost their nutrition using alternative methods (such as supplementation or processed food fortification). As a result, if not completely removed, the shortage of nutrients in the food they eat on a regular basis might be reduced (Signorell et al. 2019). Urban populations may have some access to processed and commercially fortified meals, but biofortified staples may still be more advantageous for them, particularly for the impoverished who have less access to pricey fortified and varied food.

Developing nations can benefit greatly from biofortification through a variety of techniques, such as the engineering of staple crops. Research and initiatives like HarvestPlus are concentrating on iron (Fe), zinc (Zn), and vitamin A since these micronutrients are regarded by the World Health Organization (WHO) as the scarcest (White and Broadley 2009). The goal is to enhance common crops such as maize, wheat, and rice. These widely grown crops are accessible to all people on the planet and don't require any specific care since production may be enhanced without compromising the crop's yield. It may even lead to improved development and higher yields because most of the target minerals are also crucial for the plant's own nutritional needs and may help it tolerate environmental stress. This is especially crucial in situations when new varieties outperform traditional kinds due to unfavorable agricultural environments, which are common in developing nations.

The overall goal of biofortification breeding initiatives is to match or surpass the yield level of nonfortified cultivars. Transgenic techniques were only sometimes employed, and traditional breeding has produced and mapped the majority of biofortified crops despite the difficult breeding procedure (White and Broadley 2009). For instance, conventional breeding produced wheat with greater grain Fe and Zn (Verma et al. 2016), maize enriched with provitamin A (Manjeru, Van Biljon, and Labuschagne 2019), and rice with increased Zn and Se content. Golden rice, a popular example of a genetically biofortified crop, has drawn a lot of criticism because of several social and political issues. Although most wealthy countries have not yet resolved the argument around genetically modified crops, developing nations have begun to consider the possibilities of genetically engineered nutrient-enriched crops due to the vast number of impoverished individuals who are susceptible to micronutrient deficiencies (Mangueze et al. 2018).

The adoption of biofortified food crops by consumers is a significant problem as it influences the desire for farmers to produce the crop (Manjeru, Van Biljon, and Labuschagne 2019), Considerably more work needs to go into creating a market for a biofortified crop with a greater concentration of β-carotene than a regular crop because of the color difference. Customers must be persuaded to purchase yellow maize for human use in Africa, where yellow maize is designated for animals while white maize is commonly consumed by humans. If the crops are the same color, customers need to be able to tell the difference between the regular crop and the nutrient-enriched one. Until biofortified food is widely recognized and understood, both issues must be resolved by knowledge sharing and communication between scientists, farmers, and consumers. Even yet, transportation to distant markets might be challenging, necessitating the development of local marketplaces (Signorell et al. 2019).

18.1.1 Principles of Marker-assisted Selection

The process of MAS relies on locating and applying molecular markers associated with desirable characteristics. Single-nucleotide polymorphisms (SNPs), DNA sequences, and other genetic variants linked to certain behavioral features can all be used as these markers. Breeders can forecast the existence of desired features in a breeding population by genotyping individuals, obviating the requirement for labor- and time-intensive phenotypic assessments.

18.1.2 Marker Discovery

An essential first step in using MAS to generate biofortified crops is the marker discovery process. Along the DNA sequence, genetic markers act as road signs, designating certain locations linked to desirable characteristics like elevated concentrations of vital vitamins and minerals. Several techniques are employed in the discovery of genetic markers linked to target traits.

18.1.3 Linkage Mapping

Creating genetic maps that show the relative locations of genetic markers along an organism's chromosomes is known as linkage mapping. This method is predicated on the observation of co-segregation within populations resulting from controlled crossings between molecular markers and desirable phenotypic features. Researchers can pinpoint areas of the genome that are most likely to have genes impacting the desired feature by examining inheritance patterns.

Genetic markers are chosen for linkage mapping based on their unique chromosomal positions and polymorphism properties. Simple sequence repeats (SSRs), amplified fragment length polymorphisms (AFLPs), restriction fragment length polymorphisms (RFLPs), and SNPs are a few examples of these markers. Researchers can ascertain the allelic composition at these marker loci and correlate the genetic information with observed phenotypic changes by genotyping individuals within a mapping population.

Genetic linkage, which is the phenomenon where genes near one other on the same chromosome tend to be inherited together, is the fundamental idea behind linkage mapping. Thus, in progeny resulting from crosses between parental lines with different phenotypes, genetic markers close to the gene(s) driving the target trait should co-segregate with the characteristic. Researchers (Muhammad et al. 2023; Muhammad et al. 2024; Majeed et al. 2024b) can determine the genetic distance between markers and map the target gene(s) to certain chromosomal areas by examining the frequency of recombination events between marker loci and the target gene(s). Linkage mapping is an effective method for locating genomic areas linked to complex variables, such as those pertaining to the nutritional quality of crops. Researchers can rank potential genes driving trait variation for additional validation and functional characterization by combining data on gene function and expression with genetic mapping. Linkage mapping aids in the creation of molecular markers for use in MAS and breeding initiatives, as well as offering insightful information on the genetic makeup of desired characteristics.

18.1.4 Association Studies

Association studies investigate the relationship between genetic markers and phenotypic variation in wild or breeding populations. They are sometimes referred to as genome-wide association studies (GWAS) or linkage disequilibrium mapping. As opposed to linkage mapping, which makes use of controlled crossings, association studies use the innate genetic variety seen in populations to pinpoint markers linked to desirable characteristics. In association studies, scientists examine a broad variety of people's genetic variants and phenotypic manifestations. In order to scan a large number of people from different genetic origins and provide a thorough overview of genetic variation across the genome, this frequently utilizes high-throughput genotyping methods. To link phenotypic information, including trait measures, to genetic differences, phenotypic data is methodically gathered for every person. Researchers (Majeed et al. 2023a; Majeed et al. 2023b; Majeed et al. 2023c) can find genetic markers statistically related to the desired attribute by examining genotype–phenotype connections throughout the genome. Genes that directly affect the trait or genomic areas connected to regulatory elements or quantitative trait loci (QTLs) that regulate trait variation may include these markers. For qualities affected by several genes or intricate genetic interactions, association studies are very useful. As a result of the limitations imposed by genetic linkage, association studies are able to detect markers spread throughout the genome and capture fine-scale genetic diversity, in contrast to linkage mapping, which may have low resolution or mapping precision. The ability of association studies to clarify the underlying natural allelic variations of complex characteristics is one of its main benefits. These investigations allow the discovery of genetic variations that significantly contribute to phenotypic variety by looking at a wide range of populations. This advances our knowledge of the underlying genetics of complex characteristics and facilitates the development of molecular markers that are critical to breeding initiatives. In general, association studies are an effective method for analyzing the genetic foundation of intricate features and pinpointing potential genes or genomic areas for additional functional analysis. Researchers may speed up attempts to enhance crops through MAS and breeding and obtain a thorough grasp of the genetic causes causing trait variation by combining association mapping data with other genomic and functional genomics techniques.

18.1.5 Genomic Sequencing

The technique of finding markers has been completely transformed by recent developments in DNA sequencing technology. Researchers can obtain a thorough grasp of genetic diversity inside an organism's genome through whole-genome sequencing. Researchers can find genetic variations, such as SNPs or insertions/deletions (indels), that connect with certain behaviors by examining the genomes of people with varying phenotypes. A high-resolution picture of genetic diversity across the genome is provided by whole-genome sequencing, which entails determining the nucleotide sequence of an organism's whole genome. With the use of this method, scientists may detect genetic variations down to the individual nucleotide level, shedding light on the underlying genetics of phenotypic diversity. Determining the nucleotide sequence of an organism's whole genome by whole-genome sequencing yields a high-resolution result. Whole-genome sequencing makes it easier to identify genetic variations linked to certain qualities of interest when it comes to marker discovery. Scientists can find genetic variations statistically related to a characteristic by sequencing the genomes of people with different phenotypes, such as high and low trait expression. These variations might be found in regulatory areas governing gene expression or directly altering the characteristics inside the gene. An overview of genetic diversity throughout the genome. With the use of this method, scientists may detect genetic variations down to the individual nucleotide level, shedding light on the underlying genetics of phenotypic diversity.

Whole-genome sequencing has revealed the most common kind of genetic variation: SNPs. These single-nucleotide variations (SNPs) are important markers for trait mapping and association research because they represent individual differences in one nucleotide. Furthermore, whole-genome sequencing can identify other genetic variants like insertions/deletions (indels) or structural abnormalities, allowing for their evaluation for association with desirable qualities. Whole-genome sequencing is a helpful tool for trait mapping and marker identification since it allows researchers to collect genetic diversity throughout the whole genome. Researchers can find potential genes and genetic variations underpinning complex characteristics by combining whole-genome sequencing data with other genomic and phenotypic information. This makes it easier to create molecular markers that can be used in breeding programs. All things considered, genome sequencing is a potent tool for trait mapping and marker identification in agricultural development initiatives. Researchers can find the genetic basis of trait variation and hasten the creation of biofortified crop varieties with enhanced nutritional quality and health benefits by utilizing advancements in DNA sequencing technologies.

18.1.6 Marker Validation

Marker validation is a critical stage in MAS that comes after the discovery of putative genetic markers linked to the target characteristic. The goal of marker validation is to validate the robustness and dependability of the discovered markers in predicting the presence or absence of the target trait under various environmental and genetic circumstances (Khoja et al. 2022; Khan et al. 2022; Majeed et al. 2020). In order to guarantee that MAS is useful and applicable in real-world breeding operations, this procedure is crucial. In order to evaluate the precision and consistency of marker–trait relationships, marker validation entails a number of essential stages. Verifying whether the discovered markers consistently show meaningful relationships with the desired characteristic under varied genetic backgrounds and environmental circumstances is a crucial objective (Hussain et al. 2023; Hassan et al. 2022a; Majeed et al. 2021c; Majeed et al. 2022a; Majeed et al. 2022b; Majeed et al. 2022c). This validation is crucial to guarantee the validity of markers used in breeding programs for trait prediction and selection.

To evaluate marker–trait relationships, a representative group of people from various genetic origins is usually genotyped as part of the validation process for genetic markers. Each individual's phenotypic data pertaining to the target trait are gathered, and statistical analyses are carried out to ascertain the degree and importance of correlations between marker genotypes and trait phenotypes. Testing the markers in various settings to assess their performance and stability under various growth circumstances may also be a part of marker validation. Marker validation may involve analyzing additional characteristics of the markers, such as allele frequency and linkage disequilibrium, in addition to determining marker–trait relationships. Marker-assisted breeding (MAB) tactics can be informed by knowledge of the marker allele distribution within populations and their association with neighboring genetic loci, which can shed light on the trait's genetic architecture. Genetic marker validation is a cooperative process that frequently involves contact between researchers, breeders, and other agricultural stakeholders. For MAS to be widely accepted and integrated into breeding programs, it is imperative that open access to marker data and clear communication of validation results be maintained. All things considered, marker validation is essential to guarantee the validity and efficiency of MAS for trait selection in breeding initiatives. Aiming to develop biofortified crop varieties with improved nutritional quality and agronomic performance, marker validation confirms the relationship between genetic markers and target traits across a range of genetic backgrounds and environments. This improves crop improvement efforts' precision and efficiency.

18.2 Importance of Biofortified Crops

18.2.1 Alleviating Malnutrition

Biofortified crops are essential in combating malnutrition, especially in areas with restricted access to a variety of nutrient-dense diets. The WHO estimates that over two billion people worldwide experience hidden hunger, also known as micronutrient deficiencies (WHO 2020). Numerous health issues, such as stunted growth, delayed cognitive development, and heightened vulnerability to infectious diseases, might result from these inadequacies (Bouis 2003). By offering an affordable way to increase dietary diversity and nutrient consumption, biofortified crops enriched with vital vitamins and minerals including iron, zinc, and vitamin A, offer a long-term solution to address this nutritional dilemma (Bouis and Saltzman 2017).

18.2.2 Health Benefits

It has been demonstrated that eating biofortified crops significantly improves general health and lowers the risk of illnesses linked to diet. For instance, a lack of vitamin A increases the risk of contracting diseases such as measles and diarrhea and is one of the main causes of blindness in children (Underwood 2002). According to Bouis and Saltzman (2017), biofortified crops with increased provitamin A content, such as golden rice and orange-fleshed sweet potatoes (OFSP), may be able to avert vitamin A deficiency and its related health risks (Figure 18.1). Zinc and iron deficiency are common among vulnerable groups, especially women and children, and can be addressed with zinc and iron rice and biofortified iron beans (Haas et al. 2005).

18.2.3 Economic Dividends

Growing and consuming biofortified crops not only improves health outcomes but also has major economic benefits for individuals and society. Research has demonstrated that biofortification investments may yield significant financial benefits by boosting human capital development, lowering healthcare expenses, and increasing productivity (Fan, Hazell and Haque 2012). Biofortified varieties enable farmers to attain greater revenues and lessen their reliance on expensive external inputs such as pesticides and fertilizers by improving crop yields and nutritional quality (Bouis and Saltzman 2017).

18.2.4 Environmental Sustainability

Additionally, by encouraging resource-efficient farming methods and lowering the environmental impact of food production, biofortified crops support environmental sustainability. Biofortification incorporates nutritional augmentation directly into staple crops by traditional breeding or biotechnology, in contrast to conventional efforts to treat malnutrition,

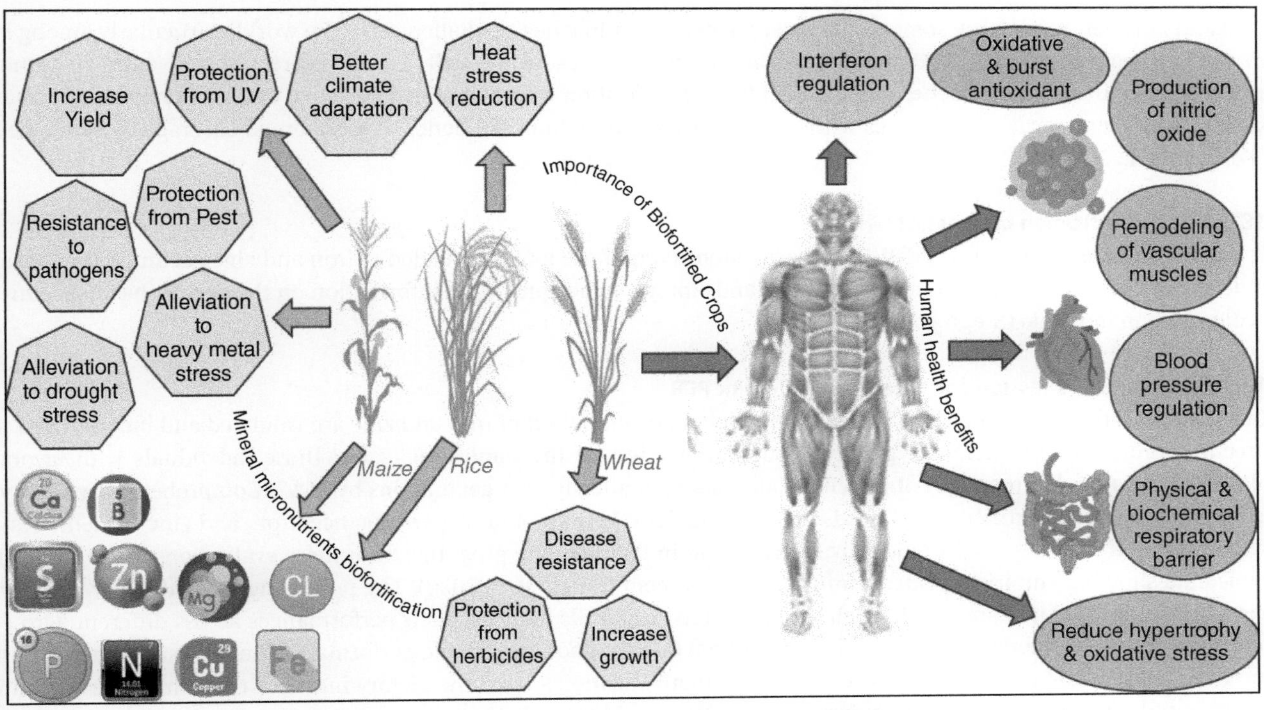

Figure 18.1 Unlocking the power of biofortified crops for a better future: addressing malnutrition, improving health, agriculture, and environmental sustainability.

which frequently rely on supplementation or fortification of processed foods (Bouis and Saltzman 2017). According to Gómez-Galera et al. (2010), this strategy lessens the need for extra inputs and lowers post-harvest losses, leading to more resilient and sustainable food systems.

18.3 Biofortification Types

Biofortification techniques come in a variety of forms, but they may be generally divided into two groups: those that utilize agronomic methods (such as applying fertilizers rich in micronutrients) and those that use genetic breading methods (Model 2). Most of the approaches that have been used thus far are covered in the following sections.

18.3.1 Biofortification Through Conventional Breeding

The traditional breeding approach is popular and has grown in significance over the past ten years. Compared to transgenics, it has been shown in recent years to be a practical, affordable, and widely accepted method for micronutrient fortification (Van Der Straeten et al. 2020). Using this method requires that the characteristics of interest have sufficient genotypic variation. Crops can benefit (Rani et al.2024; Rasool et al.2024; Tariq et al. 2023; Tassadduq et al. 2022; Shafqat et al.2024) from the use of these current variations to increase their vitamin and mineral content. Plants with desired nutrient levels and agronomic features are produced by numerous generations of crossing superior-quality parent lines with high nutrient content with recipient lines with other desirable agronomical attributes. One of the main obstacles to this strategy's success may be the absence of genetic variety in the gene pool. This may be overcome by crossing distant cousins, but doing so hinders the transmission of advantageous characteristics into the planned commercial cultivar. Mutants can also be employed to produce commercial versions. Numerous organizations worldwide have launched breeding programs to increase the micronutrient content of crops since breeding is a rapid means of improving plants. MAS has been pivotal in the development of biofortified crops with enhanced nutritional content. Below are the key applications and associated details:

18.3.1.1 Iron and Zinc Biofortification

Iron and zinc deficiencies are some of the most formidable public health challenges in the world, particularly among low- to middle-income countries. MAS in biofortification is a way that societies will save and correct the prevailing micronutrient deficiency in society for better general health. Biofortification targets biofortified staple crops like rice, wheat, maize, beans, and other cereals and legumes with an aim for large consumption; hence, it has a broad coverage of public health (Zunjare et al. 2018).

18.3.1.2 Identification of Genetic Marker

It employs linkage mapping and GWAS in easily identifying those markers related to iron and zinc accumulation, usually in loci that link with mineral transport, uptake, and storage, hence providing information on the genetic bases of nutrient accumulation in plants (Singh et al. 2020).

18.3.1.3 Marker Validation and Selection: End-point PCR

In such a way, the genetic markers linked to the higher accumulation of iron and zinc are validated and incorporated into breeding programs. Breeders can genotype populations and, at the same time, select those individuals with favorable alleles, increasing the frequency of beneficial alleles systematically over generations by MAS. Comprehensive phenotypic evaluation of the biofortified crop lines is done to establish their agronomic performance, iron, and zinc content, besides other quality traits. The results guide further selection in the breeding programs from these evaluations, with the phenotypic data showing empirical evidence of trait improvement (Singh et al. 2020). The promising biofortified crop varieties developed through MAS are put through multilocation field trials to assess their performances across different agro-ecological zones. The varieties exhibiting high nutritional quality, good agronomic performance, and consumer acceptability are then released to farmers for cultivation and commercialization so that the dietary intake of the vulnerable population for iron and zinc can be assured (Zunjare et al. 2018). In addition, the body requires it still today; vitamin A deficiency remains a problem in serious public health in areas with monotonous diets and very poor in the essential micronutrient concentrations. Provitamin A carotenoids are the precursor of vitamin A, and they occur within the biofortification carried through MAS. This approach would facilitate the efficient introgression of genes conferring carotenoid content into elite cultivars of other staple crops. Genetic marker application MAS has immensely helped in the biofortification of staples with enhanced levels of provitamin-A carotenoids. Breeders who use MAS in exploiting genetic markers linked with carotenoid biosynthesis and accumulation can considerably help in improving the nutritional quality of staple crops.

18.3.2 Transgenic Biofortification

The transgenic method can be used as an option for the creation of biofortified crops in situations where there is little to no genetic diversity (Brinch-Pedersen et al. 2007). This technique allows advantageous genes from any other plant species, regardless of evolutionary or taxonomic rank, to be transferred and expressed thanks to an infinite genetic pool. Transgenic technology seems to be the most practical option for bolstering a crop when its natural capacity to absorb a nutrient is insufficient (Pérez-Massot et al. 2013). Identification of gene functions and the use of these genes to modify plant metabolism are critical to the creation of transgenic crops (Newell-McGloughlin 2008). To find new metabolic engineering routes in crops, processes from other species, mostly bacteria, can be included (Newell-McGloughlin 2008). In the plant system, gene integration can be used simultaneously to increase micronutrient concentrations and bioavailability while reducing anti-nutrients that lower nutrient bioavailability.

Genetic modification can even transfer micronutrients from one tissue to another, enhancing micronutrient availability in edible crop tissue and raising the effectiveness of the biochemical route in edible tissues (Yang et al. 2011). Research and development for transgenic crops is expensive, time-consuming, and requires a lot of resources; but, if successful, the crop has the potential to be sustainable and cost-effective in the long term. Moreover, the transgenic method has no taxonomic limitations and allows the use of artificially produced genes. Increased micronutrient content ensures decreased micronutrient deficit among its consumers, particularly the impoverished in developing countries who cannot afford vitamin supplements (Hirschi 2009). To improve the number of micronutrients in a crop, many crops have undergone genetic modification. To increase the nutritional content of crops, many genes have been used to target specific micronutrients, including vitamins, minerals, essential fatty acids, and essential amino acids. Targets for biofortification include albumin for critical amino acids, lycopene-cyclase for vitamins, ferritin, and nicotinamide synthase for minerals, and six desaturases for essential fatty acids. Successful transgenic crop examples include high-lysine maize, high unsaturated fatty acid soybean, high provitamin A and iron-rich cassava, and provitamin A-rich golden rice (Hirschi 2009). There have been several reports of transgenic biofortified cereals, legumes, vegetables, oilseeds, fodder, and fruit crops (Khan et al.2024; Majeed et al.2024a; Waheed et al. 2022; Waheed et al. 2023; Ullah et al. 2022).

18.3.3 Employing Agronomic Techniques for Biofortification

According to Cakmak and Kutman (2017), agronomic biofortification is a quick and easy way to improve the nutritional quality of crops. Eating these crops also improves human nutrition status. This is accomplished by applying micronutrient-enriched fertilizers to plants. Mineral element solubilization, mobilization from source to sink (consumable plant parts), and fertilizer application techniques are typically used in agronomic biofortification. Higher production targets are aided by macrominerals nitrogen (N), phosphorus (P), and potassium (K). The green revolution, which prevented famine throughout the globe, especially in developing nations, was brought about by the intensified desire to make and apply macronutrient fertilizers in the 1960s, which resulted in a huge rise in agricultural yield. The current situation focuses not only on producing more with limited resources but also on enriching edible sections of the plant with micronutrients for excellent health, with a larger yield to feed around seven billion people (Graham et al. 2007; Hussain et al.2024; Nasar et al.2024). Micronutrients are typically absorbed from the soil and are present in variable amounts in different plant sections. Micronutrient deficiencies in plants and people can be addressed by applying micronutrients as fertilizers, which can also enhance the status of these nutrients in the soil. However, micronutrients that are added to the soil frequently become fixed right away and do not easily go to the sections of the plant that are consumed. It is thus advised to provide micronutrients using alternative methods, such as foliar sprays of soluble form. Agronomic biofortification is a cheap and easy technique if certain factors are taken into consideration, such as the type of fertilizer, how it is applied, and when it is applied. The efficacy of agronomic biofortification with mineral fertilizers is demonstrable by the applications of Se fertilization in Finland, Zn fertilization in Turkey (Cakmak 2010), and I fertilization in China. Agronomic biofortification has several advantages over genetic biofortification, the main one being crop-neutral fertilizer formulations and application methods. While genetic and transgenic biofortification methods are crop-specific, adding other crops to the biofortified profile

Table 18.1 Biofortification strategies for staple crops.

S.no.	Crop	Target nutrients	Biofortification strategy	Benefits
1.	Maize	Protein	Conventional breeding	Increasing protein content to address malnutrition, particularly in regions where maize is a staple food
2.	Wheat	Zinc, Iron	Genetic modification	Enhancing micronutrient content to combat deficiencies in populations heavily reliant on wheat
3.	Rice	Vitamin A	Agronomic practices	Implementing agricultural techniques to enhance vitamin A content, crucial for vision and immune function
4.	Millets	Iron, Zinc	Soil management	Improving soil health and nutrient availability to enhance iron and zinc uptake by millet plants, aiding in addressing deficiencies
5.	Sorghum	Vitamin E	Transgenic approach	Developing varieties with increased vitamin E content to support antioxidant defense mechanisms

requires a significant investment of time and resources, fertilizer application rates and procedures may be swiftly adjusted from one crop to another (Table 18.1).

18.3.4 Effects of Biofortified Variants Worldwide

The use of biofortified cultivars has enormous potential to improve human population health and well-being. Numerous investigations have exhibited the advantageous impacts of these biofortified crops on human health. A study on 246 school-age children in Maharashtra between the ages of 12 and 16 found that feeding iron-rich pearl millet was an effective way to improve the iron status of the children (Finkelstein et al. 2015). The bread, known as "bhakri," is a round, flat, unleavened bread made from conventional and iron-rich pearl millet grains. Another found that feeding provitamin-A-rich maize grains to children in Zambia considerably increased their blood xanthophylls and retinol levels (Sheftel et al. 2017). Consuming β-carotene-rich maize, as opposed to regular maize, significantly increased the blood β-carotene contents of 679 children in Zambia (Palmer et al. 2018). Additionally, QPM's positive impacts have been amply established around the globe. Compared to children who had porridge prepared from regular maize, those who consumed porridge produced from QPM had fewer sick days. Comparing the group fed simply conventional maize to the infant and young kid fed QPM, the latter showed a 9% growth rate in height and a 12% increase in weight. According to different research studies, children could satisfy their lysine needs by consuming 100 g of QPM, which led to a 40% decrease in maize compared to regular maize (Nuss and Tanumihardjo 2011). When South African schoolchildren, aged 5 to 10 years, were fed OFSP, their vitamin A status showed improvement in comparison to when they were fed the usual white variety of sweet potato (van Jaarsveld et al. 2005). Children in Mozambique who took OFSP for the course of the 2-year intervention had considerably greater blood retinol concentrations than children who were not included. Thus, tackling malnutrition and accomplishing the SDGs would benefit from the creation and promotion of biofortified cultivars (Paroda 2018).

18.4 Marker-assisted Breeding: An Overview

A potent technique in contemporary plant breeding programs, MAB provides a more accurate and effective means of producing enhanced crop varieties. This method seeks to discover and choose desired features at the DNA level by using molecular markers, such as SSRs or SNPs (Collard and Mackill). Breeders can expedite the selection process and avoid labor-intensive and time-consuming phenotypic assessments by associating distinct genes or genomic areas linked to desired characteristics with molecular markers (Varshney, Graner, and Sorrells 2005). By facilitating the introgression of advantageous alleles from various germplasm sources into top breeding lines, MAB helps to increase genetic gain and diversify the genetic basis of agricultural cultivars (Collard and Mackill 2008). Furthermore, according to Varshney, Graner, and Sorrells (2005), MAB makes it possible to precisely pyramid numerous phenotypes within a single genetic background, which facilitates the generation of customized varieties with improved biotic and abiotic stress tolerance, nutritional quality, and agronomic performance. Thus, the supply of innovative cultivars that meet the changing demands of farmers, consumers, and the environment has been accelerated by MAB, which has revolutionized crop improvement efforts globally.

18.4.1 Definition and Concept of Molecular Markers

Molecular markers and conventional breeding techniques are used in MAB, a contemporary method of plant breeding that aims to improve the accuracy and efficiency of cultivar creation. Abiotic stress tolerance, disease resistance, yield potential, and nutritional quality are just a few of the desirable features that may be linked to DNA sequences called molecular markers that are positioned at known locations across the genome. The purpose behind MAB is to identify and use these markers to select plants with desirable features more quickly during the early phases of breeding, which will cut down on the amount of time and money needed to generate new varieties (Collard and Mackill 2008). Breeders can decide which plants to move on for additional assessment and selection by genotyping individuals in breeding populations and monitoring the inheritance of target alleles connected to desirable features. By introducing advantageous alleles from various genetic backgrounds into elite cultivars, this method makes it possible to create better cultivars that are more resilient to biotic and abiotic challenges and have superior agronomic performance (Varshney, Graner, and Sorrells 2005). In summary, MAB is a potent technique that may be used to increase genetic gain, diversify crop cultivars' genetic bases, and solve changing agricultural difficulties in a fast-changing environment.

18.4.2 Marker Validation Involves Several Key Components

18.4.2.1 Genetic Diversity Assessment

The validation of the marker must be coupled with the evaluation of genetic diversity within the breeding population or germplasm collection. This ensures that the markers consistently associate with the trait of interest across different genetic backgrounds. Assessing marker–trait associations across a diverse genetic pool highlights whether these associations are universal or variable. This approach is essential for determining the strength of markers in various genetic compositions and is critical for developing strategies that are broadly applicable to genetic improvement.

18.4.2.2 Phenotypic Evaluation

Marker validation should also include phenotypic evaluations to observe the manifestation of the target trait in individuals genotyped with the identified markers. For example, conducting nutrient content analysis or specific trait assays provides concrete evidence correlating marker alleles with trait variations. Evaluating these traits in controlled environments minimizes confounding factors and enhances the reliability of the measurements, thus strengthening the validation process.

18.4.2.3 Statistical Analysis

Statistical methods play a crucial role in marker validation by quantifying the strength of associations and their significance. These methods help estimate the marker's effect size and the associated risks. Additionally, cross-validation methods are used to test the robustness of marker-based prediction models in independent validation populations, providing insights into their practical application in breeding programs.

18.4.2.4 Environmental Test

Environmental testing is critical for marker validation, considering the significant impact of environmental factors on trait expression. This testing is vital for assessing the stability and reliability of marker–trait associations under varied growing conditions. Multilocation trials and genotype-by-environment interaction studies are useful for understanding the environmental sensitivity of markers and identifying key interaction factors that may affect trait expression and selection processes.

18.4.2.5 Integration into Breeding Programs

Upon successful validation, markers are incorporated into breeding programs to facilitate marker-aided selection of the desired trait. Breeders utilize these validated markers to genotype individuals across their breeding populations, selecting favorable trait alleles over successive generations. This integration enhances the efficiency of breeding efforts, accelerating the development of biofortified crop varieties with improved nutritional quality and performance in diverse ecological settings.

18.4.2.6 Marker-assisted Selection

MAS represents a breeding revolution by providing breeders with an efficient tool to make molecular decisions in the selection of desired traits. These are breeder populations that, through informed decisions during the selection of traits, speed up the process of development of improved crop varieties by using MAS through validated genetic markers for traits. Personalized genotyping is then carried out on individuals in the breeding population and only at molecular markers that are linked to the target trait. These markers are validated using molecular techniques such as allele-specific PCR, RFLP analysis, and modern high-throughput genotyping platforms including SNP arrays or next-generation sequencing (NGS) technologies. These enable breeders to know the allelic composition at the marker loci, which is very important in selection within a breeding program. Marker analysis information will be available to empower breeders to make selections based on the expression of trait alleles that are pertinent. This focus ensures priority in breeding programs to individuals that have these favorable alleles, especially those that are enhancing nutrient content or other agronomic traits. MAS is a valuable system, which shortens the cycle durations and expenses effectively by speeding the selection and fixation of elite individuals. Thus, it mitigates the challenges of long phenotypic evaluations and labor intensity.

18.4.3 Biofortified Crops Using Marker Assistants and their Nutritional Significance

A viable way to combat global hunger and enhance human health outcomes is through the production of biofortified crops using MAB, which carries substantial nutritional relevance. Because biofortified crops have increased concentrations of vital vitamins and minerals, they are effective means of providing necessary nutrients to populations that may be deficient in them. Molecular markers are used in MAB to select and breed crops with improved nutritional profiles, increasing the accuracy and efficiency of biofortification initiatives. Research has demonstrated that the development of biofortified crops by MAB can successfully address nutritional deficits that are common in various geographic areas. In many low- and middle-income countries, vitamin A insufficiency is still a major public health problem. To address this, provitamin A has been added to golden rice, which was created by MAB (Bouis and Saltzman 2017). MAB techniques have been utilized to develop biofortified varieties of staple crops, including maize, wheat, beans, and rice. These varieties have been fortified with increased levels of iron, zinc, and other micronutrients, with the aim of addressing deficiencies that impact billions of people globally (Haas et al. 2005). The capacity of biofortified crops to sustainably supply important nutrients is what makes them nutritionally significant, especially for vulnerable people with limited access to a variety of nutrient-dense diets (Table 18.2). Biofortified crops provide a sustainable and economical means of addressing hidden hunger and enhancing the quality of food by directly incorporating increased concentrations of essential vitamins and minerals into staple crops (Bouis 2003). Furthermore, biofortification offers an all-encompassing strategy for combating malnutrition and advancing public health by enhancing other tactics like food fortification and nutritional supplements (Bouis and Saltzman 2017).

Table 18.2 Biofortification strategies for horticultural crops.

S. no.	Crop	Target nutrients	Biofortification strategy	Benefits
1.	Tomato	Vitamin C	Nutrient-fortified fertilizers	Applying fertilizers enriched with vitamin C to enhance nutritional quality of tomatoes, supporting immune health
2.	Spinach	Calcium, Iron	Biofortified varieties	Developing spinach varieties with higher calcium and iron content, addressing deficiencies prevalent in certain populations
3.	Carrot	β-Carotene	Conventional breeding	Breeding for enhanced β-carotene levels in carrots, aiding in improving vitamin A intake
4.	Broccoli	Selenium	Transgenic approach	Incorporating genes for increased selenium accumulation to boost antioxidant activity in broccoli
5.	Bell pepper	Vitamin K	Agronomic practices	Implementing agricultural techniques to improve vitamin K content in bell peppers, supporting bone health and blood clotting

18.4.4 Role of Biofortified Crops in Mitigating Malnutrition

Because biofortified crops address common deficits in critical vitamins and minerals, they are an important tool in the fight against malnutrition. Biofortified crops are generated using MAB. Higher concentrations of important elements, such as iron, zinc, vitamin A, and folate, which are frequently deficient in the diets of vulnerable groups, especially in low- and middle-income nations, are added to these crops. Biofortified crop types may be generated more quickly and effectively by using molecular markers to speed up the breeding process. This will ensure that people who are most in need of nutritious foods have timely access to them. Numerous studies have shown how biofortified crops can enhance nutritional outcomes and reduce the prevalence of malnutrition. For instance, it has been demonstrated that iron-biofortified beans greatly enhance iron status and increase iron consumption in women and children in Rwanda and Uganda (Haas et al. 2005). Similar findings have been made for zinc-biofortified wheat, which improves zinc absorption and lowers the incidence of zinc insufficiency in disadvantaged communities in Pakistan and India (Cakmak et al. 2010). The extensive use of biofortified crops has enormous potential to combat hunger worldwide. Millions of people have improved their health and well-being by having access to more nutritious meals thanks to programs such as HarvestPlus, which spearheads efforts to create and market biofortified varieties of staple crops (Bouis and Saltzman 2017). Researchers can keep improving the nutritional content of crops and reach new people and locations by combining biofortification with MAB. This will ultimately help to accomplish the global nutrition objectives and end malnutrition.

18.5 Development of Biofortified Crops through MAS

The development of biofortified crops through MAS biofortification, a process of enhancing the nutritional content of crops through plant breeding, has emerged as a cost-effective and sustainable approach to address micronutrient deficiencies globally (Gupta, Hossain, and Muthusamy 2015). MAS plays a crucial role in accelerating the development of biofortified crops by facilitating the identification and introgression of desirable traits into elite crop varieties.

18.5.1 Genetic Enhancement of Maize

Maize, a staple crop in many parts of the world, has been a focus of biofortification efforts. Researchers have successfully developed biofortified maize hybrids through the marker-assisted introgression of the opaque-2 allele, which enhances the protein content of the crop (Gupta et al. 2015). This approach has been shown to be effective in improving the nutritional quality of maize, making it a valuable tool in the fight against micronutrient deficiencies. In addition to protein content, efforts have also been made to enhance the provitamin A content in maize through MAS. The identification of favorable alleles of the crtRB1 gene, which is involved in the biosynthesis of β-carotene, has enabled the development of provitamin-A-biofortified maize hybrids (Goswami et al. 2019). This has the potential to address vitamin A deficiency, a significant public health concern in many parts of the world.

18.5.2 Iron and Zinc Biofortification

In addition to maize, efforts have been made to biofortify major crop plants with iron and zinc, two essential micronutrients (Shivay and Prasad 2012). MAS has been used to accelerate cultivar development, and the approach has been effective in other traits such as tolerance to cold, salinity, and drought. The identification of QTL for high iron and zinc content has been a significant breakthrough, enabling the development of genetic markers for MAS. One successful example of iron biofortification is the development of high-iron pearl millet hybrids through MAS (Rai, Govindaraj, and Rao 2016). The introgression of favorable alleles of the ferritin gene, which is involved in iron storage, has resulted in pearl millet varieties with up to 80% higher iron content compared to conventional varieties.

18.5.3 Marker-assisted Pyramiding

Marker-assisted pyramiding, a technique that involves the stacking of multiple genes to achieve a desired trait, has been employed to develop biofortified maize hybrids (Singh et al. 2021). This approach has been used to combine genes that enhance the content of lycopene-ε-cyclase, β-carotene hydroxylase, and opaque-2, resulting in maize hybrids with improved nutritional profiles. The pyramiding of genes for high provitamin A, high lysine, and high tryptophan has been particularly effective in developing nutritionally superior maize hybrids (Muthusamy et al. 2014). This demonstrates the power of MAS in combining multiple desirable traits to create crops with enhanced nutritional value. It is a sophisticated breeding strategy that involves the simultaneous incorporation of multiple beneficial genes into a crop variety to enhance its nutritional content and agronomic traits. In the context of biofortified maize, this approach has revolutionized the development of hybrids with improved protein quality, enhanced levels of essential nutrients like provitamin A and zinc, and superior yield potential. By leveraging the power of molecular markers and advanced breeding techniques, marker-assisted pyramiding has significantly accelerated the breeding process, reduced breeding cycles, and facilitated the creation of nutritionally enriched maize hybrids.

18.5.4 Understanding Marker-assisted Pyramiding

Marker-assisted pyramiding entails the strategic stacking of genes associated with desirable traits, such as lycopene-ε-cyclase, β-carotene hydroxylase, and opaque2, to create biofortified maize hybrids with superior nutritional profiles (Singh et al. 2021). This process involves the precise selection and introgression of target genes into elite maize lines through backcrossing and subsequent generations of progeny testing. The use of gene-specific markers enables breeders to track the presence of these genes throughout the breeding process, ensuring the efficient transfer of beneficial alleles.

18.5.5 Implementation of Marker-assisted Pyramiding

The implementation of marker-assisted pyramiding in biofortified maize breeding follows a systematic crossing scheme, involving the initial hybridization of elite maize lines and subsequent backcrossing to introgress the target genes (Singh et al. 2021). The selection of progenies at each generation based on MAS allows for the identification of individuals carrying the desired gene combinations. This iterative process of backcrossing and selection results in the development of maize hybrids with stacked traits, such as high protein content, enhanced provitamin A levels, and improved agronomic performance.

18.5.6 Advantages and Implications of Marker-assisted Pyramiding

Marker-assisted pyramiding offers several advantages in the development of biofortified maize hybrids. By combining multiple genes in a single breeding cycle, this approach accelerates the breeding process, reduces the time required for trait introgression, and enhances the efficiency of trait stacking (Singh et al. 2021). Moreover, marker-assisted pyramiding enables breeders to precisely control the genetic composition of maize hybrids, leading to the creation of nutritionally enriched varieties tailored to specific nutritional requirements.

18.5.7 Future Directions and Challenges

As marker-assisted pyramiding continues to drive advancements in biofortified maize breeding, future research efforts will focus on expanding the scope of trait stacking, optimizing gene combinations for maximum nutritional impact, and integrating novel genomic tools for enhanced precision breeding (Singh et al. 2021). Challenges such as regulatory approval, trait stability, and scalability of marker-assisted pyramiding in commercial maize production will need to be addressed to realize the full potential of this innovative breeding strategy. In conclusion, marker-assisted pyramiding represents a cutting-edge approach in the development of biofortified maize hybrids, offering a pathway to address malnutrition and enhance food security. By harnessing the power of molecular markers and advanced breeding techniques, marker-assisted pyramiding holds immense promise for the creation of nutritionally enriched maize varieties that can positively impact global health and nutrition.

18.5.8 Molecular Breeding for Nutritionally Enriched Maize

Maize is a major staple crop that plays a crucial role in food security and economic development in many regions, including sub-Saharan Africa, Latin America, and Asia (Prasanna et al. 2020). However, conventional maize varieties often lack essential amino acids, vitamins, and minerals. Biofortification, the process of enhancing the nutritional content of crops through plant breeding, has emerged as a cost-effective and sustainable approach to address these deficiencies (Goswami

Table 18.3 A snapshot of various biofortification strategies applied to different crops to enhance their nutritional value.

Sr.#	Crop	Target nutrients	Biofortification strategy	Benefits
1.	Rice	Zinc, Iron	Biofortified varieties	Improved micronutrient intake, combating iron and zinc deficiencies in populations dependent on rice as a staple food
2.	Wheat	Zinc, Iron	Agronomic practices	Enhancing soil fertility and nutrient uptake to increase zinc and iron content in wheat grains
3.	Maize	Vitamin A, Zinc	Genetic modification	Enhancing vitamin A content and zinc bioavailability through genetic engineering
4.	Sweet Potato	Vitamin A	Conventional breeding	Increasing β-carotene content through traditional breeding methods to improve vitamin A intake
5.	Cassava	Protein, Vitamin A	Transgenic approach	Developing transgenic varieties with increased protein content and enhanced vitamin A synthesis
6.	Beans	Iron, Zinc	Biofortification through soil management	Soil management techniques to improve nutrient uptake by bean plants, increasing iron and zinc content in beans
7.	Lentils	Iron, Zinc	Conventional breeding	Breeding for improved iron and zinc content in lentil varieties to address nutrient deficiencies in consuming populations

et al. 2019). Significant progress has been made in developing, testing, and deploying maize cultivars biofortified with quality protein maize (QPM), provitamin A, and kernel zinc (Prasanna et al. 2020). Molecular MAS has been particularly useful in improving these nutritional traits, as conventional breeding methods can be time-consuming and labor-intensive (Goswami et al. 2019). One successful example of MAS in maize biofortification is the development of biofortified maize hybrids through the introgression of the opaque-2 allele, which enhances the protein content of the crop (Gupta et al. 2013). Additionally, efforts have been made to biofortify maize with provitamin A and kernel zinc using MAS, with the identification of favorable alleles of the crtRB1 gene and QTLs for high iron and zinc content (Goswami et al. 2019; Suwarno et al. 2015). The integration of next-generation breeding techniques, such as gene pyramiding, transgenic breeding, and CRISPR-based gene editing, has further accelerated the development of nutritionally enriched maize (Prasanna et al. 2020). These innovative approaches have enabled the efficient stacking of multiple desirable traits, including high lysine, tryptophan, and provitamin A, in maize hybrids (Goswami et al. 2019). As biofortification efforts progress, the focus is shifting toward improving seed systems, creating demand, and enhancing marketing strategies to promote the adoption of biofortified crops (Talsma, Melse-Boonstra and Brouwer 2017). The utilization of crop wild relatives as genetic resources for biofortification and the exploration of advanced genomics techniques offer new avenues for enhancing the nutritional quality of maize and other staple crops (Guzman et al. 2014). In conclusion, the development of nutritionally enriched maize through molecular breeding represents a significant advancement in addressing global malnutrition and micronutrient deficiencies (Table 18.3). The integration of cutting-edge breeding technologies and genomic tools is poised to revolutionize the field of maize biofortification, paving the way for the creation of nutrient-rich crops that can contribute to improved public health and food security.

18.6 Vitamin A Biofortification Using MAS

18.6.1 Vitamin A Deficiency

Crop targets: The biofortification efforts may work, for instance, through a strategy where much emphasis is put on the major staples that contribute to most caloric and nutrient intakes among the population having the highest problems with Vitamin A deficiency (VAD). Notably, there are provitamin A carotenoid OFSP, maize, and cassava as a main target crop to assist effectively in tackling VAD for the improvement of the health of the population.

Orange-fleshed sweet potatoes (OFSP): The roots of the OFSP contain abundant provitamin A carotenoid β-carotene, which is further changed by the human body into vitamin A. Bright orange flesh biofortified sweet potatoes contain a very high level of β-carotene and are considered to be one of the best sources of dietary vitamin A. Focusing on OFSP for biofortification is particularly relevant since OFSP ranks among the most commonly grown and consumed root crops in areas considered to be of high prevalence to VAD, such as Sub-Saharan Africa and Southeast Asia. The household would be the primary intervention setting to increase vitamin A intake among other micronutrients and improve the nutritional status of these populations, in particular women and children.

Maize: Maize is one of the staple cereal food crops consumed by millions in many parts of the world and is particularly common in areas of the developing world blighted by VAD. Biofortified maize varieties have shown evidence of increasing the population's vitamin A intake from this staple crop. These varieties contain carotenoids in high quantity and greatly improve the nutritive outcome, in that they also enhance the livelihood of communities whose staple food is maize.

Cassava is grown on a large scale in tropical countries and serves as a major calorie provider to the millions. Although traditionally very poor in providing provitamin A carotenoids, several breeding programs aim at enhancing the nutrient through biofortification, and research has resulted in bred varieties having increased levels of β-carotene. The new varieties of cassava show great potential for enhancing dietary diversity in populations depending on the crop and, in effect, improving nutritional outcomes. This is because they greatly enhance dietary diversity and improve the nutritional outcomes of the population.

Identification and exploiting genetic markers: One of the important steps in the development of a carotenoid biosynthesis pathway-engineered biofortified crop (Aziz et al. 2023b; Bhattacharya et al. 2023) is the identification of genetic markers linked to the genes of the carotenoid biosynthesis pathway. Generally, in such studies, the research of genetic mapping and association study design is applied in the search for genomic regions that are responsible for synthesizing, regulating, and storage of carotenoids. These genes are relevant to MAS in a crop improvement program toward traits associated with the nutritional quality of staple foods.

Field trials and adoption: Biofortified varieties showing promising potential are taken up for large multilocation field trials to ascertain performance under varying environmental conditions and agronomic practices. Those showing high contents of provitamin A, high yielding potential, and good acceptability by the consumers are then recommended for release to the farmers for wider cultivation and commercialization. The extensive use of these biofortified varieties translates to a marked prevalence reduction of preventable blindness and other health-related problems of VAD.

18.7 MAS for Bioavailability Enhancement: Use of Solid Dispersion

18.7.1 Identification of Genetic Markers

MAS allows identifying markers associated with the development of the absorption, transport, and utilization of plant nutrients. These markers often correspond to genes coding for proteins that regulate nutrient uptake from the soil, internal transport, and incorporation into metabolic pathways important for nutrient storage and utilization. This will allow the breeder to understand the genetic bases that underlie these traits, identify the genes, and subsequently the genetic variants that may raise nutrient bioavailability in biofortified crops, thus alleviating cases of micronutrient malnutrition in vulnerable populations (Bouis and Welch 2010; Garcia-Casal et al. 2017).

18.7.2 Trait Selection

This will involve selecting from the population those individual animals used in breeding that possess the desired superior characteristics of nutrient utilization, recognized by the associated genetic markers. This process of selection and meticulous crossing focuses on those individuals who have favorable alleles linked to enhanced nutrient absorption and utilization in order to bring in systemic improvements in the bioavailability of essential nutrients within most of the biofortified crop varieties (Ali et al. 2023; Arshad et al. 2022; Aziz et al. 2023a; Haq et al. 2022a; Haq et al. 2022b). Moreover, MAS offers the efficient and cost-effective screening of large populations, thus favoring the selection and multiplication of elite cultivars that bear the traits of interest (Bouis and Welch 2010; Garcia-Casal et al. 2017).

18.7.3 Phenotypic Evaluation

Biofortified crop lines developed through MAS are subjected to rigorous phenotypic evaluations so that the traits enhancing nutrient bioavailability can be judged. Such evaluation does go a long way to help in establishing empirical data in relation to the efficiency of nutrient absorption and distribution within the plant tissues. This is followed by laboratory analysis and controlled feeding trials in these crop varieties to assess the bioavailability of the nutrients and whether the micronutrient content in biofortified crops can provide any human health benefits (De Moura et al. 2014; Finkelstein, Haas, and Mehta 2017).

18.7.4 Field Trials and Adoption

The promising biofortified crop varieties are taken up for multilocation field trials in different ecologies and locations to test not only their agronomic performance but also their nutritional quality and health benefits. Varieties, which express the high nutrient bioavailability trait, are tested for higher agronomic traits and are further released for cultivation and commercialization. The reasons for their adoption are meant to improve dietary intake of that nutrient and health outcomes, especially in those populations with a high vulnerability to the consequences of micronutrient malnutrition, thus improving general well-being.

18.7.5 Challenges and Future Directions

While MAS offers huge potential for the development of biofortified crops, the challenges still abound in the form of setting up robust marker–trait associations and genetic diversity impact. It is, therefore, necessary to pursue further research and development of genetic markers, improvement of marker validation procedures, and inclusion of MAS into other breeding technologies of genomic selection and gene editing to humanely overcome the identified challenges (Varshney, Terauchi, and McCouch 2014).

18.8 Conclusion

The process of improving the nutritional content of crops, grasps great ability for addressing malnutrition, predominantly in weak populations is called "Biofortification." At the molecular level, MAS has arisen as a powerful tool in biofortification efforts, facilitating the effective selection of anticipated traits. This paper provides an overview of key principles and applications of MAS in the development of biofortified crops, focusing on the identification and validation of genetic markers associated with essential nutrients such as iron, zinc, vitamin A, and carotenoids. While highlighting the importance of evaluating genetic diversity within breeding populations to ensure the universality of marker–trait associations, the process of marker discovery, validation, and selection is also discussed. Researchers can accelerate the development of biofortified crop varieties with improved nutritional quality and health benefits by leveraging genomic sequencing, linkage mapping, association studies, and genetic diversity assessment. MAS holds the potential to contribute meaningfully to global efforts to struggle with malnutrition and improve food security, through collaborative efforts between researchers, breeders, and stakeholders. MAS exemplifies a fundamental advancement in the development of biofortified crops, posing a targeted and efficient method to enhance nutritional quality. By leveraging genomic sequencing, linkage mapping, association studies, and genetic diversity assessment, MAS enables researchers to identify, validate, and select genetic markers associated with key nutrients. Through rigorous marker validation and assessment of genetic diversity within breeding populations, MAS ensures the reliability and universality of marker–trait associations across diverse genetic backgrounds and environmental conditions. The adoption of MAS in biofortification programs has the potential to significantly improve the nutritional status and health outcomes of populations affected by malnutrition. By accelerating the development of biofortified crop varieties with elevated levels of essential nutrients such as iron, zinc, vitamin A, and carotenoids, MAS contributes to efforts to address hidden hunger and promote food security in vulnerable communities. Moving forward, continued investment in research, technology, and collaboration is essential to further optimize MAS for trait improvement in crops. By integrating MAS with other breeding approaches and leveraging emerging genomic tools, researchers can enhance the precision, efficiency, and scalability of biofortification efforts. Through coordinated efforts between researchers, breeders, policymakers, and other stakeholders, MAS can play a vital role in achieving global nutrition goals and ensuring a healthier, more sustainable future for all.

References

Ali, H., Muhammad, Z., Majeed, M. et al. (2023). Vegetation diversity pattern during spring season in relation to topographic and edaphic variables in sub-tropical zone. *Botanical Studies* 64 (1): 25. doi: 10.1186/s40529-023-00398-5.

Allen, L.H., De Benoist, B., Dary, O. et al. (2006) *Guidelines on Food Fortification with Micronutrients*. Geneva, Switzerland: World Health Organization and Food and Agriculture Organization of the United Nations.

Arif, U., Bhatti, K.H., Ajaib, M. et al. (2021). Ethnobotanical indigenous knowledge of Tehsil Charhoi, District Kotli, Azad Jammu and Kashmir, Pakistan. *Ethnobotany Research and Applications* 22: 1–24. doi: 10.32859/ERA.22.50.1–24.

Arshad, F., Waheed, M., Harun, N. et al. (2022). Indigenous farmer's perception about fodder and foraging species of Semi-arid lowlands of Pakistan: A case study of District Kasur, Pakistan. *Taiwania* 67 (4): 510–523. doi: 10.6165/tai.2022.67.510.

Aziz, A., Anwar, M.M., Majeed, M. et al. (2023a). Quantifying the impacts of urbanization on urban green, evidences from Maga City, Lahore Pakistan. *Discover Sustainability 4* (1): 48. doi: 10.1007/s43621-023-00169-z.

Aziz, A., Anwar, M.M., Majeed, M. et al. (2023b). Quantifying landscape and social amenities as ecosystem services in rapidly changing peri-urban landscape. *Land* 12 (2): 477. doi: 10.3390/land12020477.

Azmach, G., Gedil, M., Menkir, A. et al. (2013). Marker-trait association analysis of functional gene markers for provitamin A levels across diverse tropical yellow maize inbred lines. *BMC Plant Biology* 13: 1–16.

Bouis, H.E. (2003). Micronutrient fortification of plants through plant breeding: Can it improve nutrition in man at low cost? *Proceedings of the Nutrition Society* 62 (2): 403–411.

Bouis, H.E. and Saltzman, A. (2017). Improving nutrition through biofortification: A review of evidence from HarvestPlus, 2003 through 2016. *Global Food Security* 12: 49–58.

Bouis, H.E. and Welch, R.M. (2010). Biofortification—A sustainable agricultural strategy for reducing micronutrient malnutrition in the global south. *Crop Science* 50: S–20.

Brinch-Pedersen, H., Borg, S., Tauris, B. et al. (2007). Molecular genetic approaches to increasing mineral availability and vitamin content of cereals. *Journal of Cereal Science* 46 (3): 308–326.

Cakmak, I. (2010). Biofortification of cereals with zinc and iron through fertilization strategy. *International Journal of Plant Production* 14 (3): 501–510.

Cakmak, I. and Kutman, U.Á. (2018). Agronomic biofortification of cereals with zinc: A review. *European Journal of Soil Science* 69 (1): 172–180.

Collard, B. C.Y. and Mackill, D.J. (2008). Marker-assisted selection: An approach for precision plant breeding in the twenty-first century. Philosophical Transactions of the Royal Society B: *Biological Sciences* 363 (1491): 557–572.

De Moura, F.F., Palmer, A.C., Finkelstein, J.L. et al. (2014). Are biofortified staple food crops improving vitamin A and iron status in women and children? New evidence from efficacy trials. *Advances in Nutrition* 5 (5): 568–570.

Fan, S., Hazell, P., and Haque, T. (2012). Targeting public investments by agro-ecological zone to achieve growth and poverty alleviation goals in rural India. *Food Policy* 37 (3): 303–315.

Finkelstein, J.L., Haas, J.D., and Mehta, S. (2017). Iron-biofortified staple food crops for improving iron status: A review of the current evidence. *Current Opinion in Biotechnology* 44: 138–145.

Finkelstein, J.L., Mehta, S., Udipi, S.A. et al. (2015). A randomized trial of iron-biofortified pearl millet in school children in India. *The Journal of Nutrition* 145 (7): 1576–1581.

Garcia-Casal, M. N., Peña-Rosas, J.P., De-Regil, L.M. et al. (2017). Fortification of maize flour with iron for preventing anaemia and iron deficiency in populations. *Cochrane Database of Systematic Reviews* Issue 11: Art. No.: CD010187.

Gómez-Galera, S., Rojas, E., Sudhakar, D. et al. (2010). Critical evaluation of strategies for mineral fortification of staple food crops. *Transgenic Research* 19 (2): 165–180.

Goswami, R., Zunjare, R., Khan, S. et al. (2019). Marker-assisted introgression of rare allele of crtRB1 gene into elite quality protein maize inbred for combining high lysine, tryptophan, and provitamin *A in maize. Plant Breeding* 138 (2): 174–183. doi: 10.1111/pbr.12676.

Graham, R.D., Welch, R.M. and Bouis, H.E. (2001). Addressing micronutrient malnutrition through enhancing the nutritional quality of staple foods: principles, perspectives, and knowledge gaps. *Advances in Agronomy* 70: 77–142.

Graham, R.L., Nelson, R., Sheehan, J. et al. (2007). Current and potential US corn stover supplies. *Agronomy Journal* 99 (1): 1–11.

Gupta, H.S., Hossain, F., Muthusamy, V. (2015). Biofortification of maize: An Indian perspective. *Indian Journal of Genetics and Plant Breeding* 75 (01): 1–22.

Gupta, H.S., Raman, B., Agrawal, P.K. et al. (2013). Accelerated development of quality protein maize hybrid through marker-assisted introgression of opaque-2 allele. Plant Breeding 132 (1): 77–82. doi: 10.1111/pbr.12009.

Guzman, C., Medina-Larqué, A.S., Velu, G. et al. (2014). Use of wheat genetic resources to develop biofortified wheat with enhanced grain zinc and iron concentrations and desirable processing quality. *Journal of Cereal Science* 60 (3): 617–622. doi: 10.1016/j.jcs.2014.07.006.

Haas, J.D., Beard, J.L., Murray-Kolb, L.E. et al. (2005). Iron-biofortified rice improves the iron stores of nonanemic Filipino women. *Journal of Nutrition* 135 (12): 2823–2830.

Haq, S.M., Tariq, A., Li, Q. et al. (2022a). Influence of Edaphic properties in determining forest community patterns of the Zabarwan mountain range in the Kashmir Himalayas. *Forests* 13 (8): 1214. doi: 10.3390/f13081214.

Haq, S.M., Yaqoob, U., Majeed, M. et al. (2022b). Quantitative ethnoveterinary study on plant resource utilization by indigenous communities in high-altitude regions. *Frontiers in Veterinary Science* 9: 944046.

Hassan, M., Haq, S.M., Ahmad, R. et al. (2022a). Traditional use of wild and domestic fauna among different ethnic groups in the Western Himalayas—A cross cultural analysis. *Animals* 12 (17): 2276. doi: 10.3390/ani12172276.

Hassan, M., Haq, S.M., Majeed, M. et al. (2022b). Traditional food and medicine: Ethno-traditional usage of fish fauna across the Valley of Kashmir: A Western Himalayan Region. *Diversity* 14 (6): 455. doi: 10.3390/d14060455.

Hirschi, K.D. (2009). Nutrient biofortification of food crops. Annual review of nutrition 29 (1): 401–421.

Hussain, F., Salam, I.U., Memon, Z.N. et al. (2024). Occurrence of fungal microbial contamination in drinking water of megacity of Karachi (Pakistan) and their physico-chemical control. *Heliyon* 10 (7): e28926. doi: 10.1016/j.heliyon.2024.e28926.

Hussain, S., Raza, A., Abdo, H.G. et al. (2023). Relation of land surface temperature with different vegetation indices using multi-temporal remote sensing data in Sahiwal region, Pakistan. *Geoscience Letters* 10 (1): 33. doi: 10.1186/s40562-023-00287-6.

Jamil, M.D., Waheed, M., Akhtar, S. et al. (2022). Invasive Plants Diversity, Ecological Status, and Distribution Pattern in Relation to Edaphic Factors in Different Habitat Types of District Mandi Bahauddin, Punjab, Pakistan. *Sustainability (Switzerland)* 14 (20): 13312. doi: 10.3390/su142013312.

Khan, A.M., Altaf, M., Hussain, T. et al. (2024). Ethnopharmacological uses of fauna among the people of central Punjab, Pakistan. *Frontiers in Veterinary Science* 11: 1351693. doi: 10.3389/fvets.2024.1351693.

Khan, A.M., Li, Q., Saqib, Z. et al. (2022). MaxEnt Modelling and Impact of Climate Change on Habitat Suitability Variations of Economically Important Chilgoza Pine (*Pinus gerardiana* Wall.) in South Asia. *Forests* 13 (5): 715. doi: 10.3390/f13050715.

Khoja, A.A., Haq, S.M., Majeed, M. et al. (2022). Diversity, Ecological and Traditional Knowledge of Pteridophytes in the Western Himalayas. *Diversity* 14 (8): 628. doi: 10.3390/d14080628.

Majeed, M., Bhatti, K.H., Amjad, M.S. et al. (2020). Ethno-veterinary uses of Poaceae in Punjab, Pakistan. *PLoS One* 15 (11): e0241705. doi: 10.1371/journal.pone.0241705.

Majeed, M., Bhatti, K.H. and Amjad, M.S. (2021a). Impact of climatic variations on the flowering phenology of plant species in Jhelum district, Punjab, Pakistan. *Applied Ecology and Environmental Research* 19 (5): 3343–3376. doi: 10.15666/aeer/1905_33433376.

Majeed, M., Bhatti, K.H., Pieroni, A., et al. (2021b). Gathered wild food plants among diverse religious groups in Jhelum District, Punjab, Pakistan. *Foods 10* (3): 593. doi: 10.3390/foods10030594.

Majeed, M., Khan, A.M., Habib, T., et al. (2022a). Vegetation analysis and environmental indicators of an arid tropical forest ecosystem of Pakistan. *Ecological Indicators* 142: 1092921. doi: 10.1016/j.ecolind.2022.109291.

Majeed, M., Lu, L., Haq, S.M. et al. (2022b). Spatiotemporal distribution patterns of climbers along an abiotic gradient in Jhelum District, Punjab, Pakistan. *Forests* 13 (8): 1244. doi: 10.3390/f13081244.

Majeed, M., Lu, L., Anwar, M.M., Tariq, A. et al. (2023a). Prediction of flash flood susceptibility using integrating analytic hierarchy process (AHP) and frequency ratio (FR) algorithms. *Frontiers in Environmental Science* 10: 1037547. doi: 10.3389/fenvs.2022.1037547.

Majeed, M., Muhammad, M., Hussain, T. et al. (2023b). Nanopriming-based management of biotic stresses for sustainable agriculture. In: *Nanopriming Approach to Sustainable Agriculture*, 263–289. Pennsylvania: IGI Global. 10.4018/978-1-6684-7232-3.ch012.

Majeed, M., Muhammad, M., Nawaz, S. et al. (2023c). Nanopriming for crop management for sustainable agriculture: An overview. *Nanopriming Approach to Sustainable Agriculture* 110–141. doi: 10.4018/978-1-6684-7232-3.ch005.

Majeed, M., Tariq, A., Anwar, M.M. et al. (2021c). Monitoring of land use–Land cover change and potential causal factors of climate change in Jhelum district, Punjab, Pakistan, through GIS and multi-temporal satellite data. *Land* 10 (10): 1026. doi: 10.3390/land10101026.

Mangueze, A.V.D.J., Pessoa, M.F., Silva, M.J. et al. (2018). Simultaneous zinc and selenium biofortification in rice. Accumulation, localization and implications on the overall mineral content of the flour. *Journal of Cereal Science* 82: 34–41.

Manjeru, P., Van Biljon, A., and Labuschagne, M. (2019). The development and release of maize fortified with provitamin A carotenoids in developing countries. *Critical Reviews in Food Science and Nutrition* 59 (8): 1284–1293.

Muhammad, M., Basit, A., Majeed, M. et al. (2024). Bacterial pigments and their applications. In: *Bacterial Secondary Metabolites*, 277–298. Amsterdam: Elsevier. doi: 10.1016/B978-0-323-95251-4.00019-3.

Muhammad, M., Waheed, A., Wahab, A. et al. (2023). Soil salinity and drought tolerance: An evaluation of plant growth, productivity, microbial diversity, and amelioration strategies. *Plant Stress* 11: 100319. doi: 10.1016/j.stress.2023.100319.

Muthusamy, V., Hossain, F., Thirunavukkarasu, N. et al. (2014). Development of β-carotene rich maize hybrids through marker-assisted introgression of β-carotene hydroxylase allele. *PLoS One* 9 (12): e113583.

Nasar, S., Muhammad, M., Siddiqi, E.H. et al. (2024). OMICS-Based knowledge for achieving food and nutritional security. *OMICs-based Techniques for Global Food Security* 67–90. doi: 10.1002/9781394209156.ch4.

Newell-McGloughlin, M. (2008). Nutritionally improved agricultural crops. *Plant Physiology* 147 (3): 939–953.

Nuss, E.T. and Tanumihardjo, S.A. (2011). Quality protein maize for Africa: Closing the protein inadequacy gap in vulnerable populations. *Advances in Nutrition* 2 (3): 217–224.

Palmer, A.C., Craft, N.E., Schulze, K.J. et al. (2018). Impact of biofortified maize consumption on serum carotenoid concentrations in Zambian children. *European Journal of Clinical Nutrition* 72 (2): 301–303.

Paroda, R.S. (Ed.). (2018). *Reorienting Indian Agriculture: Challenges and Opportunities*. Wallingford: CAB International.

Prasanna, B.M., Palacios-Rojas, N., Hossain, F. et al. (2020). Molecular breeding for nutritionally enriched maize: Status and prospects. *Frontiers in Genetics* 10: 1392. doi: 10.3389/fgene.2019.01392.

Rai, K.N., Govindaraj, M. and Rao, A.S. (2016). Genetic enhancement of grain iron and zinc content in pearl millet. *Quality Assurance and Safety of Crops & Foods* 4 (3): 119–125.

Rani, J., Rasool, A., Sher, H. et al. (2024). Precision farming with nanoscale sensors: Monitoring soil, water, and crop health. In: *Harnessing NanoOmics and Nanozymes for Sustainable Agriculture*, 78–99. Pennsylvania: IGI Global. doi: 10.4018/979-8-3693-1890-4.ch005.

Rasool, A., Tariq, M., Asif, M. et al. (2024). Nanotechnology for improved crop resilience in challenging environments. In: *Harnessing NanoOmics and Nanozymes for Sustainable Agriculture*, 149–175. Pennsylvania: IGI Global. doi: 10.4018/979-8-36 93-1890-4.ch008.

Shafqat, A., Rasool, A., Fatima, R. et al. (2024). Nanomaterials for sustainable pest and disease management in agriculture. *Harnessing NanoOmics and Nanozymes for Sustainable Agriculture* 100–125. doi: 10.4018/979-8-3693-1890-4.ch006.

Sheftel, J., Gannon, B.M., Davis, C.R. et al. (2017). Provitamin A-biofortified maize consumption increases serum xanthophylls and 13C-natural abundance of retinol in Zambian children. *Experimental Biology and Medicine* 242 (15): 1508–1514.

Shivay, Y.S. and Prasad, P. (2012). Marker-assisted selection in plant breeding. *Journal of Plant Breeding and Crop Science* 4 (2): 147–155.

Signorell, C., Zimmermann, M.B., Cakmak, I. et al. (2019). Zinc absorption from agronomically biofortified wheat is similar to post-harvest fortified wheat and is a substantial source of bioavailable zinc in humans. *The Journal of Nutrition* 149 (5): 840–846.

Singh, J., Sharma, S., Kaur, A. et al. (2021). Marker-assisted pyramiding of lycopene-ε-cyclase, β-carotene hydroxylase, and opaque2 genes for development of biofortified maize hybrids. *Scientific Reports* 11 (1): 12642.

Singh, S., Yadav, S., Singh, A. et al. (2020). Marker assisted selection and Genome editing: Potent tools for bio-fortification in Vegetable crops. *Food and Scientific Reports* 1 (12): 8.

Sinha, D., Maurya, A.K., Abdi, G. et al. (2023). Integrated genomic selection for accelerating breeding programs of climate-smart cereals. *Genes* 14 (7): 1484. doi: 10.3390/genes14071484.

Suwarno, W.B., Pixley, K.V., Palacios-Rojas, N. et al. (2015). Genome-wide association analysis reveals new targets for carotenoid biofortification in maize. *Theoretical and Applied Genetics* 128 (5): 851–864. doi: 10.1007/s00122-015-2475-3.

Talsma, E.F., Melse-Boonstra, A., and Brouwer, I.D. (2017). Acceptance and adoption of biofortified crops in low-and middle-income countries: A systematic review. *Nutrition Reviews* 75 (10): 798–829. doi: 10.1093/nutrit/nux037.

Tariq, A., Mumtaz, F., Majeed, M. et al. (2023). Spatio-temporal assessment of land use land cover based on trajectories and cellular automata Markov modelling and its impact on land surface temperature of Lahore district Pakistan. *Environmental Monitoring and Assessment* 195 (1): 114. doi: 10.1007/s10661-022-10738-w.

Tassadduq, S.S., Akhtar, S., Waheed, M. et al. (2022). Ecological distribution patterns of wild grasses and abiotic factors. *Sustainability (Switzerland)* 14 (18): 11117. doi: 10.3390/su141811117.

Tulchinsky, T. H. (2010). Micronutrient deficiency conditions: Global health issues. *Public Health Reviews* 32: 243–255.

Ullah, I., Aslam, B., Shah, S.H.I.A. et al. (2022). An integrated approach of machine learning, remote sensing, and GIS data for the landslide susceptibility mapping. *Land* 11 (8): 1265. doi: 10.3390/land11081265.

Underwood, B.A. (2002). Vitamin A deficiency disorders: International efforts to control a preventable "pandemic". *Journal of Nutrition* 132 (9): 2927S–2931S.

Van Der Straeten, D., Bhullar, N.K., De Steur, H. et al. (2020). Multiplying the efficiency and impact of biofortification through metabolic engineering. *Nature Communications* 11 (1): 5203.

van Jaarsveld, P.J., Faber, M., Tanumihardjo, S.A. et al. (2005). β-Carotene–rich orange-fleshed sweet potato improves the vitamin A status of primary school children assessed with the modified-relative-dose-response test 1–3. *The American Journal of Clinical Nutrition* 81 (5):1080–1087.

Varshney, R.K., Graner, A. and Sorrells, M.E. (2005). Genomics-assisted breeding for crop improvement. *Trends in Plant Science* 10 (12): 621–630.

Varshney, R.K., Terauchi, R. and McCouch, S.R. (2014). Harvesting the promising fruits of genomics: Applying genome sequencing technologies to crop breeding. *PLoS Biology* 12 (6): e1001883.

Verma, S.K., Kumar, S., Sheikh, I. et al. (2016). Transfer of useful variability of high grain iron and zinc from Aegilops kotschyi into wheat through seed irradiation approach. *International Journal of Radiation Biology* 92 (3): 132–139.

Waheed, M., Arshad, F., Majeed, M. et al. (2022). Community structure and distribution pattern of woody vegetation in response to soil properties in semi-arid lowland district Kasur Punjab, Pakistan. *Land* 11 (12): 2145. doi: 10.3390/land11122145.

Waheed, M., Arshad, F. and Majeed, M. et al. (2023). Potential distribution of a noxious weed (Solanum viarum Du-nal), current status, and future invasion risk based on MaxEnt modeling. *Geology, Ecology, and Landscapes* 1–16. doi: 10.1080/24749508.2023.2179752.

White, P.J. and Broadley, M.R. (2009). Biofortification of crops with seven mineral elements often lacking in human diets–iron, zinc, copper, calcium, magnesium, selenium and iodine. *New Phytologist* 182 (1): 49–84.

World Health Organization (WHO). (2020). Micronutrient deficiencies. https://www.who.int/news-room/q-a-detail/micronutrient-deficiencies.

Zunjare, R.U., Hossain, F., Muthusamy, V. et al. (2018). Development of biofortified maize hybrids through marker-assisted stacking of β-carotene hydroxylase, lycopene-ε-cyclase and Opaque2 Genes. *Frontiers in Plant Science* 9: 1785. doi: 10.3389/fpls.2018.

19

Agronomic Perspective of Improving Iodine Biofortification

Muhammad Amjad Nadim[1], *Mohammad Safdar Baloch*[1], *Qudrat Ullah Khan*[2], *Asghar Ali Khan*[1], *Muhammad Faisal Shahzad*[3], *Umar Khitab Saddozai*[1], *Imran Khan*[4], *Adnan Noor Shah*[5*]

[1] *Department of Agronomy, Gomal University, Dera Ismail Khan, Pakistan*
[2] *Department of Soil Science, Gomal University, Dera Ismail Khan, Pakistan*
[3] *Department of Entomology, Gomal University, Dera Ismail Khan, Pakistan*
[4] *Department of Forestry and Range Management, Gomal University, Dera Ismail Khan, Pakistan*
[5] *Department of Agricultural Engineering, Khwaja Fareed University of Engineering and Information Technology, Rahim Yar Khan, Pakistan*

*Corresponding author: adnan.shah@kfueit.edu.pk

19.1 Introduction

One of the deprived sources of vital mineral nutrients is our staple food crops utilized on a daily basis. The people subsisting on a monotonous cereal diet are mostly exposed to diseases caused by mineral deficiencies. In this link, different strategies have been set so far to deal with micronutrient deficiencies, such as mineral supplements, food fortification, and biofortification of crop plants with enriched mineral fertilizers. Moreover, various breeding programs have been executed followed by genetic engineering approaches to evolve new mineral-enriched genotypes of food crops (Galera et al. 2010). In the current scenario, the use of different mineral nutrients, such as iodine, iron, calcium, selenium, and zinc, has great importance in improving the quality of food grains. The enhancement in nutritional behavior of crop plants through different agronomic and breeding techniques refers to biofortification. It is comparatively the easiest and cheapest agricultural strategy emerged to improve nutrient concentrations in food grains (Hasanuzzaman et al. 2023). The biofortification of plants is an effective way of regulating the iodine content in humans. It is a set of procedures that provide iodine concentrations in plant food, which fulfill partial or entire daily intake requirements for humans. The expression of several genes is specifically regulated with iodine treatments, which are mostly involved in the defense response of plants, showing protection against both biotic and abiotic stresses. Iodine is available everywhere, but in trace amounts. Low concentrations of iodine were observed in most agricultural top soils and irrigation water, while plants don't often uptake iodine. Iodine deficiency in plants might cause yield losses, similar to deficiency of any other plant nutrient. In humans, a deficiency of iodine may cause thyroid disease, which is a serious health issue. Iodine biofortification is an effective technique to regulate it in humans (Zhang et al. 2023). Deficiency of iodine is a widespread malnutrition problem of micronutrients while adding iodine in table salt characterizes most common prophylaxes tools (Gonzali et al. 2017). To enrich the human diet, the biofortification of crops with iodine is a better strategy, which is cost-effective, well-accepted, and has good bioavailability as an iodine source (Gonzali, Kiferle, and Perata 2017). The key to establish appropriate biofortification approach is to understand iodine functioning in higher plants (Gonzali et al. 2017). Over one and half of the world's population is facing vitamin A, zinc, iron, and iodine deficiency. Various techniques have been made so far to overcome these deficiencies of micronutrients through supplements and food fortification, but new approaches are direly needed, especially to access the rural poor (Nestel et al. 2006). Scientific evidence has shown that biofortification is technically feasible without compromising agronomic productivity (Nestel et al. 2006). It is important to control micronutrient deficiencies through the biofortification of crops. The challenge is to accept biofortified crops by producers and consumers and increase the intake of target nutrients. With the advent of good seed systems, the development of markets, products, and demand creation, this

Crop Biofortification: Biotechnological Approaches for Achieving Nutritional Security under Changing Climate, First Edition. Edited by Adnan Noor Shah et al.

can be achieved (Nestel et al. 2006). Avnee et al. (2023) stated that hidden hunger (deficiency of micronutrients) prevails when the body lacks minerals and vitamins essential in trace amounts for better growth. The deficiencies are more common in developing countries where people have limited access to nutritious diets. They further reported that due to Covid-19 pandemic, mineral supplementation remained a topic of interest because of its support in improving immunity and health. Deficiencies of zinc, iron, selenium, and iodine in humans are significant global food issues. The sustainable way to reduce these deficiencies is the biofortification of food crops with essential minerals and vitamins (Avnee et al. 2023).

19.2 Essential Plant Growth Nutrients

There are a number of macronutrients and micronutrients that are utilized by plants to improve growth and development. These are chemical elements, which affect different metabolic functions. Nutrients that are considered as plant food include the following: carbon, hydrogen, oxygen, nitrogen, phosphorus, and potassium (primary major); calcium, magnesium, and sulfur (secondary major); and zinc, copper, iron, manganese, boron, chlorine, cobalt, nickel, and molybdenum (microelements).

19.3 The Use of Iodine

The least known group of plant food micronutrients are Halogens, where chlorine is the only member recognized currently in plant physiology, which might be due to its regulatory action in proton transfer functioning in the photosystem (Brahmachari et al. 2018). The use of iodine in different concentrations and forms on plant growth is of great importance, referred to as a micronutrient for plants (Kiferle et al. 2021). Different research studies focused on the increase of plant iodine content and reported iodine as a biofortificant in human and animal health (Macías et al. 2016; Gonzali et al. 2017). Generally, iodine content has been increased in different plant tissues through its exogenous administration; however, the presence of iodine might not be avoided in soil, nutrient solution, and atmosphere (Ashworth 2009). Hence, the functional role of iodine as a plant food nutrient has been masked. Macías et al. (2016) and Gonzali et al. (2017) observed that iodine is absorbed through plant roots or aboveground structures, i.e. stomata. They also reported that iodine translocated mainly through the xylem route and volatilized as methyl iodide (CH_3I) with halide-ion methyl transferase (HMT) and halide/thiol methyl transferase (HTMT) enzymes. The presence of iodine in plants is involved in different biochemical and physiological processes. The use of iodine in low concentrations showed beneficial effects on plant growth, productivity, and stress tolerance (Figure 19.1). Similarly, application at higher rates produced toxicity in plants (Incrocci et al. 2019).

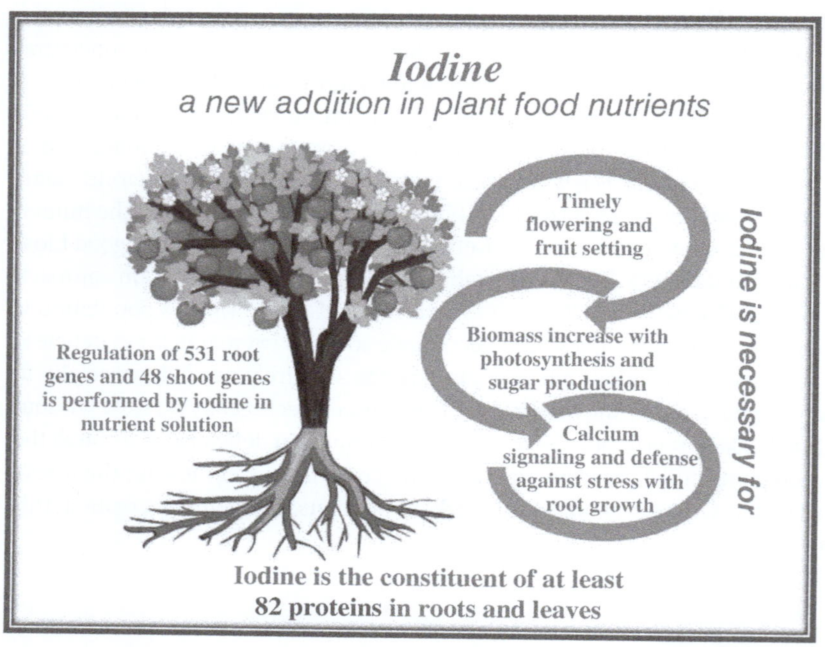

Figure 19.1 Potential role of Iodine for improving different tissues of the plants.

Regarding iodine availability in soils, it is present in small amounts in soil solutions, where the major part is associated with solid phase, i.e. organic matter and clay minerals, as well as iron and aluminum oxides (Fuge and Johnson 1986). The mobility of iodine is limited with some characteristics, such as mineral/organic composition, pH, and texture (Gonzali et al. 2017). Accordingly, the worldwide distribution of iodine is a key for better understanding its importance in living beings, from microorganisms to humans and plants. Iodine is involved in growth regulation, development, and metabolism; hence, it is essential for animals (Blasco et al. 2008) and humans as it is active in the synthesis of thyroid hormones (Landini et al. 2012). Moreover, iodine is also involved in the development and functioning of the nervous system and other organs (Koukkou et al.2017). Dai et al. (2004) reported a series of diseases and health issues such as cretinism, goiter, spontaneous abortions, and reduced intellectual capacity. The bioavailability of iodine is reported as high (99%) in food (Weng et al. 2013). Smoleń et al. (2020) and Sabatino et al. (2021) depicted that the importance of iodine in plants hasn't been fully enlightened, but the application showed greater accumulation of iodine in spinach (*Spinacia oleracea*), lettuce (*Lactuca sativa*), and curly endive (*Cichorium endivia*). Application of iodine through biofortification of wheat (*Triticum aestivum*), rice (*Oryza sativa*), and maize (*Zea mays*) is an easy way to reduce its deficiency in humans (Thai et al. 2020), particularly when used in potassium iodate (KIO_3) form (Cakmak et al. 2017).

19.4 Biofortification of Crops

Biofortification is the process of increasing the micronutrient density of multiple food crops through selective breeding, genetic modification, and by using enriched fertilizers. Multiple agricultural techniques have been developed for the crops' biofortification. Some of the important methods are selective breeding for high rates of specific nutrients, agronomic approach through the application of different fertilizer sources along with other inputs to increase iodine content in plants, fertigation, foliar spray, nutrient solution, or direct application of iodine compounds into the soil. Genetic engineering is another way to induce modification in plant structure and functioning. Modification through a genetic approach (breeding and transgenic) is a long-term and time-consuming technique to evolve new and nutrient-enriched crop varieties. Similarly, the use of different agronomic techniques is an instant and speedy solution to enhance the iodine content in different plant parts.

19.5 Biofortification Through Agronomic Techniques

A high rate of mortality (50%) across the globe has been observed due to deficient dietary nutrients. The reason behind this loss is high dependence on the use of cereal crops by the population in developing countries (Augustine and Kalyanasundaram 2020). In specified regions, the promotion of such need-based crops has declined due to non-awareness about the nutritional status of the plants. Agriculture plays a key role in food and nutritional security. More important is the bioavailability of micronutrients for different food crops, showing it a major task for agronomists and breeders to produce healthy and micronutrient-enriched food for the masses (Welch and Graham 2004). Agronomic biofortification of crops is well strategized to overcome micronutrient malnutrition in humans. Various agronomic techniques have been designed so far to improve iodine deficiency in our daily diets, such as supplementation, fortification of livestock feed, and biofortification of crop plants (Miller and Welch 2013). Apart from the iodination with salt, agronomic biofortification measures are more effective and encouraging to lessen iodine deficiency. There are several advantages of agronomic biofortification with a main focus on the production of crops containing higher concentrations of iodine as compared to traditional growing plants (Błażewicz et al. 2016) (Figure 19.2).

Gonzali et al. (2017) portrayed that the added iodine undergoes a process of incorporation into tissues, where it is available in chemical compound form and easily absorbed by the plants. Winger et al. (2008) explained "iodate" as an oxidant and "iodide" as a reducing agent, while their application in foodstuff may initiate redox reactions, and the stability, durability, and properties of food are affected. The risk of being highly concentrated foodstuff through fortification may also prevail, which might be harmful (Signore et al. 2018). It is also a fact that most genetically modified organisms (GMOs) have been opposed by consumers due to manipulation in genetic structures. However, agronomic biofortification may be easy to adopt and widely applied. The main task in iodine fortification is the fixation of dose and source fertilizer for a particular crop. The selection of fertilizer and application rate for a specific crop depends on species, soil status, growing conditions, and redox conditions (Weng et al. 2013; Gonzali et al. 2017).

Figure 19.2 The methods for agronomic biofortification of plants.

19.6 Biofortification of Crops with Iodine

Enlargement of thyroid glands and stomach cancer in humans have been reported due to the deficiency of iodine. This deficiency may be reduced or controlled by the strategic growing of cost-effective and well-accepted food crops with iodine biofortification. It is worth mentioning that one-third to one-half of the population has been suffering from iodine deficiency. The following Table 19.1 shows the daily requirements of iodine in humans.

Apart from salt iodination, the deficiency of iodine may be reduced with cost-effective and promising agronomic techniques of biofortification. Fortification of livestock feed along with biofortified crops play as corrective measures in controlling iodine deficiency. The use of iodine biofortification is more beneficial like easy uptake by plants through soil application. Application of iodine with irrigation is a much easier and more effective way to give direct benefit to plants. Iodine fertigation in leafy vegetables is more beneficial because of its quick solubility (Augustine and Kalyanasundaram 2020). Ujowundu et al. (2010) reported that the application of potassium iodate (40 μ g dm^{-3}) for four weeks at a two-day interval had a significant increase in the iodine content of tested vegetables. Leafy vegetables showed positive signs as compared to root vegetables with agronomic biofortification. Krzepiłko et al. (2019) reported that different cultivated plants have 0.1–0.15 mg kg^{-1} fresh weight as iodine, e.g. 0.12 mg kg^{-1} in spinach, 0.15 mg kg^{-1} in broccoli, and 0.14 mg kg^{-1} in pea. Iodine biofortification showed promising observations due to enhanced rates of bioactive compounds (Augustine and Kalyanasundaram 2020). High levels of iodine salts may be utilized as biofortification to evaluate the beneficial and toxin concentrations. Gonzali et al. (2017) reported that less than 50 mg kg^{-1} salt may be used to increase iodine content in plant tissues, whereas nonsignificant observations were recorded with iodine biofortification up to 50 mg kg^{-1} (Caffagni et al. 2012). It was observed in multiple trials that for efficient biofortification, foliar application of iodine salts remained much better. Lawson et al. (2016) recorded a positive impact on crop quality with foliar spraying than soil application. Foliar

Table 19.1 The daily dietary intake requirement of different age groups.

S. no.	Age group	Daily intake of iodine
1.	Young children (1–8 years)	90 μ g d^{-1}
2.	Older children (9–13 years)	120 μ g d^{-1}
3.	Adults (>14 years)	150 μ g d^{-1}
4.	Women (pregnancy and breastfeeding)	220–270 μ g d^{-1}

application of iodine significantly increased grain concentrations in brown rice, wheat, and maize (Augustine and Kalyanasundaram 2020). This enhancement was noticed in grains of wheat and polished rice. Studies on animals have confirmed that plant sources of iodine can be better absorbed, highly effective, and safe as compared to iodized salt (Krzepiłko et al. 2019). The study further revealed that iodine-enriched plant products might be safer for the consumer with low toxicity as compared to iodine salts. Crops biofortification in an appropriate manner increases iodine concentration in plants with no negative impact on plant growth and development. For better utilization, the iodine rate of application must be optimized as normal plant growth is disturbed with high doses (Signore et al. 2018).

19.7 Conclusion

Studies on the micronutrient deficiencies in crop plants have been devised more often. Different mineral nutrients have become a dire need of the population now. Agricultural techniques are the easiest and cost-effective ways to control these deficiencies. Among various microelements, iodine is directly associated with human dietary needs, which are to be supplemented through different measures. Biofortification is an effective and easy way to compensate iodine requirements of masses, as the use of iodized salt may also be harmful, especially for people having high blood pressure and/or medical treatments. Experiments carried out on agronomic biofortification must ensure good crop quality with enriched iodine content to be used by the population. This strategy should be the basic effort to lessen the iodine deficiency in the diet. Widespread iodine biofortification can be achieved by specifying detailed cultivation schemes for different crop species. Different measures for the control of contamination through environment, soil and surface water, and toxicity in living organisms should also be included in the strategy. For effective biofortification, the following measures can be adopted:

1. Standardization at post-processing stages for nutrients' availability.
2. An awareness program should be formulated for the farming community to know about the biofortification of crops.
3. New germplasm should be evolved through various techniques to be used for agronomic biofortification.
4. Research experiments on various aspects of iodine deficiency must be a part of the proposed plan in every research organization.
5. Different brochures, booklets, radio talks, extension work, etc. should be performed for the awareness of farmers and other people.

References

Augustine, R. and Kalyanasundaram, D. (2020). Agronomic biofortification of food crops with micronutrients. *Plant Archives* 20 (2): 1383–1387.

Ashworth, D.J. (2009). Transfers of iodine in soil-plant-air system: solid-liquid partitioning, migration, plant uptake and volatilization. In: *Comprehensive Handbook of Iodine* (eds. VR. Preedy, GN. Burrow and RR. Watson), 107–118. Cambridge: Academic Press.

Avnee, M., Sood, S., Chaudhary, D.R. et al. (2023). Biofortification: an approach to eradicate micronutrient deficiency. *Frontiers in Nutrition* 10. doi: 10.3389/fnut.2023.1233070.

Blasco, B., Rios, J., Cervilla, L. et al. (2008). Iodine biofortification and antioxidant capacity of lettuce: potential benefits for cultivation and human health. *Annals of Applied Biology* 152 (3): 289–299. doi: 10.1111/j.1744-7348.2008.00217.x.

Błażewicz, A., Makarewicz, A., Glowniak, I.K. et al. (2016). Iodine in autism spectrum disorders. *Journal of Trace Elements and Medical Biology* 34 (3): 32–37. doi: 10.1016/j.jtemb.2015.12.002.

Brahmachari, U., Guo, Z., Konecny, S.E. et al. (2018). Engineering proton transfer in photosynthetic oxygen evolution: chloride, nitrate, and trehalose reorganize a hydrogen-bonding network. *Journal of Physiology and Chemistry-B* 122 (26): 6702–6711. doi: 10.1021/acs.jpcb.8b02856.

Caffagni, A., Pecchioni, N., Meriggi, P. et al. (2012). Iodine uptake and distribution in horticultural and fruit tree species. *Italian Journal of Agronomy* 7 (3): e32.

Cakmak, I., Prom-u-thai, C., Guilherme, L.R.G. et al. (2017). Iodine biofortification of wheat, rice and maize through fertilizer strategy. *Plant Soil* 418: 319–335. doi: 10.1007/s11104-017-3295-9.

Dai, J.L., Zhu, Y.G., Zhang, M. et al. (2004). Selecting iodine-enriched vegetables and the residual effect of iodate application to soil. *Biology of Trace Element Research* 101 (3):265–276. doi: 10.1385/BTER:101: 3: 265.

Fuge, R. and Johnson, C.C. (1986). The geochemistry of iodine – a review. *Environmental Geochemistry and Health* 8 (2): 31–54.
Galera, S.G., Rojas, E., Sudhakar, D. et al. (2010). Critical evaluation of strategies for mineral fortification of staple food crops. *Transgenic Research* 19 (2): 165–180. doi: 10.1007/s11248-009-9311-y.
Gonzali, S., Kiferle, C. and Perata, P. (2017). Iodine biofortification of crops: agronomic biofortification, metabolic engineering and iodine bioavailability. *Current Opinion in Biotechnology* 44: 16–26. doi: 10.1016/j.copbio.2016.10.004.
Hasanuzzaman, M., Tahir, M.S., Tanveer, M. et al. (2023). *Mineral Biofortification in Crop Plants for Ensuring Food Security*. Singapore: Springer Nature Singapore Private Limited. doi: 10.1007/978-981-99-4090-5.
Incrocci, L., Carmassi, G., Maggini, R. et al. (2019). Iodine accumulation and tolerance in Sweet basil (*Ocimum basilicum* L.) with green or purple leaves grown in floating system technique. *Frontiers in Plant Sciences* 10: 1–15. doi: 10.3389/fpls.2019.01494.
Kiferle, C., Martinelli, M., Salzano, A.M. et al. (2021). Evidences for a nutritional role of iodine in plants. *Frontiers in Plant Science* 12:616868. doi: 10.3389/fpls.2021. 616868.
Koukkou, E.G., Roupas, N.D. and Markou, K.B. (2017). Effect of excess iodine intake on thyroid on human health. *Minerva Medica* 108 (2): 136–146. doi: 10.23736/S0026-4806.17.04923-0.
Krzepiłko, A., Prażak, R., Bednarz, B.S. et al. (2019). Agronomic biofortification as a means of enriching plant foodstuff with iodine. *Acta Agrobotanica* 72 (2): 1766. doi: 10.5586/aa.1766.
Landini, M., Gonzali, S., Kiferle, C. et al. (2012). Metabolic engineering of the iodine content in Arabidopsis. *Scientific Reports* 2 (1): 338. doi: 10.1038/srep00338.
Lawson, P.G., Daum, D., Czauderna, R. et al. (2016). Factors influencing the efficacy of iodine foliar sprays used for biofortifying butterhead lettuce. *Journal of Plant Nutrition and Soil Science* 179: 661–669.
Macías, J.M., Martínez, P.L., Morales, S.G. et al. (2016). Use of iodine to biofortify and promote growth and stress tolerance in crops. *Frontiers in Plant Sciences* 7: 1–20. doi: 10.3389/fpls.2016.01146.
Miller, D.D. and Welch, R.M. (2013). Food system strategies for preventing micronutrient malnutrition. *Food Policy* 42 (4): 115–128. doi: 10.1016/j.foodpol.2013.06.008.
Nestel, P., Bouis, H.E., Meenakshi, J.V. et al. (2006). Biofortification of staple food crops. *The Journal of Nutrition* 136 (4): 1064–1067. doi: 10.1093/jn/136.4.1064.
Sabatino, L., Gaudio, F.D., Consentino, B.B. et al. (2021). Iodine biofortification counters micronutrient deficiency and improve functional quality of open field grown curly endive. *Horticulturae* 7 (3): 58. doi: 10.3390/ horticulturae7030058.
Signore, A., Renna, M. and D'Imperio, M. (2018). Preliminary evidences of biofortification with iodine of "Carota di Polignano," an Italian carrot landrace. *Frontiers in Plant Sciences* 9: 170. doi: 10.3389/fpls.2018.00170.
Thai, C.P.U., Rashid, A., Ram, H. et al. (2020). Simultaneous biofortification of rice with zinc, iodine, iron and selenium through foliar treatment of a micronutrient cocktail in five countries. *Frontiers in Plant Science* 11: 589835.
Ujowundu, C.O., Ukoha, A.I. and Agha, C.N. (2010). Effects of potassium iodate application on the biomass and iodine concentration of selected indigenous Nigerian vegetables. *African Journal of Biotechnology* 9 (42): 7141–7147.
Welch, R.M. and Graham, R.D. (2004). Breeding for micronutrients in staple food crops from a human nutrition perspective. *Journal of Experimental Botany* 55 (396): 353–364.
Weng, H.X., Hong, C.L., Xia, T.H. et al. (2013). Iodine biofortification of vegetable plants – an innovative method for iodine supplementation. *Chinese Science Bulletin* 58 (17): 2066–2072. doi: 10.1007/s11434-013-5709-2.
Winger, R.J., König, J. and House, D.A. (2008). Technological issues associated with iodine fortification of foods. *Trends Food Science & Technology* 19 (2): 94–101. doi: 10.1016/j.tifs.2007.08.002.
Zhang, Y., Cao, H., Wang, M. et al. (2023). A review of iodine in plants with biofortification: uptake, accumulation, transportation, function and toxicity. *The Science of Total Environment* 878 (17): 163203. doi: 10.1016/j.scitotenv.2023.163203.

20

Applications of Nanoparticles in Biofortification of Crops: Amplifying Nutritional Quality

Zakir Ullah[1†], Javed Iqbal[2†], Banzeer Ahsan Abbasi[3], Shumaila Ijaz[4], Mamoona Munir[3], Tabassum Yaseen[2], Shobana Sampath[5], Sobia Kanwal[6], Hassan Sher[7], Zahid Ullah[7], Ahmad Ali[7], Tariq Mahmood[1]*

[1] *Department of Plant Sciences, Faculty of Biological Sciences, Quaid-i-Azam University Islamabad, Pakistan*
[2] *Department of Botany, Bacha Khan University, Charsadda, Khyber Pakhtunkhwa, Pakistan*
[3] *Department of Botany, Rawalpindi Women University, Satellite Town, Rawalpindi, Pakistan*
[4] *School of Biomedical Engineering, Shenzhen University Medical School, Shenzhen University, Shenzhen, China*
[5] *Department of Biotechnology, Vel Tech Rangarajan Dr Sagunthala R&D Institute of Science and Technology, Avadi, Chennai, India*
[6] *Department of Biology and Environmental Sciences, Allama Iqbal Open University, Islamabad, Pakistan*
[7] *Center for Plant Sciences and Biodiversity, University of Swat, Charbagh, Pakistan*
† *Equal contribution*

*Corresponding author: javed89qau@gmail.com

20.1 Introduction to Biofortification

Biofortification is the process of enhancing the nutrient content of food crops without compromising their desirable traits, achieved through conventional breeding, improved agronomic practices, or modern biotechnology (Athar et al. 2020). It has emerged as a cost-effective and promising method to deliver micronutrients to populations with limited access to diverse diets. Combining biofortification with food fortification is essential for addressing micronutrient deficiencies in low- and middle-income countries (Yadav et al. 2020). It effectively improves the micronutrient content of staple crops such as wheat, maize, and rice. Micronutrients can be efficiently transported to the endosperm using techniques such as foliar application and organic matter manipulation, ensuring sustainable agricultural production (Malik and Maqbool 2020). The United Nations Food and Agriculture Organization estimates that around 792.5 million people worldwide suffer from malnutrition, with the majority living in developing countries (Athar et al. 2020). Additionally, approximately two billion people experience "hidden hunger" due to insufficient intake of essential micronutrients despite increased food production (Yadav et al. 2020; Malik and Maqbool 2020). Overnutrition is also becoming a growing concern (Malik and Maqbool 2020).

Historically, our agricultural system prioritized maximizing grain yield and crop productivity over promoting human health, leading to a rise in micronutrient deficiencies in food grains and subsequently, among consumers (Ijaz et al. 2023a). However, there is now a paradigm shift in agriculture toward producing nutrient-rich food crops alongside ensuring adequate quantity. This shift aims to combat "hidden hunger" or micronutrient malnutrition, particularly prevalent in impoverished and developing regions where diets heavily rely on staple crops lacking in micronutrients (Lal et al. 2020). Traditionally, mass vitamin and mineral supplementation programs have been employed, but these often fall short of international health organizations' targets due to unreliable external funding, limited purchasing power, market access, healthcare infrastructure, and insufficient awareness of long-term health benefits (Virk et al. 2021; Ahmad et al. 2022). Biofortification of diverse crop varieties emerges as a sustainable, long-term solution to providing nutrient-rich crops. These biofortified crops, with enhanced bioavailable micronutrient concentrations, can reach undernourished and low-income families through traditional agricultural and food trade channels, circumventing the need for dietary supplements and fortified foods (Avnee Sood et al. 2023). Economically, biofortification represents a one-time investment, offering a cost-effective, sustainable approach to addressing hidden hunger. Once biofortified crops are developed, there are no ongoing costs associated with purchasing and adding fortificants during food processing (Avnee Sood et al. 2023; Avellan et al. 2021; Kiran et al. 2022). Looking ahead, with projected population growth in the developing world and changing

Crop Biofortification: Biotechnological Approaches for Achieving Nutritional Security under Changing Climate, First Edition. Edited by Adnan Noor Shah et al.

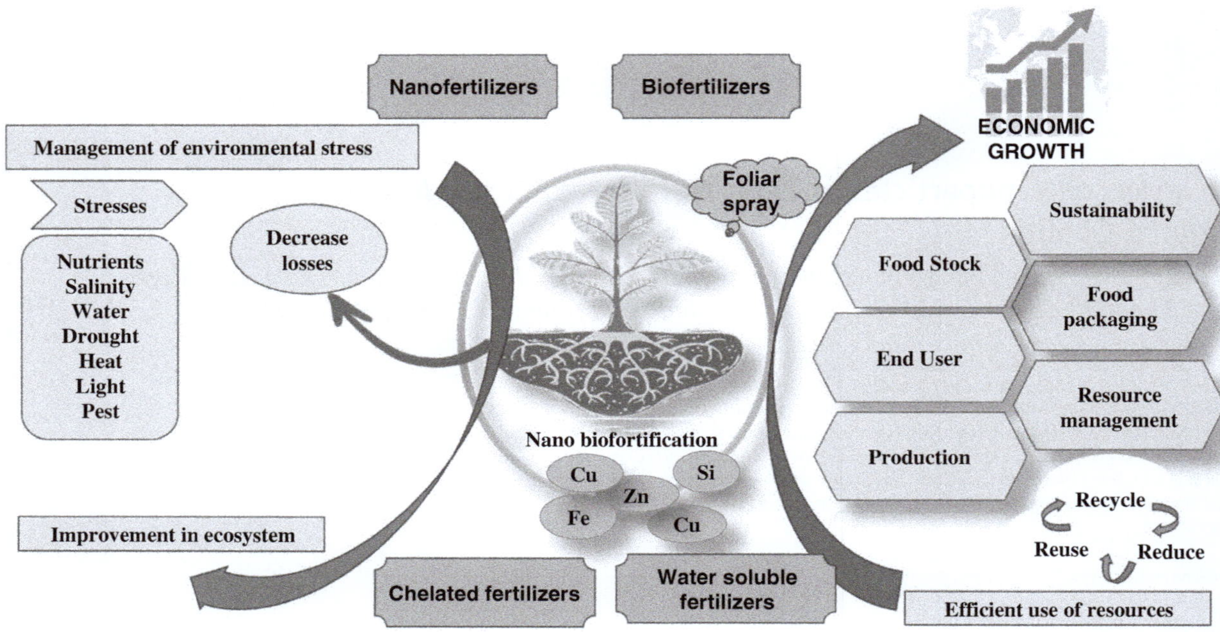

Figure 20.1 Biofortification of plants and their applications.

climatic conditions, achieving food security will become an even greater challenge (Avnee Sood et al. 2023; Masuda et al. 2020). Consequently, organizations such as the World Health Organization and the Consultative Group on International Agricultural Research have prioritized the development of nutritionally enhanced, high-yielding biofortified crops as a key objective (Figure 20.1).

20.1.1 Importance of Biofortification in Addressing Malnutrition

By raising the concentrations of vital vitamins and minerals in staple crops, biofortification can contribute to a more cost-effective and environmentally friendly solution to this issue (Kiran et al. 2022). To give nutrient-rich food to rural resource-poor people who do not have access to diverse food, supplements, or industrially fortified food is the process of enriching the bioavailable concentration in edible sections of food (Praharaj et al. 2021). Biofortification can be an affordable and sustainable solution to treat micronutrient deficiencies at the population level and enhance public health, especially for communities that primarily depend on a single staple crop for their daily calorie intake (Ramadas et al. 2020). Unlike supplements, it needs a one-time expenditure, addresses the undernourished population, and produces food of higher-quality without sacrificing yield. In developing countries where micronutrients prevail and can have detrimental effects on health, this can be particularly significant (Bouis et al. 2019). The most common deficiencies are those in iron, zinc, iodine, and selenium, which collectively make up around 60% of the total, 30% of zinc and iodine, and 15% of selenium (Athar et al. 2020). However, eliminating hunger requires more than just biofortification. It does not provide as much nutrition as supplements or food that has been fortified, but it can increase daily dietary consumption (Ahmad et al. 2022). Given the predicted decline in dietary micronutrients brought on by climate change, biofortification becomes even more crucial for vulnerable populations to preserve their health. To address the prevalent nutritional deficits, the idea of concurrently biofortifying crops with numerous critical micronutrients is novel and promising and improves overall human nutrition (Lodhi et al. 2023). Known as "multinutrient biofortification" or "combinatorial biofortification," this strategy seeks to produce crops with a well-balanced assortment of different vital vitamins and minerals (Okwuonu et al. 2021). By treating several dietary deficits at once, this technique can offer a more all-encompassing and holistic approach to fighting malnutrition. The total bioavailability and health effects of several nutrients in crops can be enhanced by mixing them. People who eat a larger variety of nutrients from staple foods might benefit from multinutrient biofortified crops, which can promote dietary diversification (Ahmad et al.

2022). Multinutrient biofortification can lead to much larger market coverage by minimizing rivalry between several single-nutrient biofortified varieties, in addition to cost savings from development to distribution and synergies through aggregated health effects (Lodhi et al. 2023).

20.1.2 Traditional Approaches to Biofortification

20.1.2.1 Agronomic Biofortification

Agronomic biofortification involves enhancing crop nutritional value through fertilization and soil management, offering a rapid and cost-effective means of producing nutrient-rich food, albeit with short-term effects (Okwuonu et al. 2021). Most crops can be biofortified with essential micronutrients such as iron, zinc, iodine, and selenium using micronutrient-enriched organic/inorganic fertilizers or biofertilizers applied via foliar or soil methods (Szerement et al. 2022). Micronutrient levels in plant edible parts depend on the fertilizer type, application method and rate, growth stage, and nutrient translocation within plants. However, due to local variations in plant species, soil chemistry, and mineral mobility, agronomic biofortification efficiency differs (Okwuonu et al. 2021; Szerement et al. 2022). The efficacy of agronomic biofortification is increased by developments in specialized fertilizers, such as water-soluble, chelated, and nanofertilizers, which increase nutrient absorption efficiency and translocation to edible plant parts. Due to their distinct characteristics that stem from their size and structure, nanoparticles (NPs) have become more attractive as biofortification instruments (Okwuonu et al. 2021; Shahane and Shivay 2022). A subset of agricultural nanotechnology known as nanofertilizers (NFs) has the potential to revolutionize traditional techniques for raising crop nutrient content and improving overall nutritional quality (Figure 20.2). Employing "nanofertilizer-assisted biofortification," NFs make use of their tiny size and larger surface area to improve plant uptake of nutrients (Ahmad et al. 2022; Lodhi et al. 2023; Okwuonu et al. 2021).

20.1.2.2 Iron Biofortification

A significant health consequence of iron deficiency, which is common in underdeveloped nations, is anemia, which affects more than half of India's under-five-year-old children and pregnant women (Szerement et al. 2022). By boosting crop iron content, which is essential for processes including the synthesis of red blood cells, the movement of oxygen throughout the body, the maintenance of healthy skin, hair, and nails, and immune system support, biofortification might lessen this (Shahane and Shivay 2022). Infants, young children, and pregnant women have higher demands for iron. Iron shortage can

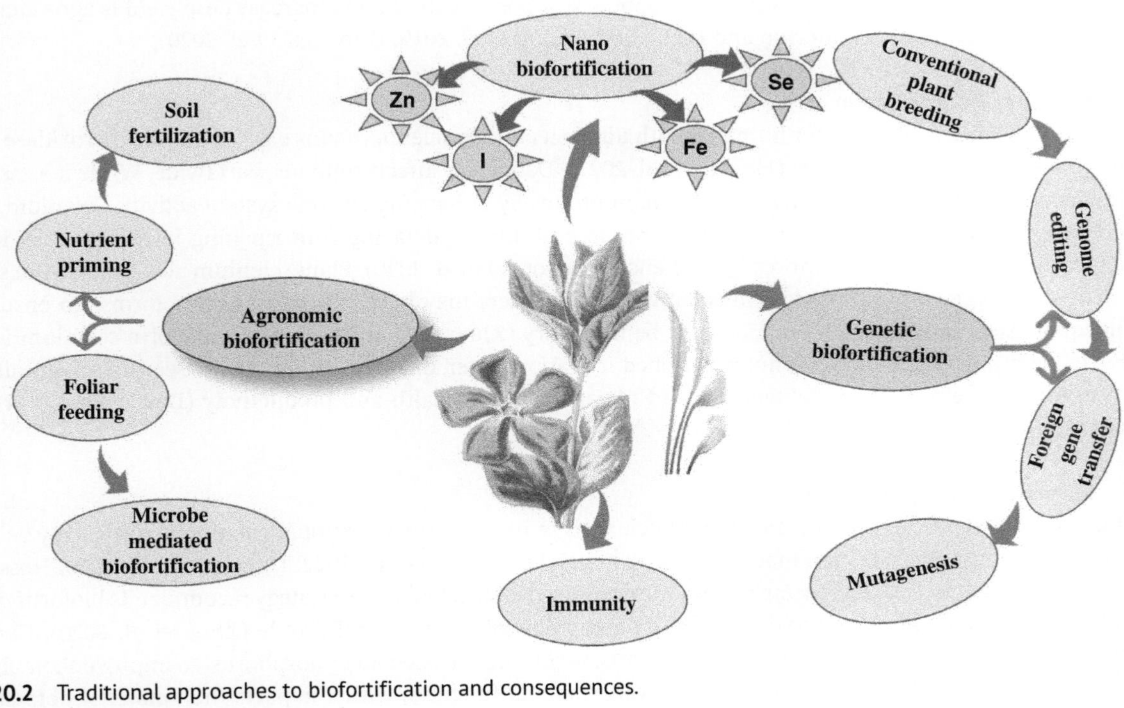

Figure 20.2 Traditional approaches to biofortification and consequences.

be characterized by anemia, fatigue, and immune system dysfunction, particularly in countries where plant-based diets are the primary source of iron (Connorton and Balk 2019). Crop absorption of iron is improved by iron-rich fertilizers and good soil management practices, such as preserving pH and nutrient balance. However, crops and growth environments have an impact on effectiveness. Notably, compared to meat eaters, vegans need 1.8 times the daily recommended amount of iron. Improving dietary iron through biofortified crops is crucial for combating anemia and related health issues, especially in regions where plant-based foods are primary iron sources (Szerement et al. 2022; Shahane and Shivay 2022; Connorton and Balk 2019).

20.1.2.3 Zinc Biofortification

Zinc is an essential micronutrient that is necessary for human and plant health. It is involved in wound healing, immunological response, protein synthesis, DNA synthesis, and the metabolism of carbohydrates, proteins, and lipids (Kumar et al. 2019). About 30% of people worldwide suffer from zinc deficiency, which makes them more susceptible to conditions including increased risk of infection, altered taste and smell, growth retardation, DNA damage, and maternal mortality. In areas where zinc insufficiency is common, biofortification provides a means of enhancing crop nutritional value while also combating the deficit (Duborská et al. 2020). Human zinc insufficiency and soil zinc deficiency overlap geographically. About 36.5% of India's agricultural land is zinc deficient, which exacerbates zinc depletion in sections of crops that are edible (Shahane and Shivay 2022). This shortage is especially widespread in underdeveloped nations with poor incomes when plant-based diets are the norm (Connorton and Balk 2019). Agronomic biofortification of zinc has shown success in various crops, offering promise for addressing this pressing nutritional challenge (Kumar et al. 2019).

20.1.2.4 Iodine Biofortification

Iodine controls body temperature and energy generation. It is essential for human metabolism and thyroid gland function. Iodine treatment has been linked to increased yields and iodine-rich crops, even though plants do not need to survive (Duborská et al. 2020). Globally, 30–38% of individuals consume insufficient amounts of iodine, which can result in health problems such as goiter, infertility, stunting, hypothyroidism, and intellectual disability (Dávila-Rangel et al. 2019). In areas where iodine insufficiency is frequent and plant-based foods are popular, iodine biofortification is beneficial for both improving individual nutrition and community iodine consumption (Szerement et al. 2022; Shahane and Shivay 2022). Iodine needs for adults range from 150 to 290 μg per day, with a tolerated maximum of 1100 μg. A popular fortification technique is iodized salt, although its effectiveness varies and it might increase blood pressure, which is a risk factor for heart disease and stroke, as well as exacerbate osteoporosis. A possible method to increase crop yield is agronomic biofortification crop iodine content (Connorton and Balk 2019; Kumar et al. 2019; Duborská et al. 2020).

20.1.2.5 Selenium Biofortification

Selenium, a trace element essential for human health due to its role in selenoproteins e.g. glutathione peroxidase (GPX), is important for proper immune function (Hossain et al. 2021). Deficiency affects millions worldwide. While not essential for plants, low doses of selenium can benefit certain plant groups by enhancing enzyme system activity. Selenium, alone or combined with iodine, can enhance plant concentration and quality, delaying fruit ripening in tomatoes by inhibiting ethylene biosynthesis and boosting antioxidant defenses (Duborská et al. 2020). Plant selenium absorption from soil varies widely, with some regions naturally low in selenium while others may have it in inaccessible forms. To ensure plants receive adequate selenium, supplementation may be necessary (Zhou et al. 2020). This can involve selenium fertilizers, selenium-rich compost or manure, selenium-enriched irrigation water, or selenium-enhanced seeds. Strategically adding selenium to the soil helps address deficiencies and maximize plant health and productivity (Dávila-Rangel et al. 2019; Hossain et al. 2021; Zhou et al. 2020).

20.1.2.6 Interaction Among Nutrients

The optimization of agronomic biofortification, which aims to improve the nutritional content of crops, may be achieved by considering the interplay between macro and micronutrients (Ahmad et al. 2022; Duborská et al. 2020; Hossain et al. 2021). Research has focused on increasing certain nutrients in the past, but a new strategy recommends biofortifying many important micronutrients at once, providing a more comprehensive nutritional profile (Zhou et al. 2020). Certain elements, such as zinc, work in concert with other nutrients, including nitrogen and phosphorus, to improve their absorption and utilization (Ullah 2017); additionally, zinc controls stomatal function, which impacts the uptake of potassium. The production of selenium-containing phytochemicals, which have important antioxidant qualities for human health, is enhanced by selenium biofortification. Plant growth-promoting bacterium (PGPR) and arbuscular mycorrhizal fungus may be combined with zinc, selenium, and iron to create environmentally acceptable biofertilizers that enrich crops with microelements (Kumar et al. 2019; Duborská et al. 2020; Dávila-Rangel et al. 2019; Hossain et al. 2021). Competition for absorption is reduced when nutrients are kept in an appropriate proportion. Zinc, manganese, and magnesium are essential nutrients for the synthesis of vitamin C and the enzyme systems that produce ascorbic acid. Copper and manganese serve as cofactors for antioxidant and flavonoid synthesis enzymes, positively influencing compound production (Zhou et al. 2020; Ullah 2017).

Plant nitrogen and phosphorus status impacts root formation, shoot movement, and nutrient redistribution to seeds, enhancing micronutrient absorption and concentration in edible parts (Ogino et al. 2019). Environmental stresses such as salt, temperature extremes, heavy metals (HMs), and drought induce oxidative stress in plants, leading to reactive oxygen species (ROS) overproduction (Dundore-Arias et al. 2019). Oxidative stress damage is decreased by biofortification with zinc, selenium, and iron utilizing a variety of fertilizer kinds and forms. Superoxide dismutase (SOD), ascorbate peroxidase, GPX, monodehydroascorbate reductase, dehydroascorbate reductase, glutathione reductase, glutathione S-transferase, and peroxiredoxin (PRX) are among the ROS-scavenging enzyme content that is increased in different plant cell sites. This coordinated strategy can reduce damage caused by oxidative stress and improve crop nutritional quality (Zhou et al. 2020; Dundore-Arias et al. 2019; Ullah et al. 2023).

20.1.2.7 Biofortification Through Conventional Breeding

Through carefully regulated crossings between various plant kinds, traditional plant breeding provides a way to selectively breed for desirable features, increasing crop nutritional value (Thind et al. 2020). To produce hybrids with enhanced attributes, this repeated procedure entails choosing plants with advantageous properties – such as a greater micronutrient content – and crossing them (Ullah et al. 2023; Thind et al. 2020). This procedure is continued over several generations, and the progeny are tested for desired features, resulting in the creation of crop types that have been biofortified. Biofortification through plant breeding targets staple food crops to address nutritional deficiencies, particularly in low-income households (Ahmad et al. 2024a). This approach has gained traction due to its sustainability; once nutritionally improved varieties are developed, they can be grown and consumed year after year without reliance on ongoing external interventions. Despite the dominance of yield and resistance breeding in the past decades, which often led to lower nutrient levels in crops, biofortification through breeding offers a promising solution (Ullah et al. 2023; Thind et al. 2020; Ahmad et al. 2024a).

Breeding efforts have focused on iron-rich beans, zinc-rich rice, selenium-rich rice, wheat, maize, iodine-rich cassava, maize, sweet potatoes, and vitamin A-rich sweet potatoes as crops to be biofortified (Thind et al. 2020). Conventional breeding is generally regarded as safe and does not give rise to the same safety issues as genetic modification. However, creating crops with higher nutritional content may be a labor-intensive and laborious process that sometimes takes years (Dundore-Arias et al. 2019). Fortification and biofortification are crucial issues, particularly in the context of rising global hunger and malnutrition (Thind et al. 2020). In 2017, hunger increased for the first time in a decade, with 815 million people suffering from hunger and 155 million children experiencing stunted growth due to poor nutrition. Known as "hidden hunger," micronutrient deficiencies impact two billion people globally (Ahmad et al. 2024a; Haider et al. 2021a). Urgent action is needed to address these challenges and prevent further escalation. Sustainable agricultural strategies and policies are essential to combat micronutrient deficiencies effectively. The Green Revolution, while successful in increasing crop production, often prioritized a narrow range of crops at the expense of more nutritious options, contributing to widespread malnutrition. Similarly, agricultural policies in West Africa have sometimes limited access to nutritious crops, exacerbating deficiencies (Dundore-Arias et al. 2019; Ullah et al. 2023). Since the 1990s, micronutrient fortification and supplementation programs have become key strategies for addressing population-wide deficiencies in low- and middle-income countries. These approaches are supported by national and international health agencies, NGOs, philanthropic organizations, and donors. Food fortification and biofortification are promoted as effective means to reach at-risk populations without requiring significant changes to existing diets or food consumption patterns (Ahmad et al. 2022; Ullah 2017).

Advocates argue that existing food supply chains can efficiently deliver fortified foods to many consumers, particularly in at-risk populations (Thind et al. 2020). Furthermore, fortification and biofortification are seen as relevant strategies for addressing deficiencies in situations where access to diverse and nutritious foods is limited, such as during emergencies or in areas with inadequate infrastructure. In summary, fortification and biofortification are essential tools in the fight against malnutrition, offering sustainable solutions to address widespread nutrient deficiencies (Duborská et al. 2020). By leveraging traditional plant breeding techniques and targeted interventions, we can work toward a future where all individuals have access to nutritious and fortified food options (Thind et al. 2020).

20.2 Nanotechnology and Nanoparticles

For over two decades, nanosciences and nanotechnologies have been extensively explored across various scientific domains, including physics, chemistry, biology, engineering, and social sciences. The allure of nanotechnology lies in the unique properties exhibited by materials at the nanoscale, opening possibilities for unprecedented functionalities (Bouis et al. 2019; Lodhi et al. 2023). Developing, monitoring, and controlling nano-objects, comprehending their characteristics and interactions with the environment, particularly living things, and incorporating them into useful systems are all extremely difficult tasks for researchers. The goal is to harness these capabilities for a multitude of applications while ensuring control and safety (Szerement et al. 2022; Shahane and Shivay 2022; Connorton and Balk 2019). Nanotechnology applications are increasingly pervasive in everyday life (Figure 20.3), spanning industries, commerce, healthcare, and societal domains. Energy, chemistry, sensors, materials, information and communication technology, biology, medicine, and environmental applications are the main areas of current research and development. Amidst the promise of nanotechnology, it is crucial not to overlook associated risks, such as potential health and environmental hazards, privacy concerns, and broader societal impacts, including evolutionary changes. Meeting these challenges requires immense effort, and competition among nations in this arena is rising (Ramadas et al. 2020; Ahmad et al. 2022; Connorton and Balk 2019; Duborská et al. 2020).

20.2.1 Two Attempts to Nano

Between the 1930s and 1980s, the advent of powerful imaging techniques enabled the exploration and manipulation of matter at scales previously unimaginable with conventional light microscopes, sparking great excitement across scientific communities (Haider et al. 2021a; Shelenga et al. 2021). A significant challenge emerged in fabricating and studying small objects composed of a few tens to thousands of atoms, pushing the boundaries of scientific inquiry (Okwuonu et al. 2021). While chemistry and biology had long been capable of synthesizing nanoscale objects using organic synthesis and peptide synthesis techniques, primarily carbon-based materials, the focus shifted when scientists began developing tools to miniaturize traditional macroscopic matter (Lodhi et al. 2023). This "top-down" approach involved creating nanometric objects, such as small clusters of metallic atoms, and then imaging and manipulating them using cutting-edge technological tools such as electron microscopes, tunneling microscopes, and nanolithography (Shahane and Shivay 2022). This combination of direct imaging and remote-controlled manipulation at the atomic level made the concept of NPs tangible, fueling the imagination of a generation with visions of nanorobots and machines operating in everyday environments, including within the human body (Connorton and Balk 2019). While the miniaturization of printed circuits, leading to exponential increases in memory and processor capacities, stands as a remarkable technological achievement of these techniques, everyday materials often require more conventional methods (Shelenga et al. 2021).

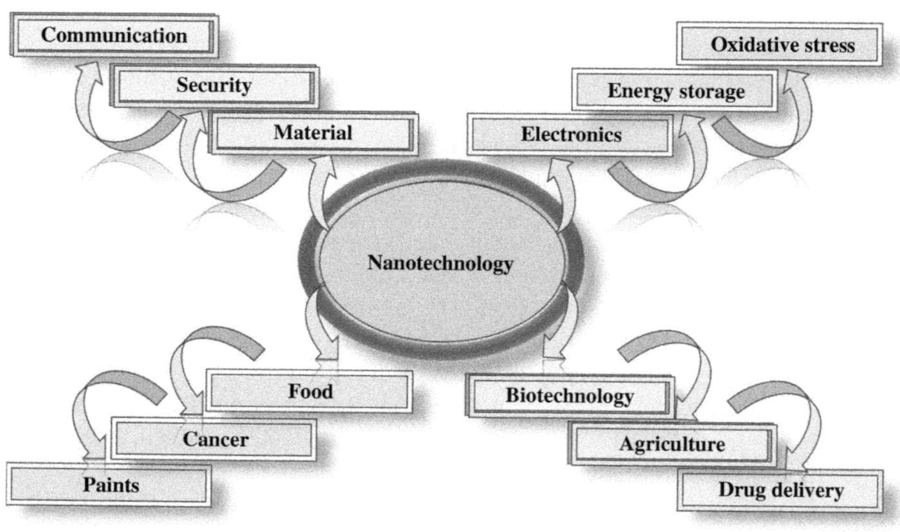

Figure 20.3 Nanotechnology and their applications.

Despite immense progress in atomic manipulation over the past few decades, exemplified by feats like creating the world's smallest film ("A Boy and His Atom"), mass production of functional nanorobots using these tools remains elusive for several reasons. Atomic manipulation is slow and labor-intensive, yielding only one object at a time, and typically requires specialized conditions, such as low temperature and pressure, for stability (Ullah et al. 2023). Thus, while impressive as a proof of concept, it falls short as a practical technological solution. The two paths to the nanoworld – top down and bottom up – sometimes converge to create innovative structures. For instance, in the fabrication of nanocars and molecular machines, organic chemistry is employed to synthesize the building blocks (e.g. wheels and axles) from small molecules (Ahmad et al. 2019). These components are then assembled into "supramolecular" objects, a technique pioneered by Nobel Prize-winning chemists J.M. Lehn and J.P. Sauvage. These assemblies are subsequently studied using top-down imaging tools such as atomic force microscopy, occasionally leading to unconventional events like nanocar races, much to the delight of enthusiasts. Overall, the journey into the nanoworld involves a blend of innovative approaches and interdisciplinary collaboration, with each path offering unique insights and challenges (Duborská et al. 2020; Dávila-Rangel et al. 2019; Ullah 2017; Thind et al. 2020).

20.2.2 Properties of Nanoparticles Relevant to Biofortification

Nanotechnology harnesses the unique properties of materials at the nanoscale (<100 nm), particularly NPs, which boast high surface-area-to-volume ratios and customizable dissolution profiles (Athar et al. 2020; Lal et al. 2020; Ahmad et al. 2022). In precision agriculture, nanotechnology shows promise for enhancing staple crop quantity and quality through biofortification methods. Applying NPs via foliar spray or soil treatment can significantly increase crop growth, production, and micronutrient levels in edible tissues while managing nutrient availability through controlled release, thereby minimizing leaching and optimizing nutrient accumulation (Avnee Sood et al. 2023; Kiran et al. 2022). However, NPs, when encountering cells, can disrupt their intracellular homeostasis by acting as oxidizing agents, potentially leading to oxidative stress. Various in vitro models have elucidated nanoparticle toxicity mechanisms, particularly through the generation of ROS (Ramadas et al. 2020; Zhou et al. 2020). To address this, models based on nanoparticle band energies and their comparison with cellular redox potential have been proposed to predict oxidative stress induction accurately. High-throughput screening platforms have been utilized to predict oxide nanoparticle toxicity profiles at the cellular level, paving the way for the development of safer nanomaterials that maintain cellular homeostasis and equilibrium in redox reactions. By modifying specific nanomaterial properties such as size and biocompatible coatings while preserving functionality, researchers aim to create "safe-by-design" nanomaterials that pose no hazards to the environment. Overall, nanotechnology-based approaches hold significant potential for revolutionizing agricultural practices, promoting sustainable crop production, and mitigating nutrient deficiencies (Ullah 2017; Ogino et al. 2019; Ullah et al. 2023).

20.2.3 Zinc

Around one-third of the global population faces Zn deficiency in their diet. Zinc plays an important role in enzyme function and hormonal regulation of carbohydrate metabolism (Miyazawa et al. 2021). However, soil Zn deficiency leads to poor Zn absorption by plants. Zinc oxide (ZnO) NPs offer a solution due to their efficient absorption, accumulation, and metabolism in plants, attributed to their high surface area-to-volume ratio. Foliar application of ZnO NPs has demonstrated remarkable results in increasing Zn concentration in maize compared to conventional zinc sulfate ($ZnSO_4$) fertilizers (Duborská et al. 2020). Surprisingly, lower concentrations of ZnO NPs were found to facilitate greater Zn accumulation in maize grains compared to higher concentrations. Similarly, wheat germinated in ZnO NPs in amended soil exhibited enhanced grain yield and Zn content. The foliar application of ZnO NPs in wheat not only increased grain Zn content but also improved plant growth and antioxidant enzyme activities (Ramadas et al. 2020; Lodhi et al. 2023; Shahane and Shivay 2022).

Various studies have reported the efficacy of ZnO NPs in enhancing Zn uptake and grain yield in wheat, rice, common bean, sorghum, and finger millet without causing toxicity (Miyazawa et al. 2021). Soil application of urea coated ZnO NPs significantly increased Zn uptake and wheat grain yield. Additionally, ZnO nanoparticle applications via foliar spray, soil application, or seed priming have led to increased Zn concentration in grains, improved protein content, decreased cadmium levels, and enhanced nutrient content in rice (Ogino et al. 2019). Seed priming of rice with ZnO NPs has shown increased Zn acquisition in seeds, along with improved antioxidant activities. Similarly, foliar application of ZnO NPs on common beans significantly increased Zn content in seeds and enhanced essential amino acid levels (Thind et al. 2020). Importantly, ZnO NPs did not pose nanoscale-specific risks, making them a safe option for agricultural use. In rice, foliar application of ZnO NPs significantly enhanced Zn concentration in grains, as well as nutrient uptake when applied in combination with basal $ZnSO_4$ (Shelenga et al. 2021). Overall, ZnO NPs show significant potential as a fertilizer to increase grain Zn content, serving as an effective biofortification strategy to combat zinc deficiency in crops (Miyazawa et al. 2021).

20.2.4 Iron

Different types of iron (Fe) may be found in soil, including magnetite (Fe_3O_4), hematite (α-Fe_2O_3), and maghemite (γ-Fe_2O_3). However, since the iron in the soil can change into forms that are inaccessible to plants, they frequently have difficulty absorbing it (Khan et al. 2019). Enough iron is essential for plant development, germination of seeds, and chlorophyll synthesis (Dundore-Arias et al. 2019). Conventional chelated iron fertilizers are widely used to correct iron shortage. However, compared to bulk iron oxide, γ-Fe_2O_3 NPs demonstrate superior translocation from roots to other plant parts due to their nanosize and low volatilization (Thind et al. 2020). Seed priming with iron oxide NPs has shown significant increases in the iron content in wheat grains compared to untreated seeds. Similarly, foliar application of iron oxide NPs has led to enhanced iron content in wheat leaves compared to conventional Fe-EDTA (Shelenga et al. 2021). Moreover, wheat seed priming with iron NPs has resulted in increased nutrient content in grains. Soil application of iron oxide NPs has also been effective in enhancing iron uptake by wheat, along with nitrogen, phosphorus, and potassium concentrations, while reducing cadmium uptake (Ahmad et al. 2019).

Fe_2O_3 NPs applied to soil have not only increased the wheat iron content and biomass but also improved the amino acid content. Additionally, soil application of nano-Fe_2O_3 has significantly increased tomato yield and fruit quality compared to conventional iron sources (Dávila-Rangel et al. 2019; Shelenga et al. 2021). Similarly, foliar application of nanoscale chelated iron on cumin has increased iron concentration in seeds and enhanced plant growth and yield (Miyazawa et al. 2021). In comparison to traditional iron sources, finger millet grain iron content has been risen because of seed priming with iron oxide NPs (Shelenga et al. 2021). Grain nutritional content, photosynthetic efficiency, and yield have all increased using nanoscale zero-valent iron (nZVI) priming of rice seeds. Additionally, nZVI has a low negative influence on the soil environment and may be used commercially as a "pro-fertilizer" for seed treatment, encouraging plant development, and yield (Ahmad et al. 2024a; Shelenga et al. 2021). Compared to traditional iron sources, biosynthesized orthorhombic Fe-oxalate-capped-Fe-oxide NPs have demonstrated efficacy in elevating soil enzymatic activity and mitigating iron deficits in tomato plants. When compared to traditional priming techniques, rice seeds that have been nanoprimed with FeO NPs capped with phytochemicals have had higher iron contents, improved germination, vigor, and antioxidant enzyme activity (Miyazawa et al. 2021). These studies highlight the potential of nanoscale iron materials as sustainable sources of iron for plants, offering tunable release properties at the nanoscale for optimal plant nutrition (Khan et al. 2019).

20.2.5 Copper

Copper oxide NPs, or CuO NPs, are used as soil additions for cleanup, plant growth regulators, and fertilizers (Ahmad et al. 2024b). CuO NPs applied topically or subsurface has both shown promise in raising the amount of copper present in food crops when compared to their untreated counterparts. However, a variety of factors, including plant species, soil properties, and environmental circumstances, affect the rate and amount of copper absorption (Khan et al. 2024). Studies have shown that Cu^{2+} released from copper NPs at a certain concentration can promote plant growth without causing phytotoxicity. For instance, foliar application of Cu NPs on tomato plants not only increased the vitamin C content but also enhanced β-carotenoid levels. Similarly, soil application of Cu NPs increased iron and copper content in sugarcane plants (Ramadas et al. 2020; Bouis et al. 2019). Furthermore, CuO NPs applied to soil enhanced the allicin content in onion leaves and increased the copper, calcium, and magnesium content in onion bulbs. In soybean plants, soil application of CuO NPs resulted in significant copper accumulation in various plant parts, particularly in the roots and seeds. Additionally, a study on zucchini plants revealed that soil application of CuO NPs triggered specific gene expression patterns related to chloroplast and mitochondrial function modulation (Ramadas et al. 2020; Ahmad et al. 2022).

Recent research has highlighted the significance of plant copper's nutritional status for crop disease prevention. For example, *Fusarium oxysporum* f. sp. niveum-induced root fungal disease in watermelon plants was greatly inhibited by applying $Cu_3(PO_4)_2{\cdot}3H_2O$ nanosheets (Okwuonu et al. 2021). Comparing treated versus untreated infected controls, foliar application of $Cu_3(PO_4)_2{\cdot}3H_2O$ nanosheets to tomato seedlings decreased the disease caused by *F. oxysporum* f. sp. lycopersici and enhanced plant biomass and nutrient levels (Duborská et al. 2020). It is noteworthy that whereas nanoscale platforms have the potential to provide essential micronutrients for crop disease response, elevated CuO concentrations may cause phytotoxicity, which in turn diminishes plant development and water content. The disruption of redox processes in plant cells by liberated copper ions might be the cause of this toxicity (Shelenga et al. 2021). Overall, the use of CuO NPs in agriculture offers opportunities for enhancing plant growth, nutrient content, and disease resistance. However, careful consideration of application methods and concentrations is essential to maximize benefits while minimizing potential risks of phytotoxicity (Ahmad et al. 2024b; Khan et al. 2024).

20.2.6 Selenium

Although selenium (Se) is essential for both human and animal health, more than a billion individuals worldwide are selenium deficient (Ali et al. 2022). Se is present in soil and biotic systems, such as plants, in a variety of forms. Compared to bulk Se, Se NPs have improved antioxidant, antibacterial, and anticancer activities and have a greater bioavailability when utilized as NFs. By scavenging free radicals, they easily bind to ligands and function as heavy metal detoxifiers. Research indicates that applying Se NPs topically to coffee grains enhances their Se content more than applying sodium selenite topically (Ali et al. 2022). In a similar vein, Se NPs applied to cucumbers improve nutrient absorption, raise nutrient concentration inside the plant, and increase crop production. Se content of rice seedlings is greatly increased by hydroponically applying Se NPs, and brown rice's foliar spray application enriches the Se level. Thus, Se NPs can serve as cost-effective micronutrients for crop biofortification and enhance the production of proteins, amino acids, phenolics, and glucosinolates (Ahmad et al. 2024a; Shelenga et al. 2021; Ali et al. 2022).

20.3 Mechanisms of Nanoparticle Uptake in Plants

20.3.1 Process of Nanoparticle Uptake, Accumulation, and Translocation in Plants

Particle size, chemical makeup, stability, functionality, and plant species all affect how successfully NPs are absorbed by plants (Lv et al. 2019). Research investigated how NP size affects wheat plants and found that Fe_2O_3 NPs that are smaller (8–20 nm) may easily penetrate roots and reach leaves. Furthermore, NP concentration and surface area-to-charge ratio are important factors in plant absorption and translocation (Shelenga et al. 2021; Lv et al. 2019). NPs can more easily penetrate the surfaces of roots and leaves when their surface-area-to-volume ratios are higher. In soybean plants, maximum uptake of Zn (at 8 nm) was observed at a concentration of 500 mg L^{-1}, with a reduction in uptake observed at higher concentrations (1000–4000 mg L^{-1}) of ZnO NPs. This decrease in uptake may be attributed to the formation of aggregates at higher concentrations, hindering passage through cell wall pores (Shelenga et al. 2021; Khan et al. 2024; Ali et al. 2022). Furthermore, nanopriming of seeds has been shown to enhance micronutrient uptake in various studies, although the specific mechanisms underlying this effect remain unclear (Lv et al. 2019).

20.3.2 Foliar Uptake and Translocation of NPs

One direct way to directly reinforce plants with NPs is by foliar treatment. NPs are taken up by stomatal and cuticular routes from the leaf surface (Banerjee et al. 2019). While polar or ionic molecules enter through stomatal holes, which generally have a diameter of 0.6–4.8 nm, lipophilic substances use the cuticular route. While bigger NPs (>5 nm) may use other channels for foliar absorption, smaller NPs (less than 4.8 nm) can probably pass through the cuticular pathway (Avellan et al. 2021). Hydrophilic compounds can enter the stomatal route through holes that have a minimum diameter of 40 nm (Athar et al. 2020). Because cuticular pores are tiny, studies often support the stomatal route for NPs absorption in leaves. After entering the leaf apoplast, NPs go, mostly through the phloem, to different sections of the plant, such as the grains, fruits, stems, and roots (Duborská et al. 2020). Research indicates that Fe_2O_3, magnesium oxide (MgO), and ZnO NPs with sizes ranging from 24 to 47 nm enter watermelon plants by their stomata and then go through the phloem to their roots. Like this, coffee plants' foliar absorption of ZnO NPs led to a notable buildup of Zn NPs in treated leaves as compared to control plants (Khan et al. 2024). The foliar absorption of zinc NPs involves both stomatal and cuticular routes although the relative importance of each channel varies. Research has documented the foliar absorption and translocation of Fe_3O_4 NPs in wheat seedlings from leaves to stems and roots through the phloem (Khan et al. 2024). Furthermore, only a tiny portion of the Cu accumulated in the leaves following the foliar application of CuO NPs in cabbage and Cu $(OH)_2$ NPs in lettuce, with the majority going to the roots. According to a different study, foliar spray increased γ-Fe_2O_3 NP uptake more than soil application (Banerjee et al. 2019).

20.3.3 Root Uptake and Translocation of NPs

Since NPs are more persistent in soil than in leaves, most research on NP absorption has concentrated on root uptake as opposed to foliar exposure (Banerjee et al. 2019). Compared to foliar absorption, the interactions of NPs in soil and their subsequent uptake and translocation in roots are more intricate. To get to the shoots through the xylem, NPs must pass through several barriers, including the root cuticle, epidermis, cortex, endodermis, and Casparian strip (Avellan et al. 2021). When added to the soil, NPs first stick to the root surface and subsequently enter the roots via a variety of processes, including endocytosis, aquaporins, ion channels, and carrier proteins. The most well-known method of uptake is endocytosis; however, aquaporin transporters are also involved in the absorption of NPs (Lv et al. 2019). Because organic acids are secreted from root hairs, the root surface is mostly negatively charged, which facilitates the adsorption and buildup of positively charged NPs. Because the cuticle of the root hair is still developing, NPs can pass through it straight and reach the epidermis (Avellan et al. 2019). Further translocation of NPs has been seen via both apoplastic and symplastic mechanisms. The symplastic pathway comprises translocation from cell to cell via plasmodesmata, while the apoplastic pathway encompasses movement via intercellular gaps. The widespread acceptance of NP apoplastic mobility is demonstrated by the numerous investigations that have found NPs in the intercellular space of root tissues (Ali et al. 2022; Lv et al. 2019). NPs may enter the xylem by the root tip or junctions in the lateral root area, where the Casparian strip is not fully grown or detached, even though the strip acts as a barrier to the apoplastic movement (Figure 20.4). One example of the significance of this junction in the apoplastic transport of NPs into the xylem is the buildup of ZnO NPs in the xylem and lateral root junction of maize (Avellan et al. 2021; 2019).

CuO NPs were shown to be absorbed and translocated across the xylem and phloem in maize and rice, as well as to accumulate in intercellular gaps, xylem sap, and leaves in endosomes harboring the particles (Lv et al. 2019). CuO NPs can enter plant cells via endocytosis, proceed to the xylem via the apoplast, and then be transported to various plant sections. Studies on Fe_2O_3 NPs in maize revealed apoplastic migration from the root epidermis to the endodermis, although there was little investigation into translocation to other plant tissues (Avellan et al. 2019). When metal acquisition is inadequate in higher plants, two main tactics are used: for nongraminaceous plants, use technique I; for graminaceous plants, use strategy II. In Strategy I, metal reduction oxidase reduces metal chelates at the root surface, and then metal-regulated transporter genes assist absorb metal ions across the plasma membrane (Su et al. 2019). As part of Strategy II, on the other hand, plants may absorb metal by secreting phytosiderophores that chelate metal and produce soluble complexes that are then absorbed by root cells (Wang et al. 2022). Research has shown that $FePO_4$ NPs are useful for raising phosphorus levels in both cucumber and maize plants. Cucumber plants use $FePO_4$ NPs as a source of phosphorus (Strategy I species), whereas maize plants use them as a source of iron (Strategy II species). Particularly during the seedling stage, the mechanism behind the absorption, transport, and storage of NPs in plants is still poorly known (Ullah et al. 2020). To comprehend these mechanisms and

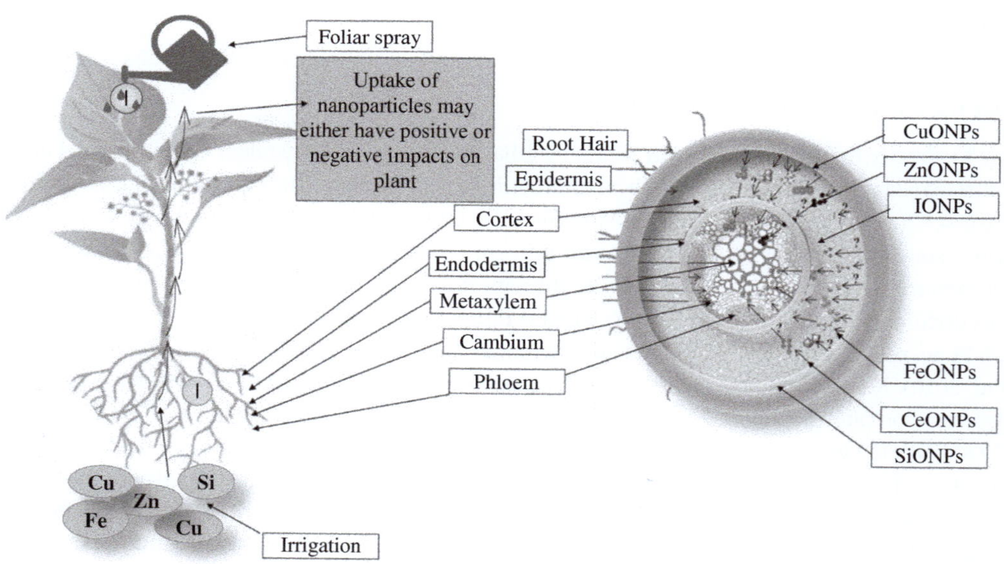

Figure 20.4 Process of different nanoparticles uptake, accumulation, and translocation in plants.

assess if NPs are appropriate for use in agriculture, more study is necessary. NPs may be more suited for soil or foliar treatment in agricultural techniques, depending on the primary transport mechanism (xylem vs. phloem) (Avellan et al. 2019).

20.4 Factors Influencing Nanoparticle Uptake in Crops

Studies show that a variety of factors, including the characteristics of NPs, plant species, and environmental conditions, influence how successfully plants absorb NPs (Ulhassan et al. 2022). The chemical composition, surface charge, surface modification, and particle size are important factors that greatly influence how much NP is absorbed by plant leaves (Avellan et al. 2019). Interestingly, NP size has become a significant factor: bigger particles are less efficiently absorbed, while smaller metal-based NPs (diameter <50 nm) can penetrate leaves through stomata. For example, ZnO NPs (30 nm) labeled with fluorescein isothiocyanate (FITC) were shown to penetrate wheat leaves mostly through stomata and accumulate in chloroplasts (Wang et al. 2022). Similarly, disruption and/or diffusion between the cuticle layer and stomata allowed wheat leaves to absorb gold NPs of different sizes (3, 10, and 50 nm) covered with different materials (Su et al. 2019).

Additionally, transmission electron microscopy revealed that watermelon leaves absorbed MgO NPs (size range: 27–35 nm) (Ullah et al. 2020). Plant leaves may also absorb NPs made of silica, polymers, and natural materials in addition to metal-based NPs; the critical absorption sizes of each NP type differ. In *Arabidopsis thaliana* leaves, for instance, SiO_2 NPs (54 nm) were shown to enter through stomata, whereas mesoporous silica nanoparticles (MSNs) (200–300 nm) were absorbed by the leaves (Ulhassan et al. 2022). NPs' entrance into plant mesophyll tissue is also influenced by their shape and surface charge. Distinct interfacial characteristics of NPs with varying shapes influence their absorption. Despite having equal diameters, *A. thaliana* leaves were able to absorb rod-shaped gold NPs more easily than spherical ones. Both positively and negatively charged NPs can be absorbed; on the stomata of wheat leaves, positively charged NPs exhibit greater adsorption (Sun et al. 2021). Plant species are particularly important because differences in the leaf pore size, density, and distribution impact the efficiency of NP absorption. Furthermore, stomatal opening and shutting are influenced by temperature, humidity, light, and the stage of plant development, all of which affect net primary productivity absorption (Avellan et al. 2019). Research comparing several species has demonstrated differences in the effectiveness of NP absorption, with dicotyledonous pumpkin demonstrating a more efficient absorption of CeO_2 NPs than monocotyledonous wheat (Zhang et al. 2019).

20.4.1 Key Factors That Affect the Uptake of Nanoparticles via Roots

The size, chemical makeup, and surface charge of NPs have a substantial impact on how well they are absorbed by plant roots (Avellan et al. 2019). Research has revealed that plant roots can absorb NPs smaller than 10 nm, such as gold NPs (3.5 nm) and CeO_2 NPs (8 ± 1 nm), as evidenced by *Vicia faba* L. and maize roots. However, absorption decreases as NP size increases, with metal-based NPs larger than 100 nm facing challenges in root absorption due to size limitations. It is interesting to note that NPs larger than 100 nm can be absorbed; these NPs are based on silicon and natural polymers (Sun et al. 2021; Zhang et al. 2019). For instance, the roots of Arabidopsis were found to absorb Si NPs measuring 200 nm, but the roots of sugarcane were shown to absorb zein NPs measuring 135 ± 3 nm. Plant absorption of NPs is also influenced by their surface charge (Avellan et al. 2021). Due to their electrostatic attraction to negatively charged cell walls, positively charged NPs are more likely to concentrate on the root surface but have difficulty penetrating the root tissues (Avellan et al. 2019). Studies using different surface charges and sizes of NPs (20–100 nm) showed that uncharged and negatively charged NPs (22 nm) may be taken up by *A. thaliana* root cells and penetrate the root xylem (Ullah et al. 2020). On the other hand, positively charged NPs did not penetrate the root tissue of Arabidopsis; instead, they simply gathered at the root epidermis (Zhang et al. 2019).

20.4.2 Key Factors That Affect the Transport of Nanoparticles Absorbed via Leaves

The size, chemical makeup, and surface charge of NPs all affect how they are transported into plants (Ramadas et al. 2020; Okwuonu et al. 2021). Smaller metal-based NPs are more likely than bigger nonmetal-based NPs to be transferred to various areas of the plant (Ullah 2017). For example, NPs with metal bases that are present in the phloem or other plant components usually have diameters between 30 and 50 nm (Avellan et al. 2019). On the other hand, it has been noted that bigger NPs composed of nonmetal elements have been translocated in plant leaves. In one experiment, after just four hours, cucumber leaves treated with 200–300 nm-sized FITC-labeled MSNs showed the presence of MSNs in petioles, stems, untreated leaves, and flowers, indicating the translocation of MSNs from leaves to other cucumber tissues (Shelenga et al. 2021; Avellan et al. 2019; Ulhassan et al. 2022). Furthermore, it was shown that leaves absorb and transfer porous hollow silica nanoparticles, which have an average particle size of 253 ± 73 nm, to stems, roots, and young leaves. The extracellular route trafficking of NPs is also influenced by their surface charge (Zhang et al. 2019). Because plant cell walls contain a lot of pectin with carboxyl groups, which have a negative charge, positively charged NPs are drawn to them. Positively charged NPs are more easily transported across the extracellular route thanks to this attraction. When carboxyl and amino-modified polystyrene NPs were sprayed on maize seedlings, for example, the NPs effectively accumulated on the leaves of the plants. However, the accumulation of amino-modified positively charged NPs was greater than that of carboxyl-modified negatively charged NPs (Azeem et al. 2021). The negatively charged NPs showed more efficient transfer to the plant's vascular system and subsequent transport to the roots through the vasculature (Wang et al. 2022). Regarding the transport of NPs absorbed through roots, their chemical composition plays a significant role (Ullah et al. 2020). Small-sized NPs such as TiO_2 NPs and Cu NPs can be absorbed by cucumber and corn roots and subsequently detected in various aboveground plant parts, including leaves and fruits (Sun et al. 2021). Surface charge also influences NP transport in roots. Root-secreted mucus and secretions serve as the initial barrier for NP entry into plant roots, inhibiting the absorption of positively charged NPs outside the cell wall. Positively charged NPs tend to accumulate at the cell wall due to electrostatic attraction, while negatively charged NPs transport within the root more efficiently (Lv et al. 2019; Avellan et al. 2019; Ulhassan et al. 2022). Studies using CeO_2 NPs with different charges demonstrated that positively charged CeO_2 NPs were more abundant on the root surface, whereas negatively charged CeO_2 NPs were transported to different plant parts. Similar results were observed in the absorption and transport of gold NPs with different charges in Arabidopsis roots (Zhang et al. 2019).

20.5 Role of Nanoparticles in Enhancing Nutrient Uptake

Plant development is hampered by drought stress (DS), which upsets the balance of nutrients. NPs are essential for preserving the balance of nutrients in plants because they improve nutrient distribution, translocation, and absorption in different areas of the plant (Zhang et al. 2019). Due to reduced nutrient absorption, transpiration, and membrane stability, water scarcity lowers the amounts of nitrogen (N), potassium (K), manganese (Mn), and zinc (Zn) in plants. On the other hand, ZnO NPs sprayed topically or applied to the soil increase the absorption of N, P, K, and Zn, hence mitigating the impacts of DS (Azeem et al. 2021). While Si-NPs in conjunction with PGPR boost photosynthesis, relative water content (RWC), antioxidant activity, and nutrient absorption in maize under DS, Fe- and Hg-NPs considerably enhance N, P, K, calcium (Ca), and magnesium (Mg) intake (Zhang et al. 2019; Azeem et al. 2021). Similarly, NP and biofertilizer combinations enhance nutrient uptake, nodule formation, and nitrogenase activity in *Phaseolus vulgaris*. Zn-NP treatment enhances wheat and sorghum productivity and nutrient uptake. NPs also enhance nitrate reductase activity, N assimilation, protein, and amino acid synthesis while increasing nutrient sequestration in plant roots and overall uptake (Lv et al. 2019; Avellan et al. 2019). Additionally, NPs improve water use efficiency, further enhancing nutrient absorption and plant growth. In summary, NPs play a crucial role in bolstering nutrient uptake, plant growth, and DS tolerance. Future research should delve into NPs' role in nutrient signaling and channels for potential advancements (Zhang et al. 2019; Azeem et al. 2021).

20.5.1 Nanoparticle-based Delivery Systems for Essential Nutrients

Nanoparticle-based delivery systems have emerged as a groundbreaking solution in agricultural science, offering significant potential to tackle key challenges in nutrient management for crops (Haider et al. 2021b). These nanocarriers encompass a wide array of materials, including metallic NPs such as iron, zinc, and copper, carbon-based nanomaterials like graphene and carbon nanotubes (CNTs), as well as liposomes and polymeric NPs (Avellan et al. 2019). Each nanoparticle type is selected for its distinct properties, such as high surface area, reactivity, or encapsulation abilities, allowing for customized approaches to nutrient delivery (Wang et al. 2022). A major advantage of nanoparticle-based systems is their capacity to safeguard essential nutrients from environmental degradation, serving as a protective barrier against leaching, microbial degradation, and oxidation. This protective function ensures a consistent nutrient supply in the soil, fostering an environment conducive to enhanced plant uptake. Moreover, encapsulation within NPs enables controlled release, facilitating gradual and targeted nutrient delivery based on plant-specific needs (Sun et al. 2021). The interaction between NPs and plant physiology is a complex yet crucial aspect of this technology. NPs enhance nutrient solubility and availability in the rhizosphere, facilitating improved uptake by plant roots (Haider et al. 2021b). Additionally, they play a vital role in transporting nutrients across cellular membranes, leading to more efficient assimilation within plant tissues. Experimental studies across various crops have demonstrated that nanoparticle-mediated nutrient delivery positively influences plant growth, development, and ultimately, crop yield. Controlled-release mechanisms are a key feature of nanoparticle-based delivery systems, allowing for finely tuned nutrient release over time to ensure optimal plant nutrition while minimizing waste (Yusefi-Tanha et al. 2020). This controlled release is particularly advantageous in nutrient-stressed environments, contributing to improved crop resilience and adaptability to adverse conditions such as drought or nutrient deficiency (Avellan et al. 2019; Su et al. 2019). However, the application of nanoparticle-based delivery systems in agriculture presents challenges, including potential environmental contamination, underscoring the need for sustainable formulations and application practices to mitigate adverse ecological impacts. Moving forward, ongoing research aims to refine formulations, optimize delivery systems, and address long-term impacts on soil health and ecosystems (Yusefi-Tanha et al. 2020). It is crucial to navigate regulatory frameworks and consider public perceptions to ensure the responsible and ethical implementation of this innovative technology for the advancement of global food security (Wang et al. 2022; Ullah et al. 2020).

20.5.2 Improving Nutrient Availability and Absorption Efficiency in Plants

Enhancing nutrient availability and absorption efficiency in plants is essential for boosting crop productivity, nutritional quality, and the sustainability of agricultural systems (Ullah et al. 2023). One effective method involves utilizing nanotechnology to formulate and deliver NPs loaded with essential nutrients such as nitrogen, phosphorus, potassium, and micronutrients. NPs offer advantages such as their small size, large surface area, and ability to penetrate plant tissues, facilitating efficient nutrient delivery and uptake. Applying NPs through methods such as foliar spraying, seed coating, or soil application can significantly improve nutrient availability and absorption efficiency, ultimately leading to enhanced crop growth and yields (Avellan et al. 2019; Ullah et al. 2020; Zhang et al. 2019; Azeem et al. 2021).

Moreover, advancements in soil and nutrient management practices, including precision agriculture, integrated nutrient management, and soil enrichment with organic matter and biochar, play a crucial role in optimizing nutrient availability and uptake by plants (Yusefi-Tanha et al. 2020). These practices aim to balance soil nutrient levels, optimize fertilizer application rates, and enhance soil structure and fertility, thereby promoting root development, nutrient uptake, and overall nutrient use efficiency in crops. Furthermore, harnessing the potential of PGPR and mycorrhizal fungi can further enhance nutrient availability in the rhizosphere, facilitate nutrient uptake by plant roots, and improve overall plant growth and health (Li et al. 2020). Additionally, breeding programs focused on developing nutrient-efficient crop varieties with improved root architecture, nutrient transporters, and metabolic pathways can contribute significantly to enhancing nutrient uptake and utilization efficiency in plants (Luo et al. 2022). By selecting traits related to nutrient acquisition and utilization, breeders can develop crops better adapted to nutrient-limited environments and capable of performing well under stress conditions. Genetic engineering techniques can also be employed to introduce genes encoding nutrient transporters and metabolic enzymes, further enhancing nutrient uptake and utilization in crops. In summary, optimizing nutrient availability and absorption efficiency in plants requires a comprehensive approach that integrates nanotechnology, soil and nutrient management practices, microbial symbiosis, and genetic improvement strategies (Su et al. 2019; Zhang et al. 2019; Azeem et al. 2021). By adopting these integrated approaches, it is possible to enhance nutrient use efficiency in agriculture, reduce reliance on fertilizers, minimize environmental pollution, and ensure sustainable food production for growing populations (Li et al. 2020; Luo et al. 2022).

20.5.3 Types of Nanoparticles Used in Crop Biofortification

Nanoparticles play a pivotal role in crop biofortification, offering a wide array of materials with distinct properties and functionalities aimed at enhancing nutrient uptake and bolstering plant health. Among the commonly utilized variants are metal NPs, such as iron (Fe), Zn, and selenium (Se) NPs, which are employed to fortify staple crops with vital micronutrients (Ramadas et al. 2020; Bouis et al. 2019; Ahmad et al. 2022; Okwuonu et al. 2021; Connorton and Balk 2019). For instance, iron NPs are applied to boost iron levels in edible plant parts, tackling iron deficiency anemia prevalent in many regions. Similarly, zinc NPs enhance zinc absorption in plants, improving crop nutritional value and combating zinc deficiency in diets. Selenium NPs, known for their antioxidant properties, fortify crops with selenium, vital for human health and disease prevention (Kumar et al. 2019; Duborská et al. 2020; Ullah et al. 2023). Carbon-based NPs, such as CNTs and graphene oxide, are gaining traction for crop biofortification due to their unique physicochemical properties. These materials offer high surface area, electrical conductivity, and biocompatibility, all conducive to enhancing nutrient delivery and plant growth. CNTs are explored for their potential to augment nutrient uptake and water retention in plants, fostering growth and stress resilience (Luo et al. 2022). Meanwhile, graphene oxide shows promise in bolstering nutrient absorption and photosynthetic efficiency, contributing to improved crop productivity and nutritional quality (Ullah et al. 2023; Khan et al. 2019). Additionally, lipid-based NPs, including liposomes and nanoemulsions, are utilized for encapsulating and delivering hydrophobic nutrients, vitamins, and phytochemicals to plants. These nanocarriers shield bioactive compounds, ensuring stability and bioavailability post-application. Liposomal formulations, for instance, deliver fat-soluble vitamins and antioxidants to plant tissues, enhancing nutritional content and health benefits. Nanoemulsions improve the solubility and absorption of lipophilic nutrients and bioactives in plants, aiding their uptake for growth and development (Ali et al. 2022; Lv et al. 2019).

In essence, leveraging various nanoparticle types in crop biofortification holds promise for addressing malnutrition, boosting crop yields, and advancing sustainable agriculture. By harnessing nanoparticles properties, such as small size and tailored surface chemistry, innovative strategies for nutrient, vitamin, and bioactive compound delivery to plants can be developed, enhancing their nutritional quality and resilience to environmental stressors (Umar et al. 2021). However, ongoing research is crucial to optimize nanoparticle formulations, application methods, and dosages to maximize efficacy while minimizing potential risks to human health and the environment (Yusefi-Tanha et al. 2020).

20.6 Techniques for Nanoparticle Application in Agriculture

Nanotechnology has gained significant traction in agriculture over the past decade, spurred by ample public funding and a growing array of methods falling under its purview (Ullah et al. 2024). However, its development in agriculture has faced unique challenges due to the nature of farm production, which operates as an open system with large-scale demand for input materials and limited control over the input of nanomaterials (Singh et al. 2021). Nevertheless, nanotechnology holds immense promise for improving crop productivity and reducing pesticide applications. Its applications span various areas, including nanoformulations of agrochemicals for enhanced crop improvement, nanosensors for disease identification and agrochemical residue detection, nanodevices for genetic engineering of plants, and precision farming techniques aimed at maximizing yields while minimizing environmental impact (Luo et al. 2022; Umar et al. 2021). Additionally, nanotechnology offers opportunities in animal health, breeding, and postharvest management, along with potential benefits in biomass-to-fuel production and food processing (Ramadas et al. 2020; Duborská et al. 2020; Shelenga et al. 2021). However, experts emphasize the need to balance the potential benefits of nanotechnology with concerns for soil, water, environment, and occupational health (Singh et al. 2021).

Nanomaterials, with their specific properties, are being explored for various agricultural applications, including utilizing agricultural waste products as renewable raw materials (Figure 20.5). Moreover, nanotechnology holds promise in addressing challenges such as insecticide resistance through targeted approaches (Pramanik et al. 2020). While nanotechnology applications in agriculture are being researched and tested, concerns regarding risk assessment and potential environmental and health impacts persist. Efforts are underway to develop nanoparticle attractants derived from biopolymers with minimal effects on human health and the environment (Li et al. 2020; Luo et al. 2022; Singh et al. 2021). Overall, nanotechnology is poised to revolutionize agriculture and the food industry by enhancing farming techniques, nutrient absorption in plants, disease detection, and pest control. However, addressing remaining challenges is crucial to realizing its full potential while ensuring sustainability and safety in agricultural practices (Pramanik et al. 2020).

20.6.1 Nanotechnology in Pesticides and Fertilizers

In today's context, the necessity for sustainable agriculture is paramount, representing an approach that fosters ecosystem health for the long-term (Manjunatha et al. 2019). Practices that could potentially inflict lasting harm on soil include excessive tilling, leading to erosion, and inadequate drainage in irrigation, resulting in salinization (Pramanik et al. 2020). These methods are used to satisfy human demands for food, animal feed, and fiber. To understand the impact of different practices on soil qualities that are critical for sustainability and to collect relevant data toward this goal, long-term studies are required. A government agency in the United States is leading the way in the creation of nanochemicals, which have potential as agents for pest control and plant growth. Fertilizers play a vital role in plant growth, and nanomaterials acting as fertilizers may possess properties conducive to crop improvement with reduced ecotoxicity (Avellan et al. 2019; Li et al. 2020). However,

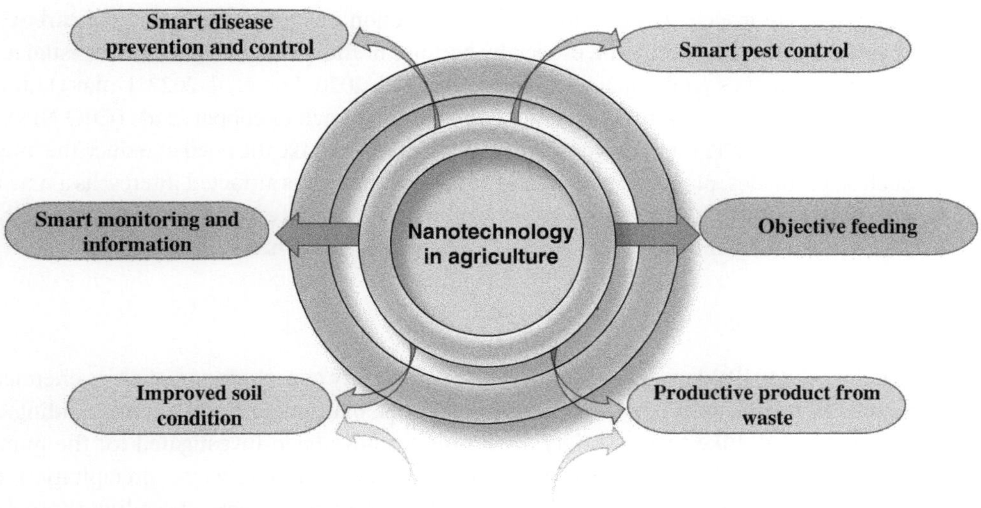

Figure 20.5 Different techniques for nanoparticle application in agriculture.

there is a concern regarding the bioaccumulation of nanomaterials in the food chain through plants. Advances in plant care recently have included the use of NPs to improve the safety and effectiveness of chemical treatments (Pramanik et al. 2020). Numerous studies has examined the impact of various NPs, such as magnetite (Fe_3O_4) NPs, alumina, zinc, zinc oxide, silver NPs, sulfur NPs, and NPs of AlO, CuO, FeO, MnO, NiO, and ZnO, on plant development and phytotoxicity. It is widely acknowledged that zinc is a micronutrient that is necessary for plant metabolism (Pramanik et al. 2020). It is also important for controlling ROS and protecting plant cells from oxidative damage. Zinc is also vital for the synthesis of auxin, chlorophyll formation, and carbohydrate metabolism. Zinc deficiency can have a major effect on agricultural productivity and quality. The increasing issue of pest insects developing resistance to insecticides is a serious threat to both human health and agriculture. Numerous methods and pathways have been investigated for the synthesis of magnesium oxide (MgO), an essential inorganic element with a wide range of uses. Green approaches utilizing nontoxic extracts from neem leaves, Citrus limon leaves, and acacia gum are among the methods that have been researched (Avellan et al. 2019; Li et al. 2020; Pramanik et al. 2020; Manjunatha et al. 2019, 2020).

20.6.2 Control of Plant Pests

Fusarium wilt poses a significant threat to tomato and lettuce production in numerous countries due to substantial yield losses, prolonged soil survival of the fungus, and the emergence of resistant strains (Riseh et al. 2022). Although resistant cultivars and chemical treatments can mitigate the disease to some extent, the constant evolution of new pathogenic races and the limited effectiveness of chemicals present ongoing challenges (Ullah et al. 2023). In recent years, nanomaterials have emerged as a potential alternative for plant pathogen control. In a greenhouse setting, Ghidan et al. synthesized magnesium oxide (MgO) NPs and examined their effects on the green peach aphid (GPA). Furthermore, aqueous extracts of *Punica granatum* peels, *Olea europaea* leaves, and *Chamaemelum nobile* flowers were used to successfully produce NPs of copper oxide (CuO), zinc oxide (ZnO), magnesium hydroxide (MgOH), and magnesium oxide (MgO) (Wytinck et al. 2020). Screening of these synthesized bio NPs demonstrated their efficacy in increasing GPA mortality rates. There was no apparent metal accumulation in the plant fruits after glasshouse studies, according to an analysis of the accumulation of metal oxide NPs in green sweet pepper fruits and leaves (Romman et al. 2020). In comparison to the control, foliar spraying green pepper leaves with artificial MgOH NPs at concentrations ranging from 100 to 800 ppm produced better plant development, greener leaves, and higher-quality fruit (Li et al. 2020). Significant progress has been achieved in the creation of NPs by researchers using a variety of techniques, such as chemical, biological, and physical methods (Avellan et al. 2019). Green synthesis methods utilizing plant extracts offer advantages such as simplicity, convenience, environmental friendliness, and shorter reaction times. Nanomaterials synthesized through eco-friendly and green methods hold potential for enhancing agricultural practices, including fertilization processes, plant growth, and pest control, while minimizing environmental pollution (Li et al. 2020; Romman et al. 2020). The GPA is recognized as a significant pest on peach trees and is considered one of the most important agricultural pests globally (Ramadas et al. 2020; Li et al. 2020; Romman et al. 2020). This pest exhibits resistance to organophosphorus and carbamate insecticides due to the overproduction of insecticide-degrading carboxyl esterases. As a result, managing this pest has grown more difficult, especially because of the quick emergence of resistance to chemical pesticides such as pyrethroids, carbamates, and organophosphates (Li et al. 2020; Luo et al. 2022; Umar et al. 2021). Numerous physical and chemical techniques have been used to create nanomaterials such as copper oxide (CuO NPs), zinc oxide (ZnO NPs), magnesium hydroxide (MgOH NPs), and magnesium oxide (MgO NPs). Like the need to reduce the usage of ecologically hazardous materials such as pesticides, plant-based nanoparticle biosynthesis has attracted interest as a new field at the nexus of nanotechnology and biotechnology. It has been discovered that plants reduce metal ions more quickly than microbes, which results in the development of stable NPs (Pramanik et al. 2020; 2020; Wytinck et al. 2020).

20.6.3 Nanoinsecticidal Potential

Copper oxide nanoparticles (CuO NPs) have been synthesized via a variety of techniques, such as chemical reduction and precipitation (Brown and Morra 1997). In these synthesis procedures, plant aqueous extracts including carob leaves and citrus lemon juice have been used. Like this, a variety of techniques have been investigated for the manufacture of zinc oxide nanoparticles (ZnO NPs), including microwave synthesis, chemical pathways, precipitation techniques, and hydrolysis in polar organic solvents. It has been established that some plant extracts, including those from *Azadirachta indica, O. europaea*, and *Solanum nigrum* leaves, contribute to the green synthesis of ZnO NPs (Haider et al. 2022b). The production of magnesium hydroxide nanoparticles (MgOH NPs) and magnesium oxide nanoparticles (MgO NPs) has been described using a variety of methods, such as the hydrothermal approach, water-in-oil microemulsion, and microwave processes (Romman et al. 2020). Environmentally friendly techniques have been used to synthesize MgOH from nontoxic materials such as acacia gum, neem leaf extract, citrus limon leaf extract, *Brassica oleracea*, and *P. granatum* peels. Nanotechnology has several uses in agriculture, including fertilizers and insecticides based on nanotechnology that have measurable effects on plant development (Brown and Morra 1997). Furthermore, a viable substitute for viral vectors in molecular farming is the use of nanovectors (Haider et al. 2022b).

20.6.4 Antimicrobial Activity

Considering their unique properties and purposes, several nanomaterials, such as silver NPs, titanium dioxide (TiO_2), zinc oxide (ZnO), silicon oxide (SiO_2), magnesium oxide (MgO), gold, and silver, are used as antimicrobial agents in food packaging. Silver NPs, for instance, have been extensively used for their disinfectant and sterilizing properties, with applications even in space missions (Palaniswamy and Divyabharathi 2023). High-temperature stability and antibacterial activity against a wide range of microorganisms are demonstrated by gold NPs. Since silver NPs are efficient against bacteria such as *Staphylococcus aureus, L. monocytogenes*, and *E. coli*, the FDA has allowed their direct use as a disinfectant in commercial water (Ullah et al. 2019). Furthermore, antibacterial action is demonstrated by NPs such as zinc oxide and titanium dioxide, especially in packaging materials. Utilizing aqueous extracts from *P. granatum* peels, zinc oxide NPs have demonstrated effectiveness as antibacterial agents against a range of bacterial species. Overall, these nanomaterials offer promising antimicrobial characteristics, enhancing the safety and shelf life of food products (Haider et al. 2022b; Palaniswamy and Divyabharathi 2023).

20.6.5 Nanotechnology Application as Nano Fungicides

The application of nanosilver against the phytopathogen *Colletotrichum gloeosporioides* has been investigated in recent research (Ullah et al. 2019). In addition to their antibacterial qualities, additional NPs such as CNTs, Fe, Cu, Si, Al, Zn, ZnO, TiO_2, CeO_2, and Al_2O_3 have also been discovered to have the ability to impact plant development (Ijaz et al. 2023b). Under some circumstances, NPs may also affect the growth of *Pseudomonas putida* KT2440 and other beneficial soil bacteria. The creation of environmentally friendly pesticides is a growing area of research attention, with herbicides and pesticides based on NPs showing promise as viable substitutes for conventional chemical pesticides in protecting crops from disease (Farooq et al. 2021). Extensive investigations into nanoparticle-based systems could potentially reduce the need for intensive pesticide use in agriculture (Pérez-de-Luque and Rubiales 2009). When producing insecticides with NPs, the antifungal characteristics of the particles are very helpful. Because silver has so many benefits over other inorganic nanoparticle-based antibacterial agents – such as copper, zinc, gold, ZnO, Al_2O_3, and TiO_2 – researchers have given it a lot of attention (Carvalho et al. 2022).

20.6.6 Nanotechnology for Controlling Plant Virus

Naturally existing nanomaterials are plant viruses, especially spherical viruses (Shafie and Abdelkader 2023). The satellite tobacco necrosis virus is one of the tiniest plant viruses yet identified; its diameter is about 18 nm. The genomes of these viruses are made up of single- or double-stranded RNA or DNA that are covered in a protein sheath (Ullah et al. 2024). Their use in nanotechnology stems from their capacity to multiply, transport nucleic acid genomes to precise locations within host cells, and leave host cells in an orderly fashion. Recently, Young et al. conducted a comprehensive review on the use of plant viruses as bio-templates for nanomaterials and their applications (Xiang et al. 2023).

20.6.7 Nanotechnology in Food Packaging

Food businesses are paving the way in producing food items with improved nutritional value (Bibi et al. 2023). For example, they use highly impermeable packaging nanomaterials to strengthen the packaging and protect food from UV rays, prolonging the shelf life of food goods. Nanosensors play a crucial role in detecting chemicals, gases, and pathogens in food, leading to the emergence of "smart packaging" in modern parlance. However, some studies have indicated that there is reluctance among consumers to accept direct nanoparticle involvement in food production, primarily due to perceived risks. Hence, it is imperative to implement safety measures to mitigate these risks and ensure human safety (Ullah et al. 2019; Ijaz et al. 2023c; Xiang et al. 2023; Bibi et al. 2023).

20.7 Various Methods for Applying Nanoparticles to Crops

Nanoparticles are gaining traction in agriculture for diverse purposes, ranging from enhancing crop yield and nutrient uptake to bolstering disease resistance and soil health (Vuong 2019). Various methods exist for applying NPs to crops, each offering distinct advantages and considerations. Here are some commonly used methods:

- Foliar spraying: NPs are dispersed in water and directly applied to plant leaves. This approach facilitates rapid nanoparticle uptake through stomata and leaf cuticles.
- Seed coating: NPs are coated onto seeds to improve germination, seedling vigor, and nutrient absorption. Seed coating allows precise nanoparticle dosage control, uniform distribution, and protection against environmental degradation.
- Soil application: NPs can be mixed directly into the soil or applied as soil amendments. This method enables sustained nanoparticle release, offering long-term benefits to plants. However, higher nanoparticle concentrations may be necessary for effective root uptake.
- Hydroponic systems: Plants are grown in nutrient solutions instead of soil, with NPs added directly to the solution. This method provides precise control over nanoparticle concentration and plant uptake, commonly used in research settings.
- Nanofertilizers: NPs are combined with traditional fertilizers to create nanofertilizers, providing controlled nutrient release. These can be applied through foliar spraying, seed coating, or soil application, depending on the formulation.
- Biological methods: Some researchers explore biological approaches for nanoparticle delivery, leveraging beneficial microorganisms or plant-microbe interactions to facilitate uptake. These methods aim to improve delivery efficiency and minimize environmental impact.
- Electrostatic and magnetic field-assisted application: NPs are applied to crops with electrostatic or magnetic field assistance, enhancing adhesion to plant surfaces, and uptake efficiency.
- Nanoencapsulation: NPs are encapsulated within biocompatible materials to protect them from degradation and control release. This method allows targeted delivery to specific plant tissues or root zones through various application routes.

Each method presents unique advantages and limitations, influenced by factors such as nanoparticle type, crop species, and environmental considerations (Singh et al. 2021). Additionally, it is crucial to assess potential risks associated with nanoparticle use, including environmental impact and human/nontarget organism toxicity, and implement appropriate risk mitigation measures. Overall, the diverse array of nanoparticle application methods offers promising avenues for enhancing agricultural productivity while minimizing adverse effects. Continued research and careful consideration of application methods are essential for maximizing the benefits of nanoparticle technology in agriculture (Herdiana et al. 2022).

20.8 Challenges and Considerations in Nanoparticle Application on a Large Scale

Scaling up nanoparticle applications in agriculture poses several challenges and considerations that must be addressed to ensure effective and sustainable implementation:

Cost-effectiveness: One major challenge is ensuring the economic viability of nanoparticle-based agricultural technologies for widespread adoption. The production of NPs can be costly, and the expenses associated with application methods such as foliar spraying or seed coating must be carefully weighed against potential benefits in crop yield, nutrient efficiency, and disease management.

Environmental impact: Unintended environmental consequences can arise from nanoparticle release into soil, water, and air if not properly managed. It is crucial to assess potential risks to ecosystems, nontarget organisms, and human health. Strategies for minimizing environmental exposure and mitigating risks need to be developed and implemented.

Regulatory approval: Regulatory approval may be required for nanoparticle-based agricultural products before they can be commercialized and used extensively. Regulatory agencies typically demand comprehensive safety and efficacy data to evaluate the risks and benefits of nanoparticle technologies. Streamlining regulatory processes and ensuring transparent risk assessment are essential for facilitating market access.

Nanoparticle stability and compatibility: NPs must maintain stability and functional properties under diverse environmental conditions, including temperature, pH, and humidity. Compatibility with other agricultural inputs such as fertilizers, pesticides, and herbicides is crucial to ensure synergistic effects and minimize adverse interactions.

Uniform application: Achieving uniform distribution and uptake of NPs across large agricultural fields is vital for maximizing efficacy and minimizing waste. Developing efficient application methods that ensure consistent nanoparticle delivery to target plants or soil zones is essential.

Long-term effects: The long-term effects of nanoparticle exposure on soil health, plant physiology, and ecosystem functioning must be thoroughly studied to assess the sustainability of nanoparticle-based agricultural practices. Research should focus on understanding nanoparticle fate and transport in the environment and their potential accumulation in soil and plants over time.

Public acceptance and perception: Public perception of nanoparticle use in agriculture can influence consumer acceptance and market demand for nanoparticle-treated crops. Addressing concerns about safety, environmental impact, and ethical considerations through transparent communication and education initiatives is crucial for building trust and fostering acceptance.

Technology transfer and adoption: Facilitating technology transfer and adoption of nanoparticle-based agricultural innovations by farmers, particularly in developing countries, requires addressing barriers such as access to knowledge, infrastructure, and financial resources. Capacity-building initiatives, extension services, and public–private partnerships can help accelerate technology uptake and maximize its impact.

Addressing these challenges and considerations will necessitate interdisciplinary collaboration among scientists, policymakers, industry stakeholders, and civil society to ensure the responsible and sustainable deployment of nanoparticle technologies in agriculture on a large scale (Babakhani et al. 2022; Yang et al. 2021).

20.9 Impact of Nanoparticles on Crop Nutritional Quality

Ongoing research is investigating the impact of NPs on crop nutritional quality, balancing potential benefits with concerns. The key points on this topic include:

Nutrient uptake enhancement: NPs, especially those containing vital nutrients such as nitrogen, phosphorus, potassium, and micronutrients, are being studied for their potential to improve nutrient absorption by plants. Delivering nutrients in nanosized particles may increase efficiency, potentially boosting crop yield and nutritional value.

Micronutrient fortification: NPs offer a means to fortify crops with essential micronutrients such as iron, zinc, and selenium. This strategy aims to address micronutrient deficiencies in diets, particularly where staple crops lack these nutrients. Nanoparticle fortification methods show promise in enhancing crop nutritional quality without altering taste or appearance.

Antioxidant content: Some research suggests that NPs, particularly those with antioxidant properties such as selenium NPs, could enhance crop antioxidant levels. Antioxidants play a vital role in protecting plants from oxidative stress and may offer health benefits to consumers when consumed in the diet.

Potential toxicity concerns: Concerns persist regarding nanoparticle toxicity to both plants and humans. NPs may accumulate in plant tissues, affecting metabolism and nutritional composition. Furthermore, long-term effects on soil health and ecosystem dynamics are not fully understood.

Nutrient imbalance: Overuse of nanoparticle-based fertilizers or micronutrient fortification may disrupt nutrient balance in crops, affecting their nutritional quality and overall health. Careful dosage management is crucial to avoid adverse effects on growth, yield, or nutritional composition.

Environmental impact: Nanoparticle release into the environment via agricultural runoff or soil leaching could affect soil microbes, water quality, and ecosystems. Understanding nanoparticle fate and transport in agricultural systems is vital for assessing environmental impact and developing mitigation strategies.

Regulatory considerations: Regulatory frameworks for nanoparticle-based agricultural products vary, with limited guidelines on their impact on crop nutrition. Robust risk assessment protocols are necessary to evaluate safety and efficacy, ensuring compliance with regulatory standards.

In conclusion, while nanoparticle-based approaches show potential for enhancing crop nutrition, a thorough evaluation of benefits, risks, and unintended consequences is essential for safe and sustainable implementation in agriculture. Continued research efforts are needed to deepen our understanding of nanoparticle-plant interactions and their implications for crop and human health (Singh et al. 2021; Herdiana et al. 2022; Babakhani et al. 2022; Yang et al. 2021; Gomez et al. 2021).

20.10 Studies Demonstrating the Effectiveness of Nanoparticle-based Biofortification

Nanoparticle-based biofortification offers a promising avenue for enhancing the nutritional value of crops, particularly in combatting mineral deficiencies prevalent in human diets (Kapoor et al. 2022; Sanadya et al. 2023). Numerous studies have underscored the efficacy of this approach in augmenting the nutrient content of crops by precisely delivering essential micronutrients. A noteworthy application involves the utilization of iron (Fe) NPs to fortify staple food crops such as rice and wheat. Iron deficiency is a widespread nutritional concern globally, especially in regions heavily reliant on cereal-based diets (Singh et al. 2023). Research has indicated that foliar application of iron NPs can significantly boost iron accumulation in grains without compromising crop yield or quality. For instance, a study by Zohra et al. (2021) demonstrated that foliar spraying of iron NPs led to a remarkable 50% increase in iron concentration in rice grains compared to untreated plants, thereby enhancing iron bioavailability for human consumption. Likewise, Zn NPs have been explored for crop biofortification to address zinc deficiency prevalent in many developing nations. Various application methods, such as foliar spraying, seed priming, or soil application, can deliver zinc NPs to crops. Studies have revealed that zinc NPs facilitate zinc uptake by plants, resulting in elevated zinc levels in edible plant parts. For instance, (Bhat et al. 2018) found that foliar application of zinc NPs boosted zinc concentration in wheat grains by 30–40% compared to untreated plants, thereby enhancing the nutritional quality of wheat flour. Selenium (Se) NPs have also gained attention for their potential in biofortification due to their superior bioavailability relative to traditional selenium sources. Selenium is a crucial micronutrient with antioxidant properties essential for human health. Several studies have showcased the effective absorption of selenium NPs by plants, leading to heightened selenium content in crops. For example, research by Haider et al. (2022a) revealed that foliar application of selenium NPs resulted in a threefold increase in selenium concentration in lettuce leaves compared to untreated plants, highlighting the promise of selenium NPs for biofortification of leafy greens (Singh et al. 2021; Yang et al. 2021). These studies underscore the effectiveness of nanoparticle-based biofortification in enhancing crop nutritional quality. However, further research is warranted to optimize nanoparticle formulations, application methods, and dosages to maximize biofortification efficiency while minimizing potential environmental impacts and ensuring food safety (Khoshgoftarmanesh et al. 2010). Additionally, comprehensive field trials and long-term monitoring are imperative to assess the efficacy and sustainability of nanoparticle-based biofortification strategies under real-world agricultural conditions (Singh et al. 2023; Zohra et al. 2021).

20.11 Enhancement of Micronutrient Content in Crops and its Significance

Augmenting the micronutrient content in crops stands as a critical endeavor in combating global malnutrition and fostering better public health outcomes, particularly in regions grappling with widespread dietary deficiencies (Fageria et al. 2008). Micronutrients, encompassing vital vitamins and minerals such as iron, zinc, selenium, and vitamin A, play pivotal roles in numerous physiological functions and are indispensable for sustaining overall health and vitality. Iron deficiency ranks among the most prevalent nutritional insufficiencies worldwide, affecting millions, especially women and children. Iron is indispensable for oxygen transportation, energy metabolism, and immune system function. Inadequate dietary iron intake can precipitate anemia, impede cognitive development, and curtail work productivity (Singh et al. 2023; Zohra et al. 2021; Bhat et al. 2018). Through strategies such as biofortification, bolstering the iron content in staple crops such as rice, wheat, and maize holds promise for ameliorating the nutritional status of populations vulnerable to iron deficiency, thereby mitigating the burden of anemia and its attendant health ramifications (Fageria et al. 2008).

Similarly, zinc deficiency is pervasive, particularly in low-income settings where diets often lack zinc-rich fare. Zinc is crucial for immune function, growth, and enzyme-mediated metabolic processes. Insufficient zinc intake can compromise immune response, heighten susceptibility to infections, and impede physical and cognitive growth in children (Sheoran et al. 2021). By augmenting the zinc content in staple crops such as rice, wheat, and legumes via biofortification, it becomes feasible to combat zinc deficiency and mitigate its adverse health effects, thus fostering improved overall health outcomes and reducing healthcare expenditures. Selenium, with its antioxidant properties and involvement in immune function and thyroid hormone metabolism, emerges as another micronutrient of interest (Trenkel 2021). Selenium deficiency has been linked to elevated risks of chronic ailments such as cancer, cardiovascular disease, and cognitive decline (Singh et al. 2023). By enhancing the selenium content in crops through biofortification, it becomes plausible to bolster selenium levels in populations grappling with inadequate selenium intake, thereby curtailing the risk of chronic diseases and promoting enhanced overall health and longevity (Zohra et al. 2021). Vitamin A deficiency poses a substantial public health challenge in numerous developing nations, particularly in regions such as sub-Saharan Africa and South Asia, where diets often lack vitamin A-rich foods such as fruits, vegetables, and animal products. Vitamin A is indispensable for vision, immune function, and growth, and its dearth can precipitate vision impairment, heightened vulnerability to infections, and stunted growth and development in children. Biofortifying staple crops such as maize, sweet potato, and cassava with provitamin A carotenoids presents a viable approach to addressing vitamin A deficiency and its attendant health implications, thereby fostering enhanced overall health and well-being in at-risk populations (Singh et al. 2023; Zohra et al. 2021; Bhat et al. 2018; Haider et al. 2022a). In summation, augmenting the micronutrient content in crops through biofortification holds immense promise for tackling global malnutrition and advancing public health outcomes, particularly among vulnerable demographics. By bolstering the availability and consumption of nutrient-dense foods, biofortification can help alleviate the burden of nutritional deficiencies and contribute to the realization of global nutrition objectives, including the eradication of hunger and malnutrition. Nonetheless, comprehensive strategies that integrate biofortification with other nutrition interventions, alongside endeavors to promote dietary diversity and enhance food systems, are imperative for maximizing the impact of biofortification on population health and well-being (Haider et al. 2022a; Khoshgoftarmanesh et al. 2010; Fageria et al. 2008; Sheoran et al. 2021).

20.12 Nanoparticles and Stress Tolerance in Plants

20.12.1 NPs Mitigate Abiotic Stress Response

Nanoparticles serve a multifaceted role in agriculture, not only promoting plant growth but also safeguarding them against abiotic stressors (Trenkel 2021). Their small size and large surface area make them effective in binding toxic metals, reducing their availability and mitigating their harmful effects on plants. Abiotic stress includes a range of environmental conditions, such as salinity, alkalinity, temperature swings, mineral and metal toxicity, and drought (Zulfiqar and Ashraf 2021). By emulating the actions of antioxidant enzymes, NPs can behave as nanoenzymes, scavenging oxidative stress and preserving the regular operation of essential metabolic processes such as photosynthesis. Photosynthesis, pivotal for plant metabolism, is highly susceptible to stress conditions. Under stress, ATP, Rubisco, and photosystem II (PS II) become prime targets for photosynthetic machinery components (Singh et al. 2023). SiO_2 NPs improve plant transpiration rate, water usage efficiency, total chlorophyll content, and carbonic anhydrase activity in response to salt stress, according to research on defense mechanisms in plants against abiotic stress. In a similar vein, TiO_2 (anatase) decreases the rate of oxygen evolution in chloroplasts while changing photoreduction activity and inhibiting linolenic acid in the electron transport chain (Singh et al. 2021; Haider et al. 2022a; Fageria et al. 2008; Trenkel 2021).

Reactive oxygen species are produced by cell organelles in response to abiotic stress, which indicates that plants are under stress. Plants have enzymatic defense mechanisms against oxidative stress, but when they are overwhelmed by stress, negative effects manifest. By collecting osmolytes, activating certain genes, and supplying free nutrients and amino acids, NPs help to mitigate this type of stress (Ramadas et al. 2020; Shelenga et al. 2021; Su et al. 2019; Li et al. 2020).

Depending on variables such as exposure concentration, coating, surface charge, plant species, and growing circumstances, cerium nanoparticles (CeO NPs), sometimes referred to as nanoceria, have a variety of impacts on plant health (Xiang et al. 2023; Singh et al. 2023). Although nanoceria is widely used in the biomedical sector as an antioxidant, its effects on plant health – both beneficial and detrimental – have been well researched. Despite the potential adverse effects of NPs on plants, their positive aspects outweigh the negatives, offering health benefits to plants. Nanoceria can augment plant tolerance against various abiotic stressors by regulating essential pathways. For instance, nanoceria mitigate excessive ROS production under stress by leveraging its unique redox potential, transitioning between Ce^{3+} and Ce^{4+} oxidation states to act as an ROS scavenger. Furthermore, nanoceria mimic the activity of SOD and catalase, thereby scavenging hydrogen peroxide and other ROS. Its antioxidative scavenging effect extends to various stressors such as excess light, heat, and chilling (Xiang et al. 2023; Singh et al. 2023; Sheoran et al. 2021; Trenkel 2021; Zulfiqar and Ashraf 2021).

Nanoceria's ameliorative effect against abiotic stress includes excess heat, light, darkness, chilling, and salinity stress. It alleviates oxidative stress in *Brassica napus* induced by NaCl, enhancing physiological and biochemical properties such as plant biomass, chlorophyll content, and photosystem efficiency (Yang et al. 2021). Nanoceria maintains cytosolic Na^+/K^+ ratios, crucial for salinity stress tolerance, by modulating hydroxyl radical generation and potassium fluxes across the plasma membrane (Singh et al. 2023). Additionally, nanoceria exhibit effective drought-resistance activity by reducing the proportion of ROS and lipid peroxidation, which enhances pollen germination and carbon absorption rates (Gomez et al. 2021; Singh et al. 2023; Zohra et al. 2021). The heightened root development and viability of seedlings under salt stress in seeds coated with nanoceria is ascribed to disruptions in antioxidative enzyme pathways, ion binding, Ca^{2+} signaling, and terpene production. These results highlight the potential of nanoceria as a flexible tool to improve plant resistance to a range of abiotic stresses, supporting food security and sustainable agriculture (Bhat et al. 2018; Haider et al. 2022a).

Silicon nanoparticles (Si NPs) are also integral to enhancing plant resilience to stressors. While silicon is not classified as an essential element for plant survival, its presence can confer various benefits to plants, making it an asset in enhancing plant resilience to stressors (Fageria et al. 2008; Sheoran et al. 2021; Trenkel 2021). NPs containing silicon can interact directly or indirectly with plants, inducing morphological and physiological changes that promote stress tolerance and improve overall plant health. These changes include increased biomass, alterations in anatomy and physiology, tissue differentiation, activation of defense mechanisms, and better adaptation to stressful conditions. Studies have demonstrated the antistress effects of Si NPs on plants exposed to drought and salt stress (Trenkel 2021; Zulfiqar and Ashraf 2021). SiNPs application in Hawthorns under DS led to enhanced photosynthetic ability, RWC, and levels of chlorophylls, carotenoids, and proline, thus improving tolerance to moderate to severe drought conditions. Similarly, nanoSiO_2 application in saline environments enhanced osmotic adjustments by accumulating proline, maintaining ionic balance, enhancing the antioxidant system, and increasing phytopropanoid levels (Zulfiqar and Ashraf 2021; Kandhol et al. 2022). Under salt stress, SiO_2 NPs increased water use efficiency, stomatal conductance, and transpiration rate while reducing chlorophyll degradation, thereby conferring tolerance to salt stress. Moreover, SiNPs application resulted in improved epicuticular wax structure and thickness in salt-stressed strawberry plants and enhanced leaf area and fresh weight in sweet pepper plants. Concentration-dependent effects of SiO_2 NPs were observed in potato plants under salt stress, with lower concentrations exhibiting more effective stress amelioration compared to higher concentrations. These findings challenge the assumption of NPs solely having toxic effects and underscore the importance of understanding NP characteristics before their application in agriculture (Kandhol et al. 2022). Plant hormones, such as gibberellin, play a crucial role in plant responses to abiotic stress. Mesoporous silica nanoparticles (MSNs) loaded with abscisic acid demonstrated prolonged hormone release and improved drought-resistance in *A. thaliana* seedlings. Harnessing the potential of NPs to deliver plant hormones offers promising avenues for enhancing plant stress tolerance and resilience to changing environmental conditions (Singh et al. 2023; Bhat et al. 2018; Sheoran et al. 2021; Trenkel 2021; Zulfiqar and Ashraf 2021).

Titanium dioxide nanoparticles (TiO_2 NPs) function through different mechanisms than antioxidative properties. For instance, exposure to TiO_2 NPs increased chlorophyll content and biomass by enhancing antioxidant enzyme activity, reducing malondialdehyde (MDA) and hydrogen peroxide levels, and inducing the production of osmolytes such as proline and soluble sugars to maintain osmotic balance (Kandhol et al. 2022). NanoTiO_2 also triggered the expression of essential noncoding RNA implicated in abiotic stress resistance, as observed in tobacco seedlings (Ijaz et al. 2023a). Water scarcity poses a significant challenge to crop cultivation, but NanoTiO_2 can ameliorate hydration status by enhancing nitrate reductase (NR) enzyme activity, leading to increased osmolyte accumulation. This heightened NR activity promotes nitric oxide (NO) production, stimulating proline and glycine betaine synthesis. TiO_2 NPs exhibit both enzymatic and nonenzymatic defense mechanisms against stress, regulating enzymes such as glutamate hydrogenase and glutamine synthase to boost nutrient accumulation and essential oil production (Bhat et al. 2018; Haider et al. 2022a; Sheoran et al. 2021). In a greenhouse experiment, TiO_2-treated Moldavian balm plants under exacerbated salinity stress displayed enhanced essential oil content, suggesting protection against stress by modifying essential oil production profiles and compositions (Zulfiqar and Ashraf 2021; Kandhol et al. 2022).

20.12.2 Role of Nanoparticles in Enhancing Plant Resilience to Environmental Stressors

Nanoparticles are minute particles that can penetrate plant cells through various organs, including the cuticle, stomata, and root tips (Ijaz et al. 2023b). Factors such as plant species, developmental stage, and exposure conditions influence how NPs affect plant physiology and morphology. Once inside plants, NPs can alter processes such as germination, antioxidant activity, chlorophyll levels, and photosynthesis (Vyas et al. 2024). Transportation within plants occurs through passive and active mechanisms, involving movement through the endodermis or utilizing osmotic pressure and cell membrane interactions. In agriculture, nanomaterials offer diverse functions and applications (Figure 20.6). Nanofertilizers, typically around 30–40 nm in size, slowly release nutrients, enhance growth hormone production, and improve resistance to stresses (Singh et al. 2023; Zohra et al. 2021; Bhat et al. 2018). Nanopesticides containing metals such as silver and copper exhibit enhanced efficacy in pest and disease control, positively impacting crop yield. Chitosan-metal oxide NPs can enhance fertilizer uptake and enzyme activity related to nitrogen metabolism. While some NPs promote seed germination, shoot and root growth, and essential oil production, others may inhibit growth, with effects varying across plant species and NP types. Sustainable

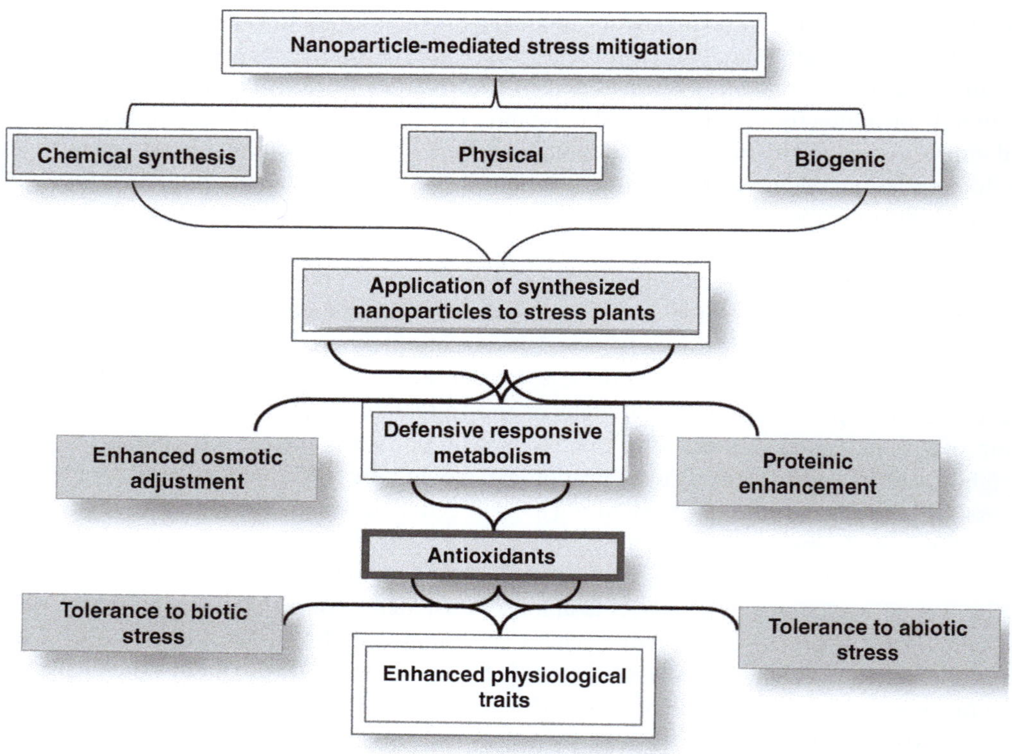

Figure 20.6 Role of nanoparticles in enhancing plant resilience to environmental stressors.

and responsible use of NPs in agriculture requires careful consideration of potential ecological and health implications (Haider et al. 2022a; Fageria et al. 2008; Trenkel 2021; Kandhol et al. 2022).

20.12.3 Application of NPs Under Salinity Conditions

The application of NPs is crucial for mitigating the adverse effects of salinity-induced abiotic stress on plants (Vyas et al. 2024). Studies have shown that the use of NPs such as Ag, Cu, ZnO, and CeO can enhance plant development and yield under salt stress conditions. For instance, Ag NPs have been found to improve germination percentage, shoot and root length, and seedling fresh and dry weight in *Lathyrus sativus*, thereby enhancing osmotic regulation and reducing the negative impacts of salinity (Vyas et al. 2024). Similarly, Cu NPs applied to soil have been shown to decrease oxidative stress in wheat, leading to significant improvements in plant growth and yield. Preapplication of Ag NPs to wheat seeds alters antioxidant enzyme activities, reducing oxidative damage and increasing salt stress tolerance (Behboudi et al. 2018). Additionally, ZnO NPs increase the dry weight of sunflowers under salt stress, while CeO NPs enhance physiological responses and biomass in salt-stressed canola. Ag NPs have also been effective in promoting seed germination and improving salt resistance in basil, *Satureja hortensis*, and cumin plants. Finally, Fe_3O_4 NPs have been reported to protect mint plants from oxidative stress-induced by high NaCl content (Vyas et al. 2024).

20.12.4 Application of NPs Under Drought Conditions

Drought poses a significant threat to crop productivity, but NPs offer a promising solution. Studies show that various NPs enhance antioxidant enzyme activity, modulate phytohormone levels, and improve physiological characteristics, mitigating DS in plants (Behboudi et al. 2018). For example, analcite NPs promote wheat germination and growth in hot, dry conditions, while ZnO NPs increase soybean seed germination rates in arid environments. Cu and Zn NPs boost antioxidant enzyme activity and relative moisture content in wheat, stabilizing photosynthetic pigment levels (Li et al. 2020; Singh et al. 2023; Zohra et al. 2021). SiO_2 NPs application in barley increases shoot length, water content, and reduces superoxide radical formation under DS. TiO_2 NPs foliar application in wheat counters drought-induced yield reduction effectively. Cu NPs enhance leaf water content, biomass, and pigment content in maize under arid conditions (Bhat et al. 2018; Haider et al. 2022a; Khoshgoftarmanesh et al. 2010). Si NPs show promise in mitigating DS in various plants, such as bananas, coriander, and cotton, as observed in increased moisture content in chickpea plants. Combined Si and selenium NPs improve growth, ion selectivity, and rice yield in saline conditions. Additionally, ZnO NPs reduce both Cd and DS in wheat, highlighting their potential to address multiple stressors simultaneously (Vyas et al. 2024; Behboudi et al. 2018).

20.12.5 Application of NPs Under Heavy Metal Stress Conditions

In environments contaminated with HMs, NPs application via soil or foliar methods shows promise in mitigating stress effects, enhancing plant development, and reducing oxidative stress-induced toxicity (Sheikhalipour et al. 2023). This approach significantly contributes to HM-contaminated environment remediation through various mechanisms. NP application reduces HM concentrations in soil, regulates genes involved in HM transfer within plants, enhances plant antioxidant systems, improves physiological functions, and stimulates protective substance production (Brown and Morra 1997; Xiang et al. 2023). For instance, silicon (Si) NPs application to maize under arsenic (As) stress reduces As's adverse effects on chlorophyll, protein content, and photosystem II quenching. Similarly, soil-applied titanium dioxide (TiO_2) NPs mitigate cadmium (Cd) toxicity in soybeans by enhancing physiological parameters and photosynthetic rates. Pea seedlings under chromium (Cr) stress benefit from silicon dioxide (SiO_2) NPs, increasing antioxidant enzyme activities and reducing oxidative stress effects (Singh et al. 2023; Zohra et al. 2021; Bhat et al. 2018). Foliar selenium (Se) NPs on Chinese cabbage under Cd stress increase biomass, chlorophyll, and antioxidant enzyme levels while reducing Cd and MDA in leaves (Haider et al. 2022a). Si NPs alleviate Cd stress in rice, and combining foliar ZnO NPs with soil biochar or using iron (Fe) NPs and biochar proves more effective against Cd stress, reducing Cd accumulation and improving nutrient uptake (Fageria et al. 2008). Additionally, foliar Fe NPs reduce Cd accumulation and toxicity while enhancing nutrient uptake in tomato plants. These findings underscore the potential of NPs in mitigating HM stress in plants and highlight their role in environmental remediation (Trenkel 2021; Sheikhalipour et al. 2023; Zahedi et al. 2023).

20.13 Regulatory and Ethical Considerations

The latter half of the twentieth century witnessed a rapid surge in technological advancements, marking the onset of the twenty-first century as the era of biotechnology, genetic engineering, and nanotechnology. While the development of biotechnology has stirred considerable debate, particularly concerning genetic research, there remains a lack of consistent legal frameworks governing this field. Responding to this, the EU has introduced patent protection for biotechnological inventions to regulate the sector. In contrast, nanotechnology innovations have not sparked as much controversy as biotechnology. Nevertheless, it is worth considering whether similar patent law solutions should be extended to nanotechnology in the future. Scholars argue that these two domains share common ground, as nanotechnology manipulates matter on a nanoscale, irrespective of whether it is organic or inanimate. This convergence has led to the emergence of nanobiotechnology, combining elements from both fields. Public acceptance plays a pivotal role in the advancement of new technologies. In Europe, recent debates have centered on nerve implants, offering new prospects for individuals with conditions such as Parkinson's disease or vision impairment.

However, concerns linger regarding the potential misuse of implants for nonmedical purposes. Recognizing these concerns, the European Commission has undertaken efforts to gauge societal expectations and apprehensions, emphasizing the importance of public awareness and engagement. Safety, utility, and public support are imperative for the success of nanotechnology. Thus, educating the public about its benefits, risks, and necessary precautions is paramount. Scientists play a crucial role in this endeavor by elucidating the principles and applications of nanotechnology to the public. In 2008, the European Commission introduced a Code of Conduct for Responsible Nanosciences and Nanotechnologies Research, encouraging adherence to ethical and sustainable practices in nanotechnology research. Several EU member states have initiated initiatives aimed at informing the public about nanotechnology developments. Projects like DEEPEN in the United Kingdom, Nanopodium in the Netherlands, and Nanosoc in Belgium aim to foster social acceptance of nanotechnology. Additionally, programs such as Nanologue in Germany strive to elucidate the benefits and ethical implications of nanotechnology while promoting dialogue among stakeholders. Addressing concerns about the long-term impact of NPs on health is crucial for garnering societal acceptance of nanotechnology. Social dialogue involving civil society representatives, scientists, and stakeholders is vital for addressing these concerns and fostering informed decision-making regarding nanotechnology development. Ultimately, social acceptance and dialogue are integral to shaping the future of nanotechnology.

20.14 Nanotechnology-based Agriculture Product

In recent times, nanotechnology has introduced various tools, such as nanodevices and nanocapsules, benefiting agriculture by improving disease diagnosis and treatment, targeted delivery of active ingredients, wastewater purification, and enhancing nutrient absorption in plants (Kumari et al. 2023). As the global population grows and agricultural challenges intensify due to climate change and pest pressures, ensuring food security becomes paramount, particularly in developing nations developing and disseminating fertilizers, insecticides, plant growth regulators, and genetically engineered plants with improved disease resistance, greater yields, and increased nutritional value, nano-based agricultural products seek to allay these worries (He et al. 2019; Ijaz et al. 2023c).

20.14.1 Nanofertilizers

Conventional chemical fertilizers are extensively used in agriculture to meet the growing demand for food due to population expansion, despite their limited nutrient uptake and high losses (Sharma et al. 2021a). Nanofertilizers, with particles smaller than 100 nm, offer a promising alternative by reducing nutrient loss and fertilizer application rates. The focus of nanofertilizer research is on micronutrients, which include copper, nickel, molybdenum, iron, zinc, manganese, and zinc. It has been demonstrated that nanocarbons such as chitosan nanoparticles (CS NPs), carbon nano-onions, and biochars improve crop growth and quality (Jyothi and Hebsur 2017). For example, applying 250 mg L^{-1} of Cu NPs to tomato plants improved fruit quality and bioactive chemicals considerably, while larger doses had negative effects. Combined application of silica NPs at concentrations improved marigold plant development and blooming traits. Similarly, ZnO NPs improved photosynthesis in cilantro at lower concentrations but affected nutritional components at higher doses, with lower toxicity compared to bulk and ionic forms (Kumar et al. 2023). However, CuO and ZnO NPs can pose risks to plant cells, soil biota, nitrogen fixation, and human health if not managed properly. Thus, while nanofertilizers hold transformative potential for agriculture, thorough studies on nanoparticle toxicity, accumulative effects, and biosafety are imperative before their commercialization (Campos et al. 2023; Camara et al. 2019).

20.14.2 Nano-based Plant Growth Regulators

Auxins, cytokinins, gibberellins, nitric oxide, abscisic acid, and ethylene are examples of plant growth hormones that are used in agriculture to increase crop productivity (Bibi et al. 2023). To provide optimum and prolonged distribution of these hormones while guarding against degradation processes, several nanoparticle systems are used. For example, CS NPs produce nitric oxide (NO) and containing the NO donor S-nitroso-mercaptosuccinic acid (S-nitroso-MSA) enhance NO bioactivity in maize plants under salinity stress (Hassanisaadi et al. 2022). Furthermore, tomato plants supplied with multi-walled CNTs in their soil produce twice as many flowers and fruits as control plants, suggesting that CNTs function as growth regulators for plants. Furthermore, compared to free GA3, gibberellic acid (GA3)-encapsulated poly (γ-glutamic acid) (γ-PGA) and chitosan (CS) polymer NPs exhibit enhanced biological activity, seed germination rates, and leaf area in *P. vulgaris* (Kumar et al. 2019; Bogue 2009).

20.14.3 Nanosensors

Among the many advantages that nanosensors provide to agriculture are real-time monitoring of crop growth, illnesses, insect infestations, and fertilizer efficiency. By enabling real-time monitoring of pesticides and fertilizers and so decreasing their excessive usage, their development makes a substantial contribution to sustainable agriculture (Javaid et al. 2021). Many nanomaterials have been used for pesticide detection, including quantum dots, CNT, gold nanoparticles (Au NP), and nanocomposites with polymers. Highly sensitive nanosensors that are based on enzymes such as choline oxidase (ChOx) and/or acetylcholinesterase (AChE) may identify minuscule levels of carbamate and organophosphorus pesticides (Riu et al. 2006). For soil nutrient detection, nanosensors utilizing materials like cysteamine-modified Au NP and graphene oxide have successfully detected nitrate in lower concentrations in field settings. Graphene-based nanoantennas integrated with CNTs can sense volatile organic compounds emitted by plants during insect attacks, facilitating insect attack monitoring. Nanosensors also detect soil water tension, pH, and nutrients, predict nitrogen intake, and identify pathogens in the soil in real-time. The advancement of nanosensor technology in agriculture holds promise for enhancing productivity while minimizing environmental impact and resource wastage (Sharma et al. 2021b; Wujcik et al. 2014; Yonzon et al. 2005; Cash and Clark 2010).

20.15 Future Directions and Conclusion

The application of NPs in crop biofortification is a significant advancement in agricultural biotechnology, offering innovative ways to enhance the nutritional content of food crops and combat widespread nutritional deficiencies. Key advancements include enhanced targeting and delivery systems, which improve nutrient assimilation by directing nutrients precisely where needed, overcoming physiological barriers effectively. Nanoparticle safety and regulation are imperative, requiring thorough investigations into long-term impacts on the environment and human health and strong regulatory frameworks to ensure responsible deployment. Integration with precision agriculture optimizes nutrient delivery through drones and IoT-based sensors, reducing waste and environmental footprint. Smart NPs respond to environmental stimuli, enhancing biofortification efficiency by releasing nutrients at the optimal time. Synergy with genetic and omic technologies aligns NP design with crop genetics and molecular profiles, customizing nutrient enhancement to meet precise goals. Sustainable production of NPs emphasizes eco-friendly synthesis methods, including biological production using plants, fungi, or bacteria. This venture holds the potential to significantly impact public health, especially in regions with limited access to diverse diets. Collaboration among scientists, policymakers, and stakeholders is vital to address ethical, environmental, and health-related concerns, ensuring NPs contribute to global malnutrition eradication.

References

Ahmad, S., Munir, S., Zeb, N. et al. (2019). Green nanotechnology: a review on green synthesis of silver nanoparticles—an ecofriendly approach. *International Journal of Nanomedicine* 14: 5087–5107.

Ahmad, Y., Haider, S., Iqbal, J. et al. (2022). The mechanisms of genome editing technologies in crop plants. In: *Principles and Practices of OMICS and Genome Editing for Crop Improvement*, 295–313. Cham: Springer International Publishing.

Ahmad, Y., Haakim, Z., Iqbal, J. et al. (2024a). Technological innovations for abiotic stress resistance in horticultural crops. In: *OMICs-based Techniques for Global Food Security* 233–244.

Ahmad, Y., Haider, S., Iqbal, J. et al. (2024b). In-silico analysis and transformation of OsMYB48 transcription factor driven by CaMV35S promoter in model plant—*Nicotiana tabacum* L. conferring abiotic stress tolerance. *GM Crops & Food* 15 (1): 130–149.

Ali, I., Salah, K.B.H., Sher, H. et al. (2022). Drought stress enhances the efficiency of floral dip method of Agrobacterium-mediated transformation in *Arabidopsis thaliana*. *Brazilian Journal of Biology* 84 (1): e259326.

Athar, T., Khan, M.K., Pandey, A. et al. (2020). Biofortification and the involved modern approaches. *Journal of Elementology* 25 (2): 717–731.

Avellan, A., Yun, J., Morais, B.P. et al. (2021). Critical review: role of inorganic nanoparticle properties on their foliar uptake and in *planta* translocation. *Environmental Science & Technology* 55 (20): 13417–13431.

Avellan, A., Yun, J., Zhang, Y. et al. (2019). Nanoparticle size and coating chemistry control foliar uptake pathways, translocation, and leaf-to-rhizosphere transport in wheat. *ACS Nano* 13 (5): 5291–5305.

Avnee Sood, S., Chaudhary, D.R., Jhorar, P. et al. (2023). Biofortification: an approach to eradicate micronutrient deficiency. *Frontiers in Nutrition* 10: 1233070.

Azeem, I., Adeel, M., Ahmad, M.A. et al. (2021). Uptake and accumulation of nano/microplastics in plants: a critical review. *Nanomaterials* 11 (11): 2935.

Babakhani, P., Phenrat, T., Baalousha, M. et al. (2022). Potential use of engineered nanoparticles in ocean fertilization for large-scale atmospheric carbon dioxide removal. *Nature Nanotechnology* 17 (12): 1342–1351.

Banerjee, K., Pramanik, P., Maity, A. et al. (2019). Methods of using nanomaterials to plant systems and their delivery to plants (mode of entry, uptake, translocation, accumulation, biotransformation and barriers). In: *Advances in Phytonanotechnology*, 123–152. Massachusetts: Academic Press.

Behboudi, F., Tahmasebi Sarvestani, Z., Kassaee, M.Z. et al. (2018). Improving growth and yield of wheat under drought stress via application of SiO_2 nanoparticles. *Journal of Agricultural Science and Technology* 20 (7): 1479–1492.

Bhat, J.S., Patil, B.S., Hariprasanna, K. et al. (2018). Genetic enhancement of micronutrient content in cereals. *Sabrao Journal of Breeding and Genetics* 50 (3): 373–429.

Bibi, S., Ullah, R., Burni, T. et al. (2023). Impact of resorcinol and biochar application on the growth attributes, metabolite contents, and antioxidant systems of tomato (Lycopersicon esculentum Mill.). *ACS Omega* 8 (48): 45750–45762.

Bogue, R. (2009). Nanosensors: a review of recent research. *Sensor Review* 29 (4): 310–315.

Bouis, H.E., Saltzman, A., and Birol, E. (2019). Improving nutrition through biofortification. In: *Agriculture for Improved Nutrition: Seizing the Momentum* 47–57. Wallingford UK: CAB International.

Brown, P.D. and Morra, M.J. (1997). Control of soil-borne plant pests using glucosinolate-containing plants. *Advances in Agronomy* 61: 167–231.

Camara, M.C., Campos, E.V.R., Monteiro, R.A et al. (2019). Development of stimuli-responsive nano-based pesticides: emerging opportunities for agriculture. *Journal of Nanobiotechnology* 17 (1): 1–19.

Campos, E.V., do ES Pereira, A., Aleksieienko, I. et al. (2023). Encapsulated plant growth regulators and associative microorganisms: nature-based solutions to mitigate the effects of climate change on plants. *Plant Science* 331; 111688.

Carvalho, R., Jones, J.B. and Paret, M.L. (2022). Utility of nanoparticles in management of plant viruses. *Nanotechnology-Based Sustainable Alternatives for the Management of Plant Diseases* 233–241.

Cash, K.J. and Clark, H.A. (2010). Nanosensors and nanomaterials for monitoring glucose in diabetes. *Trends in Molecular Medicine* 16 (12): 584–593.

Connorton, J.M. and Balk, J. (2019). Iron biofortification of staple crops: lessons and challenges in plant genetics. *Plant and Cell Physiology* 60 (7): 1447–1456.

Dávila-Rangel, I.E., Leija-Martínez, P., Medrano-Macías, J. et al. (2019). Iodine biofortification of crops. *Nutritional Quality Improvement in Plants* 79–113.

Duborská, E., Urík, M., and Šeda, M. (2020). Iodine biofortification of vegetables could improve iodine supplementation status. *Agronomy* 10 (10): 1574.

Dundore-Arias, J.P., Felice, L., Dill-Macky, R. et al. (2019). Carbon amendments induce shifts in nutrient use, inhibitory, and resistance phenotypes among soilborne Streptomyces. *Frontiers in Microbiology* 10: 420796.

Fageria, N.K., Baligar, V.C., and Li, Y.C. (2008). The role of nutrient efficient plants in improving crop yields in the twenty first century. *Journal of Plant Nutrition* 31 (6): 1121–1157.

Farooq, T., Adeel, M., He, Z. et al. (2021). Nanotechnology and plant viruses: an emerging disease management approach for resistant pathogens. *ACS Nano* 15 (4): 6030–6037.

Gomez, A., Narayan, M., Zhao, L. et al. (2021). Effects of nano-enabled agricultural strategies on food quality: current knowledge and future research needs. *Journal of Hazardous Materials* 401: 123385.

Haider, S., Iqbal, J., Naseer, S. et al. (2021a). Molecular mechanisms of plant tolerance to heat stress: current landscape and future perspectives. *Plant Cell Reports* 40: 2247–2271.

Haider, S., Iqbal, J., Shaukat, M. et al. (2021b). The epigenetic chromatin-based regulation of somatic heat stress memory in plants. *Plant Gene* 27: 100318.

Haider, S., Iqbal, J., Naseer, S. et al. (2022a). Unfolding molecular switches in plant heat stress resistance: a comprehensive review. *Plant Cell Reports* 41 (3): 775–798.

Haider, S., Raza, A., Iqbal, J. et al. (2022b). Analyzing the regulatory role of heat shock transcription factors in plant heat stress tolerance: a brief appraisal. *Molecular Biology Reports* 49 (6): 5771–5785.

Hassanisaadi, M., Barani, M., Rahdar, A. et al. (2022). Role of agrochemical-based nanomaterials in plants: biotic and abiotic stress with germination improvement of seeds. *Plant Growth Regulation* 97 (2): 375–418.

He, X., Deng, H., and Hwang, H.M. (2019). The current application of nanotechnology in food and agriculture. *Journal of Food and Drug Analysis* 27 (1): 1–21.

Herdiana, Y., Wathoni, N., Shamsuddin, S. et al. (2022). Scale-up polymeric-based nanoparticles drug delivery systems: development and challenges. *OpenNano* 7: 100048.

Hossain, A., Skalicky, M., Brestic, M. et al. (2021). Selenium biofortification: roles, mechanisms, responses and prospects. *Molecules* 26 (4): 881.

Ijaz, M., Khan, F., Ahmed, T. et al. (2023a). Nanobiotechnology to advance stress resilience in plants: current opportunities and challenges. *Materials Today Bio* 22: 100759.

Ijaz, S., Iqbal, J., Abbasi, B. A. et al. (2023b). Biofortification: lessons from the past and strategies for future food security. In: *Legumes Biofortification*, 521–545. Cham: Springer International Publishing.

Ijaz, S., Iqbal, J., Abbasi, B.A. et al. (2023c). Rosmarinic acid and its derivatives: current insights on anticancer potential and other biomedical applications. *Biomedicine & Pharmacotherapy* 162: 114687.

Javaid, M., Haleem, A., Singh, R.P. et al. (2021). Exploring the potential of nanosensors: a brief overview. *Sensors International* 2 (3): 100130.

Jyothi, T.V. and Hebsur, N.S. (2017). Effect of nanofertilizers on growth and yield of selected cereals—a review. *Agricultural Reviews* 38 (2): 112–120.

Kandhol, N., Jain, M. and Tripathi, D.K. (2022). Nanoparticles as potential hallmarks of drought stress tolerance in plants. *Physiologia Plantarum* 174 (2): e13665.

Kapoor, P., Dhaka, R.K., Sihag, P. et al. (2022). Nanotechnology-enabled biofortification strategies for micronutrients enrichment of food crops: current understanding and future scope. *NanoImpact* 26: 100407.

Khan, A., Ahmad, M., Sultan, A. et al. (2024). Contribution to the pollen morphology of Astragalus L. section Aegacantha Bunge (Galegeae-Fabaceae) and its systematic significance. *Palynology* 48 (1): 2264352.

Khan, N.A., Khan, S.U., Ahmed, S. et al. (2019). Applications of nanotechnology in water and wastewater treatment: a review. *Asian Journal of Water, Environment and Pollution* 16 (4): 81–86.

Khoshgoftarmanesh, A.H., Schulin, R., Chaney, R.L. et al. (2010). Micronutrient-efficient genotypes for crop yield and nutritional quality in sustainable agriculture. A review. *Agronomy for Sustainable Development* 30 (1): 83–107.

Kiran, A., Wakeel, A., Mahmood, K. et al. (2022). Biofortification of staple crops to alleviate human malnutrition: contributions and potential in developing countries. *Agronomy* 12 (2): 452.

Kumar, S., Nehra, M., Dilbaghi, N. et al. (2019a). Nano-based smart pesticide formulations: emerging opportunities for agriculture. *Journal of Controlled Release* 294: 131–153.

Kumar, S., Palve, A., Joshi, C. et al. (2019b). Crop biofortification for iron (Fe), zinc (Zn) and vitamin A with transgenic approaches. *Heliyon* 5 (6): e01914.

Kumar, N., Samota, S.R., Venkatesh, K. et al. (2023a). Global trends in use of nano-fertilizers for crop production: advantages and constraints–a review. *Soil and Tillage Research* 228: 105645.

Kumari, R., Suman, K., Karmakar, S. et al. (2023b). Regulation and safety measures for nanotechnology-based agri-products. *Frontiers in Genome Editing* 5: 1200987.

Lal, M.K., Kumar, A., Kardile, H.B. et al. (2020). Biofortification of vegetables. In: *Advances in Agri-Food Biotechnology*, 105–129. Singapore: Springer.

Li, Y., Zhu, N., Liang, X. et al. (2020). A comparative study on the accumulation, translocation and transformation of selenite, selenate, and SeNPs in a hydroponic-plant system. *Ecotoxicology and Environmental Safety* 189: 109955.

Lodhi, B.S., Dhakad, A., Parte, D. et al. (2023). Harnessing nature's bounty: the promise of bio-fortification in battling global malnutrition. *The Science World* 3 (11): 2804–2809.

Luo, Y., Li, L., Feng, Y. et al. (2022). Quantitative tracing of uptake and transport of submicrometre plastics in crop plants using lanthanide chelates as a dual-functional tracer. *Nature Nanotechnology* 17 (4): 424–431.

Lv, J., Christie, P. and Zhang, S. (2019). Uptake, translocation, and transformation of metal-based nanoparticles in plants: recent advances and methodological challenges. *Environmental Science: Nano* 6 (1): 41–59.

Malik, K.A. and Maqbool, A. (2020). Transgenic crops for biofortification. *Frontiers in Sustainable Food Systems* 4: 571402.

Manjunatha, R.L., Naik, D., and Usharani, K.V. (2019). Nanotechnology application in agriculture: a review. *Journal of Pharmacognosy and Phytochemistry* 8 (3): 1073–1083.

Masuda, H., Aung, M.S., Kobayashi, T. et al. (2020). Iron biofortification: the gateway to overcoming hidden hunger. In: *The Future of Rice Demand: Quality Beyond Productivity*, 149–177. Cham: Springer.

Miyazawa, T., Itaya, M., Burdeos, G.C. et al. (2021). A critical review of the use of surfactant-coated nanoparticles in nanomedicine and food nanotechnology. *International Journal of Nanomedicine* 16: 3937–3999.

Ogino, S., Nowak, J.A., Hamada, T. et al. (2019). Insights into pathogenic interactions among environment, host, and tumor at the crossroads of molecular pathology and epidemiology. *Annual Review of Pathology: Mechanisms of Disease* 14: 83–103.

Okwuonu, I.C., Narayanan, N.N., Egesi, C.N. et al. (2021). Opportunities and challenges for biofortification of cassava to address iron and zinc deficiency in Nigeria. *Global Food Security* 28 (1): 100478.

Palaniswamy, R. and Divyabharathi, R. (2023). Phytochemicals and pharmacological activities in *Abutilon indicum* L. plant a review. *International Journal of Pharmaceutical Sciences* 1 (12): 1–1.

Pérez-de-Luque, A. and Rubiales, D. (2009). Nanotechnology for parasitic plant control. *Pest Management Science: Formerly Pesticide Science* 65 (5): 540–545.

Praharaj, S., Skalicky, M., Maitra, S. et al. (2021). Zinc biofortification in food crops could alleviate the zinc malnutrition in human health. *Molecules* 26 (12): 3509.

Pramanik, P., Krishnan, P., Maity, A. et al. (2020). Application of nanotechnology in agriculture. *Environmental Nanotechnology* 4: 317–348.

Ramadas, S., Vellaichamy, S., Ramasundaram, P. et al. (2020). Biofortification for enhancing nutritional outcomes and policy imperatives. In: *Wheat and Barley Grain Biofortification*, 309–327. Darya Ganj: Woodhead Publishing.

Riseh, R.S., Tamanadar, E., Hajabdollahi, N. et al. (2022). Chitosan microencapsulation of rhizobacteria for biological control of plant pests and diseases: recent advances and applications. *Rhizosphere* 23: 100565.

Riu, J., Maroto, A., and Rius, F.X. (2006). Nanosensors in environmental analysis. *Talanta* 69 (2): 288–301.

Romman, M., Rainaz, P., Wali, K. et al. (2020). Hypoglycemic activity and ALP lowering role of Hedera nepalensis K. Koch leaves in Alloxane monohydrate induced diabetic rabbits. *Bioscience Research* 17: 2264–2268.

Sanadya, S.K., Sharma, R., Manuja, S. et al. (2023). Nanotechnology-mediated agronomic biofortification: a way forward. In: *Advances in Nanotechnology for Smart Agriculture* 135–159. Florida: CRC Press.

Shafie, R.M. and Abdelkader, H.S. (2023). Viral disease control in strawberry plants: a potential impact of green nanotechnology. *Malaysian Journal of Microbiology* 19 (3): 222.

Shahane, A.A. and Shivay, Y.S. (2022). Agronomic biofortification of crops: current research status and future needs. *Indian Journal of Fertilisers* 18 (2): 164–179.

Sharma, P., Pandey, V., Sharma, M.M.M. et al. (2021a). A review on biosensors and nanosensors application in agroecosystems. *Nanoscale Research Letters* 16: 1–24.

Sharma, S., Rana, V.S., Pawar, R. et al. (2021b). Nanofertilizers for sustainable fruit production: a review. *Environmental Chemistry Letters* 19 (2): 1693–1714.

Sheikhalipour, M., Gohari, G., Esmaielpour, B. et al. (2023). Melatonin and TiO_2 NPs application-induced changes in growth, photosynthesis, antioxidant enzymes activities and secondary metabolites in stevia (Stevia rebaudiana bertoni) under drought stress conditions. *Journal of Plant Growth Regulation* 42 (3): 2023–2040.

Shelenga, T.V., Kerv, Y.A., Perchuk, I.N. et al. (2021). The potential of small grains crops in enhancing biofortification breeding strategies for human health benefit. *Agronomy* 11 (7): 1420.

Sheoran, P., Grewal, S., Kumari, S. et al. (2021). Enhancement of growth and yield, leaching reduction in *Triticum aestivum* using biogenic synthesized zinc oxide nanofertilizer. *Biocatalysis and Agricultural Biotechnology* 32: 101938.

Singh, A., Rajput, V.D., Chakrawarti, N. et al. (2023). Nano-enabled approaches for biofortification strategies to enhance crop production with micronutrient enrichment. In: *Nano-Biofortification for Human and Environmental Health*, 1–14. Cham: Springer International Publishing.

Singh, R.P., Handa, R., and Manchanda, G. (2021). Nanoparticles in sustainable agriculture: an emerging opportunity. *Journal of Controlled Release* 329: 1234–1248.

Su, Y., Ashworth, V., Kim, C. et al. (2019). Delivery, uptake, fate, and transport of engineered nanoparticles in plants: a critical review and data analysis. *Environmental Science: Nano* 6 (8): 2311–2331.

Sun, H., Lei, C., Xu, J. et al. (2021). Foliar uptake and leaf-to-root translocation of nanoplastics with different coating charges in maize plants. *Journal of Hazardous Materials* 416: 125854.

Szerement, J., Szatanik-Kloc, A., Mokrzycki, J. et al. (2022). Agronomic biofortification with Se, Zn, and Fe: an effective strategy to enhance crop nutritional quality and stress defense—a review. *Journal of Soil Science and Plant Nutrition* 22 (1): 1129–1159.

Thind, S., Hussain, I., Ali, S., Hussain, S. et al. (2020). Physiological and biochemical bases of foliar silicon-induced alleviation of cadmium toxicity in wheat. *Journal of Soil Science and Plant Nutrition* 20: 2714–2730.

Trenkel, M.E. (2021). *Slow-and Controlled-Release and Stabilized Fertilizers: An Option for Enhancing Nutrient Use Efficiency in Agriculture*. Paris, France: International Fertilizer Industry Association (IFA).

Ulhassan, Z., Bhat, J.A., Zhou, W. et al. (2022). Attenuation mechanisms of arsenic-induced toxicity and its accumulation in plants by engineered nanoparticles: a review. *Environmental Pollution* 302: 119038.

Ullah, H., Li, X., Peng, L. et al. (2020). In vivo phytotoxicity, uptake, and translocation of PbS nanoparticles in maize (*Zea mays* L.) plants. *Science of the Total Environment* 737: 139558.

Ullah, S., Begam, L., Ullah, Z. et al. (2019). Ecological study of different community site from district Karak Khyber Pakhtun Khwa Pakistan. *International Journal of Research Agronomy* 2 (1): 12–16.

Ullah, S., Shakir, L., Subhan, G. et al. (2024). Phytodiversity and conservation assessment of ethnobotanically significant flora in Khall Hagram Dara, Lower Dir, Khyber Pakhtunkhwa, Pakistan. *Plant Protection* 8 (1): 143–162.

Ullah, Z. (2017). Estimation of genetic diversity in the germplasm of *Brassica Campestris* (L.) using morphological and biochemical markers. *Journal of Botany* 1 (1): 45–59.

Ullah, Z., Gul, F., Iqbal, J. et al. (2023a). Biogenic synthesis of multifunctional silver oxide nanoparticles (Ag_2ONPs) using parieteria alsinaefolia delile aqueous extract and assessment of their diverse biological applications. *Microorganisms* 11 (4): 1069.

Ullah, Z., Iqbal, J., Abbasi, B.A. et al. (2023b). Assessment of Gus expression induced by anti-sense Os PPO gene promoter and antioxidant enzymatic assays in response to drought and heavy metal stress in transgenic *Arabidopsis thaliana*. *Sustainability* 15 (17): 12783.

Ullah, Z., Iqbal, J., Abbasi, B.A. et al. (2023c). How genome editing can be helpful in the biofortification of legumes. In: *Legumes Biofortification*, 207–232. Cham: Springer International Publishing.

Ullah, R., Ullah, Z., Iqbal, J. et al. (2023d). Aspartic acid-based nano-copper induces resilience in *Zea mays* to applied lead stress via conserving photosynthetic pigments and triggering the antioxidant biosystem. *Sustainability* 15 (16): 12186.

Umar, W., Hameed, M.K., Aziz, T. et al. (2021). Synthesis, characterization and application of ZnO nanoparticles for improved growth and Zn biofortification in maize. *Archives of Agronomy and Soil Science* 67 (9): 1164–1176.

Virk, P.S., Andersson, M.S., Arcos, J. et al. (2021). Transition from targeted breeding to mainstreaming of biofortification traits in crop improvement programs. *Frontiers in Plant Science* 12: 703990.

Vuong, L.D. (2019). Nanoparticles for the improved crop production. In: *Nanotechnology for Agriculture: Crop Production & Protection*, 85–106. Singapore: Springer.

Vyas, T.K., More, B., and Mehta, M.P. (2024). Improving stress resilience in plants by nanoparticles. In: *Improving Stress Resilience in Plants* 73–96. Massachusetts: Academic Press.

Wang, F., Feng, X., Liu, Y. et al. (2022). Micro (nano) plastics and terrestrial plants: up-to-date knowledge on uptake, translocation, and phytotoxicity. *Resources, Conservation and Recycling* 185: 106503.

Wujcik, E.K., Wei, H., Zhang, X. et al. (2014). Antibody nanosensors: a detailed review. *RSC Advances* 4 (82): 43725–43745.

Wytinck, N., Manchur, C. L., Li, V.H. et al. (2020). dsRNA uptake in plant pests and pathogens: insights into RNAi-based insect and fungal control technology. *Plants* 9 (12): 1780.

Xiang, H., Meng, J., Shao, W. et al. (2023). Plant protein-based self-assembling core–shell nanocarrier for effectively controlling plant viruses: evidence for nanoparticle delivery behavior, plant growth promotion, and plant resistance induction. *Chemical Engineering Journal* 464: 142432.

Yadav, D.N., Bansal, S., Tushir, S. et al. (2020). Advantage of biofortification over fortification technologies. In: *Wheat and Barley Grain Biofortification*, 257–273. Darya Ganj: Woodhead Publishing.

Yang, G., Yuan, H., Ji, H. et al. (2021). Effect of ZnO nanoparticles on the productivity, Zn biofortification, and nutritional quality of rice in a life cycle study. *Plant Physiology and Biochemistry* 163: 87–94.

Yonzon, C.R., Stuart, D.A., Zhang, X. et al. (2005). Towards advanced chemical and biological nanosensors—an overview. *Talanta* 67 (3): 438–448.

Yusefi-Tanha, E., Fallah, S., Rostamnejadi, A. et al. (2020). Root system architecture, copper uptake and tissue distribution in soybean (*Glycine max* (L.) Merr.) grown in copper oxide nanoparticle (CuONP)-amended soil and implications for human nutrition. *Plants* 9 (10): 1326.

Zahedi, S.M., Hosseini, M.S., Hoveizeh, N.F. et al. (2023). Comparative morphological, physiological and molecular analyses of drought-stressed strawberry plants affected by SiO_2 and SiO_2-NPs foliar spray. *Scientia Horticulturae* 309 (21): 111686.

Zhang, P., Ma, Y., Xie, C. et al. (2019). Plant species-dependent transformation and translocation of ceria nanoparticles. *Environmental Science: Nano* 6 (1): 60–67.

Zhou, X., Yang, J., Kronzucker, H.J. et al. (2020). Selenium biofortification and interaction with other elements in plants: a review. *Frontiers in Plant Science* 11: 586421.

Zohra, E., Ikram, M., Omar, A.A. et al. (2021). Potential applications of biogenic selenium nanoparticles in alleviating biotic and abiotic stresses in plants: a comprehensive insight on the mechanistic approach and future perspectives. *Green Processing and Synthesis* 10 (1): 456–475.

Zulfiqar, F. and Ashraf, M. (2021). Nanoparticles potentially mediate salt stress tolerance in plants. *Plant Physiology and Biochemistry* 160: 257–268.

21

Zinc Biofortification in Rice – From Conventional Breeding to Biotechnological Approaches

Swapan K. Tripathy[1*]

[1] *Department of Agricultural Biotechnology, College of Agriculture, OUAT, Bhubaneswar, India*

*Corresponding author: swapankumartripathy@gmail.com

21.1 Introduction

Rice (*Oryza sativa* L., 2n = 24, family Poaceae) is grown in 120 countries covering 162.06 million hectares with a production of 496.1 million tons in 2019–2020 (Shahbandeh 2021). It feeds over half of the global population but serves as a poor source of essential micronutrients, including zinc (Zn) (Bouis and Welch 2010). The present global population is 7.8 billion (double that in 1970), and it is expected to be 9.7 billion by 2050. Globally, about 1 billion people are chronically hungry (FAO 2017), and more than one-quarter of the world population is at a risk of Zn deficiency (Gernand et al. 2016) which includes 17.3% of the population with acute low intake of Zn due to undernutrition. The situation is more alarming in developing countries than their developed counterparts. Therefore, increasing the production of nutritious (biofortified) rice seems to be a lasting solution to food and nutritional security. HarvestPlus, a joint venture of the International Center for Tropical Agriculture (CIAT) and the International Food Policy Research Institute (IFPRI), works closely with 40 partner countries to develop biofortified crops to address micronutrient deficiencies, including Zn. In addition, several national and international research centers are in place to combat the hidden hunger from nutritional perspectives.

Zn is a trace element, and it is neither synthesized by the body nor can be stored. It serves as a cofactor of more than 300 enzymes (Roohani et al. 2013) and acts as a coordinating element for more than 1000 transcription factors involved in cellular metabolism. The biggest class of zinc-requiring proteins is the zinc-finger transcription factors. In animals, Zn enhances the body's capacity to absorb iron (Fe) (Chang et al. 2010). Zn deficiency leads to loss of immunity to a variety of diseases, stunted growth, hair loss, skin lesions, loss of appetite and taste, impaired learning ability (brain function), reduced sexual appetite, improper wound healing, DNA damage, and even cancer. Zn also reduces the severity of diarrhea, colds (pneumonia), diabetes, sickle cell anemia, and kidney disorders (Figure 21.1). Low intake of Zn is also linked to pregnancy and childbirth complications, lower birth weight, and other fetal effects. Globally, 45% of child deaths below five years are associated with micronutrient deficiency (WHO 2020). Vegetarians and specifically children, pregnant and breastfeeding women, and patients after surgery need Zn supplements. Rice being the staple food, Zn biofortified rice in the food chain can highly contribute to human nutrition. Besides, Zn biofortification in rice is reported to increase seed yield (Kumar et al. 2009), support plant growth, and improve immunity to a variety of biotic and abiotic stresses, including Zn deficiency (Figure 21.1). Most of the rice fields have low available Zn. Zn-efficient rice varieties grown under such conditions can accumulate more Zn in rice grains than sensitive genotypes.

Rice has poor bioavailability of Zn compared to other cereals (Myers et al. 2014). Low phytic acid content is indeed needed to improve the bioavailability of micronutrients, including Zn in rice. Bran, along with the aleurone cell layer, and part of the embryo are removed during milling and polishing, resulting in a loss of about 40% of Zn in brown rice. Thus, the endosperm being the major source of Zn (~57%) in milled rice, the higher accumulation of endosperm-specific Zn in rice grain is always a preferred option. The estimated average requirement (EAR) of Zn is 12 mg day^{-1} for women, 4 mg day^{-1} for four to six years old children, and 10 mg day^{-1} for adults. With 25% Zn absorption and 90% retention in our blood serum, at least 25% of the daily EAR can be met by feeding 422 gm of biofortified rice, which has a Zn concentration of 28–30 mg kg^{-1} in milled kernel (HarvestPlus 2014). This means the breeding target needs to be fixed at 28–30 mg Zn kg^{-1}

Crop Biofortification: Biotechnological Approaches for Achieving Nutritional Security under Changing Climate, First Edition. Edited by Adnan Noor Shah et al.

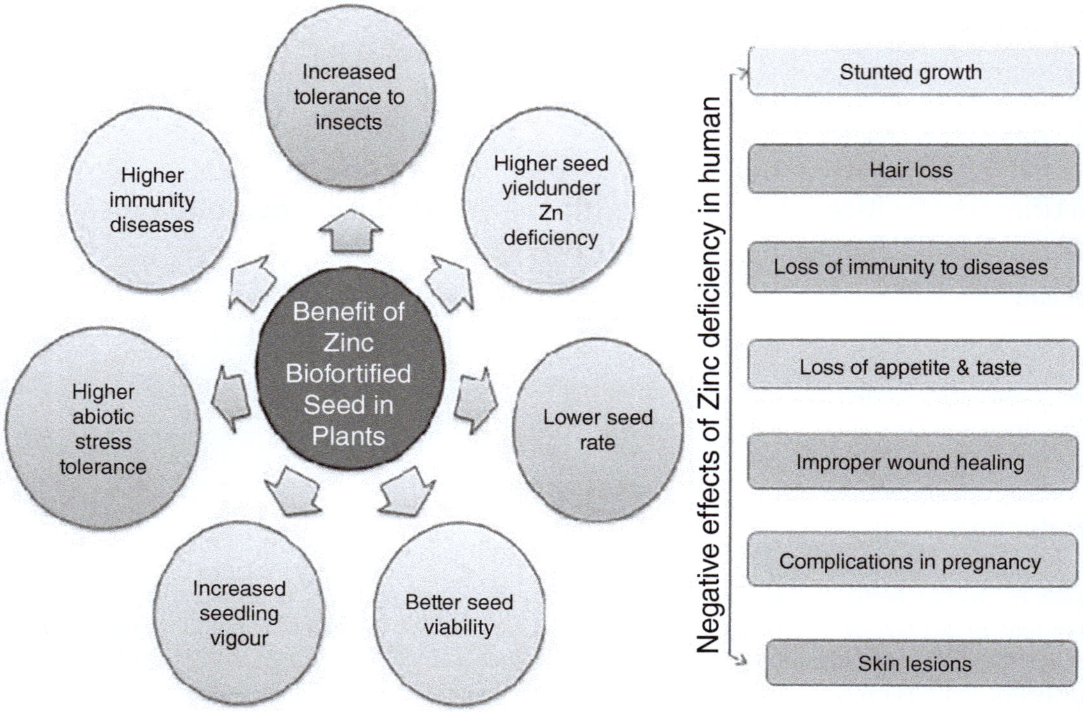

Figure 21.1 Benefit of zinc-biofortified seeds in plants and the negative effects of zinc deficiency in humans.

(Shahzad et al. 2014) which is double the present status of Zn (15–16 mg kg^{-1}) in rice (Trijatmiko et al. 2016). It is indeed a challenging task. Hence, rice breeding needs to be reoriented to improve the status of grain Zn content while increasing the yield potential. This chapter is a detailed review on phenotyping techniques, available genetic variation, association of grain Zn content with important agronomic traits, molecular basis of Zn transport and loading to grain, progress of conventional breeding, and prospect of innovative breeding strategy coupled with biotechnological approaches for the development of high Zn rice.

21.2 High-throughput Phenotyping and Exploring High Zn Donors

Grain Zn content is a highly variable trait, and hence, its precise phenotyping is an essential step for Zn biofortification in rice. The estimation technique(s) should ideally be cost-effective and rapid for high-throughput phenotyping. Colorimetric approaches, though used for different cereal crops, are laborious and not much sensitive (Choi et al. 2007). A preliminary qualitative assessment of Zn content can be done using dithizone (diphenylthiocarbazone) stain (1.56 mM) for mass screening of germplasm and initial breeding materials. Subsequently, quantification of Zn by image analysis with Adobe Photoshop (Choi et al. 2007) can effectively separate low, moderate, and high Zn rice genotypes. However, the estimation of Zn using atomic absorption spectrophotometer or more specifically by inductive coupled plasma-optical emission spectrophotometer (ICP-OES) and inductive coupled plasma-mass spectrometry (ICP-MS) analysis is more reliable owing to high accuracy (Beauchemin 2017), and high-throughput assay (around 500 samples per day). These techniques clearly identify the elements in the matter and their concentrations but require expensive equipment, highly trained manpower, contamination-free reagents, and extensive sample preparation (Velu et al. 2013). Similarly, the particle-induced X-ray emission (PIXE) is also a very powerful technique for the elemental study to detect the trace elements present in rice grains. Near infra-red spectroscopy and X-ray fluorescence (XRF) spectrometry (either handheld or benchtop XRF) are very fast and useful for nondestructive determination of Zn in rice grains (Stangoulis 2010). These are reported to offer multielemental (Na, Mg, P, S, Cl, K, Ca, Mn, Fe, Cu, Zn, Br, Rb, and Sr) analytical advantages in rice plant samples (Bado et al. 2016). Arnold et al. (2015) used the isotope fractionation technique to elucidate Fe and Zn homeostasis in different tissues and organs in rice. However, a more robust technique is needed for use in biofortification to minimize harvest and postharvest contaminations (Stangoulis et al. 2010).

As a crop, rice is the most genetically diverse food crop (Sandeep and Godi 2023) with the largest world germplasm resources amenable for identification of potential high Zn donors and harnessing valuable genes/quantitative trait locus (QTLs) for Zn biofortification. Broadly, germplasm lines can be grouped into three categories: low (<14 mg kg^{-1}), moderate (14–24 mg kg^{-1}), and high (>24 mg kg^{-1}) Zn rice. Modern high-yielding rice varieties are deficient in Zn and iron (Fe). Some landraces (Maganti et al. 2020), basmati types (Brar et al. 2011), wild species (*Oryza nivara* and *Oryza rufipogon*) (Sarla et al. 2012) and the African cultivated rice *O. glaberrima* (Ohmori et al. 2016) retain high grain Zn content, but japonica rice harbors the least Zn (Anuradha et al. 2012a). Similarly, some aromatic rice (Gregorio 2002), local upland rice (cv. Nam Roo: 31 mg kg^{-1}) (Jaksomsak et al. 2015), deepwater rice, and colored rice (Swamy et al. 2016) are reported to have higher Zn status in rice grains. Brown rice of local landraces shows considerable variation for Zn content (14.5–35.3 mg kg^{-1}) (Maganti et al. 2020), but it is reported to be two- to threefold in *O. nivara* (Acc IRGC81848 and IRGC81832) (Swamy et al. 2018a) and *O. rufipogon* (Sarla et al. 2012) compared to the mega variety Swarna (11.0 mg kg^{-1}). Similarly, many researchers revealed wide variations in the Zn content (13.0 mg kg^{-1} to maximum 40.0 mg kg^{-1}) in brown rice (Rao et al. 2014; Patil et al. 2015), suggesting a tremendous scope for its enrichment in rice grains. A quest for Zn-dense genotype can help breeders to breed high-yielding biofortified rice. The top Zn-dense (>24 mg kg^{-1}) genotypes identified by the author are Nagina 22 (28.1 mg kg^{-1}), BG 102 (25.8 mg kg^{-1}), Dudh Kandar (25.6 mg kg^{-1}), R-RHZ-7 (24.7 mg kg^{-1}), and IR 85850-AC157-1 (24.0 mg kg^{-1}) compared to Zn biofortified released varieties of Bangladesh (BRRI Dhan 62:19.6 mg kg^{-1}, BRRI Dhan 64:24.0 mg kg^{-1}, BRRI Dhan 72:23.0 mg kg^{-1}) and India (DRR Dhan 45:22.3 mg kg^{-1}, DRR Dhan 48:20.9 mg kg^{-1}, DRR Dhan 49:26.1 mg kg^{-1}, CGZR 1:21.7 mg kg^{-1}, CGZR 2:21.0 mg kg^{-1}, Hurz 1:20.0 mg kg^{-1} and Hurz 3:20.5 mg kg^{-1}), and popular mega varieties, e.g. MTU 1010 (15.1 mg kg^{-1}) (Tripathy Swapan et al. 2019a) and Swarna (11.0 mg kg^{-1}) (Tripathy Swapan 2020). Among these, two high Zn breeding lines, IR 85850-AC157-1 and R-RHZ-7 showed moderately high seed yield potential. A few local landraces, Basudha, Malliphulajhuli, Tikimahsuri, and Nikipankhia, revealed higher grain Zn content (>40.0 mg kg^{-1}) in addition to Fe (Tripathy Swapan 2020). Swamy et al. (2016) revealed significant variation for grain Zn in IRRI rice germplasm. The local landrace "Chittimuthyalu" of India (Av. Zn: 23.45 mg kg^{-1}) (ICAR-IIRR 2016; ICAR-IIRR 2017) and the popular Indonesian variety "Chiherang" (Av. Zn: 23.04 mg kg^{-1}) (Barokah et al. 2018) maintain stable grain Zn content with acceptable seed yield and hence serve as a quality check in Zn biofortification trials.

Recently, the author estimated grain Zn content of brown rice from a set of 518 rice genotypes using ICP-OES, which varied from 1.5 mg kg^{-1} in Kathia to 73.8 mg kg^{-1} in Asumakanda (a local landrace) with an overall mean Zn content of 14.297 mg kg^{-1}. Local landraces, notably Asumakanda (73.8 mg kg^{-1}), Basudha (44.4 mg kg^{-1}), Malliphulajhuli (43.8 mg kg^{-1}), Nikipankhia (42.8 mg kg^{-1}), and Tikimahsuri (41.5 mg kg^{-1}) harbor high status of Zn in rice grains.

Crossing between diverse genotypes offers scope for the recovery of transgressive segregants (Saxesena et al. 2013). Rathod et al. (2017) studied genetic diversity of 56 high Fe and Zn genotypes of rice and revealed distinct genotypic groups for high micronutrient contents. Tripathy Swapan (2020) identified a high Zn and high Fe genotypic group (≥40 mg kg^{-1}) that comprising the erstwhile mentioned Zn-dense local landraces (Tikimahsuri, Nikipankhia, Basudha, and Malliphulajhuli). Besides, "Nagina 22 (28.1 mg kg^{-1})" and "Dudh Kandar (25.6 mg kg^{-1})" are reported to be highly divergent though with poor yield potential (Tripathy Swapan et al. 2019a). Inter-Se hybridization involving such genotypes could be an excellent approach for the development of high-yielding Zn-dense rice.

21.3 Association of Grain Zn with Yield and Quality Traits

Zn concentration in rice grains is reported to maintain a positive relationship with maturity duration, plant height, and seed yield (Sri Devi et al. 2019). In general, high Fe and Zn lines bred through biofortification do not show any yield penalty (Kumar et al. 2009). But os-NAS/IR 64 transgenic lines with Fe and Zn beyond 23 mg kg^{-1} and 40 mg kg^{-1} respectively, showed a significant reduction in yield and ancillary traits (Moreno-Moyano et al. 2016). In contrast, Trijatmiko et al. (2016) selected a transgenic event with a single insert NASFer-274 (containing rice *OsNAS2* and soybean ferritin *SferH-1* genes) showing the highest reported Fe and Zn concentration (15 mg kg^{-1} and 45.7 mg kg^{-1}) without yield penalty under field evaluation in the Philippines and Cambodia. This also did not reveal any difference in protein content, amylose content, gel consistency, seed size, or chalkiness.

A number of researchers reported a strong positive relationship of grain Zn content with Fe (Maganti et al. 2020; Kumar et al. 2014) possibly due to the co-location of QTLs for Fe and Zn in chromosome 7 and chromosome 12 (Anuradha et al. 2012b). Similar is the case among 1138 IRRI germplasm lines (Gregorio 2002) and 85 introgression lines derived from Teqing × *O. rufipogon* tested over two seasons (Garcia-Oliveira et al. 2008). This means that the varieties that have efficient transport of Zn would also have higher translocation of Fe and loading of the same in the grain owing to similar genetic control. However, no such strong relationship is realized between these trace elements in polished rice (Ajmera et al. 2017; Sala et al. 2013) possibly due to loss of the major fraction of Fe during milling compared to Zn. High grain Zn trait is also shown to be linked with aroma (Gregorio 2002) and phytic acid content (Stangoulis et al. 2007).

A positive, significant association of Zn content with grain breadth (GB), but no or even an inverse relationship with grain length (GL) indicating efficient Zn transport and loading to sink in some of the varieties, resulting in better grain filling and bold grains (Tripathy Swapan et al. 2020). Such a typical relationship was also associated with Zn-dense cv. BRRI 64, traditional short bold landraces (Anandan et al. 2011) and wild rice (Gregorio et al. 2000) harboring high grain Zn content. Bekele et al. (2013) revealed favorable relationship between Zn content and grain weight coupled with GL. Further, Tripathy Swapan et al. (2020) revealed better scope for selection of medium duration semi-dwarf plant types with more number of fertile bold grains/panicles coupled with enriched Zn content in rice.

21.4 Molecular Basis of Zn Uptake and Transport in Rice

Zn accumulation in rice grain is a complex polygenic trait. Several genes are involved in metal uptake, translocation, sequestration, and loading to grains. Their coordinated interactions play a key role in the regulation of metal homeostasis in rice. In fact, Zn stored in the endosperm is transported to the embryo at the onset of germination, resulting in a high Zn localization in the radicle and leaf primordium (Takahashi et al. 2009), followed by in the vascular bundle of the scutellum. Most of the varieties also match the co-localization of Zn with Fe (Ishimaru et al. 2011).

Plants suffer from Zn deficiency in flooded soil (Impa et al. 2013) as Zn is immobilized under submergence. Besides, a higher concentration of iron (Fe^{2+}) in the rhizosphere is reported to have an antagonistic effect on the uptake of many nutrients, including Zn (Fageria et al. 2008). Zn acquisition from soil occurs in the form of both Zn^{2+} ion and Zn-DMA (zinc-deoxymugineic acid complex) (Figure 21.2). Zn-efficient genotypes absorb a higher amount of Zn owing to their more and finer small root hairs and release of Zn chelating DMA family phytosiderophores (PS) possibly as an adaptive response to Zn deficiency (Mingotte et al. 2018). Once Zn gets entry into the root, it follows symplastic transport across the plasmamembrane through the use of apoplastic spaces (Olsen and Palmgren 2014). In Zn deficient soil (≤ 0.5 mg kg^{-1}), the element is absorbed into the root hair cells mostly by diffusion (passive transport) (Reddy et al. 2018). Besides, intracellular Zn concentration is finely regulated for its bioavailability to cellular metabolism. Genotypic differences do exist in the loading of Zn to xylem and the phloem and it's unloading into different tissues (Johnson-Beebout et al. 2016). Once minerals are transported in the shoot xylem, transpirational pull can carry micronutrients to the leaves where they can be taken through the symplast system. A greater proportion (>100 mg kg^{-1}) of this is retained in leaves with limited loading (12–16 mg kg^{-1}) to rice kernel (Yin et al. 2016). A direct uptake pathway (mass flow) of root-stem-rachis and vascular bundle of grains through the xylem works in Zn sufficient soil (Jiang et al. 2008) as there is no xylem discontinuity in rice (Krishnan and Dayanandan 2003). In addition to root uptake by xylem (Impa et al. 2013), some varieties (Zn-efficient) also show phloem remobilization of Zn from root and shoot to grain (Hefferon 2019) in Zn deficient soil, possibly triggered by Zn stress-induced signaling (Kabir et al. 2017). As a result, these varieties show greater seedling vigor and increased crop yields (Sadeghzadeh 2013). Therefore, identification of genotypes with enhanced Zn loading efficiency can pave the way for genetic biofortification in rice. Within the rice grain, the apoplastic pathway plays an important role in Zn transport from the nucellar epidermis and aleurone cells to the endosperm (Jiang et al. 2008).

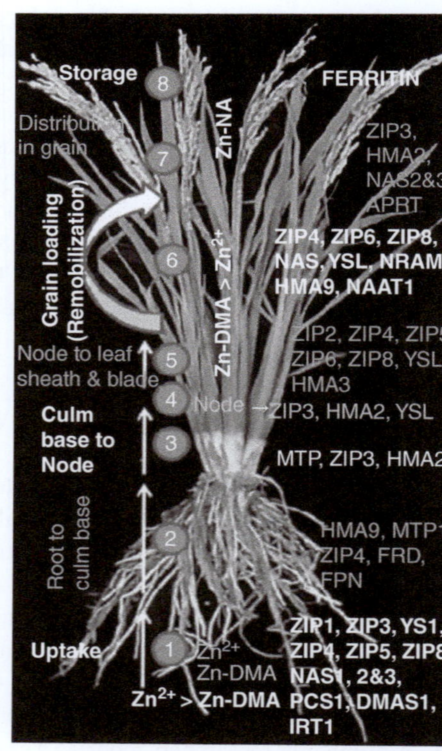

Figure 21.2 Candidate genes involved in uptake, transport, and grain loading of Zn in rice.

There exists nearly a fourfold difference in grain Zn content in rice germplasm, indicating the presence of potential inherent genetic factors amenable for Zn biofortification (Brar et al. 2014). Several transporter genes from different families, such as Zn-regulated transporter (ZRT)-like protein (ZIP), yellow stripe 1-like (YSL), nicotianamine synthase (NAS), natural resistance-associated macrophage protein (NRAMP), ZRT, Fe-regulated transporter (IRT)-like proteins, deoxymugineic acid synthase (DMAS), ferritin protein (FER), metal transport proteins (MTP), oligopeptide transporters (OPT), NAC transcriptional factors, cation diffusion facilitator (CDF), vacuolar iron transporter (VIT), and aldo-ketoreductase are known to take part in Zn uptake, transport, and accumulation. Several members of *ZIP* gene family, first reported in plants, are involved in the uptake and transport of metal cations (Zn, Fe, Cd, Mn, etc.) in rice (Bashir et al. 2012) possibly due to their unique conserved histidines or certain adjacent residues. OsZIP1, OsZIP3, OsZIP4, OsZIP5, OsZIP8 (Lee et al. 2010a), and OsZIP9 (Yang et al. 2020) are reported to overexpress during Zn deficiency. OsZIP1 (Ramesh et al. 2003), OsZIP5 (Lee et al. 2010a), OsZIP8 (Lee et al. 2010b) and OsIRT1 (Kawakami and Bhullar 2018) seem to be important for Zn^{2+} uptake from soil under Zn deficiency, while OsHMA2 (Yamaji et al. 2013), OsZIP4, OsZIP5, and OsZIP8 are meant for root-to-shoot translocation (Lee et al. 2010b). However, little is known about the expression of these genes through different growth stages of rice. Real-time PCR is reported to reveal pronounced upregulation of *OsZIP4, OsDMAS1, OsNAS2*, and *OsPCS1* in Zn-deficient roots of Pokkali compared to the sensitive variety BRRI 33 (Kabir et al. 2017).

Zn in the cell sap exists both in a free ion form or soluble form once bound with the ligand nicotianamine (NA). NA is a metal chelator and also acts as the precursor of DMA (Kawakami and Bhullar 2018). Biosynthesis of NA is mediated by *NA* genes. *OsNAS1* and *OsNAS2* are expressed in companion and pericycle cells adjacent to xylem in rice roots. While OsNAS3 is upregulated for Zn uptake in rice roots under Zn deficiency (Suzuki et al. 2008a). The function of osZIP1 is restricted to the uptake of Zn at the rhizosphere (roots) of plants, and it also has role in metal homeostasis. In fact, osZIP 1 continues to be at low and constant expression in the endoplasmic reticulum (ER) to ensure proper allocation of Zn in the cells. But, at the instance of a high concentration of Zn in the cell sap, it overexpresses to turn back the Zn concentration to normal by Zn partitioning to vacuoles. Similarly, MTP1 and OsHMA3 (localized at tonoplast) help in Zn sequestration to leaf and root vacuoles, respectively (Cai et al. 2019). OsHMA9 and OsZIP4 are reported to be involved in the loading of Zn to xylem and phloem in root respectively, under Zn deficiency (Ishimaru et al. 2005). Thereafter, Zn is transported to the node, where intervascular Zn transfer is mediated by OsZIP3 and OsHMA2 (Sasaki et al. 2015). The stem node serves as a hub to deliver Zn to different above-ground organs (Yamaji and Ma 2014). Zn-efficient rice genotypes containing OsHMA9 are potent enough to retranslocate Zn from older to emerging leaves under Zn deficiency (Lee et al. 2007). Zinc plays an important role in photosystem II in leaf mesophyll cells, and excess of it is stored in vacuoles. However, the sequestration of the excess Zn to vacuoles is restricted in flag leaf due to repression of the tonoplast Zn transporters (OsVIT1 and OsVIT2) (Zhang et al. 2012) and the Zn pool of the cell sap is loaded to phloem, though how Zn travels out of mesophyll cells is not yet clear. The pH of the phloem is slightly alkaline, and Zn is often bound to a metal chelator, NA for loading to the grain. In cereals, the remobilization of Zn during the period of grain loading also occurs from stems, peduncles, florets and rachis (Walker and Waters 2011). OsZIP4 and OsZIP8 are reported to be important for Zn remobilization from leaves to seed (Nakandalage et al. 2016). While OsZIP3 and OsHMA2 are involved in Zn distribution to the grain (Che et al. 2019). NAM-B1, a transcriptional factor that promotes senescence during grain filling and increases the remobilization of Fe and Zn to the grains in wheat (Uauy et al. 2006), can also play a key role in rice. Besides, the NA and DMA biosynthesis genes, e.g. *OsNAS2, OsNAS3, OsDMAS1*, and *OsNAAT1*, are also shown to be overexpressed at flowering, which evidences their definitive role in Zn homeostasis in the developing rice grain (Hefferon 2019).

21.5 Progress in Conventional Breeding

The idea of biofortification in staple crops for iron, zinc, and vitamin A dates back to the 1990s by the lead young Consultative Group on International Agricultural Research (CGIAR) researcher, Dr. Howarth Bouis (who won the World Food Prize in 2016), to address the world crisis of nutritional insecurity. Enriching the Zn status of rice grains along with high yield potential, acceptable eating quality, and tolerance to biotic and abiotic stresses is indeed a challenging task, as all such traits are highly complex with high G×E interaction (Sadeghzadeh et al. 2013). A build-up of high Zn donors by harnessing crop biodiversity using either phenomics and/or genomics-assisted prebreeding is the stepping stone for the Zn bio-fortification program.

Conventional breeding approaches, e.g. pedigree, backcross, heterosis, and mutation breeding followed for other traits, are also important for the development of high Zn rice (Figure 21.3). Zn content in rice grains is reported to have moderate

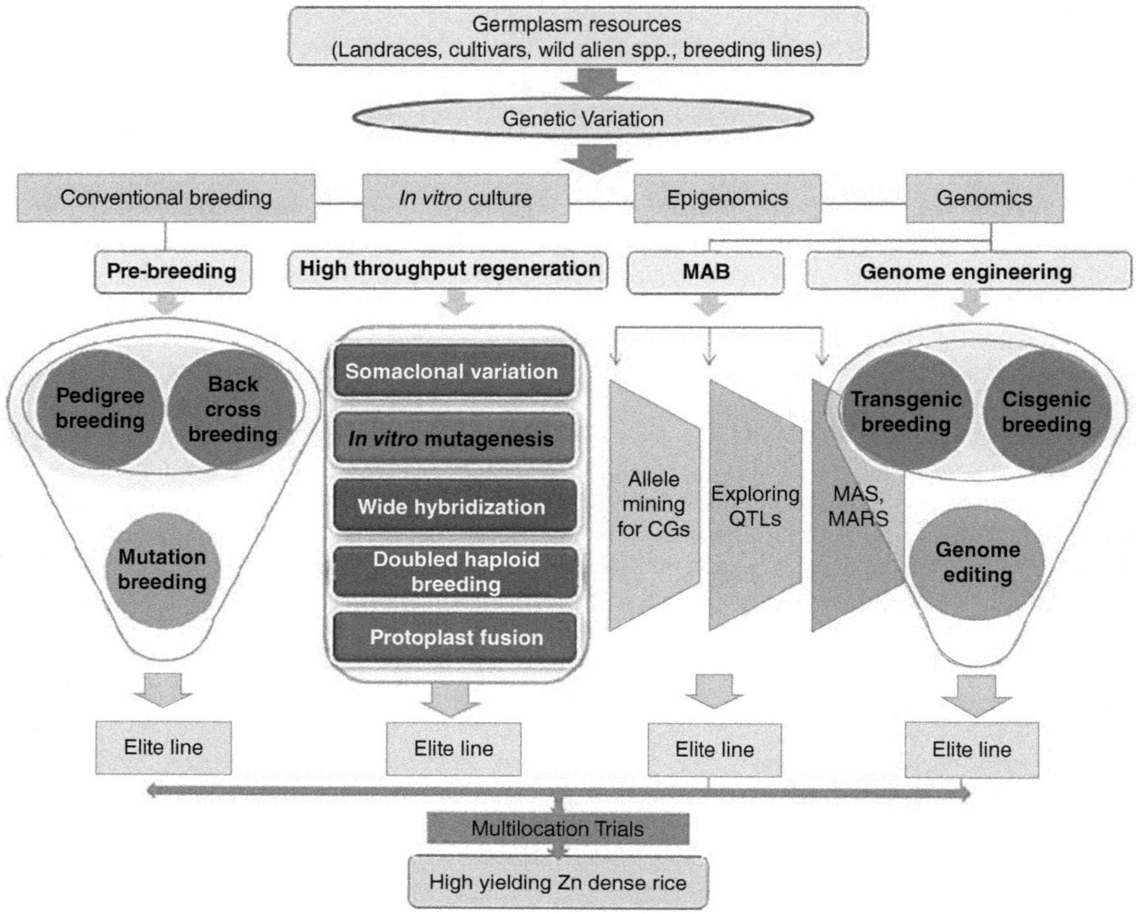

Figure 21.3 Schematic breeding strategies for development of Zn-dense rice varieties in rice.

heritability with a preponderance of nonadditive gene interaction (Anusha et al. 2021). Therefore, appropriate choice of parents, crossing designs, and follow-up selection strategies, and field evaluation are crucial for high Zn rice development. Selection of desirable plant type with improved agronomic traits even if starts from F2 itself, but screening of Zn-dense pedigrees needs to be started not earlier than F4 generation to hasten Zn status (Swamy et al. 2016).

Biparental crosses or even three-way or four-way multiple crosses involving several high Zn donors and popular high-yielding varieties can enhance the grain Zn content and yield potential. Several researchers recovered superior high-yielding transgressive lines with high grain Zn content (>40 mg kg^{-1}) among RIL populations of Azucena × Moromutan (Bekele et al. 2013), Madhukar × Swarna, Jalmagna × Swarna (Anuradha et al. 2012b; Sarla et al. 2012), BPT 5204 × Chittimutyalu (Babu et al. 2013), PAU 201 × Palman 579 (Kumar, Jain, and Jain 2014), IRRI38 × Jeerigesanna (Gande et al. 2014). Susanto et al. (2017) identified three high-yielding (˃5.0 t/ha) Zn-dense(˃30 mg kg^{-1}) rice breeding lines (BR7840-54-2-5-1, IR84020-84-2-3-2, and IR68144-2B-2-2-3-1-166) out of 22 test genotypes under wet season in Indonesia. One elite IRRI breeding line, IR68144-4B-2-2-3 had about 80% more Fe than IR64 (check), with a higher status of Zn (Gande et al. 2014). The high Zn trait of several IRRI donors and breeding materials has been attempted for transfer to the background of popular rice varieties (IR64, Swarna, Swarna Sub1, NSICRc222, PSBRc82, BR28, BR29, BR11, and Ciherang). Besides, multiparent advanced generation inter-cross (MAGIC) populations are the potential breeding resources for isolation of high Zn pyramided RILs (Cavanagh et al. 2008). At IRRI, several high Zn transgressive lines were recovered from several MAGIC-indica, MAGIC-japonica and MAGIC-global populations (Bandillo et al. 2013). A global MAGIC population comprising eight indica and eight japonica elite parents was used to identify genomic determinants for Zn content in milled rice to support the selection process in Zn biofortification (Zaw et al. 2019). A national network of rice biofortification trials across India since 2013 was shown to increase the Zn content of polished rice from 16.25 mg kg^{-1} (baseline) to 24 mg kg^{-1} till 2015 (Sanjeeva Rao et al. 2020). Out of 170 entries, 40 out yielded the best checks (IR 64, BPT 5204) in addition to high Zn content (≥24 mg kg^{-1}) and desirable cooking qualities.

The development of hybrids is an important option for Zn biofortification. Important candidate genes (CGs)/QTLs for elevated grain Zn can be stacked into the parents to develop "all in one" hybrid rice varieties. But progress in this regard is very slow (Kumar 2020). Babu et al. (2012) have identified two heterotic crosses (PR116 × Chittimutyalu and Mandya Vijay × Jalamagna) with 50% higher Zn over the standard check "Chittimutyalu." Besides, grain Zn and Fe content are also reported to be higher in some hybrids than parents in a 4 × 4 line (CMS) × tester (restorer) design under aerobic soil (Anusha et al. 2021).

Few researchers isolated high-yielding Zn-rich (>26 mg kg^{-1}) mutants from cv.IR 64 (Jeng et al. 2012) and P44 (Tripathy Swapan 2020) following treatment with sodium azide (SA) and 0.5% ethyl methanesulfonate (EMS), respectively. A few low phytate mutants have also been reported (Jeng et al. 2012), which may offer better bioavailability of Zn for human nutrition. Agarwal et al. (2018) identified two EMS-induced mutants (NH686 and NH787) of an aus variety of rice (Nagina22) having high Zn in grains when tested across several locations. Wang et al. (2017) characterized mutants of a rice variety Lilizhi and showed 35% higher Zn concentration in grains than in the wild type. These mutants had shown efficient Zn transport to the grains than storing in roots.

Almost all Zn biofortified rice varieties so far released are the products of conventional breeding (Table 21.1). Bangladesh (BRRI) in collaboration with HarvestPlus was able to release the world's first Zn-rich rice BRRI 62 (19–20 mg kg^{-1}) in 2013 (Josh 2013). Subsequently, BRRI 64 (24 mg kg^{-1}) and BRRI 72 (22.8 mg kg^{-1}) and BRRI Dhan 84 (>24 mg kg^{-1}) were released by the country. BRRI Dhan 100 is the most recent high-yielding Zn-dense (25.6 mg kg^{-1}) rice variety released by BRRI, Bangladesh in 2021. The first high Zn rice "DRR Dhan 45" was released in 2016 through conventional breeding by IIRR, Hyderabad, in India, which is a medium duration (125 days) semi-dwarf high-yielding (>5.0 t ha^{-1}) variety suitable for irrigated conditions. It has an average Zn content of 22.6 mg kg^{-1} (twice that of IR64 in polished rice) (Sanjeeva Rao et al. 2020), good cooking quality, moderate amylose content (20.7%), moderate resistance to blast, sheath rot, and rice tungro virus. Other Zn-rich varieties developed by IIRR were DRR Dhan 48 (20.91 mg kg^{-1}) and DRR Dhan 49 (25.2 mg kg^{-1}), all released in 2017. The recent high zinc and protein variety DRR Dhan 63 of IIRR has an average zinc content of 24.2 ppm and 10% protein in polished rice (ICAR-IIRR 2023). Another Zn-rich rice variety, "Surabhi" (22.8 mg kg^{-1}) was a successful venture in 2017 by Nuziveedu Seeds Ltd., India. The technology efficacy of the above four Indian Zn biofortified varieties

Table 21.1 Zn biofortified rice varieties released in target countries.

Variety	Source	Year of release	Av. grain yield (t ha^{-1})	Grain Zn (mg kg^{-1})
BRRI Dhan 62	BRRI, Gazipur, Bangladesh	2013	4.2	19–20
BRRI Dhan 64	-do-	2014	6.5	24.0
BRRI Dhan 72	-do-	2015	5.7	22.8
BRRI Dhan 84	-do-	2020	6.0–6.5	>24.0
BRRI Dhan 100	-do-	2021	7.7	25.6
CR Dhan 315	NRRI, Cuttack	2020	5.0	24.9
CGZR 1	IGKVV, Raipur, India	2015	4.5	>22.0
CGZR 2	-do-	-	4.2	21.0
HURZ 1	IRRI South Asia Hub, Hyderabad, India	-	5.74	22.94
HURZ 3	-do-	-	5.68	23.86
DRR Dhan 45	IIRR, Hyderabad, India	2016	>5.0	22.6
DRR Dhan 48	-do-	2017	5.0–5.5	20.9
DRR Dhan 49	-do-	2017	5.0–5.5	25.2
Impari IR nutri zinc	Indonesian Centre for Rice Research, Indonesia	2018	6.2	29.5
Surabhi	Nuziveedu Seeds Ltd., India	2017	4.0	22.8
NSIC Rc 460	IRRI, Philippines	2016	4.0	19.0
Chiherang	Indonesian Center of Rice Research-IRRI	2000	4.07	23.04
Swarna (non-biofortified check)	ANGRAU, AP, India	1982	5.5–6.0	11.0

ranged from 48 to 75%, with Zn intake of 38–47% and 46–57% of the recommended dietary allowances for males and females, respectively (Sanjeeva Rao et al. 2020). Besides, IGKV, Raipur, India (CGZR 1 and CGZR 2) also released two rice varieties each with grain Zn content of 21–22 mg kg^{-1} compared to the mega variety Swarna (11 mg kg^{-1}) and MTU 1010 (15.1 mg kg^{-1}). The first Zn-rich rice released by the Philippines is NSIC Rc 460 (19.0 mg kg^{-1}) with an average Zn increment of 4.5 mg kg^{-1} over the popular check variety IRRI 123. A popular protein-rich (10.1%) rice variety CR Dhan 311 (Mukul) bred by NRRI, Cuttack (India) in the "cv. Naveen" background using ARC 10075 as the donor was also reported to harbor moderately high grain Zn (20 mg kg^{-1}) (Chattopadhyay et al. 2019). Recently, NRRI, Cuttack also released another Zn-rich (25 mg kg^{-1}) rice variety, CR Dhan 315 (ICAR-NRRI 2022).

21.6 Prospect of Biotechnological Approaches for Development of High Zn Rice

Rice is a model cereal crop with a small genome size (430 Mb). It is the first crop whose genome is completely sequenced, and the genomic databases are freely available in the public domain for breeding purposes. Besides, the availability of high-throughput phenotyping techniques, proven Zn transporter genes, high-density genetic linkage maps, optimized in vitro culture protocols, high-throughput genome-wide sequencing platforms, advanced transgenic techniques, and precise gene editing tools have widened the scope for the development of Zn biofortified rice for the future. The schematic flow chart for innovative breeding strategy is given in Figure 21.3.

21.6.1 Somaclonal Variation

Somaclonal variation (SV) can be induced in the culture process using hormonal combinations (Della Costa et al. 2017) and more specifically long-term callus cultures. A few mutational events that survived the in vitro stress are passed on to the regenerants due to morphogenetic selectivity (Tripathy Swapan et al. 2016). A high-throughput regeneration system with optimized culture variables in the Zn-rich cv. Chittimuthyalu (Tripathy Swapan 2021) can be used to recover SV for a number of nutritional quality traits in rice grain (Ferdausi et al. 2009). Chakravorty (2020) reported a good response for recovery of somaclonal variation in a promising aromatic variety, "Gandhibiroin" for yield and nutritional traits. Al-Zahrani et al. (2018) successfully induced callus in three nutrient-rich (Fe and Zn) genotypes (Eg-N-1, Eg-N-2, and Eg-N-9) using MS +3 mg L^{-1} 2,4-D under a biofortification program and recovered somaclonal variants following regeneration in MS +2 mg L^{-1} BAP. The SV lines varied in quantitative traits, e.g. duration, leaf blast, plant height, number of panicles/plants, number of grains/panicles, test weight, fertility percentage, and grain yield in SC1 generation compared to the parents. Al-Noor et al. (2019) assessed induced variability among somaclonal variants derived from Iratom 24, BRRI Dhan 58, BRRI Dhan 62, and Binnatoa. A SV derived from Binnatoa had the highest Fe (26.93 and 21.51 mg kg^{-1}) and Zn (42.67 and 38.84 mg kg^{-1}) contents in brown and polished rice grain, respectively, as compared to Fe (14.2 and 9.45 mg kg^{-1}) and Zn (36.5 and 27.17 mg kg^{-1}) in the parent variety. This envisaged that viable genetic changes in the Fe and Zn transporter genes or their regulatory elements have been induced in callus culture and preferentially carried over to regenerants without affecting growth and survival. Such biofortified somaclones could be the potential breeding material for developing Fe and Zn enriched rice cultivars.

21.6.2 In Vitro Mutagenesis

The induced mutations play a key role in the improvement of plant varieties. The requirement of a large size mutagenic population for screening and the methods of selection are still the most stumbling blocks for success in conventional mutation breeding in rice. However, the transcriptomic studies revealed that in vitro mutagenesis of plant tissues can be a highly preferable approach to induce significant changes in gene expression patterns (Batista et al. 2008) throughout the genome within a manageable population size, and a single plant may carry a number of mutations in different genes (Serrat-Gurrera 2016). Other advantages include high mutation frequency, uniform mutagen treatment, application of selective agents to homogeneous cell populations, use of a single-cell system (somatic embryos) to avoid chimeras, the requirement of less space and time to handle large population, and keeping the plant material disease-free (Jain 2010).

Success has been achieved to generate variability for enhancing nutrition (Jain 2010), salt tolerance (Saleem et al. 2005), and yield-related traits (Chen et al. 2012). Recently, the author attempted in vitro mutagenesis of embryogenic calli of high Zn rice cv. "Chittimuthyalu" (23.5 mg kg^{-1}) with 0.005% EMS. The regeneration medium (MS +2 mg L^{-1} BAP +0.5 mg L^{-1} NAA) was supplemented with 125 mg L^{-1} L-ascorbic acid and 125 mg L^{-1} citric acid to prevent callus oxidation during mutagenesis. The follow-up phenotyping for grain Zn concentration of regenerated plants using ICP-OES led to the isolation of five Zn-dense mutants (>30 mg kg^{-1} in brown rice) without yield penalty.

In vitro mutagenesis can also be integrated with molecular marker techniques for high-throughput screening of mutants (Shu 2009). EMS mainly induces C-to-T substitutions, resulting in point mutations with C/G to T/A transitions (IRGSP 2005). EMS mutagenesis of mature seed-derived rice calli is reported to be a new method for rapidly obtaining TILLING mutant populations (Serrat-Gurrera 2016) amenable for direct screening of in vitro regenerated plants. Such a system saves time and reduces the greenhouse space and work load during the generation of mutant plant populations, besides ensuring a high mutagenesis rate.

21.6.3 Wide Hybridization

The wild forms of rice are in fact very rich sources for micronutrients. *O. nivara, O. rufipogon, O. latifolia, O. officinalis*, and *O. granulata* retain around two- to threefold higher grain Zn content than in the cultivated rice (Tripathy Swapan 2020). Grain Zn content in brown rice of *O. nivara* and *O. rufipogon* can be as high as 72 and 67 mg kg^{-1}, respectively (Sarla et al. 2012; Swamy et al. 2011) indicating enough scope for the transfer of the high grain Zn trait to the cultivated varieties through wide hybridization. However, wild species have been less utilized in rice breeding due to crossing barriers in addition to their poor phenotypic appearance, less seed-setting rate, and segregation distortions. But it can be made possible to recover distant hybrids by embryo culture (embryo rescue).

The "embryo rescue" technique allows recovery of inherently weak immature hybrid embryos into fertile plants that are otherwise subject to degeneration due to genetic disharmony. Once the hybrid embryo is rescued, the fertile progenies are obtained either through backcrossing with *O. sativa* or double haploid production. Sarla et al. (2012) recovered a set of backcross inbred lines (BILs) from three crosses (Swarna × *O. nivara* IRGC81832, Swarna × *O. nivara* IRGC81848, and Samba Mahsuri × *O. rufipogon* WR119). A wide range of Zn concentrations (0.4–104 mg kg^{-1}) was revealed in the above BIL populations, which paved the way for the isolation of 20 high Zn (>50 mg kg^{-1}) fixed breeding lines. Besides, wide hybridization between *O. nivara* (Accession IRGC 81832 and IRGC 81848) with Swarna (Swamy et al. 2018a) and *O. rufipogon* with Teqing (Indica rice) (Garcia-Oliveira et al. 2008) has been reported for the identification of large effect QTLs to support Zn biofortification in rice. In fact, a number of high Zn introgression lines are now available for use as potential donors, but yet there is no technical publication about the direct product development for Zn biofortification in rice using wide hybridization.

21.6.4 Double Haploid Breeding

Doubled haploid (DH) breeding starts with anther/pollen culture of F1 hybrids to form stable DHs within a short time (Tripathy Swapan 2018; Tripathy Swapan et al. 2019b). Several researchers characterized double haploid lines (DHLs) for yield, yield components, and grain mineral contents (Du et al. 2013; Calayugan et al. 2020; Descalsota-Empleo et al. 2019). Most of these were derived from japonica × indica crosses. Lee et al. (2020) developed a DH population consisting of 123 lines derived from a cross between the rice cultivars 93–11 (Chinese indica, low Zn and Fe) and Milyang 352 (Korean japonica, high Zn and Fe) for identification of QTLs governing Fe and Zn content in brown rice to support DH breeding. They detected qFe3-1 and qZn3-1 on chromosome 3 under early transplanting conditions, and another two QTLs, qFe3-2 and qZn3-2 on chromosome 3 and qZn4-2 on chromosome 4 under late transplanting conditions. Some of the DHLs had shown Zn content as high as 31.5 and 28 mg kg^{-1} compared to the 30 and 17 mg kg^{-1} in the donor (Milyang 352) under early and late transplanting conditions, respectively.

Faza et al. (2016) developed DHLs from four biparental crosses comprising five parents with contrasting grain Zn content. Zn concentration varied appreciably, and they isolated a few high Zn DHLs. Recently, a set of 148 DHLs derived from a cross IR05F102 (Indica) × IR69428 (Indica) showed wide variability in all three seasons for almost all agronomic traits along with micronutrient (Fe and Zn) concentrations (Calayugan et al. 2020). Similarly, Descalsota-Empleo et al. (2019) developed two DH populations comprising 111 DHLs and 146 DHLs derived from the crosses IR64 × IR69428 and BR29 × IR75862 respectively, using high Zn IRRI lines. A four-QTL DH line (qZn2.1 + qZn5.1 + qZn5.1 + qZn11.1) showed the highest grain Zn of 28.2 and 24.3 mg kg^{-1} in dry and wet seasons, respectively. Besides, Swamy et al. (2018b) identified a number of nutrient-dense DHLs in polished rice in two DH populations derived from PSBRc82 × Joryeongbyeo and PSBRc82 × IR69428. On the other hand, Stangoulis et al. (2007) recovered a few low phytic acid DHLs from a cross IR64 × Azucena, which may offer increased bioavailability of Zn.

21.6.5 Protoplast System for Breeding

In rice, protoplasts are isolated from suspension-cultured cells or etiolated seedlings. However, few studies have shown successful isolation of protoplasts from green tissues (stem and leaf sheath) and their suitable use (Zhang et al. 2011a) in investigations of plant signal transduction pathways. Besides, protoplasts are considered better candidates for somatic hybridization (protoplast fusion) and vector-independent gene transfer (Potrykus 2015) resulting in rapid analysis of transient gene expression as well as the generation of stably transformed transgenic plants. The use of protoplast fusion for transfer of salt tolerance (Finch et al. 1990), bacterial blight resistance (Yan et al. 2004) and cytoplasmic male sterility (Kyozuka et al. 1989) have been demonstrated. However, such a technique has not been so far attempted for Zn biofortification in rice. However, there is ample scope for transfer of the high Zn trait of the erstwhile mentioned wild species, e.g. *O. nivara* and *O. rufipogon*, and the African cultivated rice *O. glaberrima* (Garcia-Oliveira et al. 2008; Sarla et al. 2012) to the popular cultivated varieties of indica rice for Zn-dense product development through protoplast fusion.

Protoplasts allow free uptake of DNA sequence through transient pores on plasma membrane (PM) stimulated by various treatments, including polyethylene glycol (PEG), polyvinyl alcohol, calcium phosphate precipitation, and electroporation (Potrykus 1991). The PEG-mediated direct gene transfer technique was established originally for tobacco protoplasts by Negrutiu et al. (1987) and later extended to rice and different other crop plants owing to its ease in genetic transformation. OsZIP 1-GFP (a Zn transporter + green fluorescence protein) fusion was transformed and expressed in rice protoplasts (cv. Nipponbare, Kitaake, and Dong Jing). Transient expression of GFP in the ER and PM suggested localization of OsZIP 1 function in ER and PM (Liu et al. 2019). Similarly, OsZIP7-YFP (a Zn transporter + yellow fluorescence protein) construct in *Arabidopsis thaliana* protoplasts, either alone or co-transfected with AHA2-RFP (red fluorescence protein) revealed localization of OsZIP 7 in PM under a fluorescence microscope (Ricachenevsky et al. 2018). The above protoplast-derived transformed lines were shown to harbor higher Zn content than control varieties under Zn deficiency.

Epigenetic modification of DNA: The overall chromatin structure determined by histone variants and their posttranslational modifications, and DNA methylation greatly regulates the cellular functions. Nutritional stress signals can induce epigenetic modifications in the DNA sequence, leading to regulation of gene expression (Chinnusamy and Zhu 2009). They influence transcriptional responses to stress and could thus confer an adaptive advantage for the plant (Finatto et al. 2015). It would be interesting to study the role of epigenetic regulators in Fe and Zn homeostasis in rice. Thus, epigenetic regulation of genes associated with nutritional stress, e.g. Zn deficiency in soil can unravel their role in Zn biofortification in rice. With emerging data on epigenomics and the role of DNA methylation in plant stress response, it would be worth exploring to harness the epigenetic variations for biofortification.

21.6.6 Molecular Marker-assisted Breeding

Molecular breeding for Zn biofortification essentially starts with conventional biparental or multiple crosses (three-way and double-crosses) or MAGIC to unlock genetic variability. Besides, the approach is equally fit to the recurrent backcrosses meant for converging genetic variability to recover fixed breeding introgression lines with the maximum genetic background of the recurrent parent. Single plant selections are made even at the seedling stage based on validated molecular markers tightly linked with genes/QTLs. Many often CGs are found to be located within the QTLs, which offer the marker aided selection much effective and simpler using a limited number of molecular markers. MAS for introgression of major QTLs may be rewarding for increasing grain Zn content in rice.

The success of Zn biofortification depends upon the choice of key CGs involved in efficient Zn uptake, transport, loading to grains, and storage with minimum hindrance due to phytate for bioavailability. Each Zn-dense donor is expected to harbor different sets of CGs and/or QTLs, and therefore, Zn accumulation in such varieties might be under different genetic control. Transcriptome and in silico analysis can be the methods of choice for gene discovery. Judicious selection of parents with complementary sets of Zn transporter genes and/or QTLs can pave the way for raising potential crosses and follow-up isolation of transgressive segregates containing all such genetic factors pooled together from both the parents (gene pyramiding). Allele mining and the marker-trait association (MTA) can accomplish the purpose to supporting marker assisted selection for Zn biofortification, though progress for high Zn rice development is indeed slow.

Allele mining for exploring candidate genes: A number of CGs are responsible for the natural variation of Fe and Zn concentrations (Jin et al. 2015). Allele mining can be precisely performed for identification of diverse and valuable genetic resources (Banerjee et al. 2010). The sequences of erstwhile mentioned known candidate gene families (*ZIP, YSL, NRAMP, FER, TFs* of *NAC* genes, *IRO2, DMAS, MTP, OPT*, and *VIT*) are publicly available in NCBI databases. Sequence information may be utilized for designing gene-specific markers. However, the report on identification of Zn transporter candidate gene(s) is indeed scanty (Kumar et al. 2019; Patil and Shashidhar 2016). The polymorphic nature of OsZip 1 specific designed primer was demonstrated in a set of rice genotypes (Patil and Shashidhar 2016). PCR analysis of 17 Zn-rich rice genotypes using gene-specific primers (Sahoo 2019) revealed the presence of *OsZIP4c* in all test genotypes, *OsNAS2* in 10 genotypes, and OsZIP 7d in only five test genotypes. Further, five high Zn (>22 mg kg^{-1}) varieties (Chittimuthyalu, IR 85850-AC-157-1, IR 91143-AC239-1, Kalanamak, and Nagina 22) were shown to carry all the above CGs (*OsZIP4c, OsZIP7d* and *OsNAS2*). Neeraja et al. (2018) identified six CGs (in a RIL population of BPT 5204/Chittimuthyalu) using simple sequence repeats (SSR) markers at grain filling for their significant association with grain Zn content in brown and polished rice. Patil et al. (2015) validated 20 Zn transporter gene-specific primers, out of which nine showed polymorphism. Among these, only two primers for OsZIP21 and OsZIP32 showed significant association with grain Zn content with 6.1 and 10% phenotypic variation (PV), respectively. In another study, 17 polymorphic gene-specific markers were shown to be related to gene families *OsZIP, OsYSL, OsNRAMP, OsNAAT, OsFRO, OsFDH, OsGSTU*, and *OsPDR*, which are involved in metal (Fe and Zn) transport and homeostasis in rice (Chandu et al. 2020). Similarly, Gande et al. (2014) revealed significant association of four candidate gene-specific markers (OsNAC, OsZIP8a, OsZIP8c, and OsZIP4b) for grain Zn content with PV of 4.5, 19.0, 5.1, and 10.2%, respectively using single marker analysis in a set of RILs derived from IRRI38 × Jeerigesanna. In addition, they validated three markers (OsZIP8a, OsNAC, and OsZIP4b) for grain Zn content with PV of 11.0, 5.8, and 4.8% respectively in 96 germplasm accessions of rice.

Transcriptome and in silico analysis for exploring differentially expressed genes (DEGs): Functional annotation of high priority DEGs a in few elite high Zn varieties compared to low Zn genotypes using transcriptomic analysis at varying stages of the rice plant can unravel the genes accountable for mineral uptake and transport to grains (Sauer et al. 2007). This can be helpful to develop functional markers (to be used for genomic-assisted breeding [GAB]) besides understanding the underlying molecular mechanisms. *OsZIP1* gene was shown to be upregulated under Zn-deficient conditions, whereas *OsZIP3* was upregulated under both Zn available and deficient conditions in rice (Ramesh et al. 2003). The integrated analyses of miRNAs and transcriptome data revealed expression of 12 DEGs regulated by Zn-responsive miRNAs (miR171g-5p, miR397b-5p, miR398a-5p, and miR528-5p) under Zn deficiency (Zeng et al. 2019).

From flag leaf transcriptome, Roda et al. (2020) identified 13 DEGs (on chromosomes 1 and 4) having their role in binding metal ion, cofactor and coenzyme for effective use in the Zn biofortification program. Neeraja et al. (2018) revealed high Zn status in the straw of cv. BPT 5204, but a high concentration of Zn in polished rice of landraces under Zn sufficient soil, indicating efficient transport system to grain in the later. They identified eight DEGs associated with high Zn in Chittimuthyalu and Kala Jeera Joha through a panicle transcriptome study. One transcript for NAS3 viz. OS07G0689600-NAS 3 was shown to be threefold upregulated in cv. "Kala Jeera Joha" compared to the Zn-rich cv. "Chittimuthyalu."

Further, transcriptome study of different regions of the developing rice grains revealed expression of tissue-specific DEGs having roles in metal binding, transporter, transcriptional factor activity, and phytohormone biosynthesis and signaling (Wu et al. 2020). The majority of the ovular vascular trace (OVT) tissue-specific TFs belong to bHLH and MYB families that regulate mobilization and loading of metabolites to endosperm and the developing embryo. Besides, Zn and Fe transporters (NRAMP, ABC, HMA, and ZIP) are also preferentially expressed in OVT on the fourth and eighth days after flowering (Wu et al. 2020). Recently, in silico analysis in rice led to identifying as many as 260 CGs which were regulated in response to Fe and Zn stresses. Among these, at least 37 CGs were reported to be differentially expressed under both Fe and Zn deficiency (Raza et al. 2019). Several candidate genes, e.g. *osIRT, osZIP8a, osZip4b*, and *APRT* involved in enhanced loading of Zn to sink, have been detected, and these may be transferred to elite rice genotypes using marker-assisted selection. Besides, Ricachenevsky et al. (2018) annotated several rice genes concerning to the characterized genes of *A. thaliana* for each family related to Zn homeostasis.

QTL mapping and MTA: QTLs can be identified with the help of association mapping (AM), viz., genome-wide AM, candidate gene based AM, and bulk segregant analysis. The complex polygenic inheritance of the high Zn trait often relies on marker-assisted breeding (MAB) using major effect QTLs to precisely develop Zn-dense rice (Swamy et al. 2016). Microsatellites (SSRs) and single-nucleotide polymorphisms (SNPs) are the markers of choice for genetic dissection of complex traits. Until now, several mapping populations, e.g. recombinant inbred lines (Anuradha et al. 2012b), double haploids (Du et al. 2013; Zhang et al. 2011), BILs (Ishikawa et al. 2010), and introgression lines (Zhang et al. 2014) have been used in QTL mapping studies for grain Zn content. However, DH populations are considered ideal for mapping QTLs owing to their low frequency of deleterious alleles and negligible background noise (Calayugan et al. 2020). Several major QTLs with high PV (>10%) and CGs for grain Zn content have been mapped covering the whole genome (Zhang et al. 2011b; Norton et al. 2010) of rice using various germplasm resources (Table 21.2) including landraces, indica, japonica, aus

Table 21.2 QTLs identified for grain Zn content in different mapping population of rice.

Chromosome	QTLs	Markers	Source of mapping population	Type (population size)	R^2 (%)	References
1	qZn1	RM34-RM237	IR 64/Azucena	DHs (129)	15.0	James et al. (2007)
1	qZn1.1	id1008679–439764	IR05F102 (Indica) × IR69428 (Indica)	DHs (148)	8.96	Calayugan et al. (2020)
10	qZn10.1	RM474-RM184	PAU201(Indica)/Palman 579(Indica)	F_2 (247)	19.1	Kumar, Jain and Jain (2014)
2	qZn2.1	RM573	Nibare/*O. meridionalis* (W1627)	BILs (151)	15.2	Ishikawa et al. (2017)
2	qZn2.1	LM2110566-RMid2009463	PSBRc82/Joryeongbyeo and PSBRc82/IR69428	DHs (130 and 97)	17.3	Swamy et al. (2018b)
3	qZn3.1	RM7-RM517	Madhukar/Swarna	RILs (168)	31.0	Anuradha et al. (2012b)
4	qZn.4	CT206-G177	JYQ8/(Indica)/ JX17(Japonica)	DHs (127)	10.83	Zhang et al. (2011b)
5	qZn5.1	RM1089	Teqing(Indica)/*O. rufipogon*	ILs (85)	5.0	Garcia-oliveira et al. (2008)
5	qZn5.1	4904312–4908650	IR05F102 (Indica) × IR69428 (Indica)	DHs (148)	12.15	Calayugan et al. (2020)
5	qZn5	R3166-RG360	Zhenshan 97/Minghui 63	RILs (241)	12.34	Lu et al. (2008)
6	qZn.6	RZ516-G30	JYQ8/(Indica)/ JX17(Japonica)	DHs (127)	12.38	Zhang et al. (2011b)
6	qZn.6	RG172-RM340	*O. sativa*/*O. rufipogon*	BILs (202)	2.2	Hu et al. (2016)
6	qZn6.1	RM585-RM3	PAU201(Indica)/Palman 579 (Indica)	F_4 (106)	25.0	Kumar, Jain and Chowdhury (2019)
7	qZn7.1	RM234-RM248	Madhukar/Swarna	RILs (168)	35.0	Anuradha et al. (2012b)
7	qZn7.2	RM248-RM8007	Madhukar/Swarna	RILs (168)	35.0	Anuradha et al. (2012b)
7	qZn7.3	RM501-OsZIP2	Madhukar/Swarna	RILs (168)	29.0	Anuradha et al. (2012b)
7	qZn.8	RM234-R1789	Zhenshan 97/Minghui 63	RILs (241)	5.3	Lu et al. (2008)
8	qZn8.1	RM152	Teqing(Indica)/*O. rufipogon*	ILs (85)	19.0	Garcia-oliveira et al. (2008)
9	qZn9.1	9809545–9819278	IR05F102 (Indica) × IR69428 (Indica)	DHs (148)	13.79	Calayugan et al. (2020)
10	qZn10	RM1125-RM6704	*O. sativa*/*O. rufipogon*	BILs (202)	5.0	Hu et al. (2016)
12	qZn12.1	c12p4887439–12172332	IR05F102 (Indica) × IR69428 (Indica)	DHs (148)	15.26	Calayugan et al. (2020)
12	qZn12.1	RM3331	Teqing (Indica)/*O. rufipogon*	ILs (85)	9.0	Garcia-oliveira *et al.* (2008)
12	qZn12.1	RM17-RM260	Madhukar/Swarna	RILs (168)	35.0	Anuradha et al. (2012b)
12	qZn12.2	RM260-RM7102	Madhukar/Swarna	RILs (168)	34.0	Anuradha et al. (2012b)

accessions, and wild species (*O. nivara* and *O. rufipogon*). But yet their use for Zn biofortification is indeed scanty. In this context, QTL pyramiding, marker-assisted recurrent selection and GAB approaches are worth trying to develop high-Zn rice.

At least 10 major QTLs determine mineral element (Cu, Ca, Zn, Mn, and Fe) accumulation in rice grains (Norton et al. 2010). QTL mapping of two BC2F3 populations from crosses of *O. sativa* cv. Swarna with two different accessions of *O. nivara* revealed most of the QTLs for Fe and Zn from *O. nivara* origin (Swamy et al. 2018a). Besides, two co-located Fe and Zn QTLs were detected closely linked to yield QTLs on chromosome 2 (RM6-RM535) and chromosome 12 (RM415-RM247). In another study, 26 SSR markers on a set of 176 RILs of Azucena × Moromutant showed close association of RM21 with grain Zn content, explaining 11% PV (Bekele et al. 2013) consistently over two years. QTL analysis in Madhukar × Swarna F7 RILs using 110 SSR markers and nine gene-specific markers revealed seven QTLs for Fe and six QTLs for Zn, each explaining >30% PV. These QTLs for Fe and Zn concentration were shown to be co-located on chromosomes 7 and 12 (Sarla et al. 2012). Recently, it was confirmed that Fe and Zn homeostasis genes *OsMTP6, OsNAS3, OsMT2D, OsVIT*, and *OsNRAMP7* were co-located with QTLs (Majumder et al. 2019). In another study, six and five QTLs on chromosomes 7 and 12 respectively, were found to be more consistent across the genetic backgrounds and environments. Two such QTLs (qZn7.2 and qZn12) that revealed PV as high as 35.0% and an additive effect of more than 10 mg kg^{-1} (Anuradha et al. 2012b), seem to be good targets for MAS. Garcia-Oliveira et al. (2008) identified two grain Zn QTLs such as qZn5-1 and qZn8-1, consistently over two years. Zn content in leaf and grain maintained a feeble relationship, possibly due to the positioning of their respective QTLs on different chromosomal locations. Further, it envisages different mechanisms underlying Zn accumulation in vegetative and reproductive tissues (Norton et al. 2010), and it is less likely that they are co-inherited.

Brar et al. (2015) validated 48 SSR markers in 14 diverse rice varieties and revealed significant associations of RM 152, RM 201, RM 205, RM 23, RM 300, RM 447, RM 585, and RM 6641 with grain Zn content. Marker analysis in 13 black rice genotypes revealed one QTL each for Zn on chromosomes 5 and 8 near SSR marker RM 153 and RM 152, respectively (Faza et al. 2016). In a panel of 85 introgression lines (ILs) derived from a cross between an elite indica cv. Teqing and the wild rice (*O. rufipogon*), Garcia-Oliveira et al. (2008) also reported one major effect QTL (qZn8-1) with the maximum PV (19%), in the vicinity of SSR marker RM152 on chromosome 8. Recently, three novel QTLs, e.g. qZn2.2, qZn8.3, and qZn12.3 for Zn content were detected by Pradhan et al. (2020) in a panel of 120 germplasm lines. SSR markers, e.g. RM 1132 and RM 339, were commonly associated with grain Fe, Zn and yield QTLs indicating their usefulness in breeding high-yielding Zn and Fe-rich rice. Genome-wide association study (GWAS) in a set of 192 Indian rice germplasm revealed 11 and 5 significant MTAs for Zn content in brown rice and milled rice, respectively (Bollinedi et al. 2020). Further, the MTAs for Zn in milled rice closely correspond to Zn in brown rice, primarily due to less loss of Zn compared to Fe during milling.

More than 20 million SNPs are now available from a panel of 3000 varieties (Alexandrov et al. 2015). Use of in silico and transcriptome analysis led to the discovery of 176 novel SNPs and 39 SSRs within CGs underlying Zn and Fe homeostasis in rice (Banerjee et al. 2010). Babu et al. (2020) generated a huge number (39 137) of polymorphic SNPs in 40 rice genotypes through double digest restriction site associated DNA (dd-RAD) sequencing. Once the SNPs linked with grain Zn content and seed yield are identified, these could be of direct use in GAB to develop Zn-dense, high-yielding varieties. Significant SNPs were identified at 22.26 Mb on chromosome 3 and these were shown to be consistently associated with grain Zn over four out of five locations using genome-wide AM in 421 accessions (Norton et al. 2014). Further, the study of GWAS revealed associations of Zn content in polished rice with stable SNPs located in three putative CGs (*LOC_Os03g47980, LOC_Os07g47950*, and *LOC_Os07g48050*) on chromosomes 3 and 7. The last two CGs for Zn content on chromosome 7 co-localized with the earlier reported genomic region (rMQTL 7.1) (Raza et al. 2019) and OsNAS3 gene. Swamy et al. (2018b) identified several CGs (*OsNRAMP, OsNAS, OsZIP, OsYSL, OsFER*, and *OsZIFL* family) near QTLs for grain Zn and grain yield using 6K SNP chip analysis in two DH populations. Eight QTLs were detected on chromosomes 2, 3, 6, 8, 11, and 12 with 7.5–22.8% PV, and among this, qZn2.1 was consistent over two seasons. However, no QTL that commonly shares both yield and grain Zn was identified in either of the DH populations. In another study, OsNAS3 locus was consistently identified for association with Zn content in milled rice in two dry seasons following GWAS as well as interval mapping of the global MAGIC population comprising eight indica and eight japonica elite parents (Zaw et al. 2019). For this, the sequence reads obtained following Illumina's GBS protocol (Elshire et al. 2011) were aligned to the reference genome Nipponbare sequence MSUv7 to derive the physical positions of SNP markers and extracted for genetic mapping at no closer than 0.1 cM (Zaw et al. 2019). Detection of above Zn-related loci (QTL/candidate genes) flanked by stable SNP markers can support Zn biofortification using marker-assisted selection.

21.6.7 Genome Engineering (GE)

Transgenic breeding: Conventional breeding for grain Zn content seems to be not much rewarding as genetic factors for the trait cover the whole genome of rice and the contribution of each is minimum. However, the transfer of foreign candidate gene(s) (Table 21.3) into farmer-preferred genotypes made it possible to enhance uptake, transport, mobilization of Zn, and other micronutrients to grains without yield penalty, and even with increased bioavailability (Trijatmiko et al. 2016). This can be accomplished by manipulation of transport proteins in root plasma membrane and tonoplast, overexpression of genes for increased PS synthesis, and enhanced activity of transporter genes, transcription factors (ZIP19 and ZIP23), and

NAS2 gene for efficient transport and accumulation of Zn in developing rice grains (Johnson et al. 2011). Bio-fortified transgenic lines have been reported to harbor 36 and 23% enhanced Zn and Fe concentrations in unpolished rice grains, respectively (Laura et al. 2016). Insertion of a novel functional allele from osNAS family (Swamy et al. 2018b) together with the *ZIP* gene family (Descalsota et al. 2018) can be a potential strategy for enhancing grain Zn content. In this context, novel alleles of heavy metal ATPase OsHMA3 from rice and ZIF1 homologue from Arabidopsis meant for Zn sequestration to vacuoles are also considered as potential candidates for Zn biofortification. However, "marker free" transgenic rice (Baisakh et al. 2006) and more specifically, targeted DNA insertion at suitable genomic sites would be more preferable.

OsIRT and *MxIRT* family genes are reported to be better candidates for increased Fe and Zn concentration in rice grains (Tan et al. 2015). *OsIRT1* was shown to be overexpressed under the promoter pZmUbq1 resulting, 13 and 12% increase in Fe and Zn content in rice grains, respectively (Lee and An 2009). Transgenic plants overexpressing *MxIRT1* exhibited a threefold increase in Fe and Zn accumulation coupled with a decrease in cadmium (Hefferon 2019). Concomitant Zn accumulation with Fe is quite common in transgenic ferritin-rich rice grains. For instance, the endogenous ferritin gene *OsFer2* is overexpressed in rice, resulting in a 2.09-fold and 1.37-fold increase of Fe and Zn, respectively (Paul et al. 2012). One of the high-yielding indica T4 transgenic lines expressed over 2.6-fold higher levels of ferritin with concomitant 2.54-fold and 1.54-fold increases in Fe and Zn content, respectively in polished rice (Paul et al. 2014). Besides, *NAS* genes from *A. thaliana* (Singh et al. 2017; Wu et al. 2018) and *Hordeum vulgare* (Suzuki et al. 2008b) have been overexpressed and verified for Zn and Fe content in rice grains. *OsZIP7* is a Zn transporter gene localized in PM, and it seems to be a good candidate for biofortification with overexpression in two transgenic lines of rice (Ricachenevsky et al. 2018).

Transgenic approach to deliver multiple transgenes into the rice genome (Table 21.3) was attempted by a few researchers (Banakar et al. 2017). Recently, five Nipponbare multi-nutrient transgenic lines (CP22, CP87, CP97, CP101, and CP105) have shown enhanced Fe, Zn, and β-carotene content in polished rice grains by expressing Arabidopsis AtNAS1, bean

Table 21.3 Candidate genes used for development of transgenic Zn-rich rice.

SL. No.	Gene	Level of expression	References
1	*OsNas1*	1.3-fold Zn and 1.0-fold Fe	Zheng et al. (2010)
2	*HvNas1*	1.5-fold Zn and 2.5-fold Fe	Higuchi et al. (2001)
3	*OsNas2*	2.7-fold Zn and 3.0-fold Fe	Lee et al. (2012)
4	*OsNas2*	2.2-fold Zn	Johnson et al. (2011)
5	Activation tagging of *OsNas2*	2.9-fold Zn	Lee et al. (2011)
6	*OsNas1, OsNas2,* and *OsNas3*	2.0-fold Zn and 2.0-fold Fe	Johnson et al. (2011)
7	*SoyFerH1*	1.5-fold Zn and 2.5-fold Fe	Paul et al. (2014)
8	*SoyFerH1*	1.1-fold Zn and 3.0-fold Fe	Qu et al. (2005)
9	*OSFer1*	Low expression than OsFer2	Stein et al. (2009)
10	*OsFer2*	1.37-fold Zn and 2.9-fold Fe	Soumitra et al. (2012)
11	*OsIRO2*	1.4-fold Zn and 2.8-fold Fe	Ogo et al. (2011)
12	Barley genes	1.3-fold Zn and 1.4-fold Fe	Masuda et al. (2008)
13	*OsYSL15*	1.0-fold Zn and 1.1-fold Fe	Lee et al. (2009)
14	*HvNas1, HvNAAT,* and *IDS3*	1.4-fold Zn and 1.2-fold Fe	Suzuki et al. (2008b)
15	*OsNas1. PvFer* and *Afphytase*	1.6-fold Zn and 6.3-fold Fe	Wirth et al. (2009)
16	*SoyFerH2, HvNas1,* and *OsYSL2*	1.3-fold Zn and 3.4-fold Fe	Aung et al. (2013)
17	*AtFRD3 + AtNAS1 + PvFER*	2.4-fold Zn and 5.4-fold Fe	Wu et al. (2018)
18	*OsNAS1 + HvNAATb*	2.2-fold Zn and 4.3-fold Fe	Banakar et al. (2017)
19	*OsNAS2 + GmFER*	2.8-fold Zn and 7.5-fold Fe	Trijatmiko et al. (2016)
20	*PaCRT1 + ZmPSY + AtNAS1 + PvFER*	1.3-fold Zn and 3.3-fold Fe	Singh et al. (2017)
21	*AtIRT1 + AtNAS1 + PvFER*	1.8-fold Zn and 4.7-fold Fe	Boonyaves et al. (2017)

ferritin (Pvferritin), bacterial CRTI, and corn ZmPSY in a single genetic locus (Singh et al. 2017). Zn and Fe content was also shown to be increased up to 2.4- and 5.4-fold in polished rice grains using a combination of AtFRD3 under pZmUBQ1 along with pCaMV35S:AtNAS1 and pOsGlob1:PvFER (Wu et al. 2018). However, combining candidate gene cascade pAt-IRT1: AtIRT1 with pCaMV35S:AtNAS1 + pOsGlob1: PvFER did not reveal much dividend, which led to 1.8 and 4.7-fold increase in Zn and Fe content (Boonyaves et al. 2017). Trijatmiko et al. (2016) were able to achieve 15 mg kg^{-1} Fe and 45.7 mg kg^{-1} Zn in rice grain without yield penalty and altered grain quality in a transgenic event NASFer-274 having single locus insertion of rice *OsNAS2* and soybean ferritin (*SferH-1*) genes following rigorous selection of 1689 IR64 transgenic events. However, it is inferred that expression of nonrice ferritin, especially from legumes (Pvferritin, SferH) (Trijatmiko et al. 2016; Wu et al. 2018) in rice endosperm is more effective than expression of OsFer (Oliva et al. 2014). Other approaches combined the overexpression of the natural resistance-associated macrophage protein (AtNRAMP3) gene and ferric reductase defective 3 (FRD3) gene from Arabidopsis with OsNAS and also reached 11–13 ppm for Fe and 33–48 ppm for Zn (Wu et al. 2018).

The RNAi technique has been proven to be a successful attempt for the development of lysine-rich biotech rice (Galili 2002). It also seems to be an ideal biotech tool for increasing the bioavailable Fe and Zn content in rice grains. In fact, *OsHRZ 1* and *OsHRZ 2* in rice and their close homologue BTS in Arabidopsis bind Fe and Zn under Fe deficiency, resulting in reduced availabilities of these elements for growth and development as well as decreased accumulation of Fe and Zn in rice grains. However, RNAi-mediated silencing of *OsHRZ 2* (a negative regulator of iron) was shown to increase Fe and Zn concomitantly in rice endosperm (Kobayashi et al. 2013). Phytate (InsP6) is the major anti-nutrient phosphorous component in rice grains that reduces the bioavailability of Zn. RNAi-mediated silencing of *MIPS* (Ali et al. 2013a), *IPK1* (Ali et al. 2013b), and *OsITP/6K-1* gene (Karmakar et al. 2019) of the phytic acid biosynthesis pathway increased the bioavailable Zn content along with Fe, Ca, and Mg content in milled rice grain. Kuwano et al. (2009) showed a reduction in seed phytic acid content up to 68% without any deleterious effects on dry seed weight, seed viability, or plant growth in transgenic rice developed by silencing RINO1 gene.

Besides functional regions of the genome, the small (~22nts) regulatory RNAs (miRNAs) play a pivotal role in almost all biological and metabolic processes (Mangrauthia et al. 2017) including uptake and transport of minerals in plants (Kehr 2013). MiRNAs play an important role in the adaptive mechanisms of plants by regulating the expression of various stress-related genes (Tang and Chu 2017). Downregulation of miRNAs and consequent upregulation of their putative target genes indicated the involvement of miRNA-mediated regulation of Zn homeostasis. Zn deficiency in Sorghum bicolor increased the upregulation of several miRNA families, such as miR166, miR171, miR172, miR398, miR399, and miR319, which in turn target many gene family members, including transporters (Li et al. 2013). However, engineering miRNA expression levels could provide an effective alternate strategy for improving rice grain Zn concentration. The candidate miRNAs involved in Fe and Zn accumulation in rice grain can be engineered.

Cisgenesis and gene stacking: Transgenic high Zn rice using foreign genes is a sure shot to achieve the target, but it still faces strict regulatory hurdles and a lack of consumer acceptance. In contrast, cisgenic breeding involves the transfer of genes to a variety from the crossable plants of the same species or closely related species using a similar strategy as in transgenic breeding. This offers a unique advantage of overexpression of a cisgene under a preferred promoter and also to overcome the problem of linkage drag (Telem et al. 2013) compared to conventional breeding. Nicotianamine synthase genes from rice produced several fold nutrient (Fe and Zn) dense rice (HarvestPlus 2014; Johnson et al. 2010; Lee et al. 2012). *OsNAS2* gene under the control of 2xpCaMV35S promoter is reported to enhance 2.2-fold Zn concentration in rice seed (Johnson et al. 2011). Lee et al. (2011) were able to breed NA-rich NAS2 activation-tagged transgenic lines with more exudation of PS from roots and greater Zn uptake, translocation, and loading to grains. One such transgenic product contained as high as 20-fold NA and 2.7-fold more Zn in the rice endosperm.

Recombinase technology offers site-specific integration of a gene. The targeted gene insertion could facilitate the stacking of multiple genes (gene pyramiding) for desired traits in a specific location in the genome, analogous to that of multiple gene introgressions by conventional backcross breeding. Cisgenic barley with one to six extra phytase gene (*HvPAPhy_a*) inserts has shown a perfect positive linear correlation with the level of phytase activity in mature grains of barley (Holme et al. 2020). Similarly, the multivitamin corn inbred "M37W" is a product of five stacked genes (Naqvi et al. 2009). Until now, the highest increase in grain Zn content has been achieved through OsNAS2 overexpression (Trijatmiko et al. 2016). Therefore, cisgene stacking of additional copies of the rice *OsNAS2* gene for Zn biofortification in rice is worth trying. Cisgene staking may also be relevant to biofortification for multiple micronutrients and to offer tolerance to a number of stresses. Masuda et al. (2012) developed GM Tsukino Hikari rice through the introduction of *SoyferH2, HvNAS1*, and *OsYSL2* genes and observed 4.4-fold and 1.6-fold increased iron and zinc accumulation in brown rice grain. Similarly,

Aung et al. (2013) achieved GM Paw San Yin rice variety with an iron and zinc increase of 3.4-fold and 1.3-fold in polished grain using the above combination of candidate genes. Besides, multi-biofortified rice with enhanced concentrations of provitamin A, Zn, Fe, and folate has been reported to have potential health benefits and as such, be a cost-effective strategy for nutritional security (De Steur et al. 2011).

Genome editing: Conventional transgenic approaches often lead to off-target insertion, resulting in undesirable effects. However, genome editing offers efficient, precise, and targeted modifications at the genomic loci. Several functional and regulatory genes are involved in the key steps, including uptake, transport, and loading to grains and storage of Zn in the seed. These genes can be inserted (Lee et al. 2019) or even modulated (Qi et al. 2013) by available genome editing tools, e.g. clustered regulatory interspaced short palindromic repeat-associated protein 9 (CRISPR/Cas9), oligonucleotide site-directed mutagenesis, Zn finger nucleases (ZFNs) and transcription activator-like effector nucleases (TALENs) keeping the rest of the genome intact (Yin et al. 2017).

Editing of key genes has been successful for nutritional and other important traits in rice, e.g. OsPDS for carotenoid biosynthesis, GBSS II for amylose content (Zhang et al. 2018), OsSBEIIb for resistant starch content (Sun et al. 2017), BADH2 for fragrance (Shan et al. 2015), Rc for anthocyanin content (Zhu et al. 2019), LOX 3 for seed viability (Ma et al. 2015); *SD1, OsDEP1, OsIPA1, OsGn1a, OsGS3*, and *DST* genes for semi-dwarf plant type, dense erect panicle, panicle architecture, grain number, grain size, and drought tolerance (Li et al. 2016), OsRR22 for salt tolerance (Zhang et al. 2019) and OsMPK5 for disease resistance (Xie and Yang 2013). However, there is indeed limited progress for grain zinc improvement in rice, possibly due to the fact that the mutable locations in the DNA are limited due to the requirement of the appropriate protospacer adjacent motif (PAM) for editing (Zhang et al. 2019). Recently, variants of Cas have been developed that recognize varied PAM sites (Hu et al. 2016; Zhang et al. 2019). Moreover, a recently engineered efficient SpRY Cas9 can target PAM-less sequences.

Phytase can catalyze the hydrolysis of phytic acid (phytate) and release the chelated minerals (i.e. Fe^{2+} and Zn^{2+}). In the genetically modified Taipei-309 rice, the phytase level was 130-fold increased (in the grains). Conversely, the phytate biosynthesis gene(s) can be knocked out by the CRISPR-Cas9 system to increase the bioavailability of Zn in brown rice (Waltz 2016). Jiang et al. (2019) successfully demonstrated significant decrease in phytic acid content (−10.1 and −32.1%) in two rice mutant lines (ositpk6_1 and ositpk6_2) developed by targeted mutagenesis using the CRISPR/Cas9 method. OsVMT (vacuolar mugineic acid transporter) gene is involved in sequestering 2-deoxymugineic acid (DMA) into the vacuoles. Two CRISPR/Cas9 knock-out mutant lines (vmt-a and vmt-t) of cv. Nipponbare have been reported using *Agrobacterium tumefaciens*-mediated transformation (Che et al. 2019). Knockout of this gene enhances the accumulation of Fe and Zn in polished rice grains (without yield penalty) through DMA-increased solubilization of Fe and Zn deposited in the node (Che et al. 2019).

CRISPR/Cas9 involves NHEJ and HDR mechanisms for DNA damage repair. Its use is limited owing to undesired rearrangements due to double dtrand breaks (DSBs) at the targeted loci. While CRISPR/Cas9-based base editor tool brings about a sure shot permanent alteration of a targeted base without involving a DSB (Komor et al. 2016; Nishida et al. 2016). This comprises a fusion of the mutated Cas9 domain with a base-modifying deaminase enzyme. The guide RNA-mutated Cas 9 binds to the target DNA, making a short segment ss-DNA bubble, while the deaminase enzyme modifies the targeted base of the ssDNA (Nishida et al. 2016). Such a gene editing approach may be used for site-specific alteration in Zn transporter genes to recover novel alleles for Zn transport and loading to grains.

21.7 Conclusion

Zinc is essential for normal growth, and it bolsters immunity to viruses and other health threats. Rice grains are deficient in Zn. Agronomic biofortification seed priming and direct intake of Zn tablets cannot be sustainable approaches to meet our body's Zn requirement. However, developing Zn-rich rice and getting it on everyone's plate seems to be a game-changer to fight hidden hunger. A few high Zn rice varieties have been released in the target countries, but Zn content is sufficiently below the breeding target (28–30 mg kg^{-1}). Plants acquire Zn from soil and transport it to grains through the root, culm base, node, leaves, and finally remobilize to develop seeds. Until now, the knowledge of transcriptomics, haplotype analysis using SNPs, and in silico studies has greatly helped to explore key players (genes/QTLs) for micronutrient uptake, transport, and loading to the sink. Pyramiding the large effect candidate genes/QTLs by MAB using multiple crosses or more appropriately, the MAGIC populations seems to be a better alternative to biparental pedigree breeding. Wide hybridization and protoplast fusion techniques can also harness valuable genes related to high nutritional status that are available in wild relatives. However, cutting edge technologies, e.g. targeted mutagenesis (genome editing), recombinase technology (for site-specific gene replacement), as well as multiple cisgene stacking by GE, seem to have great promise for Zn biofortification and consumer's acceptance in the context of strict regulatory clearance for GM crops. It is expected that a new way for product development would provide sustainable nutritional security.

Declaration and Competing Interest

The author declares that there is no conflict of interest.

Funding Statement

Financial assistance was provided by CIAT and IFPRI through HarvestPlus Project on rice vide Grant No. 2018H8503. OUA.

Acknowledgments

The author is very thankful to the Central Instrumentation Facility (CIF), OUAT for providing ICP-OES facility. Besides, financial assistance provided by CIAT and IFPRI through HarvestPlus Project is highly acknowledged.

References

Agarwal, S., Mangrauthia, S.K. and Sarla, N. (2018). Genomic approaches for micronutrients biofortification of rice (Chapter 14). In: *Plant Micronutrient Use Efficiency*. Netherlands: Elsevier. doi: 10.1016/B978-0-12-812104-7.00014-9.

Ajmera, S., Kumar, S.S. and Ravindrababu, V. (2017). Character association analysis for grain iron and zinc concentrations and grain yield components in rice genotypes. *Bulletin of Environment, Pharmacology and Life Sciences* 6 (1): 177–181.

Alexandrov, N., Tai, S., Wang, W. et al. (2015). SNP-Seek database of SNPs derived from 3000 rice genomes. *Nucleic Acids Research* 43: 1023–1027.

Ali, N., Paul, S., Gayen, D. et al. (2013a). RNAi mediated down regulation of myo-inositol-3-phosphate synthase to generate low phytate rice. *Rice* 6 (1): 1–12.

Ali, N., Paul, S., Gayen, D. et al. (2013b). Development of low phytate rice by RNAi-mediated seed-specific silencing of inositol 1,3,4,5,6-pentakisphosphate 2-kinase gene (IPK1). *PLoS One* 8: e68161.

Al-Noor, M.M., Quadir, Q.F., Naher, J. et al. (2019). Inducing variability in rice for enriched iron and zinc content through *in vitro* culture. *Plant Tissue Culture and Biotech* 29 (2): 161–174.

Al-Zahrani, H.S.M., Almaghrabi, O.A., El-Malky, M.M. et al. (2018). Development of nutrient-rich and blast-resistant rice cultivars through tissue culture and monogenic lines. *International Journal of Agriculture and Biology* 20 (11): 2508–2520. doi: 10.17957/IJAB/15.0798.

Anandan, A., Rajiv, G., Eswaran, R. et al. (2011). Genetic variation and relationship between quality traits and trace elements in traditional and improved rice(*Oryza sativa* L.) genotypes. *Journal of Food Sciences* 76 (4): H122–H130.

Anuradha, K., Agarwal, S., Batchu, A.K. et al. (2012a). Evaluating rice germplasm for iron and zinc concentration in brown rice and seed dimensions. *Journal of Phytology* 4 (1): 19–25.

Anuradha, K., Agarwal, S., Rao, V.Y. et al. (2012b). Mapping QTLs and candidate genes for iron and zinc concentrations in unpolished rice of Madhukar × Swarna RILs. *Gene* 508: 233–240.

Anusha, G., Rao, D.S., Jaldhani, V. et al. (2021). Grain Fe and Zn content, heterosis, combining ability and its association with grain yield in irrigated and aerobic rice. *Scientific Reports* 11 (1): 10579.

Arnold, T., Markovic, T., Kir, G.J.D. et al. (2015). Iron and zinc isotope fractionation during uptake and translocation in rice (*Oryza sativa*) grown in oxic and anoxic soils. *Comptes Rendus Geoscience* 347 (7–8): 397–404. doi: 10.1016/j.crte.2015.05.005.

Aung, M.S., Masuda, H., Kobayashi, T. et al. (2013). Iron biofortification of Myanmar rice. *Frontiers in Plant Science* 4: 158.

Babu, P.M., Neeraja, C.N., Rathod, S. et al. (2020). Stable SNP allele associations with high grain zinc content in polished rice (*Oryza sativa* L.) identified based on ddRAD sequencing. *Frontier in Genetics* 11: 763. doi: 10.3389/fgene.2020.00763.

Babu, V.R., Shreya, K., Dangi, K.S. et al. (2012). Genetic variability studies for qualitative and quantitative traits in popular rice (*Oryza sativa* L) hybrids of India. *International Journal of Scientific and Research Publications* 2: 2250–3153.

Babu, V.R., Shreya, K., Kuldeep, S. D. et al. (2013). Evaluation of popular rice (*Oryza sativa* L.) hybrids for quantitative, qualitative and nutritional aspects. *International Journal of Scientific and Research Publications* 3 (1): 1–8.

Bado, S., Padilla-Alvarez, R., Migliori, A. et al. (2016). The application of XRF and PIXE in the analysis of rice shoot and compositional screening of genotypes. In: 22nd International conference on ion beam analysis. Special issue: *Nuclear Instruments and Methods in Physics Research, Section B: Beam Interactions With Materials and Atoms* 371: 407–412. doi: 10.1016/j.nimb.2015.08.081.

Baisakh, N., Rehana, S., Rai, M. et al. (2006). Marker-free transgenic (MFT) near-isogenic introgression lines (NIILs) of 'golden' indica rice (cv. IR64) with accumulation of provitamin A in the endosperm tissue. *Plant Biotechnology Journal* 4: 467–475. doi: 10.1111/j.1467-7652.2006.00196.x.

Banakar, R., Alvarez, F. A., Díaz-Benito, P. et al. (2017). Phytosiderophores determine thresholds for iron and zinc accumulation in biofortified rice endosperm while inhibiting the accumulation of cadmium. *Journal of Experimental Botany* 68: 4983–4995.

Bandillo, N., Raghava, C., Muyco, P.A. et al. (2013). Multi-parent advanced generation inter-cross (MAGIC) populations in rice: progress and potential for genetics research and breeding. *Rice* 6 (1): 1–15.

Banerjee, S., Sharma, D.J., Verulkar, S.B. et al. (2010). Use of *in silico* and semiquantitative RT-PCR approaches to develop nutrient rich rice (*Oryza sativa* L.). *Indian Journal of Biotechnology* 9 (2): 203–212.

Barokah, U., Susanto, U., Swamy, M.B.P. et al. (2018). High-zinc rice as a breakthrough for high nutritional rice breeding program. *IOP Conference Series Earth and Environmental Science* 129 (1): 012004.

Bashir, K., Ishimaru, Y. and Nishizawa, N.K. (2012). Molecular mechanisms of zinc uptake and translocation in rice. *Plant and Soil* 129 (1–2):189–201.

Batista, R., Saibo, N., Lourenco, T. et al. (2008). Microarray analyses reveal that plant mutagenesis may induce more transcriptomic changes than transgene insertion. *Proceedings of the National Academy of Sciences* 105 (9): 3640–3645.

Beauchemin, D. (2017). Inductively coupled plasma mass spectrometry methods. In: *Encyclopedia of Spectroscopy and Spectrometry*, 3e (eds. J.C. Lindon. G.E. Tranter and D.W. Koppenaal), 236–245. Massachusetts: Academic Press. doi: 10.1016/B978-0-12-409547-2.11222-3.

Bekele, B.D., Naveen, G.K., Rakhi, S. et al. (2013). Genetic evaluation of recombinant inbred lines of rice (*Oryza sativa* L.) for grain zinc concentrations, yield related traits and identification of associated SSR markers. *Pakistan Journal of Biological Sciences* 16 (23): 1714–1721. doi: 10.3923/pjbs.2013.1714.1721.

Bollinedi, H., Yadav, A.K., Vinod, K.K. et al. (2020). Genome-wide association study reveals novel marker-trait associations (MTAs) governing the localization of Fe and Zn in the rice grain. *Frontiers Genetics* 11: 213. doi: 10.3389/fgene.2020.00213.

Boonyaves, K., Wu, T.Y., Gruissem, W. et al. (2017). Enhanced grain iron levels in rice expressing an iron regulated metal transporter, nicotianamine synthase, and ferritin gene cassette. *Frontier in Plant Science* 8: 130.

Bouis, H. and Welch, R. (2010). Biofortification-A Sustainable agricultural strategy for reducing micronutrient malnutrition in the global south. *Crop Science* 50: 20–32.

Brar, B., Jain, R.K. and Jain, S. (2015). Correlation of molecular marker allele size with physio-morphological and micronutrient (Zn, Fe) traits among rice genotypes. *International Journal of Current Science* 15: E 42– 50.

Brar, B., Jain, S. and Jain, R.K. (2014). Molecular profiling of rice (*Oryza sativa* L.) genotypes differing in micronutrients (iron and zinc) content. *Indian Journal of Genetics and Plant Breeding* 74 (1): 81–85.

Brar, B., Jain, S., Singh, R. et al. (2011). Genetic diversity for iron and zinc contents in a collection of 220 rice (*Oryza sativa* L.) genotypes. Indian *Journal of Genetics and Plant Breeding* 71 (1): 67–73.

Cai, H., Huang, S., Che, J. et al. (2019). The tonoplast-localized transporter OsHMA3 plays an important role in maintaining Zn homeostasis in rice. *Journal of Experimental Botany* 70 (10): 2717–2725.

Calayugan, M.I.C., Formantes, A.K., Amparado, A. et al. (2020). Genetic analysis of agronomic traits and grain iron and zinc concentrations in a doubled haploid population of rice (*Oryza sativa* L.). *Scientific Reports* 10 (1): 2283. doi: 10.1038/s41598-020-59184-z.

Cavanagh, C., Morell, M., Mackay, I. et al. (2008). From mutations to magic: resources for gene discovery, validation and delivery in crop plants. *Current Opinion in Plant Biology* 11: 215–221. doi: 10.1016/j.pbi.2008.01.002.

Chakravorty, A. (2020). The role of variability in biofortification and *in vitro* conservation of diversed rice landraces of West Bengal. *Journal of Advanced Scientific Research* 11 (6): 208–212.

Chandu, G., Addanki, K.R., Balakrishnan, D. et al. (2020). SSR markers for grain iron zinc and yield-related traits polymorphic between Samba Mahsuri (BPT 5204) and a wild rice *Oryza rufipogon*. *Electronic Journal of Plant Breeding* 11 (3): 841–847. doi: 10.37992/2020.1103.138.

Chang, S., El Arifeen, S., Bari, S. et al. (2010). Supplementing iron and zinc: Double blind, randomized evaluation of separate or combined delivery. *European Journal of Clinical Nutrition* 64 (2): 153–160.

Chattopadhyay, K., Sharma, S.G., Bagchi, T.B. et al. (2019). CR Dhan 310 and CR Dhan 311 high protein, nutrient-rich rice varieties of NRRI. NRRI, Cuttack (India). https://icar-nrri.in/wp-content/uploads/2019/06/2.-leaflet_highprotein_final.pdf.

Che, J., Yokosho, K., Yamaji, N. et al. (2019). A Vacuolar phytosiderophore transporter alters iron and zinc accumulation in polished rice grains. *Plant Physiology* 181: 276–288. doi: 10.1104/pp.19.00598.

Chen, Y.L., Liang, L., Ma, X.L. et al. (2012). An efficient rice mutagenesis based on suspension –cultured cells. *Journal of Integrative Plant Biology* 55 (2): 122–130. doi: 10.1111/jipb.12000.

Chinnusamy, V. and Zhu, J.K. (2009). Epigenetic regulation of stress responses in plants. *Current Opinion in Plant Biology* 12 (2):133–139.

Choi, E.Y., Graham, R.and Stangoulis, J. (2007). Semi-quantitative analysis for selecting Fe- and Zn-dense genotypes of staple food crops. *Journal of Food Composition and Analysis* 20 (7): 496–505.

De Steur, H., Gellynck, X., Straeten, D.V.D. et al. (2011). Potential impact and cost effectiveness of multi-biofortified rice in China. *New Biotechnology* 29: 432–442.

Descalsota-Empleo, G.I., Amparado, A., Inabangan-Asilo, M.A. et al. (2019). Genetic mapping of QTL for agronomic traits and grain mineral elements in rice. *The Crop Journal* 7 (4): 560–572. doi: 10.1016/j.cj.2019.03.002.

Descalsota, G., Swamy, B., Zaw, H. et al. (2018). Genome-wide association mapping in a rice MAGIC plus population detects QTLs and genes useful for biofortification. *Frontiers in Plant Science* 9: 1347. doi: 10.3389/fpls.2018.01347.

Du, J., Zeng, D., Wang, B. et al. (2013). Environmental effect on mineral accumulation in rice grains and identification of ecological specific QTLs. *Environmental Geochemistry and Health* 35 (2): 161–170.

Elshire, R.J., Glaubitz, J.C., Sun, Q. et al. (2011). A robust, simple genotyping-by-sequencing (GBS) approach for high diversity species. *PLoS One* 6 (5): e19379.

FAO (2017). The future of food and agriculture-Trends and challenges. http://www.fao.org/3/a-i6583e.pdf.

Fageria, N.K., Santos, A.B., BarbosaFilho, M.P. et al. (2008). Iron toxicity in low land rice. *Journal of Plant Nutrition* 31 (9): 1676–1697.

Faza, M., Anis, G., Mazal, T.M. et al. (2016). Genetic identification of quantitative trait loci (QTL) for high nutritional value traits in DHLs derived from black rice. *Journal of Agriculture Biotechnology* 1 (2): 59–67. doi: 10.20936/JAB/160203.

Ferdausi, A., Nath, U.K., Das, B.L. et al. (2009). *In vitro* regeneration system in brinjal (*Solanum melongena* L.) for stress tolerant somaclone selection. *Journal of the Bangladesh Agricultural University* 7 (2): 253–258.

Finch, R.P., Slamet, H. and Cocking, E.C. (1990). Production of heterokaryons by the fusion of mesophyll protoplasts of porteresia coarctata and cell suspension-derived protoplasts of *Oryza sativa*: a new approach to somatic hybridization in rice. *Journal of Plant Physiology* 136 (5): 592–598.

Finatto, T., de Oliveira, A.C. and Chaparro, C. (2015). Abiotic stress and genome dynamics: specific genes and transposable elements response to iron excess in rice. *Rice* 8: 13.

Galili, G. (2002). New insights into the regulation and functional significance of lysine metabolism in plants. *Annual Review of Plant Biology* 53 (1): 27–43.

Gande, N.K., Kundur, P.J., Soman, R. et al. (2014). Identification of putative candidate gene markers for grain zinc content using recombinant inbred lines (RIL) population of IRRI38 X Jeerigesanna. *African Journal of Biotechnology*: 13 (5): 657–663. doi: 10.5897/AJB2013.13334.

Garcia-Oliveira, A.L., Tan, L., Fu, Y. et al. (2008). Genetic identification of quantitative trait loci for contents of mineral nutrients in rice grain. *Journal of Integrative Plant. Biology* 51 (1): 84–92. doi: 10.1111/j.1744-7909.2008.00730.x.

Gernand, A.D., Schulze, K.J., Stewart, C.P. et al. (2016). Micronutrient deficiencies in pregnancy worldwide: health effects and prevention. *Nature Reviews Endocrinology* 12 (5): 274–289. doi: 10.1038/nrendo.2016.37.

Gregorio, G.B. (2002). A new tools for fetching micronutrient malnutrition. *Journal of Nutrition* 132: 500–502.

Gregorio, G.B., Senadhira, D., Htut, H. et al. (2000). Breeding for trace mineral density in rice. *Food and Nutrition Bulletin* 21 (4): 382–386. doi: 10.1177/156482650002100407.

HarvestPlus (2014). Biofortification progress briefs. http://www.harvestplus.org/sites /default/files/Biofortification _Progress _Briefs_August2014_WEB_0.pdfHarvest plus 2014.

Hefferon, K. (2019). Biotechnological approaches for generating zinc-enriched crops to combat malnutrition. *Nutrients* 11 (2): 253. doi: 10.3390/nu11020253.

Higuchi, K., Watanabe, S., Takahashi, M. et al. (2001). Nicotianamine synthase gene expression differs in barley and rice under Fe-deficient conditions. *Plant Journal* 25 (2): 159–167.

Holme, I.B., Madsen, C.K., Wendt, T. et al. (2020). Horizontal stacking of PAPhy_a cisgenes in barley is a potent strategy for increasing mature grain phytase activity. *Frontiers Plant Science* 11: 592139. doi: 10.3389/fpls.2020.592139.

Hu, B.L., Huang, D.R., Xiao, Y.Q. et al. (2016). Mapping QTLs for mineral element contents in brown and milled rice using an *Oryza sativa* × *Oryza rufipogon* backcross inbred line population. *Cereal Research Communications* 44: 57–68.

ICAR-IIRR (2016). Annual progress report, varietal improvement, All India Co-ordinated rice improvement project, ICAR-Indian Institute of Rice Research (IIRR), Rajendranagar, Hyderabad, Telangana, India 1:1.420–1.447.

ICAR-IIRR (2017). Annual progress report, varietal improvement, All India Co-ordinated Rice Improvement Project, ICAR-Indian Institute of Rice Research (IIRR), Rajendranagar, Hyderabad, Telangana, India 1:1.313–1.337.

ICAR-IIRR (2023). Dhan 63: IIRR develops biofortified rice rich in zinc, other minerals. The New Indian Express. https://www.newindianexpress.com/states/telangana/2023/mar/31/dhan-63-iirr-develops-biofortified-rice-rich-in-zinc-other-minerals-2561181.html.

ICAR-NRRI (2022). ICAR-NRRI develops bio-fortified rice varieties. The Statesman. https://www.thestatesman.com/india/icar-nrri-develops-bio-fortified-rice-varieties-1503137049.html.

Impa, S.M., Morete, M.J., Ismail, A.M. et al. (2013). Zn uptake, translocation and grain Zn loading in rice (*Oryza sativa* L.) genotypes selected for Zn deficiency tolerance and high grain Zn. *Journal of Experimental Botany* 64 (10): 2739–2751. doi: 10.1093/jxb/ert118.

International Rice Genome Sequencing Project (IRGSP) (2005). The map-based sequence of the rice genome. *Nature* 436 (7052): 793–800.

Ishikawa, R., Iwata, M., Taniko, K. et al. (2017). Detection of quantitative trait loci controlling grain zinc concentration using Australian wild rice (*Oryza meridionalis*) a potential genetic resource for biofortification of rice. *PLoS One* 12: e0187224.

Ishikawa, S., Abe, T., Kuramata, M. et al. (2010). A quantitative trait locus for increasing cadmium-specific concentration in rice grain is located on short arm of chromosome 7. *Journal of Experimental Botany* 613: 923–934.

Ishimaru, Y., Bashir, K. and Nishizawa, N.K. (2011). Zn uptake and translocation in rice plants. *Rice* 4: 21–27. doi: 10.1007/s12284-011-9061-3.

Ishimaru, Y., Suzuki, M., Kobayashi, T. et al. (2005). OsZIP4, a novel zinc regulated zinc transporter in rice. *Journal of Experimental Botany* 56: 3207–3214.

Jain, S.M. (2010). Induced mutation for enhancing nutrition for food production. *NJF Rep.* 6 (2):64–67.

Jaksomsak, P., Jaksomsak, N., Dell, B. et al. (2015). Variation of seed zinc in a local upland rice germplasm from Thailand. *Plant Genetic Resources* 13 (2): 168–175.

James, C.R., Huynh, B.L., Welch, R.M. et al. (2007). Quantitative trait loci for phytate in rice grain and their relationship with grain micronutrient content. *Euphytica* 154 (3): 289–294.

Jeng, T.L., Lin, Y.W., Wang, C.S. et al. (2012). Comparisons and selection of rice mutants with high iron and zinc contents in their polished grains that were mutated from the indica type cultivar IR64. *Journal of Food Composition and Analysis* 28 (2):149–154.

Jiang, M., Liu, Y., Liu, Y. et al. (2019). Mutation of inositol 1, 3, 4-trisphosphate 5/6-kinase6 impairs plant growth and phytic acid synthesis in rice. *Plants* 8: 114.

Jiang, W., Struik, P.C., Van Keulen, H. et al. (2008). Does increased zinc uptake enhance grain zinc mass concentration in rice? *Annals of Applied Biology* 153: 135–147. doi: 10.1111/j.1744-7348.2008.00243.x.

Jin, T., Chen, J. and Zhu, L. (2015). Comparative mapping combined with homology based cloning of the rice genome reveals candidate genes for grain zinc and iron concentration in maize. *BMC Genetics* 16: 17.

Johnson, A.A.T., Kyriacou, B., Callahan, D.L. et al. (2011). Constitutive overexpression of the OsNAS gene family reveals singlegene strategies for effective iron- and zinc-biofortification of rice endosperm. *PLoS One* 6: e24476. doi: 10.1371/journal.pone.0024476.

Johnson-Beebout, S.E., Goloran, J.B., Rubianes, F.H. et al. (2016). Zn uptake behavior of rice genotypes and its implication on grain Zn biofortification. *Scientific Reports*. 6: 38301.

Josh, J. (2013). Bangladesh developed world's first zinc-enriched rice variety. https://www.jagranjosh.com/current-affairs/bangladesh-developed-worlds-first-zincenriched-rice-variety-1377523748-1.

Kabir, A.H., Hossain, M.M., Khatun, M.A. et al. (2017). Biochemical and molecular mechanisms associated with Zn deficiency tolerance and signaling in rice (*Oryza sativa* L.). *Journal of Plant Interactions* 12 (1): 447–456. doi: 10.1080/17429145.2017.1392626.

Karmakar, A., Bhattacharya, S., Sengupta, S. et al. (2019). RNAi-mediated silencing of ITPK gene reduces phytic acid content, alters transcripts of phytic acid biosynthetic genes, and modulates mineral distribution in rice seeds. *Rice Science* 27 (4): 315–328. http://www.ricescience.org/EN/abstract/abstract9861.shtml.

Kawakami, Y. and Bhullar, N.K. (2018). Molecular processes in iron and zinc homeostasis and their modulation for biofortification in rice. *Journal of Integrative Plant Biology* 60 (12): 1181–1198.

Kehr, J. (2013). Systemic regulation of mineral homeostasis by microRNAs. *Frontiers Plant Science* 4:145.

Kobayashi, T., Nagasaka, S., Senoura, T. et al. (2013). Iron-binding haemerythrin RING ubiquitin ligases regulate plant iron responses and accumulation. *Nature Communications* 4: 2792.

Krishnan, S. and Dayanandan, P. (2003). Structuralandhisto-chemicalstudieson grain-filling inthecaryopsisofrice(*Oryza sativa* L.). *Journal of Biosciences* 28: 455–469. doi: 10.1007/BF02705120.

Kumar, A. (2020). Genetic analysis in relation to iron and zinc accumulation in rice. Ph. D. thesis. Department of Plant Breed and Genetics, India: Dr. Rajendra Prasad Central Agril. University, Pusa. http://krishikosh.egranth.ac.in/handle/1/5810150573.

Kumar, A.A., Reddy, B.V.S., Ramaiah, B. et al. (2009). Genetic variability and plant character association of grain Fe and Zn in selected core collection accessions of *sorghum* germplasm and breeding lines. *SAT eJournal* 7: 1–4.

Kumar, J., Jain, S. and Jain, R.K. (2014). Linkage mapping for grain iron and zinc content in F_2 population derived from the cross between PAU201 and Palman 579 in rice (*Oryza sativa* L.). *Cereal Research Communications* 42 (3): 389–400. doi: 10.1556/CRC.42.2014.3.3.

Kumar, N, Jain, R.K. and Chowdhury, V.K. (2019). Linkage mapping of QTLs for grain minerals (iron and zinc) and physio-morphological traits for development of mineral rich rice (*Oryza sativa* L.). *Indian Journal of Biotechnology* 18: 69–80.

Kumari, K., Kumar, P., Sharma, V.K. et al. (2019). Genomic marker assisted identification of genetic loci and genes associated with variation of grain zinc concentration in rice. *Journal of Genetics* 98: 111. doi: 10.1007/s12041-019-1144-8.

Komor, A.C., Kim, Y.B., Packer, M.S. et al. (2016). Programmable editing of a target base in genomic DNA without double-stranded DNA cleavage. *Nature* 533: 420–424.

Kuwano, M., Mimura, T., Takaiwa, F. et al. (2009). Generation of stable 'low phytic acid' transgenic rice through antisense repression of the 1d-myo-inositol 3-phosphate synthase gene (RINO1) using the 18-kDa oleosin promoter. *Plant Biotechnology Journal* 7: 96–105.

Kyozuka, J., Kaneda, T. and Shimamoto, K. (1989). Production of cytoplasmic male sterile rice (*Oryza sativa* L.) by cell fusion. *Nature Biotechnology* 7: 1171–1174. doi: 10.1038/nbt1189–1171.

Laura, T.M., Julien, P., Jose, T. et al. (2016). Association of increased grain iron and zinc concentrations with Agro-morphological traits of biofortified rice. *Plant Science* 7: 1463.

Lee, K., Eggenberger, A.L., Banakar, R. et al. (2019). CRISPR/Cas9-mediated targeted T-DNA integration in rice. *Plant Molecular Biology* 99: 317–328. doi: 10.1007/s11103-018-00819-1.

Lee, S. and An, G. (2009). Over-expression of OsIRT1 leads to increased iron and zinc accumulations in rice. *Plant Cell and Environment* 32: 408–416. doi: 10.1111/j.1365-3040.2009.01935.x.

Lee, S., Jeong, H.J., Kim, S.A. et al. (2010a). OsZIP5 is a plasma membrane zinc transporter in rice. *Plant Molecular Biology* 73: 507–517.

Lee, S., Jeon, U.S., Lee, S.J. et al. (2009). Iron fortification of rice seeds through activation of the nicotianamine synthase gene. *Journal Proceedings of the National Academy of Science of the United States of America* 106: 22014–22019.

Lee, S., Kim, S.A., Lee, J. et al. (2010b). Zinc deficiency-inducible OsZIP8 encodes a plasma membrane-localized zinc transporter in rice. *Molecules and Cells* 29: 551–558.

Lee, S., Kim, Y.S., Jeon, U.S. et al. (2012). Activation of rice nicotianamine synthase 2 (OsNAS2) enhances iron availability for biofortification. *Molecules and Cells* 33: 269–275.

Lee, S., Kim, Y.Y., Lee, Y. et al. (2007). Rice P1B-type heavy-metal ATPase, OsHMA9, is a metal efflux protein. *Plant Physiology* 145: 831–842.

Lee, S., Persson, D.P., Hansen, T.H. et al. (2011). Bio-available zinc in rice seeds is increased by activation tagging of nicotianamine synthase. *Plant Biotechnology Journal* 9 (8): 865–873.

Lee, S.M., Kang, J.W., Lee, J.Y. et al. (2020). QTL Analysis for Fe and Zn concentrations in rice grains using a doubled haploid population derived from a cross between rice (*Oryza sativa*) cultivar 93-11 and Milyang 352. *Plant Breeding and Biotechnology* 8 (1): 69–76. doi: 10.9787/PBB.2020.8.1.69.

Li, J., Meng, X., Zong, Y. et al. (2016). Gene replacements and insertions in rice by intron targeting using CRISPR-Cas9. *Nature Plants* 2 16139.

Li, Y., Zhang, Y. and Shi, D. (2013). Spatial-temporal analysis of zinc homeostasis reveals the response mechanisms to acute zinc deficiency in Sorghum bicolor. *New Phytologist* 200: 1102–1115.

Liu, X.S., Feng, S.J., Zhang, B.Q. et al. (2019). OsZIP1 functions as a metal efflux transporter limiting excess zinc, copper and cadmium accumulation in rice. *BMC Plant Biology* 19: 283. doi: 10.1186/s12870-019-1899-3.

Lu, K., Li, L., Zheng, X. et al. (2008). Quantitative trait loci controlling Cu Ca Zn Mn and Fe content in rice grains. *Journal of Genetics* 87 (3): 305–310.

Ma, L., Zhu, F., Li, Z. et al. (2015). TALEN-based mutagenesis of lipoxygenase LOX_3 enhances the storage tolerance of rice (*Oryza sativa*) seeds. *PLoS One* 10: e0143877.

Maganti, S., Swaminathan, R. and Parida, A. (2020). Variation in iron and zinc content in traditional rice genotypes. *Agricultural Research* 9: 316–328. doi: 10.1007/s40003-019-00429-3.

Majumder, S., Datta, K. and Datta, S.K. (2019). Rice biofortification: High iron, zinc, and vitamin-A to fight against "hidden hunger". Agronomy 9 (12): 803. doi:10.3390/agronomy9120803.

Mangrauthia, S.K., Bhogireddy, S. and Agarwal, S. (2017). Genome-wide changes in microRNA expression during short and prolonged heat stress and recovery in contrasting rice cultivars. *Journal of Experimental Botany* 68 (9): 2399–2412.

Masuda, H., Ishimaru, Y., Aung, M.S. et al. (2012). Iron biofortification in rice by the introduction of multiple genes involved in iron nutrition. *Scientific Reports* 2: 543.

Masuda, H., Suzuki, M., Morikawa, K.C. et al. (2008). Increase in iron and zinc concentrations in rice grains via the introduction of barley genes involved in phytosiderophore synthesis. *Rice* 1: 100–108.

Mingotte, F.L.C., Revolti, L.T.M., Unêda-Trevisoli, S.H. et al. (2018). Rice (Oryza sativa) breeding strategies for grain biofortification. *African Journal of Biotechnology* 17 (14): 466–477. doi: 10.5897/AJB2017.16329.

Moreno-Moyano, L.T., Bonneau, J.P., Sánchez-Palacios. et al. (2016). Association of increased grain iron and zinc concentrations with agro-morphological traits of biofortified rice. *Frontiers Plant Science* 7: 1463. doi: 10.3389/fpls.2016.01463.

Myers, S.S., Zanobetti, A., Kloog, I. et al. (2014). Increasing CO_2 threatens human nutrition. *Nature* 510: 139– 142.doi:10.1038/nature13179.

Nakandalage, N., Nicolas, M., Norton, R.M. et al. (2016). Improving rice zinc biofortification success rates through genetic and crop management approaches in a changing environment. *Frontier Plant Science* 7: 764. doi: 10.3389/fpls.2016.00764.

Naqvi, S., Zhu, C., Farre, G. et al. (2009). Transgenic multivitamin corn through biofortification of endosperm with three vitamins representing three distinct metabolic pathways. In: *Proceedings of the National Academy of Sciences USA* 106: 7762–7767.

Neeraja, C.N., Kulkarni, K.S., Madhu Babu, P. et al. (2018). Transporter genes identified in landraces associated with high zinc in polished rice through panicle transcriptome for biofortification. *PLoS One* 13 (2): e0192362. doi: 10.1371/journal.pone.0192362.

Nishida, K., Arazoe, T., Yachie, N. et al. (2016). Targeted nucleotide editing using hybrid prokaryotic and vertebrate adaptive immune systems. *Science* 102: 553–563.

Norton, G.J., Deacon, C.M., Xiong, L. et al. (2010). Genetic mapping of the rice ionome in leaves and grain: identification of QTLs for 17 elements including arsenic cadmium iron and selenium. *Plant Soil* 329: 139–153.

Norton, G.J., Douglas, A., Lahner, B. et al. (2014). Genome wide association mapping of grain arsenic copper molybdenum and zinc in rice (*Oryza sativa* L) grown at four international field sites. *PLoS One* 9: e89685.

Ogo, Y., Itai, R.N., Kobayashi, T. et al. (2011). $OsIRO_2$ is responsible for iron utilization in rice and improves growth and yield in calcareous soil. *Plant Molecular Biology* 75: 593–605.

Ohmori, Y., Sotta, N. and Fujiwara, T. (2016). Identification of introgression lines of *Oryza glaberrima* Steud. with high mineral content in grains. *Soil Science and Plant Nutrition* 62 (5–6): 456–464. doi: 10.1080/00380768.2016.1204896.

Oliva, N., Mohanty, P.C., Poletti, S. et al. (2014). Large-scale production and evaluation of marker-free indica rice IR64 expressing phytoferritin genes. *Molecular Breeding* 33: 23–37.

Olsen, L.I. and Palmgren, M. (2014). Many rivers to cross: The journey of zinc from soil to seed. *Frontiers in Plant Science* 5: 30.

Patil, A.C. and Shashidhar, H.E. (2016). Validation of Zn transporter specific primers in rice (*Oryza sativa* L.) accessions grown under aerobic condition. *Research Journal of Chemistry and Environment* 20 (5): 34–38.

Patil, A.C., Shivakumar, N., Rajanna, M.P. et al. (2015). Micronutrient (zinc and iron) productivity in rice (*Oryza sativa* L.). *Ecology Environment and Conservation* 21: S337–S346.

Paul, S., Ali, N., Datta, S.K. et al. (2014). Development of an iron-enriched high-yieldings indica rice cultivar by introgression of a high-iron trait from transgenic iron-biofortified rice. *Plant Foods for Human Nutrition* 69 (3): 203–208.

Paul, S., Ali, N., Gayen, D. et al. (2012). Molecular breeding of Osfer 2 gene to increase iron nutrition in rice grain. *GM Crops Food* 3: 310–316.

Potrykus, I. (1991). Gene transfer to plants: assessment of published approaches and results. *Annual Review of Plant Physiology and Plant Molecular Biology* 42: 205–225.

Potrykus, I. (2015). From the concept of totipotency to biofortified cereals. *Annual Review of Plant Biology* 66, 1–22.

Pradhan, S.K., Pandit, E., Pawar, S. et al. (2020). Linkage disequilibrium mapping for grain Fe and Zn enhancing QTLs useful for nutrient dense rice breeding. *BMC Plant Biology* 20: 57.

Qi, L.S., Larson, M.H., Gilbert, L.A. et al. (2013). Repurposing CRISPR as an RNA guided platform for sequence-specific control of gene expression. *Cell* 152: 1173–1183.

Qu, L.Q., Yoshihara, T., Ooyama, A. et al. (2005). Iron accumulation does not parallel the high expression level of ferritin in transgenic rice seeds. *Planta* 222 (2): 225–233.

Ramesh, S.A., Shin, R., Eide, D.J. et al. (2003). Differential metal selectivity and gene expression of two zinc transporters from rice. *Plant Physiology* 133: 126–134.

Rathod, R., Pulagam, M.B., Sanjeeva Rao, D. et al. (2017). Genetic divergence in high iron and zinc genotypes of rice (*Oryza sativa* L). *Electronic Journal of Plant Breeding* 7 (3): 638–640.

Raza, Q., Riaz, A., Sabar, M. et al. (2019). Meta-analysis of grain iron and zinc associated QTLs identified hotspot chromosomal regions and positional candidate genes for breeding biofortified rice. *Plant Science* 288: 110214. doi: 10.1016/j.plantsci.2019.110214.

Reddy, M., Turaidar, V., Krupa Anantapur, K.N.R. et al. (2018). Enhancement of iron and zinc in rice grain through biofortification approach. *International Journal of Current Microbiology and Applied Sciences* (7): 628–637.

Ricachenevsky, F.K., Punshon, T., Lee, S. et al. (2018). Elemental profiling of rice FOX lines leads to characterization of a new Zn plasma membrane transporter, OsZIP7. *Frontier Plant Science* 9: 865. doi: 10.3389/fpls.2018.00865.

Roda, F.A., Marques, I., Batista-Santos, P. et al. (2020). Rice biofortification with zinc and selenium: a transcriptomic approach to understand mineral accumulation in flag leaves. *Frontiers in Genetics* 11: 543.doi: 10.3389/fgene.2020.00543.

Roohani, N., Hurrell, R., Kelishadi, R. et al. (2013). Zinc and its importance for human health: an integrative review. *Journal of Research in Medical Sciences* 18: 144–157.

Sadeghzadeh, B. (2013). A review of zinc nutrition and plant breeding. *Journal of Soil Science and Plant Nutrition* 13 (4): 905–927.

Sahoo, B. (2019). Exploring candidate genes and in vitro mutagenesis in zinc rich rice. Bhubaneswar, India: M.Sc. (Agril. Biotechnology) thesis, Odisha University of Agril. And Technology.

Sala, M., Anandakumar, C.R., Geetha, S., Gnanamalar, R.P. et al. (2013). Variability studies for iron and zinc content on segregating population of rice. *Electronic Journal of Plant Breeding* 4 (3): 1201–1204.

Saleem, M.Y., Mukhtar, Z., Cheema, A.A. et al. (2005). Induced mutation and *in vitro* techniques as a method to induce salt tolerance in Basmati rice (*Oryza sativa* L.). *International Journal of Environmental Science and Technology* 2 (2): 141–145.

Sandeep, T.S.R.S. and Godi, S. (2023). Insights into the biofortification of zinc and iron in rice. *Nutrition. and Food Science International Journal* 12 (1): 1–3. doi: 10.19080/NFSIJ.2023.12.555827.

Sanjeeva Rao, D., Neeraja, C.N., Madhu Babu, P. et al. (2020). Zinc biofortified rice varieties: challenges, possibilities, and progress in India. *Frontiers in Nutrition* 7: 26. doi: 10.3389/fnut.2020.00026.

Sarla, N., Swamy, B.P.M., Kaladhar, K. et al. (2012). Increasing iron and zinc in rice grains using deep water rices and wild species–identifying genomic segments and candidate genes. *Quality Assurance and Safty of Crops & Food* 4: 138.

Sasaki, A., Yamaji, N., Mitani-Ueno, N. et al. (2015). A node-localized transporter OsZIP3 is responsible for the preferential distribution of Zn to developing tissues in rice. *Plant Journal* 84: 374–384.

Sauer, U., Heinemann, M. and Zamboni, N. (2007). Genetics: getting closer to the whole picture. *Science* 316: 550–551. doi: 10.1126/science.1142502.

Saxesena, R.R., Lal, G.M., Yadav, P.S. et al. (2013). Diversity analysis and identification of promising lines for hybridization in field pea (*Pisum sativum* L.). *The Bioscan* 8 (4): 1437–1440.

Serrat-Gurrera, X. (2016). *Applied Biotechnology to Improve Mediterranean Rice Varieties*. Bercelona: A doctoral thesis submitted to University of Barcelona.

Shahbandeh (2021). World rice acreage 2010–2019. Statista. https://www.statista.com/statistics/271969/world-rice-acreage-since-2008/.

Shahzad, Z., Rouached, H. and Rakha, A. (2014). Combating mineral malnutrition through iron and zinc biofortification of cereals. *Comprehensive Reviews in Food Science and Food Safety* 13 (3): 329–346. doi: 10.1111/1541-4337.12063.

Shan, Q., Zhang. Y., Chen, K. et al. (2015). Creation of fragrant rice by targeted knockout of the OsBADH2 gene using TALEN technology. *Plant Biotechnology Journal* 13 (6): 791–800.

Shu, Q.Y. (2009). A summary of the international symposium on induced mutations in plants. In: *Shu QY (Ed) Induced Plant Mutations in the Genomics Era*, 15–18. Rome: Food and Agriculture Organization of the United Nations.

Singh, S.P., Gruissem, W. and Bhullar, N.K. (2017). Single genetic locus improvement of iron, zinc and-carotene content in rice grains. *Scientific Reports.* 7 (1): 6883.

Soumitra, P., Nusrat, A., Dipak, G. et al. (2012). Molecular breeding of Osfer 2 gene to increase iron nutrition in rice grain. *GM Crops Food Biotechnology. Agriculture and the Food Chain* 3 (4): 310–316.

Sri Devi, P., Krishna Veni, B., Jyothula, D.P.B. et al. (2019). Association studies for yield components, physico-chemical and nutritional quality parameters in colored rice (*Oryza sativa* L.). *International Journal of Current Microbiology and Applied Sciences* 8 (6): 359–365.

Stangoulis, J. (2010). Technical aspects of zinc and iron analysis in biofortification of the staple food crops, wheat and rice. In: *Proceeding of the 19th World Congress of Soil Science Soil Solutions for a Changing World*, 42–44. Australia: Brisbane.

Stangoulis, J., Huynh, B.L., Welch, R.M. et al. (2007). Quantitative trait loci for phytate in rice grain and their relationship with grain micronutrient content. *Euphytica* 154: 289–294.

Stein, R.J., Ricachenevsky, F.K. and Fett, J.P. (2009). Differential regulation of the two rice ferritin genes (OsFER1 and OsFER2). *Plant Science* 177: 563–569. doi: 10.1016/j.plantsci.2009.08.001.

Sun, Y., Jiao, G., Liu, Z. et al. (2017). Generation of high-amylose rice through CRISPR/Cas9-mediated targeted mutagenesis of starch branching enzymes. *Frontier Plant Science* 8: 298.

Susanto, U., Barokah, U., Hidayatullah, A. et al. (2017). Yield and Zn content of biofortified rice genotypes in an Indonesian rice agro-ecosystem. *Nusantara Bioscience* 9 (3): 288–294.

Suzuki, M., Morikawa, K.C., Nakanishi, H. et al. (2008b). Transgenic rice lines that include barley genes have increased tolerance to low iron availability in a calcareous paddy soil. *Soil Science. and Plant Nutrition* 54: 77–85.

Suzuki, M., Tsukamoto, T., Inoue, H. et al. (2008a). Deoxymugineic acid increases Zn translocation in Zn-deficient rice plants. *Plant Molecular Biology* 66: 609–617.

Swamy, B.P.M., Kaladhar, K., Anuradha, K. et al. (2018a). QTL analysis for grain iron and zinc concentrations in two *O. nivara* derived backcross populations. *Rice Science* 25 (4): 197–207.

Swamy, B.P.M., Descalsota, G.I.L., Nha, C.T. et al. (2018b). Identification of genomic regions associated with agronomic and biofortification traits in DH populations of rice. *PLoS One* 13 (8): e0201756. doi: 10.1371/journal.pone.0201756.

Swamy, B.P.M., Kaladhar, K., Anuradha, K. et al. (2011). Enhancing iron and zinc concentration in rice grains using wild species. In: *ADNAT Convention and International*, 71. Hyderabad: Symposium on Genomics and Biodiversity, CCMB, (23–25th Feb 2011).

Swamy, B.P.M., Rahman, M.A., Inabangan-Asilo, M.A. et al. (2016). Advances in breeding for high grain zinc in rice. *Rice* 9: 49.

Takahashi, M., Nozoye, T., Kitajima, N. et al. (2009) *In vivo* analysis of metal distribution and expression of metal transporters in rice seed during germination process by microarray and x-ray fluorescence imaging of Fe, Zn, Mn, and Cu. *Plant Soil* 325: 39–51. doi: 10.1007/s11104-009-0045-7.

Tan, S., Han, R., Li, P. et al. (2015). Over-expression of the MxIRT1 gene increases iron and zinc content in rice seeds. *Transgenic Research* 24: 109–122.

Tang, J. and Chu, C. (2017). MicroRNAs in crop improvement: fine-tuners for complex traits. *Nature Plants* 3 (7): 17077. doi: 10.1038/nplants.2017.77.

Telem, R.S., Wani, S.H., Singh, N.B. et al. (2013). Cisgenics—a sustainable approach for crop improvement. *Current Genomics* 14 (7): 468–476. doi: 10.2174/13892029113146660013.

Trijatmiko, K.R., Dueñas, C., Tsakirpaloglou, N. et al. (2016) Biofortified indica rice attains iron and zinc nutrition dietary targets in the field. *Scientific Report* 6: 19792. doi: 10.1038/srep19792.

Tripathy Swapan, K. (2018). Anther culture for double haploid breeding in rice-A way forward. *Rice Genomics and Genetics* 9 (1): 1–6.

Tripathy Swapan, K. (2020). Genetic variation for micronutrients and study of genetic diversity in diverse germplasm of rice. *Journal of Crops and Weeds* 16 (1): 101–109.

Tripathy Swapan, K. (2021). Optimization of culture variables for efficient callus induction and rapid plant regeneration in zinc rich rice (*Oryza sativa* L.) cv. "Chittimuthyalu". *Journal of Applied Biology and Biotechnology* 9 (3): 1–9.

Tripathy Swapan, K., Behera, S. and Sahu, D. (2020). Phenotyping and association of grain zinc content with agromorphological traits in a core rice germplasm. *International Journal of Chemical Studies* 8 (2): 685–691.

Tripathy Swapan, K., Behera, S.K., Bastia, D.N. et al. (2019a). Exploring genetic variation and identification of divergent zinc rich donors in a core rice germplasm. *The Bioscan* 14 (4): 233–237.

Tripathy Swapan, K., Lenka, D., Prusti, A.M. et al. (2019b). Anther culture in rice: progress and breeding perspective. *Applied Biological Research* 21 (2): 87–104.

Tripathy Swapan, K., Panda, A. and Lenka, D. (2016). Cytological elucidation of somaclonal variation in grasspea (*Lathyrus sativus* L.). *Legume Research* 39 (2): 177–182.

Uauy, C., Distelfeld, A., Fahima, T. et al. (2006). A NAC gene regulating senescence improves grain protein, Zn, and Fe content in wheat. *Science* 314: 1298–1301

Velu, G., Ortiz-Monasterio, I., Cakmak, I. et al. (2013). Biofortification strategies to increase grain zinc and iron concentrations in wheat. *Journal of Cereal Science* 59 (3): 365–372.

Waltz, E. (2016). CRISPR-edited crops free to enter market: skip regulation. *Nature Biotechnology* 34 (6): 582.

Wang, Y., Mei, S. and Wang, Z. (2017). Metabolite profiling of a zinc-accumulating rice mutant. *Journal of Agricultural and Food Chemistry* 65 (18): 3775–3782. doi: 10.1021/acs.jafc.7b00105.

WHO (2020). Malnutrition-Key facts. 1st April, 2020. https://www.who.int/news-room/fact-sheets/detail/malnutrition.

Wirth, J., Poletti, S., Aeschlimann, B. et al. (2009). Rice endosperm iron biofortification by targeted and synergistic action of nicotianamine synthase and ferritin. *Plant Biotechnology Journal* 7: 631–644.

Wu, T.Y., Gruissem, W. and Bhullar, N.K. (2018). Facilitated citrate dependent iron translocation increases rice endosperm iron and zinc concentrations. *Plant Science* 270: 13–22.

Wu, T.Y., Muller, M., Gruissem, W. et al. (2020). Genome wide analysis of the transcriptional profiles in different regions of the developing rice grains. *Rice* 13: 62. doi: 10.21203/rs.3.rs-23378/v1.

Xie, K. and Yang, Y. (2013). RNA-guided genome editing in plants using a CRISPR-Cas system. *Molecular Plant* 6 (6): 1975–1983. doi: 10.1093/mp/sst119. PMID: 23956122.

Yamaji, N. and Ma, J.F. (2014). The node, a hub for mineral nutrient distribution in graminaceous plants. *Trends in Plant Science* 19: 556–563.

Yamaji, N., Xia, J., Mitani-Ueno, N. et al. (2013). Preferential delivery of zinc to developing tissues in rice is mediated by P-type heavy metal ATPase OsHMA2. *Plant Physiology* 162: 927–939.

Yan, C., Qian, K., Xue, G. et al. (2004). Production of bacterial blight resistant lines from somatic hybridization between *Oryza sativa* L. and *Oryza meyeriana* L. *Journal of Zhejiang University - Science A: Applied Physics and Engineering* 5 (10): 1199–1205. doi: 10.1631/jzus.2004.1199.

Yang, M., Li, Y., Liu, Z. et al. (2020). A high activity zinc transporter OsZIP9 mediates zinc uptake in rice. *The Plant Journal* 103: 1695–1709. doi: 10.1111/tpj.14855.

Yin, H., Kauffman, K.J. and Anderson, D.G. (2017). Delivery technologies for genome editing. *Nature Reviews Drug Discovery* 16 (6): 387–399.

Yin, H.J., Gao, X.P., Stomph, T. et al. (2016). Zinc concentration in rice (*Oryza sativa* L.) grains and allocation in plants as affected by different zinc fertilization strategies. *Communications in Soil Science and Plant Analysis* 47: 761–768.

Zaw, H., Raghavan, C., Pocsedio, A. et al. (2019). Exploring genetic architecture of grain yield and quality traits in a 16-way indica by japonica rice MAGIC global population. *Scientific Report* 9: 19605. doi: 10.1038/s41598-019-55357-7.

Zeng, H., Zhang, X., Ding, M. et al. (2019). Integrated analyses of miRNAome and transcriptome reveal zinc deficiency responses in rice seedlings. *BMC Plant Biology* 19: 585. doi: 10.1186/s12870-019-2203-2.

Zhang, A., Liu, Y., Wang, F. et al. (2019). Enhanced rice salinity tolerance via CRISPR/Cas9-targeted mutagenesis of the OsRR 22 gene. *Molecular Breeding* 39 (3): 47. doi: 10.1007/s11032-019-0954-y.

Zhang, J., Zhang, H., Botella, J.R. et al. (2018). Generation of new glutinous rice by CRISPR/Cas9-targeted mutagenesis of the Waxy gene in elite rice varieties. *Journal of Integrative Plant Biology* 60: 369–375.

Zhang, M., Pinson, S.R.M., Tarpley, L. et al. (2014). Mapping and validation of quantitative trait loci associated with concentration of 16 elements in un milled rice grain. *Theoretical and Applied Genetics* 127: 137–165.

Zhang, X., Zhang, G., Guo, L. et al. (2011b). Identification of quantitative trait loci for Cd and Zn concentrations of brown rice grown in Cd-polluted soils. *Euphytica* 180: 173–179.

Zhang, Y., Su, J., Duan, S. et al. (2011a). A highly efficient rice green tissue protoplast system for transient gene expression and studying light/chloroplast-related processes. *Plant Methods* 7 (1): 30. doi: 10.1186/1746-4811-7-30.

Zhang, Y., Xu, Y.H., Yi, H.Y. et al. (2012). Vacuolar membrane transporters OsVIT1 and OsVIT2 modulate iron translocation between flag leaves and seeds in rice. *Plant Journal* 72 (3): 400–410. doi: 10.1111/j.1365-313X.2012.05088.x.

Zheng, L., Cheng, Z., Ai, C. et al. (2010). Nicotianamine, a novel enhancer of rice iron bioavailability to humans. *PLoS One* 5: e10190.

Zhu, Y., Lin, Y., Chen, S. et al. (2019). CRISPR/Cas9-mediated functional recovery of the recessive rc allele to develop red rice. *Plant Biotechnology Journal* 17: 2096–2105.

22

Modification in Conventional Methods and Modern Plant Breeding Techniques to Enhance Genetic Gain for Future Food Security

Afifa Younas[1*], *Nadia Riaz*[2], *Madiha Rashid*[1], *Sajid Fiaz*[3], *Aasma Tufail*[1], *Zahra Noreen*[1], *Muhammad Aslam*[1], *Mehnoob Ullah Khan*[1], *Mehwish Tabassum*[1]

[1] *Department of Botany, Division of Science and Technology, University of Education, Lahore, Pakistan*
[2] *Department of Botany, Lahore College for Women University, Lahore, Pakistan*
[3] *Department of Plant Breeding and Genetics, The University of Haripur, Haripur, Pakistan*

*Corresponding author: botanistafifa@yahoo.com

22.1 Introduction

The average food production rate is 32 million metric tons per hectare which is far from global demand (Alston et al. 2009; Sharma et al. 2017). To feed billions of people worldwide, which is expected to double by 2050, 70% more food is required, which means 44 million metric tons per hectare should be persistently harvested according to the World Food Security Summit held in 2009 in Italy. Meeting this demand (zero hunger and improved nutrition) is challenging and could be achieved only through exercising different breeding techniques followed by the distribution of modern technologies to developing countries and focusing on yield gain traits (Tester and Langridge 2010). Although novel high-yield varieties developed through breeding techniques contributed much to hunger reduction, chronic hunger is still persistent. Hunger proportion which was 50% in mid-twentieth century (1900–1945) has been reduced to 11% today (FAO 2019).

Plant breeding means "Crossing the best with the best and hopping for the best" Acquaah (2015). Plant breeding leads to the development of new varieties which provide assurance to food security. Long-term efforts have been made to enhance different plant features such as yield, quality of crops, resistivity of plants against diseases, and resistivity against biotic and abiotic stresses through traditional and modern breeding technology. Crossing different plants acts as a source of genetic diversity. Whereas equipping a single variety with all the desirable traits through modern breeding techniques reduce diversity within a variety. This leads to uniformity in the farmer's field Louwaars (2018).

Plant breeding started in synchronization with agricultural society when people initiated farming instead of hunting. Plant breeding started 9000–11 000 years ago in the form of "Crop-domestication" to meet the demand of an ever-increasing population. Following this, farmers bred plants to aggrandize yield through selection techniques based on appealing phenotype. In addition, farmers imported plants from different geographical regions to increase diversity and paved a gateway for breeding (Doebley et al. 2006). Mendel's law of inheritance was discovered in 1865. The Rediscovery of Mendel's law (1900) acted as an inception point for plant breeding resulting in hybrid production (Hickey et al. 2017).

The average time duration for plant breeding phenomenon ranges between 4 and 10 years (Poehlman and Sleper 1995a), but it also depends on the number of genes involved (qualitative or quantitative), plant life cycle (annual, biennial, or perennial), type of trait (dominant or recessive), location of breeding trait (epigeal or hypogeal), and trait to be transfer, i.e. biotic stress–resistant trait or abiotic stress–resistant trait (Chahal and Gosal 2002; Acquaah 2009). More time is required for tree breeding as compared to annual crops. Qualitative traits are easy to breed as compared to quantitative traits. Recessive traits take more time because an additional step of self-fertilization has to be done before each backcross. Whereas dominant traits take less time. Breeding of air-borne-disease-resistant traits is easy as compared to soil-borne-disease-resistant traits.

Crop Biofortification: Biotechnological Approaches for Achieving Nutritional Security under Changing Climate, First Edition. Edited by Adnan Noor Shah et al.

22.2 Conventional Breeding Techniques for Self-pollinated Crops

Conventional breeding is a selective technique that has been under human practice since thousands of years ago to develop cultivars for desirable characteristics using old tools. The selection of plants is based on appealing phenotypic traits. Selection technique is the oldest but is still mostly used in collaboration with modern breeding techniques for crop improvement. Conventional breeding techniques include pure-line selections, mass selection, backcross breeding, recurrent selection, and hybridization, which are the most famous traditional breeding methods.

Following are some of the common techniques for self-pollinated crops, i.e. pure-line selection, mass selection, and hybridization.

22.2.1 Pure-line Selection

One of the top-ranking primitive techniques involves the selection of true breeding (homozygous) parental plants followed by self-fertilization. The pure-line selection is common to self-pollinated plants like wheat and rice (Breseghello and Coelho 2013). Johannsen's theory of pure line in 1903 opened a gateway to pure-line selection practices. He suggested that a pure line of beans (*Phaseolus vulgaris*) can be sorted out from a mixed population of self-pollinated species (Poehlman 2012). Experimental study reveals that the yield of wheat has been aggrandized by 15–20% from its high efficient parent plant through pure-line selection (Lane 1981).

1. Lines having different genotypes can be isolated from a population of mixed genotypes.
2. Genetic variation within a pure line is not heritable, but environmental factors cause variation.

Homozygous plants with desirable traits are self-pollinated. The progenies are assessed phenotypically. The high-performing progenies are selected to be grown next year and offtype progenies are excluded. Next year selected high-performing progenies are grown followed by self-pollination. The same process is repeated multiple times until desirable traits and uniformity are attained. This is followed by a multilocation trial in the presence of check varieties. After six to seven years, the superior genotype is released as a variety (Lamichhane and Thapa 2022). EDOA1 and DOA2 are japonica varieties of Rice. DOA2 was selected to be passed through pure-line selection technique, and as a result, five candidate lines (Tana 1–Tana 5) were obtained. The agronomic traits and grain yield of the candidate line were compared with the DOA2 (original), DOA 1, Koshihikari, Akitakomachi, and Sasanishiki. Agronomic trait and grain yield of Tana 1 were the highest. Tana 1 was obtained through pure-line selection of DOA2 (Nakwilai et al. 2020).

22.2.2 Mass Selection

It is the oldest and most commonly used method of plant selection over time but with a poor selection efficiency rate (Abreu et al. 2010). This technique is applicable to self-pollinated as well as cross-pollinated crops (Allard 1961) Mass selection technique is important for traits having a high heritable rate (Brown and Caligari 2011). Seven years after the successful generation, the plants are declared as a variety followed by market distribution. This technique involves the selection and cultivation of landraces (native plants having intact genomes) over many generations, offtype plants are excluded, and desirable plants are allowed to produce the next generation (Acquaah 2015)

In 1958–1961 for four consecutive years, a research experiment was conducted on a heterogeneous (different heights) oat population with the aim to reduce the population height to uniform. The mass selection technique remains effective for oat height, and uniform height was obtained after a few generation. The height of mass-selected population decreased by 0.45 inch per generation and the height of nonselected population increased by 0.21 inch per generation, so the height was reduced to uniform (Romero and Frey 1966).

22.2.3 Backcross Breeding

This technique involves producing cultivars identical to one of the two parents containing favorable traits called recurring parent by incorporating desirable traits from less viable (donor plants) plants to well-established recurring plants providing the other traits remain intact. That is why backcross breeding is also called recurring selection (Fujimak 1978; Aleksoski 2018). For cereals, crops' backcross breeding is a well-appropriate technique firstly suggested by Harlan and Pope in 1922. The backcrossed progenies produced after successive three to four generations are 98% genetically identical to recurring parents.

Or we can say that the recurring parent conserved its 98% genome in backcrossed progenies (Vogel 2009; Briggs 1938). The genetic contribution between parents (donor and recurring plants) to produce backcrossed progenies is not equal. Backcrossed progenies received genes mostly from recurring plants (Singh 1982). Recurrent selection is a time-taken technique as compared to the backcross breeding technique to obtain stable maize inbred lines (Hallauer et al. 2010; Hallauer et al. 1992; Pandey and Gardner 1992). Backcross breeding can be made more efficient through the knockout technique (Meredith 1977).

1. A plant lacking in one or a few desirable traits is called a recurrent parent.
2. Nonrecurring plant or donor plant having desirable trait to be transferred to recurrent parent.
3. The trait to be transferred must be highly heritable.
4. Cross must be continued until donor trait is obtained in recurrent plant.

Donor and recipient parent are crossed to produce F1 hybrid. This F1 hybrid is backcrossed with recurrent parent to produce F2 hybrid followed by scrutiny for desirable traits. F2 hybrids with desirable traits are again backcrossed to recurrent parent. This step is repeated for six to eight generations until desirable trait is obtained in recurrent parent.

Backcross method for recessive traits is a little bit different compared to the backcross for dominant traits in the sense that an additional step of self-fertilization after each backcross is necessary to produce homozygous plant. After each backcross, desirable traits lie in the heterozygous plant which is to be self-fertilized to produce homozygous recessive plant followed by backcross. Tropical maize line (low grain yields) was crossed with temperate maize line (high grain yield) for three consecutive years (2003–2006) with the aim to increase grain yield and to check the transfer ability of multiple quantitative traits from exotic (temperate) to local (tropical or elite maize). Multiple quantitative traits (row number per ear, ear length, kernel number per row, and kernel weight) were transferred from exotic to elite class successfully also with high grain yield through backcross breeding (Chen et al. 2010).

Two different varieties of cotton *Gossypium hirsutum* L. (DPL 16 and FTA 263-20) were backcrossed with the aim to increase the fiber strength. DPL being the recurrent parent with low fiber strength was crossed with the donor FTA 263-20 having high fiber strength. Fiber strength was successfully transferred from FTA 263-20 to DPL 16 through backcross breeding. The cultivars produced as a result of backcross breeding were found to have greater fiber strength (Meredith 1977).

22.3 Conventional Breeding Techniques for Cross-pollinated Crops

Following are the techniques for self-pollinated crops, i.e. recurrent selection and hybridization.

22.3.1 Recurrent Selection

Recurrent selection is a term coined by Hull (1945). This process involves crosses between high-performing plants (heterozygous recurrent parent and inbred parent) followed by selecting and reselecting desirable progenies generation after generation. This technique is valid for cross-pollinated crops. This technique is important in maintaining variability and increasing the frequency of desirable traits (Khadr 1964). This technique was initially applied and practiced for maize and later exercised for rice, millet, wheat, and soybeans (Ramya et al. 2016). There are three different types of recurrent selections (Luckett and Halloran 2003), which are simple recurrent selection, recurrent selection for combining ability, and reciprocal recurrent selection.

22.3.2 Hybridization

Hybridization is the genetically superior hybrid (progenies) production, which involves natural or artificial process of mating genetically different plants belonging to the same species, different species, or even genera. This technique was first practiced by Joseph Koerauter German Botanist for crop improvement (Begna 2021). Hybridization is safe from genetic alteration, but recombinant DNA formation is common which leads to favorable novel traits (Mwangangi et al. 2019). Hybridization has two common types: (1) intraspecific hybridization and (2) interspecific hybridization.

22.3.2.1 Interspecific Hybridization

Crossing of parent plants belonging to different genera or even species that are sexually incompatible is called intraspecific hybridization also termed as wide cross (Lidder and Sonnino 2012). The success rate of wide crosses is not always positive.

This phenomenon of incompatibility between distantly related species is called incongruity (Hoggenboom 1973). Interspecific hybrids are important because of their novel genetic combination and to understand barriers in hybridization (Dandekar et al. 2005). The factors that lead to sexual incompatibility are pre- and postfertilizations.

Prefertilization or prezygotic barriers are also called morphological barriers, which include different positions of anthers and stigmas in space and time and failure of the pollen tube to develop through stigma, style, and ovary (e.g. solanaceae and brassicaceae; Harlan and De Wet 1971). Postfertilization or postzygotic barriers are also called physiological barriers, which include hybrid sterility, embryo deterioration, seed lethality, and seed abortion (e.g. *Oryza sativa* and *Oryza meyeriana*; Murray 2016).

1. **Prezygotic Barrier in Interspecific Hybridization of *Juglans regia* and *Juglans nigra*** In 2010 and 2011, experiment was conducted on two genetically different walnut species *Juglans regia* and *Juglans nigra*. The hybridization remained unsuccessful due to a prezygotic barrier, i.e. the flowering time of the two species was different. The flowering time recorded for *J. regia* ranges from 9 April to 6 May while for *J. nigra* it ranges from 3 May to 6 June. So, the flowering overlapping time recorded was negligible (Pollegioni et al. 2013).
2. **Postzygotic Barrier in the Interspecific Hybridization Between *O. sativa* and *O. meyeriana*** A research was conducted to develop an interspecific hybrid of *O. sativa* and *O. meyeriana*. Prezygotic barrier was negligible but the hybrid formation was restricted by postzygotic barrier, i.e. less viability of the embryo. Three days after pollination, 73.33% of embryos failed and were restricted to the globular stage while only 20% of embryos developed to pear shape (mature form) which later turned brown and shrined. The endosperm nuclei did not develop into endosperm cells and deteriorated followed by embryo degeneration due to starvation, which was the main reason for the lower viability of the embryo. Later, through a nonconventional method, i.e. asymmetric somatic hybridization, a hybrid was developed which was unfortunately highly sterile (Fu et al. 2009).
3. **Interspecific Hybridization in *P. vulgaris* (L.)** In 2019, an experiment was conducted on *P. vulgaris* (L.) with the actual aim to increase the yield. The parental organisms with contrasting features, i.e. high yield and resistance to pathogenic damage, are selected for crossbreeding. After hybridization, the parental and hybrid plants were compared by the following parameters: "number of beans per plant," "number of seeds from the plant," and "mass from the plant." The coefficient of variation for "the number of beans per plant" is 16.3–58.3% for the parents and 19.8–81.2% for F1–F2 hybrids, respectively. The coefficient of variation for "the number of seeds from the plant" was 23.6–76.6% for the parents and 30.3–84.6% for F1–F2 hybrids, respectively. The coefficient of variation for "the seed mass from the plant" was 19.6–76.1% for the parents and 24.1–90.3% for F1–F2 hybrids, respectively). The result clearly indicates that the yield of the hybrid was greater as compared to parent plants (Golovan et al. 2019)
4. **Interspecific Hybridization in *Paspalum dilatatum* Poir.** *Paspalum dilatatum* Poir., commonly called as dallisgrass, is a forage crop that can stand under high grazing. It has two varieties sexual biotype ♀ with high fertility and less forage characteristics and apomictic biotype ♂ with less fertility and high forage characteristics. These two varieties were used as a parental generation to obtain hybrid with the aim to increase fertility and forage traits. All the hybrid progenies obtained were sexual and high fruit set, i.e. highly fertile. The fruit set increased progressively, i.e. 40%, 70%, 80%, and 90% in F2, F3, F4, and F5, respectively (Table 22.1; Bennett et al. 1969).

22.3.2.2 Intraspecific Hybridization

Mostly, this type of hybridization gives positive results due to lacking of barriers. This is the leading source of commercial plant breeding. It is the process of creating varieties by synchronization of favorable traits of plants belonging to the same species (Thomas 2016). Intraspecific hybridization can be simple or complex based on the number of parents involved. If two parents are involved, it is called simple intraspecific hybridization. It involves crossing two plants to

Table 22.1 The list of interspecific hybridization in which embryos deteriorate at an early stage.

Plant	References
Vigna unguiculata and *Vigna vexillata*	Barone et al. (1992)
Asparagus officinalis and *Asparagus densiflorus*	Marcellan and Camadro (1999)
Cicer arietinum and its related annual wild species	Ahmad and Slinkard (2004)

produce F1 followed by self-fertilization to produce F2 or backcross. If more than two parents are involved, it is called complex intraspecific hybridization. It involves the accumulation of desirable traits from more than two parents into a single hybrid offspring.

22.4 Modern Plant Breeding Technology

Modern plant breeding involves modern technology aided by traditional breeding practices for the past few years (Fu 2015). Some of the most common modern breeding practices include genomic selection (GS), marker-assisted selection (MAS), high-throughput phenotyping (HTP), and clustered regularly interspaced short palindromic repeats (CRISPR)-Cas9. Modern interdisciplinary approaches have been exercised to undergo the breeding process efficiently, which leads to sustainable agriculture development and increased food security (Qaim 2020). Cultivation of genetically modified (GM) plants aggrandized yield by 22% and reduced pesticide usage by 33% (Klümper and Qaim 2014). The cornerstones of modern plant breeding were laid by Darwin and Mendel in the late nineteenth century (Borlaug, 1983) (Figure 22.1).

1. High cost value and regulatory complexity of modern technology restricted it to developed countries only.
2. Breeding program is based on GM varieties, whereas non-GM varieties have limited access to advanced technology.
3. Many people, especially Europeans, have serious concerns regarding the use of genetically modified organisms (GMOs) as food because it is a common perception that GMOs could have serious health effects.
4. Many countries have strict policies for the first open-field trial of GMO.
5. Many developing countries still sow low-yield seeds because of the high cost of GM seeds.

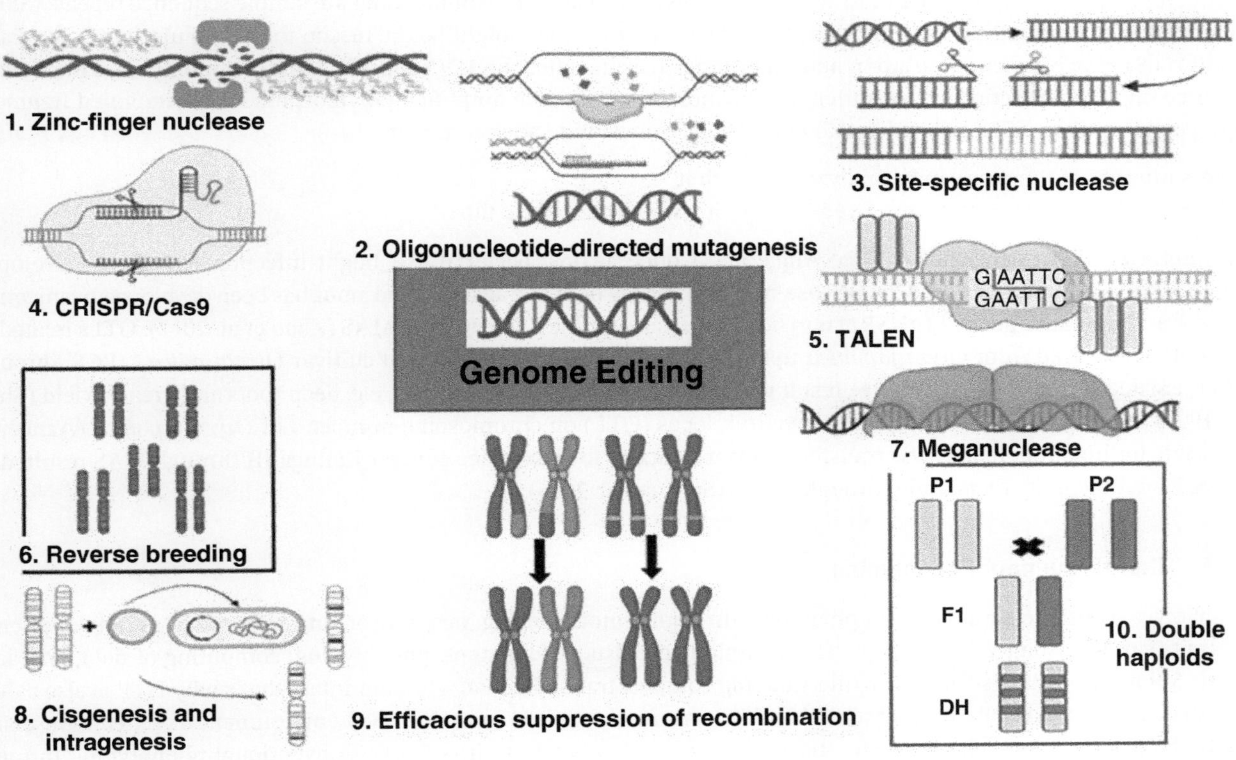

Figure 22.1 Modern plant breeding techniques such as marker-assisted selection, high-throughput phenotyping (HTP), CRISPR-Cas9, site-specific nucleases (SSNs), zinc-finger nucleases (ZFNs), transcription activator–like effector nucleases (TALENs), clustered regularly interspaced short palindromic repeat, meganucleases (homing endonucleases), oligonucleotide-directed mutagenesis (ODM), cisgenesis and intragenesis, reverse breeding (RB), efficacious suppression of recombination, and double haploids (DH).

22.4.1 Genomic Selection

GS is more efficient in the sense that it gives more yield in less time. It ranks high as compared to MAS because it uses genome-wide markers to speculate the effect of quantitative gene loci, thereby calculating the estimated breeding value (Wang et al. 2018). Research conducted by the International Maize and Wheat Improvement Center (CIMMYT) showed that in plants like maize, the breeding interval could be as low as half (Crossa et al. 2017). GS does not involve detecting like in linkage analysis and genome-wide association studies but make assumptions on the performances of future offspring based on DNAs collected. At present, penalized regression method and Bayesian method can be used to predict the breeding value. However, the quality of prediction can be gauged by the correlation between estimated and actual breeding values (Rabier et al. 2016). Two types of data training population and validation population are used in GS to calculate the estimated and actual breeding value. The already phenotyped and genotyped training population can be a good source of predicting breeding values in the validation population. Small genes that are located at multiple loci but have less effect can be easily mapped through GS. These genes are important because economically and agronomically important traits are under the control of these small effective genes. With an increase in the price of phenotyping of crops, quantitative traits that are difficult to pinpoint through traditional phenotyping can be successfully detected through GS.

22.4.2 Marker-assisted Selection

The detection of genes responsible for desirable traits through DNA markers is called MAS (Das et al. 2017). The easy availability and variety of DNA markers made breeding phenomena effortless. Molecular marker technology initiated in 1980s contributed much towards crop improvement (Zadokar et al. 2023). DNA markers detect allelic variation of genes in the same gene pool (Nogoy et al. 2016). For a single gene transfer, marker selection is helpful, but for polygene transfer, it is less useful. The success rate for single gene transfer through MAS has been reported for several plants, e.g. transfer of *Bt* genes (Ribaut and Hoisington 1998). The most widely used DNA markers for plant breeding are simple sequence repeats (SSRs) also termed as microsatellites. It is inexpensive and easy to use which might be the reason for its popularity (Collard and Mackill 2008). Environmental variation and inheritance have no effect on DNA marker. Most commonly used markers in plant breeding are restriction fragment length polymorphism, random amplified polymorphic DNA, amplified fragment length polymorphism, SSRs or microsatellites and single-nucleotide polymorphism

1. Less number of samples can be analyzed through it
2. Less number of lines can be improved through it within the specific time

Two disease-resistant varieties of Rice Angle and Conde against bacterial leaf blight infection have been developed through MAS (Bustamam et al. 2002). Disease-resistant variety of maize against head smut has been developed by injection of head smut resistance genes 1 (*qHSR1*) into head smut susceptible lines through MAS (Zhao et al. 2012). QTLs related to deep roots transferred from *Oryza japonica* upland cultivar (Azucena) to low land cultivar *Oryza indica* "*IR64*" through marker-assisted technique. The positive result was obtained in mass-selected lines, e.g. deep roots and greater yield (Shen et al. 1999). Five (RM242-RM201) quantitative trait locus (QTL) on chromosome number 9 of *Oryza japonica* (Azucena) responsible for high root length was transferred from Azucena to Indian rice cultivar Kalinga III through MAS resulted in Kalinga III with lengthy roots under drought stress (Steele et al. 2006).

22.4.3 High-throughput Phenotyping

HTP involves plant selection based on phenotype through remote sensing, new sensor, and software. A variety of phenotypic traits can be visualized through it. This technique consists of four steps: phenotyping, computing of data, and data analysis followed by plant selection. While studying complex traits, the plants remain intact that's why HTP is also called nondestructive method (Pabuayon et al. 2019). HTP was first exercised in a controlled environment like greenhouse and growth chamber to check plant growth and performance (Araus and Cairns 2014). Conventional phenotyping through visual observation is time-consuming while HTP is time-saving technique because phenotypic alteration in plants due to genetic variations and environmental factors of hundreds of thousands of plants can be recorded per day (Rebetzke et al. 2019). HTP involves modern automated equipment which reduces human efforts, large population can be studied in less time with more accuracy. HTP is aided by various imaging techniques that take images of crops time to time (Pasala and

Pandey 2020). Conventional phenotyping was a common method for superior variety selection before genomic selection (GS) and MAS. GS and MAS are great fit for superior variety selection but still modern technique of phenotyping HTP is necessary for selection (Jannick et al. 2010). Accurate and fast phenotyping through HTP aided in GS of superior variety.

HTP can be done at different platforms, i.e. ground-based and aerial phenotyping.

1. **Ground-based Phenotyping** It is done by phenomobiles (a modified agriculture harvester) equipped with sensors and GPS. The measurement rate of ground-based phenotyping is 0.8 hectares per hour. This is limited to a single plot, i.e. multiple plots cannot be measured through it and also the data received required further processing through different software.
2. **Aerial Phenotyping** It is faster than ground-based phenotyping. Multiple plots can be measured through it simultaneously. It is done by small airplanes, phenotowers, blimps, polycopters, etc.

HTP for different traits (yield gain, adaptation to different biotic and abiotic stresses, leaf morphology, chlorophyll content, plant physiology, surface temperature, opening and closing of stomata, and chlorophyll senescence) can be measured accurately in control environment like pots, but HTP cannot be exercised accurately in field due to heterogeneous environmental condition like low soil, water, and nutrients contents in pots as compared to field, and plants in pots grow solely but in the field it grows in canopy where it is under the effect of neighbor plants; moreover, environmental conditions are not measured, such as soil acidity, basicity, salinity, and moisture content (Passioura 2006; Poorter et al. 2012; Reynolds et al. 2012)

Following are the different imaging techniques:

1. Visible Light Imaging: It is used to study plant structures like leaf index, biomass, and yield characters. It is popular because it is easy to handle and less expensive. It uses visible light to take pictures, and the collected data are expressed with intensity values of blue, green, or red bands (Li et al. 2014).
2. Fluorescence Imaging: This technique is used to analyze metabolic activity within plants. It is popular for plant physiology, plant health, and disease-resistant traits (Zhang and Zhang 2018). Bombardment of plants by blue or actinic lights results in re-emission of the absorbed shorter wavelength lights, which is called fluorescence. This re-emitted light acts as a good indicator of photoassimilation in plants. Fluorescence imaging is done with charge-coupled device cameras.
3. Thermal Imaging: This technique works by using infrared light within a range of 3–4 μm. It is sensitive to surface temperature of objects (plants; Li et al. 2014). It is popular to study water and plant relations, biotic and abiotic stresses, and the opening and closing of stomata. High or low temperature reflects the opening or closing of stomata. Images are taken through thermal cameras. It measures temperature differences in plants' canopy.
4. Spectral Imaging: It is used in remote sensing of vegetation. It is popular to study plant's water content, plant pathological condition, and chlorophyll senescence. It records plants' radiation to study vegetation indices of plants. Solar radiation will be more from whole plants as compared to single leaf (Li et al. 2014).

22.4.4 Clustered Regularly Interspaced Short Palindromic Repeats-Cas9

These are DNA sequences aided by Cas9 protein (endonucleases) in prokaryotic organisms against invading phages (Chen et al. 2019). Breeding through CRISPR was initiated in 2013 for crops like rice and tobacco. Genetic modification through CRISPR is done either through intrusion or extrusion of genes. CRISPR-Cas genomic sequence editing technique was first exercised in *Arabidopsis* (Feng et al. 2013). Till 2017, about 20 crops have been made diseases resistant and tolerant against biotic and abiotic stresses through CRISPR-Cas9 (Ricroch et al. 2017). Introgression of CRISPR complex into the plant is done by three methods.

1. Agrobacterium-mediated transfer: T-DNA insertion into plants via living organisms (Bacteria) is called agrobacterium-mediated transfer, which is also called a biological method. T-DNA provides assurance of CRISPR complex expression in plants.
2. Polyethylene glycol: The only parallel of agrobacterium-mediated transfer was polyethylene glycol characterized by the direct uptake of DNA in protoplast, efficiently also called as electroporation-mediated transfer. This method was first exercised for the successful transfer of plasmid DNA into the protoplast of tobacco and petunia.
3. Particle gun method: It is a widely accepted method due to introgression of wide-range molecules. Genetic modification in maize, wheat, and soybean has been done through it.

22.4.5 CRISPR History

The history of CRISPR related to chemistry is divided into two periods given as follows:

1. **Detection Period (1987–1993)** During 1987, process of detecting the gene causing the transformation of alkaline phosphatase isozyme in bacteria (*Escherichia coli*), the five direct repeats comprising 24 nucleotide repetitious sequences of the CRISPR-Cas system were first encountered. Another existence was observed in *Haloferax mediterranei*, where direct repeats were separated by spacer sequences, which are 35 bp long and possess 30–34 sequences of nucleotides. Inverted and short repeats resembling those in the earlier sequence also occurred in the direct repeats. Scientists were fascinated by the varying sequences that lay between the direct repeats and spacers of the CRISPR-Cas system (CCS) during the period ranging from 1987 to 1993.
2. **Functional and Structural Characterization Period (1993–2011)** Subsequent research teams studying bacteria and archaea persisted in finding similar enigmatic patterns of consistently spaced repetitive sequences, which eventually resulted in the realization of their chemical and biological relevance. Ruud Jansen and associates identified the repetitive DNA sequences from 1993 to 2005 as CRISPRs in 2002. These repeats are situated in the genome close to the "DNA repair system." The term "CRISPR-associated" (Cas) genes was coined to refer to these close genes, which were previously thought to be a component of the DNA repair system. These genes were primarily located to accompany CRISPR genes. Two separate investigations performed in 2005 connected bacteriophages to the spacer's origin. After performing a computational investigation in 2006, Eugene V. Koonin and his associates discovered a resemblance between the functions of the Cas genes and CRISPR as a system and an immune system based on RNA interference in prokaryotes. After a viral attack, the CCS endows resistance to viruses due to chemical changes. The phage genomic sequences are converted into spacers and kept in the host genome. A revised investigation of the evolutionary associations between Cas proteins and CRISPR was introduced in 2011, and this work was followed by multiple other papers.

Following that, CRISPR investigations blossomed annually in thousands of labs worldwide; a few articles provide a schematic illustration of this innovative technology's success stories. After the CRISPR array and its allies were discovered to be effective defense means for protection and survival against viral attack in the genomes of bacteria and archaea, they were repurposed as a gene-editing strategy to alter the genomes of eukaryotic organisms. This strategy has massive applications in a variety of chemical and biological areas, ranging from medicine to agriculture. During this period, the CRISPR-Cas system underwent extensive reprogramming and transfer. According to an investigation, *E. coli* acquired the CRISPR-Cas9 system from *S. thermophilus*, which offered the bacteria resistance against phage and plasmid invasion. Researchers David Bikard and colleagues engineered Cas9 as a transcriptional repressor that inhibits or prevents RNA polymerase from binding to sequences of promoter. In 2012, the first report described in vitro reconstructions of functional CRISPR-Cas systems and outlined their possible use as RNA-programmable genome editing instruments. In 2013, the first publication demonstrated how to effectively edit the genome of mammalian cells employing CRISPR-Cas instruments. Jennifer Doudna (USA) and Emmanuelle Charpentier (Sweden) have been granted the Nobel Prize during year 2020 in chemistry for their novel research on CRISPR-Cas9—an approach to editing DNA. The Nobel Committee honored these two scientists for finding that it is possible to modify genomes with extreme precision through the modification of a microbial immune mechanism. In 2013, the first report of effective genome engineering in mice employing CRISPR-Cas instruments was made at numerous loci.

Xanthomonas oryzae pv. oryzae (bacterial species) is a causative agent of blight disease in one of the common host rice leading to yield reduction by 10–20% (Zhang and Wang 2013). UPT box (promoter for transcription-activator-like effectors [TAL]) was GM through CRISPR/Cas12a technique that retard protein binding of TAL and gene activation. So GM rice was developed that was resistant to blight disease (Yu et al. 2021). This was further validated by modification in *Oryza sativa Photochromic and Flowering Time 1* (*OsPFT1*) gene through CRISPR-Cas9. The mutant plant developed was resistant against sheath blight (Shah et al. 2019).

SEMI-ROLLED LEAF 1 and *2* (*SRL1* and *SRL2*) genes which regulate leaf rolling by encoding glycosylphosphatidylinositol-anchored protein located on the plasma membrane. *SRL1* and *SRL2* genes modified through CRISPR-Cas technology resulted in a rolled leaf mutant type that was resistant to drought (Liao et al. 2019).

Wheat dehydration responsive element binding protein 2 (TaDREB2) and wheat ethylene-responsive factor 3 (TaERF3) are the two abiotic stress-related genes in wheat which have modified through CRISPR-Cas9 genomic sequence editing technique with the aim to make plant stress tolerant. The success rate for gene modification was about 70% (Kim et al. 2018). Multiple genomes or multiple agronomic traits can be modified simultaneously through CRISPR-Cas9 genomic sequence editing technique demonstrated by Wang et al. (2018) by modifying three wheat genes *TaGW2* (a negative regulator of grain traits), *TaLpx-1* (lipoxygenase, which provides resistance to *Fusarium graminearum*), and *TaMLO* (loss of function, confers resistance to powdery mildew resistance) simultaneously.

22.5 Genome Editing

Presently, the global population stands at approximately eight billion individuals. Based on UN statistics, around two billion people, constituting 26.4% of the world's populace, are facing either moderate or severe food shortages. Significant efforts are underway to combat hunger and malnutrition worldwide, with many of them leveraging scientific advancements in life sciences and agriculture. Despite notable progress in plant breeding, the pressing need remains for the rapid development of new high-gaining and stress-resistant crop varieties. These endeavors aim to address challenges such as crop losses attributed to climate fluctuations, diminishing cultivated areas, and the proliferation of increasingly aggressive and resilient pathogens. Equally crucial is the burgeoning global population. Currently, meeting these challenges is inconceivable without harnessing biotechnological and genetic engineering (GE) methodologies.

The primary aim of plant breeding is to cultivate crop varieties that exhibit high yields, adaptability to biotic and abiotic stresses, and superior product quality. Traditional breeding methods face limitations due to the time required to create fresh cultivars possessing the sought-after agronomic characteristics. This timeframe is influenced not solely by factors such as the length of the growing phase and the time taken for plants to reach maturity (particularly relevant for long-growth period plants like trees) but also by the necessity for copious phases of crossing, selection, and testing within the breeding process. Furthermore, typical modalities such as physical and chemical mutation lack the precision required to target specific genomic regions. As the twenty-first century began, the implementation of MAS techniques has notably enhanced the efficacy of breeding activities, leading to improved crop productivity under diverse environmental conditions. Nonetheless, it is important to acknowledge that these methods do not provide the capability to specifically target the crop genome.

Concurrently, the advancements in next-generation sequencing, coupled with escalating number of sequenced genomes for major crops and the identification of novel genes and their roles, have encouraged researchers to explore more tailored approaches in plant breeding. These developments have appreciably catalyzed the progress of targeted genome editing (GE) methodologies. Targeted genome engineering has surfaced as a substitute for conventional plant breeding and transgenic (GMO) techniques, providing a pathway to enhance crop plants and ensure the sustainability of food production. Recent advancements in genome engineering suggest that novel crops created through targeted genome modification (TGM) utilizing site-specific nucleases (SSNs) could sidestep the regulatory concerns associated with GMOs. Man-made SSNs like zinc-finger nucleases (ZFNs), transcription activator–like effector nucleases (TALENs), CRISPR paired with the endonuclease Cas (CRISPR/Cas), and meganucleases (MegaNs; homing endonucleases) have demonstrated remarkable efficacy in inducing targeted mutagenesis across various model plants and crops. Furthermore, oligonucleotide-directed mutagenesis (ODM) enables genome editing at the level of single nucleotides.

One of the primary issues among the general populace regarding transgenic crops revolves around the amalgamation of genetic materials among species incapable of natural hybridization. Addressing this apprehension, two transformation strategies, namely cisgenesis and intragenesis, were devised as substitutes for transgenesis. These methodologies advocate for the plant transformation solely with genetic material sourced from identical species or sexually concordant species. Intragenesis is distinct from cisgenesis in that it permits the utilization of novel gene combinations formed through in vitro realignment of operational genetic components. Various surveys indicate greater public approval of intragenic or cisgenic crops when compared to crops produced via transgenesis.

Plant breeding has consistently progressed to cultivate new varieties with improved quality and yield through thorough field evaluations, aiming to sustainably support the growing global population. RB represents an innovative approach to breeding that reconstructs the heterozygous genotype from complementing homozygous lines, operating in reverse compared to standard breeding. It is a quick track breeding method that gives more freedom to mix in favored parents to create heterozygous genotypes. RB significantly expedites the breeding process and expands the pool of genetic combinations, enabling breeders to promptly address the requirements of farmers and growers with enhanced varieties. Another notable benefit of RB lies in its facilitation of superior hybrid plant selection. By generating large populations of plants for screening, well-performing specimens can be perpetually regenerated without prior knowledge of their genetic makeup. This effectively mitigates the unpredictability inherent in earlier hybrid breeding approaches. Presently, RB is constrained to crops with relatively small, diploid genomes.

Genome editing, commonly mentioned as gene editing, involves altering the genetic composition of a living organism by inserting, deleting, or substituting DNA sequences. This process is typically aimed at enhancing desirable traits in crops or farm animals, or treating genetic disorders. The overarching objective of genome editing is to gain deeper insights into gene functionality and to develop practical applications, particularly in addressing inherited and acquired medical conditions.

22.5.1 Principle of Genome Editing

The fundamental principle of genome editing utilizing SSNs relies on the activation of deoxyribonucleic acid repair mechanism provoked by DSBs (DNA double-strand breaks) induced by SSNs at specific genomic locations. These DSBs can undergo repair through either NHEJ (nonhomologous end-joining) or HR (homologous recombination) pathways. Through these operations, SSNs have the capacity to create gene ablations, substitutions, incorporations, and chromosomal reconfiguration. The accuracy of these modification techniques aims to address some of the limitations and ambiguities related to conventional breeding methods and transgenic approaches. Additionally, it enables newly developed TGM crops to potentially bypass regulatory mechanisms applied to GMOs.

Genome editing can yield three sorts of modifications: Type I implies changing a few nucleotides, Type II implies supplanting an allele, and Type III involves inserting new genes into specific sections of the genome. Crops generated through gene editing may not require the strict regulating procedures linked with GM crops, as most genome-editing techniques solely alter a few nucleotides. Many scientists feel that plants modified by sophisticated gene editing approaches are likely to garner greater public acceptance compared to transgenic alternatives. Genome editing has the potential to accelerate crop development while reducing the danger of unwanted impacts. This technique can be used in any laboratory with any crop, entailing those with complex genomes that are difficult to breed using traditional approaches.

22.5.2 Mechanism of Genome Editing

22.5.2.1 Site-specific Nucleases

Genome editing involves a diverse array of techniques utilizing a SSN system, which necessitates the identification of a specific sequence. Four primary classes of SSNs have been developed for cleaving genetic sequences: meganucleases, ZFNs, TALENs, and CRISPR/Cas nuclease system. These four types of SSNs differ in mode of action, activity and structure resulting in varying target selection, efficacy, precision, and mutation patterns.

22.5.2.2 Zinc-finger Nucleases

ZFNs represent a fusion of tailored Cys2-His2 ZF protein with the cleavage domain found in restriction endonuclease FokI, marking the pioneering targeted nuclease widely adopted. ZFNs are dimers that recognize a particular half-site sequence of 9–18 bps of deoxyribonucleic acid using the ZF deoxyribonucleic acid-binding domain. The FokI cleavage domain breaks deoxyribonucleic acid inside a five to seven base pair spacer sequence that separates two adjacent binding sites of zinc-finger, facilitating ZFN protein dimerization. A typical ZFN comprises three or four ZF domains, each consisting of almost 30 amino acid residues arranged in a ββα motif. The residues responsible for deoxyribonucleic acid identification are situated within the alpha-helical domain and usually interact with 3 bps of deoxyribonucleic acid, occasionally overlapping with neighboring domains. Utilizing techniques like phage display, numerous ZF domains capable of identifying unique deoxyribonucleic acid triplets have been detected. These domains can be combined together sequentially utilizing a standard linker peptide to create polydactyl ZF proteins capable of targeting a broad spectrum of deoxyribonucleic acid sequences. Apart from this "modular assembly" method for constructing ZFs, selection-based approaches for ZF protein construction have also been documented, inclusive of those accounting for contextual interactions among neighboring zinc-finger domains, like that of oligomerized pool engineering (OPEN). Additionally, unique collections of verified 2 fingered ZF components have been employed to assemble ZF collections, inclusive of those considering context-responsive effects.

In 1996, the initial report showcased the effective utilization of ZFNs for genetic modifications in plants. This technology was adeptly employed to modify the genomes of both *Nicotiana tabacum* and *Arabidopsis*. Specifically in tobacco, ZFNs were instrumental in restoring the functionality of the impaired reporter gene *GUS:NPTII*. Meanwhile, in *Arabidopsis*, the controlled expression of ZFNs via the heat shock protein promoter resulted in 106 DNA mutations. Currently, numerous studies affirm the feasibility of GE in numerous plant species inclusive of tobacco, maize, soybean, *Arabidopsis*, canola, and others through ZFN technology. Notably, ZFNs have enabled the introduction of mutations in *CHN50* (endochitinase-50 gene) in *N. tabacum*, leading to immunity against patented herbicides. Similarly, promising outcomes were achieved through the targeted editing of the *IPK1* (inositol-1, 3, 4, 5, 6-pentakisphosphate kinase 1) genetic element accountable for the production of phytic acid in *Zea mays*. Additionally, ZFN-directed targeting has been successful in modifying various genes such as Dicer-like genes (*DCL4a and DCL4b*) of *Glycine max*, ABA Insensitive-4 (*ABI4*) of *Arabidopsis*, and genes encoding alcohol dehydrogenase and chalcone synthase of *Arabidopsis*.

Nevertheless, despite the relative success observed with ZFNs, their widespread adoption as a (GE tool has been hindered by several drawbacks. Primarily, the complexity and high cost associated with this technology pose significant barriers, particularly in constructing protein domains tailored for every particular genomic position. Moreover, there remains a risk of faulty cleavage of target deoxyribonucleic acid, stemming from single-nucleotide replacement or erroneous interactions among domains. To mitigate nontarget impacts and consequent cellular toxicity of ZFNs, heterodimer formation of the FokI nuclease is employed.

22.5.2.3 Transcription Activator-like Effector Nucleases

Comparable to ZFN, TALENs are enzymes made up of certain deoxyribonucleic acid–binding domains of extremely conserved repetitions derived from effectors like TALEs (transcription activators), which bind to FokI; 33 to 34 highly conserved sequences of amino acid are found in 15–30 copies inside TALEs domains (Jankele and Svoboda, 2014). The exclusions are the highly variable residues of the 12th and 13th amino acids (also known as repeat-variable residues, or RVD; Gaj et al. 2013; Jankele and Svoboda 2014). With the use of two of these amino acids found in recurring peptide chains of a certain protein, it is possible to determine the recognition code for particular nucleotides (Jankele and Svoboda 2014). Although certain amino acid combinations are preferred, this code is degenerate (Gaj et al. 2013; Jankele and Svoboda 2014). It makes it possible to create recombinant proteins with the ability to identify particular DNA sequences (Jankele and Svoboda 2014). The number of amino acids that separate the transcription activator domain from FokI and the number of bases that separate the binding sites determine the TALEN's activity. Every repetition in the TALEN domain identifies a single nucleotide, in contrast to ZF (Gaj et al. 2013; Jankele and Svoboda 2014).

After FokI dimerization, the TALE domains can detect 15 to 30 nucleotides or 30 to 60 nucleotides for every dimer of TALEN. Furthermore, nucleotide mismatches are more likely to occur in TAL domains even if they possess greater binding selectivity (Abdurakhmonov et al. 2016; Gaj et al. 2013; Jansing et al. 2019). In addition to ZFN, TALEN is used in the heterodimer formation of FokI to reduce off-target impacts (Guilinger et al. 2014). TALEN analysis of mutations that happened during GE reveals that deletions outnumber insertions by a significant margin (89% vs 1.6%). The longer TALEN spacers, which give the DNA fragments after DSBs more extended projecting ends, are the cause of deletions (Y. Kim et al. 2013). It is theoretically possible to induce DSB into any region of the genome by using TALEN. The TALEN nuclease recognition sites require thymidine to be present upstream of the five prime end of the target sequence, which is the only restriction. Nonetheless, selecting restriction sites is possible by varying the spacer length (Jankele and Svoboda 2014).

When it comes to genome editing, TALENs have two different advantages over ZFNs. First, TALE arrays may be designed without the necessity for selection or guided evolution, which drastically cuts down on the time and expertise required to construct a working nuclease. Second, TALENs have been found to exhibit better specificity and lower toxicity in contrast to certain ZFNs (Mussolino et al. 2014). This could be due to either a high energy penalty for interacting with base mismatches or a stronger affinity for target deoxyribonucleic acid (Meckler et al. 2013).

22.5.2.4 Clustered Regularly Interspaced Short Palindromic Repeats

Advancements in the field of targeted genome modification have progressed so swiftly that just as one was adjusting to the impact of TALENs, the remarkable CRISPR/Cas system emerged, capable of generating DSBs even more efficiently, simply, and rapidly, ushering in another captivating era in the ongoing narrative of genome engineering (Barrangou 2012; Pennisi 2013). CRISPR and CRISPR-associated systems (Cas) constitute a sophisticated mechanism that offers RNA-arbitrated adaptive protection against invasive viruses and plasmids. These systems are widely distributed among archaea and bacteria. The CRISPR systems have been categorized into three types depending on their sequences and components. Multiple proteins come together to form a huge functional Cas complex in type I and type III CRISPR systems. In contrast, the type II CRISPR system, found in *Streptococcus pyogenes*, depends on sole endonuclease (Cas9). This type II system has garnered significant attention and has been extensively utilized for genome engineering purposes. In this mechanism, when small fragments of DNA originating from a plasmid or virus are incorporated into a CRISPR locus, sections of the foreign deoxyribonucleic acid undergo conversion into CRISPR RNAs (crRNAs). These crRNAs subsequently bind with the trans-activating crRNA (tracrRNA), resulting in the formation of a double-RNA structure. This double-RNA structure serves as a guide for directing Cas9 to a specific region known as the protospacer within the invader's DNA. Once targeted, Cas9 initiates cleavage of the protospacer DNA on both strands. The complementary strand is broken by the HNH nuclease domain, while the noncomplementary strand is broken by the RuvC-like domain (Barrangou 2013).

In 2012, in vitro precise cleavage was showcased utilizing Cas9 alongside a single-guided RNA (sgRNA) molecule. This sgRNA was created through the hybridization of a double tracrRNA: crRNA complex (Jinek et al. 2012). This breakthrough marked a significant advancement and laid the groundwork for genome engineering via CRISPR. Early in 2013, two separate research groups autonomously illustrated targeted deoxyribonucleic acid cleavage in mouse and human cells using CRISPR/Cas technology (Cong et al. 2013; Mali et al. 2013). The primary limitation of CRISPR/Cas is that a 50-NGG protospacer adjacent motif must always come before recognition sites. Solely tailored sgRNA, encoded by roughly 100 nucleotide sequences, is all that is needed to target a particular sequence; Cas9 does not need to be redesigned for every new target site. For this reason, the CRISPR system is far simpler than TALENs or ZFNs (Pennisi 2013). The CASPR-Cas platform represents a cost-efficient and user-friendly technology for precise and effective targeting, editing, modification, regulation, and labeling of genomic loci across various cells and species. The advent of CRISPR-Cas9 has sparked a revolution, with laboratories worldwide harnessing this technology for new applications in the field of biology (Abdallah et al. 2015).

22.5.2.5 Meganucleases (Homing Endonucleases)

MegaNs are a class of naturally existing endonucleases first identified in the late 1980s. They belong to a group of endonucleases capable of recognizing and cleaving large DNA sequences, typically ranging from 12 to 40 bps, which are either unique or nearly so in the majority of the genomes (Gallagher et al. 2014). Additionally, MegaNs offer the advantage of exhibiting reduced cytotoxicity in cells as compared to alternative techniques like ZFNs (Julianne Smith et al. 2006). Their ability to recognize large DNA sequences makes MegaNs well-suited for genomic engineering applications. However, the availability of naturally existing MegaNs is limited and insufficient to target all potentially significant genomic loci.

Various companies have adapted MegaNs for genome editing purposes. Scientists at Cellectis modified the regions responsible for DNA cleavage and specific DNA interaction in MegaNs to recognize different DNA sequences while preserving their ability to cut DNA with high specificity. Similarly, Precision Biosciences, Inc. developed a method known as the directed nuclease editor (DNE), which employs a fully rational design process to create engineered MegaNs tailored to target and modify particular locations within the genome (Ashworth et al. 2010). Following DNA cleavage, cells undergo indigenous deoxyribonucleic acid repair mechanisms, with homologous recombination facilitating the integration of modified sequences or novel genes.

MegaNs have shown that they can produce precise genetic changes with high efficiency by targeting specific DNA sequences. These findings demonstrate MegaNs' value as a flexible tool for genetic editing in a variety of biological conditions (Periwal 2017). MegaNs have significant features and advantages over other genome editing approaches such as ZFNs and TALENs (Sharma et al. 2022). MegaNs, with their distinct structure and DNA recognition properties, provide benefits in terms of specificity and target site selection (Khalil 2020). Meganuclease-mediated genome editing mechanisms and investigating novel applications in areas such as gene therapy, biotechnology, and synthetic biology can further be clarified with thorough research. MegaNs have practical uses in a variety of domains, including targeted gene modification, gene regulation, and genome engineering (Li et al. 2020; Suzuki et al. 2020). Future meganuclease technology research should focus on addressing key difficulties and improving our understanding of genome editing mechanisms. Enhancing delivery systems, boosting target specificity, and investigating applications in developing sectors like gene drive technology are all areas that require additional research. In conclusion, MegaNs provide a precise and adaptable tool for targeted genomic alteration, with far-reaching ramifications in research, health, and biotechnology.

However, one drawback of using MegaNs is the considerable cost and time involved in constructing sequence-specific enzymes for all potential target sequences, in contrast to other SSNs. Every novel genome modification target necessitates a primary protein engineering phase to generate a tailored meganuclease, rendering MegaNs technically demanding to handle and also subject to patent disputes (Abdallah et al. 2015).

22.5.2.6 Oligonucleotide-directed Mutagenesis

ODM, a site-specific mutagenesis technique, employs mutagenic deoxyribonucleic acid pieces ranging from 20 to 200 nucleotides in size (Sauer et al. 2016). This method matches the sequence of fragments to a genome's target sequence, with the exception of one bp, which may represent a putatively introduced mutation (Sauer et al. 2016; Songstad et al. 2017).

Action Mechanism of ODM The ODM oligonucleotide in eukaryotic cells enters the cytoplasm by means of cell membrane, proceeds to the nucleus, and engages in complementary interactions with the DNA's target sequence. Here, nucleotides that do not match help to trigger a particular sequence alteration in the target gene as a result of the cell's error-correction system. The first step in this two-stage method is to anneal the targeted oligonucleotides with the target DNA. The second step is to repair the nucleotide mismatch, which causes a directed mutation (Sauer et al. 2016). Mammals were the first animals to exhibit this system, followed by plants (Beetham et al. 1999). It should be highlighted, nonetheless, that in plants, endonucleases that break down oligonucleotides and alter their 5' and 3' ends impede DNA ligation, which prevents the oligonucleotide from integrating into the genome (Songstad et al. 2017).

Single-stranded DNA (ssDNA) fragments can be used for ODM, but their short intracellular half-life restricts their application (Jansing et al. 2019). For single-stranded oligodeoxynucleotides (ssODN), stabilizing adjustments are required in order to overcome this drawback. Among them are chimeraplasts (methylated DNA/RNA duplexes), phosphorothioate modification, ssODNs with a 5'-Cy3, and modified 3'idC reverse base (Sauer et al. 2016). It should be highlighted that ODM has a rather low efficiency and a positive correlation with oligonucleotide fragment length. Research has demonstrated that lengthening the ssODN to 200 nucleotides can improve editing precision in *Arabidopsis thaliana* by up to 0.05%. Using chimeraplasts, the frequency of mutations in tobacco or rapeseed did not increase above the level of spontaneous mutations. In this sense, ODM is frequently employed in conjunction with nonspecific treatments that induce DSB, like TALENs or antibiotics and CRISPR/Cas9 in *Linum usitatissimum* and *Arabidopsis*, to boost the efficacy of mutagenesis and the frequency of target mutations.

22.5.2.7 Cisgenesis and Intragenesis

Cisgenic plants are genomically altered organisms that carry one or more genes sourced from identical species or sexually concordant species. These genes, known as cisgenes, replicate the natural genes entirely, including their endogenous promoter and terminator regions, and are arranged in the correct orientation. Unlike transgenic plants, which may contain additional genes like selection markers for antibiotic or herbicide resistance, cisgenic plants do not incorporate any foreign genes. Consequently, they closely resemble conventionally bred plants. Unlike typical breeding methods, the cisgenic technique averts the incorporation of unwanted genes associated with the target gene, thus bypassing the issue of linkage drag, which can lead to unwanted impacts.

Intragenesis distinguishes itself from cisgenesis based on the genetic construct's composition. An intragenic plant results from the introduction of a genetic construct containing regulatory and coding sequences sourced from either a sexually concordant species or an identical species. However, it is necessary to note that the intragene is not an exact replica of an endogenous gene. An intragene typically comprises a hybrid gene formed from a novel amalgamation of either a partial/full coding sequence along with promoter and terminator sequences. These sequences may originate from endogenous genes/genes of a cross-compatible species. Elements within the intragene can also be arranged in an antisense fashion or in a sense–antisense orientation (hairpin construct) to inhibit endogenous gene expression. Intragenes are inserted into vectors where the right and left borders of T-DNA of *Agrobacterium tumefaciens* are substituted with P-DNA (plant-derived transfer DNA) sequences and then integrated into the genome of the plant through genetic transformation without using markers. By emulating varieties resulting from conventional breeding techniques, plants developed via cisgenesis and intragenesis may garner greater public acceptance compared to transgenic plants. For instance, a 2010 Eurobarometer survey revealed that in European countries, cisgenically modified *Malus pumila* received more support than transgenic apples.

In the agricultural sector, a genomic modification is deemed advantageous if it brings in a beneficial phenotypic characteristic without compromising productiveness or introducing undesirable alterations. Both intragenesis and cisgenesis have the potential to meet these criteria. Cisgenesis is particularly suited for transferring phenotypic characteristics like resistance to pathogens from a wild, invulnerable, cross-compatible donor plant to a recipient one. Intragenesis offers greater flexibility compared to cisgenesis since it allows for the modification of a native target gene's spatial or temporal expression. This flexibility enables the directed or enhanced synthesis of valuable products in edible organs, such as fruits or seeds. Additionally, intragenesis enables the silencing of unwanted genes, like those liable for producing anti-nutritional compounds, through RNA interference. RNA gene silencing, commonly employed in studies of functional genomics, is increasingly recognized as a potent tool for crop improvement and breeding.

22.5.2.8 Reverse Breeding

RB is a pioneering approach in plant breeding aimed at reconstructing elite hybrids using genotypes that, while not identical to the original parents, possess complementary traits that enable the replication of the original hybrid. This method was initially trialed in the model plant *A. thaliana*, which boasts a compact genome. By inhibiting meiotic recombination, RB facilitates the generation of parental lines derived from specific heterozygous plants.

Mechanism of Reverse Breeding RB involves two fundamental stages: firstly, the inhibition of crossover recombination within a chosen plant, succeeded by the generation of DH lines from spores containing chromosomes without recombination. These DH lines serve as the basis for reproducing the elite heterozygote on a large scale for commercial purposes.

At the core of RB lies the mechanism of achiasmatic meiosis. In achiasmatic chromosomes, where crossovers fail to occur, chromosomes remain univalent. Unlike in bivalents where chiasmata facilitate the separation of homologous chromosomes to opposite poles during regular disjunction, univalents lack these chiasmata. Consequently, homologous chromosomes may separate to the identical pole (nondisjunction), resulting in aneuploidy in spores. As a result, achiasmatic plants exhibit high levels of sterility.

22.5.2.9 Efficacious Suppression of Recombination

Efficacious suppression of meiotic crossovers is necessary for RB to work. As a result, genes that are crucial for the production of crossovers but preserve chromosomal structure are very helpful. Genes can be targeted with RNA interference or siRNAs to achieve the suppression of gene expression, which is necessary for RB and will mostly result in posttranscriptional gene silencing (PTGS). For instance, in the human meiotic recombination protein DMC1, an amino-terminal deletion devoid of 81 amino acids renders it fully defective in both ssDNA and double-stranded DNA binding activities. In crops where achieving steady transformation is unfeasible, alternative methods such as virus-induced gene silencing can be employed to induce PTGS. The application of RB would proceed much more quickly if substances that impede or omit recombination during meiosis were applied exogenously. For RB to benefit, crossover suppression does not have to be fully achieved. Any chromosomal pair(s) may nevertheless have a single remaining crossover. The homologs participating in a single crossover often separate, and this raises the likelihood of producing a doubly balanced gamete. Two recombinant chromatids are also produced by a crossing, but they are useless for RB. However, only half of the bivalent pair's chromatids are impacted by a crossover, meaning the other two chromatids are nonrecombinant and functional. As a result, half of the generated spores may be beneficial to RB. Recombinant chromosomes are carried by DHs more frequently when residual crossovers (assuming there is only one per bivalent) occur, although 50% of spores with nonrecombinant chromosomes are still produced. By utilizing molecular markers, it is possible to select these nonrecombinant spores.

22.5.2.10 Double Haploids

Through achiasmatic meiosis, doubled-haploid (DH) plants can be produced from unfertilized ovules or from cultures of microspores and anther. Different species have different levels of DH production efficiency from haploid spores. The implementation of RB is confined to crops where DH technique is extensively used. This method has proven successful for the vast majority of crop species, and reputable breeding businesses regularly employ it in their breeding operations. Nonetheless, there are a few well-known exceptions, such as tomatoes, lettuce, soybeans, and cotton, where the formation of double-haploid plants is either extremely unusual to nonexistent. In modern plant breeding, genotyping DHs using molecular markers is a standard procedure and essential for RB. If meiotic recombination is entirely absent, a single polymorphic molecular marker per chromosome would be adequate for genotyping every DH plant since each chromosome would act as a unified linkage block. However, in the presence of residual crossovers, two markers per chromosome, ideally positioned distally, are necessary for accurate genotyping.

22.6 Conclusion

There are few significant obstacles, but the most important one is the availability of food. In the past, plant breeding demonstrated exceptionally high productivity. The latest plant breeding techniques participate in several areas of the plant. The recent plant breeding techniques are more filling, finding the knowledge gaps that can lead to new varieties. The previous techniques can be improved by adding new strategies. Most recent techniques like GS, MAS, HTP, CRISPR-Cas9, SSNs, ZFNs, TALENs, CRISPR, meganucleases (homing endonucleases), ODM, cisgenesis and intragenesis, RB, efficacious suppression of recombination, and DH showed significant progress in plants. CRISPR and all the latest techniques help to improve synthetic biology, and plant genetics, plant architecture, flower production, and fruit attributes provide new opportunities to improve the productivity of crops.

References

Abreu, G.B., Ramalho, M.P., Toledo, F.H.R.B. et al. (2010). Strategies to improve mass selection in maize. *Maydica 55* (3–4): 219–225.

Acquaah, G. (2009). *Principles of Plant Genetics and Breeding*. New Jersey: John Wiley & Sons.

Acquaah, G. (2015). Conventional plant breeding principles and techniques. *Advances in plant breeding strategies: Breeding, Biotechnology and Molecular Tools*, 115–158. Cham: Springer International Publishing.

Ahmad, F. and Slinkard, A.E. (2004). The extent of embryo and endosperm growth following interspecific hybridization between *Cicer arietinum L.* and related annual wild species. *Genetic Resources and Crop Evolution* 51: 765–772.

Aleksoski, J. (2018). The effect of backcross method in tobacco breeding. *Journal of Agriculture and Plant Sciences 16* (1): 9–19.

Allard, R.W. (1961). Principles of plant breeding. *Soil Science* 91 (6): 414.

Alston, J.M., Beddow, J.M., and Pardey, P.G. (2009). Agricultural research, productivity, and food prices in the long run. *Science* 325 (5945): 1209–1210.

Araus, J.L. and Cairns, J.E. (2014). Field high-throughput phenotyping: The new crop breeding frontier. *Trends in Plant Science* 19 (1): 52–61.

Barone, A., Del Giudice, A., and Ng, N.Q. (1992). Barriers to interspecific hybridization between *Vigna unguiculata* and *Vigna vexillata*. *Sexual Plant Reproduction 5*: 195–200.

Begna, T. (2021). Conventional breeding methods widely used to improve self-pollinated crops. *International Journal of Research* 7(1): 1–16.

Bennett, H. W., Burson, B. L., and Bashaw, E. C. (1969). Intraspecific hybridization in dallisgrass, *Paspalum Dilatatum Poir.* 1. *Crop Science* 9 (6): 807–809.

Borlaug, N.E. (1983). Contributions of conventional plant breeding to food production. *Science* 219 (4585): 689–693.

Breseghello, F. and Coelho, A.S.G. (2013). Traditional and modern plant breeding methods with examples in rice (*Oryza sativa L.*). *Journal of Agricultural and Food Chemistry* 61 (35): 8277–8286.

Briggs, F.N. (1938). The use of the backcross in crop improvement. *The American Naturalist* 72 (740): 285–292.

Brown, J. and Caligari, P. (2011). *An Introduction to Plant Breeding*. New Jersey: John Wiley and Sons.

Bustamam, M., Tabien, R. E., Suwarno, A. et al. (2002). Asian rice biotechnology network: Improving popular cultivars through marker-assisted backcrossing by the NARES. In: *Poster presented at the international rice congress* 16–20. China: Beijing.

Chahal, G.S. and Gosal, S.S. (2002). *Principles and Procedures of Plant Breeding: Biotechnological and Conventional Approaches*. Oxford: Alpha Science Int'l Ltd.

Chen, H.M., Zhang, Y.D., Chen, W. et al. (2010). Improving grain yield and yield components via backcross procedure. *Maydica* 55 (2): 145.

Chen, K., Wang, Y., Zhang, R. et al. (2019). CRISPR/Cas genome editing and precision plant breeding in agriculture. *Annual Review of Plant Biology* 70: 667–697.

Collard, B.C. and Mackill, D.J. (2008). Marker-assisted selection: an approach for precision plant breeding in the twenty-first century. *Philosophical Transactions of the Royal Society B: Biological Sciences* 36(1491): 557–572.

Crossa, J., Pérez-Rodríguez, P., Cuevas, J. et al. (2017). Genomic selection in plant breeding: Methods, models, and perspectives. *Trends in Plant Science* 22 (11): 961–975.

Dandekar, A., Leslie, C. and McGranahan, G. (2005). *Juglans regia* walnut. In: *Biotechnology of Fruit and Nut Crops*, 307–323. Wallingford UK: CAB International.

Das, G., Patra, J.K. and Baek, K.H. (2017). Insight into MAS: A molecular tool for development of stress resistant and quality of rice through gene stacking. *Frontiers in Plant Science* 8: 985.

Doebley, J.F., Gaut, B.S. and Smith, B.D. (2006). The molecular genetics of crop domestication. *Cell* 127 (7): 1309–1321.

FAO, I. and UNICEF. (2019). WFP, WHO, The State of food security and nutrition in the world. In: *Safeguarding Against Economic Slowdowns and Downturns*. United Nations: Food and Agriculture Organization (FAO).

Feng, Z., Zhang, B., Ding, W. et al. (2013). Efficient genome editing in plants using a CRISPR/Cas system. *Cell Research* 23 (10): 1229–1232.

Fu, X.L., Lu, Y.G., Liu, X.D. et al. (2009). Crossability barriers in the interspecific hybridization between *Oryza sativa* and *O. meyeriana*. *Journal of Integrative Plant Biology* 51 (1): 21–28.

Fu, Y.B. (2015). Understanding crop genetic diversity under modern plant breeding. *Theoretical and Applied Genetics* 128: 2131–2142.

Fujimak, H. (1978). New techniques in backcross breeding for rice improvement.

Golovan, L.V., Klymenko, I.V., Stankevych, S.V. et al. (2019). The inheritance of economically valuable features in the intraspecific hybridization of bean (*Phaseolus L*). *Ukrainian Journal of Ecology* 9 (2): 156–169.

Hallauer, A.R. (2010). Recurrent selection in maize. *Plant Breeding Reviews* 9: 115–179.

Hallauer, A.R., Carena, M.J. and Miranda Filho, J.D. (2010). *Quantitative Genetics in Maize Breeding (Vol.* 6). Netherland: Springer Science and Business Media.

Hallauer, A.R., Lamkey, K.R., Russell, W.A. et al. (1992). Registration of B95 parental inbred line of maize. *Crop Sciences* 32: 1515.

Harlan, J.R. and De Wet, J.M. (1971). Toward a rational classification of cultivated plants. *Taxon* 20 (4): 509–517.

Hickey, J.M., Chiurugwi, T., Mackay, I. et al. (2017). Genomic prediction unifies animal and plant breeding programs to form platforms for biological discovery. *Nature Genetics* 49 (9): 1297–1303.

Jannink, J.L., Lorenz, A.J., and Iwata, H. (2010). Genomic selection in plant breeding: from theory to practice. *Briefings in Functional Genomics* 9 (2): 166–177.

Khadr, F.H. (1964). *Effectiveness of Recurrent Selection and Recurrent Irradiation in Oat Breeding*. Iowa: Iowa State University.

Kim, D., Alptekin, B., and Budak, H. (2018). CRISPR/Cas9 genome editing in wheat. *Functional and Integrative Genomics* 18: 31–41.

Klümper, W. and Qaim, M. (2014). A meta-analysis of the impacts of genetically modified crops. *PLoS One* 9 (11): e111629.

Lamichhane, S. and Thapa, S. (2022). Advances from conventional to modern plant breeding methodologies. *Plant Breeding and Biotechnology* 10 (1): 1–14.

Lane, M. (1981). The achievements of conventional plant breeding. *Philosophical Transactions of the Royal Society B. Biological Sciences* 292 (1062): 441–455.

Li, L., Zhang, Q., and Huang, D. (2014). A review of imaging techniques for plant phenotyping. *Sensors* 14 (11): 20078–20111.

Liao, S., Qin, X., Luo, L. et al. (2019). CRISPR/Cas9-induced mutagenesis of semi-rolled leaf1, 2 confers curled leaf phenotype and drought tolerance by influencing protein expression patterns and ROS scavenging in rice (*Oryza sativa L.*). *Agronomy* 9 (11): 728.

Lidder, P. and Sonnino, A. (2012). Biotechnologies for the management of genetic resources for food and agriculture. *Advances in Genetics* 78: 1–167.

Louwaars, N.P. (2018). Plant breeding and diversity: A troubled relationship? *Euphytica* 214 (7): 114.

Luckett, D. and Halloran, G. (2003). Plant breeding. In: *Principles of Field Crop Production*, 4e, 1–3. New York, USA: Oxford University Press.

Marcellan, O.N. and Camadro, E.L. (1999). Formation and development of embryo and endosperm in intra-and inter-specific crosses of *Asparagus officinalis* and *A. densiflorus cv. Sprengeri. Scientia Horticulturae* 81 (1): 1–11.

Meredith Jr, W. (1977). Backcross breeding to increase fiber strength of cotton 1. *Crop Science* 17 (1): 172–175.

Murray, B.G. 2016. Hybridization and plant breeding. *Encyclopedia of Applied Plant Sciences* 2: 168–173.

Mwangangi, I.M., Muli, J.K., and Neondo, J.O. (2019). Plant hybridization as an alternative technique in plant breeding improvement. *Asian Journal of Research in Crop Science* 4 (1): 1–11.

Nakwilai, P., Cheabu, S., Narumon, P. et al. (2020). Evaluation of japonica rice (*Oryza sativa L.*) varieties and their improvement in terms of stability, yield and cooking quality by pure line selection in Thailand. *Science Asia* 46: 157–168.

Nogoy, F.M., Song, J.Y., Ouk, S. et al. (2016). Current applicable DNA markers for marker assisted breeding in abiotic and biotic stress tolerance in rice (*Oryza sativa L.*). *Plant Breeding and Biotechnology* 4 (3): 271–284.

Pabuayon, I.L.B., Sun, Y., Guo, W. et al. (2019). High-throughput phenotyping in cotton: A review. *Journal of Cotton Research* 2: 1–9.

Pandey, S. and Gardner, C.O. (1992). Recurrent selection for population, variety, and hybrid improvement in tropical maize. *Advances in Agronomy* 48: 1–87.

Pasala, R. and Pandey, B.B. (2020). Plant phenomics: high-throughput technology for accelerating genomics. *Journal of Biosciences* 45 (1): 111.

Passioura, J.B. (2006). The perils of pot experiments. *Functional Plant Biology* 33 (12): 1075–1079.

Poehlman, J.M. and Sleper, D.A. (1995a). *Breeding Field Crops*. Ames: Iowa State University Press.

Pollegioni, P., Olimpieri, I., Woeste, K. E. et al. (2013). Barriers to interspecific hybridization between *Juglans nigra L.* and *J. regia L* species. *Tree Genetics and Genomes* 9: 291–305.

Poorter, H., Bühler, J., van Dusschoten, D. et al. (2012). Pot size matters: a meta-analysis of the effects of rooting volume on plant growth. *Functional Plant Biology* 39 (11): 839–850.

Qaim, M. (2020). Role of new plant breeding technologies for food security and sustainable agricultural development. *Applied Economic Perspectives and Policy* 42 (2):129–150.

Rabier, C.E., Barre, P., Asp, T. et al. (2016). On the accuracy of genomic selection. *PLoS One* 11 (6): e0156086.

Ramya, P., Singh, G.P., Jain, N. et al. (2016). Effect of recurrent selection on drought tolerance and related morpho-physiological traits in bread wheat. *PLoS One* 11 (6): e0156869.

Rebetzke, G.J., Jimenez-Berni, J., Fischer, R.A. et al. (2019). High-throughput phenotyping to enhance the use of crop genetic resources. *Plant Science* 282: 40–48.

Reynolds, M.P., Pask, A.J.D., and Mullan, D.M. (2012). *Physiological Breeding I: Interdisciplinary Approaches to Improve Crop Adaptation*. Mexico: CIMMYT.

Ribaut, J.M. and Hoisington, D. (1998). Marker-assisted selection: new tools and strategies. *Trends in Plant Science* 3 (6): 236–239.

Ricroch, A., Clairand, P., and Harwood, W. (2017). Use of CRISPR systems in plant genome editing: Toward new opportunities in agriculture. *Emerging Topics in Life Sciences* 1 (2): 169–182.

Romero, G.E. and Frey, K.J. (1966). Mass selection for plant height in oat populations 1. *Crop Science* 6 (3): 283–287.

Shah, P.R., Varanavasiappan, S., Kokiladevi, E. et al. (2019). Genome editing of rice *PFT1* gene to study its role in rice sheath blight disease resistance. *International Journal of Current Microbiology and Applied Sciences* 8 (6): 2356–2364.

Sharma, M., Gupta, S.K., Deeba, F. et al. (2017). Effects of reactive oxygen species on crop productivity: An overview. In: *Reactive Oxygen Species in Plants: Boon Or Bane-Revisiting the Role of ROS*, 117–136. Hoboken, NJ: Wiley & Sons Ltd.

Shen, L., Courtois, B., McNally, K. et al. (1999). Developing near-isogenic lines of IR64 introgressed with QTLs for deeper and thicker roots through marker-aided selection. *Genetic Improvement of Rice for Water-Limited Environments. Los Baños (Philippines)*, 275–292. Los Banos: International Rice Research Institute.

Singh, S.P. (1982). Alternative methods to backcross breeding. *Annual report of the Bean Improvement Cooperative* 25: 11–12.

Steele, K.A., Price, A.H., Shashidhar, H.E. et al. (2006). Marker-assisted selection to introgress rice QTLs controlling root traits into an Indian upland rice variety. *Theoretical and Applied Genetics* 112: 208–221.

Tester, M. and Langridge, P. (2010). Breeding technologies to increase crop production in a changing world. *Science* 327 (5967): 818–822.

Thomas, B. (2016). *Encyclopedia of Applied Plant Sciences*. Cambridge, Massachusetts: Academic Press.

Vogel, K.E. (2009). Backcross breeding. *Transgenic Maize: Methods and Protocols*, 161–169. Totowa: Humana Press.

Wang, X., Xu, Y., Hu, Z. et al. (2018). Genomic selection methods for crop improvement: Current status and prospects. *The Crop Journal* 6 (4): 330–340.

Yu, K., Liu, Z., Gui, H. et al. (2021). Highly efficient generation of bacterial leaf blight-resistant and transgene-free rice using a genome editing and multiplexed selection system. *BMC plant biology* 21 (1): 1–10.

Zadokar, A., Kumar, P., Kumar, A. et al. (2023). Insight into the molecular breeding research status for crop improvement in India: Prospects and achievements. *Plant Molecular Biology Reporter* 41 (4): 537–558.

Zhang, H. and Wang, S. (2013). Rice versus *Xanthomonas oryzae pv. oryzae*: A unique pathosystem. *Current Opinion in Plant Biology* 16 (2): 188–195.

Zhang, Y. and Zhang, N. (2018). Imaging technologies for plant high-throughput phenotyping: A review. *Frontiers of Agricultural Science and Engineering* 5 (4): 406–419.

Zhao, X., Tan, G., Xing, Y. et al. (2012). Marker-assisted introgression of qHSR1 to improve maize resistance to head smut. *Molecular Breeding* 30: 1077–1088.

23

Biofortification of Crops and Vegetables to Achieve Food Nutritional Security

Fahad Shafiq[1*], *Sumera Anwar*[2], *Umar Farooq Awan*[1], *Zunaira Yaqoob*[1], *Sadaf Mehfooz*[1], *Maham Ishfaq*[3], *Ayesha Ali*[1], *Areej Fatima*[1], *Ruqayya Ali*[1], *Areeba Azhar*[1], *Iqra Mukhtar*[1], *Minahil Sheikh*[1], *and Muhammad Ashraf*[3]

[1] *Department of Botany, Government College University Lahore, Punjab, Pakistan*
[2] *Department of Botany, Government College Women University, Faisalabad, Pakistan*
[3] *Institute of Molecular Biology and Biotechnology, The University of Lahore, Pakistan*

*Corresponding author: fahadsheikh@gcu.edu.pk

23.1 Introduction

23.1.1 Hidden Hunger: Mineral Deficiencies Across the World

Hidden hunger is defined as "the inadequate intake of micronutrients by humans" affecting more than two billion people worldwide (IFPRI 2016; Ritchie, Reay et al. and Higgins 2018; Lowe 2021). In this context, micronutrient malnutrition affects over 150 million children worldwide (including 36 million in Asia) leading to stunted growth, weight reduction, and a 50% mortality rate (Lyons 2018; Weffort and Lamounier 2024) and the problem is more severe in developing countries (Harding, Aguayo and Webb et al. 2018). In adults, impaired physical health, higher infection rate, and progression of degenerative and cognitive diseases are common (Ibeanu et al. 2020). Other medical conditions caused by hidden hunger include cardiovascular diseases, thyroid, anemia, and osteoporosis linked with Fe, Zn, Ca, and iodine deficiencies (Tulchinsky 2010; Weffort and Lamounier 2024). Moreover, deficiencies such as vitamins A, B_{12}, and D are prevalent in the human population worldwide (Lowe 2021; Das and Padhani 2022; OMS 2023). In Pakistan, malnutrition affects individuals belonging to all age groups (Rai and Rani 2024). A large population of Pakistan is deficient in essential vitamins and minerals (Ahmed 2019; Amjad et al., 2020). As mentioned earlier, these deficiencies particularly affect women and children across all age groups leading to impaired cognitive development, higher mortality rates, and higher disease incidence rates (Coppinger and Diamond, 2001; Shimada et al., 2021; Rai and Rani 2024). Similar reports have been published in several reputed journals highlighting the need to address malnutrition in humans (Ruel et al. 2013; Black et al. 2013; Horton and Lo, 2013; Taylor et al. 2013; Hawkes et al. 2020; Popkin et al. 2020).

Although the United Nations declared 2016–2025 for an Action on Nutrition, the issues related to food security continue to increase (United Nations 2020). Micronutrient deficiencies in humans are a major issue since they go unnoticed until irreversible damage is caused and contribute to 7% of the global disease burden (Ibeanu et al.2020). Unfortunately, the prevalence of micronutrient malnutrition is increasing in the human population, especially in developing nations (FAO et al. 2019). Major factors that contribute to hidden hunger include lack of dietary diversity, reliance on high-yielding cereal crops (often micronutrient deficient), and financial resources (Von Grebmer 2018; Zikankuba et al. 2019). Moreover, an imbalanced diet, excessive consumption of fast foods, and lack of awareness contribute to hidden hunger (Mehboob 2022; Mashhadi et al. 2022). It has been reported that as children become adults, there is a prominent reduction in the quality of their diet including a reduction in fruit and vegetable consumption with an increased intake of carbonated drinks (Black et al. 2013).

Taken together, micronutrient deficiencies pose significant challenges for humans of all age groups and efforts are required to address this issue at regional and global scale. Efforts to reduce hidden hunger are the top priority of the UN and a lot of progress has been made so far through various approaches like food fortification, dietary supplements, and biofortification (Obersteiner et al. 2016; FAO 2017; Allen and de Brauw 2018).

Crop Biofortification: Biotechnological Approaches for Achieving Nutritional Security under Changing Climate, First Edition. Edited by Adnan Noor Shah et al.

23.2 Sustainable Developmental Goal 2: Nutritional Food Security

In 2015, all the UN member states adopted the 2030 Agenda for Sustainable Development and identified 17 goals (herein referred to as SDGs 1-17) to tackle climate change and a prosperous way forward to end poverty and inequality. Of these, Sustainable Developmental Goal 2 (SDG-2) primarily focuses on achieving zero hunger by 2030. The objectives of SDG-2 are to end hunger, achieve food security, improve nutrition, and promote sustainable agriculture. It is estimated that over 1.5 billion people (20% of the global population) are affected by Fe and Zn deficiency (McLean et al. 2009; Wessells and Brown 2012; Hefferon 2019) and a major fraction of this population lives in third-world countries (Ritchie et al. 2018). It is pertinent to mention that the costs associated with disease burden due to hidden hunger have progressed to over US$180 billion (Global Health 2013). Not only this, malnutrition also compromises the learning abilities of the people, negatively affecting socioeconomic development (Black et al. 2013).

Currently, it does not seem like we are on track to achieve the SDG-2 of hidden hunger (Lowe 2021). Collaboration across disciplines among academicians, legislators, and stakeholders is crucial for improving global nutrition. Moreover, determining the specific nutrients to target, and monitoring the effects of nutritionally enhanced crops on human health is also challenging. Nonetheless, improving the nutritional quality of crops and vegetables is crucial for addressing micronutrient shortages and promoting global health. Techniques such as fertilization, genetic engineering, and biofortification have been used to enhance the bioavailability and nutrient composition of food crops. To ensure the widespread adoption and effect of new technologies, it is imperative to continue studying, collaborating, and including the public (Zinkernagel et al. 2020). In this chapter, an attempt has been made to address the issue of hidden hunger through biofortification, and comprehensive insights into various biofortification avenues have been discussed in length.

23.3 Biofortification: Improving Nutritional Status in Edible Plant Parts

Biofortification is the process by which the nutritional density of food is improved either by conventional breeding, agronomic practices, or through modern biotechnological tools without compromising its characteristics important to farmers or consumers. Since low-income families mostly rely on low-cost staple food and generally lack dietary diversity in their eating habits (Popkin et al. 2020), by using this approach, the concentration of various essential micronutrients including Zn and Fe as well as the concentration of vitamins can be improved (Siwela et al. 2020). In addition, people mostly rely on cereal consumption which does not fulfill micronutrient requirements (Perez-Massot et al. 2013; Garcia-Oliveira et al. 2018). One of the main sources of vitamins, minerals, and phytochemicals that protect people from malnutrition is vegetables. Nutrient enrichment is crucial in agriculture. On the other hand, vegetables are deficient in essential mineral nutrients.

As a way forward, crop biofortification is a feasible and cost-effective process to address the issue of micronutrient deficiency, especially in poor people (Singh et al. 2021). Biofortification shows potential by enhancing the amount of essential elements like zinc, iron, iodine, and vitamin A in the edible parts of many agricultural plants. The consensus is that biofortification works better than supplements, food diversity, and fortification. The biofortification process includes enhancing mineral bioavailability, transport, and root uptake (Zulfiqar et al. 2024). To boost the nutritional quality of vegetables, micronutrient application is crucial in addition to primary nutrients (Figure 23.1).

23.4 Approaches to Enhance the Nutritional Quality of Crops and Vegetables

There are various ways to tackle malnutrition and address the issue of hidden hunger. These methods include supplementation, fortification, biofortification, and diet diversification (Gibson and Hotz 2001; Wakeel et al. 2018; Lowe 2021). Of various ways, supplementation or the use of supplements to treat food nutritional deficiencies offers a direct solution and is quite effective (Casgrain et al. 2012; Maughan et al. 2018); however, this is not practically feasible for a long-term solution and is difficult to practice in remote areas. The second possible solution is diet diversification (Maqbool and Beshir 2019). This could also be challenging due to regional food habits and due to some economic issues. In particular, poor people cannot afford a diversified diet and are mostly restricted to limited dietary habits (Meenakshi et al. 2010). Another approach is fortification by adding vitamins and minerals to increase the food's nutritional value and it can be carried out at an industrial scale (FAO 1995; Garrett 2018; Mannar and Hurrell 2018). This method is quite effective in increasing the micronutrient content of the food, however, there are certain limitations associated with this process. This

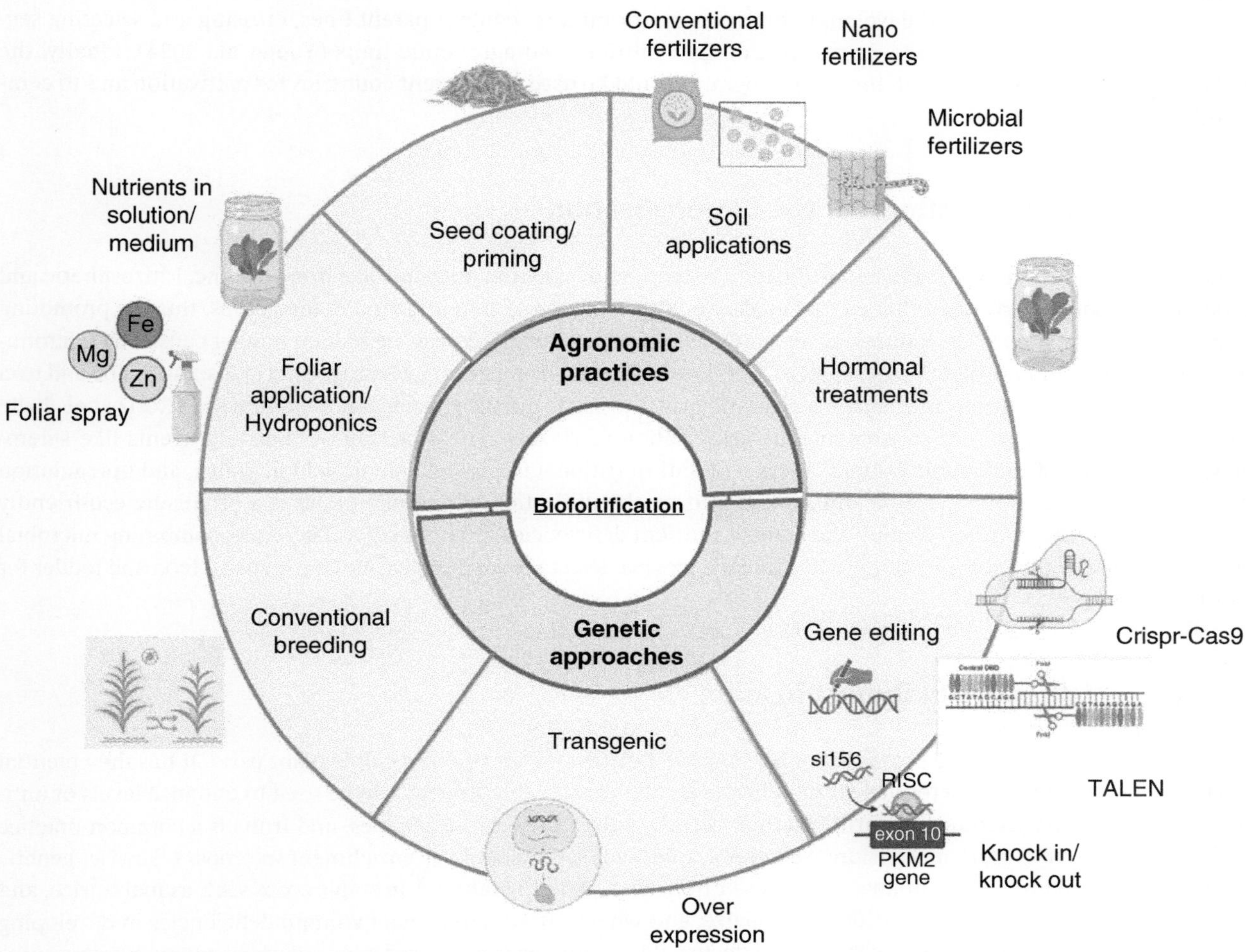

Figure 23.1 Schematic diagram showing multifaceted approaches to target and enhance biofortification of crops.

is because food fortification may change the color, appearance, taste, and shelf life of the food product (Blanco-Rojo and Vaquero 2019). Moreover, all the communities may not have equal access to fortified food (Lowe 2021) and the costs may also pose an issue. Finally, the fortified requires proper monitoring during the fortification process to ensure the quality which is also challenging.

So, what is the feasible and sustainable way to address the issue of hidden hunger? The answer lies in biofortification. Biofortification involves increasing the nutrient content of crops either through agronomic practices or breeding programs (WHO/FAO 2006; Bouis 2018). The agronomic practices may include the use of nutrient-rich fertilizers, foliar formulations, seed coating with nutrients, soil amendments to enhance crop nutrient fraction, or the use of various nanomaterials to enhance the crop mineral nutrients. It is practically feasible and better than the previously mentioned methods and offers solutions to address the problem of hidden hunger, especially among third-world countries. The biofortified crops possess greater nutrient concentrations mainly in wheat, rice, corn, potatoes, radish, pea, and beans and it is a practically feasible approach to enhance the nutritional quality of the food crops (Garg et al. 2018; Raza et al. 2022; Shafiq et al. 2023; Nazir et al. 2024; Mahmood et al. 2024).

Alternatively, the process of genetic biofortification significantly raises the available levels of micronutrients in edible crops. A key strategy in the fight against hunger is genetic biofortification, which aims to increase the nutritional content of crops through genetic engineering. This method involves the identification of key genes involved in plant micronutrient uptake. Hence, genetic biofortification could improve the uptake of essential elements like iron, zinc, and vitamin A in plant parts. In addition, conventional breeding has a considerable record of being effective in producing high yield, by using techniques including mass selection, hybridization, and mutation breeding. Conventional plant breeding involves identifying and developing high-vitamin or mineral-content parent lines, crossing and selecting segregates over generations to produce plants with desired nutrient and agronomic traits (Yue et al., 2024). Finally, the biofortified crops developed through breeding programs could be used in different countries for cultivation and to combat hidden hunger (Lowe 2021).

23.5 Role of Soil Microflora for the Biofortification

Soil microflora plays a vital role in the biofortification of crops with essential nutrients like iron and zinc. Rhizospheric and endophytic microorganisms can enhance the uptake and translocation of iron and zinc in food crops, thereby promoting plant growth (Lurthy et al. 2021; Rahman et al. 2024). This approach helps address the hidden hunger caused by micronutrient deficiencies in human diet (Verma et al. 2021). These rhizospheric microorganisms transform insoluble zinc and iron to soluble form through various mechanisms enabling plants to easily uptake these metal ions from soil (Roriz et al. 2020; Haroon et al. 2022). These mechanisms include acidification of rhizosphere, secretion of chelating agents like siderophores, modification of root architecture, reduction of anti-nutritional factors like phytic acid in grains, and upregulation of transporter genes in plants for zinc and iron. Moreover, the utilization of soil microflora is a promising eco-friendly method to improve the nutritional value and address nutrient deficiencies in human diet. Therefore, combining microbial biofortification with other sustainable agricultural practices can lead to more nutritive and inexpensive food and fodder for mankind and livestock.

23.6 Current Avenues in Biofortification

Agronomic biofortification is the process of enhancing the nutrient composition of edible plant parts. It has the potential to improve crop growth and micronutrient mobilization. Additionally, this strategy can be used to enhance levels of antioxidants, minerals, and vitamins. Biofortification of cereals, pulses, oilseeds, vegetables, and fruits is a common practice nowadays targeting zinc, iron, magnesium, selenium, folic acids, and vitamin A enrichment in crops. Likewise, genetic engineering has been utilized to enhance the levels of iron, zinc, and provitamin A in staple crops such as maize, rice, and cassava. Scientists have created biofortified rice, maize, and wheat varieties to combat vitamin deficiencies in developing countries. Agronomic biofortification can also be effective for optimizing the nutrient status of crops grown in micronutrient-deficient soils (Asif et al. 2024). A brief account is presented here.

23.6.1 Mineral Biofortification

Mineral nutrients are essentially required for normal human development. By contrast, mineral deficiencies are very prevalent and result in some serious health issues (Bailey et al. 2015). Among these, the deficiency of Fe, Zn, Ca, I, and Se is commonly associated with the residents of third-world countries. In this context, several studies highlighted the potential of mineral biofortification in crops. For instance, iodine deficiency is also a significant challenge and is currently being addressed via salt biofortification and is mandatory in various countries (Lowe 2021). In Pakistan, both Fe and Zn are commonly deficient in agricultural soils due to high pH, and cereal crops cultivated on alkaline soils generally exhibit Fe/Zn concentrations well below the recommended threshold level (Ishfaq et al. 2021; Shafiq et al. 2023). Apart from cereals, tubers, legumes, and vegetables also exhibit mineral deficiencies which also contribute to significant yield losses (Noulas et al. 2018; Joshi-Saha et al. 2022; Vélez-Bermúdez and Schmidt 2022). An attempt has been made to tabulate data from some previous studies and the benefits are reported here in Table 23.1.

23.6.2 Vitamin Biofortification

Vitamin biofortification is also a much-needed avenue to address hidden hunger. It is reported that the prevalence of vitamin deficiencies is significant (i.e. vitamin A 6%; vitamin D 29%; vitamin B_1 11%; vitamin B_2 50%; vitamin B_6 46%; folate 32% and vitamin B_{12} 46%) and has become a major concern (Bamji et al. 2021). Vitamin A is important for normal vision, gene expression, reproduction, embryonic development, growth, and immune function. It is necessary for the integrity of epithelial cells throughout the body (Weffort and Lamounier 2024).

Table 23.1 Biofortification of crops and vegetables with minerals by soil and foliar application of micronutrient formulations.

Mineral	Crop/Vegetable	Biofortification method	Benefits	References
Fe	Rice	DTPA-Fe, $FeSO_4$, EDTA-FeNa, HEDTA-Fe foliar application	DTPA-Fe promoted Fe content in grains	He et al. (2013)
Fe	Rice	Fe-EDDHA foliar application	0.8% improved PSII electron and energy transport properties	Gao et al. (2022)
Fe	Rice/Soybean/ Lettuce	Fe-citrate, Fe-EDTA, Fe-HA, $FePO_4$, nano-Fe oxide	Fe-HA showed the highest Fe mobilization and seed Fe content	Sharma et al. (2023)
Fe	Rice	Fe-amino acid foliar application	14.5% average grain Fe content	Yuan et al. (2013)
Fe	Rice	Soil (5 kg ha^{-1}), foliar (1.5 kg ha^{-1}), and combined	Combined soil + foliar application showed maximum grain Fe	Butsai et al. (2022)
Fe	Rice	$FeSO_4$ (0.5, 0.75, 1, 1.5%) foliar application	1.5% $FeSO_4$ resulted in 61.2 μg/pot Fe uptake	Kamali, Kalaiselvi and Chitdeshwari (2020)
Fe	Rice	$FeSO_4$ and $FeSO_4$ + nicotianamine foliar application	Grain Fe increased by 17% and 30%	Wei et al. (2012a)
Fe	Rice	$FeSO_4$ foliar application	Positively regulated flavonoid metabolism; Grain Fe content was not affected	Zhao et al. (2024)
Fe	Wheat	FeEDTA, $FeSO_4$, Fe-EDDHA, Fe-citrate (0.25%) foliar application	Improved grain Fe content	Aciksoz et al. (2011)
Fe	Wheat	$FeSO_4$ soil and foliar application	Foliar application is more effective in increasing grain Fe	Ramzan et al. (2020)
Fe	Wheat	Fe-EDDHA, $FeSO_4$, and nano-zero-valent Fe, foliar at a dose of 0.2% Fe	Grain Fe content was significantly increased by all Fe sources	Taskin and Gunes (2022)
Zn	Rice	Zn-amino acid and $ZnSO_4$, foliar application	Grain Zn was increased by 24% and 22%, respectively	Wei et al. (2012b)
Zn	Rice	0.5% $ZnSO_4$ foliar application	Zn sprayed after flowering increased grain Zn	Boonchuay et al. (219)
Zn	Rice	0.15–2250 μmol L^{-1} $ZnSO_4$, sand and solution	Higher grain Zn from 9 to 37 mg kg^{-1}	Jiang et al. (2008)
Zn, Fe	Wheat	$ZnSO_4$ (50 kg ha^{-1}) as soil application, 0.5% (4 kg ha^{-1}) $ZnSO_4$ as foliar; $FeSO_4$, Fe-EDTA, Fe-EDDHA, and Fe-citrate at 5 and 10 mg kg^{-1} in soil, or 0.2% Fe-EDTA as foliar	Combined soil and foliar Zn application caused more than a threefold increase in grain Zn; Fe-EDTA proved best Fe source for increasing grain Fe	Cakmak et al. (2010)
Zn	Maize	ZnO-NPs, Zn-chitosan NPS, $ZnSO_4$, Zn+Fe+Se, foliar applications 452 mg L^{-1}	Grain Zn was maximally increased by ZnSO4	Xue et al. (2023)
Zn	Maize	Zn-EDTA, Zn-EDDHA, and Zn-organic compounds, soil application	Highest Zn plant content was obtained by Zn-EDTA (20 mg Zm/kg)	Alvarez and Rico (2003)
Zn	Maize	Zn-EDTA, Zn-EDDHA, and Zn-organic complexes, soil application	Zn complexes significantly increased Zn uptake by the plant	Lopez-Valdivia et al. (2002)

(*Continued*)

Table 23.1 Biofortification of crops and vegetables with minerals by soil and foliar application of micronutrient formulations. *(continued)*

Mineral	Crop/Vegetable	Biofortification method	Benefits	References
Fe, Zn	Rice	Fe-glycine, Zn-glycine, $ZnSO_4$ (0.5, 1, 1.5 kg ha^{-1}) foliar application	Fe-glycine and Zn-glycine increased grain Fe and Zn by 33% and 22%, respectively	Soltani et al. (2023)
Se	Rice	Na-selenate (30–120 g ha^{-1}), foliar application	IRGA-426 cultivar with the dose 60 g ha^{-1}Se was appropriate combination to increase Se content in the grains	Ramos et al. (2022)
Se	Rice	Selenite and selenate (30 g ha^{-1}), application methods: soil, flood water, foliar and fluid fertilizer	Highest grain Se was obtained with selenate in flood water	Premarathna et al. (2012)
Se	Rice	Na-selenite, Na-selenate (300 g Se ha^{-1}), foliar application	Selenite and selenate application resulted in up to 884- and 347-fold increases in grain Se, respectively	Lidon et al. (2019)
Se	Alfalfa	Se-nanoparticles (30–250 mg L^{-1}), foliar application	All concentrations enhanced Se content; 50 mg $L^{-1}$was considered suitable for growth	Sun et al. (2023)
Se	Wheat	Selenate only, granular and Se-enriched fertilizer (10 g ha^{-1}), soil application; 33 µg L^{-1} Se with an without N carrier, foliar application	Grain Se ranges from 0.13 to 0.84 mg kg^{-1}; Co-application of Se and N doubled the Se concentration in grains	Ramkissoon et al. (2019)
Se	Wheat	Selenate (20 g ha^{-1}), soil application	Enhanced grain Se	Ahmad et al. (2019)
Se	Wheat	Na-selenate and Na-selenite (10, 20, 40 g ha^{-1}), foliar application	Na-selenate (10 g ha^{-1}) was found optimum in increasing grain Se according to the recommended Se dose	Poblaciones et al. (2014)
Se	Wheat	Na-selenate and Na-selenite (10 and 20 g Se ha^{-1}), foliar application	Grain Se ranges from 13 to 15% by 20 g Se ha^{-1}	Ducsay et al. (2016)
Se	Wheat	Na-selenate solution, granular Se as soil application	10 g $ha^{-1}$was found as the optimum dose; 16–26 ng Se g^{-1} fresh weight was increased with 1 g of Se application	Broadley et al. (2010)
Se	Tomato	Na-selenite (2, 4, 6, 8 mg L^{-1}), hydroponic system	2 mg $L^{-1}$was recommended as the optimum Se	Gaucin-Delgado et al. (2020)
Se	Lettuce	Na-selenite and Na-selenate (2–64 µmol L^{-1}), Hydroponic nutrient	Selenate at low concentrations proved appropriate for enhancing Se in lettuce	Ramos et al. (2010)
Se	Spinach	Se(IV) solution (120 g ha^{-1}), soil application	Se(IV) solution at 14 days after plantation caused optimum Se accumulation in leaves and stems	Nayan et al. (2020)
Se	Cucumber	Na-selenate (5 and 10 mg L^{-1}), foliar application	10 mg L^{-1}Se proved effective in enhancing Se content, salinity tolerance, and grafting efficiency	Amerian et al. (2024)
Se+Zn	Field pea	Zn (50 kg ha^{-1}), soil application; Se (10 g ha^{-1}), Zn (8 kg ha^{-1}), and Se+Zn, foliar application	Combined foliar application of Zn (8 kg ha^{-1}) and Se (10 g ha^{-1}) showed 30% and 73% increases in Zn and Se content in grains	Reynolds-Marzal et al. (2022)
I, Se	Carrot	I (5 kg ha^{-1}) in the form of KI and KIO_3; Se (1 kg ha^{-1}) in the form of Na_2SeO_4 and Na_2SeO_3, and combined I + Se	KI and Na_2SeO_4 more effectively increased the I and Se in carrot	Smoleń, Skoczylas and Ledwożyw-Smoleń (2016)
I	Tomato	$Na_{125}I$; Hydroponic and leaf applications	Hydroponic tomato uptake more I in fruit than leaf application	Landini et al. (2011)

(Continued)

Table 23.1 Biofortification of crops and vegetables with minerals by soil and foliar application of micronutrient formulations. *(continued)*

Mineral	Crop/Vegetable	Biofortification method	Benefits	References
I	Lettuce	Se (Na_2SeO_4), I (KIO_3), and combined Se + I; foliar application and nutrient application	Foliar application more effectively enhanced Se and I	Smoleń, Kowalska and Sady (2014)
I	Tomato	KI, KIO_3, and organic iodine compounds, soil applications	Influenced iodine metabolism through the HMT gene	Halka et al. (2019)
Micronutrient	Rice	Zn, I, Fe, Se, and combined foliar application	Grains Zn, I, and Se were increased; Fe was increased only at two sites	Prom-U-Thai et al. (2020)

The deficiencies of vitamin B are also very prevalent due to its water solubility, and it affects the nervous system and muscles (reviewed by Green et al. 2017; Mathew et al. 2024). Vitamin C is an antioxidant that cannot be synthesized in humans, and it must be obtained from plants. A deficiency of Vitamin C in the older population can induce nonspecific symptoms such as fatigue, anxiety, and depression and make it difficult to identify. Likewise, vitamin C deficiency is also very common and is linked to multiple medical conditions in humans including pulmonary hypertension, and kidney diseases (Rowe and Carr 2020; Doseděl et al. 2021; Hemilä and de Man 2024). Finally, vitamin D deficiency affects people resulting in certain physiological disorders including arthritis, depression, and cardiovascular diseases (SBP 2016; Sîrbe et al. 2022).

Different strategies are being used for vitamin biofortification. For instance, wheat flour is often fortified with minerals and B-complex vitamins, and WHO/FAO (2006) has provided guidelines for it. Likewise, fortification of oil with vitamins A and D is mandatory in many countries, including India (Bamji et al. 2021). Since provitamin A can be converted into vitamin A in the human body, staple crops that are biofortified with provitamin A are useful in reducing vitamin A deficiency (Rai et al. 2024). For biofortification of vitamin B_{12}, nanotechnology is used, by delivering vitamin B_{12}-coated nanoparticles to specific plant locations, which increases the plant's bioavailability (Oh, et al. 2021). Likewise, biofortification of vitamin C is necessary to ensure a consistent supply without compromising other nutrients. Different biotechnology and breeding strategies have been used largely to biofortify vitamin C. This is probably because the increase in vitamin C achieved by agronomic approaches is unstable and disappears after harvest or after 24–72 hours in mature plants (Kathi, et al. 2023).

Since various strategies have been used to perform vitamin biofortification, a summary of different studies has been presented in Table 23.2.

23.6.3 Biofortification of Essential Fatty Acids

Fatty acids are long-chain hydrocarbon molecules that have carboxylic acid groups attached to them. They prominently play their role in maintaining the health and well-being of people worldwide by regulating the membrane structure as well as functions (Jain 2020). Both omega-3 (ω-3) and omega-6 (ω-6) fatty acids are essential fatty acids and have many important structural, metabolic, and functional roles in human health. Polyunsaturated fatty acids (PUFAs) have been reported to play a role in the prevention and treatment of many other chronic diseases, such as neurological disorders, cancer, inflammatory diseases, obesity, and diabetes mellitus (Yashodhara et al. 2009). The WHO suggests the ratio to be greater than 10 (Simopoulos and DiNicolantonio 2016). Humans cannot synthesize these fatty acids in sufficient amounts. Linoleic acid and α-linolenic acid (ALA), being a precursor for the formation of ω-3 and ω-6 fatty acids, cannot be produced in the human body, and it must be taken through the diet, making it an essential fatty acid. Humans can convert the ALA into DHA to use it, but this process takes place at a very slow rate, being unable to fulfill the dietary requirements of people (Burns-Whitmore et al. 2019). Therefore, humans must get the essential fatty acids including cis-LA and ALA from their diet. However, the uptake of adequate quantity of essential fatty acids from the diet has become challenging over time because of the growing global population.

People living in poor areas are at greater risk of hidden hunger and malnutrition as they are dependent on a diet that is not considered sufficient to fulfill the nutritional requirements of the people Abdoli (2024). To fulfill this requirement, incorporation of ω-3 and ω-6-rich dietary sources in daily meals is required. Several past studies have fortified a range of foods including eggs, milk, and butter (Derbyshire et al. 2024). Biofortification is considered a novel approach to fulfill the dietary requirements of people worldwide as it adds the desired nutrients to the crops before they are harvested. In this

Table 23.2 Vitamin biofortification in crops via transgenic approaches.

Vitamin	Transgenic crop	Genetic modification method	Enhanced trait	Reference
Provitamin A	Wheat	*PSY, Crtl, CrtB*	Enhanced provitamin A and carotenoid	Wang et al. (2014)
	Wheat	Expression of phytoene synthase1 (*y1* gene) from maize and carotene desaturase (*crtI* gene) from *Erwinia uredovora*	Increased total carotenoids content in the endosperm	Cong et al. (2009)
	Maize	Overexpression of bacterial genes *crtB* (phytoene synthase) and *crtI* (phytoene desaturase and zeta-carotene desaturase) under "super gamma-zein promoter"	Enhanced provitamin A (β-carotene) content in the kernels	Aluru et al. (2008)
	Maize	Overexpression of *psy1* (phytoene synthase gene) from maize in combination with a bacterial *crtI* (desaturase gene)	Higher carotenoid biosynthesis in maize kernels; changes in metabolism of other pathways	Decourcelle et al. (2015)
	Maize	White maize variety was bombarded with five carotenogenic genes (*Zmpsy1, Pacrtl, Gllycb, Glbch, Paracrtw*)	34-fold increase in β-carotene content in kernels	Zhu et al. (2008)
	Sorghum	Biofortified lines developed for increased provitamin A carotenoids	Increased bioaccessible β-carotene content	Lipkie et al. (2013)
Vitamin B_1 (Thiamin)	Rice	Ectopic expression of thiamin biosynthesis genes *THIC, THI1*, and *TH1* under strong endosperm-specific promoters	Increased thiamin content in endosperm	Strobbe et al. (2021)
	Rice	Expression of SiTBP (thiamine binding protein from *Sesamum indicum*) under the rice endosperm-specific glutelin D-1 promoter	Increased thiamine content in polished grains	Fitzpatrick et al. (2024)
	Rice	CRISPR/Cas9 knockout of *OsPALE1* gene	Involvement in vitamin B_1 biosynthesis, increased understanding of biosynthetic pathway	Hsieh et al. (2021)
Vitamin B_2 (Riboflavin)	Rice	Insertion of six riboflavin biosynthesis genes from *Saccharomyces cerevisiae* into rice endosperm	Increased riboflavin content in rice endosperm	Tian et al. (2021)
Vitamin B_5 (Pantothenic acid)	Oilseed rape	Expression of *Escherichia coli KPHMT* and *PS* genes under CaMV35SS promoter	1.5–2.5-fold increase in pantothenate levels in various plant parts	Chakauya et al. (2008)
Vitamin B_6 (Pyridoxine)	Rice	Expression of *AtPDX1.1* and *AtPDX2* genes	Increased vitamin B_6 levels in leaves (up to 28.3-fold) and roots (up to 12-fold)	Mangel et al. (2019)
	Wheat	Overexpression of endogenous *PDX1* gene under an endosperm promoter	Up to 34.9-fold increase in vitamin B_6 content in seeds	Mohsin et al. (2022)
	Cassava	Transgenic 35S and PAT lines	Increased vitamin B_6 content in leaves and roots	Li et al. (2015)
Vitamin B_9 (Folic acid)	Rice	Overexpressing genes of the pterin (*Glb-1*) and para-aminobenzoate (*GluB1*) branches of the folate biosynthetic pathway	Enhanced vitamin B_9	Storozhenko et al. (2007)
	Rice	Complexing folate to folate-binding proteins to improve stability and biofortification	Increased folate concentration up to 150-fold higher than wild-type, improved stability for long-term storage	Blancquaert et al. (2015)
	Tomato	Overexpression of a synthetic gene for GTP cyclohydrolase I (mammalian) using fruit-specific promoters	Increased folate content in fruits by an average of twofold, with up to 10-fold increase when supplied with PABA	de la Garza et al. (2004)

(*Continued*)

Table 23.2 Vitamin biofortification in crops via transgenic approaches. *(continued)*

Vitamin	Transgenic crop	Genetic modification method	Enhanced trait	Reference
Vitamin C (Ascorbic acid)	Tobacco, Lettuce	Transformation using *Agrobacterium tumefaciens* expressing rat cDNA encoding L-gulono-γ-lactone oxidase under CaMV 35S promoter	Increased ascorbic acid content up to seven times more than untransformed plants	Jain and Nessler (2000)
	Tomato, Strawberry, Potato	Transformation to expressing GDP-L-galactose phosphorylase (*GGP* or *VTC2*) under CaMV 35S promoter or polyubiquitin promoter	Increased ascorbate content (two to sixfold) in fruits and tubers	Bulley et al. (2012)
	Tobacco and Maize	Overexpression of dehydroascorbate reductase cDNA from wheat	Increased ascorbic acid levels two- to fourfold, improved ascorbate redox state in leaves and kernels	Chen et al. (2003)
	Tomato	Overexpression of two GDP-Mannose 3',5'-epimerase (*SlGME1* and *SlGME2*) genes	Increased ascorbate accumulation in leaves and red fruits, enhanced tolerance to oxidative stress, cold, and salt stress	Zhang et al. (2011)
Vitamin D (Calciferol)	Tomato	CRISPR-Cas9 genome editing	Accumulation of provitamin D3 in leaves and fruit	Li et al. (2022)
Vitamin E (Tocopherol)	Maize	Transformation of barley *HGGT* (homogentisate phytyltransferase) gene	Tocopherol and tocotrienols increased sixfold in kernels	Cahoon et al. (2003)
	Canola	Overexpression of *γ-TMT* (γ-tocopherol methyltransferase) gene	Increased γ-tocopherol	Van Eenennaam et al. (2003)
	Soybean	Arabidopsis MPBQ/MSBQ methyltransferase (*VTE3*)	Increased vitamin E activity	Sattler, Cheng and DellaPenna (2004)
	Perilla frutescens Britt.	Transformation of *γ-TMT* gene driven by the CaMV 35S promoter	Increased α-tocopherol content	Ghimire et al. (2011)
Multivitamins	Maize	Overexpressing *Zmpsy1, Pacrtl, Osdhar*, and *EcfolE* genes	169-, 6-, and 2-fold increase in β-carotene, ascorbate, and folate	Naqvi et al. (2009)

way, the production and quality of the crop is enhanced. In the past, the first biofortification success was the production of quality protein maize, which resulted in then enhancement in the amount of essential amino acids lysine and tryptophan (Labuschagne 2023). In recent times, different strategies for the biofortification of different ranges of foods with fatty acids have been carried out; n-3-fortified eggs are produced by fortifying chicken feed with n-3 FA, as flaxseed. This process resulted in the production of eggs containing up to five to six times more n-3 FA (Marques et al. 2021). Similarly, in chickpeas, Linoleic acid (LA; ω-6) is the most abundant fatty acid while ALA (ω-3), the other essential fatty acid is less in amount. Therefore, biofortification in chickpeas to enrich the seed in ALA is predicted as a beneficial way to increase the ALA content in it (Madurapperumage et al. 2021). Furthermore, fish oil or marine algae with long-chain ω-3 PUFAs have also been used in animal feed for this purpose (Patel et al. 2022). Apart from this, biofortified crops hold a very bright future as they can eliminate micronutrient deficiency among poor people, particularly in developing countries. The health advantages of ω-3 PUFAs are widely recognized. Ultimately, dietary intake of ω-3 PUFAs through biofortification is the most efficient way to deal with the deficiency of fatty acids among people.

23.6.4 Biofortification of Vital Metabolites

Both lysine and arginine have been regarded as essential amino acids. Because neither humans nor animals can produce essential amino acids (EAAs), such as lysine and methionine, these nutrients must be obtained through diet. However, a deficiency in certain EAAs frequently results in nutritional imbalances in plants (Galili et al. 2016). The ICAAS (International Council of Amino Acid Science) held a meeting in Tokyo in 2001 to promote the vision to organize and emphasize the general question of specific Amino Acids. The 6th ICAAS workshop held in Budapest focused on basic Arginine and Lysine (Rogers and Smriga 2012).

Arginine is an essential amino acid, L-form physiologically active and the L-form of arginine helps your body to produce proteins. Natural sources of arginine include nuts, meat, and fish. Although your body can produce the amino acid, L-arginine is semi-essential (conditioned), therefore humans still need to eat certain sources of the amino acid in their diet (Lei et al. 2018). L-arginine is normally required for the synthesis of nitric oxide, a signaling molecule that is essential for many biological processes and activities, including the control of blood flow, the operation of mitochondria, and cellular communication (Farazi et al. 2015). It also serves as a basic component of several protein molecules, enzymes, receptors, and other structural proteins. Humans require lysine for a variety of processes, chief among them being proteinogenesis. It is also necessary for the crosslinking of collagen polypeptides, the absorption of vital mineral elements, and the synthesis of carnitine, which is critical for the metabolism of fatty acids (Koetzle et al. 1972).

In addition to the essential amino acids, there are several other vital metabolites that are essentially required for humans and are generally deficient in our diet. The information related to the metabolite biofortification is given in Table 23.3.

Table 23.3 Transgenic crops for metabolites biofortification.

Metabolite	Transgenic Crop	Genetic modification method	Enhanced trait	References
β-carotene	Rice	Introduction of phytoene synthase (psy), lycopene β-cyclase (lcy), and phytoene desaturase (crtI) under different promoters; selection with hygromycin and mannose	Enhanced β-carotene content in the endosperm, evident from the yellow coloration of polished rice grains	Datta et al. (2003)
β-carotene, Oleic acid	Soybean	Overexpression of seed-specific bacterial phytoene synthase gene from *Pantoea ananatis*	Enhanced β-carotene content (845 $\mu g\ g^{-1}$ dry seed weight), increased oleic acid and protein content	Schmidt et al. (2015)
β-carotene	Soybean	Phytoene synthase *crtB*	Enhanced β-carotene	Pierce et al. (2015)
β-carotene	Banana	Phytoene synthase *PSY2a*	Enhanced β-carotene	Waltz (2014)
β-carotene, Zeaxanthin	Potato	Expression of Erwinia uredovora *crtB* gene encoding phytoene synthase	Increased β-carotene (up to 11 $\mu g\ g^{-1}$ DW) and lutein (19-fold higher) in tubers	Ducreux et al. (2005)
β-carotene		Antisense RNA silencing of lycopene epsilon cyclase gene under patatin promoter	Increased β-carotene (up to 14-fold) and total carotenoids (up to 2.5-fold) in tubers	Diretto et al. (2006)
Flavonoids/ antioxidants	Rice	Expression of maize C1 and R-S regulatory genes under a rice prolamin promoter to restrict expression to endosperm	Production of various flavonoids, including dihydroquercetin, dihydroisorhamnetin, and 3'-O-methyl quercetin in the endosperm	Shin et al. (2006)
	Rice	Introduction of multiple genes encoding flavonoid biosynthesis enzymes (PAL, CHS, flavonol synthase, isoflavone synthase, flavone synthase)	Synthesis and accumulation of diverse flavonoids including naringenin, kaempferol, genistein, apigenin, and tricin in seeds	Ogo et al. (2013)
Lysine	Barley	Expression of dihydrodipicolinate synthase (*dapA* gene from *Escherichia coli*) using Agrobacterium-mediated transformation	Increased lysine content by 50% in leaves and 30% in seeds compared to wild-type barley	Ohnoutkova et al. (2012)
	Rice	Expression of bacterial aspartate kinase and dihydrodipicolinate synthase, inhibition of rice lysine ketoglutarate reductase/ saccharopine dehydrogenase activity	Increased free lysine levels in seeds up to 25-fold compared to wild type; improved total free amino acids and protein content	Yang et al. (2016)

(*Continued*)

Table 23.3 Transgenic crops for metabolites biofortification. *(continued)*

Metabolite	Transgenic Crop	Genetic modification method	Enhanced trait	References
	Rice	Expression of mutant aspartate kinase and dihydrodipicolinate synthase (*DHDPS1*) with inhibition of rice lysine ketoglutarate reductase/ saccharopine dehydrogenase activity	Up to 58.5-fold increase in free lysine compared to wild type; improved total free amino acids and protein content	Yang et al. (2021)
Cysteine	Soybean	Overexpression of cytosolic isoform of O-acetylserine sulfhydrylase	Enhanced levels of both protein-bound and free cysteine	Kim et al. (2012)
Methionine and cysteine	Soybean	Expression of feedback-insensitive Arabidopsis cystathionine γ-synthase (*AtD-CGS*) gene under a seed-specific promoter (legumin B4)	Increased methionine by 1.8- to 2.3-fold in mature seeds; increased total protein content	Song et al. (2013)
Methionine and cysteine	Soybean	Overexpression of maize 15 kDa zein protein gene behind the β-phaseolin promoter for seed-specific expression using somatic embryogenic protocols	Increased methionine by 12–20% and cysteine by 15–35%	Dinkins et al. (2001)
Methionine	Common bean	Expression of a methionine-rich storage albumin from Brazil nut using biolistic methods; driven by doubled 35S CaMV promoter and AMV enhancer sequences	Increased methionine content by 14–23%	Aragão et al. (1999)
Methionine	Soybean; Azuki bean	Expression of a mutated form of Arabidopsis cystathionine γ-synthase (AtCGS1 mto1-1) under a seed-specific glycinin promoter	Increased soluble methionine; variable effects on total methionine content	Hanafy et al. (2013)
Methionine and cysteine	Alfalfa	Overexpression of Arabidopsis cystathionine γ-synthase (AtCGS) under the Arabidopsis rubisco small subunit promoter	Increased levels of methionine, S-methylmethionine, protein-bound methionine, free cysteine, glutathione, and protein-bound cysteine	Avraham et al. (2005)
Anthocyanin	Wheat	Transient expression using tissue-specific promoters (Dhn12, Itr1, Ltp1) and anthocyanin biosynthesis genes (C1, B-peru)	Tissue-specific expression of anthocyanin in embryos	Doshi et al. (2006)
Amylose content	Wheat	Silencing of *SBEIIa* genes using RNA interference (RNAi) targeted to seed endosperm using tissue-specific promoters	Increased amylose content; altered starch granule morphology; new pasting properties in high amylose lines	Sestili et al. (2010)
Amylopectin	Potato	*nptII*	Released potato cultivar (Amflora™) with high amylopectin component of starch	Tilocca et al. (2014)
Lactoferrin	Barley	Particle bombardment of zygotic embryos with plasmids containing human lactoferrin gene and selectable marker genes	Production of recombinant human lactoferrin in seeds	Kamenarova et al. (2007)
γ-Linolenic Acid + stearidonic acid	Soybean	RNA interference (RNAi) using siRNA targeting *GmFAD3* gene family	Significant reduction in ALA levels, mimicking fad3-mutant phenotype; improved seed oil stability	Flores et al. (2008)
γ-Linolenic acid	Mustard	Expression of Δ6 desaturase gene (*PiD6*) from *Pythium irregulare* under *Brassica napus* napin promoter	High-level production of GLA (up to 40% of total seed fatty acids), mainly incorporated into triacylglycerols	Hong et al. (2002)

23.7 Potential Limitations and Challenges

Despite the feasibility and effectiveness of the biofortification process, the success in this field is still limited due to multiple reasons. One of the major limitations is that most of the studies have been conducted at a very small scale and farmer trials are only limited to the use of genetically modified cereals through conventional and molecular breeding. The application of different foliar sprays to enhance micronutrient levels still requires large-scale trials in collaboration with farmers. Likewise, vitamin biofortification also requires further validation at field scale to integrate it with local communities. Although some rice varieties have been developed with enhanced vitamin A, their acceptance among the general public is very limited. Lack of awareness is also a major issue that limits the success of biofortification. In short, the biofortification process should be coupled with awareness programs to achieve a successful transition towards sustainable food practices.

References

Abdoli, M. (2024). Stable crops biofortified with increased protein, oils, and vitamins. In: *Phytoremediation and Biofortification*, 249–276. New Jersey and Canada: Apple Academic Press.

Aciksoz, S.B., Yazici, A., Ozturk, L. et al. (2011). Biofortification of wheat with iron through soil and foliar application of nitrogen and iron fertilizers. *Plant and Soil* 349: 215–225.

Ahmad, S., Young, S.D., Bailey, E.H. et al. (2019). Selenium biofortification of a wheat crop in northeastern Pakistan using 77Se as a tracer. In: *Selenium Research for Environment and Human Health: Perspectives, Technologies and Advancements*, 107–108. Boca Raton, FL: CRC Press.

Ahmed, F. (2019). Understanding food insecurity experiences, dietary perceptions and practices in the households facing hunger and malnutrition in Rajanpur District, Punjab Pakistan. *Pakistan Perspectives* 24 (2):115–132.

Allen, S. and De Brauw, A. (2018). Nutrition sensitive value chains: Theory, progress, and open questions. *Global Food Security* 16: 22–28.

Aluru, M., Xu, Y., Guo, R. et al. (2008). Generation of transgenic maize with enhanced provitamin A content. *Journal of Experimental Botany* 59 (13): 3551–3562.

Alvarez, J.M. and Rico, M.I. (2003). Effects of zinc complexes on the distribution of zinc in calcareous soil and zinc uptake by maize. *Journal of Agricultural and Food Chemistry* 51(19): 5760–5767.

Amerian, M., Palangi, A., Gohari, G. et al. (2024). Enhancing salinity tolerance in cucumber through Selenium biofortification and grafting. *BMC Plant Biology* 24 (1): 24.

Amjad, M. and Akbar, M. (2020). Role of socioeconomic factors to overcome micronutrient malnutrition in Pakistan: Application of partial proportional odds model. *Progress in Nutrition* 22 (3): e2020021.

Aragão, F.J.L., Barros, L.M.G., De Sousa, M.V. et al. (1999). Expression of a methionine-rich storage albumin from the Brazil nut (*Bertholletia excelsa* HBK, Lecythidaceae) in transgenic bean plants (*Phaseolus vulgaris* L., Fabaceae). *Genetics and Molecular Biology* 22: 445–449.

Asif, M., Asad, M.S., Safdar, M.E. et al. (2024). Agronomic bio-fortification of zinc improves the yield and quality of fodder oat. *Journal of Ecological Engineering* 25 (6): 153–163.

Avraham, T., Badani, H., Galili, S. et al. (2005). Enhanced levels of methionine and cysteine in transgenic alfalfa (*Medicago sativa* L.) plants over-expressing the *Arabidopsis* cystathionine γ-synthase gene. *Plant Biotechnology Journal* 3 (1): 71–79.

Bailey, R.L., West Jr, K.P., and Black, R.E. (2015). The epidemiology of global micronutrient deficiencies. *Annals of Nutrition and Metabolism* 66 (Suppl. 2): 22–33.

Bamji, M.S., Murty, P.V.V.S., and Sudhir, P.D. (2021). Nutritionally sensitive agriculture – an approach to reducing hidden hunger. *European Journal of Clinical Nutrition* 75 (7): 1001–1009.

Black, R.E., Alderman, H., Bhutta, Z.A. et al. (2013). Maternal and child nutrition: Building momentum for impact. *The Lancet* 382 (9890): 372–375.

Blanco-Rojo, R. and Vaquero, M.P. (2019). Iron bioavailability from food fortification to precision nutrition. A review. *Innovative Food Science & Emerging Technologies* 51: 126–138.

Blancquaert, D., Van Daele, J., Strobbe, S. et al. (2015). Improving folate (vitamin B_9) stability in biofortified rice through metabolic engineering. *Nature Biotechnology* 33 (10): 1076–1078.

Bouis, H. (2018). Reducing mineral and vitamin deficiencies through biofortification: Progress under HarvestPlus. In: *Hidden Hunger: Strategies to Improve Nutrition Quality* 118: 112–122. Basel: Karger Publishers.

Broadley, M.R., Alcock, J., Alford, J. et al. (2010). Selenium biofortification of high-yielding winter wheat (*Triticum aestivum* L.) by liquid or granular Se fertilisation. *Plant and Soil* 332: 5–18.

Bulley, S., Wright, M., Rommens, C. et al. (2012). Enhancing ascorbate in fruits and tubers through over-expression of the l-galactose pathway gene GDP-l-galactose phosphorylase. *Plant Biotechnology Journal* 10 (4): 390–397.

Burns-Whitmore, B., Froyen, E., Heskey, C. et al. (2019). Alpha-linolenic and linoleic fatty acids in the vegan diet: do they require dietary reference intake/adequate intake special consideration? *Nutrients* 11 (10): 2365.

Butsai, W., Kaewpradit, W., Harrell, D.L. et al. (2022). Effect of iron application on rice plants in improving grain nutritional quality in Northeastern of Thailand. *Sustainability* 14 (23): 15756.

Cahoon, E.B., Hall, S.E., Ripp, K.G. et al. (2003). Metabolic redesign of vitamin E biosynthesis in plants for tocotrienol production and increased antioxidant content. *Nature Biotechnology* 21 (9): 1082–1087.

Casgrain, A., Collings, R., Harvey, L.J. et al. (2012). Effect of iron intake on iron status: a systematic review and meta-analysis of randomized controlled trials. *The American Journal of Clinical Nutrition* 96 (4): 768–780.

Cakmak, I., Kalayci, M., Kaya, Y., et al. (2010) Biofortification and localization of zinc in wheat grain. *Journal of Agricultural and Food Chemistry* 58 (16): 9092–9102.

Chakauya, E., Coxon, K.M., Wei, M. et al. (2008). Towards engineering increased pantothenate (vitamin B_5) levels in plants. *Plant Molecular Biology* 68: 493–503.

Chen, Z., Young, T.E., Ling, J. et al. (2003). Increasing vitamin C content of plants through enhanced ascorbate recycling. *Proceedings of the National Academy of Sciences* 100 (6): 3525–3530.

Cong, L., Wang, C., Chen, L. et al. (2009). Expression of phytoene synthase1 and carotene desaturase *crtI* genes result in an increase in the total carotenoids content in transgenic elite wheat (*Triticum aestivum* L.). *Journal of Agricultural and Food Chemistry* 57 (18): 8652–8660.

Coppinger, R.J. and Diamond, A.M (2001). Selenium Deficiency and Human Disease. In *Selenium: Its Molecular Biology and Role in Human Health*, 219–233. Boston, MA: Springer US.

Das, J.K. and Padhani, Z.A. (2022). Alleviating hidden hunger: An infallible bridge to improved health and nutrition. *The Lancet Global Health* 10 (11): e1539–e1540.

Datta, K., Baisakh, N., Oliva, N. et al. (2003). Bioengineered 'golden' indica rice cultivars with β-carotene metabolism in the endosperm with hygromycin and mannose selection systems. *Plant Biotechnology Journal* 1 (2): 81–90.

Decourcelle, M., Perez-Fons, L., Baulande, S., et al. (2015). Combined transcript, proteome, and metabolite analysis of transgenic maize seeds engineered for enhanced carotenoid synthesis reveals pleotropic effects in core metabolism. *Journal of Experimental Botany* 66 (11): 3141–3150.

de la Garza, R.D., Quinlivan, E.P., Klaus, S.M. et al. (2004). Folate biofortification in tomatoes by engineering the pteridine branch of folate synthesis. *Proceedings of the National Academy of Sciences* 101 (38): 13720–13725.

Derbyshire, E.J., Birch, C.S., Bonwick, G.A. et al. (2024). Optimal omegas – barriers and novel methods to narrow omega-3 gaps. A narrative review. *Frontiers in Nutrition* 11: 1325099.

Dinkins, R.D., Srinivasa Reddy, M.S., Meurer, C.A. et al. (2001). Increased sulfur amino acids in soybean plants overexpressing the maize 15 kDa zein protein. *In Vitro Cellular & Developmental Biology-Plant* 37: 742–747.

Diretto, G., Tavazza, R., Welsch, R. et al. (2006). Metabolic engineering of potato tuber carotenoids through tuber-specific silencing of lycopene epsilon cyclase. *BMC Plant Biology* 6: 1–11.

Doseděl, M., Jirkovský, E., Macáková, K. et al. (2021). Vitamin C – sources, physiological role, kinetics, deficiency, use, toxicity, and determination. *Nutrients* 13 (2): 615.

Doshi, K.M., Eudes, F., Laroche, A. et al. (2006). Transient embryo-specific expression of anthocyanin in wheat. *In Vitro Cellular & Developmental Biology-Plant* 42: 432–438.

Ducreux, L.J., Morris, W.L., Hedley, P.E. et al. (2005). Metabolic engineering of high carotenoid potato tubers containing enhanced levels of β-carotene and lutein. *Journal of Experimental Botany* 56 (409): 81–89.

Ducsay, L., Ložek, O., Marček, M. et al. (2016). Possibility of selenium biofortification of winter wheat grain. *Plant Soil and Environment* 62 (8):379–383.

FAO. (1995). Technical consultation on food fortification: Technology and quality control, 20–23. In: *Micronutrient Fortification of Food: Technology and Quality Control*. Rome, Italy.

FAO, IFAD, UNICEF, WFP W. (2017). The state of food security and nutrition in the world 2017. Building resilience for peace and food security. http://www.fao.org/policy-support/tools-andpublications/resources-details/en/c/1107528/ (accessed 13 September 2021).

FAO, IFAD, UNICEF, WFP, WHO. (2019). *The State of Food Security and Nutrition in the World 2019. Safeguarding Against Economic Slowdowns and Downturns*. Rome, Italy: FAO.

Farazi, A., Shafaat, O., Sofian, M. et al. (2015). Arginine adjunctive therapy in active tuberculosis. *Tuberculosis Research and Treatment* 2015 (1): 205016.

Fitzpatrick, T.B., Dalvit, I., Chang, F.-H., et al. (2024) Vitamin B1 enhancement in the endosperm of rice through thiamine sequestration. *Plant Biotechnology Journal* 22 (8): 2330.

Flores, T., Karpova, O., Su, X. et al. (2008). Silencing of *GmFAD3* gene by siRNA leads to low α-linolenic acids (18: 3) of fad3-mutant phenotype in soybean [Glycine max (Merr.)]. *Transgenic Research* 17 (5): 839–850.

Galili, G., Amir, R., and Fernie, A.R. (2016). The regulation of essential amino acid synthesis and accumulation in plants. *Annual Review of Plant Biology* 67 (1): 153–178.

Gao, D., Ran, C., Zhang, Y. et al. (2022). Effect of different concentrations of foliar iron fertilizer on chlorophyll fluorescence characteristics of iron-deficient rice seedlings under saline sodic conditions. *Plant Physiology and Biochemistry* 185 (4): 112–122.

Garcia-Oliveira, A.L., Chander, S., Ortiz, R. et al. (2018). Genetic basis and breeding perspectives of grain iron and zinc enrichment in cereals. *Frontiers in Plant Science* 9: 937.

Garg, M., Sharma, N., Sharma, S. et al. (2018). Biofortified crops generated by breeding, agronomy, and transgenic approaches are improving lives of millions of people around the world. *Frontiers in Nutrition* 5: 12.

Garrett, G.S. (2018). National mandated food fortification programs. In: *Food Fortification in a Globalized World*, 53–62. London: Academic Press.

Gaucin-Delgado, J.M., Hernandez-Montiel, L.G., Sanchez-Chavez, E. et al. (2020). Agronomic biofortification with selenium improves the yield and nutraceutical quality in tomato under soilless conditions. *Notulae Botanicae Horti Agrobotanici Cluj-Napoca* 48 (3): 1221–1232.

Ghimire, B.K., Seong, E.S., Lee, C.O. et al. (2011). Enhancement of alpha-tocopherol content in transgenic *Perilla frutescens* containing the *gamma-TMT* gene. *African Journal of Biotechnology* 10 (13): 2430–2439.

Gibson, R.S. and Hotz, C. (2001). Dietary diversification/modification strategies to enhance micronutrient content and bioavailability of diets in developing countries. *British Journal of Nutrition* 85 (S2): S159–S166.

Global Health (2013). Hidden hunger index: why micronutrients matter. http://www.borgenmagazine.com/hidden-hunger-indexwhy-micronutrients-matter/ (accessed 10 Jan 2020).

Green, R., Allen, L.H., Bjørke-Monsen, A.L. et al. (2017). Vitamin B_{12} deficiency. *Nature Reviews Disease Primers* 3 (1): 1–20.

Halka, M., Smoleń, S., Czernicka, M. et al. (2019). Iodine biofortification through expression of *HMT, SAMT* and *S3H* genes in *Solanum lycopersicum* L. *Plant Physiology and Biochemistry* 144: 35–48.

Harding, K.L., Aguayo, V.M., and Webb, P. (2018). Hidden hunger in South Asia: A review of recent trends and persistent challenges. *Public Health Nutrition* 21 (4): 785–795.

Haroon, M., Khan, S.T., and Malik, A. (2022). Zinc-solubilizing bacteria: An option to increase zinc uptake by plants. In: *Microbial Biofertilizers and Micronutrient Availability: The Role of Zinc in Agriculture and Human Health*, 207–238. Cham: Springer Cham.

Hawkes, C., Ruel, M.T., Salm, L. et al. (2020). Double-duty actions: Seizing programme and policy opportunities to address malnutrition in all its forms. *The Lancet* 395 (10218):142–155.

He, W., Shohag, M.J.I., Wei, Y., et al. (2013). Iron concentration, bioavailability, and nutritional quality of polished rice affected by different forms of foliar iron fertilizer. *Food Chemistry* 141(4): 4122–4126.

Hefferon, K. (2019). Biotechnological approaches for generating zinc-enriched crops to combat malnutrition. *Nutrients* 11 (2): 253.

Hemilä, H. and de Man, A.M. (2024). Vitamin C deficiency can lead to pulmonary hypertension: A systematic review of case reports. *BMC Pulmonary Medicine* 24 (1): 140.

Hong, H., Datla, N., Reed, D.W. et al. (2002). High-level production of γ-linolenic acid in *Brassica juncea* using a Δ6 desaturase from *Pythium irregulare*. *Plant Physiology* 129 (1): 354–362.

Horton, R. and Lo, S. (2013). Nutrition: A quintessential sustainable development goal. *The Lancet* 382 (9890): 371–372.

Hsieh, P.H., Chung, Y.H., Lee, K.T. et al. (2021). The rice PALE1 homolog is involved in the biosynthesis of vitamin B_1. *Plant Biotechnology Journal* 19 (2):218.

Ibeanu, V.N., Edeh, C.G., and Ani, P.N. (2020). Evidence-based strategy for prevention of hidden hunger among adolescents in a suburb of Nigeria. *BMC Public Health* 20 (1): 1–10.

IFPRI. (2016). Hidden hunger: Approaches to tackling micronutrient deficiencies. Nourishing millions stories change nutrition. https://doi.org/10.2499/9780896295889_04.

Ishfaq, M., Wakeel, A., Shahzad, M.N. et al. (2021). Severity of zinc and iron malnutrition linked to low intake through a staple crop: A case study in east-central Pakistan. *Environmental Geochemistry and Health* 43 (10): 4219–4233.

Jain, A.K. and Nessler, C.L. (2000). Metabolic engineering of an alternative pathway for ascorbic acid biosynthesis in plants. *Molecular Breeding* 6 (1): 73–78.

Jain, T. (2020). Fatty acid composition of oilseed crops: A review. In: *Emerging Technologies in Food Science: Focus on the Developing World*, 147–153.

Jiang, W., Struik, P.C., Van Keulen, H. et al. (2008). Does increased zinc uptake enhance grain zinc mass concentration in rice? *Annals of applied biology* 153 (1): 135–147.

Joshi-Saha, A., Sethy, S.K., Misra, G. et al. (2022). Biofortified legumes: Present scenario, possibilities and challenges. *Field Crops Research* 279: 108467.

Kamali, B., Kalaiselvi, T., and Chitdeshwari, T. (2020). Exogenous foliar application of $FeSO_4$ on enrichment of iron in rice grain and yield. *Journal of Pharmacognosy and Phytochemistry* 9 (4): 3344–3348.

Kamenarova, K., Gecheff, K., Stoyanova, M. et al. (2007). Production of recombinant human lactoferrin in transgenic barley. *Biotechnology & Biotechnological Equipment* 21 (1): 18–27.

Kathi, S., Laza, H., Singh, S. et al. (2023). Vitamin C biofortification of broccoli microgreens and resulting effects on nutrient composition. *Frontiers in Plant Science* 14: 1145992.

Kim, W.S., Chronis, D., Juergens, M. et al. (2012). Transgenic soybean plants overexpressing O-acetylserine sulfhydrylase accumulate enhanced levels of cysteine and Bowman-Birk protease inhibitor in seeds. *Planta* 235 (1): 13–23.

Koetzle, T.F., Lehmann, M.S., Verbist J.J. et al. (1972). Precision neutron diffraction structure determination of protein and nucleic acid components. VII. The crystal and molecular structure of the amino acid l-lysine monohydrochloride dihydrate. *Acta Crystallographica Section B: Structural Crystallography and Crystal Chemistry* 28 (11): 3207–3214.

Labuschagne, M. (2023). Biofortification to improve food security. *Emerging Topics in Life Sciences* 7 (2): 219–227.

Landini, M., Gonzali, S., and Perata, P. (2011). Iodine biofortification in tomato. *Journal of Plant Nutrition and Soil Science* 174 (3): 480–486.

Lei, H.T., Ma, J., Sanchez Martinez, S. et al. (2018). Crystal structure of arginine-bound lysosomal transporter SLC38A9 in the cytosol-open state. *Nature Structural and Molecular Biology* 25 (6): 522–527.

Li, J., Scarano, A., Gonzalez, N.M. et al. (2022). Biofortified tomatoes provide a new route to vitamin D sufficiency. *Nature Plants* 8 (6): 611–616.

Li, K.T., Moulin, M., Mangel, N. et al. (2015). Increased bioavailable vitamin B_6 in field-grown transgenic cassava for dietary sufficiency. *Nature Biotechnology* 33 (10): 1029–1032.

Lidon, F.C., Oliveira, K., Galhano, C. et al. (2019). Selenium biofortification of rice through foliar application with selenite and selenate. *Experimental Agriculture* 55 (4): 528–542.

Lipkie, T. E., De Moura, F. F., Zhao, Z. Y., Albertsen, M. C., Che, P., Glassman, K., & Ferruzzi, M. G. (2013). Bioaccessibility of carotenoids from transgenic provitamin A biofortified sorghum. *Journal of agricultural and food chemistry* 61 (24): 5764–5771.

Lopez-Valdivia, L.M., Fernandez, M.D., Obrador, A. et al. (2002). Zinc transformations in acidic soil and zinc efficiency on maize by adding six organic zinc complexes. *Journal of Agricultural and Food Chemistry* 50 (6): 1455–1460.

Lowe, N.M. (2021). The global challenge of hidden hunger: Perspectives from the field. *Proceedings of the Nutrition Society* 80 (3): 283–289.

Lurthy, T., Pivato, B., Lemanceau, P. et al. (2021). Importance of the rhizosphere microbiota in iron biofortification of plants. *Frontiers in Plant Science* 12: 744445.

Lyons, G. (2018). Biofortification of cereals with foliar selenium and iodine could reduce hypothyroidism. *Frontiers in Plant Science* 9: 730.

Madurapperumage, A., Tang, L., Thavarajah, P. et al. (2021). Chickpea (*Cicer arietinum* L.) as a source of essential fatty acids – a biofortification approach. *Frontiers in Plant Science* 12: 734980.

Mahmood, S., Ahmad, M.Q., Saleem, A. et al. (2024). Agronomic and genetic biofortification of wheat: progress and limitations. In: *Biofortification of Grain and Vegetable Crops*, 81–95. London: Academic Press.

Mangel, N., Fudge, J.B., Li, K.T. et al. (2019). Enhancement of vitamin B_6 levels in rice expressing Arabidopsis vitamin B_6 biosynthesis de novo genes. *The Plant Journal* 99 (6): 1047–1065.

Mannar, M.G.V. and Hurrell, R.F. (2018). Chapter 1 – food fortification: past experience, current status, and potential for globalization. In: *Food Fortification in a Globalized World*, 3–11. Amsterdam: Elsevier, Academic Press.

Maqbool, M.A. and Beshir, A. (2019). Zinc biofortification of maize (*Zea mays* L.): Status and challenges. *Plant Breeding* 138 (1): 1–28.

Marques, E., Darby, H., and Kraft, J. (2021). Omega-3 Fatty acid fortification of flax through nutri-priming. *Frontiers in Nutrition* 8: 715287.

Mashhadi, S.F., Khan, N., Maryam, Z. et al. (2022). Hidden hunger or knowledge hunger? An interventional study in a public school of Rawalpindi. *Pakistan Armed Forces Medical Journal* 72 (SUPPL-4): S859–63.

Mathew, A.R., Di Matteo, G., La Rosa, P. et al. (2024). Vitamin B_{12} deficiency and the nervous system: Beyond metabolic decompensation – comparing biological models and gaining new insights into molecular and cellular mechanisms. *International Journal of Molecular Sciences* 25 (1): 590.

Maughan, R.J., Burke, L.M., Dvorak, J. et al. (2018). IOC consensus statement: Dietary supplements and the high-performance athlete. *International Journal of Sport Nutrition and Exercise Metabolism* 28 (2): 104–125.

McLean, E., Cogswell, M., Egli, I. et al. (2009). Worldwide prevalence of anaemia, WHO vitamin and mineral nutrition information system, 1993–2005. *Public Health Nutrition* 12 (4): 444–454.

Meenakshi, J.V., Johnson, N.L., Manyong, V.M. et al. (2010). How cost-effective is biofortification in combating micronutrient malnutrition? An ex ante assessment. *World Development* 38 (1): 64–75.

Mehboob, R. (2022). Hidden hunger, its causes and impact on human life. *Pakistan Journal of Health Sciences* 3 (4): 01–01.

Mohsin, S., Irfan, M., Saeed, A. et al. (2022). Enhanced expression of *PDX1* accumulates vitamin B6 in transgenic wheat seeds. *Journal of Cereal Science* 107 (4): 103502.

Naqvi, S., Zhu, C., Farre, G. et al. (2009). Transgenic multivitamin corn through biofortification of endosperm with three vitamins representing three distinct metabolic pathways. *Proceedings of the National Academy of Sciences* 106 (19): 7762–7767.

Nayan, D.S., Zainol, R., Abdul Sukor, A.S. et al. (2020). Selenium biofortification of green spinach with optimum phosphorus fertilization and selenium application timing. *Agrosystems, Geosciences & Environment* 3 (1): e20089.

Nazir, T., Anwar, S., Mahmood, A. et al. (2024). Nano-zinc oxide foliar application at the tuber initiation stage enhances the starch synthesis and zinc bioaccumulation in potato tubers. *International Journal of Applied and Experimental Biology* 3 (1): 51–59.

Noulas, C., Tziouvalekas, M., and Karyotis, T. (2018). Zinc in soils, water and food crops. *Journal of Trace Elements in Medicine and Biology* 49: 252–260.

Obersteiner, M., Walsh, B., Frank, S. et al. (2016). Assessing the land resource - food price nexus of the Sustainable Development Goals. *Science Advances* 2 (9): e1501499.

Ogo, Y., Ozawa, K., Ishimaru, T. et al. (2013). Transgenic rice seed synthesizing diverse flavonoids at high levels: A new platform for flavonoid production with associated health benefits. *Plant Biotechnology Journal* 11 (6): 734–746.

Oh, S., Cave, G., and Lu, C. (2021). Vitamin B_{12} (cobalamin) and micronutrient fortification in food crops using nanoparticle technology. *Frontiers in Plant Science* 12: 668819.

Ohnoutkova, L., Zitka, O., Mrizova, K. et al. (2012). Electrophoretic and chromatographic evaluation of transgenic barley expressing a bacterial dihydrodipicolinate synthase. *Electrophoresis* 33 (15): 2365–2373.

OMS (2023). Micronutrientes. Citado em: 30/07/23. Disponível em https://www.paho.org/es/temas/micronutrientes.

Patel, A., Desai, S.S., Mane, V.K. et al. (2022). Futuristic food fortification with a balanced ratio of dietary ω-3/ω-6 omega fatty acids for the prevention of lifestyle diseases. *Trends in Food Science & Technology* 120: 140–153.

Pérez-Massot, E., Banakar, R., Gómez-Galera, S. et al. (2013). The contribution of transgenic plants to better health through improved nutrition: Opportunities and constraints. *Genes & Nutrition* 8: 29–41.

Pierce, E.C., LaFayette, P.R., Ortega, M.A. et al. (2015). Ketocarotenoid production in soybean seeds through metabolic engineering. *PLoS One* 10 (9): e0138196.

Poblaciones, M.J., Santamaría, O., García-White, T. et al. (2014). Selenium biofortification in bread-making wheat under Mediterranean conditions: influence on grain yield and quality parameters. *Crop and Pasture Science* 65 (4): 362–369.

Popkin, B.M., Corvalan, C., and Grummer-Strawn, L.M. (2020). Dynamics of the double burden of malnutrition and the changing nutrition reality. *The Lancet* 395 (10217): 65–74.

Premarathna, L., McLaughlin, M.J., Kirby, J.K. et al. (2012). Selenate-enriched urea granules are a highly effective fertilizer for selenium biofortification of paddy rice grain. *Journal of agricultural and food chemistry* 60 (23): 6037–6044.

Prom-U-Thai, C., Rashid, A., Ram, H. et al. (2020). Simultaneous biofortification of rice with zinc, iodine, iron and selenium through foliar treatment of a micronutrient cocktail in five countries. *Frontiers in Plant Science* 11: 589835.

Rahman, A., Ahmad, M.A., Mehmood, S. et al. (2024). Isolation and screening of Zn (Zn) solubilizing rhizosphere bacteria from different vegetations for their ability to improve growth, Zn uptake, and expression of Zn transporter genes in tomato. *Current Microbiology* 81 (3): 83.

Rai, D., Chaudhary, C., Khatak, A. et al. (2024). A sustainable approach to combat micronutrient deficiencies and ensure global food security through biofortification. *European Journal of Nutrition & Food Safety* 16 (4): 15–30.

Ramkissoon, C., Degryse, F., da Silva, R.C. et al. (2019). Improving the efficacy of selenium fertilizers for wheat biofortification. *Scientific Reports* 9 (1):19520.

Ramos, D.P., Chan, G.A.H., Tavares, T.C. de O. et al. (2022). Impacts of selenium foliar biofortification on the biochemical composition and grain yield of rice cultivars. *American Scientific Research Journal for Engineering, Technology, and Sciences* 88 (1): 1–14.

Ramos, S.J., Faquin, V., Guilherme, L.R.G. et al. (2010). Selenium biofortification and antioxidant activity in lettuce plants fed with selenate and selenite. *Plant Soil Environment* 56 (12): 584–588.

Ramzan, Y., Hafeez, M.B., Khan, S. et al. (2020). Biofortification with zinc and iron improves the grain quality and yield of wheat crop. *International Journal of Plant Production* 14 (3): 501–510.

Raza, S.H., Shahzadi, A., Iqbal, M. et al. (2022). Foliar application of nano-zinc oxide crystals improved zinc biofortification in cauliflower (*Brassica oleracea* L. var. botrytis). *Applied Nanoscience* 12 (6): 1803–1813.

Reynolds-Marzal, M.D., Rivera-Martín, A.M., Pinheiro, N.M. et al. (2022). Combined zinc and selenium biofortification in field peas under Mediterranean conditions. *Crop and Pasture Science*. https://doi.org/10.1071/CP21711.

Ritchie, H., Reay, D.S., and Higgins, P. (2018). Quantifying, projecting, and addressing India's hidden hunger. *Frontiers in Sustainable Food Systems* 2: 11.

Rogers, M.D. and Smriga, M. (2012). Improving regulatory decisions through targeted research: A case study concerning amino acids. *European Journal of Risk Regulation* 3 (2): 161–168.

Roriz, M., Barros, M., Castro, P.M. et al. (2020). Improving iron nutrition in plant foods: The role of legumes and soil microbes. In: *Vitamins and Minerals Biofortification of Edible Plants* 103–122. Hoboken, NJ: John Wiley & Sons Ltd.

Rowe, S. and Carr, A.C. (2020). Global vitamin C status and prevalence of deficiency: A cause for concern?. *Nutrients* 12 (7): 2008.

Ruel, M.T. and Alderman, H. (2013). Nutrition-sensitive interventions and programmes: How can they help to accelerate progress in improving maternal and child nutrition? *The Lancet* 382 (9891): 536–551.

Sattler, S.E., Cheng, Z., and DellaPenna, D. (2004). From *Arabidopsis* to agriculture: Engineering improved vitamin E content in soybean. *Trends in Plant Science* 9 (8): 365–367.

SBP (2016). Consenso sobre anemia ferropriva: atualizac¸ ao: destaques ~ 2021. Departamenos científicos de hematologia e nutrologia. [Citado em: 30/07/23]. Disponível em https://www.sbp.com.br/fileadmin/user_upload/23172d-Diretrizes-Consenso_sobre_Anemia_Ferropriva-OK.pdf.

Schmidt, M.A., Parrott, W.A., Hildebrand, D.F. et al. (2015). Transgenic soya bean seeds accumulating β-carotene exhibit the collateral enhancements of oleate and protein content traits. *Plant Biotechnology Journal* 13 (4): 590–600.

Sestili, F., Janni, M., Doherty, A. et al. (2010). Increasing the amylose content of durum wheat through silencing of the *SBEIIa* genes. *BMC Plant Biology* 10: 1–12.

Shafiq, F., Ahmad, A., Anwar, S. et al. (2023). Spinel nanocomposite ($nMnZnFe_2O_4$) synchronously promotes grain yield and Fe-Zn biofortification in non-aromatic rice. *Plant Physiology and Biochemistry* 201: 107830.

Sharma, S., Anand, N., Bindraban, P.S. et al. (2023). Foliar application of humic acid with Fe supplement improved rice, soybean, and lettuce iron fortification. *Agriculture* 13 (1): 132.

Shimada, B.K., Alfulaij, N., and Seale, L.A. (2021). The impact of selenium deficiency on cardiovascular function. *International Journal of Molecular Sciences* 22 (19): 10713.

Shin, Y.M., Park, H.J., Yim, S.D. et al. (2006). Transgenic rice lines expressing maize C1 and R-S regulatory genes produce various flavonoids in the endosperm. *Plant Biotechnology Journal* 4 (3): 303–315.

Simopoulos, A.P. and DiNicolantonio, J.J. (2016). The importance of a balanced ω-6 to ω-3 ratio in the prevention and management of obesity. *Open Heart* 3 (2): e000385.

Singh, B., Goutam, U., Kukreja, S. et al. (2021). Potato biofortification: An effective way to fight global hidden hunger. *Physiology and Molecular Biology of Plants* 27 (10): 2297–2313.

Sîrbe, C., Rednic, S., Grama, A. et al. (2022). An update on the effects of vitamin D on the immune system and autoimmune diseases. *International Journal of Molecular Sciences* 23 (17): 9784.

Siwela, M., Pillay, K., Govender, L. et al. (2020). Biofortified crops for combating hidden hunger in South Africa: Availability, acceptability, micronutrient retention and bioavailability. *Foods* 9 (6): 815.

Smoleń, S., Kowalska, I., and Sady, W. (2014). Assessment of biofortification with iodine and selenium of lettuce cultivated in the NFT hydroponic system. *Scientia Horticulturae* 166: 9–16.

Smoleń, S., Skoczylas, Ł., Ledwożyw-Smoleń, I. et al. (2016). Biofortification of carrot (*Daucus carota* L.) with iodine and selenium in a field experiment. *Frontiers in Plant Science* 7: 730.

Soltani, S.M., Hossieni Chaleshtori, M., Tajaddodi Talab, K. et al. (2023). Rice growth improvement, bio-fortification, and mitigation of macronutrient requirements through foliar application of zinc and iron-glycine chelate and zinc sulfate. *Journal of Plant Nutrition* 46 (8): 1777–1786.

Song, S., Hou, W., Godo, I. et al. (2013). Soybean seeds expressing feedback-insensitive cystathionine γ-synthase exhibit a higher content of methionine. *Journal of Experimental Botany* 64 (7): 1917–1926.

Storozhenko, S., De Brouwer, V., Volckaert, M. et al. (2007). Folate fortification of rice by metabolic engineering. *Nature Biotechnology* 25 (11): 1277–1279.

Strobbe, S., Verstraete, J., Stove, C. et al. (2021). Metabolic engineering of rice endosperm towards higher vitamin B_1 accumulation. *Plant Biotechnology Journal* 19 (6): 1253–1267.

Sun, P., Wang, Z., Yuan, N. et al. (2023). Evaluation of nano-selenium biofortification characteristics of alfalfa (*Medicago sativa* L.). *Green Processing and Synthesis* 12 (1): 20228121.

Taskin, M.B. and Gunes, A. (2022). Iron biofortification of wheat grains by foliar application of nano zero-valent iron (nZVI) and other iron sources with urea. *Journal of Soil Science and Plant Nutrition* 22 (4): 4642–4652.

Taylor, A., Dangour, A.D., and Reddy, K.S. (2013). Only collective action will end undernutrition. *The Lancet* 382 (9891): 490–491.

Tian, Y.S., Xu, J., Wang, B. et al. (2021). Riboflavin fortification of rice endosperm by metabolic engineering. *Plant Biotechnology Journal* 19 (8): 1483.

Tilocca, M.G., Serratrice, G., Oggiano, M.A. et al. (2014). Monitoring the presence of genetically modified potato EH92-527-1 (BPS-25271-9) in commercial processed food. *Italian Journal of Food Safety* 3 (1): 1628.

Tulchinsky, T.H. (2010). Micronutrient deficiency conditions: Global health issues. *Public Health Reviews* 32 (1): 243–255.

United Nations (2020). The Sustainable Development Goals Report 2020. https://unstats.un.org/sdgs/report/2020/TheSustainable-Development-Goals-Report-2020.pdf (accessed February 2021).

Van Eenennaam, A.L., Lincoln, K., Durrett, T.P. et al. (2003). Engineering vitamin E content: From *Arabidopsis* mutant to soy oil. *The Plant Cell* 15 (12): 3007–3019.

Vélez-Bermúdez, I.C. and Schmidt, W. (2022). How plants recalibrate cellular iron homeostasis. *Plant and Cell Physiology* 63 (2): 154–162.

Verma, S., Chakdar, H., Kumar, M. et al. (2021). Microorganisms as a sustainable alternative to traditional biofortification of iron and zinc: Status and prospect to combat hidden hunger. *Journal of Soil Science and Plant Nutrition* 21 (2): 1700–1717.

Rai, V.R., and Rani, B. (2024). Hidden Hunger: A looming crisis in Pakistan's nutrition landscape. *Liaquat National Journal of Primary Care 2024* 6 (1): 1–2.

von Grebmer, K. (2018). How to accelerate the end of hunger and undernutrition. *Hidden Hunger: Strategies to Improve Nutrition Quality* 118: 93–101.

Wakeel, A., Farooq, M., Bashir, K. et al. (2018). Micronutrient malnutrition and biofortification: Recent advances and future perspectives. In: *Plant Micronutrient Use Efficiency* 225–243. Amsterdam: Academic Press.

Waltz, E. (2014). Vitamin A super banana in human trials. *Nature Biotechnology* 32 (9): 857.

Wang, C., Zeng, J., Li, Y., et al. (2014). Enrichment of provitamin A content in wheat (Triticum aestivum L.) by introduction of the bacterial carotenoid biosynthetic genes CrtB and CrtI. *Journal of Experimental Botany* 65 (9): 2545–2556.

Weffort, V.R.S. and Lamounier, J.A. (2024). Hidden hunger – a narrative review. *Jornal de Pediatria* 100 (suppl 1): S10–S17.

Wei, Y., Shohag, M.J.I., and Yang, X. (2012). Biofortification and bioavailability of rice grain zinc as affected by different forms of foliar zinc fertilization. *PLoS One* 7 (9): e45428.

Wei, Y., Shohag, M.J.I., Yang, X. et al. (2012). Effects of foliar iron application on iron concentration in polished rice grain and its bioavailability. *Journal of Agricultural and Food Chemistry* 60 (45): 11433–11439.

Wessells, K.R. and Brown, K.H. (2012). Estimating the global prevalence of zinc deficiency: Results based on zinc availability in national food supplies and the prevalence of stunting. *PLoS One* 7 (11): e50568.

WHO/ FAO, Allen, L., de Benoist, B., Dary O, R Hurrell, (eds.) (2006). *Guidelines on Food Fortification with Micronutrients.* Geneva, Switzerland: World Health Organization/Food and Agriculture Organization of UN.

Xue, Y.F., Li, X.J., Yan, W. et al. (2023). Biofortification of different maize cultivars with zinc, iron and selenium by foliar fertilizer applications. *Frontiers in Plant Science* 14: 1144514.

Yang, Q.Q., Yu, W.H., Wu, H.Y. et al. (2021). Lysine biofortification in rice by modulating feedback inhibition of aspartate kinase and dihydrodipicolinate synthase. *Plant Biotechnology Journal* 19 (3): 490–501.

Yang, Q.Q., Zhang, C.Q., Chan, M.L. et al. (2016). Biofortification of rice with the essential amino acid lysine: molecular characterization, nutritional evaluation, and field performance. *Journal of Experimental Botany* 67 (14): 4285–4296.

Yashodhara, B.M., Umakanth, S., Pappachan, J.M. et al. (2009). Omega-3 fatty acids: a comprehensive review of their role in health and disease. *Postgraduate Medical Journal* 85 (1000): 84–90.

Yuan, L., Wu, L., Yang, C. et al. (2013). Effects of iron and zinc foliar applications on rice plants and their grain accumulation and grain nutritional quality. *Journal of the Science of Food and Agriculture* 93 (2): 254–261.

Yue, G.H., Ma, K.Y., and Xia, J.H. (2024). Status of conventional and molecular breeding of salinity-tolerant tilapia. *Reviews in Aquaculture* 16 (1): 271–286.

Zhang, C., Liu, J., Zhang, Y. et al. (2011). Overexpression of SlGMEs leads to ascorbate accumulation with enhanced oxidative stress, cold, and salt tolerance in tomato. *Plant Cell Reports* 30: 389–398.

Zhao, X., Zhang, X., Wang, L. et al. (2024). Foliar application of iron impacts flavonoid glycosylation and promotes flavonoid metabolism in coloured rice. *Food Chemistry* 444: 138454.

Zhu, C., Naqvi, S., Breitenbach, J. et al. (2008). Combinatorial genetic transformation generates a library of metabolic phenotypes for the carotenoid pathway in maize. *Proceedings of the National Academy of Sciences* 105 (47): 18232–18237.

Zikankuba, V.L., Mteremko, D., and James, A. (2019). Staple crops biofortification linking agriculture, food and nutrition towards eliminating hidden hunger. *European Journal of Nutrition & Food Safety* 9 (2): 112–121.

Zinkernagel, J., Maestre-Valero, J.F., Seresti, S.Y. et al. (2020). New technologies and practical approaches to improve irrigation management of open field vegetable crops. *Agricultural Water Management* 242: 106404.

Zulfiqar, U., Khokhar, A., Maqsood, M.F. et al. (2024). Genetic biofortification: Advancing crop nutrition to tackle hidden hunger. *Functional & Integrative Genomics* 24 (2): 34.

24

Genetic Diversity and Crop Genome-wide Association Studies to Identify Biofortified Traits for Micronutrients

Praviinkumar R[1*]

[1] *Department of Genetics and Plant Breeding (GPB), Faculty of Agriculture, Annamalai University, Annamalai Nagar, Chidambaram, Tamil Nadu, India*

*Corresponding author: praviinkumar21@gmail.com

24.1 Introduction

Global attention has been drawn to biofortification, the process of increasing the nutrient content of crops, as a sustainable way to combat malnutrition. In the early days of biofortification, more emphasis was placed on ideals to improve the plant ideotype, but this new era of genetics is moving rapidly to decipher the pool of genomes with the ultimate aim of adapting and sustaining the modern changing environment without any changes in quality. Malnutrition in the form of hidden hunger happens when a person consumes a diet high in energy but low in nutrients, leading to several micronutrient deficiencies. Apart from modern transgenic techniques, the data from diversity analysis plays a significant role in crop improvement. Genome-wide association studies (GWAS) is one of the most effective methods for utilizing both biometrical and molecular fields to pinpoint the target loci linked to vital nutrients such as zinc (Zn) and iron (Fe) (Morris et al. 2012). In context, GWAS uses high-throughput technologies to scan the entire genome, which is correlated with the genetic variants influencing nutrient content in crops by utilizing large-scale genomic data to discover several loci significantly linked with trait variations (Stangoulis and Knez 2022). GWAS in biofortification focused on breeding efforts by identifying critical genetic loci associated with essential micronutrient elements, primarily iron (Fe) and zinc (Zn). The combo of GWAS and cutting-edge technologies has the potential to expedite biofortification and guarantee nutritional security for people on a global scale.

Micronutrients are just as important as macronutrients, but they are required only in trace amounts for normal human health functions such as digestion, hormone production, and brain function. Ineffective utilization of micronutrients from modern diet supplements tends to increase disease risk (Berti, Faber, and Smuts 2014). The GWAS based on large-scale resequencing has been demonstrated to be a reliable platform for identifying micronutrient genetic variants that can directly contribute to biofortified crop improvement (Huang et al. 2011).

24.1.1 Genetic Diversity

According to Hughes et al. (2008), genetic variety is the range of distinct features found in members of a species as a result of differences in their genetic composition. It is a crucial measure of the health of evolution because it offers a range of phenotypic variants that can be selected artificially by humans, serving as the basis for crop improvement, or naturally by environmental factors, allowing natural populations to adapt to changing conditions and colonize new habitats. Moreover, genetic diversity is essential to agriculture since it provides the basis for robust crop varieties and agricultural innovation (Jovovic et al. 2020). GWAS is an effective method for taking advantage of a panel of germplasm's inherent variety. Furthermore, gene banks act as essential storage facilities for this priceless genetic variety, guaranteeing its survival for upcoming generations. By using GWAS, we can discover the mysteries of germplasm, which utilize diversity panels that are frequently used by genetics researchers to identify genotypes associated with particular traits, as in quantitative trait loci (QTLs) mapping with GWAS.

Crop Biofortification: Biotechnological Approaches for Achieving Nutritional Security under Changing Climate, First Edition. Edited by Adnan Noor Shah et al.

24.1.2 Micronutrient Malnutrition and Hidden Hunger

Malnutrition and dietary deficiencies remain a pressing concern, particularly in regions with limited access to diverse food sources. Climate change further threatens agricultural ecosystems, jeopardizing future food security. The primary goal is to combat malnutrition and "hidden hunger" by enhancing the micronutrient content of staple crops (Singh et al. 2016). Fortunately, advancements in germplasm analysis offer a powerful tool to combat these challenges. By identifying germplasm with high nutrient content alongside desirable traits like drought tolerance, researchers are now engaging in breeding new crops that are both resilient and nutritious (da Silva et al. 2011). Association mapping (AM) serves as a substitute for traditional linkage mapping, providing a detailed understanding of the genetic underpinnings of complex characteristics (Abdurakhmonov et al. 2012). This new era of GWAS-bio fortification tackles malnutrition through a creative blend of both traditional plant breeding and modern biotechnology, with the aim of identifying the candidate gene responsive for high quality improvement in nutrition such as protein and vital nutrients.

This chapter delves into the exciting potential of germplasm analysis. The application of candidate gene detection in nutrigenomics offers opportunities to mitigate malnutrition (Dable-Tupas et al. 2023). This approach holds promise for preventing both acute and chronic sequelae (long-term effects) associated with poor dietary intake. Moreover, the application of genome-wide AM in sorghum has uncovered substantial correlations between genetic markers and numerous traits that are advantageous for agriculture. This effectively illustrates how germplasm analysis can be employed to develop new crop varieties that are both hardy and nutrient-rich (Upadhyaya et al. 2014).

24.2 Genomic Biofortification Strategies

The agronomic strategies used in conventional biofortification procedures, such as fertilizer dosage compensation and foliar application, are merely nutrient management systems and are not much efficient than breeding strategies. Subsequently, plant breeding strategies were presented, which show great promise for enhancing nutritional improvement through three key methods: (i) Conventional breeding: Enhance potential cultivars through crossbreeding; (ii) Transgenic biofortification: genetic modification through the transfer of the desired target gene; and (iii) Gene editing biofortification: CRISPR-Cas9 tools to trigger the nutritional trait (Kiran 2020). The nutritional gain on cultivars has been significantly increased by the above areas, but difficulties continue to identify the target traits that relate to nutritional traits for biofortification improvement. As a result, researchers are currently engaging in GWAS to identify the associated marker traits so that the above breeding methods can be used efficiently.

24.2.1 Significant Drift Towards GWAS Biofortification and Methodologies

The transition in plant breeding from a design-centric methodology to genome-wide strategies has resulted in the emergence of GWAS. GWAS has emerged as a favored approach for accelerating breeding procedures, with choices being influenced by the correlations between marker-trait associations (MTAs), which correspond to the combined impact of all advantageous alleles. Initially, linkage mapping is the tool to transfer the trait of interest, such as QTLs, by developing a biparental population as the source to create genetic diversity. But it offers a low transfer rate and resolution. This limits researchers to have wider genetic variations and utilize the variation present in those specific parents only. Alternatively, AM in diversity panels has accelerated the identification of genomic regions associated with phenotypic traits by the historical recombination accumulated in natural populations (meiotic events) that caused the nonrandom association of alleles at different loci across the genome (Zhu et al. 2008). Thus, association mapping enables a higher mapping resolution than biparental linkage analysis.

The GWAS and candidate gene approach are two kinds of approaches. While the latter provides phenotypic associations between prespecified genes of interest, the former examines the entire genome for such associations. The candidate gene approach has also increased statistical power to detect differences due to limiting the analysis to a relatively few number of genes (Modena et al. 2019).

24.3 Capturing the Common Variation in Genome

24.3.1 Genotyping Through Genetic Markers

Utilizing unlinked and putatively neutral markers to assess genetic diversity in the accessions used in the mapping study and account for population structure is a frequent approach in many association genetic studies (Hall et al. 2010). There are many markers available to mitigate the effects of population structure (for more on this, see below), but some have their own limitations. Amplified fragment length polymorphism (AFLP) markers are commonly used because they can be easily obtained in almost any organism, even without existing genomics data. However, AFLP markers have limitations, such as being dominant, meaning that the heterozygous genotype cannot be distinguished from one of the homozygous genotypes. This lack of allelic information introduces challenges when estimating population structure or using it directly in mapping studies. SSR (simple sequence repeat) markers (microsatellite markers) are highly polymorphic and provide valuable genetic information. However, they require significant effort to isolate and are rarely transferable between species unless they are closely related. Despite these challenges, SSR markers have been popular for estimating population structure and assessing pairwise relatedness among individuals. In summary, both AFLP and SSR markers play essential roles in association studies, but researchers must carefully consider their advantages and limitations when designing experiments and interpreting results (Hasan et al. 2021).

The integration of next-generation sequencing (NGS) technologies with high-throughput phenotyping, AM, and comprehensive analyses of the proteome and metabolome has revolutionized the identification of candidate genes. Single-nucleotide polymorphisms – relatively high polymorphic marker, frequently called SNPs (pronounced as “snips”) are stably inherited, distributed, and highly abundant throughout the genome. This tool enables the rapid and comprehensive identification of a vast array of genetic variants. The mutational rate is smaller when compared to SSR, SNPs are easily incorporated with high-throughput genotyping in multiplex or microarray. Thus, in presumably any population, a handful of common SNPs chosen from each segment would be sufficient to determine the appropriate haplotypes (genetic polymorphisms that tend to be inherited together from a single parent). Therefore, Linkage disequilibrium (LD) patterns can be predicted, which are nothing more than discrete haplotype blocks (Wall and Pritchard 2003). Nowadays, SNPs are favorable markers to be used on genotyping association studies, and thousands of SNP polymorphisms were identified for a single phenotype. According to Liu and Zhang (2006), in the rice genome, a total of 80 127 SNP polymorphic sites were identified. According to (Bukomarhe et al. 2023) following the GWAS on *Oryza sativa*, the Rice Genome Annotation Project (RGAP) database can be used to annotate potential candidate genes for the Fe and Zn concentration.

24.3.2 Minor Allele Frequency

Minor allele frequency (MAF) denotes the frequency at which the second most prevalent allele is present within a specified population. Imperfect LD between SNPs and QTLs is the cause of the genetic variance that SNPs are unable to detect. GWAS SNP arrays include a wide range of MAFs, from nearly monomorphic (MAF $< 0.5\%$) to very common (MAF $\approx 50\%$). Rare variants (MAF < 0.05) may have a more significant impact on heritability than common variants (MAF > 0.05). After the elimination of SNP markers with MAF below 0.05 and above 0.95, missing data exceeding 30%, and heterozygosity surpassing 20% from a total of 35 153 markers that underwent processing for the procurement of high-quality informative markers, a cohort of 9503 high-quality SNP markers remained for utilization in GWAS to analyze candidate genes for wheat grain micronutrients (Rathan et al. 2022). The extent of LD between SNPs and QTLs depends on different minor allele frequencies (MAF) between them.

24.3.3 Linkage Disequilibrium

The basic principle behind the key success of AM is LD, the nonrandom association of alleles of different loci. Simply, LD doesn’t follow one of the three (law of independent assortment) Mendel’s laws of inheritance. Linkage describes the nonrandom coinheritance of genetic loci due to their physical proximity on the same chromosome. In contrast, LD refers to the nonrandom association of alleles at different loci within a population, not necessarily because they are physically close. LD is also present on distant loci (or) loci located on different chromosomes. The significant relation is that high linkage can directly proportion with the high LD, and in contrast, more recombination between loci lowers the LD between alleles. The idea of LD was initially introduced by Jennings in 1917, and the method to quantify it (D) was established by Lewontin in 1964 (Abdurakhmonov and Abdukarimov 2008). The simplified explanation of the commonly used LD measure, D or D' (standardized version of D), is the difference between the observed gametic frequencies of haplotypes and the expected gametic frequencies of haplotype under linkage equilibrium as $D = P_{AB} - P_{APB}$. Where P_A and P_B are the product of the frequencies of the alleles A and B, respectively; PAB is the frequency of gametes carrying alleles A and B at two loci. There are two methods to quantify LD, such as r^2 and D'. While r^2 encapsulates both the history of recombination and mutation and D' solely accounts for the history of recombination, making it a more precise metric for gauging differences in recombination. A more detailed review of LD is given by Flint-Garcia et al. (2003).

24.3.4 LD Visualization and Interpretation

The interpretation of LD analysis (or) haplotype blocks found to be in significant LD can be interpreted as two basic models. (1) LD-decay scatter plots and (2) disequilibrium matrices (or) LD heat maps. The former gives a graphical idea of how LD declines exponentially as the physical or genetic distance between the markers increases, and these scatterplots help establish the average effective distance threshold beyond which significant LD, typically 0.5 for *D*' and 0.1 for *D*, is anticipated based on the curve of a nonlinear logarithmic trend that is plotted through the data points on the scatterplot (Breseghello and Sorrells 2006). There are tools available for creating LD decay plots. For instance, "PopLDdecay" is a software tool that allows users to calculate LD decay statistics and plot variant call format (VCF) files directly (Zhang et al. 2019). The latter is used to interpret disequilibrium matrices, also known as LD heat maps, which are graphical representations used to visualize pairwise LD measurements between SNPs within a genomic region. In an LD heat map, the color intensity of each cell corresponds to the strength of LD between a pair of SNPs. This heat map is typically a false-color image in the upper left diagonal of a square plot. The "LDheatmap()" function in R is commonly used to generate these heat maps. This function takes an input data set that provides information on pairwise LD between SNPs in a genomic region, plots color-coded values of the pairwise LD measurements, and returns an object of class "LDheatmap" containing a number of components (Figure 24.1).

24.4 Micronutrient Phenotyping

Phenotyping forms the foundation of successful GWAS studies by ensuring accuracy, power, and specificity. Scientists can leverage GWAS to unlock the secrets of how genes influence nutritional breeding traits. Large-scale field measurements offer the statistical power needed to detect subtle genetic effects on these broad traits. Manipulating the large population can account for environmental variations that might show true underlying genetic effects. Phenotyping typically includes measuring field data, which is crucial for GWAS investigations. However, biofortification procedures need phenotyping of individual nutrient estimation, and it demands higher specificity and precision.

The grain micronutrient phenotyping such as zinc, iron, and manganese concentrations (GZnC, GFeC, and GMnC). Inductively coupled plasma (ICP) spectrometry is a commonly used instrument that allows for the analysis of trace elements in a variety of grain samples. By utilizing a precise electromagnetic array along with a plasma torch and an integrated spectro-analytical detector, ICP solutions can accurately identify a wide range of chemical elements within a sample, even at trace and ultratrace levels. A standard ICP array can be integrated with a selection of detectors, predominantly an atomic emission spectrometer or a mass spectrometer. Both methods enable operators to generate reliable analytical reports for multielement analysis within the sub-parts per million (ppm) range for nearly all elements listed on the periodic table.

Unlike traditional methods such as inductively coupled plasma-optical emission spectrometry (ICP-OES), which require expensive equipment, extensive sample preparation, and trained analysts. ICP-MS allows for the efficient quantification of macrolevel nutritional elements alongside lower level elements in a single analysis, such as mineral concentrations for Ca, Cd, Co, Cu, Fe, Li, Mg, Mn, Ni, and Zn, eliminating the need for sample dilution.

Atomic emission spectrometry (AES) is a technique that involves analyzing the light emitted by atoms transitioning from an excited state to a lower energy state. These specific wavelengths are unique to individual elements found on the periodic table. In a study conducted by (Jaiswal et al. in 2019), a total of 10 elements, including potassium (K), nickel (Ni), calcium (Ca), boron (B), magnesium (Mg), phosphorus (P), sulfur (S), zinc (Zn), manganese (Mn), and iron (Fe), were phenotyped using ICP-AES to identify associations with marker traits.

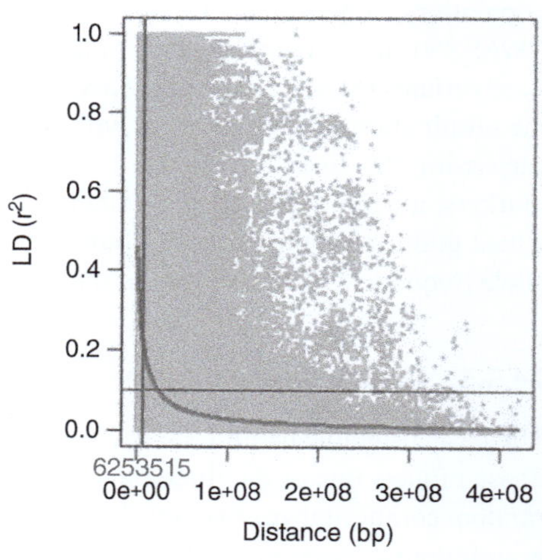

Figure 24.1 A plot illustrating the decay of linkage disequilibrium (LD) derived from the LD values of a theoretical marker dataset is presented to illustrate the assessment of average genome-wide LD block sizes. The pairwise LD values (r^2) are graphed against genetic distance using R software using TASSEL LD results. *Note:* This presentation is purely for illustrative purposes and does not hold to any specific genomic segment of an organism.

Modern energy-dispersive X-ray fluorescence spectrometry (ED-XRF) provides a convenient and economical alternative for assessing Zn and Fe concentrations in whole grain rice and pearl millet, particularly in the context of biofortification breeding programs (Paltridge et al. 2012). ED-XRF offers a streamlined approach. X-rays are emitted from the sample following excitation by energetic X-rays, revealing characteristic energy levels associated with specific elements. By calibrating the EDXRF instrument using reference, both grain zinc concentration (GZnC) and grain iron concentration (GFeC) concentrations obtained via ICP-OES. Remarkably, the ED-XRF results exhibited a strong correlation with the ICP-OES measurements for both GZnC and GFeC in both species (Paltridge et al. 2012). Likewise, ANOVA was performed to analyze the differences between group means and their associated procedures. Through the segregation of the complete phenotypic variance into separate components, ANOVA offers researchers insight into how genetic, environmental, and interactional factors contribute to the variability in traits. ANOVA showed significant effects of the genotypes and the year on micronutrient concentration in grains (Arora et al. 2019).

24.5 Genomic Traits Associated with Micronutrients

In recent years, particularly within this decade, numerous researchers have been engaged in genome-wide studies. A significant portion of these studies has focused on the resilience approach. A substantial amount of literature reveals the effectiveness of micronutrient GWAS in various cereals such as rice, wheat, barley, chickpea, and some other legumes. These examinations are mainly focused on uncovering the genes/QTLs associated with genetic markers, i.e. MTA. The assessment scale covers the analysis of one nutrient to almost 10 nutrients in Rice (Bhatta et al. 2018) and Foxtail millet (Jaiswal et al. 2024). The table below collectively depicts the recent studies on various crops in terms of improvement in Zn and Fe with the ultimate aim of finding MTA (Table 24.1).

Table 24.1 Biofortification of crops and vegetables with minerals by soil and foliar application of micronutrient formulations.

Plant species	Grain micronutrients	Sample size	Type of markers	Trait association	Reference
Wheat	Zn, Fe, and Mn	161 Wild Emmer lines	13 116 DArT	14 MTA	Liu et al. (2021)
	Zn and Fe	166 Diverse Lines	373 106 SNPs	41 MTA	Tong et al. (2022)
	Zn, Fe, Cu, and Mn	167 *Aegilops tauschii* (wild progenitor of the bread wheat)	5249 SNPs	19 MTA	Arora et al. (2019)
	Zn, Fe, and Quality traits	184 Diverse lines	35 153 SNPs	55 MTA	Rathan et al. (2022)
	Zn and Fe content mainly	205 Chinese lines	24 355 SNPs	101 MTA	Wang et al. (2021)
	Fe only	369 European lines	15 523 SNPs	41 MTA	Alomari et al. (2018)
	Zn and Fe	161 Aegilops substitution and addition lines	10 825 SNPs	111 MTA	Kaur et al. (2023)
	Zn, Fe, and Kernel traits	280 Diverse lines	14 790 SNPs	17 MTA	Krishnappa et al. (2022)
	10 Grain Nutrients (Ca, Cd, Cu, Co, Fe, Li, Mg, Mn, Ni, and Zn)	123 Synthetic hexaploid wheat	35 648 SNPs	92 MTA	Bhatta et al. (2018)
Rice	Zn and Fe	190 Diverse lines	148 SSRs	22 MTA	Sahu et al. (2020)
	Zn and Fe	144 MAGIC lines	14 242 SNP	57 MTA	Descalsota et al. (2018)
	Zn and Fe	192 diverse lines	50k Illumina SNP chip	29 MTA	Bollinedi et al. (2020)
	Zn, Fe, and Quality traits	300 indica accessions	1.02 million SNPs	109 MTA	Pasion et al. (2023)
	Zn, Fe, and Quality traits	96 Diverse lines	SSR	78 MTA	Chattopadhyay et al. (2023)
	Na, Mo and agronomic traits	529 Diverse lines	6.4 million SNPs	72 MTA	Yang et al. (2018)

(*Continued*)

Table 24.1 Biofortification of crops and vegetables with minerals by soil and foliar application of micronutrient formulations. *(continued)*

Plant species	Grain micronutrients	Sample size	Type of markers	Trait association	Reference
	Zn, Fe, and other	219 Diverse lines	155 SSR	60 MTA	Nawaz et al. (2015)
	Six mineral elements (Cu, Fe, K, Mg, Mn, and Zn)	174 Diverse lines	6565 SNPs	14 MTA	Islam et al. (2022)
	12 Grain Nutrients (B, Ca, Co, Cu, Fe, K, Mg, Mn, Mo, Na, P, and Zn)	Diverse indica accessions	5.2 million SNPs	57 MTA	Cu et al. (2021)
Barley	Zn	298 Diverse lines	7842 SNPs	4 MTA (Zn)	Mamo et al. (2014)
	B, Cu, Fe, Mn, and Zn	232 Diverse lines	62 654 SNPs	37 MTA	Abendroth et al. (2022)
	Zn and Fe	496 Diverse lines	50k Ilumnia SNP chip	18 MTA	Nyiraguhirwa, et al. (2022)
	Zn, Fe, and Se	216 spring lines	9k Ilumnia SNP chip	222 MTA	Thabet et al. (2022)
Maize	Zn and Cd	430 inbred lines	147 762 SNPs	191 MTA	Baseggio et al. (2021)
	Zn and Fe	923 inbreds	955 690 SNPs	46 MTA	Hindu et al. (2018)
Pearl Millet	Zn and Fe	130 Diverse lines	114 SSR	16 MTA	Anuradha et al. (2017)
	Zn and Fe and Protein Contents	281 inbreds	58 719 SNPs	78 MTA	Pujar et al. (2020)
Foxtail millet	Ten Grain Nutrients (K, Ni, Ca, B, Mg, P, S, Zn, Mn, and Fe)	93 Diverse lines	10 000 SNPs	74 MTA	Jaiswal et al. (2024)
Finger Millet	Zn and Fe	202 Diverse lines	2977 SNPs	12 MTA	Chandra et al. (2024)
	Zn,Fe, Ca, Mg, Na	190 Diverse lines	169 365 SNPs	18 MTA	Puranik et al. (2020)
Lentils	Zn and Fe	95 Diverse lines	33 745 SNPs	14 MTA	Singh et al. (2023)
		96 Diverse lines	80 SSR	5 MTA	Kumar, et al. (2019)
		96 Diverse lines	73 SSRs	4 MTA	Singh et al. (2017)
Chickpea	Zn, Cu, Fe, and Mn	147 Diverse lines	50k Ilumnia SNP chip	6 MTA	Fayaz et al. (2022)
	Fe and Zn	280 Diverse lines	4995 SNPs	13 MTA	Srungarapu et al. (2022)
	Fe and Zn	92 Diverse lines	24 620 SNPs	11 MTA	Upadhyaya et al. (2016)
Mung bean	Seven Grain Nutrients (Ca, Fe, K, Mn, P, S, and Zn)	95 Diverse lines	6486 SNPs	43 MTA	Wu et al. (2020)
	Nine Grain Nutrients (K, P, Ca, Mg, Cu, Mn, S, Zn, and Fe)	178 Mesoamerican lines	25 011 SNPs	48 MTA	Delfini et al. (2021)
	Zn and Fe	145 diverse	14 447 SNPs	20 MTA	Sinha et al. (2023)
Common Bean	Eight Grain Nutrients (N, P, K, Ca, Mg, Fe, Zn, and Mn)	174 Croatian landraces	6311 SNPs	2 MTA (Zn)	Gunjača et al. (2021)
	Zn and Fe	289 Diverse lines	11 480 SNPs	5 MTA	Gelaw et al. (2023)
Field Pea	Zn and Fe	135 diverse lines	16 877 SNPs	5 MTA	Dissanayaka et al. (2020)

Similarly, Rathan et al. (2022) also conducted a GWAS on wheat to identify loci and candidate genes for grain micronutrients and quality traits. Additionally, Zhong et al. (2022) investigated zinc toxicity tolerance in rice through a GWAS within a rice core collection, highlighting the importance of genetic variations in response to high zinc concentrations. These studies collectively demonstrate the utility of GWAS in identifying genetic markers associated with micronutrient concentrations in various crops, providing valuable insights for biofortification and crop improvement efforts.

24.6 Population Structure

Population structure is an essential aspect of GWAS as it can significantly impact the results and interpretation of genetic associations (Zhou and Stephens 2012). These studies aim to identify genetic variants linked to a specific trait or disease by comparing the genomes of individuals with and without the trait (Bush and Moore 2012). However, the presence of population structure, referring to the existence of subgroups within a population with varying allele frequencies due to factors like ancestry or geographic location, can introduce confounding biases in GWAS (Novembre and Stephens 2008).

In order to alleviate the impacts of population structure, it is advisable to utilize a range of statistical techniques to adjust for ancestry or other variables that contribute to population stratification. (Price et al. 2006). Common approaches include principal component analysis (PCA), mixed linear models (MLMs), and genomic control (GC) (Kang et al. 2008). Furthermore, investigating population structure can provide valuable insights into the genetic ancestry of groups and individuals (Rosenberg et al. 2002). The issue of population structure arises due to the correlation between any phenotypic trait and the underlying population structure at neutral loci, resulting in an increased number of positive associations.

GC is a method used to correct for population stratification. It aims to adjust for confounding biases caused by population structure, ensuring accurate results (Devlin and Roeder 1999). GC assumes that a majority of genetic variants in the genome are not associated with the trait under investigation, and their distribution can be used to estimate the underlying population structure. However, GC has limitations, particularly when there are structured associations between subpopulations and environmental factors that affect the trait of interest. GC may not capture all the underlying population structure, leading to residual confounding biases in GWAS results. The kinship matrix (K) can be derived from pedigree data or calculated using relative kinship coefficients obtained from genetic marker data. For example, in a study of genome-wide association in *Arabidopsis thaliana*, the mixed model demonstrated superior control over false-positive rates compared to other tested methods, despite the intricate sub-structuring of the association population (Figure 24.2). Estimation of the kinship matrix can be achieved using various methods, such as the method-of-moments estimator or the maximum likelihood estimator (Astle and Balding 2009). In addition, there are different types of kinship matrices that can be used depending on the context, such as the additive genetic relationship matrix or the dominance genetic relationship matrix (Wolak et al. 2012). PCA method reduces the dimensionality of a dataset by transforming it into a new set of uncorrelated variables called principal components (PCs) applied to genetic markers across the genome to capture the underlying genetic diversity and identify the major axes of variation in the data (Novembre and Stephens 2008). PCA was employed to examine the population structure concerning GZnC, GFeC, grain protein content, test weight, and thousand kernel weight in a collection of 184 varied bread wheat genotypes through the utilization of the GAPIT tool (Rathan et al. 2022).

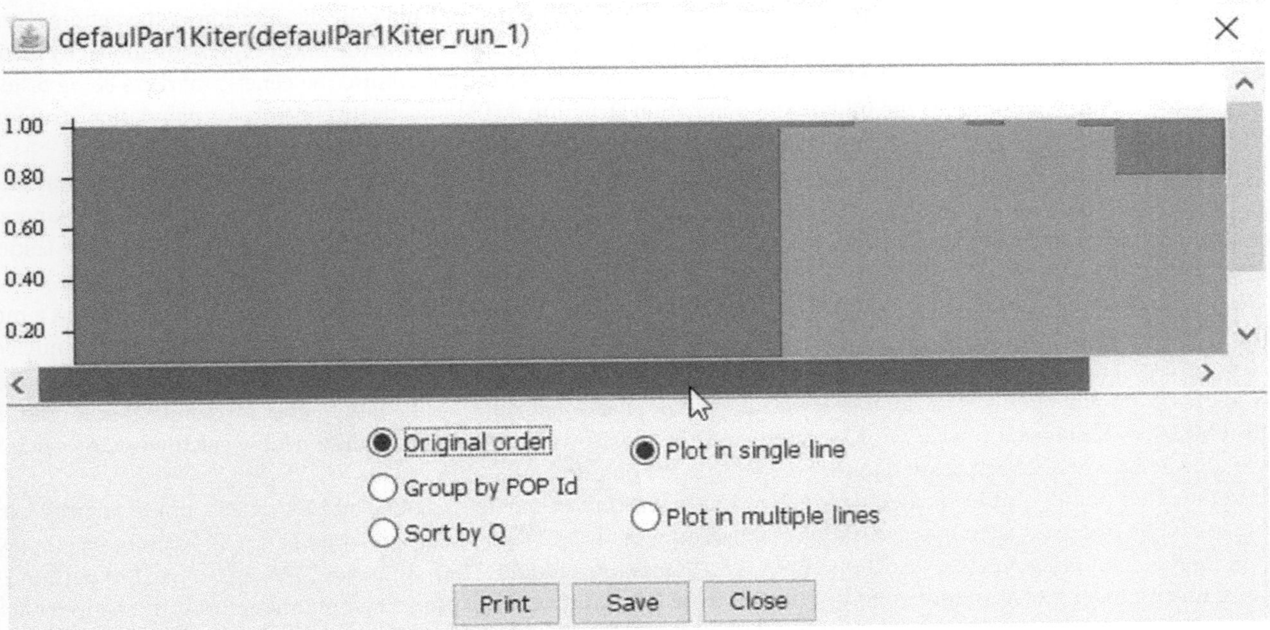

Figure 24.2 STRUCTURE software sample illustrates the population genetic structure using four distinct colors ($K = 2$).

MLMs have become powerful in addressing (Zhou and Stephens 2012). MLMs are an extension of linear regression models that account for both fixed and random effects (Pinheiro and Bates 2000). In the context of GWAS, fixed effects represent the genetic markers being tested for association with a trait, while random effects capture the underlying population structure (Kang et al. 2008). By incorporating random effects, MLMs can effectively control for the influence of population structure on genetic associations. Trait analysis by association, evolution, and linkage (TASSEL) (Bradbury et al. 2007) package tool used for the purpose of examining the association between 1233 SNP markers and the concentrations of iron, zinc, and selenium, respectively.

The program ADMIXTURE version 1.3.0 generates a Q matrix with ancestor estimates for every tested person. At $K^{1/4}$ 4, which corresponds to four different subpopulations, the matching Q matrix with the lowest cross-validation error was determined to be the most representative of the research population.

The advantage of MLMs is that they have the ability to account for multiple sources of heterogeneity, such as additive and nonadditive genetic variance, spatial heterogeneity, and genotype-environment interactions (Desaint, Hereil, and Causse 2023). This makes them particularly suitable for analyzing complex traits and controlling for structured associations in GWAS. While MLM-based methods have proven useful for controlling population structure, they can be computationally challenging for large datasets (Lippert et al. 2011).

24.6.1 Software Used for Population Stratification

Software	Website	Reference
R Package	http://www.r-project.org/	R Development Team
STRUCTURE	https://web.stanford.edu/group/pritchardlab/structure.html	Pritchard, Stephens and Donnelly (2000)
AWclust	https://awclust.sourceforge.net/	Gao and Starmer (2008)
AdMixture	https://dalexander.github.io/admixture/index.html	Alexander, Novembre and Lange (2009)

24.7 Marker Trait Analysis of Biofortified Traits

On GWAS of nutritional traits, the Bayesian-information and linkage-disequilibrium iteratively nested keyway (BLINK) method was utilized due to its high statistical power and nonassumption of normal gene distribution in the genome, preventing false positives and exclusion of causal genes (Sinha et al. 2023). MLMs extend the linear regression model by incorporating both fixed and random effects (Pinheiro and Bates 2000). Fixed effects capture the genetic markers being tested, while random effects account for the underlying population structure. MLM effectively controls for population structure effects but can be computationally intensive for large datasets (Lippert et al. 2011).

Jaiswal et al. (2019) utilized the fixed and random model circulating probability unification (FarmCPU) approach in their multilocus GWAS, a method originally developed by Liu et al. (2016). The FarmCPU technique is renowned for its superior computational efficiency and its capability to tackle issues pertaining to population structure, kinship, and multiple testing corrections in the investigation of the correlations among ten nutritional elements in Foxtail millet (*Setaria italica L.*).

Huang et al. (2019) discussed two methods of introducing cryptic associations. Including every genetic marker as a random effect is one approach. Through LD, some of these markers are able to capture the impacts of quantitative trait nucleotides (QTNs). The alternative is to use all genetic markers to determine kinship among individuals first. The variance structure of individual effects as random effects is then defined by kinship. In the latter case, a fixed effect and random effect MLM can include both population structure and kinship. However, the MLM requires a lot of calculation. As a result, several techniques have been created to shorten MLM's computation durations.

On the other hand, multilocus methods such as multilocus mixed-model (MLMM) and FarmCPU aim to identify variants to explain observed phenotypic variation (Tibbs Cortes et al. 2021). The freshly devised FarmCPU technique employs a bin method assuming an even distribution of QTNs across the genome. The estimated QTNs are utilized to partition a mixed linear model into a computationally effective fixed effect model (FEM) and a computationally costly random effect model (REM), which are subsequently employed iteratively. In order to entirely eradicate the computationally expensive

REM, we substituted REM with FEM utilizing Bayesian information criteria (Huang et al. 2019). The bin method was replaced with LD information to remove the necessity for QTNs to be evenly distributed across the genome. This novel approach is termed BLINK. The analyses of both actual and simulated data revealed that BLINK enhances statistical power in comparison to FarmCPU, while also significantly reducing computational time. The BLINK model, an improvement over FarmCPU, uses a multilocus approach for better accuracy. This model eliminates the need for evenly distributed genes and uses Bayesian information content (BIC) for faster analysis. BIC analysis showed no models required principal components (PCs). The BLINK method was chosen for its high statistical power and ability to avoid false positives by not assuming normal gene distribution. BLINK stands out by avoiding the assumption of causal genes being normally distributed across the genome, thereby reducing the risk of false positives and excluding potential causal genes. This makes BLINK an effective approach for accurately identifying genetic associations related to nutritional traits (Huang et al. 2019). The mixed linear models were fitted using the following formula:

$$y = X\beta + Z\mu + e$$

In this formula, y represents the observed phenotypes, β is the coefficient of the SNP marker the effect being estimated, population structure (Q), and intercept as fixed effects, μ is an unknown vector of random additive effects from background QTLs and individuals, X and Z are the known design matrices, and e is an unobserved vector of residuals, as explained in the GAPIT user manual. In MLM ($X\beta$) represents a fixed effect, and relative kinship and residuals represent random effects. This program creates a "bar graph" as a GWAS result at the conclusion of the study, showing the significant markers with a threshold line (FDR = Q-value = 0.05) and −log10(P-values). Then, the false discovery rate (FDR) adjusted marker scores were outputted, from which P-values were calculated (Mamo, Barber, and Steffenson 2014).

24.8 Result and Validation – Genome-wide Significance

In GWAS, the P and Q-values are general notations for validating the significance of genome-wide studies. The importance and interpretation of these values in understanding the results of GWAS for grain micronutrients are used to assess the bounded significance of estimated micronutrients. The P-value is the probability that data could occur without any difference, while the Q-value is a measure of significance based on the false discovery rate (FDR).

24.8.1 *P*-values: False-Positive Rate

A significant P-value for GWAS threshold may vary depending on the analysis, but it is often less than 1×10^{-5} (or even less than 5×10^{-8}) indicating natural evidence of marker trait association towards grain micronutrients. However, in GWAS thousands to millions of SNPs are tested for association, and the threshold for significance may be much smaller than 0.05. A Bonferroni correction is often applied to control for multiple testing and minimize false positives. For example, in a GWAS study of *Aegilops tauschii* (diploid progenitor of the D-genome of hexaploid bread wheat). A Bonferroni-corrected threshold of −log10 (P-value) ≥ 4.68 was used to detect 19 MTAs with significant SNPs for various traits (Arora et al. 2019). To account for multiple testing in GWAS, a fixed P-value threshold of 5×10^{-8} is widely used to identify the association between a common genetic variant and a trait of interest having universal significance commonly, known as genome-wide significance (GWS).

24.8.2 *Q*-values: Controlling False Discovery Rate

The Q-value is a measure of significance based on the FDR an alternative approach for controlling the rate of false discoveries in multiple testing scenarios. The significance of each feature is measured by the q value, which automatically accounts for the fact that thousands of features are being tested at once. Q-values estimate the proportion of false positives among all rejections of the null hypothesis, allowing researchers to directly control the desired FDR level. Small Q-values indicate that the result remains statistically significant after correcting for multiple tests, increasing confidence in the identified associations. In this case, Q-values were employed to ensure that the identified associations were robust to multiple testing. Although the false positive rate and FDR are sometimes confused, there is a significant difference between the two.

Visualization methods such as Manhattan plots and quantile–quantile (Q–Q) plots play essential roles in interpreting and communicating the results of these studies. Manhattan Plots: Manhattan plots provide a graphical representation of the association between genetic markers and traits, with the significance of the association (typically represented as −log10 [*P*-value]) plotted against the chromosomal position of markers (Dudbridge 2013). For example, in a GWAS of a simulated trait, GAPIT3 (genome association and prediction integrated tool) generated interactive Manhattan plots to extract information on significant markers. QQ Plots: QQ plots help assess the deviation of observed *P*-values from the null hypothesis, which states that there is no association between genetic markers and the trait of interest. In a QQ plot, the observed *P*-values are sorted, and their quantiles are plotted against the quantiles of expected *P*-values under the null hypothesis (Dudbridge 2013). If the observed *P*-values follow the null distribution, the points in the QQ plot will align closely with the diagonal line. Deviations from this line, particularly at the tail end, indicate potential genuine associations between markers and traits.

24.9 Future Prospects

As genome sequencing technologies become more accessible and affordable, the scope and impact of GWAS-driven biofortification are expected to grow exponentially. The integration of genomic data with other cutting-edge approaches, such as gene editing and high-throughput phenotyping, can further accelerate the development and deployment of nutrient-rich, climate-resilient crop varieties (McMahon et al. 2021). Ultimately, the successful implementation of GWAS-driven biofortification strategies will require a collaborative and multidisciplinary approach, involving plant breeders, geneticists, agronomists, nutritionists, and policymakers. By working together, these stakeholders can harness the power of genomics to create lasting solutions for global food security, improve human health, and ensure the sustainability of agricultural systems worldwide (Aryee et al. 2021).

The integration of GWAS with other advanced techniques such as CRISPR-Cas9 gene editing and genomic selection could revolutionize the field of biofortification. It could lead to the development of new crop varieties with enhanced nutritional profiles, thereby contributing to the alleviation of micronutrient deficiencies worldwide. Moreover, the use of GWAS in biofortification could also help in adapting our food systems to climate change. By identifying the genes associated with stress tolerance, we could develop crops that not only have improved nutritional content but are also resilient to changing environmental conditions (Figure 24.3).

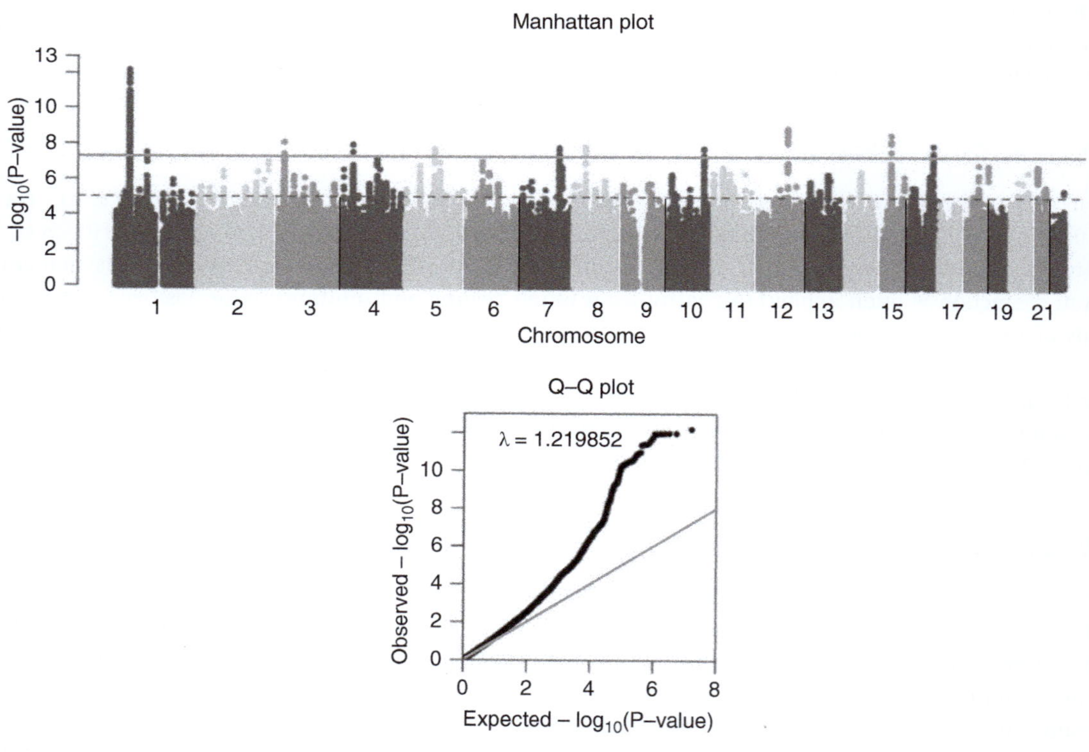

Figure 24.3 Manhattan Plots and QQ plots are generated using R software with GWAS summary statistics data. Left: Manhattan plots depicting 21 chromosomes carrying the important markers identified by MLM models are displayed. Right: Quantile–quantile (Q–Q) plots for the distribution of *P*-values observed using a mixed linear model (MLM) as black dots, and the expected null distribution of *P*-values, assuming no relationships, as a red solid line.

24.10 Conclusion

In conclusion, the potential impact of GWAS biofortification on global food security cannot be overstated. It represents a powerful tool in our fight against malnutrition and hunger. By improving the nutritional quality of our crops, we can ensure that every individual, regardless of where they live, has access to healthy and nutritious food. However, it's important to remember that biofortification is just one piece of the puzzle. To truly achieve global food security, we must also address other critical issues such as poverty, inequality, and unsustainable agricultural practices. Only then we can hope to create a world where everyone can lead healthy and productive lives.

References

Abdurakhmonov, I.Y. and Abdukarimov, A. (2008). Application of association mapping to understanding the genetic diversity of plant germplasm resources. *International Journal of Plant Genomics* 2008: 574927.

Abendroth, J.A., Sallam, A.H., Steffenson, B.J. et al. (2022). Identification of genomic loci controlling grain macro and micronutrient variation in a wild barley (*Hordeum vulgare spp. spontaneum*) diversity panel. *Agronomy* 12 (11): 2839. doi: 10.3390/agronomy12112839.

Alexander, D.H., Novembre, J., and Lange, K. (2009). Fast model-based estimation of ancestry in unrelated individuals. *Genome Research* 19 (9): 1655–1664.

Alomari, D.Z., Eggert, K., Von Wirén, N. et al. (2018). Whole-genome association mapping and genomic prediction for iron concentration in wheat grains. *International Journal of Molecular Sciences* 20 (1): 76.

Anuradha, N., Satyavathi, C.T., Bharadwaj, C. et al. (2017). Deciphering genomic regions for high grain iron and zinc content using association mapping in pearl millet. *Frontiers in Plant Science* 8: 238474.

Arora, S., Cheema, J., Poland, J. et al. (2019). Genome-wide association mapping of grain micronutrients concentration in *Aegilops tauschii*. *Frontiers in Plant Science* 10: 422013.

Aryee, S.N.D., Owusu-Adjei, D., Osei-Amponsah, R. et al. (2021). Sustainable genomic research for food security in Sub-Saharan Africa. *Agriculture & Food Security* 10: 1–12.

Astle, W. and Balding, D.J. (2009). Population structure and cryptic relatedness in genetic association studies. *Statistical Science* 24 (4): 451–471.

Baseggio, M., Murray, M., Wu, D. et al. (2021). Genome-wide association study suggests an independent genetic basis of zinc and cadmium concentrations in fresh sweet corn kernels. *G3* 11 (8): jkab186.

Berti, C., Faber, M., and Smuts, C.M. (2014). Prevention and control of micronutrient deficiencies in developing countries: Current perspectives. *Nutrition and Dietary Supplements* 6: 41–57.

Bhatta, M., Baenziger, P.S., Waters, B.M. et al. (2018). Genome-wide association study reveals novel genomic regions associated with 10 grain minerals in synthetic hexaploid wheat. *International Journal of Molecular Sciences* 19 (10): 3237.

Bollinedi, H., Yadav, A.K., Vinod, K.K. et al. (2020). Genome-wide association study reveals novel marker-trait associations (MTAs) governing the localization of Fe and Zn in the rice grain. *Frontiers in genetics* 11: 497543.

Bradbury, P.J., Zhang, Z., Kroon, D.E. et al. (2007). TASSEL: Software for association mapping of complex traits in diverse samples. *Bioinformatics* 23 (19): 2633–2635.

Breseghello, F. and Sorrells, M.E. (2006). Association analysis as a strategy for improvement of quantitative traits in plants. *Crop Science* 46 (3): 1323–1330.

Breseghello, F. and Sorrells, M.E. (2006). Association mapping of kernel size and milling quality in wheat (*Triticum aestivum L.*) cultivars. *Genetics* 172 (2): 1165–1177.

Bukomarhe, C.B., Kimwemwe, P.K., Githiri, S.M. et al. (2023). Association mapping of candidate genes associated with iron and zinc content in rice (*Oryza sativa L.*) grains. *Genes* 14 (9): 1815.

Bush, W.S. and Moore, J.H. (2012). Chapter 11: Genome-wide association studies. *PLoS Computational Biology* 8 (12): e1002822.

Chandra, A.K., Pandey, D., Sood, S. et al. (2024). Uncovering the genomic regions underlying grain iron and zinc content using genome-wide association mapping in finger millet. *3 Biotech* 14: 47. doi: 10.1007/s13205-023-03889-1.

Chattopadhyay, K., Bagchi, T.B., Sanghamitra, P. et al. (2023). Mapping genetic determinants for grain physicochemical and nutritional traits in brown and pigmented rice using genome-wide association analysis. *Euphytica* 219 (5): 57.

Cu, S.T., Warnock, N.I. and Pasuquin, J. (2021). A high-resolution genome-wide association study of the grain ionome and agronomic traits in rice *Oryza sativa subsp. indica*. *Scientific Reports* 11 (1): 19230.

da Silva, E.C., Nogueira, R.J.M.C., da Silva, M.A. et al. (2011). Drought stress and plant nutrition. *Plant Stress* 5 (1): 32–41.

Dable-Tupas, G., Talampas-Abundo, M.D., Abundo, I.C.S. et al. (2023). Nutrigenomics in the management and prevention of malnutrition, stunting, and other nutritional disorders. In: *Role of Nutrigenomics in Modern-day Healthcare and Drug Discovery,* 147–175. Amsterdam, The Netherlands: Elsevier. doi: 10.1016/B978-0-12-824412-8.00005-9.

Delfini, J., Moda-Cirino, V., dos Santos Neto, J. et al. (2021). Genome-wide association study for grain mineral content in a Brazilian common bean diversity panel. *Theoretical and Applied Genetics* 134: 2795–2811.

Desaint, H., Hereil, A., and Causse, M. (2023). Genome-wide association study: A powerful approach to map QTLs in crop plants. In: *Advanced Crop Improvement 1: Theory and Practice*, 379–455. Cham: Springer International Publishing.

Descalsota, G.I.L., Swamy, B.M., Zaw, H. et al. (2018). Genome-wide association mapping in a rice MAGIC plus population detects QTLs and genes useful for biofortification. *Frontiers in Plant Science* 9: 1347.

Devlin, B. and Roeder, K. (1999). Genomic control for association studies. *Biometrics* 55 (4): 997–1004.

Dissanayaka D.N., Gali K.K., Jha A.B. et al. (2020). Genome-wide association study to identify single nucleotide polymorphisms associated with Fe, Zn and Se concentration in field pea. *Crop Science* 60: 2070–2084. doi: 10.1002/csc2.20161.

Dudbridge, F. (2013). Power and predictive accuracy of polygenic risk scores. *PLoS Genetics* 9 (3): e1003348.

Fayaz, H., Tyagi, S., Wani, A.A. et al. (2022). Genome-wide association analysis to delineate high-quality SNPs for seed micronutrient density in chickpea (*Cicer arietinum* L.). *Scientific Reports* 12 (1): 11357.

Flint-Garcia, S.A., Thornsberry, J.M., and Buckler I.V.E.S. (2003). Structure of linkage disequilibrium in plants. *Annual Review of Plant Biology* 54 (1): 357–374.

Gao, X. and Starmer, J.D. (2008). AWclust: Point-and-click software for non-parametric population structure analysis. *BMC Bioinformatics* 9: 77. doi: 10.1186/1471-2105-9-77.

Gelaw, Y.M., Eleblu, J.S.Y., Ofori, K. et al. (2023). Genome-wide association study of grain iron and zinc concentration in common bean (*Phaseolus vulgaris*). *Plant Breeding* 142 (3): 357–371. doi: 10.1111/pbr.13090.

Gunjača J, Carović-Stanko K, Lazarević B. et al. (2021). Genome-wide association studies of mineral content in common bean. *Frontiers of plant science* 12: 636484. doi: https://pubmed.ncbi.nlm.nih.gov/33763096/.

Hall, D., Tegström, C. and Ingvarsson, P.K. (March 2010). Using association mapping to dissect the genetic basis of complex traits in plants. *Briefings in Functional Genomics* 9 (2): 157–165. doi: 10.1093/bfgp/elp048.

Hartl, D.L. and Orel, V. (1992). What did Gregor Mendel think he discovered? *Genetics* 131 (2): 245.

Hasan, N., Choudhary, S., Naaz, N. et al. (2021). Recent advancements in molecular marker-assisted selection and applications in plant breeding programmes. *Journal, Genetic Engineering & Biotechnology* 19 (1): 128. doi: 10.1186/s43141-021-00231-1.

Hindu, V., Palacios-Rojas, N., Babu, R. et al. (2018). Identification and validation of genomic regions influencing kernel zinc and iron in maize. *Theoretical and Applied Genetics* 131: 1443–1457.

Huang, M., Liu, X., Zhou, Y. et al. (2019). BLINK: A package for the next level of genome-wide association studies with both individuals and markers in the millions. *GigaScience* 8 (2): giy154.

Huang, X., Zhao, Y., Wei, X. et al. (2011). Genome-wide association study of flowering time and grain yield traits in a worldwide collection of rice germplasm. *Nature Genetics* 44 (1): 32–39. doi: 10.1038/ng.1018.

Hughes, A.R., Inouye, B.D., Johnson, M.T. et al. (2008). Ecological consequences of genetic diversity. *Ecology Letters* 11 (6): 609–623.

Islam, A.F., Mustahsan, W., Tabien, R. et al. (2022). Identifying the genetic basis of mineral elements in rice grain using genome-wide association mapping. *Genes* 13 (12): 2330.

Jaiswal, V., Bandyopadhyay, T., Singh, R.K. et al. (2024). Multi-environment GWAS identifies genomic regions underlying grain nutrient traits in foxtail millet (*Setaria italica*). *Plant Cell Reports* 43 (1): 1–13.

Jaiswal, V., Gupta, S., Gahlaut, V. et al. (2019). Genome-wide association study of major agronomic traits in foxtail millet (*Setaria italica L.*) using ddRAD sequencing. *Scientific Reports* 9 (1): 5020.

Jovovic, Z., Andjelkovic, V., Przulj, N. et al. (2020). Untapped genetic diversity of wild relatives for crop improvement. *Rediscovery of Genetic and Genomic Resources for Future Food Security* 25–65.

Kang, H.M., Zaitlen, N.A., Wade, C.M. et al. (2008). Efficient control of population structure in model organism association mapping. *Genetics* 178 (3): 1709–1723.

Kaur, H., Sharma, P., Kumar, J. et al. (2023). Genetic analysis of iron, zinc and grain yield in wheat – *Aegilops* derivatives using multi-locus GWAS. *Molecular Biology Reports* 50: 9191–9202. doi: 10.1007/s11033-023-08800-y.

Kiran, K. (2020). Advanced approaches for biofortification. *Advances in Agri-Food Biotechnology* 29–55.

Krishnappa, G., Khan, H., Krishna, H. et al. (2022). Genetic dissection of grain iron and zinc, and thousand kernel weight in wheat (*Triticum aestivum L.*) using genome-wide association study. *Scientific Reports* 12 (1): 12444.

Kumar, H., Singh, A., Dikshit, H.K. et al. (2019). Genetic dissection of grain iron and zinc concentrations in lentil (*Lens culinaris Medik.*). *Journal of Genetics* 98 (66). doi: 10.1007/s12041-019-1112-3.

Lippert, C., Listgarten, J., Liu, Y. et al. (2011). FaST linear mixed models for genome-wide association studies. *Nature Methods* 8 (10): 833–835.

Liu, C.G. and Zhang, G.Q. (2006). Single nucleotide polymorphism (SNP) and its application in rice. *Yi Chuan= Hereditas* 28 (6): 737–744.

Liu, J., Huang, L., Li, T. et al. (2021). Genome-wide association study for grain micronutrient concentrations in wheat advanced lines derived from wild emmer. *Frontiers in Plant Science* 12: 651283.

Liu, X., Huang, M., Fan, B. et al. (2016). Iterative usage of fixed and random effect models for powerful and efficient genome-wide association studies. *PLoS Genetics* 12 (2): e1005767.

Mamo, B.E., Barber, B.L., and Steffenson, B.J. (2014). Genome-wide association mapping of zinc and iron concentration in barley landraces from Ethiopia and Eritrea. *Journal of Cereal Science* 60 (3): 497–506.

McMahon, A., Lewis, E., Buniello, A. et al. (2021). Sequencing-based genome-wide association studies reporting standards. *Cell Genomics* 1(1): 100005.

Modena, B.D., Doroudchi, A., Patel, P. et al. (2019). Leveraging genomics to uncover the genetic, environmental and age-related factors leading to asthma. In: *Genomic and Precision Medicine,* 331–381. Cambridge: Academic Press.

Morris, G., Ramu, P., Deshpande, S. et al. (2012). Population genomic and genome-wide association studies of agroclimatic traits in Sorghum. *Proceedings of the National Academy of Sciences* 110 (2): 453–458. doi: 10.1073/pnas.1215985110.

Nawaz, Z., Kakar, K.U., Li, X.B. et al. (2015). Genome-wide association mapping of quantitative trait loci (QTLs) for contents of eight elements in brown rice (*Oryza sativa L.*). *Journal of Agricultural and Food Chemistry* 63 (36): 8008–8016.

Novembre, J. and Stephens, M. (2008). Interpreting principal component analyses of spatial population genetic variation. *Nature Genetics* 40 (5): 646–649.

Nyiraguhirwa, S., Grana, Z., Ouabbou, H. et al. (2022). A genome-wide association study identifying single-nucleotide polymorphisms for iron and zinc biofortification in a worldwide barley collection. *Plants* 11 (10): 1349.

Paltridge, N.G., Milham, P.J., Ortiz-Monasterio, J.I. et al. (2012). Energy-dispersive X-ray fluorescence spectrometry as a tool for zinc, iron and selenium analysis in whole grain wheat. *Plant and Soil* 361: 261–269.

Pasion, E.A., Misra, G., Kohli, A. et al. (2023). Unraveling the genetics underlying micronutrient signatures of diversity panel present in brown rice through genome – ionome linkages. *The Plant Journal* 113 (4): 749–771.

Pinheiro, J.C. and Bates, D.M. (2000). Linear mixed-effects models: Basic concepts and examples. *Mixed-Effects Models in S* and *S-Plus* 3–56.

Price, A.L., Patterson, N.J., Plenge, R.M. et al. (2006). Principal components analysis corrects for stratification in genome-wide association studies. *Nature Genetics* 38 (8): 904–909.

Pritchard, J.K., Stephens, M., and Donnelly, P. (2000). Inference of population structure using multilocus genotype data. *Genetics* 155 (2): 945–959.

Pujar, M., Gangaprasad, S., Govindaraj, M. et al. (2020). Genome-wide association study uncovers genomic regions associated with grain iron, zinc and protein content in pearl millet. *Scientific Reports* 10: 19473. doi: 10.1038/s41598-020-76230-y.

Puranik, S., Sahu, P.P., Beynon, S. et al. (2020). Genome-wide association mapping and comparative genomics identifies genomic regions governing grain nutritional traits in finger millet (*Eleusine coracana L. Gaertn.*). *Plants, People, Planet* 2 (6): 649–662.

Rathan, N.D., Krishna, H., Ellur, R.K. et al. (2022). Genome-wide association study identifies loci and candidate genes for grain micronutrients and quality traits in wheat (*Triticum aestivum L.*). *Scientific Reports* 12 (1): 7037. doi: 10.1038/s41598-022-10618-w.

Rosenberg, N.A., Pritchard, J.K., Weber, J.L. et al. (2002). Genetic structure of human populations. *Science* 298 (5602): 2381–2385.

Sahu, P.K., Mondal, S., Sao, R. et al. (2020). Genome-wide association mapping revealed numerous novel genomic loci for grain nutritional and yield-related traits in rice (*Oryza sativa* L.) landraces. *3 Biotech* 10: 1–22.

Singh, A., Sharma, V., Dikshit, H.K. et al. (2017). Association mapping unveils favorable alleles for grain iron and zinc concentrations in lentil (*Lens culinaris subsp. culinaris*). *PLoS One* 12 (11): e0188296.

Singh, B., Singh, S., Mahato, A.K. et al. (2023). Delineation of novel genomic loci and putative candidate genes associated with seed iron and zinc content in lentil (*Lens culinaris* Medik.). *Plant Science* 335: 111787.

Singh, U., Praharaj, C.S., Chaturvedi, S.K. et al. (2016). Biofortification: Introduction, approaches, limitations, and challenges, (eds. U. Singh, C. Praharaj, S. Singh, N. Singh). In: *Biofortification of Food Crops*. New Delhi: Springer. doi: 10.1007/978-81-32 2-2716-8_1.

Sinha M.K., Aski M.S., Mishra G.P. et al. (2023 Feb 7). Genome wide association analysis for grain micronutrients and anti-nutritional traits in mungbean [*Vigna radiata* (L.) R.Wilczek] using SNP markers. *Frontiers in Nutrition* 10: 1099004. doi: 10.3389/fnut.2023.1099004. PMID: 36824166; PMCID: PMC9941709.

Srungarapu, R., Mahendrakar, M.D., Mohammad, L.A. et al (2022). Genome-wide association analysis reveals trait-linked markers for grain nutrient and agronomic traits in diverse set of chickpea germplasm. *Cells* 11 (15): 2457.

Stangoulis, J. and Knez, M. (2022). Biofortification of major crop plants with iron and zinc – achievements and future directions. *Plant and Soil* 474 (1–2): 57–76. doi: 10.1007/s11104-022-05330-7.

Thabet, S.G., Alomari, D.Z., Brinch-Pedersen, H. et al. (2022). Genetic analysis toward more nutritious barley grains for a food secure world. *Botanical Studies* 63 (1): 6.

Tibbs Cortes, L., Zhang, Z., and Yu, J. (2021). Status and prospects of genome-wide association studies in plants. *The Plant Genome* 14 (1): e20077.

Tong, J., Zhao, C., Sun, M., et al. (2022). High resolution genome wide association studies reveal rich genetic architectures of grain zinc and iron in common wheat (*Triticum aestivum* L.). *Frontiers in Plant Science* 13: 840614.

Upadhyaya, H.D., Bajaj, D., Das, S. et al. (2016). Genetic dissection of seed-iron and zinc concentrations in chickpea. *Scientific Reports* 6 (1): 24050.

Upadhyaya, H.D., Dwivedi, S.L., Sharma, S. et al. (2014). Enhancement of the use and impact of germplasm in crop improvement. *Plant Genetic Resources* 12 (S1): S155–S159. https://doi:10.1017/S1479262114000458.

Wall, J. and Pritchard, J. (2003). Haplotype blocks and linkage disequilibrium in the human genome. *Nature Reviews Genetics* 4: 587–597. doi: 10.1038/nrg1123.

Wang, W., Guo, H., Wu, C. et al. (2021). Identification of novel genomic regions associated with nine mineral elements in Chinese winter wheat grain. *BMC Plant Biology* 21 (1): 311.

Wolak, M.E. (2012). Nadiv: An R package to create relatedness matrices for estimating non-additive genetic variances in animal models. *Methods in Ecology and Evolution* 3: 792–796.

Wu, X., Islam, A.F., Limpot, N. et al. (2020). Genome-wide SNP identification and association mapping for seed mineral concentration in mung bean (*Vigna radiata* L.). *Frontiers in Genetics* 11: 656.

Yang, M., Lu, K., Zhao, F.J. et al. (2018). Genome-wide association studies reveal the genetic basis of ionomic variation in rice. *The Plant Cell* 30 (11): 2720–2740.

Zhang, C., Dong, S.S., Xu. J.Y. et al. (2019). PopLDdecay: A fast and effective tool for linkage disequilibrium decay analysis based on variant call format files. *Bioinformatics* 35 (10): 1786–1788. doi: 10.1093/bioinformatics/bty875.

Zhong, K., Xie, L., Hu, S. et al. (2022). Genome-Wide Association Study of Zinc Toxicity Tolerance within a Rice Core Collection (*Oryza sativa* L.). *Plants* 11 (22): 3138.

Zhou, X. and Stephens, M. (2012). Genome-wide efficient mixed-model analysis for association studies. *Nature Genetics* 44 (7): 821–824.

Zhu, C., Gore, M., Buckler, E.S. et al. (2008). Status and prospects of association mapping in plants. *The Plant Genome* 1 (1). doi: 10.3835/plantgenome2008.02.0089.

25

Modification of Conventional Methods and Modern Plant Breeding Techniques to Enhance Genetic Gain for Future Food Security

Saira Sattar[1], *Razia Sultana*[2], *Hiddayatullah*[1], *Ikram Ullah*[3*]

[1] *Horticultural Research Institute, National Agricultural Research Center (NARC) Park Road, Islamabad, Pakistan*
[2] *Wheat Research Institute, Ayub Agricultural Research Institute, Faisalabad, Pakistan*
[3] *College of Horticulture, Northwest A&F University, Yangling 712100, China*

*Corresponding author: ikram@nwafu.edu.cn

25.1 Objective of Study

New and improved crop varieties can be developed through innovative plant breeding techniques and by their modifications with time to acquire our objectives within a short period to cope with world hunger. The objective of this chapter is to know about the latest scientific work and modifications of previous techniques to increase crop productions.

25.2 Introduction

Dated back to 10 000 BCE, ancient human civilizations depended on bygone methods of hunting and gathering for their need of food. Gradually, with the passage of time, primordial humans started the domestication of crops like wheat, maize, and potatoes (Agronomy and Agriculture 2021). In the next 50 years, agricultural production is supposed to double up to meet the requirement of 9 billion people (Food and Agriculture Organization [FAO] 2009, Searchinger et al. 2014). According to FAO, in 2023, the global food bill will increase to US$1.98 trillion, which is 1.5% higher as compared to 2022.

World cereals requirements need to be increased by 38–67% because of the rising population (Ricroch et al. 2014). For plant ideotype, the integration of plant genetics is required for the selection of desirable traits as well as novel techniques (tissue culture, genomic biotechnology, epigenetics, next-generation sequencing (NGS), genome editing, and nanobiotechnology) to enhance the efficiency of selection and breeding processes. Plant breeding techniques have been evolving with the passage of time to shorten the plant life cycle (Ahmar et al. 2020a). Modification in conventional breeding will shorten the longer process of varietal development which usually takes more than 10 years (Bharti and Chimata 2019).

Conventional breeding is a prerequisite for the creation of variation, but it relies on phenotypic expressions that are affected by environmental fluctuations (Lema 2018). Hence, genomic selection is more reliable and fruitful to get maximum genetic gain of cultivar within less time; however, it requires diverse technical skills to operate marker-assisted selection, CRISPR-Cas9, phenomics, etc. (Wang et al. 2018; Wang 2022, Crossa et al. 2017). With the aid of modern breeding technologies, crop breeders and researchers try to explore more diverse characters that have high adaptability to distant locations as well as diverse climatic changes (Ewing et al. 2019; Wang 2022).

25.2.1 Conventional Breeding Techniques and Their Modifications

Since agricultural practices started about 10 000 years ago, farmers have been domesticating different crops, distinguishing the best traits from the wild relatives of crop populations, landraces, and improved species. Earlier, improvement in domesticated plants was due to panoramic vital forces of nature that interact along with farmers' endeavors to modify the economic traits (Sweeney and McCouch 2007) (Figure 25.1). With the passage of time, evolution in morphological and physiological functions of traditional crop plants eventually led to the dramatic change in some crop species that laid the foundation of

Crop Biofortification: Biotechnological Approaches for Achieving Nutritional Security under Changing Climate, First Edition. Edited by Adnan Noor Shah et al.

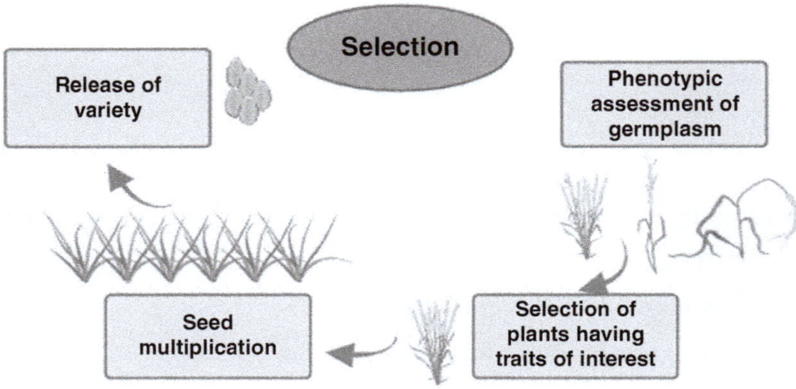

Figure 25.1 Conventional breeding methods for crop improvement.

current cultivars, and along with advanced technology and crop engineering, crop breeders upgraded yield with minimal impact on the environment; scientists referred it as "Super Domestication" (Vaughan, Balázs, and Heslop-Harrison 2007).

Variations in the genetic architecture of species not only enhance crop germplasm but also aid in providing well and widely characterized genetic resources to cumulate superior characters through suitable breeding strategies for economic purposes (Acquaah 2007). Conscious selection by experts/breeders as well as autonomic environmental variations play an imperative role in developing competitive and highly productive crops. Selection of desirable plants could be done effectively from a heterogenous gene pool to maximize the genetic gain. Earlier, farmers used to select individuals on the basis of morphological parameters, but with the help of other indispensable tools like DNA markers, DNA fingerprinting, and enzyme-linked immunosorbent assay (ELISA) techniques, selection has been made more efficient to develop precision plants (Joshi 2017).

Hybridization is a practice used to create heritable genetic variations by mating between genetically dissimilar individuals. Hybridization is a prerequisite to converge valuable alleles, break unsuitable linages, and diversify germplasm (Mwangangi, Kiilu Muli, and Neondo 2019). Hybridization can occur naturally in self-pollinated as well as cross-pollinated crops depending on the mode of pollination (maize = wind pollinated, sunflower and cotton = insect pollinated, wheat and barley = self-pollinated). To get novel characters, mating can be done between individuals of unrelated species of the same genera (interspecific hybridization) and between individuals of the same species (intraspecific hybridization) (Figure 25.2).

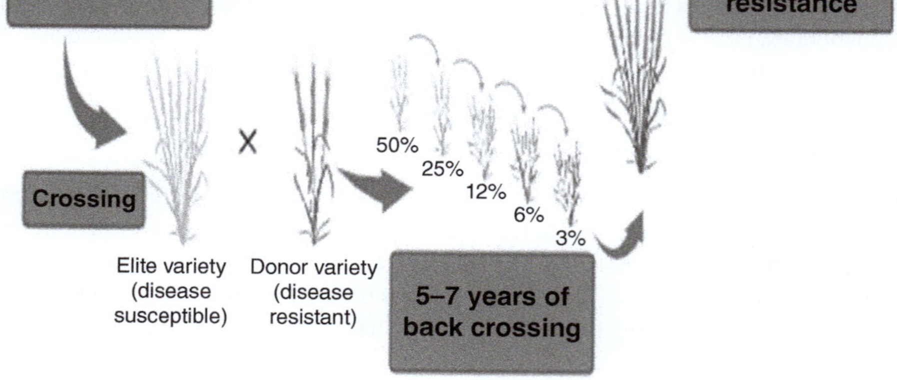

Figure 25.2 The process of hybridization for crop improvement.

The first generation produced by mating the contrasting genotypes is called hybrid/F1 generation. Hybrid breeding was first started in maize in 1909 by G. H. Shull (1908); he proposed the term "heterosis" to find out the vigor of hybrid as well as the expression of heterozygosis. In cross-pollinated species, inbred lines can be combined in many ways to develop single cross, double cross, and triple cross.

Ideotype breeding has gained emphasis because of changing climate patterns; therefore, a conceptual model plant needs to be developed to ensure wider adaptability and accumulation of desirable alleles (Arnaud 2018). Donald (1967) first proposed the term "Ideotype" to define ideotype (model plant development) in wheat. There are multifarious aspects of model plant development depending upon breeding goals, i.e. the term "Isolation Ideotype" is often used in cereals when ideotype plants are required to enhance the quality and yield of oil- and grain-related parameters. "Competitive Ideotype" is mostly used for annual plants and heterogenous populations, and the parameters accentuated in this ideotype include the speed of germination, leaf canopy, various characteristics of roots, seed size, and growing habits of the plant. In "Market Ideotype," the focus is to enhance economic market traits like baking quality, cooking quality, and color of grain. "Climate Ideotype" is used to develop model plants to adapt best under changing climatic patterns, i.e. smog, drought, and heat stresses. However, there are some drawbacks; this is a slow process and due to linkage, it is difficult to accumulate all desirable traits in our ideotype plant (Humera et al. 2018) (Figure 25.3).

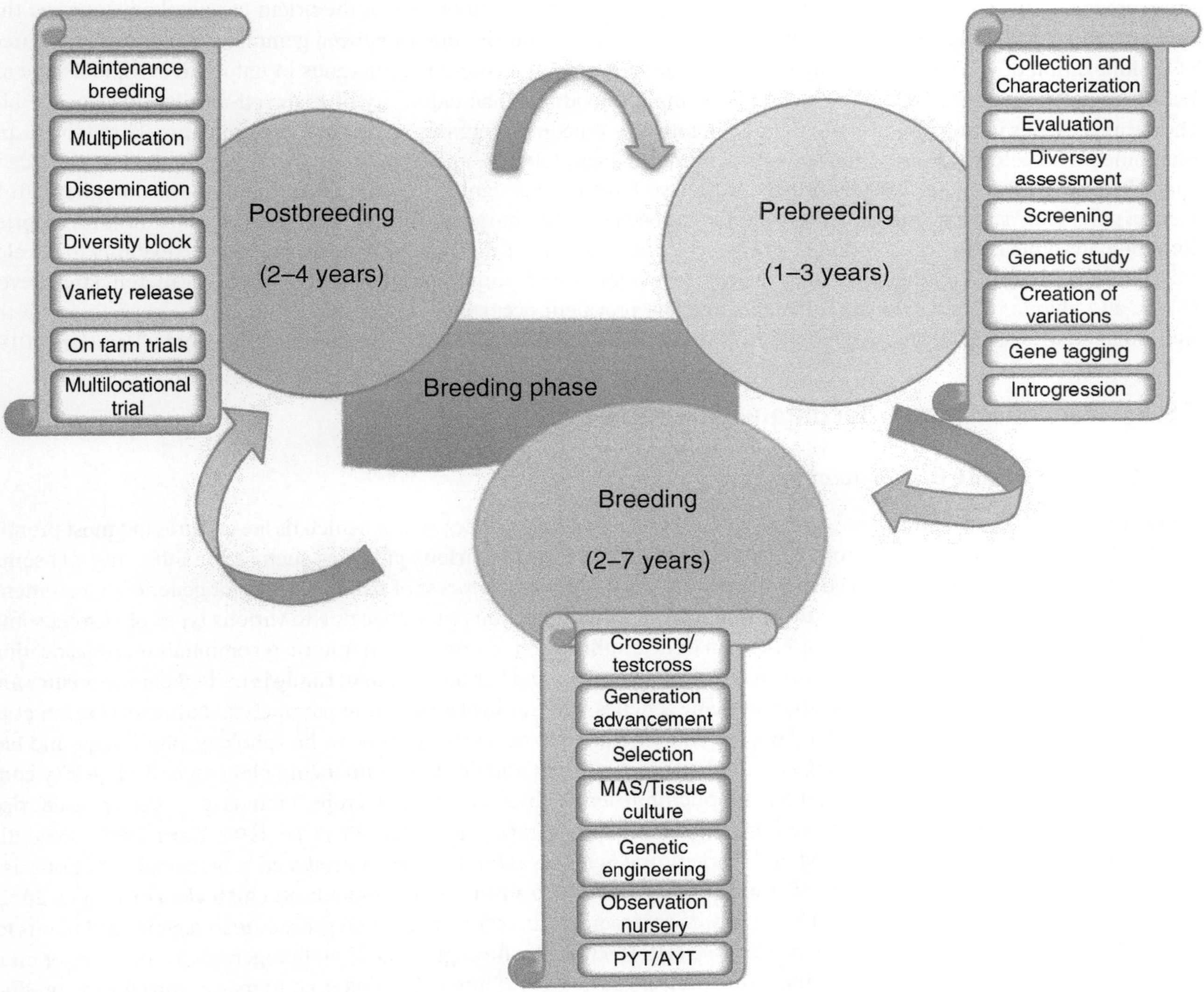

Figure 25.3 Post and pre-breeding methods for crop improvement.

25.2.2 Mutation Breeding and Crop Improvement

Mutation is one of the heritable but slow processes to create genetic variability. Cosmic rays, X-rays, gamma rays, and fast neutrons are ionizing elements of the electromagnetic spectrum that are capable of inducing variations by altering the genetic architecture of the specimen. Variations induced through mutation can either be beneficial or deleterious; variations may change the nature of chromosomes by altering their polyploidy level and induce structural changes in chromosomes, i.e. translocation, insertion, and duplication (Jiang and Ramachandran 2010). Another reversal technique "TILLING (Targeting-induced local lesions in the genome)" helps to track down the point mutation on the target gene. The technique was first proffered in *Arabidopsis thaliana* in 2000 and used to identify the induced mutations in the mutagenized population. "Eco-tilling" is the modified form of tilling that can also determine the natural mutation (Till, Comai, and Henikoff 2007). Gel analyzer "L1-COR" is used to determine the mutation and Coddle (web program) helps to find out functions of mutant genes and their functions in mutagenized populations.

25.2.3 Polyploidy and Crop Improvements

In plants, the phenomenon of polyploidy plays a significant role in evolution, and ancient duplication in *Zea mays* (Kuwada's hypothesis) also directs attention toward plant polyploidy. The whole-genome duplication can occur either through autopolyploid or allopolyploid. According to the evolutionary timeline of plant species, ancient polyploids are categorized as paleopolyploid, mid-ancient evolved polyploids in the historic time are classified as "mesopolyploid," and those emanated in recent years are known as "Neopolyploids". Many studies about the origin of polyploids reported that mitotic nondisjunction, polyspermy, doubling of somatic cells in mitosis, and unreduced gemmates in meiosis are sources of diploidization of chromosomes. Polyploid plants are reported to be more heterozygous in nature than diploids and are hence considered more adaptive and good performing individuals. The capability of high self-fertility and flexible biochemical activities in polyploids made them preferable for breeding programs in stressed environments. Through natural phenomenon of allopolyploid, achievements in cotton, wheat, tobacco, and brassica crops have been made. Triploids in sugar beet and watermelon and tetraploids in the rye have gained significant importance (Simon and Jonathan 2014). Hence, polyploidy is the pivotal pathway to generate speciation and burgeoning biodiversity in "Plantae." Studies reported that the polysomic nature of polyploids occurs recurrently in most of the plant species that have the capability to develop inimitable gene regulation that diversifies the expression levels and strengthens the plant to adapt to different stress levels (biotic and abiotic). Despite all these benefits and the prevalent occurrence of polyploids, there is a need to revise the molecular studies using advanced genomic tools to understand the regulation and processes of polyploids (Van Hieu 2019).

25.3 Nonconventional Techniques in Crop Development

25.3.1 Tissue Culture Biotechnology

Biotechnology has provided various technologies to assist breeding practices among which tissue cultures are most promising, enabling breeders to exploit crop plants in in-vitro conditions for various purposes such as the following: (1) somaclonal variations (variations produced in regenerated plants during the process of tissue culture) for genetic improvements (Larkin and Scowcroft 1981). Tissue culture–induced variability in plants may arise due to various types of chromosomal and DNA rearrangements like gene amplification or deamplification, nonreciprocal mitotic recombination, translocation, point mutations, activation of silent genes in multigene families, and variation in maternally inherited characteristics and in epigenome creating additional variation in plant genetics without hybridization in commercial cultivars (Larkin et al. 1985; Scowcroft et al. 1987; Karp 1994). These variations lead to desirable traits in plant morphology, physiology, and biochemical processes, enabling plant resistance to various biotic and abiotic stress, enhancing plant vigor and quality compounds like essential oils, macronutrients, and micronutrients, observed in many crops, including wheat, triticale, rice, oats, maize, sugarcane, alfalfa, tobacco, tomato, potato, oilseed rape, and celery (Thorpe 1990; Karp 1994). Recently, University of Florida's Institute of Food and Agricultural Sciences (UF/IFAS) has introduced somaclonal variations as a breeding method for the improvement of strawberries' resilience to adverse environments and high yield (Yoo et al. 2022). However, this technology has produced few cultivars as it requires an efficient selection procedure in regenerated plants for desired traits. (2) Haploids (Hs) and double-haploid (DH) technology through gametic embryogenesis for the development of a completely homozygous line in a single step (Maria 2011). DH technique is being used in many crops for the production of DH inbred lines in winter rapeseed, and it is estimated that 50% of the contemporary barley breeding program involves DH breeding method in Europe (Mollers and Iqbal 2009; Dunwell 2010). Studies showed that heritability and estimates of genetic variations are higher in DH-produced inbred lines as compared to recombinant inbred lines (Lemos et al. 2023). Moreover, DH technique is also practiced for gene expression of recessive alleles in a homozygous state that may remain unexpressed in a disomic state and also for enhancing expression of genes in multigenic traits or production of seedless fruits as in triple haploids (Humphreys et al. 2007; Germanà 2006, 2009). Previously, haploids were produced via culturing of anther and pollen, a latest advancement has been occurred in which mutation in the genes encoding egg-cell-specific aspartic endopeptidases (ECSs) induces haploid plants (Zhang et al. 2023). (3) Micropropagation, for regeneration of plantlets from ex-plant or callus on a commercial scale for the production of true-to-type in asexually propagated crops (Debergh and Zimmerman 1991). Despite of massive multiplication of plants, micropropagation technology is extensively used for the creation of somaclonal variations and for the conservation of specific genotypes (Bautista-Montes, Hernández-Soriano, and Simpson 2022). As in micropropagation, thousands of plants are produced under controlled conditions and their harvesting, processing, and marketing can be monetarized using artificial intelligence (AI) technology (Malabadi et al. 2023). (4) Protoplast fusion – production of hybrid plants through protoplast fusion that cannot be produced through conventional breeding (Feher and Dudits 1994). When the protoplast of two different species are fused together through physical or chemicals means, only few of hybrid of protoplasts regenerate or produce viable seed. Potential example of crop improvement through protoplast fusion was practiced in *Nicotiana*, in which various disease-resistant traits and alkaloid compounds were modified (Pandeya et al. 1986; Evans and Bravo 1988). To overcome the challenges associated with protoplast fusion and protoplast-based genetic improvements in crop plants, a new method has been developed. The method involves efficient transfection of potato protoplasts using linearized DNA and utilizing UBIQUITIN10 promoters from various plants for effective transgene expression (Rather et al. 2022) (Figure 25.4).

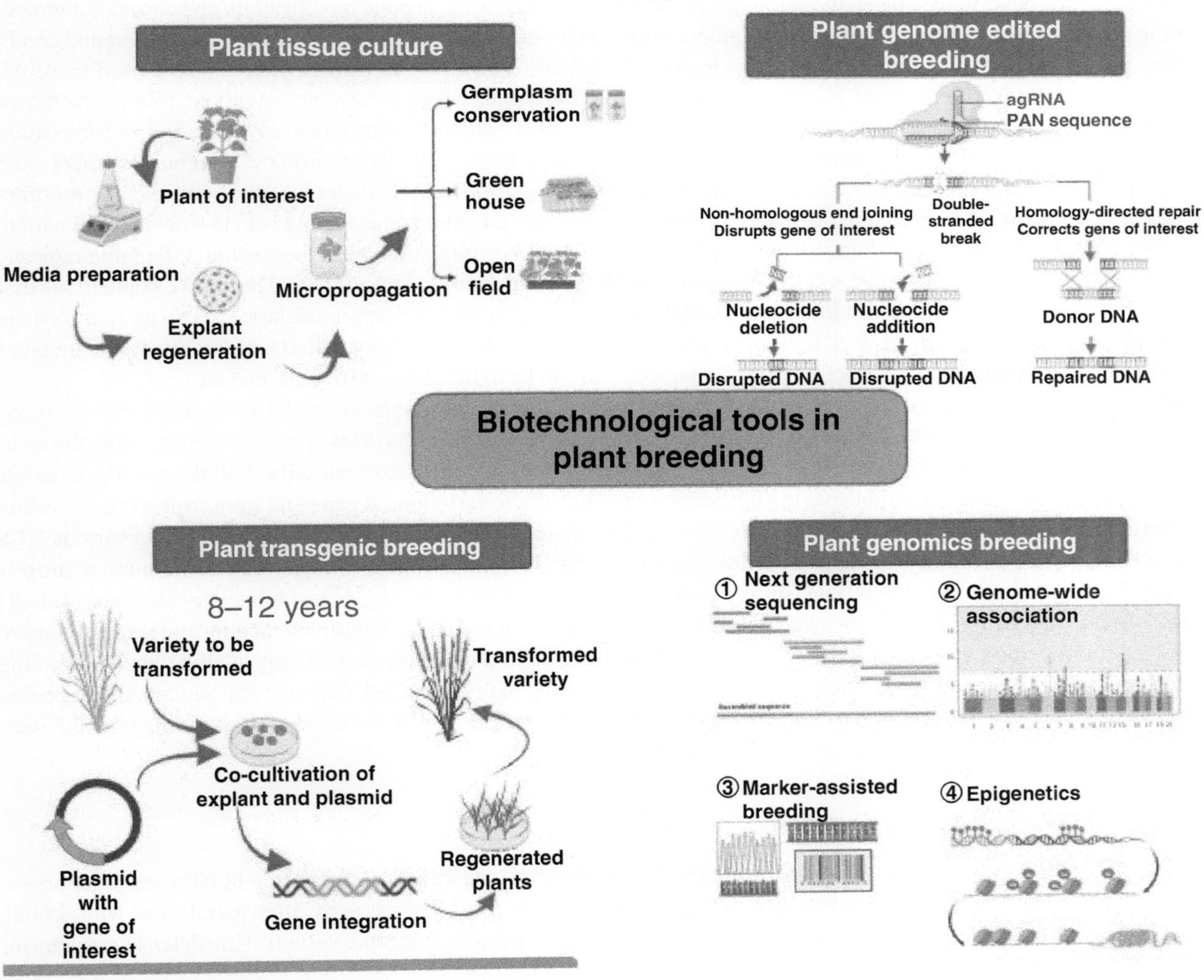

Figure 25.4 Biotechnological tools for plant breeding

25.3.2 Plant Genomic Breeding

Marker-assisted technologies have evolved in the past few years from morphological, physiological, and biochemical to DNA-based markers. Through these techniques, we can overcome the problems of environmental influence on the morphology in conventional breeding techniques. Recent advancements in this field are genome-wide association study (GWAS), QTL mapping, and analysis of genetic diversity, germplasm characterization, and haplotype determination by using various types of markers such as restriction fragment length polymorphism (RFLP), amplified fragment length polymorphism (AFLP), simple sequence repeats (SSRs), and single-nucleotide polymorphism (SNP). Molecular markers assist in the determination of changes in nucleotide (duplication, inversion, transition, and transversion).

1. **Genomic selection (GS)**: It is a kind of marker-assisted selection in which selection markers are distributed over the whole genome of desired genotypes using SNPs as a selected marker that covers all the QTL of the targeted traits (Zargaret al. 2015). SNPs are the most efficient markers being biallelic, but become more expensive as they require genotyping using various NGS technologies (Metzker 2010; Varshney et al. 2009). The other SNP genotyping technologies like genotyping by sequencing (GBS; Elshire et al. 2011; Sonah et al. 2013) can recognize and genotype the SNP markers at the same time or the identified markers help to detect more markers to genotyping assay (Davey et al. 2011). GS is seen as a key component of the variety of developmental procedures, capable of being integrated into various stages and providing predictive power within and across breeding cycles for selection of diverse traits in hybrid development and identification of traits that can play a key role in resilience of crops against adverse environmental conditions (Merrick et al. 2022).
2. **Epigenomics:** Both direct (genomic) and indirect (epigenomic) variation contributes to plant adaption under different environmental conditions. Epigenetic variations like DNA (de)methylation, histone modification, and chromatin remodeling produced by reprograming the transcriptome during plant development and in response to environmental conditions provide sources of variation for breeders to exploit (Gallusci et al. 2017). DNA methylation would need to exhibit substantial natural variation and be able to influence important plant traits to have an important role in traditional approaches to crop improvement. Scans of diverse accessions of *A. thaliana*, maize, rice, soybean, and *Brachypodium distachyon* have revealed substantial levels of natural variation of DNA methylation profiles (Vaughn, Tanurdzić, and Lippman 2007; Garg et al. 2015). Researchers have discovered several potential epialleles that can provide major morphological variation. It is observed that the association of epialleles with the phenotype of plant is a slow process and hard to understand. However, the development and study of epiRILs (epi recombinant inbred lines) using *A. thaliana* radically changed the rate at which epialleles could be created and linked with morphological traits. The epiRIL populations that have been generated exhibit broad phenotypic variation, ranging from altered disease resistance to variation in biomass (Reinders et al. 2009; Johannes et al. 2009). The science of epigenome is like an iceberg that requires the development of new methods to understand indirect variations produced and can be induced for the trait of interest.
3. **Next-generation Sequencing:** In recent years, genomic-assisted breeding approaches have become more advanced due to the availability of whole genome and transcriptome data of several crops (Varshney et al. 2009). Using the technologies of NGS, it is possible to sequence and resequence genomic and transcriptomic data of all the genetic materials present in a crop species. Thereby making it possible to predict all the variations at gene and transcript levels providing information to exploit gene regulation, expression, and biological circuits with great precision. Among various NGS technologies, commercially available technologies are Roche/454 (http://www.454.com/), Solexa/Illumina (http://www.illumina.com/), and AB SOLiD (http://www3.appliedbiosystems.com/AB_Home/applicationstechnologies/SOLiDSystemSequencing/index.htm) rather than the Sanger sequencing which is laborious, robotic, and space-intensive for larger operation (Smailus et al. 2005). Moreover, NGS technologies facilitate the breeding of crop plants by making marker-assisted selection more efficient through linkage mapping in the marker-deficient crops. Sequencing of pooled amplicons effectuated for numerous candidate genes of the entire population gives more understanding of population biology and to study genome-wide association genetics (Ronaghi et al. 1996).

25.3.3 Plant Genome Editing Breeding

Crop improvement through changing the genetic architecture of the plant according to desired trait by editing the genome is becoming revolutionized after the discovery of a mutagenesis technique, clustered regularly interspaced short palindromic repeat (CRISPR)/CRISPR-Cas9 system naturally present in bacteria s adaptive immune system. This defense mechanism is like RNA interference which involves the identification and breakdown of foreign DNA. Genome editing modification by CRISPR involves a guide RNA that is complementary to a target gene and brings double-strand breaks using a Cas9 nuclease, followed by joining of end through a homology-directed repair mechanism (Jinek et al. 2012).

25.4 Nanobiotechnology

Nanotechnology is another promising and emerging tool with little efforts and environmental friendly way to overcome potential problems of agriculture (Sugunan and Dutta 2008) via the following:

1. **Crop genetics improvement:** Plants' genetic could be modified through bombardment of NPs having gene of interest or nanomaterials-based targeted delivery of CRISPR mRNA and sgRNA (Ran, Liang, and Gao 2017; Miller et al. 2017; Kim, Gopal, and Sivanesan 2017).
2. **Crop growth improvement:** Several research works on nanoparticles (NPs) with both metal-based nanoparticles (MBNPs) and carbon-based nanoparticles (CBNPs) in combination with irradiation (Al Sherbini et al. 2015) have shown significant role in crop growth and yield improvement via enhancing seed germination percentage (Lu et al. 2002; Khodakovskaya et al. 2009; Nair et al. 2010; Gopinath et al. 2014), root and shoot length, and increased fruit yield (Kole et al. 2013) in wheat, lettuce, soybean, and melon.
3. **Biotic and abiotic stress tolerance:** Crop resistance to various biotic and abiotic stress could be achieved through exposure to NPs (Afsharinejad et al. 2016; Kwak et al. 2017).
4. **Advancing agronomic practices:** Using nanotechnologies input costs can be reduced through the application of engineered nanomaterials (ENMs) by exploiting various agronomic practices reducing the leaching of plant protection chemicals and fertilizer through slow availability of nanopesticides and fertilizers (Chen et al. 2013; Gardea-Torresdey, Rico, and White 2014). Therefore, advancement in nanotechnology techniques in combination with other emerging breeding approaches has significant potential for unsustainable agriculture growth.

25.5 Conclusion

Ancient breeding techniques are useful in crop improvement by introducing desirable characteristics in plants that lead to increases in genetic resources. But due to unpredictable climatic changes, some varieties get vulnerable to different environmental stresses; hence, to cope with the leading environmental issues, modification in traditional techniques, marker-assisted selection, and modern techniques (CRISPER-Cas, nanobiotechnology, mutation breeding, tissue culture, genetic engineering, etc.) are key to get quantity and quality of useful agronomic traits. Hence, emerging techniques are required to enhance traits under different conditions, and modern emerging technologies coupled with modified traditional techniques are crucial to achieve sustainable goals of zero hunger and future food security.

References

Acquaah, G. (2007). *Principles of Plant Genetics and Breeding*. USA: Blackwell Publishing.

Afsharinejad, A., Davy, A., Jennings, B. et al. (2016). Performance analysis of plant monitoring nanosensor networks at THz frequencies. *IEEE Internet of Things Journal* 3 (1): 59–69.

Agronomy, P. and Agriculture, I. (2021). *Important Events in Early History of Agriculture*. agriinfo.in. (accessed 1 October 2021).

Ahmar, S., Gill, R.A., Jung, K.H. et al. (2020a). Conventional and molecular techniques from simple breeding to speed breeding in crop plants: recent advances and future outlook. *International Journal of Molecular Sciences* 21 (7): 2590.

Ahmar, S., Saeed, S., Hafeez, M. et al. (2020b). A revolution toward gene-editing technology and its application to crop a revolution toward gene-editing technology and its application to crop improvement. *International Journal of Molecular Sciences*. 21 (16): 5665.

Al Sherbini, A., Abd El-Gawad, HG., Kamal, MA. et al. (2015). Potential of He-Ne laser irradiation and iron nanoparticles to increase growth and yield of pea. *American-Eurasian Journal Of Agricultural & Environmental Sciences* 15 (7): 1435–1446.

Arnaud, G. (2018). Using ideotypes to support selection and recommendation of varieties. *OCL* 25 (6): 1–9.

Bautista-Montes, E., Hernández-Soriano, L. and Simpson, J. (2022). Advances in the micropropagation and genetic transformation of Agave species. *Plants* 11 (13): 1757.

Bharti, G. and Chimata, M.K. (2019). Review on new plant breeding techniques. *International Journal of Science and Research* 8 (4): 723–730.

Chen, H.C., Roco, M.C., Son, J.B., Jiang, S. et al. (2013). Global nanotechnology development from 1991 to 2012: patents, scientific publications, and effect of NSF funding. *Journal of Nanoparticle Research* 15 (9): 1951.

Crossa, J., Pérez-Rodríguez, P., Cuevas, J. et al. (2017). Genomic selection in plant breeding: methods, models, and perspectives. *Trends in Plant Science* 22 (11): 961–975.

Davey, J.W., Hohenlohe, P.A, Etter, P.D. et al. (2011). Genome-wide genetic marker discovery and genotyping using next-generation sequencing. *Nature Reviews*. 12 (7): 499–510.

Debergh, P.C. and Zimmerman, R.H. (eds.) (1991). *Micropropagation*. Dordrecht: Kluwer Academic.

Donald, C.M. (1967). The breeding of crop ideotypes. *Euphytica* 17: 385–403.

Dunwell, J.M. (2010). Haploids in flowering plants: origins and exploitation. *Plant Biotechnology Journal* 8 (4): 377–424.

Elshire, R.J., Glaubitz, J.C., Sun, Q. et al. (2011). A robust, simple genotyping-by-sequencing (GBS) approach for high diversity species. *PLoS One* 6 (5): e19379.

Evans, D.A. and Bravo, J.E. (1988). Agricultural applications of protoplast fusion. In: *Plant Biotechnology* (ed. T.J. Marby), 51–91. Austin: IC' Institute, University of Texas.

Ewing, P.M., Runck, B.C., Kono, T.Y.J. et al. (2019). The home field advantage of modern plant breeding. *PLoS One* 14 (12): e0227079.

Feher, A. and Dudits, D. (1994). Plant protoplasts and cell fusion and direct DNA uptake: Culture and regeneration systems. In: *Plant Cell and Tissue Cube* (eds. I.K. Vasil and T.A. Thorpe), 7X–118. Dordrecht: Kluwer Academic.

Food and Agriculture Organization of the United Nations (FAO). High level expert forum—how to feed the world in 2050. 2009, 1–4. https://www.fao.org/wsfs/forum2050/wsfs-forum/en/ (accessed on 8 December 2021).

Gallusci, P., Dai, Z., Ge´nard, M. et al. (2017). Epigenetics for plant improvement: current knowledge and modeling avenues. *Trends in Plant Science* 22 (7): 610–623.

Gardea-Torresdey, J.L., Rico, C.M. and White, J.C. (2014). Trophic transfer, transformation, and impact of engineered nanomaterials in terrestrial environments. *Environmental Science and Technology* 48 (5): 2526–2540.

Garg, R., Narayana Chevala, V., Shankar, R. et al. (2015). Divergent DNA methylation patterns associated with gene expression in rice cultivars with contrasting drought and salinity stress response. *Scientific Reports* 5: 14922.

Germanà, M.A. (2006). Doubled haploid production in fruit crops. *Plant Cell Tissue and Organ Culture* 86 (2): 131–146.

Germanà, M.A. (2009). Haploid and doubled haploids in fruit trees. In: *Advances in Haploid Production in Higher Plants* (eds. A. Touraev, B. Forster and M. Jain), 241–263. Heidelberg: Springer.

Gopinath, K., Gowri, S., Karthika, V. et al. (2014). Green synthesis of gold nanoparticles from fruit extract of Terminaliaarjuna, for the enhanced seed germination activity of Gloriosa superba. *Journal of Nanostructure in Chemistry* 4: 1–11.

Humphreys, M.W., Gasior, D., Lesniewska-Bocianowska, A. et al. (2007). Androgenesis as a means of dissecting complex genetic and physiological controls: Selecting useful gene combinations for breeding freezing tolerant grasses. *Euphytica* 158: 337–345.

Jiang, S.Y. and Ramachandran, S. (2010). Natural and artificial mutants as valuable resources for functional genomics and molecular breeding. *International Journal of Biological Sciences* 6 (3): 228–251.

Jinek, M., Chylinski, K., Fonfara, I. et al. (2012). A programmable dual-RNA–guided DNA endonuclease in adaptive bacterial immunity. *Science* 337: 816–821.

Johannes, F., Porcher, E., Teixeira, F.K. et al. (2009). Assessing the impact of transgenerational epigenetic variation on complex traits. *PLoS Genetics* 5 (6): e1000530.

Joshi B.K. (2017). Plant breeding in Nepal: past, present and future. *Journal of Agriculture and Forestry University* 1: 1–33.

Karp, A. (1994). Origins, causes and uses of variation in plant tissue cultures. In: *Plant Cell und Tissue Culture* (eds. 1.K. Vasil and T.A. Thorpe), X39–151. Dordrecht: Kluwer Academic.

Khodakovskaya, M., Dervishi, E., Mahmood, M. et al. (2009). Carbon nanotubes are able to penetrate plant seed coat and dramatically affect seed germination and plant growth. *ACS Nano* 3 (10): 3221–3227.

Kim, D.H., Gopal, J. and Sivanesan, I. (2017). Nanomaterials in plant tissue culture: the disclosed and undisclosed. *RSC Advances* 7 (58): 36492–36505.

Kole, C., Kole, P., Randunu, K.M. et al. (2013). Nanobiotechnology can boost crop production and quality: first evidence from increased plant biomass, fruit yield and phytomedicine content in bitter melon (Momordica charantia). *BMC Biotechnology* 13: 37.

Kwak, S.-Y., Wong, M.H., Lew, T.T.S. et al. (2017). Nanosensor technology applied to living plant systems. *Annual Review of Analytical Chemistry* 10 (1): 113–140.

Larkin, P.J. and Scowcroft, W.R. (1981). Somaclonal variation – a novel source of variability from cell culture for plant improvement. *Theoretical and Applied Genetics* 60 (4): 197–214.

Larkin, P.J., Brettell, R.I.S., Ryan, S.A. et al. (1985). Somaclonal variation: Impact on plant biology and breeding strategies. In: *Biotechnology in Plant Science* (eds. P. Day, M. Zaitlin and A. Hollaender), 83–100. New York: Academic Press.

Lee, J., Chin, J.H., Ahn, S.N. et al. (2015). Brief History and Prespectives on plant breeding. *Current Technologies in Plant Molecular Breeding* 1–14.

Lema, M. (2018). Marker assisted selection in comparison to conventional plant breeding: Review article. *Agricultural Research & Technology* 14: 555914.

Lemos, R.D.C., Pádua, J.M.V., Bruzi, A.T. et al. (2023). Comparison between doubled haploid lines and lines obtained via the bulk method in tobacco. *Crop Breeding and Applied Biotechnology* 22 (4): e42992249.

Lu, CM., Zhang, CY., Wen, JQ. et al. (2002). Research of the effect of nanometer materials on germination and growth enhancement of Glycine max and its mechanism. *Soybean Science* 21: 168–172.

Malabadi, R.B., Nethravathi, T.L., Kolkar, K.P. et al. (2023). Cannabis sativa: applications of artificial intelligence (AI) and plant tissue culture for micropropagation. *International Journal of Research and Innovations in Applied Science (IJRIAS)* 8 (6): 117–142.

Merrick, L.F., Herr, A.W., Sandhu, K.S. et al. (2022). Utilizing genomic selection for wheat population development and improvement. *Agronomy* 12 (2): 522.

Metzker, M.L. (2010). Sequencing technologies – the next generation. *Nature reviews Genetics* 11 (1): 31–46.

Miller, J.B., Zhang, S., Kos, P. et al. (2017). Non-viral CRISPR/Cas gene editing in vitro and in vivo enabled by synthetic nanoparticle co-delivery of Cas9 mRNA and sgRNA. *Angewandte Chemie International Edition* 56 (4): 1059–1063.

Mollers, C. and Iqbal, MCM. (2009). Doubled haploids in breeding winter oilseed rape. In: *Advances in Haploid Production in Higher Plants* (eds. A. Touraev, B.P. Forster and S.M. Jain), 161–170. Heidelberg: Springer.

Nair, R., Varghese, S.H., Nair, B.G. et al. (2010). Nanoparticulate material delivery to plants. *Plant science* 179: 154–163.

Pandeya, R.S., Douglas, G.C., Keller, W.A. et al. (1986). Somatic hybridization between Nicotiunu rustica and N. fabucum: development of tobacco breeding strains with disease resistance and elevated nicotine content. *Zeifschriff fir Pflanzenziichtung* 96,34&352.

Ran, Y., Liang, Z. and Gao, C. (2017). Current and future editing reagent delivery systems for plant genome editing. *Science China Life sciences* 60 (5): 490–505.

Rather, G.A., Ayzenshtat, D., Teper-Bamnolker, P. et al. (2022). Advances in protoplast transfection promote efficient CRISPR/Cas9-mediated genome editing in tetraploid potato. *Planta* 256 (1): 14.

Reinders, J., Wulff, B.B., Mirouze M. et al. (2009). Compromised stability of DNA methylation and transposon immobilization in mosaic Arabidopsis epigenomes. *Genes & Development* 23 (8): 939–950.

Ricroch, A., Chopra, S. and Fleischer, S.J. (2014). *Plant Biotechnology: Experience and Future Prospects*. Switzerland: Springer.

Ronaghi, M., Karamohamed, S., Pettersson, B. et al. (1996). Real-time DNA sequencing using detection of pyrophosphate release. *Analytical Biochemistry* 242 (1): 84–89.

Scowcroft, W.R., Brettell, R.I.S., Ryan, S.A. et al. (1987). Somaclonal variation and genomic flux. In: *Plant Tissue and Cell Culture* (eds. C.E. Green, D.A. Somers, W.P. Hackett. et al.), 275–286. New York: A.R. Liss.

Searchinger, T., Hanson, C., Ranganathan, J. et al. (2014). *Creating a Sustainable Food Future. A Menu of Solutions to Sustainably Feed More Than 9 Billion People by 2050; World resources report 2013–14: Interim findings 2014*. Washington, DC, USA: World Resources Institute.

Shull, G.H. (1908). The composition of a field of maize. *Reports of the American Breeders Association* 4: 296–301.

Simon, R.B. and Jonathan, F.W. (2014). Doubling down on genomes: Polyploidy and crop plants. *Journal of Botany* 101 (10): 1711–1725.

Smailus, D.E., Marziali, A., Dextras, P. et al. (2005). Simple, robust methods for high-throughput nanoliter-scale DNA sequencing. *Genome Research* 15 (10): 1447–1450.

Sonah, H., Bastien, M., Iquira, E. et al. (2013). An improved genotyping by sequencing (GBS) approach offering increased versatility and efficiency of SNP discovery and genotyping. *PLoS One* 8 (1): e54603.

Sugunan, A. and Dutta, J. (2008). Pollution treatment, remediation and sensing. In: *Nanotechnology* (ed. K. Harald), vol. 3, 125–143. Weinheim: Wiley-VCH.

Sweeney, M. and McCouch, S. (2007). The complex history of the domestication of rice. *Annals of Botany* 100 (5): 951–957.

Thorpe, T.A. (1990). The current status of plant tissue culture. In: *Plunt Tissue Culture; Applications and Limifafions* (ed. S.S. Bhojwani), l–33. Amsterdam: Elsevier.
Till, B.J., Comai, L. and Henikoff, S. (2007). Tilling and ecotilling for crop improvement. In: *Genomics-Assisted Crop Improvement* (eds. R.K. Varshney and R. Tuberosa), 333–349. Dordrecht: Springer.
Van Hieu, P. (2019). Polyploid gene expression and regulation in polysomic polyploids. *American Journal of Plant Sciences* 10:1409–1443.
Varshney, R.K., Nayak, S.N., May, G.D. et al. (2009). Nextgeneration sequencing technologies and their implications for crop genetics and breeding. *Trends in Biotechnology*. 27 (9): 522–530.
Vaughan, D.A., Balázs, E. and Heslop-Harrison, J.S. (2007). From crop domestication to super-domestication. *Annals of Botany* 100 (5): 893–901.
Vaughn, M.W., Tanurdzić, M. and Lippman, Z. (2007). Epigenetic natural variation in Arabidopsis thaliana. *PLoS Biology* 5 (7):e174.
Wang, X. (2022). Managing land carrying capacity: key to achieving sustainable production systems for food security. *Land* 11 (4):484.
Wang, X., Xu, Y., Hu, Z. et al. (2018). Genomic selection methods for crop improvement: Current status and prospects. *The Crop Journal* 6 (4): 330–340.
Yoo, C.M., Dalid, C., Moyer, C. et al. (2022). Improving strawberry varieties by somaclonal variation: HS1448, 10/2022. *Electronic Data Information Source* 2022 (5): 4.
Zargar, S.M., Raatz, B., Sonah, H. et al. (2015). Recent advances in molecular marker techniques: insight into QTL mapping, GWAS and genomic selection in plants. *Journal of Crop Science and Biotechnology* 18 (5): 293–308.
Zhang, X., Shi, C., Li, S. et al. (2023). A female in vivo haploid-induction system via mutagenesis of egg cell-specific peptidases. *Molecular Plant* 16 (2): 471–480.

26

Nanofertilizers for Growing Fortified Crops: A Need of the Day

Umair Ashraf[1*], *Muhammad Naveed Shahid*[1], *Shiza Ayaz Khilji*[1], *Munazza Kiran*[1], *Sammina Mahmood*[1], *Iqra Yousaf*[1], *Shakeel Ahmad Anjum*[2], *Lin Li*[3]

[1] *Department of Botany, Division of Science and Technology, University of Education, Lahore, Punjab, Pakistan*
[2] *Department of Agronomy, University of Agriculture, Faisalabad, Punjab, Pakistan*
[3] *School of Breeding and Multiplication, Sanya Institute of Breeding and Multiplication, Hainan University, Sanya, Hainan, China*

*Corresponding author: umair.ashraf@ue.edu.pk; umairashraf2056@gmail.com

26.1 Introduction

The human population is increasing very rapidly, which demands more food. According to the population survey, there will be about 9.7 billion people on the planet by the year 2050 (DESA 2015). Farm lands are becoming less fertile due to societal changes in lifestyle and other human activities, which inevitably impact crop productivity and may cause famine and hunger. Therefore, coordinated efforts are required to improve crop productivity and resource use efficiency (Elemike et al. 2019). Humans need about 40 essential nutrients, including about 20 minerals for a healthy life; however, the daily diet mostly contains insufficient amounts of the required nutrients, which results in malnutrition (WHO 2000). Malnutrition or micronutrient deficiency results in poor mental health, stumpy educational abilities, and low intellectual capacity, as well as lowers one's earning potential and job capacity, which has far-reaching effects on long-term national development. Therefore, fortifying crops with essential nutrients is necessary to tackle this worldwide threat. A fortified food minimizes the risk of anemia, cancer, heart disease, blindness, and early death (Hossain and Mohiuddin 2012).

One of the promising strategies to deal with malnutrition is the biofortification of food crops. Biofortification refers to growing vitamin- and mineral-rich crop varieties on a large scale, which may ensure that more people have access to healthier and nutrient-rich food (Ashraf et al. 2023). The goal of biofortification is to address micronutrient deficiencies by producing cheap, highly nutritious, and sustainable staple foods. Nevertheless, biofortified staple crops provide a lower daily dose of vital minerals and vitamins than supplements or industrially enriched foods (Bouis, McEwan, and Tanumihardjo 2013). Unlike commercial or industrial fortified foods, biofortified crops allow rural customers to access nutrient-dense food locally. According to Stein et al. (2007), agronomic and breeding techniques are important strategies that can be employed to increase the nutritional content of crops. Agronomic interventions emphasize the use of various fertilizers, such as organic, inorganic, and biofertilizers. On the other hand, inorganic fertilizers require a high rate of application with more nutrient loss, whereas organic fertilizers have low mineral nutrient content and release nutrients slowly. Scientists made lots of attempts to improve the nutrient use efficiency of conventional fertilizers; nevertheless, fertilizer application rates and demands are still high with less use efficiencies. Nanotechnology is a promising technique used for the targeted application of plant nutrition (Subramanian and Tarafdar 2011; Elemike et al. 2019); however, there are concerns related to its adoption, application, and management in the field of agriculture (Xiong et al. 2017; Thakur et al. 2018).

Nanofertilizers (NFs) emerge as a key innovation in the field of nanotechnology in the agricultural context (Ashraf et al. 2023). They are manufactured by incorporating nanoscale materials that efficiently regulate plant growth by releasing nutrients on crop demand (DeRosa et al. 2010; Nair et al. 2010). The NFs comprise various components such as zinc oxide NPs, quantum dots, iron, silica, titanium dioxide, gold nanorods, and others (Parasad and Bhattacharyya 2017). The success of employing nanomaterials as fertilizers depends on factors such as plant type, nanomaterial characteristics (size,

Crop Biofortification: Biotechnological Approaches for Achieving Nutritional Security under Changing Climate, First Edition. Edited by Adnan Noor Shah et al.

composition, concentration, chemical properties), and application methods (Thakur et al. 2018). The utilization of NFs offers a promising avenue to improve crop quality and essential nutrient content, thereby contributing to addressing the malnutrition problems of a growing population.

26.2 Why Nanofertilizers?

The application of fertilizers determines crop productivity; however, agricultural soils have become depleted due to intensive cropping systems and excessive application rates of synthetic fertilizers (Adeyemi et al. 2020; Diatta et al. 2020; Kumar et al. 2019a; Kumar et al. 2019b; Seleiman and Abdelaal 2018). Some recent estimates suggested that the usage of fertilizers may contribute roughly 30% of the total crop production, whereas the remaining 70% relies on the effective use of agricultural inputs and other external factors (Li et al. 2014). However, the efficiency of synthetic fertilizers is drastically reduced owing to their fixation in the soil or lost to the environment through leaching, volatilization, and runoff (Eid et al. 2020). For example, N (nitrogen), P (phosphorus), and K (potassium) have nutrient use efficiency (NUE) around 35%, 20%, and 40%, respectively (Husen and Iqbal 2019; Subramanian et al. 2015), thus a small portion of applied fertilizers can be taken up by the plant (Tariq et al. 2020). In order to achieve high yields, farmers apply synthetic fertilizers in higher quantities than recommended dosages, which raise salt concentrations and modify the natural nutritional equilibrium of soils; both cause negative impacts on plant productivity. Therefore, to boost crop output, enhance crop quality, and improve the overall sustainability of agricultural systems, it is crucial to design novel fertilizers that can release their nutrients gradually in synchronization with plant growth stage and demands (Seleiman et al. 2020a). In this regard, the NFs are found to release nutrients sustainably with maximum utilization (Qureshi et al. 2018).

Interestingly, the NFs can release nutrients for a longer period of time, i.e. 40–50 days as compared to 4–10 days for synthetic fertilizers. For example, synthetic urea fertilizer can quickly lose more than 70% of its N content after field application, leaving less than 20% of the N content easily accessible to plants (Kahrl et al. 2010). On the other hand, urea coating with hydroxyapatite NPs resulted in a gradual release of N. Similarly, the application of nanohybrid of urea (i.e. a modified form of hydroxyapatite) in rice fields can release N up to 12 times slower than synthetic urea and can boost grain yields by 50% when applied in conjunction with conventional urea (Kottegoda et al. 2017). Moreover, synthetic P fertilizers have low rates of absorption and high rates of fixation in the soil (Shenoy and Kalagudi 2005), but nanoformulations of P can lower nutrient losses by directly internalizing the crops (Poudel et al. 2023). For instance, it has been found that the application of porous nanomaterials, such as zeolites and chitosan, significantly increases absorption efficiency by reducing N loss. Furthermore, despite the fact that a number of methods have lately been developed to improve phosphorus absorption (Seleiman and Abdelaal 2018), however, their effectiveness has been somewhat restricted. Conversely, when compared to plants grown with synthetic P fertilizers, the application of P-enriched hydroxyapatite NPs was observed to significantly improve the growth and yield of soybeans (Liu and Lal 2014). Similarly, carbon-based nanomaterials (i.e. graphene oxide sheets) may prolong the potassium nitrate release mechanism, hence reducing leaching losses (Shalaby et al. 2016). Additionally, some plant metabolic processes that affect a plant's ability to mobilize nutrients like P are influenced by NPs (Zahra et al. 2015). For example, zinc NPs are utilized to enhance the absorption capacity of synthetic P (Seleiman and Abdelaal 2018). Additionally, biosensors can be added to NFs to regulate the release of nutrients and their bioavailability according to crop needs (León-Silva et al. 2018). The NFs are only required in small amounts; therefore, their application rates are lower than synthetic fertilizers (León-Silva et al. 2018). Overall, NFs are more advantageous than synthetic fertilizers due to their unique properties (Figure 26.1) as they have small particle sizes (<100 nm) and, therefore, can easily enter into the plants when applied either as foliar or soil amendments (Liscano et al. 2000). They have large surface areas that maximize the reactivity and enhance the nutrient availability, as well as plant NUE (Liscano et al. 2000; Siddiqi and Husen 2017). Synthetic fertilizers have large particle sizes and high adsorption and fixation to soil particles, which accounts for their low solubility. Unlike synthetic fertilizers, NFs are soluble in water, thus they are readily available to plants. In addition, to reduce bulk requirements and to enhance plant absorption and availability, fertilizers are encapsulated in NPs (Chhipa 2017). For instance, using zeolite-based NFs makes it possible for plants to have high nutrient availability throughout the growing period. Moreover, the targeted and gradual nutrient release (Singh and Kumar 2017) of NFs (Siddiqi and Husen 2017) reduces their toxicity to plants (Sohair et al. 2018) and lowers the N losses through leaching, denitrification, volatilization, fixation, as well as salt accumulation in soil. The multidimensional roles of nanofertilizers have been summarized in Figure 26.1.

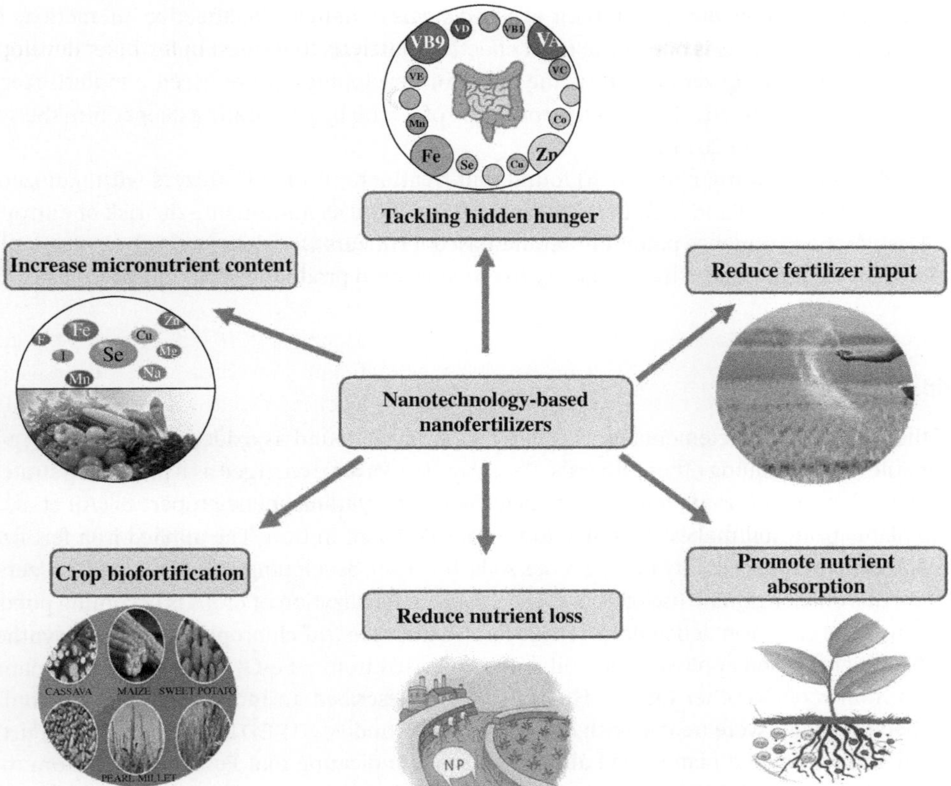

Figure 26.1 Multidimensional roles on nanotechnology-based nanofertilizers.

26.3 Role of Different Nanofertilizers

26.3.1 Zeolite-based Nanofertilizers

Zeolites, a family of microporous aluminosilicate minerals, have been utilized as a soil supplement for many years owing to their capacity to absorb and hold onto water, nutrients, and other organic molecules (Jakkula and Wani 2018). Zeolites are renowned for their soil-friendly properties and are important for improving crop productivity while mitigating environmental risks (Morales-Díaz et al. 2017). In recent years, natural zeolite, a mixture of more than 50 minerals primarily alkali and alkaline earth aluminosilicates, has been converted into NFs owing to its availability and excellent properties (Eroglu et al. 2017; Suppan 2017). It not only provides plants with some nutrients but is also a smart carrier of important minerals for manufacturing fertilizers. Moreover, zeolites help in preventing nutrient loss by progressively releasing nutrients into the soil and facilitating uptake by plants (Guo 2004). Zeolites are now extensively used in food and agriculture after their declaration as nontoxic by the Food and Drug Administration and the International Agency for Research on Cancer (Eroglu et al. 2017; Suppan 2017). The top-down method of ball milling can be used to convert zeolites to nanoforms by decreasing their diameters and increasing their surface area for appropriate ion adsorption and desorption (Yuvaraj and Subramanian 2018). These characteristics make it possible to hold some ions and release others in a mechanistic way that is suitable for crop improvement. Nanozeolite, therefore, is designed to have better qualities, which encourages the efficient use of more macronutrients and micronutrients and can serve as pesticides, improve soil nutrients, regulate soil acidity, and improve seed germination (Eroglu et al. 2017; Suppan 2017).

Zeolite-based nitrogenous and potassium fertilizers resulted in lower application amounts with higher productivity (Polat et al. 2004). In addition, zeolite nanocomposites of potassium, nitrogen, phosphorus, calcium, zinc, and iron improved nutrient uptake, growth, and crop productivity (Harper 2015); however, zeolites poorly absorb and adsorb anions (Morales-Díaz et al. 2017). Owing to its large surface area, three-dimensional crystal structure, and high porosity, zeolite allows high cationic exchange and can hold onto both positive and negative nutritional ions for an extended period of time.

The hydrated negative inorganic ions present on their exposed surfaces help in the effective interactions of these surfaces with polar molecules and cations. This is one of the most effective materials to be used in fertilizer development and nano-formulations. The process of creating zeolite NPs is the basis for developing zeolite-based nanofertilizers. Because these fertilizers are tiny, they can more effectively reach the root zone of a crop by penetrating deeper into the soil (Sharma et al. 2022; Manikandan and Subramanian 2016).

Moreover, Manikandan and Subramanian (2016) found that treatment of urea fertilizers with nanozeolite (nano-zeourea) resulted in slow nutrient release and high uptake of nitrogen, as well as minimizing the risk of eutrophication and the greenhouse effect. Despite their immense potential, zeolite-based fertilizers were less focused. However, the exploration of the potential of nanozeolites underscores their capacity to enhance crop productivity and nutrient use efficiencies in crop production systems.

26.3.2 Iron Oxide Nanofertilizers

Iron (Fe) is one of the most common elements on earth and is primarily found as oxides of maghemite (g-Fe_2O_3), hematite (α-Fe_2O_3), and magnetite (Fe_3O_4) among other minerals. Recently, Fe NPs have emerged as a promising strategy for enhancing agricultural productivity due to their availability, low toxicity, and superparamagnetic properties (Ali et al. 2016). Less bioavailability of iron to plants from soil makes the staple food crops deficient in iron. The applied iron fertilizers might not be adsorbed by plants due to their unavailability in calcareous soils; however, developing iron-chelated fertilizers can be prepared and applied to address this challenge. The use of γ-Fe_2O_3 NPs for iron fertilization of crops is becoming popular as a potential solution to minimize soil and crop iron deficiencies. These Fe NPs improve leaf chlorophyll and photosynthesis, root growth, and crop stand establishment. When applied to the soil, iron is released from γ-Fe_2O_3 NPs due to the dynamic nature of NPs and then translocates from roots to other tissues. Hu et al. (2017) described an increase in the iron content of *Cucurbita maxima* L. shoots when the plants were treated with both γ-Fe_2O_3 NPs and Fe (II)-EDTA-treated plants. Furthermore, root Fe contents in Fe-treated and nontreated plants were almost the same, indicating that Fe had moved from root to other plant parts (Corredor et al. 2009; Hu et al. 2017). In sum, the mechanism underscores the potential of Fe-NPs in facilitating nutrient uptake and distribution within plants, thereby enhancing overall crop productivity and nutritional value.

26.3.3 Zinc Oxide Nanofertilizers

Zinc oxides (ZnO) and zinc sulfates ($ZnSO_4H_2O$ or $ZnSO_4.7H_2O$) serve as primary sources of zinc micronutrients for fertilizer fortification. Zinc plays a crucial role in living organisms as a component of enzymes, essential for hormone production, and vital for the regulation of chlorophyll and carbohydrate metabolism. When in the form of NPs, zinc oxide can be absorbed, metabolized, and stored in plant systems. However, excessive quantities of zinc oxide nanoparticles (ZnONPs) can disrupt plant development, leading to intrinsic abnormalities in plants (Singh et al. 2013). However, its deficiency could also impair plant growth and development. For example, zinc-deficient soils result in poor Zn concentrations in the edible plant parts of plants. According to the World Health Organization (WHO), one-third of the world's population suffers from zinc deficiency, which is due to the zinc-deficient foods, particularly cereals (Biesalski and Biesalski 2013; Guilbert 2003; Sadeghzadeh 2013). Hidden hunger diseases are the result of zinc and other mineral deficiencies in our diet. Micronutrients are required daily in a specific amount, if not met for an extended period of time, might result in illness. Therefore, by fertilizing crops with enough micronutrients (in general) and Zn (in particular), the crops can be fed with the required Zn levels. On the other hand, the availability of Zn in alkaline soils is poor and challenging (Sadeghzadeh 2013). While zinc oxide and zinc sulfate alone can be somewhat effective as micronutrients, unlike simply zinc, they can be used effectively in blends with macronutrients that can serve as carriers (Guilbert 2003), whereas the dissolution and absorption of micronutrients depends on the characteristics of carriers. For example, 2.0% of zinc may dissolve in polyphosphate, ammonia, and anhydrous ammonium nitrate, while only 0.05% of zinc can be absorbed in orthophosphate (Guilbert 2003). Moreover, the ionic strength, mineral content, and organic matter of the soil affect nutrient absorption and bioavailability. High pH and calcium carbonate content of calcareous soils have great impacts on the solubility and amount of zinc absorbed into the soil (Milani et al. 2015).

Zinc carbonate is precipitated when zinc is exchanged for calcium in the presence of calcium carbonate, which affects availability and reactivity in the soil. The nature of the soil also determines nutrient availability to plants. For instance, sandy soils are particularly poor at retaining nutrients, thus preventing plants from absorbing micronutrients. The availability of micronutrients in the soil is also determined by soil pH. The availability of micronutrients decreases with increase in soil pH except for molybdenum. Regarding zinc, it gets adsorbed on the soil surface along with other clay and inorganic minerals as the soil pH rises, and it can precipitate as $ZnCO_3$, $Zn(OH)_2$, and Zn_2SiO_4 (Biesalski and Biesalski 2013), which results in ineffective mineral desorption for plant uptake in the soil. Nevertheless, Zn-based nanofertilizers with targeted delivery proved much more effective than conventional Zn fertilizers. For instance, the application of nano-Zn fertilizer improved the photosynthesis, plant growth, antioxidant activity, and yield of hybrid maize (Azam et al. 2022).

Moreover, the application of urea coated with nano-Zn substantially improved nutrient utilization and Zn bioaccumulation in wheat (Dimkpa et al. 2022). The application of ZnO nanofertilizer improved the fruit Zn contents and nutrient distribution pattern in cherry tomatoes (Almendros et al. 2022). Overall, Zn-based nanofertilizers could help to fulfill the Zn deficiencies in staple foods, especially cereals, pulses, legumes, and vegetables.

26.3.4 Titanium Dioxide Nanofertilizers

Titanium dioxide nanoparticles (TiO_2 NPs) have shown promising potential in improving nitrogen fixation and stimulating photosynthesis in plants and overall plant growth (Zheng et al. 2005), however, their uptake and efficacy depend on the particle size. According to Asli and Neumann (2009), the TiO_2 NPs applied to maize at a size of around 30 nm did not translocate owing to the bigger size than pore diameter (6.6 nm) of the root cells. Similarly, Du et al. (2011) Dağhan et al. (2020) revealed that the application of TiO_2 NPs improved plant growth and nutrient uptake, i.e. N, P, Zn, and Cu in wheat. Overdose of TiO_2 NPs may also alter enzyme activity and cause toxicity in plants, whereas nontranslocation of TiO_2 NPs within the soil can have significant impacts on the soil ecosystem (Elemike et al. 2019). Therefore, understanding the behavior of TiO_2 NPs in soil and plants and their effects on enzyme activity and soil ecosystems is crucial for assessing their potential as a nutrient fortification strategy and ensuring their safe and sustainable application in agriculture.

26.4 Role of NFs of Major Nutrients to Enhance Crop Productivity

Fertilizers play a crucial role in supplying essential nutrients for plant growth and enhancing crop productivity and quality. Plants require high amounts of macronutrients such as N, P, and K (Adeyemi et al. 2020; Czymmek et al. 2020; Kumar et al. 2019a, b). Farmers typically employ huge quantities of fertilizers to partially remediate low NUE because most of these nutrients are not efficiently absorbed by plants, which could have a negative effect on soil, water, and the environment as a whole (Chhipa 2017; Czymmek et al. 2020; Seleiman et al. 2013; Seleiman et al. 2020b). The negative effects of synthetic fertilizers can be reduced by applying NFs that can help to improve the NUE of fertilizers and increase crop productivity (Liu and Lal 2014; Seleiman et al. 2020c). The type and efficacy of NFs generally depend on the carrier present in them, e.g. mesoporous silica NPs, hydroxyapatite NPs, copper, silica, zinc, carbon, nitrogen, and polymeric NPs (Guo et al. 2018; Liscano et al. 2000).

26.4.1 Role of Nitrogen NFs

Synthetic N fertilizers undergo leaching and volatilization losses during and immediately after their application to the field. To address this problem, N-based NFs could be used that have a slow-release rate than conventional nitrogenous fertilizers. Higher nutrient uptake, growth rate, yield, and better grain quality of maize were observed when treated with zeolite-urea NF as compared to synthetic urea fertilizers (Manikandan and Subramanian 2016). Similarly, significant improvements in growth and oil contents of starflower (*Borago officinalis* L.) were reported when applied with nitrogen-based NFs (Mahmoodi et al. 2017). Moreover, the seed yield of soybeans was increased over synthetic fertilizers when treated with urea-modified zeolites (Liu and Lal 2015). Similarly, potato yield was improved with reduced nitrate leaching with the application of nano-N and chelated nano-N (Zareabyaneh and Bayatvarkeshi 2015). Overall, nitrogen in the form of NFs can be highly useful because it decreases the rates of leaching and volatilization helps in slow nutrient release, and thus increases the growth and productivity of crops.

26.4.2 Role of Phosphatic NFs

Phosphorus (P) is considered to be the second most vital nutrient for plant growth after nitrogen. It has both structural and functional roles in plants (Soliman et al. 2016). Phosphorus is poorly available in synthetic fertilizers owing to its slow release and high fixation with soil colloids. Liu and Lal (2015) reported that P-based NFs can deliver P gradually for up to

40–50 days after application. Thus, P-based NFs not only improve phosphorous use efficiency value but also significantly improve the growth and yield of *Vigna radiata* (Patra et al. 2013). The growth and antioxidant activity in *Adansonia digitata* L. were improved with the application of nanosized hydroxyapatite (Soliman et al. 2016), whereas the application of nano-P in place of diammonium phosphate provided effective nutrient use efficiency, grain yield, and nutritional quality of wheat (Poudal et al. 2023). In summary, P-based NFs are suitable options in agriculture due to poor fixation with soil colloids with better use efficiency and enhancing the productivity and quality of crops.

26.4.3 Role of Potassium NFs

Potassium (K) is the third most important macronutrient for plants following, N and P (Preetha and Balakrishnan 2017). The maximum use efficiency of K fertilizer ranges from 30% to 50%, whereas almost 50–70% of applied K fertilizer can be lost, which is causing serious threats to soil health, water quality, and the economy (Czymmek, et al. 2020). It was observed that K losses in soil could be reduced by applying nano-K fertilizer (Kubavat et al. 2020). For instance, the application of K-loaded zeolites improved the chlorophyll content, yield, and harvest index of hot pepper (*Capsicum annuum* L.) (Li, Wee, and Sohn 2010). Similarly, biomass, quality, and growth of *Cucurbita pepo* L. were improved with the foliar application of nano-K fertilizer, whereas nano-K application improved the drought tolerance in soybean (Rostami Ajirloo and Amiri 2022). Thus, the application of nano-K-based fertilizers can help plants to improve their growth and yield under normal and stressful conditions.

26.5 Nanofertilizers for Stress Management

Biotic and abiotic stresses pose significant threats to crop productivity and global food security (Seleiman and Kheir 2018; Seleiman et al. 2020a). Therefore, it is imperative to develop innovative stress management technologies to improve crop resilience against stress conditions. Plant nutrition management can be employed to regulate the physiological functions of plants under stress conditions. The NFs or NPs improve the antioxidant defense system in plants, thus helping them to combat environmental stresses (Farhangi-Abriz and Torabian 2018; Seleiman et al. 2020b). Application of Si NPs improved drought stress tolerance by modulating the physiological and biochemical functions in hawthorn (*Crataegus* sp.) (Ashkavand et al. 2015). Application of foliar ZnO NPs improved the early growth and low-temperature stress tolerance in fragrant rice (Mai et al. 2024), whereas nano-priming with La_2O_3 improved the morpho-physiological traits, antioxidant defense system, and Cd stress tolerance in rice (Sun et al. 2023). Seed priming with multi-walled carbon nanotubes promoted the seed germination potential and root-shoot growth of maize under Cd toxic conditions.

Moreover, the yield and oil percentage of safflower (*Carthamus tinctorius* L.) grown under water stress were improved with the foliar spray of Fe NPs (Davar et al. 2014), whereas seed germination and growth of bean plants (*Phaseolus vulgaris* L.) were substantially improved with nano-chitosan under salt stress (Zayed et al. 2017). Application of ZnO NPs was also reported to be effective against *Clostridium perfringens, Bacillus subtilis, and Escherichia coli* (Nawaz et al. 2011).

26.6 Efficacy of Different Nanomaterial-based Nanofertilizers

The use of NPs to regulate the delivery of nutrients makes nanofertilizers more effective and economical than conventional fertilizers. Among the nanomaterials employed in the development of nanofertilizers, carbon nanotubes, graphene, and quantum dots are quite common. The synthesis of nanofertilizers uses cutting-edge technology to offer a productive and environment-friendly way to fertilize crops efficiently (Yadav et al. 2023). Three primary categories can be used to classify nanofertilizers, i.e. nanoscale nutrient particles, nanocoated fertilizers, and nanocomposites, which provide special benefits such as reduced environmental footprints, enhanced nutritional availability, and controlled release of nutrients. The NFs are synthesized via physical, chemical, and biological techniques. Physical methods are simpler and more scalable than chemical procedures, while chemical methods provide exact control over particle shape and size. On the other hand, biological approaches, which make use of microbes and plant extracts, offer environmentally beneficial options. Whatever technique is utilized for the synthesis of nanofertilizers, the size, shape, surface characteristics, and interactions with plant cells substantially affect the efficacy of NFs (Monreal et al. 2016; Padhan et al. 2023).

In addition, the development of carbon-based nanofertilizers also offers great potential to improve agricultural productivity and profitability (Singh et al. 2015), whereas carbon nanotubes (CNTs) are one of the most studied carbon-based nanofertilizers, known for their high mechanical strength and unique chemical properties (Safdar et al. 2022). Moreover, chitosan, a biopolymer, is recognized as an ideal choice among agriculture, food, and health experts. It is a deacetylated form of chitin that comes from the shells of arthropods. Chitosan is a cationic polymer that interacts with negatively charged objects owing to its structural properties. Fertilizer nutrients interact with chitosan to generate complexes that increase the availability of nutrients to plants (Ashraf et al. 2022). In agricultural applications, chitosan NPs can act as a vehicle to regulate the nutrient release from NPK fertilizers. Fertilizers can be dissolved in the NPs solution and magnetically agitated, making the process reasonably simple (Corradini et al. 2010). Chitosan molecules can be easily modified to absorb and release fertilizers, herbicides, insecticides, and plant growth regulators (Mujtaba et al. 2020).

Clay-based NPs can be used to create controlled release fertilizers (CRF) owing to their wide surface area and nanolayer reactivity. Nanoclay is an essential part of the synthesis of CRF as it offers an active surface for several physicochemical and biological processes (Basak et al. 2012). Fertilizers can be encapsulated and delivered via nanocapsules, which are minuscule capsules made of either organic or inorganic material (Petosa et al. 2017), whereas carbon nanotubes, lipids, metal oxides, silica, and biopolymers are the materials used to develop nanotubes (Kottegoda et al. 2017).

Furthermore, nanogels are hydrophilic porous structures that can absorb fertilizers or other compounds and release them gradually overtime. The major nutrients, i.e. nitrogen, phosphorus, and potassium, can be impregnated into the nanogels (Bindra and Singh 2021). Nanogels are also used in water filtration, medication delivery, and manufacturing of fertilizers. To enhance water retention and lessen soil erosion, the nanogel particles can absorb and hold water in the soil. Additionally, compared to conventional fertilizers, nanogels may contain more nutrients due to their high surface area:volume ratio. Moreover, nanogels are safe for the environment owing to their nontoxic and biodegradable properties (Madzokere et al. 2021). In addition, polyurethane-based nanofertilizers are made of a polymer made of organic units joined by urethane linkages. Typically, ring-opening metathesis polymerization (ROMP), which involves generating a monomer from two distinct molecules with a double bond and polymerizing is used to create polyurethane-based nanofertilizers. The resultant polymer produces NPs that release nutrients into the soil efficiently (Thirugnanasambandan 2021). Nanocrystals made of starch make up starch-based nanofertilizers that improve soil nutrient status and plant nutrient use efficiency when applied to plants in liquid or aerosol form (Lin et al. 2011).

Certain nanofertilizers, such as CNTs, work with plant root systems to enhance hormone synthesis, promote plant growth, and improve raised soil carbon and other nutrients. CNTs are made of rolled up sheets of carbon atoms and have special qualities that make them excellent fertilizers. Nutrients can be absorbed and released by CNTs, which can also strengthen soil structure, retain more water, and promote plant growth. Low carbon nanotube concentrations may help seed germination, root development, and water transport with no phytotoxicity, in contrast to the unfavorable effects of high fertilizer doses (Mondal et al. 2011). Overall, nutrient-based nanofertilizers are important to address the challenge of poor nutrient availability, uptake, and utilization in plants (Mondal et al. 2011).

In addition, inorganic nanofertilizers include metals, metalloids, and nonmetallic NPs, and they can provide essential nutrients, such as nitrogen, phosphorus, and potassium, to plants. These fertilizers are designed to improve the efficiency of nutrient uptake by plants and can be used to improve yields in agriculture (Kalia and Sharma 2019). The choice of material depends on the desired properties of the fertilizer. Inorganic nanofertilizers offer a unique advantage due to their ability to be customized to the specific nutrient requirements of targeted plants, thereby allowing targeted applications to enhance yield. Inorganic nanofertilizers are already being used in agriculture, and their use is expected to increase as more farmers adopt precision agriculture practices. In a study on safflower plant growth, the foliar application of silicon dioxide NPs ($nSiO_2$) improved canopy spread, stem diameter, plant height, and ground cover (Janmohammadi et al. 2016).

The NPs in organic nanofertilizers are made of organic ingredients and are intended to progressively release nutrients into the soil. They are natural, organic, and eco-friendly and can also change the pH of the soil and retain soil moisture, which helps plants absorb essential nutrients more effectively. According to Soman and Balachandran (2021), organic NPs can be made from atoms or molecules and can take on a variety of shapes, including capsules, vesicles, micelles, liposomes, polymersomes, dendrimers, and polymeric NPs. Nutrient-loaded nanofertilizers, such as nanoporous zeolites, are an efficient way to supply nutrients to crops. Zeolite fertilizers are made of minerals that are created when volcanic ash and alkaline lake water react. The result is a structure analogous to a honeycomb that has a high porosity, which makes it easier for moisture, nutrients, and other substances to be absorbed and retained. To create a nanofertilizer tailored to different crop needs, the natural zeolite is ground into a powder mixed with nitrogen, phosphorus, and potassium, increasing the zeolite particles' surface area and improving the soil's ability to absorb and retain nutrients. Engineered nanoporous zeolites, also known as aluminosilicates, are distinguished by the presence of micropores (2 nm), mesopores (2–50 nm), and macropores (>50 nm) that may be used as fertilizer carriers owing to their increased ion-exchange and adsorption capacities. Nanoporous zeolite fertilizers improve soil fertility, making it more suitable for crop growth (Sharma et al. 2022). Furthermore, the zeolite particles' water retention capability helps reduce the need for frequent irrigation, which can benefit areas where water resources are scarce or expensive. Additionally, zeolite fertilizers can help to reduce soil erosion due to the increased water retention capabilities of the zeolite particles (Sharma et al. 2022).

Consistency-based nanofertilizers, which offer advantages including enhanced nutrient absorption, decreased nutrient loss, and limited environmental impact, comprise surface-coated and nanocarrier-based technologies. They represent a viable alternative for sustainable agriculture (Liu and Lal 2015). Synthetic polymer- and biological product-coated NPs make up surface-coated nanofertilizers (Vejan et al. 2021). Crop yields can be increased by using nano-carrier-based nanofertilizers, which use mesoporous silica NPs as carriers to promote nutrient solubility, stability, and availability (Kong et al. 2021). Surface-coated nanofertilizers comprise NPs of silica, iron, and other nutrient-containing materials, including established fertilizers with nanoscale additives, nanoscale coatings, and fertilizers coated with nanoscale materials (Solanki et al. 2015).

To increase the effectiveness of conventional fertilizers, nanotechnology-based fertilizers made of NPs of a specific nutrient have been developed (Raliya et al. 2017). Plant availability of NPs is increased when they are coated with a protective layer, which regulates their release. Combining traditional and nanotechnology-based fertilizers, hybrid NFs improve nutrient availability and release nutrients gradually and continuously (Tarafder et al. 2020). According to Kah et al. (2013), these nanofertilizers increase fertilizer efficiency and lessen the fertilizer's negative effects on the environment. Fertilizers based on nanocarriers are used to transfer and supply nutrients to plants and soil. These nanofertilizers can improve plant growth, lower nutrient losses, and boost nutrient availability. Controlled release capabilities are essential for efficient nutrient dissemination and the reduction of detrimental impacts on the environment and human health when designing next-generation nanocarriers for nutrient biofortification. The benefits of nanocarrier-based fertilizers include reduced environmental impact, improved nutrient uptake, and regulated nutrient release (Liu and Lal 2015; Kah et al. 2018).

26.7 Biofortification of Food Crops from Conventional to Modern Approaches

As the global population aims to reach 9.6 billion by 2050, efforts are underway to enhance crop production strategies to ensure food security. However, the persistent challenge of hidden hunger, characterized by micronutrient deficiencies resulting from inadequate vitamins and minerals in diets, remains a significant global concern. This deficiency is associated with severe health issues such as intellectual disabilities, stunted growth, premature mortality, and an elevated risk of chronic diseases like cardiovascular diseases and cancers (Jha and Warkentin 2020). Approximately one-third of the global population is suffering from hidden hunger, with the most prevalent deficiencies being in vitamin A, iron, and zinc (Bouis, McEwan, and Tanumihardjo 2013). This issue poses a significant public health challenge, particularly in regions like Sub-Saharan Africa, the Caribbean, and East and West Asia (Siwela et al. 2020).

To combat hidden hunger, diverse fortification strategies are being explored, primarily through two main approaches: direct intervention and indirect intervention. Direct intervention focuses on nutrition-specific measures and addresses food consumption behavior through methods like food supplementation and promoting dietary diversity. On the other hand, indirect intervention, which is nutrition-sensitive, involves techniques such as biofortification (de Valença et al. 2017). The suitability of a particular strategy for a given location depends on factors like the prevalence of specific deficiencies, the availability of food sources, infrastructure, and government regulations (Olson et al. 2021). The rural poor population is disproportionately affected by micronutrient malnutrition, and direct intervention strategies are not readily available (Bouis, McEwan, and Tanumihardjo 2013).

Biofortification involves enhancing the quantity or accessibility of essential nutrients in crops during their growth (de Valença et al. 2017). Unlike other fortification methods that involve supplement addition during food processing, biofortification directly targets the crop itself (Malik and Maqbool 2020). The three primary techniques in biofortification include conventional plant breeding, application of soil and foliar fertilizers, and genetic engineering (Garcia-Casal et al. 2016). The primary objective is to elevate the nutritional content of crops with favorable agronomic traits (Singh et al. 2016) while meeting the agricultural needs (yield) and cultural preferences (color, taste, and cooking time) of farmers. The resulting product must be physically and economically accessible to consumers while preserving its nutritional quality during preparation (Siwela et al. 2020).

In biofortification through conventional plant breeding, parent lines containing high concentrations of desired micronutrients can be crossed for offspring rich in micronutrients. Alternatively, crossing parent lines with reduced anti-nutrient levels, compounds hindering nutrient absorption, enhances bioavailability (Campos-Bowers and Wittenmyer 2007). Bioavailability, crucial for physiological functions, aids smallholder farmers in improving micronutrient density sustainability. Despite its long-term benefits in addressing health concerns in low-income areas (Jha and Warkentin 2020), conventional breeding faces challenges, such as time-consuming trait identification and potential issues like low heritability and linkage drag postgene incorporation (Garcia-Casal et al. 2016; Malik and Maqbool 2020).

Another method of biofortification involves soil and foliar micronutrient applications. Depleted soil minerals can be replenished using mineral fertilizers, increasing the transfer of minerals to edible plant portions (Jha and Warkentin 2020). Foliar fertilization, achieved by spraying fertilizers on aboveground plant parts, enriches plants with micronutrients and macronutrients; however, the effectiveness of foliar applications depends on environmental factors and plant tissue permeability. Although foliar application is beneficial for immediate crop response and controlling nutrient loss, however, high doses may have adverse effects on plant health. On the other hand, geographic variations in soil composition and micronutrient deficiencies pose challenges to nutrient-enriched crop production (Alshaal and El-Ramady 2017).

Genetic engineering is one of the modern approaches, which utilizes genes from diverse sources to introduce them directly into crops (Singh et al. 2016) to produce nutrient-enriched crop cultivars. Genetic engineering, successful in crops like wheat and soybeans, can accumulate significant amounts of vitamins and minerals in edible parts (Campos-Bowers and Wittenmyer 2007). However, cultural, ethical, and scientific barriers, including cost and technology accessibility, genomic understanding, and concerns about safety, labeling, and intellectual property, must be addressed for the successful integration of genetic engineering in biofortification (Garcia-Casal et al. 2016; Siwela et al. 2020).

The development of highly specialized fertilizers, such as water-soluble, chelated, and nanofertilizers, with high nutrient uptake efficiency and greater nutrient translocation to the consumable sections of a crop plant, has increased the effectiveness of agronomic biofortification (Bhardwaj et al. 2022). Biofortification has various applications, particularly in addressing nutritional deficiencies in remote populations with limited access to fortified foods (Olson et al. 2021). Unlike other strategies focusing on populous areas, biofortification uniquely targets rural regions, where micronutrient deficiencies are prevalent, with the expectation that surpluses will extend to urban communities. This approach is crucial, as rural areas often experience the most significant micronutrient gaps. Biofortification can enhance multiple nutrients, including iron, iodine, zinc, calcium, selenium, and vitamins A, B, C, and E (Alshaal and El-Ramady 2017).

The ability to enrich crops with a diverse range of vitamins and minerals addresses deficiencies in heavily consumed crops across various cultures, boosting their nutritional value. Genetic engineering, a component of biofortification, enables the incorporation of multiple micronutrients into the same crop, enhancing nutritional quality and economic benefits (Olson et al. 2021). By the end of 2016, over 20 million people in 30 countries were consuming biofortified crops, with 150 varieties across 10 crops available, showcasing the widespread impact of biofortification efforts (Jha and Warkentin 2020).

26.8 Nanonutrition for Biofortification in Crops

Cereals are the seeds or grains that belong to the Poaceae family, a common family of plants grown for their fruit, known as "caryopsis." Cereals are a cost-effective and abundant source of carbohydrates, vitamins, minerals, fats, oils, and proteins. They provide more food energy to humans than any other crop group. Cereals are not just packed with essential nutrients, but they also contain bioactive substances that are beneficial for human health. The application of NPs in agriculture is providing a promising alternative to traditional breeding and agronomic practices.

NPs, due to their small size and large surface area, can significantly improve the bioavailability of essential nutrients in cereals. For instance, zinc oxide NPs have demonstrated an increase in zinc uptake in rice, contributing to higher yield and better nutritional quality (Rameshraddy et al. 2017). The application of biogenic ZnO NPs improved germination indices and growth morphology of rice (Cyriac et al. 2020). The combination of K fertilizer and $ZnSO_4$ NPs improved the proximate composition and growth of maize in Cd-contaminated soil (Umair et al. 2023). Exogenous nanochelated zinc–boron–silicon fertilizers improved the antioxidant defense system, water use efficiency, and yield and protein contents in wheat under drought conditions (Ahmadian et al. 2021). In addition, soil mineral fertilizer application in integration with foliar nanofertilizer application (P and K) improved the agronomic and yield attributes of maize as well as nutrition contents in aboveground plant parts (MA et al. 2017). Coating nanozeolite with conventional urea fertilizer substantially improved the morphological, yield, and quality attributes as well as nutrient uptake in maize (Manikandan and Subramanian 2016).

Nanochelated Fe fertilizer enhanced the grain NPK by 46%, 43%, and 41% and Fe and Zn concentrations by 25% and 50%, respectively (Fakharzadeh et al. 2020). Vattani et al. (2012) reported that applying nanochelated iron fertilizer enhanced the accumulation of iron and potassium but decreased sodium and nitrate contents in spinach leaves. Application of nano-Zn fertilizer improved boron concentration, whereas Zn concentration ranged from 4.5 to 4.8 and 2.5–3.5 mg Zn kg^{-1} in the acidic and calcareous soil, respectively, which were reported as adequate for human consumption (Almendros et al. 2022).

Moreover, differences were noticed in the uptake and accumulation patterns of micronutrients, i.e. Cu, B, Mn, Fe and macronutrients, i.e. Ca, Mg, S, and P in relation to different forms of Zn fertilizers, i.e. ZnO NPS, $ZnCl_2$ and bulk ZnO application in kidney beans (Medina-Velo et al. 2017). The maximum Zn, P, and Ca concentrations were found in tomatoes applied with nano-Zn application as compared to conventional $ZnSO_4$ application (Parra-Torrejón et al. 2023). Application of micronutrient nanoparticles MN-NPs (MnO_2, ZnO, and MoO_3) at 40 mg L^{-1} improved the morphological growth, flowering, and yield traits of common beans (Salama et al. 2022), whereas application of Mo-NF improved the biomass, yield, and nitrogen use efficiency in green beans (Muñoz-Márquez et al. 2022). In addition, the application of ZnO NPs at 5 ppm enriched the grain Zn contents of sorghum and finger millet by 94% and 14%, respectively (Dimkpa et al. 2019; Kumar et al. 2019a, b). Nanochelated Fe application at 0.5–1.0 g L^{-1} enhanced the growth, yield, and Fe concentrations in the grains of Cumin (Sabet and Mortazaeinezhad 2018). Similarly, foliar sprays of Se NPs ranged from 10 to 160 mg L^{-1} increased the coffee grain Se contents up to 4.84–5.82 ppm (Mateus et al. 2021). Furthermore, foliar Se NPs application at 25 mgL^{-1} improved the NPK uptake as well as the crop yield of cucumber (Shalaby et al. 2021). The Cu, Mg, and Ca contents in onion bulbs as well as leaf allicin contents were increased substantially with the soil application of CuO NPs (Wang et al. 2020). Overall, the application of nanofertilizers/NPs is important to improve the crop growth, yield, and nutrient enrichment of crop plants; however, care must be taken regarding the application method, dose, and type of nanofertilizers applied to avoid the toxic effects and pollution-related issues. The application of nanofertilizers and their impacts on crops have been summarized in Table 26.1.

Table 26.1 Effect of different nanofertilizers on growth, yield, and mineral nutrients of different crops.

S. no.#	Nanofertilizers	Application dose	Application method	Crop	Outcomes	Reference
1	Nanochelated iron fertilizer	2.5 g L^{-1}	Foliar application	Rice	Increased biological yield by 27% and decreased hollow. Grain number by 25%; raised protein content by 13%. Increase in nitrogen, phosphorus, potassium, iron, and zinc concentrations in white rice by 46%, 43%, 41%, 25%, and 50%, respectively.	Fakharzadeh et al. (2020)
2	Nanofertilizers of potassium and boron	K = 0, 1, 2 K L^{-1} B = 0, 1.5, 3 mL^{-1}	Foliar application	Wheat	Increased chlorophyll content, plant height, number of spikes, and grains per spike. The grain and biological yield was increased by 17.94% and biological yield by 14.31%, respectively.	Noaema and Alhasany (2020)
3	Nano-N	0, 15, and 30 L ha^{-1}	Spray	Wheat	Application of Nano-N at 30 L ha−1 resulted in the highest flag leaf area, number of spikes, grains per spike, 1000 grain weight, and grain yield.	Al-Shamary and Al-Ansari (2022)
4	Nano-urea	No nano-urea, 100% N through nano-urea, 90% Urea + 10% N through Nano-urea, 80% + 20%, 70% +30%, 60% +40%, 50% +50%, 40% + 60%	Spray	Rice	The maximum plant height, number of tillers, grain yield, and straw yield were found in the treatment 50% RDN + 50% Nano-N.	Midde et al. (2022)

(*Continued*)

Table 26.1 Effect of different nanofertilizers on growth, yield, and mineral nutrients of different crops. *(continued)*

S. no.#	Nanofertilizers	Application dose	Application method	Crop	Outcomes	Reference
5	NPK nanofertilizer	N>1.2%; P2O5>0.001%; K2O>0.0001%	Foliar	Rice	Growth and antioxidant activity were improved	Benzon et al. (2015)
6	Zinc nanofertilizer	5 kg Zn ha^{-1}	Exogenous	Rice	Improved plant height and root length	Yuvaraj and Subramanian (2014)
7	NPK NPs	3,4.5, 6, and 9 mL	Foliar	Cucumber	Increase in plant height, chlorophyll content, NPK% in leaves and fruits.	Merghany et al. (2019)
8	P and K as nanofertilizes	P at 0, 250, and 500 ppm and K at 0, 150, and 300 ppm	Foliar	Broad bean	Significantly increased the proteins, straw, seeds, and biological yield of broad bean	El-Azizy et al. (2021)
9	Nanofertilizers of zinc, iron, and manganese	6 g L^{-1}	Foliar	Wheat	Leaf area, spikelets in spike, and grain yield were increased up to 13.8%, 20%, and 32%	Jassim et al. (2023)
10	NPK nanofertilizer	(0, 2, and 4 g L^{-1})	Foliar spray	Maize	Increase in plant growth rates increase chlorophyll content and dry matter accumulation	Hussein and Ahmed (2023)
11	Nanofertilizer of NPK	Nano (N+P, N+K, P+K) 50 + 50 and 75 + 75 mL Nano (N+P+K) 33.3 + 33.3 + 33.3 50 + 50 + 50 mL Nano-super micro plus (SMP) 100 150 mL	Spray	Wheat	All nanofertilizers improved the growth parameters and nutrient content, i.e. NP and K in wheat, whereas SMP nanofertilizer was found superior to others.	Al-Juthery et al. (2018)
12	Hybrid nanocomposite (HNCSF)	Applied at five different growth stages by using 5 mg as granular form and 5 mg as a foliar spray.	Foliar + soil application	Strawberry	Ash, protein, and fat content of strawberry fruits were significantly increased	Rahman et al. (2021)
13	Nano NPK	Nano-micro elements 2 g L^{-1} + nano-seaweed extract 2 mL L^{-1} + NPK nano-concentration 2 g L^{-1}	Spraying	Fenugreek	Fresh and dry biomass, seed, and oil yield was improved	Al-Saidi et al. (2022)
14	Nanofertilizer of Zn and Cu with normal NPK	Soil application of 50% N + 100% PK + 50% Zn along with 1st foliar spray with Nano-N, 2nd spray with Nano-Zn and 3rd spray with Nano-Cu	Spray	Tomato	Plant height, number of branches, fruit length, fruit girth, number of fruits per plant, fruit weight, and overall yield were improved.	Mishra et al. (2020)
5	Nanofertilizer N2O P2O K2O	3 mg L^{-1}	Spray	Cucumber	Vegetative growth and yield of cucumber were substantially improved	Al-Bayati et al. (2023)
16	Chitosan-silicon nanofertilizer (CS-Si NF)	0.04–0.12% (w/v)	Foliar + seed priming	maize	Improved seedling vigor, maintained cellular-redox homeostasis, improved leaf area, and grain yield	Kumaraswamy et al. (2021)

(*Continued*)

Table 26.1 Effect of different nanofertilizers on growth, yield, and mineral nutrients of different crops. *(continued)*

S. no.#	Nanofertilizers	Application dose	Application method	Crop	Outcomes	Reference
17	Nanofertilizers NPK	0, 20, and 30g L^{-1}	Spray	Brassica	Leaf morphological characteristics and chlorophyll contents were improved	Al-Jubouri and Abdulrahman. (2023)
18	ZnO NPs	0, 25, 50,100, and 200 mg L^{-1}	Foliar application	Wheat	Increase growth, yield, and Zn biofortification in wheat plants	Nazir et al. (2024)
19	NPK Nanofertilizer	1, 2, and 3 g L^{-1}	Foliar spraying	Wheat	Increase in spike length, number of grains per spike, 1000- grain weight, biological yield, straw yield and harvest index	Alqasim and Al-Ghazal (2023)
20	Potassium nanofertilizers	1, 0, and 2g L^{-1}	Spraying	Oat	Significant increase in plant height, flag leaf area, index of chlorophyll pigments in flag leaf, tillers per plant, panicles per plant, and grains per panicle were noticed	Al-Sultani et al. (2023)

26.9 Benefits and Challenges of Nanofertilizer-based Crop Biofortification

No doubt, synthetic chemical fertilization has successfully met the agricultural fertilizer needs of the world despite a number of challenges, such as poor uptake and significant losses – particularly of up to 70% for nitrogen and potassium and 90% for phosphorus. The risk of important mineral and nutrient deficiencies affects about two-third of the global population (Alemu 2020). Globally, there are currently between 720 and 811 million undernourished individuals, and that number is predicted to rise to 840 million by 2030 (World Health Organization 2020).

Conventional fertilizers are becoming outdated owing to their own application and environmental challenges; therefore, nanotechnology has emerged as a promising improvement pathway (Christen et al. 2013). Designing incredibly small objects, ranging in size from 1 to 100 nm, that stand in between individual atoms or molecules and bulk materials is the focus of nanotechnology. When compared to their bulk counterparts, these nano-sized materials show remarkable qualities that affect reactivity and other physicochemical characteristics (Bhardwaj et al. 2019).

Enduring the apparent benefits of nanofertilizers, limited findings have raised concerns regarding nanotoxicity and genotoxicity, and there have been inconsistent results from comparable formulations on different plant species (Koo et al. 2015). Global interest is growing in the application of fertilizer to crops in the form of nanofertilizers. Application of hybrid nanofertilizer including nano-urea-modified hydroxyapatite NPs improved the availability of essential plant nutrients, plant yield, and nutrient use efficiency (Tarafder et al. 2020).

The amount of fertilizer needed for crop production can be significantly decreased with the use of nanofertilizers as compared to conventional fertilizers. For example, in contrast to conventional fertilizer, Abdel-Aziz, Hasaneen, and Omer (2016) showed that a lower concentration (10%) of chitosan-NPK nanofertilizer greatly boosted the harvest index, crop index, and nutrient mobilization in wheat. On the other hand, it is essential to minimize runoff and the leaching of mineral nutrients into surface and subsurface aquifers in order to stop the buildup of pollutants and the contamination of drinking water, which can result in eutrophication and algal bloom, with overuse of phosphate fertilizers being a major cause of eutrophication (Hazra 2016). Nanofertilizers, especially those based on phosphorus, have the potential to reduce global water pollution and enhance water quality. Thus, nanotechnology has great potential in precision agriculture to improve the quality and quantity of crops through biofortification (Datta and Vitolins 2016; Sharma et al. 2017; Xiong et al. 2017). The application of nanofertilizer can increase growth, crop production, and plant micronutrient status (Deepa and Ganesan 2015; Sharma et al. 2014). Engineered NPs with the desired structure and physicochemical properties can be a safe alternative to facilitate the biofortification of food crops (Elemike et al. 2019). The application of nanofertilizers also improved crop growth and crop nutrient status under saline conditions (Siddiqi and Al-Whaibi 2014).

No doubt, nanofertilizers may be the future fertilizers owing to their superiority over conventional fertilizers; however, still there is a long way to go to achieve the desired goals. In addition to their potential benefits, nanofertilizers have their own implications and challenges. For example, the genotoxicity of NPs is directly elated with exposure time and concentration and inversely related to size factors that promote NPs accumulation in plants, whereas it seems that the toxicity rises with a decrease in size (Chen 2018). Moreover, the NPs/NFs can damage DNA chemically or physically through direct effect and/or can also cause indirect genotoxicity through interactions with nuclear proteins, decreased DNA repair, and the production of reactive oxygen species (Mehrian and De Lima 2016). Another mechanism of nanotoxicity is the disruption of cell wall membranes caused by the sharp edges of NPs, which can cause leakage of cellular content and ultimately result in cell death. For example, it was found that Ag NPs caused damage to the root cell walls and vacuoles of rice, with the extent of the damage related to the concentration of Ag NPs applied (Mahakham et al. 2017). Severe damage was observed as a result of fullerene NPs obstructing the apertures of root cells. Spermatozoa were physically bound by graphene oxide nanosheets, which caused deactivation and death (Hashemi et al. 2014). In addition, crop biofortification has also some challenges over other biofortified food supplements. For example, biofortification through NFs is unable to match the micronutrient amounts offered by supplements or foods with industrial fortification that may be easier for rural populations to obtain (Singh et al. 2016). Even though biofortification has improved, without higher nutritional levels, it may not be able to completely eradicate micronutrient shortages. Concerns regarding crop reliance and decreased environmental and dietary diversity are raised by the fact that biofortification currently concentrates on a small number of staple crops (Johns and Eyzaguirre 2007). Variety in the diet is important since even biofortified staple crops may not contain all the nutrients found in other foods. This constraint highlights how crucial it is to take into account how normal foods might help correct micronutrient shortages in order to ensure the long-term viability of biofortification initiatives.

26.10 Summary and Future Perspectives

In summary, the global increase in population necessitates a corresponding rise in fortified crop production, a challenge that NFs can meet in the future. The NFs offer significant potential for enhancing crop productivity while promoting sustainable agriculture practices. The large surface area of NFs facilitates the controlled release of nutrients, resulting in rapid prolonged nutrient availability and efficient use than conventional fertilizers. Moreover, their small particle size enables efficient root penetration and uptake by plants, enhancing nutrient absorption and NUE. The NFs contribute to sustainable agriculture by minimizing fertilizer losses from volatilization, runoff, and leaching during production, reducing environmental impacts. The NFs offer a smart delivery system, reducing the quantity of fertilizers needed while maximizing plant growth and nutrient uptake. The NFs can be applied via foliar application or soil amendments; however, foliar application shows promising results in enhancing plant morphological, physiological, biochemical, and yield traits. The nanofertilizers can be synthesized through physical, chemical, or biological methods, with green synthesis emerging as the most effective and environmentally friendly approach. Beyond enhancing agricultural productivity, the NFs confer stress resistance to crops, further enhancing their suitability for agricultural applications. In conclusion, the unique characteristics of NFs position them as valuable tools for addressing the global food security challenge, offering a sustainable and efficient approach to enhancing crop yields and resilience. Continued research and development in this field hold promise for further optimizing NFs for widespread adoption in agriculture.

References

Abdel-Aziz, H.M., Hasaneen, M.N. and Omer, A.M. (2016). Nano chitosan-NPK fertilizer enhances the growth and productivity of wheat plants grown in sandy soil. *Spanish Journal of Agricultural Research* 14 (1): 0902–0902.

Adeyemi, O., Keshavarz-Afshar, R., Jahanzad, E. et al. (2020). Effect of wheat cover crop and split nitrogen application on corn yield and nitrogen use efficiency. *Agronomy* 10 (8): 1081.

Ahmadian, K., Jalilian J., Pirzad A. (2021). Nanofertilizers improved drought tolerance in wheat under deficit irrigation. *Agricultural Water Management* 244, 106544

Al-Bayati, H.J., Al-Harbawy, K.A., Allela, W.B. et al. (2023). Response growth and yield of three cucumber hybrids (*Cucumis sativus* L.) to nano fertilizer N 20 P 20 K 20 under protected House. *IOP Conference Series: Earth and Environmental Science* 1213 (1): 012051.

Alemu, E.A. (2020). Malnutrition and its implications on food security. In: *Zero Hunger Encyclopedia of the UN Sustainable Development Goals* (eds. W. Leal Filho, A.M. Azul, L. Brandli et al.). Cham: Springer.

Ali, A., Zafar, H., Zia, M. et al. (2016). Synthesis, characterization, applications, and challenges of iron oxide nanoparticles. *Nanotechnology, Science and Applications* 9 (9): 49–67.

Al-Jubouri, A.H.A. and Abdulrahman, H.B.A. (2023). Effect of nano-fertilizer and boron on vegetative growth characteristics of two hybrids of cabbage plant (*Brassica oleracea* var. Capitata L.). *IOP Conference Series: Earth and Environmental Science* 1158 (4): 042019.

Al-Juthery, H.W., Habeeb, K.H., Altaee, F.J.K. et al. (2018). Effect of foliar application of different sources of nanofertilizers on growth and yield of wheat. *Bioscience Research* 15 (4): 3976–3985.

Almendros, P., González, D., Fernández, MD. et al. (2022). Both Zn biofortification and nutrient distribution pattern in cherry tomato plants are influenced by the application of ZnO nanofertilizer. *Heliyon* 8 (3): e09130.

Alqasim, Y.F.Y. and Al-Ghazal, S.A.Y. (2023). Impact foliar Spraying of NPK Nano fertilizer on yield traits of two varieties of bread wheat (*Triticum aestivum* L.). *IOP Conference Series: Earth and Environmental Science* 1213 (1): 012006.

Al-Saidi, A.A.R.N., Al-Mohammad, M.H. and Al-Juthery, H.W. (2022). Effect of spraying some nanofertilizers and their combinations on the growth and yield of fenugreek (*Trigonella foenum-graecum* L.). *IOP Conference Series: Earth and Environmental Science* 1060 (1): 012041.

Alshaal, T. and El-Ramady, H. (2017). Foliar application: From plant nutrition to biofortification. *Environment Biodiversity and Soil Security* 1: 71–83.

Al-Shamary, A.J.Z. and Al-Ansari, A.M. S. (2022). Response of growth and productivity of wheat cultivars (*Triticum aestivum* L.) to Nano-N and Urea Fertilizer. *IOP Conference Series: Earth and Environmental Science* 1060 (1): 012040.

Al-Sultani, A.J.K., Al-Mohammad, M.H. and Lehmood, A.M. (2023). Response of two oat cultivars to spraying nano potassium in some growth parameters and yield components. In: *IOP Conference Series: Earth and Environmental Science* 1262 (5): 052029.

Ashkavand, P., Tabari, M., Zarafshar, M. et al. (2015). Effect of SiO_2 nanoparticles on drought resistance in hawthorn seedlings. *Leśne Prace Badawcze* 76 (4): 350–359.

Ashraf, U., Kiran, M., Shahid, M.N. (2023). Biofortification through seed priming in food crops: potential benefits and future scope. In: *Mineral Biofortification in Crop Plants for Ensuring Food Security*, 261–296. Singapore: Springer Nature.

Ashraf, U., Shahid, M.N., Batool, F. et al. (2023). Application of nanomaterials in agriculture. In: *The Impact of Nanoparticles on Agriculture and Soil*, 259–283. Academic Press.

Ashraf, U., Zafar, S., Ghaffar, R. et al. (2022). Impact of nano chitosan-NPK fertilizer on field crops. In: *Role of Chitosan and Chitosan-Based Nanomaterials in Plant Sciences*, 165–183. Cambridge: Academic Press.

Asli, S. and Neumann, P.M. (2009). Colloidal suspensions of clay or titanium dioxide nanoparticles can inhibit leaf growth and transpiration via physical effects on root water transport. *Plant Cell and Environment* 32 (5): 577–584.

Azam, M., Bhatti, H.N., Khan, A. et al. (2022). Zinc oxide nano-fertilizer application (*foliar and soil*) effect on the growth, photosynthetic pigments and antioxidant system of maize cultivar. *Biocatalysis and Agricultural Biotechnology* 42: 102343.

Basak, B.B., Pal, S. and Datta, S.C. (2012). Use of modified clays for retention and supply of water and nutrients. *Current Science* 102 (9): 1272–1278.

Benzon, H.R.L., Rubenecia, M.R.U., Ultra Jr, V.U. et al. (2015). Nano-fertilizer affects the growth, development, and chemical properties of rice. *International Journal of Agronomy and Agricultural Research* 7 (1): 105–117.

Bhardwaj, A.K., Chejara, S., Malik, K. et al. (2022). Agronomic biofortification of food crops: An emerging opportunity for global food and nutritional security. *Frontiers in Plant Science* 13: 1055278.

Bhardwaj, A.K., Hamed, L.M., Sharma, N. et al. (2019). Engineered polymeric and nano-materials for taming salty soils and waters used for crop production. In: *Research Developments in Saline Agriculture* (eds J. Dagar, R. Yadav and P. Sharma). Singapore: Springer.

Bhargava, A., Srivastava, S., Bhargava, A. et al. (2019). Cereals. In: *Participatory Plant Breeding: Concept and Applications*, 129–173. New York: Springer.

Biesalski, H.K. and Biesalski, H.K., (2013). Trapped on the hunger carousel: Generation after generation. *Hidden Hunger* 65–130.

Bindra, H.S. and Singh, B. (2021). Nanofertilizers and nanopesticides: Future of plant protection. In: *Advances in Nanofertilizers and Nano-pesticides in Agriculture*, 57–84. Delhi: Woodhead Publishing.

Bouis, H., McEwan, M. and Tanumihardjo, S. (2013). ICN2 Second international conference on nutrition. In: *Biofortification: Evidence and Lessons Learned Linking Agriculture and Nutrition*, 1–23. Rome: Food and Agriculture Organization of the United Nations.

Campos-Bowers, M.H. and Wittenmyer, B.F. (2007). Biofortification in China: Policy and practice. *Health Research Policy and Systems* 5 (1): 1–7.

Chen, H. (2018). Metal based nanoparticles in agricultural system: behavior, transport, and interaction with plants. *Chemical Speciation and Bioavailability* 30 (1): 123–134.

Chhipa, H. (2017). Nanofertilizers and nanopesticides for agriculture. *Environmental chemistry letters* 15 (1): 15–22.

Christen, V., Capelle, M. and Fent, K. (2013). Silver nanoparticles induce endoplasmatic reticulum stress response in zebrafish. *Toxicology and Applied Pharmacology* 272 (2): 519–528.

Corradini, E., De Moura, M.R. and Mattoso, L.H.C. (2010). A preliminary study of the incorporation of NPK fertilizer into chitosan nanoparticles. *Express Polymer Letters* 4 (8): 509–515.

Corredor, E., Testillano, P.S., Coronado, M.J. et al, (2009). Nanoparticle penetration and transport in living pumpkin plants: In situ subcellular identification. *BMC Plant Biology* 9 (1): 1–11.

Cyriac, J., Melethil, K., Thomas, B. et al. (2020). Synthesis of biogenic ZnO nanoparticles and its impact on seed germination and root growth of *Oryza sativa* L. and *Vigna unguiculata* L. *Materials Today: Proceedings* 25: 224–229.

Czymmek, K., Ketterings, Q., Ros, M. et al. (2020). The New York phosphorus index 2.0. In: *Agronomy Fact Sheet Series. Fact Sheet #110*; New York, NY, USA: Cornell University Cooperative Extension.

Datta, M. and Vitolins, M.Z. (2016). Food fortification and supplement use – are there health implications? *Critical Reviews in Food Science and Nutrition* 56 (13): 2149–2159.

Davar, Z.F., Roozbahani, A. and Hosnamidi, A. (2014). Evaluation the effect of water stress and foliar application of Fe nanoparticles on yield, yield components and oil percentage of safflower (*Carthamus tinctorious L.*). *International Journal of Advanced Biological and Biomedical Research* 2: 1150–1159.

De França., Bettencourt, G.M., Degenhardt, J., Torres, L.A.Z. et al. (2020). Green biosynthesis of single and bimetallic nanoparticles of iron and manganese using bacterial auxin complex to act as plant bio-fertilizer. *Biocatalysis and Agricultural Biotechnology 30*: 101822.

De Valença, A.W., Bake, A., Brouwer, I.D. et al. (2017). Agronomic biofortification of crops to fight hidden hunger in sub-Saharan Africa. *Global Food Security* 12: 8–14.

Deepa, B. and Ganesan, V. (2015). Bioinspired synthesis of selenium nanoparticles using flowers of *Catharanthus roseus* (L.) G. Don. and *Peltophorum pterocarpum* (DC.) Backer ex Heyne – a comparison. *International Journal of Chemical Technology Research* 77 (22): 725–733.

DeRosa, M.C., Monreal, C., Schnitzer, M. et al. (2010). Nanotechnology in fertilizers. *Nature Nanotechnology* 5 (2): 91.

DESA, (2015). *World Population Projected to Reach 9.7 Billion by 2050*. New York: United Nations Department of Economic and Social Affairs.

Diatta, A.A., Fike, J.H., Battaglia, M.L. et al. (2020). Effects of biochar on soil fertility and crop productivity in arid regions: A review. *Arabian Journal of Geosciences* 13 (14): 1–17.

Dimkpa, C.O., Campos, M.G., Fugice, J. et al. (2022). Synthesis and characterization of novel dual-capped Zn–urea nanofertilizers and application in nutrient delivery in wheat. *Environmental Science: Advances* 1 (1): 47–58.

Dimkpa, C.O., Singh, U., Bindraban, P.S. et al. (2019). Zinc oxide nanoparticles alleviate drought-induced alterations in sorghum performance, nutrient acquisition, and grain fortification. *Science Total Environ* 688 (9): 926–934.

Du, W., Sun, Y., Ji, R. et al. (2011). TiO_2 and ZnO nanoparticles negatively affect wheat growth and soil enzyme activities in agricultural soil. *Journal of Environmental Monitoring* 13 (4): 822–828.

Eid, M.A., Abdel-Salam, A.A., Salem, H.M. et al. (2020). Interaction effects of nitrogen source and irrigation regime on tuber quality, yield, and water use efficiency of *Solanum tuberosum L. Plants* 9 (1): 110.

El-Azizy, F.A. and Habib, A., A M. (2021). Effect of nano phosphorus and potassium fertilizers on productivity and mineral content of broad bean in North Sinai. *Journal of Soil Sciences and Agricultural Engineering* 12 (4): 239–246.

Elemike, E.E., Uzoh, I.M., Onwudiwe, D.C. et al. (2019). The role of nanotechnology in the fortification of plant nutrients and improvement of crop production. *Applied Sciences* 9 (3): 499.

Eroglu, N., Emekci, M. and Athanassiou, C.G. (2017). Applications of natural zeolites on agriculture and food production. *Journal of the Science of Food and Agriculture* 97 (11): 3487–3499.

Fakharzadeh S., Hafizi M., Baghaei MA. et al. (2020). Using nanochelating technology for biofortification and yield increase in rice. *Scientific Reports* 10 (1): 4351.

Farhangi-Abriz, S. and Torabian, S. (2018). Nano-silicon alters antioxidant activities of soybean seedlings under salt toxicity. *Protoplasma* 255 (4): 953–962.

Garcia-Casal, M.N., Peña-Rosas, J.P., Giyose, B. et al. (2017). Staple crops biofortified with increased vitamins and minerals: Considerations for a public health. *Strategy Annals of the New York Academy of Sciences* 1390 (1): 3–13.

Guilbert, J.J. (2003). The world health report 2002 – reducing risks, promoting healthy life 1. *Education for Health* 16 (2): 230.

Guo, H., White, J.C., Wang, Z. et al. (2018). Nano-enabled fertilizers to control the release and use efficiency of nutrients. *Current Opinion in Environmental Science and Health* 6: 77–83.

Guo, J. (2004). Synchrotron radiation, soft-X-ray spectroscopy and nanomaterials. *International Journal of Nanotechnology* 1 (1–2): 193–225.

Harper, T. (2015). *The Year of the Trillion Dollar Nanotechnology Market*. UK Manchester: AZoNetwork UK Ltd.

Hashemi, E., Akhavan, O., Shamsara, M. et al. (2014). Cyto and genotoxicities of graphene oxide and reduced graphene oxide sheets on spermatozoa. *RSC Advances* 4 (52): 27213–27223.

Hazra, G. (2016). Different types of eco-friendly fertilizers: An overview. *Sustainability In: Environment* 1 (1): 54.

Hossain, S.M. and Mohiuddin, A.K.M. (2012). Study on biofortification of rice by targeted genetic engineering. *International Journal Agriculture Research Innovation and Technology* 2 (2): 25–35.

Hu, J., Guo, H., Li, J. et al. (2017). Interaction of γ-Fe_2O_3 nanoparticles with *Citrus maxima* leaves and the corresponding physiological effects via foliar application. *Journal of Nanobiotechnology* 15 (1): 1–12.

Husen, A. and Iqbal, M. (2019). *Nanomaterials and Plant Potential: An Overview*. New York: Springer International Publishing 3–29.

Hussein, S.I. and Ahmed, K.K. (2023). Response of yield traits and its components to nano-fertilization in maize (*Zea mays* L.). *IOP Conference Series: Earth and Environmental Science* 1262 (5): 052048.

Jakkula, V.S. and Wani, S.P. (2018). Zeolites: Potential soil amendments for improving nutrient and water use efficiency and agriculture productivity. *Scientific Reviews and Chemical Communications* 8 (1): 1–15.

Janmohammadi, M., Amanzadeh, T., Sabaghnia, N. and. (2016). Effect of nano-silicon foliar application on safflower growth under organic and inorganic fertilizer regimes. *Botanica Lithuanica* 22 (1): 53–64.

Jassim, R.A.H., Barkat, H.N. and Nassir, M.A. (2023). The effect of the level and methods of adding nano-fertilizer on the absorption of some elements and the growth and yield of wheat (*Triticum aestivum* L.). *IOP Conference Series: Earth and Environmental Science* 1158 (6): 062023.

Jha, A.B. and Warkentin, T.D. (2020). Biofortification of pulse crops: Status and future perspectives. *Plants* 9 (1): 73.

Johns, T. and Eyzaguirre, P.B. (2007). Biofortification, biodiversity and diet: A search for complementary applications against poverty and malnutrition. *Food Policy* 32 (1): 1–24.

Kah, M., Beulke, S., Tiede, K. et. al. (2013). Nanopesticides: state of knowledge, environmental fate, and exposure modeling. *Critical Reviews in Environmental Science and Technology* 43 (16): 1823–1867.

Kah, M., Kookana, R.S., Gogos, A. et al. (2018). A critical evaluation of nanopesticides and nanofertilizers against their conventional analogues. *Nature Nanotechnology* 13 (8): 677–684.

Kahrl, F., Li, Y., Su, Y. et al. (2010). Greenhouse gas emissions from nitrogen fertilizer use in China. *Environmental Science and Policy* 13 (8): 688–694.

Kalia, A. and Sharma, S.P. (2019). Nanomaterials and vegetable crops: realizing the concept of sustainable production. In: *Nanoscience for Sustainable Agriculture* (eds. R. Pudake, N. Chauhan and C. Kole), 323–353. Cham: Springer.

Kong, X.P., Zhang, B.H. and Wang, J. (2021). Multiple roles of mesoporous silica in safe pesticide application by nanotechnology: A review. *Journal of Agricultural and Food Chemistry* 69 (24): 6735–6754.

Koo, Y., Wang, J., Zhang, Q. et al. (2015). Fluorescence reports intact quantum dot uptake into roots and translocation to leaves of *Arabidopsis thaliana* and subsequent ingestion by insect herbivores. *Environmental Science and Technology* 49 (1): 626–632.

Kottegoda, N., Sandaruwan, C., Priyadarshana, G. et al. (2017). Urea-hydroxyapatite nanohybrids for slow release of nitrogen. *ACS Nano* 11 (2): 1214–1221.

Kubavat, J., Sarvaiya, J., Tyagi, T. et al. (2023). Development of Levan capped silver nanoparticles based product and its effect on wound healing. *Burns Open* 7 (1): 8–21.

Kumar, P., Lai, L., Battaglia, M.L. et al. (2019a). Impacts of nitrogen fertilization rate and landscape position on select soil properties in switchgrass field at four sites in the USA. *Catena* 180 (1): 183–193.

Kumar, S., Lai, L., Kumar, P. et al. (2019b). Impacts of nitrogen rate and landscape position on soils and switchgrass root growth parameters. *Agronomy Journal* 111 (3): 1046–1059.

Kumaraswamy, R.V., Saharan, V., Kumari. S. et al. (2021). Chitosan-silicon nanofertilizer to enhance plant growth and yield in maize (*Zea mays* L.). *Plant Physiology and Biochemistry* 159 (24): 53–66.

León-Silva, S., Arrieta-Cortes, R., Fernández-Luqueño, F. et al. (2018). Design and production of nanofertilizers. In: *Agricultural Nanobiotechnology* (eds F. López-Valdez, F. Fernández-Luqueño). Cham: Springer. doi: 10.1007/978-3-319-96719-6_2.

Li, J.X., Wee, C.D. and Sohn, B.K. (2010). Growth Response of Hot Pepper Applicated with Ammonium (NH_4^+) and Potassium (K^+)-Loaded Zeolite. *Korean Journal of Soil Science and Fertilizer* 43 (5): 741–747.

Li, S.X., Wang, Z.H., Miao, Y.F. et al. (2014). Soil organic nitrogen and its contribution to crop production. *Journal of Integrative Agriculture* 13 (10): 2061–2080.

Li, Y., Liang, L., Li, W. et al. (2021). ZnO nanoparticle-based seed priming modulates early growth and enhances physio-biochemical and metabolic profiles of fragrant rice against cadmium toxicity. *Journal of Nanobiotechnology* 19: 1–19.

Lin, N., Huang, J., Chang, P.R. et al. (2011). Preparation, modification, and application of starch nanocrystals in nanomaterials: A review. *Journal of Nanomaterials* 2011 (1): 1–13.

Liscano, J.F., Wilson, C.E., Norman-Jr, R.J. et al. (2000). *Zinc availability to rice from seven granular fertilizers*, 963. Fayetteville, CA, USA: Arkansas Agricultural Experiment Station.

Liu, R. and Lal, R. (2014). Synthetic apatite nanoparticles as a phosphorus fertilizer for soybean (*Glycine max*). *Scientific Reports* 4 (1): 5686.

Liu, R. and Lal, R. (2015). Potentials of engineered nanoparticles as fertilizers for increasing agronomic productions. *Science of the Total Environment* 514: 131–139.

MA, G., Radwan, F.I., Kandil, E.E. et al. (2017). Comparison of some new maize hybrids response to mineral fertilization and some nanofertilizers. *Alexandria Science Exchange Journal* 38: (July–September) 506–514.

Madzokere, T.C., Murombo, L.T. and Chiririwa, H. (2021). Nano-based slow releasing fertilizers for enhanced agricultural productivity. *Materials Today: Proceedings* 45 (6): 3709–3715.

Mahakham, W., Sarmah, A.K., Maensiri, S. et al. (2017). Nanopriming technology for enhancing germination and starch metabolism of aged rice seeds using phytosynthesized silver nanoparticles. *Scientific Reports* 7 (1): 8263.

Mahmoodi, P., Yarnia, M., Amirnia, R. et al. (2017). Comparison of the effect of nano urea and nono iron fertilizers with common chemical fertilizers on some growth traits and essential oil production of Borago officinalis L. *Agricultural and Food Sciences* 2: 1–4.

Mai, Y., Ren, Y., Deng, S. et al. (2024). Influence of ZnO nanoparticles on early growth stage of fragrant rice at low temperature (LT) stress. *Journal of Soil Science and Plant Nutrition* 24 (1): 1301–1317.

Malik, K.A. and Maqbool, A. (2020). Transgenic crops for biofortification. *Frontiers in: Sustainable Food Systems* 4: 571402.

Manikandan, A. and Subramanian KS. (2016). Evaluation of zeolite based nitrogen nanofertilizers on maize growth, yield and quality on inceptisols and alfisols. *International Journal Plant Soil Science* 9 (4): 1–9

Mateus, M.P., Tavanti, R.F.R., Tavanti, T.R. et al. (2021). Selenium biofortification enhances ROS scavenge system increasing yield of coffee plants. *Ecotoxicol Environ Safety* 209 (209): 111772 doi: org/10.1016/j.ecoenv.2020.111772.

Medina-Velo, I.A., Dominguez, O.E., Ochoa, L. et al. (2017). Nutritional quality of bean seeds harvested from plants grown in different soils amended with coated and uncoated zinc oxide nanomaterials. *Environmental Science Nano* 4 (12): 2336–2347.

Mehrian, S.K. and De Lima, R. (2016). Nanoparticles cyto and genotoxicity in plants: Mechanisms and abnormalities. *Environmental Nanotechnology, Monitoring and Management* 6: 184–193.

Merghany, M.M., Shahein, M.M., Sliem, M.A. et al. (2019). Effect of nanofertilizers on cucumber plant growth, fruit yield and it's quality. *Plant Archives* 19 (2): 165–172.

Midde, S.K., Perumal, M.S., Murugan, G. et al. (2022). Evaluation of nano urea on growth and yield attributes of rice (*Oryza sativa* L.). *Chemical Science Review and Letters* 11 (42): 211–214.

Milani, N., Hettiarachchi, G.M., Kirby, J.K. et al. (2015). Fate of zinc oxide nanoparticles coated onto macronutrient fertilizers in an alkaline calcareous soil. *PLoS One* 10 (5): e0126275.

Mishra, B., Sahu, G.S., Mohanty, L.K. et al. (2020). Effect of nanofertilizers on growth, yield and economics of tomato variety Arka Rakshak. *Indian Journal of Pure and Applied Biosciencesis* 8: 200–204.

Mondal, A., Basu, R., Das, S. et al. (2011). Beneficial role of carbon nanotubes on mustard plant growth: An agricultural prospect. *Journal of Nanoparticle Research* 13 (10): 4519–4528.

Monreal, C.M., DeRosa, M., Mallubhotla, S.C. et al. (2016). Nanotechnologies for increasing the crop use efficiency of fertilizer-micronutrients. *Biology and Fertility of Soils* 52 (3): 423–437.

Morales-Díaz, A.B., Ortega-Ortíz, H., Juárez-Maldonado, A. et al. (2017). Application of nanoelements in plant nutrition and its impact in ecosystems. *Advances in Natural Sciences: Nanoscience and Nanotechnology* 8 (1): 013001.

Mujtaba, M., Khawar, K.M., Camara, M.C. et al. (2020). Chitosan-based delivery systems for plants: A brief overview of recent advances and future directions. *International Journal of Biological Macromolecules* 154: 683–697.

Muñoz-Márquez, E., Soto-Parra, J.M., Noperi-Mosqueda, L.C. et al. (2022). Application of molybdenum nanofertilizer on the nitrogen use efficiency, growth and yield in green beans. *Agronomy* 12 (12): 3163.

Nair, R., Varghese, S.H., Nair, B.G. et al. (2010). Nanoparticulate material delivery to plants. *Plant Science* 179 (3): 154–163.

Nawaz, H.R., Solangi, B.A., Zehra, B. et al. (2011). Preparation of nano zinc oxide and its application in leather as a retanning and antibacterial agent. *Canadian Journal on Scientific and Industrial Research* 2 (4): 164–170.

Nazir, M.A., Hasan, M., Mustafa, G. et al. (2024). Zinc oxide nano-fertilizer differentially effect on morphological and physiological identity of redox-enzymes and biochemical attributes in wheat (Triticum *aestivum* L.). *Scientific Reports* 14 (1): 13091.

Noaema, A.H. and Alhasany, A.R. (2020). Effect of spraying nanofertilizers of potassium and boron on growth and yield of wheat (*Triticum aestivum* L.). In: *IOP Conference Series: Materials Science and Engineering* 871 (1): 012012.

Olson, R., Gavin-Smith, B., Ferraboschi, C. et al. (2021). Food fortification: The advantages, disadvantages and lessons from sight and life programs. *Nutrients* 13 (4): 1118.

Padhan, S.R., Kar, I., Mohanty, A. et al. (2023). Nanofertilizers: Types, synthesis, methods, and mechanisms. In: *Nanofertilizers for Sustainable Agroecosystems: Recent Advances and Future Trends*, 61–98. Cham: Springer Nature Switzerland.

Pal, M. and Molnár, J. (2021). Growing importance of fruits and vegetables in human health. *International Journal of Food Science and Agriculture* 5 (4): 567–569.

Parra-Torrejón, B., Cáceres, A., Sánchez, M. et al. (2023). Multifunctional nanomaterials for biofortification and protection of tomato plants. *Environmental Science and Technology* 57 (40): 14950–14960.

Patra, P., Choudhury, S.R., Mandal, S. et al. (2013). Effect sulfur and ZnO nanoparticles on stress physiology and plant (*Vigna radiata*) nutrition. In: *Advanced Nanomaterials and Nanotechnology: Proceedings of the 2nd International Conference on Advanced Nanomaterials and Nanotechnology*, Dec 8–10, 2011, *Guwahati, India*, 301–309. Berlin: Springer Heidelberg.

Petosa, A.R., Rajput, F., Selvam, O. et al. (2017). Assessing the transport potential of polymeric nanocapsules developed for crop protection. *Water Research* 111 (13): 10–17.

Polat, E., Karaca, M., Demir, H. et al. (2004). Use of natural zeolite (clinoptilolite) in agriculture. *Journal of Fruit and Ornamental Plant Research* 12 (1): 183–189.

Preetha, P.S. and Balakrishnan, N. (2017). A review of nanofertilizers and their use and functions in soil. *International Journal of Current Microbiology Applied. Science* 6 (12): 3117–3133.

Qureshi, A., Singh, D.K. and Dwivedi, S. (2018). Nanofertilizers: A novel way for enhancing nutrient use efficiency and crop productivity. *Internation Journal of Current Microbiology Applied Science* 7 (2): 3325–3335.

Rahman, M.H., Hasan, M.N. and Khan, M.Z.H. (2021). Study on different nanofertilizers influencing the growth, proximate composition and antioxidant properties of strawberry fruits. *Journal of Agriculture and Food Research* 6: 100246.

Raliya, R., Saharan, V., Dimkpa, C. et al. (2017). Nanofertilizer for precision and sustainable agriculture: current state and future perspectives. *Journal of Agricultural and Food Chemistry* 66 (26): 6487–6503.

Rameshraddy, Pavithra, G.J., Rajashekar Reddy, B.H., Salimath, M. et al. (2017). Zinc oxide nano particles increases Zn uptake, translocation in rice with positive effect on growth, yield and moisture stress tolerance. *Indian Journal of Plant Physiology* 22 (372): 287–294.

Rostami Ajirloo, A.A. and Amiri, E. (2022). Effects of nano-potassium fertilizer on yield and water use efficiency of soybean under water deficit conditions (case study: Moghan plain, Iran). *Communications in Soil Science and Plant Analysis* 53 (12): 1542–1551.

Sabet, H., Mortazaeinezhad, F. (2018). Yield, growth and Fe uptake of cumin (*Cuminum cyminum* L.) affected by Fe-nano, Fe-chelated and Fe-siderophore fertilization in the calcareous soils. *Journal Trace Element Medicine Biology* 50: 154–160. doi: org/10.1016/j. jtemb.2018.06.020.

Sadeghzadeh, B., (2013). A review of zinc nutrition and plant breeding. *Journal of Soil Science and Plant Nutrition*, 13 (4): 905–927.

Safdar, M., Kim, W., Park, S. et al. (2022). Engineering plants with carbon nanotubes: A sustainable agriculture approach. *Journal of Nanobiotechnology* 20 (1): 1–30.

Salama, D.M., Abd El-Aziz, M.E., Shaaban, E.A. et al. (2022). The impact of nanofertilizer on agro-morphological criteria, yield, and genomic stability of common bean (*Phaseolus vulgaris* L.). *Scientific Reports* 12 (1): 18552.

Seleiman, M.F. and Abdelaal, M.S. (2018). Effect of organic, inorganic and bio-fertilization on growth, yield and quality traits of some chickpea (*Cicer arietinum* L.) varieties. *Egyptian Journal of Agronomy* 40 (1): 105–117.

Seleiman, M.F. and Kheir, A.M. (2018). Maize productivity, heavy metals uptake and their availability in contaminated clay and sandy alkaline soils as affected by inorganic and organic amendments. *Chemosphere* 204 (1-2): 514–522.

Seleiman, M.F., Ali, S., Refay, Y. et al. (2020a). Chromium resistant microbes and melatonin reduced Cr uptake and toxicity, improved physio-biochemical traits and yield of wheat in contaminated soil. *Chemosphere* 250: 126239.

Seleiman, M.F., Almutairi, K.F., Alotaibi, M. et al. (2020b). Nano-fertilization as an emerging fertilization technique: Why can modern agriculture benefit from its use? *Plants* 10 (1): 2.

Seleiman, M.F., Alotaibi, M.A., Alhammad, B.A. et al. (2020c). Effects of ZnO nanoparticles and biochar of rice straw and cow manure on characteristics of contaminated soil and sunflower productivity, oil quality, and heavy metals uptake. *Agronomy* 10 (6): 790.

Seleiman, M.F., Santanen, A. and Mäkelä, P.S. (2020). Recycling sludge on cropland as fertilizer – advantages and risks. *Resources, Conservation and Recycling* 155: 104647.

Seleiman, M.F., Santanen, A., Jaakkola, S. et al. (2013). Biomass yield and quality of bioenergy crops grown with synthetic and organic fertilizers. *Biomass and Bioenergy 59*: 477–485.

Shalaby, T.A., Abd-Alkarim, E., El-Aidy, F. et al. (2021). Nano-selenium, silicon and H_2O_2 boost growth and productivity of cucumber under combined salinity and heat stress. *Ecotoxicology and Environmental Safety* 212: 111926. doi: 10.1016/j.ecoenv.2021.111962.

Shalaby, T.A., Bayoumi, Y., Abdalla, N. et al. (2016). Nanoparticles, soils, plants and sustainable agriculture. *Nanoscience in Food and Agriculture* 1: 283–312.

Sharma, A.K., Tiwari, A.K., Dixit, A.R. et al. (2017). Investigation into performance of SiO_2 nanoparticle based cutting fluid in machining process. *Materials Today: Proceedings* 4 (2): 133–141.

Sharma, G., Sharma, A.R., Bhavesh, R. et al. (2014). Biomolecule-mediated synthesis of selenium nanoparticles using dried *Vitis vinifera* (raisin) extract. *Molecules* 19 (3): 2761–2770.

Sharma, V., Javed, B., Byrne, H. et al. (2022). Zeolites as carriers of nanofertilizers: From structures and principles to prospects and challenges. *Applied Nano* 3 (3): 163–186.

Shenoy, V.V. and Kalagudi, G.M. (2005). Enhancing plant phosphorus use efficiency for sustainable cropping. *Biotechnology Advances* 23 (7-8): 501–513.

Siddiqi, K.S. and Husen, A. (2017). Plant response to engineered metal oxide nanoparticles. *Nanoscale Research Letters* 12 (1): 1–18.

Siddiqui, M.H. and Al-Whaibi, M. H. (2014). Role of nano-SiO_2 in germination of tomato (*Lycopersicum esculentum seeds Mill.*). *Saudi Journal of Biological Sciences* 21 (1): 13–17.

Singh, A., Singh, N.B., Hussain, I. et al. (2015). Plant-nanoparticle interaction: an approach to improve agricultural practices and plant productivity. *Internation Journal Pharm Science Invent* 4 (8): 25–40.

Singh, M.D. and Kumar, B. (2017). Bio efficacy of nano zinc sulphide (ZnS) on growth and yield of sunflower (*Helianthus annuus* L.) and nutrient status in the soil. *International Journal Agriculture Science* 9: 3795–3798.

Singh, N.B., Amist, N., Yadav, K. et al. (2013). Zinc oxide nanoparticles as fertilizer for the germination, growth and metabolism of vegetable crops. *Journal of Nanoengineering and Nanomanufacturing* 3 (4): 353–364.

Singh, U., Praharaj, C.S., Chaturvedi, S.K. et al. (2016). Biofortification: Introduction, approaches, limitations, and challenges. *Biofortification of Food Crops* 3–18.

Siwela, M., Pillay, K., Govender, L. et al. (2020). Biofortified crops for combating hidden hunger in South Africa: Availability, acceptability, micronutrient retention and bioavailability. *Foods* 9 (6): 815. doi: 10.3390/foods9060815.

Sohair, E.E.D., Abdall, A.A., Amany, A.M. et al. (2018). Evaluation of nitrogen, phosphorus and potassium nanofertilizers on yield, yield components and fiber properties of Egyptian cotton (*Gossypium barbadense* L.). *Journal Plant Science Crop Protection* 1 (2): 208.

Solanki, P., Bhargava, A., Chhipa, H. et al. (2015). Nanofertilizers and their smart delivery system. In: *Nanotechnologies In Food and Agriculture*. (eds. M. Rai, C. Ribeiro, L. Mattoso et al.), 81–101. Cham: Springer.

Soliman, A.S., Hassan, M., Abou-Elella, F. et al. (2016). Effect of nano and molecular phosphorus fertilizers on growth and chemical composition of Baobab (*Adansonia digitata L.*). *Journal of Plant Sciences* 11 (4): 52–60.

Soman, R. and Balachandran, S. (2021). Trends and technologies behind controlled-release fertilizers. In: *Controlled Release Fertilizers for Sustainable Agriculture*, 155–168. Massachusetts: Academic Press.

Stein, A.J., Nestel, P., Meenakshi, J.V. et al. (2007). Plant breeding to control zinc deficiency in India: how cost-effective is biofortification? *Public Health Nutrition* 10 (5): 492–501.

Subramanian, K.S. and Tarafdar, J.C. (2011). Prospects of nanotechnology in Indian farming. *Indian Journal Agriculture Science* 81 (10): 887–893.

Subramanian, K.S., Manikandan, A., Thirunavukkarasu, M. et al. (2015). Nanofertilizers for balanced crop nutrition. In: (eds. M. Rai, C. Ribeiro, and L. Mattoso). Duran *Nanotechnologies in Food and Agriculture*. Cham: Springer. doi: org/10.1007/978-3-319-14024-7_3.

Sun, F., Chen, W., Ren, Y. et al. (2023). Nano-priming with La_2O_3 improves early growth and regulates physio-biochemical mechanisms in fragrant rice against cadmium toxicity. *Journal of Soil Science and Plant Nutrition* 23 (3): 4004–4020.

Suppan, S. (2017). Applying nanotechnology to fertilizer: Rationales, research, risks and regulatory challenges. *The Institute for Agriculture and Trade Policy Works Locally and globally, Brazil.*

Tarafder, C., Daizy, M., Alam, M.M. et al. (2020). Formulation of a hybrid nanofertilizer for slow and sustainable release of micronutrients. *ACS Omega* 5 (37): 23960–23966.

Tariq, M., Khan, F., Shah, A.H. et al. (2020). Effect of micronutrients foliar supplementation on the production and eminence of plum (*Prunus domestica* L.). *Quality Assurance and Safety of Crops and Foods* 12: 32–40.

Thakur, S., Thakur, S. and Kumar, R. (2018). Bio-nanotechnology and its role in agriculture and food industry. *Journal Molecular Genetic Medicine* 12 (324): 1747–0862.

Thirugnanasambandan, T. (2021). Advances of engineered nanofertilizers for modern agriculture. In: *Plant-Microbes-Engineered Nano-particles (PM-ENPs) Nexus in Agro-Ecosystems: Understanding the Interaction of Plant, Microbes and Engineered Nano-particles (ENPS*), 131–152.

Umair, M., Zafar, SH., Cheema, M. et al. (2023). Unraveling the effects of zinc sulfate nanoparticles and potassium fertilizers on quality of maize and associated health risks in Cd contaminated soils under different moisture regimes. *Science Total Environment* 896 (4): 165147. doi: org/10.1016/j.scitotenv.2023.165147.

Vattani, H., Keshavarz, N., Baghaei, N., (2012). Effect of sprayed soluble different levels of iron chelate nano fertilizer on nutrient uptake efficiency in two varieties of spinach (*Varamin 88 and Virofly*). *International Research Journal of Applied and Basic Sciences* 3 (13): 2651–56.

Vejan, P., Khadiran, T., Abdullah, R. et al. (2021). Controlled release fertilizer: A review on developments, applications and potential in agriculture. *Journal of Controlled Release* 339 321–334.

Wang, Y., Deng, C., Cota-Ruiz, K. et al. (2020). Improvement of nutrient elements and allicin content in green onion (*Allium fistulosum*) plants exposed to CuO nanoparticles. *Science of Total Environment* 725: 138387. doi: org/10.1016/j.scitotenv.2020.138387.

WHO. (2000). *Nutrition for Health and Development*. Switzerland: Geneva: WHO.

World Health Organization. (2020). *The State of Food Security and Nutrition in the World 2020: Transforming Food Systems for Affordable Healthy Diets.* Rome: FAO. doi: org/10.4060/ca9692en.

Xiong, T., Dumat, C., Dappe, V. et al. (2017). Copper oxide nanoparticle foliar uptake, phytotoxicity, and consequences for sustainable urban agriculture. *Environmental Science and Technology* 51 (9): 5242–5251.

Yadav, A., Yadav, K. and Abd-Elsalam, K.A. (2023). Nanofertilizers: types, delivery and advantages in agricultural sustainability. *Agrochemicals* 2 (2): 296–336.

Yuvaraj, M. and Subramanian, K.S. (2018). Development of slow release Zn fertilizer using nano-zeolite as carrier. *Journal of Plant Nutrition* 41 (3): 311–320.

Yuvaraj, M. and Subramanian, K.S. (2014). Fabrication of zinc nano fertilizer on growth parameter of rice. *Trends Bioscience* 7 (17): 2564–2565.

Zahra, Z., Arshad, M., Rafique, R. et al. (2015). Metallic nanoparticle (TiO_2 and Fe_3O_4) application modifies rhizosphere phosphorus availability and uptake by *Lactuca sativa*. *Journal of Agricultural and Food Chemistry* 63 (31): 6876–6882.

Zareabyaneh, H. and Bayatvarkeshi, M. (2015). Effects of slow-release fertilizers on nitrate leaching, its distribution in soil profile, N-use efficiency, and yield in potato crop. *Environmental Earth Sciences* 74: 3385–3393.

Zayed, M.M., Elkafafi, S.H., Zedan, A.M. et al. (2017). Effect of nano chitosan on growth, physiological and biochemical parameters of *Phaseolus vulgaris* under salt stress. *Journal of Plant Production* 8 (5): 577–585.

Zheng, L., Hong, F., Lu, S. et al. (2005). Effect of nano-TiO_2 on strength of naturally aged seeds and growth of spinach. *Biological Trace Element Research* 104 (1): 83–91.

Index

Note: Page numbers in *italics* and **bold** refers to figures and tables respectively.

Crop Biofortification: Biotechnological Approaches for Achieving Nutritional Security under Changing Climate, First Edition. Edited by Adnan Noor Shah et al.

n

o

p